Covers Previous Years Questions Fully Solved

disha
Nurturing Ambitions

ALGEBRA Vol. 1

for Boards/ JEE main/ Advanced/ KVPY/ Olympiad

Covers

- Linear Equation
- Logarithm
- Piecewise Functions
- Quadratic Equation
- Sequence & Series
- Complex Number

2000+ Solved Questions

5 Difficulty Level Questions

Concept Applicator(CA)

Concept Builder (CB)

Concept Cracker (CC)

Concept Deviator (CD)

Concept Eliminator (CE)

Gajendra Kumar

PREFACE

It gives me immense pleasure to introduce a comprehensive book on Algebra "Theory of Algebra"

The book "Theory of Algebra" delves deep into the basics of the topic and then advances into the IIT JEE level and ends up high with the International Maths Olympiad level. Each chapter has been well explained and illustrated with numerous examples for the better understanding of the reader.

The entire algebra is divided in two volumes

Volume 1	Volume 2
(1) Linear Equation	(1) Introduction of Factorial (!)
(2) Logarithm	(2) Binomial Theorem
(3) Piece-wise defined Functions	(3) Set Theory
(4) Quadratic Equation	(4) Permutation & Combination
(5) Sequence and Series	(5) Factor theory
(6) Complex Numbers	(6) Probability
	(7) Determinant & Matrices

The unique features of this book will be:

(1) Comprehensive theory given along with numerous illustrations.

(2) Questions divided into 5 different difficulty levels.

(3) Board level questions are given to emphasize on Board exam.

(4) Previous 38 years IIT JEE/ AIEEE questions are included.

(5) Advance level theory provided for those, who aim for IIT JEE Advanced/ Olympiad.

(6) Questions provided are in MCQ form (including all variety of IIT JEE questions) along with Subjective Questions.

(7) Olympiad level questions are added in different section (Only for extra-ordinary students)

(8) Each and Every question is given with detailed solution

(9) Theory starts from the scratch and goes upto Olympiad level

Each Chapter is divided in two part -

Part A: Basic, concepts and detailed theory with numerous examples and exercise in three different difficulty levels.

Part B: Advance level theory required for Olympiad and IIT JEE Advanced with 2 level of exercises.

It is advisable to students that before attempting Part B of any Chapter please solve Part A of all the chapters.

Quantitative problems in this book have been divided into 5 difficulty levels viz.

- **Concept Applicator**: Basic concept testing questions.
- **Concept Builder**: Concept strengthening questions.
- **Concept Cracker**: IIT JEE Mains / State level exams
- **Concept Deviator**: IIT JEE - Advanced equivalent level questions.
- **Concept Eliminator**: KVPY and International Maths Olympiad standard questions.

The best bet is that this book is self-sufficient in itself. For the aspirants of any competitive examination like IIT JEE Main & Advanced, RMO and also Board exams etc. Their quest for an overall adaptive book ends here. Once the reader is able to successfully complete this book, he can be rest assured of his excellent performance on the exam day.

I would like to express my sincere thanks to Mr. Deepak Agarwal, Founder – Director of Disha Publication for consistently guiding me to bring this book as immaculate.

To understand the fact that a book is not an individual work, it incorporates the efforts of other people, I am thankful to all my students who have helped me throughout the compiling of this book and also proof-reading it. These include but not limited **to Argha Kamal Das (JU), Rohit Srivastava (IIT Kgp), Atreya Roy.**

Last but not the least, I would like to express my sincere gratitude to my father (my only Maths teacher), mother, siblings and have no words for my better half 'Suvra', who is my inspiration and partner in all ventures. Finally, thanks to the Hero of my life, my son – 'Sanyukt' (meaning summation in Math). He is the great source of inspiration for this book.

I have taken extreme care while writing the theories, compiling the questions and preparing the solutions, however, some errors may have crept in, so hereby I invite all readers to bring it to our notice.

Gajendra Kumar

- **Head Office :** B-32, Shivalik Main Road, Malviya Nagar, New Delhi-110017
- **Sales Office :** B-48, Shivalik Main Road, Malviya Nagar, New Delhi-110017
  Tel. : 011-26691021 / 26691713

Typeset by Disha DTP Team

Printed at Repro Knowledgecast Limited, Thane

DISHA PUBLICATION

For further information about the books and ebooks from DISHA,

Log on to **www.dishapublication.com**

or email to **info@dishapublication.com**

CHAPTER-4 : BASIC MATHS (PIECEWISE DEFINED FUNCTION) Page : 139 to 288

CHAPTER-5 : SEQUENCE & SERIES **Page : 289 to 396**

Graph of $a_1x + b_1y + c_1 = 0$

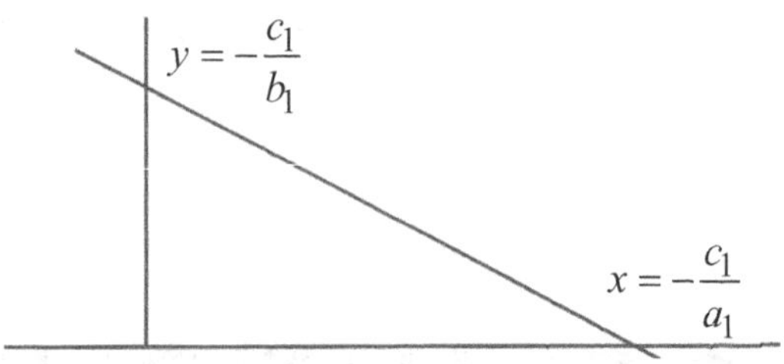

To solve an equation with two variables we need two equations. The equation $a_1x + b_1y + c_1 = 0$ represents *a* straight line. And solution of the above two equations graphically means point of intersection of these two straight lines. But if two straight lines are parallel to each other then these two straight lines will never cut each other and we will get no solution, while if both the equation represents same straight line then we will get infinite many solutions. There are some situations described here.

Case (i) $\frac{a_1}{a_2} \neq \frac{b_1}{b_2}$

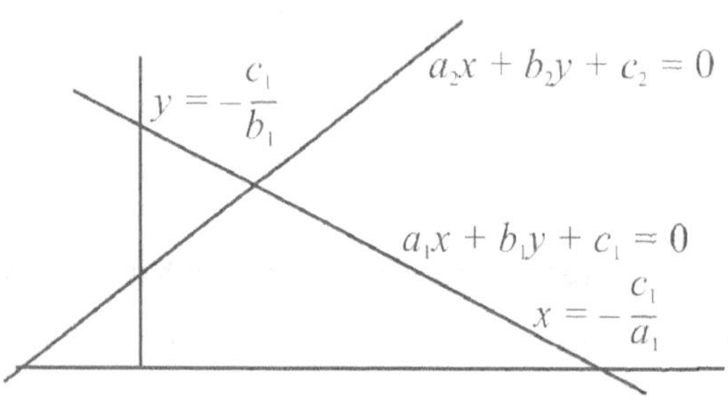

As per the condition if $\frac{a_1}{a_2} \neq \frac{b_1}{b_2}$ the two straight lines will intersect each other once and co-ordinate of that point is called the solution of the simultaneous equations $a_1x + b_1y + c_1 = 0$ and $a_2x + b_2y + c_2 = 0$

Case (ii) $\frac{a_1}{a_2} = \frac{b_1}{b_2} \neq \frac{c_1}{c_2}$

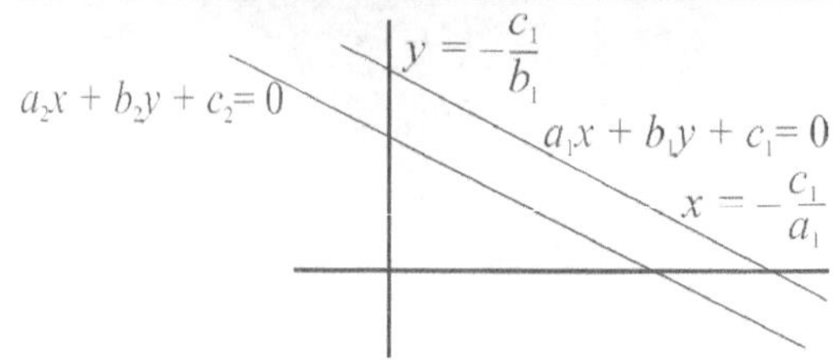

As per the condition if $\frac{a_1}{a_2} = \frac{b_1}{b_2} \neq \frac{c_1}{c_2}$ the two straight lines are parallel to each other so they never intersect so in this case there exist no solution of simultaneous equations $a_1x + b_1y + c_1 = 0$ and $a_2x + b_2y + c_2 = 0$

Case (iii) $\frac{a_1}{a_2} = \frac{b_1}{b_2} = \frac{c_1}{c_2}$

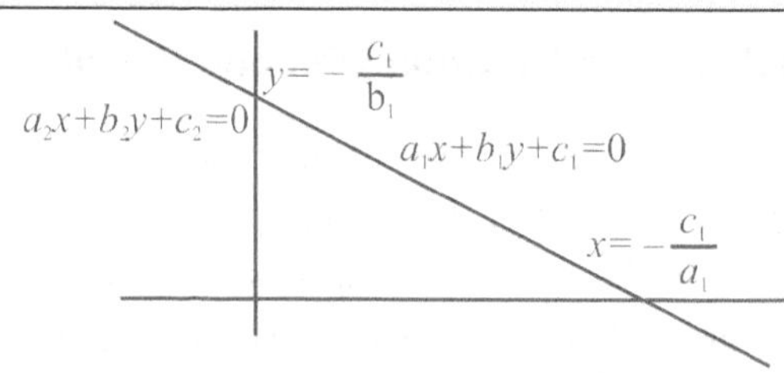

As per the condition if $\frac{a_1}{a_2} = \frac{b_1}{b_2} = \frac{c_1}{c_2}$ the two equations represent same straight line and hence there exist infinite number of solutions of simultaneous equations $a_1x + b_1y + c_1 = 0$ and $a_2x + b_2y + c_2 = 0$

Summary of above discussion is as follows:

- There is *a* unique solution if $\frac{a_1}{a_2} \neq \frac{b_1}{b_2}$
- There is no solution if $\frac{a_1}{a_2} = \frac{b_1}{b_2} \neq \frac{c_1}{c_2}$

  (the two straight lines are parallel to each other)
- There is infinite many solutions if $\frac{a_1}{a_2} = \frac{b_1}{b_2} = \frac{c_1}{c_2}$

  (two equation represents same straight line)

1.1.3 Solution of Simultaneous Equation

Consider two simultaneous equation with two variables:

$$a_1x + b_1y + c_1 = 0 \quad \text{... (i)}$$

And $a_2x + b_2y + c_2 = 0$ ...(ii)

We have to solve this system of equation means we need to find the value of 'x' and 'y' which satisfy both the equation. In order to solve these two equations just eliminate one of the variable by multiplying one or both the equations by *a* constant.

Illustration 3. Solve $3x + 4y = 11$ and $4x + 3y = 10$

Solution: Consider the given equations

$$3x + 4y = 11 \quad \text{...(i)}$$

$$4x + 3y = 10 \quad \text{...(ii)}$$

Let we want to eliminate y, whose co-efficients are 4 and 3. We know that LCM of 4 and 3 is 12 so we have to make both the coefficients 12, in order to do so multiply eq.(i) by 3 and eq (ii) by 4, we get

$$9x + 12y = 33 \quad \text{... (iii)}$$

$$16x + 12y = 40 \quad \text{... (iv)}$$

Subtract eq. (iii) from eq. (iv), we get $7x = 7$ or $x = 1$ putting this value of x in eq.(i) we get y =2

Illustration 4. Solve $\sqrt{2}x - \sqrt{3}y = 1$

and $\sqrt{8}x + \sqrt{27}y = 17$

Solution: Given equations are

$$\sqrt{2}x - \sqrt{3}y = 1 \quad \text{... (i)}$$

$$2\sqrt{2}x + 3\sqrt{3}y = 17 \quad \text{... (ii)}$$

Multiply 1st equation by 3 and add in 2nd equation we get $5\sqrt{2}x = 20$ or $x = 2\sqrt{2}$ substitute this value of x and we get $y = \sqrt{3}$.

Illustration 5. Solve $\dfrac{15}{x+2y} - \dfrac{11}{3x+4y} = 2$

and $\dfrac{5}{2(x+2y)} + \dfrac{22}{3x+4y} = \dfrac{5}{2}$

Solution: Given that $\dfrac{15}{x+2y} - \dfrac{11}{3x+4y} = 2$ ... (i)

$$\frac{5}{2(x+2y)} + \frac{22}{3x+4y} = \frac{5}{2} \quad \text{...(ii)}$$

Multiply equation (i) by 2 and add in equation (ii) we get

$$\frac{65}{2(x+2y)} = \frac{13}{2}$$

or $x + 2y = 5$ ...(iii)

Multiply equation (ii) by 6 and then subtract equation (i) we get

$$\frac{143}{3x+4y} = 13$$

or $3x + 4y = 11$ ...(iv)

On solving equations (iii) and (iv) we get $x = 1, y = 2$

1.1.4 Simultaneous Equation with Three Variables

In this type of problem since 3 variables are given. so in order to solve it completely (means to find the value of each variable) we require three distinct equations.

Methods that can be used to solve these type of problems:

- **Elimination Process:** In this process, take any 2 pair of equations out of 3 given equations, eliminate one of the variable. As *a* result we will get two equations with two variables now this can be dealt easily.
- **Substitution process**: In this process, take one variable, find its value in terms of other two variables and substitute it in remaining two equations. Again as *a* result we will get two equations with two variables which can be dealt easily.
- **By determinant (Cramer's Rule)**: Consider *a* system of equations:

$$a_1x + b_1y + c_1z = d_1$$

$$a_2x + b_2y + c_2z = d_2$$

$$a_3x + b_3y + c_3z = d_3$$

Solution of this system of equation is:

$$x = \frac{D_1}{D},\ y = \frac{D_2}{D},\ z = \frac{D_3}{D}$$

Here $D = \begin{vmatrix} a_1 & b_1 & c_1 \\ a_2 & b_2 & c_2 \\ a_3 & b_3 & c_3 \end{vmatrix}$, $D_1 = \begin{vmatrix} d_1 & b_1 & c_1 \\ d_2 & b_2 & c_2 \\ d_3 & b_3 & c_3 \end{vmatrix}$,

$$D_2 = \begin{vmatrix} a_1 & d_1 & c_1 \\ a_2 & d_2 & c_2 \\ a_3 & d_3 & c_3 \end{vmatrix},\ D_3 = \begin{vmatrix} a_1 & b_1 & d_1 \\ a_2 & b_2 & d_2 \\ a_3 & b_3 & d_3 \end{vmatrix}$$

Conditions for Consistency:

We have following cases:

(i) If $D \neq 0$, the system of equation is consistent and has *a* unique solution.

(ii) If $D = 0$ and $D_1 = D_2 = D_3 = 0$ the system of equation has infinite solution

(iii) If $D = 0$ and at least one of D_1, D_2, D_3 is non-zero the system of equation is inconsistent.

Illustration 6. Solve the system of linear equation:

$$x + 2y + 3z = 14$$
$$2x + y + 3z = 13$$
$$3x + 2y + z = 10$$

Solution: Given equations are:

$$x + 2y + 3z = 14 \quad \text{... (1)}$$
$$2x + y + 3z = 13 \quad \text{... (2)}$$
$$3x + 2y + z = 10 \quad \text{... (3)}$$

From equations (1) and (2) if we eliminate z, then we will get

$$-x + y = 1 \quad \text{... (4)}$$

Similarly from equations (2) and (3). If we eliminate z there we will get

$$7x + 5y = 17 \quad \text{... (5)}$$

Now, after solving equations (4) and (5) we will get $x = 1, y= 2$, and when we substitute this value of x and y in any of the equation we will get $z = 3$.

$x =1, y = 2, z =3$.

Illustration 7. Solve the system of linear equation

$$3x + 2y + 4z = 16$$
$$x + 4y + 2z = 12$$
$$5x + 4y + 6z = 26$$

Solution. From the given system of linear equation:

$$D = \begin{vmatrix} 3 & 2 & 4 \\ 1 & 4 & 2 \\ 5 & 4 & 6 \end{vmatrix} = -8$$

$$D_1 = \begin{vmatrix} 16 & 2 & 4 \\ 12 & 4 & 2 \\ 26 & 4 & 6 \end{vmatrix} = -8$$

$$D_2 = \begin{vmatrix} 3 & 16 & 4 \\ 1 & 12 & 2 \\ 5 & 26 & 6 \end{vmatrix} = -12$$

$$D_3 = \begin{vmatrix} 3 & 2 & 16 \\ 1 & 4 & 12 \\ 5 & 4 & 26 \end{vmatrix} = -20$$

Then, $x = \dfrac{D_1}{D} = \dfrac{-8}{-8} = 1,\ y = \dfrac{D_2}{D} = 1.5,\ z = \dfrac{D_3}{D} = 2.5$

1.1.5 Partial Solution of Simultaneous Equation with Three Variables

In this section, we will learn how to solve equation partially or to find value of x or y or z (if possible) from set of 2 equations with 3 variables.

Let the given equations are:

$$a_1x + b_1y + c_1 z = d_1 \quad \text{.... (i)}$$
$$a_2x + b_2y + c_2 z = d_2 \quad \text{.... (ii)}$$

From the set of above two equation it may be possible to find one of the variable x or y or z.

We can find value of x when we are able to eliminate y and z together and it is possible when their co-efficients are in the same ratio.

Equation is solvable for x when, $\left(\dfrac{b_1}{b_2} = \dfrac{c_1}{c_2}\right)$

Equation is solvable for y when, $\left(\dfrac{b_1}{b_2} = \dfrac{c_1}{c_2}\right)$

Equation is solvable for z when, $\left(\dfrac{a_1}{a_2} = \dfrac{b_1}{b_2}\right)$

Method to solve: Let from the given equations (i) and (ii), we are able to find the value of x (or we can solve for x) then $\dfrac{b_1}{b_2} = \dfrac{c_1}{c_2} = \dfrac{p}{q}$ (say)

Here p/q is in the simplest from then multiply equation (i) by q and equation (ii) by p, we get

$$a_1\, qx + b_1\, qy + c_1\, qz = qd_1 \quad \text{... (i)}$$
$$a_2\, px + b_2\, py + c_2\, pz = pd_2 \quad \text{... (ii)}$$

Subtracting these two we will get, $x = \frac{d_1q - d_2p}{a_1q - a_2p}$

Since $b_1q = b_2p$ and $c_1q = c_2p$.

And terms containing y and z will get eliminated simultaneously

Illustration 8. Find the value of x (if it is possible) if:

$$3x + 6y + 8z = 15,$$
$$11x + 21y + 28z = 55$$

Solution: Equation is solvable for x if co-efficient of y and z are in the same ratio:

Since $\frac{6}{21} = \frac{8}{28} = \frac{2}{7}$ so, if we multiply eq. (i) by 7 and eq (ii) by 2 than we can eliminate y and z together

$$21x + 42y + 56z = 105... \text{(i)}$$
$$22x + 42y + 56z = 110...\text{(ii)}$$

Subtracting equation (i) from eq (ii), we get $x = 5$

Here we can not find the value of y and z.

1.1.6 Establish a New Relation between given Variables

In this section we will learn how to form (or get value) of 3rd relation from two given relations; let the two given relations are:

$$a_1x + b_1y + c_1z = d_1 \quad ...\text{(i)}$$
$$a_2x + b_2y + c_2z = d_2 \quad ... \text{(ii)}$$

then find the value of $a_3x + b_3y + c_3z$ ... (iii)

here 3rd equation is dependent on equations (i) and (ii) there are two methods by which we can solve these type of question:

(i) by observation and (ii) by calculation.

(i) **Observation Method:** In this case, equations are closely related and final result can be calculated by some trial and error.

Illustration 9. If $x + 2y + 3z = 10$ and $2x + 3y + 4z = 18$.

Then find the value of $3x + 5y + 7z$

Solution: Here just by observation we find that by adding eqs. (i) and (ii), we get

$$3x + 5y + 7z = 10 + 18 = 28 \quad ...\text{(ii)}$$

Calculation Method: We will try to understand this method by taking and example.

Illustration 10. If

$$x + 2y + 3z = 10 \quad ...\text{(i)}$$
$$2x + 3y + 4z = 15 \quad ...\text{(ii)}$$

Then find the value of $5x + 6y + 7z = ?$

Solution: Here, let eq (i) is multiplied by (p)and eq. (ii) is multiplied by (q) and then added to get the final result.

$$px + 2py + 3pz = 10P$$
$$2qx + 3qy + 4qz = 15q$$

Adding these two equations, we get

$(p + 2q)x + (2p + 3q)y + (3p + 4q)z = (10p + 15q)$

And this result is $5x + 6y + 7z$

Comparing the co-efficients, we get

$$p + 2q = 5 \quad ....\text{(i)}$$
$$2p + 3q = 6 \quad ...\text{(ii)}$$
$$3p + 4q = 7 \quad ... \text{(iii)}$$

If we solve eqs. (i) and (ii) we will get $p = -3$, $q = 4$

This value of p and q should satisfy 3rd equation, $3p + 4q = 7$

So, $p = -3$ and $q = 4$

Then substituting this value of p and q we get

$$5x + 6y + 7z = 10(-3) + 15(4) = 30$$

Concept Applicator (CA)-1

1. Solve for x and y if $3x + 4y = 25$ and $4x + 5y = 32$
2. Solve for x and y if $3x-2y = 11$ and $2x-3y = 4$
3. Solve for x and y if $2x + 3y = 7$ and $6x + 9y = 21$
4. Solve for x and y if $2x+ 4y =7$ and $7x + 14y = 21$
5. Solve $6x - 8y = 21$ and $39x - 52x = 111$
6. Solve for x and y if $\frac{5}{2x+6y} + \frac{15}{3x+4y} = 2$ and $\frac{30}{x+3y} - \frac{20}{3x+4y} = 4$
7. Find the value of x, y and z if $3x + 2y + 4z = 16$; $5x + 3y + z = 12$; $2x + y + 3z = 11$
8. Find the value of y if $6x + 7y + 8z = 10$ and $27x + 27y + 36z = 18$
9. Find the value x,y and z if it is possible, if $3x + 4y + 6z = 6$ and $12x + 14y + 21z = 24$
10. Find the value of $x + y + z$ if $3x+2y + 4z = 16$; and $5x + 3y + 7z = 27$.

1.2 INEQUATIONS

1.2.1 Introduction

If any statements/expression, having some variable or variables, contains the sign less than (<), greater than (>), less than or equal to $(\leq)$ or greater than or equal $(\geq)$ then such statement is called as inequation or inequality.

Example of inequation with 1 variable

$$x > 5, y < 8, z \geq 1, w \leq -5$$

Example of inequation with 2 variable

$$x + 2y \geq 3, 3x + 4y < 5$$

Example of quadratic inequality

$$x^2 + 3x + 2 \leq 0, x^2 - x - 6 > 0$$

Open and Closed Interval

In open interval end points are not included in the solution set while in closed interval, end points are included in the solution set.

Open interval is denoted by (.) while closed interval by [.]

If $a \leq x \leq b$ then it is denoted by $x \in [a,b]$

If $a < x < b$ then it is denoted by $x \in (a,b)$

If $a < x \leq b$ then it is denoted by $x \in (a,b]$

If $a \leq x < b$, then it is denoted by $x \in [a,b)$

1.2.2 Linear Inequation in One Variable

In this case inequality must involve only one variable. For example; $3x - 5 \geq 23$, $x - 5 \geq 2$, it involves only one variable x., another example, $2y < 7$, it involves only one variable y.

We have to find the value or interval of x (variable) for which given inequality holds true.

Let us see how to solve inequation step by step

Step 1: Write the inequation

Step 2: Bring all the variable terms (in general term containing x) on one side and constant terms on other side.

Step 3: Do the simplification.

Step 4: Mark the inequation on number line.

Step 5: Write the solution set.

Let us understand with some examples

Illustration 1. Solve $3x - 9 \geq 0$

Solution. Given in equation is $3x - 9 \geq 0$

or $3x \geq 9$ [Bringing 9 on right side]

Thus, $x \geq 3$

This can also be represented as $x \in [3, \infty)$

The same thing can be represented on number line as

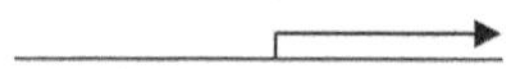

x=3

Thus, the solution set is $[3, \infty)$.

Illustration 2. Solve $-2x + 5 < 3x - 20$

Solution: Bring all terms containing variable x on one side and all the constants on other side.

$$5 + 20 < 2x + 3x$$

$$25 < 5x$$

Thus, $x > 5$

This can also be represented as $x \in (5, \infty)$

Now use number line

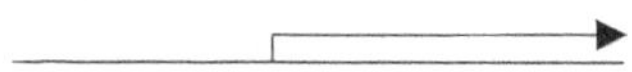

$x = 5$

Illustration 3. Solve $x + 9 \leq 3(1 - x)$

Solution: Given $\quad x + 9 \leq 3(1 - x)$

$$x + 9 \leq 3 - 3x$$

Transpose $-3x$ on L. H. S. and 5 on R. H. S.

$$x + 3x \leq 3 - 9 = -6$$

$4x \leq -6$ (Now divide by 4)

$$x \leq -\frac{6}{4} = -\frac{3}{2}$$

Mark $x = -\frac{3}{2}$ on number line and show that $x \leq -\frac{3}{2}$

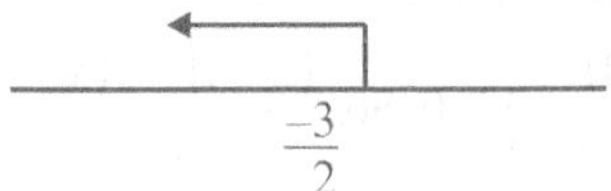

Thus, solution set is $\left(-\infty, -\frac{3}{2}\right]$

Illustration 4. Solve $\frac{3(x-5)}{5} \geq \frac{5(5-x)}{3}$

Solution: Given that $\frac{3(x-5)}{5} \geq \frac{5(5-x)}{3}$

Multiply both sides by 15, we get

$$3\times3\ (x-5) \geq 5\times5\ (5-x)$$

$$9x - 45 \geq 125 - 25x$$

Transpose all the variable terms one side and constant terms other side

$$9x + 25x \geq 125 + 45$$

$$34x \geq 170$$

$$x \geq \frac{170}{34} = 5$$

Mark $x = 5$ on number line and show $x \geq 5$

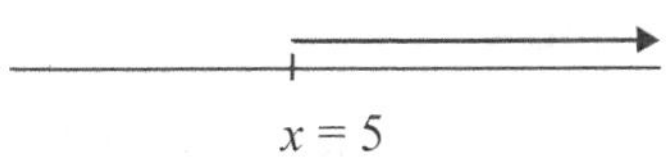

Thus solution set is $[5, \infty)$

1.2.3 Linear Rational Inequation

Type 1: Inequality of type $\frac{1}{ax+b} \leq 0$ **or** $\frac{1}{ax+b} < 0$

Both the inequalities $\frac{1}{ax+b} \leq 0$ and $\frac{1}{ax+b} < 0$ are same as $\frac{1}{ax+b} \neq 0$. The solution set of $\frac{1}{ax+b} < 0$ and $(ax+b) < 0$ are same

So $\frac{1}{ax+b} < 0 \Rightarrow (ax+b) < 0 \Rightarrow x < -\frac{b}{a}$

Type 2: Inequality of type $\frac{ax+b}{cx+d} \leq 0$ **or** $\frac{ax+b}{cx+d} \leq 0$

Sign scheme of

$\frac{ax+b}{cx+d}$ & $(ax + b)(cx + d)$ is same.

So if $\frac{ax+b}{cx+d} \leq 0$,

then $(ax + b)(cx + d) \leq 0$, but $x \neq -\frac{d}{c}$

Type 3: Inequality of type $\frac{1}{ax+b} \leq k$ or $\frac{1}{ax+b} < k$

Simplify the given expression as

$$\frac{1}{ax+b} - k \leq 0 \Rightarrow \frac{1-kb-kax}{ax+b} \leq 0,$$

then solve

it as explained in type 2.

First of all assure that the coefficient of variable x is positive if it is negative then multiply it with -1 to make variable x positive.

Let us see with some example

Illustration 5. Solve $\frac{1}{x-5} \leq 0$

Solution: Given $\frac{1}{(x-5)} \leq 0$

Solution set of $\frac{1}{(x-5)} \leq 0$ and $\frac{1}{(x-5)} < 0$ is same

For this inequality to hold true $x - 5 < 0$ so $x < 5$.

Mark $x = 5$ on number line and show $x < 5$.

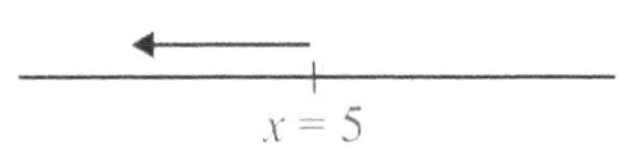

Thus, solution set is $(-\infty, 5)$

Illustration 6. Solve $\frac{x+2}{x+3} \geq 1$

1.3 DIOPHANTINE EQUATIONS

1.3.1 Diophantine Equation

We know that for a system of equations, to solve it completely we require number of independent equations equal to number of variables. If we have three variables then we require three independent equation and so on.

In a system of Diophantine equations, the number of unknown variables is more than the number of equations, so the solutions may not be unique / finite, or say, the number of solutions is more than one as usual. Because of the uncertainty of solutions, this kind of equations are also called **indefinite equations**. But if put restriction that solution must be integer then we may have finite solution.

Diophantine Equation is called the problem for finding **integer solutions of equations.**

Here the discussion is confined to linear equations and the systems of linear equations. The simplest equation is $ax + by = c$ ...(i)

where a, b, c are constant integers.

Theorem 1: The equation has no integer solution if *gcd* (*a*, *b*) is not divisible by *c*.

Proof: Let HCF of 'a' and 'b' is H then $ax + by$ is divisible by H hence 'c' must be divisible by H.

Theorem 2: When HCF (*a*, *b*) =1, then the equation (i) always has at least one integer solution.

Theorem 3: If x_0, y_0 is a special integer solution of the Equation (i), then the general solution of (i) is given by

$$\begin{cases} x = x_0 + bt \\ y = y_0 - at, \end{cases} \quad \forall t \in z \qquad ...(ii)$$

Proof : Let (x,y) be any solution of (i) differ from the given solution (x_0, y_0).

Then $$ax + by = c \qquad ...(i)$$

$$ax_0 + by_0 = c \qquad ...(ii)$$

Subtracting equation (ii) from equation (i) gives $a(x - x_0) = -b(y - y_0)$.

So $$\frac{x - x_0}{b} = \frac{y - y_0}{-a}.$$

Since, $b \mid a(x - x_0)$ and $a \mid b(y - y_0)$ implies that $b \mid (x - x_0)$ and $a \mid (y - y_0)$.So each side of the last equality is an integer. Hence $x - x_0 = bt$ and $y - y_0 = -$ at, Hence proved

For solving *a* Diophantine equation theorem III is very important, We can find the general solution if *a* special solution is given or can be found easily, then determine the range or values of 't' according to the given conditions in question.

Illustration 1: Find the number of non-negative integral solution of equation $5x + 7y = 100$

Solution : We will use Theorem III to solve this question. From observation we can say that $(x, y) =$ (20, 0) is one root of the equation.

In order to get the next value of x and y subtract 7 (co-efficient of y) from 20 and add 5 (co-efficient of x) so next solution is $(x,y) = (13, 5)$, similarly next solution is $(x, y) = (13 - 7, 5 + 5) = (6,10)$. If we apply the same concept then next integral solution will be (–1, 15) but as per the condition we need only non negative integral solution hence ruled out.

Required set of solution is (20,0),(13, 5) and (6, 15)

Illustration 2. Find all non-negative integer solutions to the equation $12x + 5y = 125$.

Solution: We will use Theorem III to solve this question. From observation we can say that $(x, y) =$ (0, 25) is one root of the equation.

In order to get the next value of y, and x, subtract 12 (co-efficient of x) from 25 and add 5 (co-efficient of y) so next solution is $(x,y) =$ (5,13) , similarly next solution is $(x, y) = (5 + 5, 13 - 12) = (10, 1)$. If we apply the same concept then next integral solution will be (15, –11) but as per the condition we need only non-negative integral solution hence ruled out.

Required set of solution is (0, 25), (5, 13) and (10, 1)

Illustration 3: Given that the positive integers $x > 1$ and $y > 4$ satisfies the equation $2007x - 21y =$ 1923. Find the minimum value of $2x + 3y$

Solution: $x - 1 > 0$, so $y > 4$

Let $x-1 = a$ or $x = a+1$

And $y-4 = b$ or $y = b+4$

Then $2007a + 2007 - 21b - 84 = 1923$

$$2007a - 21b = 2007 - 2007 = 0$$

or $2007\,a = 21b$

So least value of $a = 7$, $b = 669$ and minimum value of $x = 8$ and y is 673

$$2x_{min} + 3y_{min} = 2 \times 8 + 3 \times 673 = 2035$$

1.3.2 Special Equation

TYPE 1: Find the total number of natural number solutions of $x^2 - y^2 = k$.

Case 1: $x^2 - y^2 = p$, where p = prime number

In this case total number of solutions is only one pair of solution, lets understand this with one example

Illustration 4: Find the positive integral solution of the equation $x^2 - y^2 = 37$.

Solution: Since 37 is a prime number so only way to write is $37 = 37 \times 1$

$$(x+y)(x-y) = 37 \times 1$$

$$x + y = 37$$

$$x - y = 1$$

$$x = 19, y = 18$$

So only one pair of solution exist.

Case 2: $x^2 - y^2 = N$, where N is an odd natural number other than prime number and not a perfect square total number of solutions is same as (total number of factors of N)/2

Illustration 5: Find the positive integral solution of the equation $x^2 - y^2 = 15$

Solution: 15 can be written as 15×1 or 5×3

$$(x+y)(x-y) = 15*1$$

$$x + y = 15$$

$$x - y = 1$$

$$x = 8, y = 7$$

or $(x+y)(x-y) = 5*3.$

$$x + y = 5$$

$$x - y = 3$$

$$x = 4, y = 1$$

or total number of factors of 15 is 4, so total number of solutions $= 4/2 = 2$.

Case 3: $x^2 - y^2 = N$, where N is odd and perfect square.

Total number of integral solutions is given by (total number of factors of $(N-1)$)/2

Illustration 6: Find the positive integral solution of the equation $x^2 - y^2 = 49$

Solution: 49 can be written as 49×1 or 7×7

$$(x+y)(x-y) = 49*1$$

$$x + y = 49$$

$$x - y = 1$$

$x = 25, y = 24$

or total number of factors of 49 is 3, so total number of solutions $= (3-1)/2$

Case 4: $x^2 - y^2 = k$, where k is an even number divisible by 2 but not by 4.

Total number of solution = NO SOLUTION.

Illustration 7: Find the integral solution of the equation $x^2 - y^2 = 10$

Solution: $(x+y)(x-y) = 10*1$

$$x + y = 10$$

$$x - y = 1$$

$$2x = 11$$

$$x = 5.5$$

which does not an integer.

or $(x+y)(x-y) = 5*2$

$$x + y = 5$$

$$x - y = 2$$

$$2x = 7 \text{ or } x = 3.5$$

which again not an integer. Therefore no possible solutions.

Case 5: $x^2 - y^2 = k$, where k is an even natural number divisible by 4 and is not a perfect square.

Total number of solutions = (total number of factors of $k/4$)/2.

Illustration 8 : Find the integral solution of the equation $x^2 - y^2 = 8$.

Solution: $(x+y)(x-y) = 4*2$

$x + y = 4$ and $x - y = 2$

$2x = 6$ or $x = 3$ and $y = 1$.

or $(x+y)(x-y) = 8*1$

$x + y = 8$ and $x - y = 1$

$2x = 9$ or $x = 4.5$ which is not an integer.

Therefore only one solution

Case 6: $x^2 - y^2 = k$, where k is an even number multiple of 4 and *a* perfect square.

Total number of solutions= (total number of factors of $k/4 - 1)/2$

Type 2: Find the number of integral solution of $\frac{1}{x}+\frac{1}{y}=\frac{1}{k}$ Number of natural number solutions = total number of factors of k^2.

Case 1 : If $\frac{1}{x}+\frac{1}{y}=\frac{1}{p}$

where p is a prime number.

Total number of solutions = 3.

We will understand this by one example

Illustration 9: Find the number of integral solution of $\frac{1}{x}+\frac{1}{y}=\frac{1}{7}$

Solution : The given equation can be written as

$$7x + 7y - xy = 0$$

$$x(7-y) - 7(7-y) + 49 = 0$$

$$(x-7)(7-y) = -49$$

$$(x-7)(y-7) = 49$$

Since $49 = 1 \times 49 = 7 \times 7 = 49 \times 1$

So $x - 7 = 7$ or $x = 14$ and $y = 14$

Similarly, we will get two more pair of solutions

Number of ways in which 49 can be written as *a* product of 2 numbers is equal to the total number of factors of 49.

Illustration 10. The total number of integral solutions if $xy = 15$, $xz = 18$, $yz = 30$.

Solution. Multiplying $(xy)(xz)(yz)$

we get $x^2y^2z^2 = 15 \times 18 \times 30$.

$$xyz = \pm\sqrt{15 \times 18 \times 30}$$

$$xyz = \pm(90)$$

$$z = xyz/xy = \pm 6$$

$$y = xyz/xz = \pm 5$$

$$x = xyz/yz = \pm 3$$

Concept Applicator (CA)-3

1. Find the possible number of combinations of x, y that satisfies $3x + 7y = 37$ where $x, y \in z^+$

   (a) 1 (b) 2

   (c) 3 (d) 0

2. If $5x + 16y = 100$ where $x, y \in z^+$ then

   (a) y must be a multiple of 10

   (b) x must be a multiple of 4

   (c) y must be a multiple of 5

   (d) None of these

3. Rahul purchased some pens and copy. The price of each pen is ₹ 10 and the price of each copy is ₹ 11. If Rahul spends a total amount of ₹ 130, then how many copies did he purchase?

   (a) 2 (b) 8

   (c) 10 (d) 12

4. The number 284 has to be divided into two parts such that one part is a multiple of 9 and the other is a multiple of 17. In how many ways can the number be divided?

   (a) 4 (b) 3

   (c) 2 (d) 1

5. Rapti appeared in TAAT examination where *a* question paper contains a certain number of 10-mark, 5-mark and 2-mark questions with a total of 35 questions in it. The maximum marks that a student can secure in the paper is 100. How many 10-mark questions are there in the paper?

   (a) 3 (b) 5

   (c) 6 (d) 4

6. Given that the equation $\frac{4}{3}x - a = \frac{2}{5}x + 140$ has a positive integer solution, where a is a parameter . Find the minimum positive integer value of a.

7. If k is a positive integer such that the equation $kx - 12 = 3k$ has an integer root, then the number of such k is
   (a) 3 (b) 4
   (c) 5 (d) 6

8. Find two positive integers A and B which satisfy $\frac{A}{11} + \frac{B}{3} = \frac{17}{33}$

9. Find the integer solutions of the equation $13x - 7y = 0$ satisfying the condition $80 < x + y < 120$.

10. Find the number of ordered pairs of positive integers (x, y) that satisfy the equation $\frac{1}{x} + \frac{1}{y} = \frac{1}{2004}$.
   (a) 32 (b) 36
   (c) 45 (d) None of these

Exercise A

Concept Builder (CB)

1. Sayani, an over-confident girl, had to do a multiplication. Instead of taking 35 as one of the multipliers, she took 53. As a result, the product went up by 540. What is the new product?
   (a) 1050 (b) 540
   (c) 1440 (d) 1590

2. "Shubham" car rental agency has the following terms. If a car is rented for 5 hours or less the charge is 60 per hour or ₹12 per kilometer whichever is more. On the other hand, if the car is rented for more than 5 hours, the charge is ₹50 per hour or ₹7.50 per Kilometer whichever is more. Ankur rented a car from this agency, drove it for 30 kilometers and ended up paying ₹300. For how many hours did he rent the Car?
   (a) 4 (b) 5
   (c) 6 (d) None of these

3. A rich merchant "Utsa Roy" had collected many gold coins. He did not want anybody to know about them. One day, his wife asked, "How many gold coins do we have?" After pausing a moment, he replied, "Well! If I divide the coins into two unequal numbers, then 48 times the difference between the two numbers equals the difference between the squares of the two numbers." The wife looked puzzled. Can you help the merchant's wife by finding out how many coins the merchant has?
   (a) 96 (b) 53
   (c) 43 (d) None of these

4. A piece of string is 40 centimeters long. Aritra, cut it into three pieces. The longest piece is 3 times as long as the middle-sized piece and the shortest piece is 23 centimeters shorter than the longest piece. Find the length of the shortest piece.
   (a) 27 (b) 5
   (c) 4 (d) 9

5. The number of solutions of the equation 2x + $y = 40$ where both x and y are positive integers and $x \leq y$ is:
   (a) 7 (b) 13
   (c) 14 (d) 18

6. In a certain factory, each day the expected number of accidents is related to the number of overtime hours by a linear equation. Suppose that on one day there were 1000 overtime hours logged and 8 accidents reported, and on another day there were 400 overtime hours logged and 5 accidents. What are the expected

numbers of accidents when no overtime hours are logged?

(a) 2 (b) 3

(c) 4 (d) 5

7. "Rahul" brokerage house offers 3 stock portfolios. Portfolio I consists of 2 blocks of common stock and 1 municipal bond. Portfolio II consists of 4 blocks of common stock, 2 municipal bonds and 3 blocks of preferred stock. Portfolio III consists of 2 blocks of common stock, 2 municipal bond and 3 blocks preferred stock. A customer wants 12 blocks of common stock, 6 municipal bonds and 6 preferred stocks. How many portfolio III should be offered?

(a) 1 (b) 2

(c) 3 (d) None of these

8. Find the value of $x : y : z$ if

$x + 2y + 3z = 14$ ...(i)

$2x + 3y + z = 11$ ...(ii)

$3x + y + 2z = 11$ ...(iii)

(a) 1:2:3 (b) 2:3:1

(c) 2:1:3 (d) None of these

9. A computer program was tested 300 times before its release. The testing was done in three stages of 100 tests each. The software failed 15 times in Stage I, 12 times in Stage II, 8 times in Stage III, 6 times in both Stage I and Stage II, 7 times in both Stage II and Stage III, 4 times in both Stage I and Stage III, and 4 times in all the three stages. How many times the software failed in a single stage only?

(a) 10 (b) 13

(c) 15 (d) 17

10. The taxis plying in wasseypur have the following fare structure: ₹ 20 for the first two kilometers, ₹ 5 for every km in excess of 2 km and up to 10 km, and Rs 8 for every km in excess of 10 km. Bullock carts on the other hand charge ₹ 2 per km. Sardar Khan takes *a* taxi from the Wasseypur railways station to his home. On the ways, at a distance of 14 km from the railway station, he meets Faizal Khan, and gets down from the taxi to talk to him. Later he takes a bullock cart to reach his home. He spends a total of ₹ 102 to reach his home from the railway station. How far is his home from the railways station?

(*a*) 17 (*b*) 18

(*c*) 19 (d) 20

Concept Cracker (CC)

1. If $2x + 3y + 4z = 10$ ...(i)

$3x + 5y + 7z = 5$ .... (ii)

Then find the value of $3x + 4y + 6z$

(a) 18 (b) 25

(c) 21 (d) None of these

2. If $x + 2y + 3z = 5$ and $3x + 5y + kz = 10$. What should be the value of K such that $3x + 4y + 3z$ can be calculated.

(a) 5 (b) 7

(c) 6 (d) None of these

3. What could be the value of K such that value of y can be calculated from these two equations .....

$$2x + 3y + 4z = 10$$

$$3x + 5y + kz = 15$$

(a) 5 (b) 6

(c) 4 (d) None of these

4. If $9x + 12y + 24z = 20$

$8x + 6y + 20z = 30$

Then which one of the following can be calculated

(a) $x + 6y + 5z$ (b) $x + 5y + 4z$

(c) $x + 6y + 4z$ (d) none of these

5. The number of positive integers valued pairs (x, y) satisfying $4x - 17y = 1$ and $x \leq 1000$ is

(a) 59 (b) 57

(c) 55 (d) 58

6. Three labeled boxes containing red and white cricket balls are all mislabeled. It is known that one of the boxes contains only white balls and one only red balls. The third contains a mixture of red and white balls. You are required to cor-

rectly label the boxes with the labels red, white and red and white by picking a sample of one ball from only one box. What is the label on the box you should sample?

(a) White (b) Red

(c) Red and White

(d) Not possible to determine from a sample of one ball

7. If $x > 2$ and $y > -1$, then which of the following statements is necessarily true?

(a) $xy > -2$ (b) $-x < 2y$

(c) $xy < -2$ (d) $-x > 2y$

8. If $x > 5$ and $y < -1$, then which of the following statements is true?

(a) $(x + 4y) > 1$ (b) $x > -4y$

(c) $-4x < 5y$ (d) None of these

9. Three friends returning from a movie, stopped to eat at a restaurant. After dinner, they paid their bill and noticed a bowl of mints at the front counter. Sita took 1/3 of the mints, but returned four because she had *a* momentary pang of guilt. Fatima then took 1/4th of what was left but returned three for similar reasons. Eswari then took half of the remainder but threw two back into the bowl. The bowl had only 17 mints left when the raid was over. How many mints were originally in the bowl?

(a) 38 (b) 31

(c) 41 (d) None of these

10. All the page numbers from a book are added, beginning at page 1. However, one page number was mistakenly added twice. The sum obtained was 1000. Which page number was added twice?

(a) 44 (b) 45

(c) 10 (d) 12

11. At a certain fast food restaurant, Brain can buy 3 burgers, 7 shakes and one order of fries for ₹120 exactly. At the same place it would cost ₹164.5 for 4 burgers, 10 shakes and one order of fries. How much would it cost for an ordinary meal of one burger, one shake and one order of fries?

(a) ₹31

(b) ₹41

(c) ₹21

(d) Cannot be determined

12. A set of consecutive positive integers beginning with 1 is written on the blackboard. A student came along and erased one number. The average of the remaining numbers is $35\frac{7}{17}$. What was the number erased?

(a) 7 (b) 8

(c) 9 (d) None of these

13. A change making machine contains 1 rupee, 2 rupees and 5 rupee coins. The total number of coins is 300. The amount is ₹ 960. If the number of 1 rupee coins and number of 2 rupee coins are interchanged, the value comes down by ₹40. The total number of 5 rupee coins is

(a) 100 (b) 140

(c) 60 (d) 150

14. I have a total of ₹1000. Item A costs ₹110, item B costs ₹90, item C costs ₹70, item D cost ₹40 and item E costs ₹45. For every item D that I purchase, I must also buy two of item B. For every item A, I must buy one of item C. For every two items of E, I must also buy one of item C. For every item purchased I earn 1000 points and for every rupee not spent I earn a penalty of 1500 points. My objective is to maximise the points I earn. What is the number of items that I must purchase to maximise my points?

(a) 13 (b) 14

(c) 15 (d) 16

15. A child was asked to add first few natural numbers (that is, $1 + 2 + 3 \ldots$) so long his patience permitted. As he stopped, he gave the sum as 575. When the teacher declared the result wrong the child discovered he had missed one number in the sequence during addition. The number he missed was:

(a) less than 10 (b) 10

(c) 15 (d) more than 15

16. Three travelers are sitting around *a* fire, and are about to eat *a* meal. One of them has five small loaves of bread, the second has three small loaves of bread. The third has no food, but has eight coins. He offers to pay for some bread. They agree to share the eight loaves equally among the three travelers, and the third traveler will pay eight coins for his share of the eight loaves. All loaves were the same size. The second traveler (who had three loaves) suggests that he be paid three coins and that the first traveler be paid five coins. The first traveler says that he should get more than five coins. How much the first traveler should get?

(a) 5 (b) 7
(c) 1 (d) one of these

17. Mayank, Mirza, Little and Jaspal bought a motorbike for \$60.00. Mayank paid one half of the sum of the amounts paid by the other boys. Mirza paid one third of the sum of the amounts paid by the other boys; and Little paid one fourth of the sum of the amounts paid by the other boys. How much did Jaspal have to pay?

(a) 15 (b) 13
(c) 17 (d) none of these

18. The owner of *a* local jewellery store hired 3 watchmen to guard his diamonds, but *a* thief still got in and stole some diamonds. On the way out, the thief met each watchman, one at *a* time. To each he gave ½ of the diamonds he had then, and 2 more besides. He escaped with one diamond. How many did he steal originally?

(a) 40 (b) 36
(c) 25 (d) None of these

19. A test has 50 questions. A student scores 1 mark for *a* correct answer, –1/3 for *a* wrong answer, and –1/6 for not attempting *a* question. If the net score of *a* student is 32, the number of questions answered wrongly by that student cannot be less than

(a) 6 (b) 12
(c) 3 (d) 9

20. Each family in a locality has at most two adults, and no family has fewer than 3 children. Considering all the families together, there are more adults than boys, more boys than girls, and more girls than families. Then the minimum possible number of families in the locality is

(a) 4 (b) 5
(c) 2 (d) 3

21. The sum of four consecutive two digit odd numbers, when divided by 10, becomes a perfect square. Which of the following can possibly be one of these four numbers?

(a) 21 (b) 25
(c) 41 (d) 67

22. When you reverse the digits of the number 13, the number increases by 18. How many other two digit numbers increase by 18 when their digits are reversed?

(a) 5 (b) 6
(c) 7 (d) None of these

23. Consider the set S = {2, 3, 4, ..., 2n + 1), where n is *a* positive integer larger than 2007. Define, X as the average of the odd integers in S and, Y as the average of the even integers in S. What is the value of $X - Y$?

(a) 1 (b) n/2
(c) (n + 1)/2n (d) 2008

24. Ten years ago, the ages of the members of *a* joint family of eight people added up to 231 years. Three years later, one member died at the age of 60 years and *a* child was born during the same year. After another three years, one more member died, again at 60, and a child was born during the same year. The current average age of this eight-member joint is nearest to

(a) 22 years (b) 21 years
(c) 25 years (d) 24 years

25. A confused bank teller transposed the rupees and paise when he cashed a cheque for Shailaja, giving her rupees instead of paise and paise

instead of rupees. After buying a toffee for 50 paise, Shailaja noticed that she was left with exactly three times as much as the amount on the cheque. Which of the following is a valid statement about the cheque amount?

(a) Over Rupees 7 but less than Rupees 8
(b) Over Rupees 22 but less than Rupees 23
(c) Over Rupees 18 but less than Rupees 19
(d) Over Rupees 4 but less than Rupees 5

26. A shop stores *x* kg of rice. The first customer buys half this amount plus half a kg of rice. The second customer buys half the remaining amount plus half a kg of rice. Then the third customer also buys half the remaining amount plus half a kg of rice. Thereafter, no rice is left in the shop. Which of the following best describes the value of *x*?

(a) $2 \leq x \leq 6$ (b) $5 \leq x \leq 8$
(c) $9 \leq x \leq 12$ (d) $11 \leq x \leq 14$

27. The integers 1, 2, …, 40 are written on *a* blackboard. The following operation is then repeated 39 times: In each repetition, any two numbers, say *a* and *b,* currently on the blackboard are erased and *a* new number $a + b - 1$ is written. What will be the number left on the board at the end?

(a) 820 (b) 821
(c) 781 (d) 819

28. In a rainy season, Jamshedpur experienced some rain for thirteen days. A morning rain was always followed by a clear evening. An evening rain was always preceded by *a* clear morning. A total of 9 mornings and 12 evenings were clear. How many rainless days did Jameshedpur have in the period?

(a) 7 (b) 6
(c) 5 (d) 3

29. Consider the system of linear equations.

$$2x + 3y + 4z = 16$$
$$4x + 4y + 5z = 26$$
$$ax + by + cz = r$$

For r = 5 and $a = 1$ then the system of linear equation will have infinite number of solution if c = A. Then the value of A is

(a) 3/2 (b) 1
(c) ½ (d) 0

30. In an examination there are 30 question. 1 mark is given for each correct answer and 0.25 is deducted for every incorrect answer. Ankur attempted all the questions and scored 13.75. How many incorrect answers did he have?

(a) 10 (b) 11
(c) 12 (d) 13

31. A rural child specialist has to determine the weight of five children of different ages. He knows from his past experience that each of the children would weigh less than 30 kg and each of them would have different weights. Unfortunately, the scale available in the village can measure weight only over 30 kg. The doctor decides to weigh the children in pairs. However his new assistant weighted the children without noting down the names. The weights were: 35, 36, 37, 39, 40, 41, 42, 45, 46 and 47 kg. The weight of the lightest child is:

(a) 15 kg. (b) 16 kg.
(c) 17 kg. (d) 18 kg.

32. Nikhil's mother asks him to buy 100 pieces of sweets worth ₹ 100/-. The sweet shop has 3 kinds of sweets, *kajubarfi, gulabjamun* and *sandesh.Kajubarfi* costs ₹ 10/- per piece, *gulabjamun costs* ₹ 3/- per piece and sandesh costs 50 paise per piece. If Nikhil decides to buy at least one sweet of each type, how many *gulabjamuns* should he buy?

(a) 1 (b) 2
(c) 3 (d) 4

33. Three truck drivers, Amar, Akbar and Anthony stop at *a* road side eating joint. Amar orders 10 *rotis*, 4 plates of *tadka*, and *a* cup of tea. Akbar orders 7 *rotis*, 3 plates of *tadka*, and *a* cup of tea. Amar pays ₹ 80 for the meal and Akbar pays ₹ 60. Meanwhile, Anthony orders

5 *rotis*, 5 plates of *tadka* and 5 cups of tea. How much (in ₹) will Anthony pay?

(a) 75 (b) 80

(c) 95 (d) 100

34. Rakesh has four weights of different weights; he measured three weights at a time and got result as 13, 14, 16 and 17 kgs. Find the value of smallest weights?

(a) 4kg (b) 3kg

(c) 2kg (d) 5 kg

35. A unique bar was there in *a* city. In the bar both entry and exit fees was \$10 per person. Pankaj went to bar played in casino and doubled his amount and came out of the bar, again he went in the bar and played the casino and doubled his amount and came out of the bar and then he has finally no amount left with him, so with how much amount he entered the bar?

(a) 12.75 (b) 11.25

(c) 12.25 (d) 12.50

36. A and B have some amount with them. B gave some amount to A hence due to this change now A and B both have as much rupee as much they had paisa and as much paisa as much they had rupee. What can be the maximum total sum of money with them?

(a) 199.98 (b) 188.98

(c) 198.97 (d) 198.98

37. Five friends A, B, C, D and E had some amount with them. A gave half of his amount to B, B gave half of his total to C, C again give half of his total to D and so on D gave to E. if at the end the sum of the amount with A and D is equal to that with E then find the ratio of the amount with A to that with E?

(a) 1:2 (b) 2:1

(c) 3:1 (d) 1:3

38. The ratio of father's age when his son will attain mother's present age to the father's age when son will attain father's present age is 11:12 then what is the ratio of mother's age for the same time period.

(a) 12:13 (b) 11:13

(c) 10:11 (d) 10:13

39. There are 10 students standing in a row with each of them have some chocolates. If 10th student gave away his all chocolates equally to others then average number of chocolates for all students from 2nd student to 9th student becomes 5. If 1st student gave away all his chocolates to others equally then the average number of chocolates from 2nd student to 9th student becomes 4. If total number of chocolates with all of them is 61 then find out how many chocolates was with 1st student?

(a) 27 (b) 18

(c) 9 (d) 36

40. In *a* family mother's age is equal to sum of the ages of son and her daughter. The average age of all four members in the family is equal to average of son and mother, while average of father and son's age is 33, and then find the age of son?

(a) 28years (b) 21years

(c) 22years (d) 24years

41. A and B are standing on a ladder such that A know how many steps are behind him and the number of steps between them while B knows the number of steps ahead of him and the number of steps between them. A says to B if we both approaches to each other 2 steps then the number of steps behind me and the number of steps between us will be equal. In reply B say if we both move away from each other by 2 steps then the number of steps between us will be equal to number of steps ahead of me. If total numbers of steps are equal to 23 then find the number of steps between them?

(a) 7 (b) 12

(c) 10 (d) 15

42. In the 1st day of college 1st year students are standing in such a way that total number of rows are equal to total number of column (i.e. they form a square). Next day minimum number of students joined such that again they form a square. On 3rd day again minimum number of students joined so that again they can stand in a square. If the ratio of the number of students joined on 2nd day to that of 3rd day is 25:27 then find the total number of students on the 1st day?

(a) 121 (b) 144

(c) 169 (d) 196

43. Three friends A,B and C have some amounts with them while their friend D has no amount, they shared the amount in specific manner. A divided half of his amount equally among B, C and D. similarly B divided half of his total in between C and D equally, and C gave half of his total to D. After all these changes at the end if amount with A, B and C are equal then by what percentage D has more amount than A?

(a) 66.67% (b) 70%

(c) 75% (d) 83.33%

44. Three friends A, B and C have some gold coins with them such that total number of gold coins are 24. A gave away some of his coins to C and double of that to B then they all have equal number of coins with them. How many maximum possible number of coins A had?

(a) 12 (b) 14

(c) 17 (d) 16

45. When son will attain his mother's present age then the average age of son, mother and father will become 160/3 and when he will attain his father's present age then the average will becomes 172/3, by how many years father is older than mother?

(a) 6years (b) 5years

(c) 4years (d) 7 years

46. In the above question if when mother was as old as her son is now then the ratio of age of mother to father was 3:4 then find the present age of son.

(a) 10 years (b) 12 years

(c) 11 years (d) 8 years

47. Three friends A, B and C bought *a* cycle for \$80 with each one of them have some share in that. If B reduces his contribution by 20% then C has to increase his contribution by 25% to compensate that. When A reduces his contribution by 2/7 then C has to increase his contribution by 1/2. Find the share of C.

(a) 20 (b) 25

(c) 30 (d) 24

48. There are two groups of 4 students each. Average number of coins with group *a* is 12 while that with group B is 15. 1st student of group 1 gave half of his coins to 1st student of group B while 1st student of group B half of his coins to 1st student of group A, similarly 2nd student, 3rd student and 4th student interchanged half of their coins. After all these interchanges find the total number of coins with students of group 1.

(a) 44 (b) 34

(c) 64 (d) 54

49. In a class if all students are present then four students don't get chair, while if 50% of the students are absent then 33.33% of the chairs are vacant. Find the number of chairs in the class.

(a) 14 (b) 18

(c) 12 (d) 10

50. If cost price of 40 pens is more than cost of 15 copies but less than cost of 16 copies. What can be the minimum total cost of a pen and a copy if price of both pen and copy is an integer

(a) 16 (b) 18

(c) 20 (d) 21

PART-B

1.4 ADVANCE MATERIAL FOR JEE ADVANCED, KVPY AND OLYMPIAD

1.4.1 Chicken McNugget Theorem

The Chicken McNugget Theorem (also known as Postage Stamp Problem or Frobenius coin problem). The name of this theorem may surprise you but if go back to its origin then we can relate with its name. In McDonalds, chicken McNuggets were available in 9 and 20 piece per package. Somebody wondered what the largest amount you could never buy was, assuming that you can not buy half or any fraction of a package. It was then found that 151 is the largest McNuggets that one can not buy. And thus creating the "Chicken McNugget" theorem.

The theorem says that for any two relatively prime positive integer '*p*' and '*q*' the greatest integer that can not be written in the form of 'px+qy' for non negative integer '*x*' and '*y*' is '*pq-p-q*'

Further it can be derived from the above theorem that there are exactly $\frac{(p-1)(q-1)}{2}$ positive integers which can not be expressed in the form of '$px+qy$'.

For positive integers $a_1, a_2, \ldots, a_n$ such that gcd $(a_1, a_2, \ldots, a_n) = 1$, then the largest integer that *cannot* be expressed as an integer conical combination of these numbers, i.e., as a sum $k_1a_1 + k_2a_2 + \cdots + k_na_n$, where $k_1, k_2, \ldots, k_n$ are non-negative integers is called the **Frobenius number** of the set $\{a_1, a_2, \ldots, a_n\}$, and is usually denoted by $g(a_1, a_2, \ldots, a_n)$.

If the GCD were not 1, every integer that is not a multiple of the GCD would be inexpressible as a linear combination of $\{a_1, a_2, \ldots, a_n\}$,

Now we have different cases-

Case (i) When n= 2 then, we have already discussed this above i.e for any two relatively prime positive integer 'p' and 'q' the greatest integer that can not be written in the form of 'px+qy' for non negative integer 'x' and 'y' is 'pq-p-q'

Case (ii) When n =3 then, finding the largest number (Frobenius Number) is a bit tedious work but we can find its range as

$$g(a_1, a_2, a_3) \geq (3a_1a_2a_3)^{\frac{1}{3}} - a_1 - a_2 - a_3$$

Case (iii) When a_is are in AP

Consider set

$$\{a_1, a_2, \ldots a_n\} = \{a, a+d, a+2d, \ldots, a+nd\}$$

here a and d are co-prime to each other.

Then its Frobenius number is defined as

$$g\{a, a+d, a+2d, \ldots, a+nd\}$$

$$= \left(\left[\frac{a-2}{n}\right]+1\right)a + (d-1)(a-1) - 1$$

Case (iv) When a_is are in GP

Consider set

$$\{a_1, a_2, \ldots a_n\} = \{m^k, m^{k-1}n, m^{k-2}n^2, \ldots n^k\}$$

here m and n are co-prime to each other.

Then its Frobenius number is defined as

$$g\{m^k, m^{k-1}n, m^{k-2}n^2, \ldots n^k\}$$

$$= n^{k-1}(mn - m - n) + \frac{\{m^2(n-1)(m^{k-1} - n^{k-1})\}}{m-1}$$

Illustration 1. What is the largest amount that can not be paid by using notes of denomination Rs5 and Rs7.

Solution: Let number of Rs 5 notes is '*x*' and that of Rs 7 notes is '*y*' then the amount that can be paid is in the form of $5x + 7y$, and the largest number which can not be achieved by $5x + 7y$ is $5 \times 7 - 5 - 7 = 23$

Illustration 2. What is the largest number cannot be achieved by $5x+9y$. Here ***x*** and ***y*** are positive integers.

Solution: Largest number that cannot be achieved by $5x + 9y$ is $5 \times 9 - 5 - 9 = 31$

Illustration 3. How many bills of distinct integral amount cannot be paid by using notes of denomination ₹ 5 and ₹ 7.

Solution: Let number of ₹ 5 notes is 'x' and that of ₹ 7 notes is 'y' then there are exactly $\frac{(5-1)(7-1)}{2}$ = 12 positive integers which cannot be expressed in the form of '$5x+7y$'.

Illustration 4. Find the largest integer that can not be expressed as $21x + 35y$

Solution: Since 21 and 35 are not co-prime to each other, and their HCF is 7 so, any number which is not a multiple of 7 can not be expressed as $21x+35y$

Illustration 5. Find the largest integer that can not be expressed as $4x + 7y + 10z$.

Solution: Here 4, 7 and 10 are in AP with a= 4 and common difference =3 and also 3 and 4 are co prime to each other, we know that

$$g\{a, a+d, a+2d, \ldots, a+nd\}$$

$$= \left(\left[\frac{a-2}{n}\right]+1\right)a+(d-1)(a-1)-1$$

So $g(4, 7, 10)$

$$= \left(\frac{4-2}{2}+1\right)4+(3-1)(4-1)-1 = 131$$

1.4.2 Pigeonhole principle

Principle I: When $(m + 1)$ pigeons enter m pigeonholes (m is a positive integer), there must be at least one hole having more than 1 pigeon.

Principle II: When $m + 1$ pigeons enter n pigeonholes, there must be one hole having atleast $\left[\frac{m}{n}\right]+1$ pigeons.

Principle III: When infinitely many elements are partitioned into finitely many sets, there must be at least one set containing infinitely many elements.

Pigeonhole Principle is easy to understand, and can be proven at once by contradiction. It has various applications. The main problem for applying the principle is doing classification of the possible cases, i.e., to find out different but appropriate " pigeonholes " for different problems

Illustration 6. A bag contains stones of 7 colours and there are 50 stones of each colour. Minimum how many stones we have to pick so that we have at least 2 stones of same colour.

Solution: Let each colour be one pigeonhole, and a ball drawn be a pigeon then from Principle I, We have to draw at least 7+1 = 8 stones.

Illustration 7. A bag contains stones of 7 colours and there are 50 stones of each colour. Minimum how many stones we have to pick so that we have at least 7 stones of same colour.

Solution: Let each colour be one pigeonhole, and a ball drawn be a pigeon then from Principle II, We have to draw at least 6x7+1 = 43 stones. Since if only 42 stones are picked out, there may be exactly 6 for each colour. By pigeonhole principle, there must be one colour such that at least [42/7] + 1 = 7 drawn balls have this color.

Illustration 8. A bag contains stones of 7 colours and there are 50 stones of each colour. Minimum how many stones we have to pick so that we have at least 7 stones of each colour.

Solution: From previous solution, we need to have at least 6x7+1 = 43 stones such that we have at least one colour has 7 stones. Now for 2nd colour we have to draw another 6 stones and so on. Repeating this process for 6 times, the 7 groups of 7 same colour stones are obtained. Thus, the least number of drawn stones is $43 + 6 \times 7 = 85$.

Illustration 9. A bag contains 200 marbles. There are 60 red ones, 60 blue ones, 60 green ones and the remaining 20 consist of yellow and white ones. If marbles are chosen from the bag without looking, what is the smallest number of marbles one must pick in order to ensure that, among the chosen marbles, at least 20 are of the same color ?

Solution: Consider when 77 marbles are chosen, there may be 19 red, 19 blue, 19 green, and 20 yellow and white.

So, if 78 marbles are chosen at random, the number of yellow and white ones among them is at most 20.

Illustration 10. Determine the maximum number of elements of a subset S of $\{1, 2, 3, \ldots, 1999\}$ such that the difference of any two distinct elements of S is not equal to 4.

Solution: Given that $S = \{1, 2, 3, \ldots, 1999\}$. At first,by modulo 4, we can partition A into four subsets S_0, S_1, S_2 and S_3, where

$$S_i = \{n \in A: n \equiv i (\text{mod } 4)\}, \text{ for } i = 0, 1, 2, 3.$$

So, there are 500 numbers in each of S_1, S_2, S_3 but 499 numbers in S_0.

Now, consider for the set S_1, arrange all the numbers in an ascending order and consider the 250 pairs formed by its numbers (1, 5), (9, 13), (17, 21), ..., (1993, 1997).

By pigeonhole principle if more than 250 numbers in S_1 are selected out as part of S, then there must be some two numbers coming from one of the above pairs, so their difference is 4. So from S_1 we can select 250 elements similarly we can select 250 elements from each of remaining three groups.

Therefore, the maximum number of the elements is 1000.

Exercise B

Concept Deviator (CD)

(Single Correct MCQs)

1. Each of the numbers $x_1, x_2, \ldots, x_n$, $n > 4$, is equal to 1 or –1. Suppose, $x_1x_2x_3x_4 + x_2x_3x_4x_5 + x_3x_4x_5x_6 + \ldots + x_{n-3}x_{n-2}x_{n-1}x_n + x_{n-2}x_{n-1}x_nx_1 + x_{n-1}x_nx_1x_2 + x_nx_1x_2x_3 = 0$, then,
   (a) n is even
   (b) n is odd
   (c) n is an odd multiple of 3
   (d) n is prime
2. A shipping clerk has five boxes of different but unknown weights each weighing less than 100 kgs. The clerk weighs the boxes in pairs. The weights obtained are 110, 112, 113, 114, 115, 116, 117, 118, 120 and 121 kgs. What is the weight, in kgs, of the heaviest box?
   (a) 60
   (b) 62
   (c) 64
   (d) Cannot be determined
3. Of 128 boxes of oranges, each box contains at least 120 and at most 144 oranges. The number of boxes containing the same number of oranges is at least
   (a) 5
   (b) 103
   (c) 6
   (d) Cannot be determined
4. Shyam visited Ram on vacation. In the mornings, they both would go for yoga. In the evenings they Would play tennis. To have more fun, they indulge only in one actively per day, i.e., either they went for yoga or played tennis each day. There were days when they were lazy and stayed home all day long.

   There were 24 mornings when they did nothing, 14 evenings when they stayed at home, and a total of 22 days when they did yoga or played tennis. For how many days Shyam stayed with Ram?

(a) 32 (b) 24

(c) 30 (d) None of these

5. Three Englishmen and three Frenchmen work for the same company. Each of them knows a secret not known to others. They need to exchange these secrets over person-to-person phone calls so that eventually each person knows all six secrets. None of the Frenchmen knows English, and only one Englishmen knows French. What is the minimum number of phone calls needed for the above purpose?

(a) 5 (b) 10

(c) 9 (d) 15

6. A rectangular floor is fully covered with square tiles of identical size. The tiles on the edges are white and the tiles in the interior are red. The number white tiles is the same as the number of red tiles. A possible value of the number of tiles along one edge of the floor is

(a) 10 (b) 12

(c) 14 (d) 16

7. In a tournament, there are n team T_1, T_2, ..., T_n, with $n > 5$. Each team consists of k players, $k > 3$. The following pairs of teams have one player in common:

T_1 & T_2, T_2 & T_3,....., T_{n-1} & T_n, and T_n & T_1.No other pair of teams has any player in common. How many players are participating in the tournament, considering all the n teams together?

(a) k (n – 1) (b) n (k – 2)

(c) k (n – 2) (d) n (k – 1)

8. Suppose you have a currency, named Miso, in three denominaitons: 1 Miso, 10 Misos and 50 Misos. In how many ways can you pay a bill of 107 Misos?

(a) 16 (b) 18

(c) 15 (d) 19

9. How many pairs of positive integers m, n satisfy $\frac{1}{m}+\frac{4}{n}=\frac{1}{12}$ where n is an odd integer less than 60?

(a) 4 (b) 7

(c) 5 (d) 3

10. Suppose, the seed of any positive integer n is defined as follows:

Seed(n) = n, if $n < 10$ = seed($s(n)$), otherwise, where $s(n)$ indicates the sum of digits of n. For example, seed(7) = 7, seed(248) = seed(2 + 4 + 8) = seed(14) = seed(1 + 4) = seed(5) = 5 etc. How many positive integers n, such that $n < 500$, will have seed(n) = 9?

(a) 39 (b) 72

(c) 81 (d) 55

11. Given five points A = (7, 4), B = (–10, 0), C = (–10, 3), D = (0, 10) and E = (7, 7). Every second all the points move by halving their abscissas and by doubling their ordinates. This process continues for 500 years. After 500 years, which two points are closest?

(a) A and B (b) B and C

(c) A and E (d) A and C

12. Little Pika who is five and half years old has just learnt addition. However, he does not know how to carry. For example, he can add 14 and 5, but he does not know how to add 14 and 7. How many pairs of consecutive integers between 1000 and 2000 (both 1000 and 2000 included) can Little Pika add?

(a) 150 (b) 155

(c) 156 (d) 258

13. A potter asked his two sons to sell some pots in the market. The amount received for each pot was same as the number of pots sold. The two brothers spent the entire amount on some packets of potato chips and one packet of banana chips. One brother had the packet of banana

chips along with some packets of potato chips, while the other brother just had potato chips. Each packet of potato chips costs ₹ 10/- and the packet of banana chips costs less than ₹ 10/- . The packets of chips were divided between the two brothers so each brother received equal number of packets. How much money should one brother give to the other to make the division financially equitable?

(a) 1 (b) 2
(c) 4 (d) 5

14. m, n are integers satisfying $3m + 2 = 5n + 3$ and $30 < 5n + 2 < 40$. Find the value of $m + n$.

(a) 17 (b) 21
(c) 13 (d) None of these

15. Given that the integers solutions of the inequality $0 \leq ax + 5 \leq 4$ for x are 1, 2, 3, 4. Find the range of the constant a.

(a) $-\frac{5}{4} \leq a \leq -1$ (b) $-\frac{5}{4} \leq a \leq 1$
(c) $-\frac{5}{4} \leq a \leq 0$ (d) None of these

16. Suppose x, y and z are positive integers such that $x > y > z > 663$ and x, y and z satisfy the following:

$$x + y + z = 1998$$
$$2x + 3y + 4z = 5992.$$

Find the value of $y - x$

(a) 1 (b) 2
(c) 3 (d) None of these

17. Let x be a 3-digit number such that the sum of the digits equals 21. If the digits of x are reversed, the number thus formed exceeds x by 495. What is product of digits of x?

(a) 144 (b) 288
(c) 240 (d) None of these

18. Consider a four digit number such that when its units digit and hundreds digit are interchanged, and so does the tens digit and thousands digit, then the value of the number increases 5940 and remainder is 8 when it is divided by 9. If K is the minimum odd number satisfying these conditions then find the sum of digits of K. Given that unit digit and hundred's digit is same

(a) 18 (b) 21
(c) 24 (d) None of these

19. The number of positive integer solutions (x, y, z) for the system of simultaneous equations $\begin{cases} xy + yz = 63, \\ xz + yz = 23 \end{cases}$ is

(a) 1; (b) 2;
(c) 3; (d) 4;

20. Find the number of positive integer solutions of the equation $\frac{x}{3} + \frac{14}{y} = 3$.

(a) 2 (b) 3
(c) 5 (d) None of these

21. Find the number of positive integer solutions of the equation $\frac{2}{x} - \frac{3}{y} = \frac{1}{4}$.

(a) 8 (b) 5
(c) 7 (d) None of these

22. The number of positive integer solutions (x, y, z) for the system of simultaneous equations $\begin{cases} xy + xz = 255, \\ xy + yz = 31 \end{cases}$ is

(a) 3 (b) 2
(c) 1 (d) 0

23. Given real numbers and b, if $a = \frac{x+3}{4}$, $b = \frac{2x+1}{3}$, $b < \frac{7}{3} < 2a$, find the range of the value of x.

(a) $\frac{5}{3} < x < 3$ (b) $\frac{5}{3} < x < 5$

(c) $1\frac{1}{3} < x < 3$ (d) None of these

24. Given that the solution set for x of the inequality $\frac{2m + x}{3} \leq \frac{4mx - 1}{2}$ is $x \geq \frac{3}{4}$, find the value of the parameter m.

(a) $\frac{7}{10}$ (b) $\frac{9}{10}$

(c) $\frac{9}{11}$ (d) None of these

25. Given that a, b are two integers such that the positive integer solutions of the system of inequalities $9x - a \geq 0$, and $8x - b < 0$ are 1, 2, 3. Find the number of the ordered pairs (a, b).

(a) 72 (b) 75

(c) 81 (d) None of these

26. Given $0 \leq a - b \leq 1$, $1 \leq a + b \leq 4$. Find the value of $8a + 2016b$ when the value of $a - 2b$ is maximum.

(a) 6 (b) 8

(c) 10 (d) None of these

27. Find all the positive integer valued solutions (x, y, z) of the systems of inequalities

$$\begin{cases} 3x + 2y - z = 4, \\ 2x - y + 2z = 6, \\ x + y + z < 7. \end{cases}$$

Determine how many such triplets exist?

(a) 0 (b) 1

(c) 3 (d) None of these

28. Given that x 1-cent coins, y 2-cent, and z 5- cent coins have a total value of 10 dollars. How many triplets (x, y, z) exist?

(a) 51 (b) 37

(c) 541 (d) None of these

29. N is the least natural number, such that its value increases by 5 times when its last digit is moved to the first place then find the sum of digits of N

(a) 27 (b) 18

(c) 36 (d) None of these

30. In an examination, two types of questions are asked: one mark questions and two marks questions. For each wrong answer, of one mark question, the deduction is 1/4 of a mark and for each wrong answer, of two marks question, the deduction is 1/3 of a mark. Moreover, 1/2 of a mark is deducted for any unanswered question. The question paper has 10 one mark questions and 10 two marks questions. In the examination, students got all possible marks between 25 and 30 and every student had different marks. What would be the rank of a student, who scores a total of 27.5 marks?

(a) 5 (b) 6

(c) 7 (d) 8

(Passage Type)

Passage-I

Directions for questions 1 to 3:: These questions are based on the situation given below:

Ten coins are distributed among four people P, Q, R, S such that one of them gets one coin, another gets two coins, the third gets three coins and the fourth gets four coins. It is known that Q gets more coins than P, and S gets fewer coins than R.

31. If the number of coins distributed to Q is twice the number distributed to P then which one of the following is necessarily true?

(a) R gets an even number of coins.

(b) R gets an odd number of coins.

(c) S gets an even number of coins.

(d) S gets an odd number of coins.

32. If R gets at least two more coins than S, then which one of the following is necessarily true?

(a) Q gets at least two more coins than S.

(b) Q gets more coins than P.

(c) P gets more coins than S.

(d) P and Q together get at least five coins.

33. If Q gets fewer coins than R, then which one of the following is not necessarily true?

(a) P and Q together get at least four coins.

(b) Q and S together get at least four coins.

(c) R and S together get at least five coins.

(d) P and R together get at least five coins.

Passage-II

Directions for questions 4 to 6: These questions are based on the situation given below:

A young girl Roopa leaves home with x flowers, goes to the bank of a near by river. On the bank of the river, there are four places of worship standing in a row. She dips all the x flowers into the river. The number of flowers doubles. Then she enters the first place of worship, offers y flowers to the deity. She dips the remaining flowers into the river, and again the number of flowers doubles. She goes to the second place of worship, offers y flowers to the deity. She dips the remaining flowers into the river and again the number of flowers doubles. She goes to the third place of worship and offers y flowers to the deity. She dips the remaining flowers into the river again the number of flowers doubles. She goes to the fourth place of worship, offers y flowers to the deity. Now she is left with no flowers in hand.

34. If Roopa leaves home with thirty flowers, the number of flowers that she offers to each deity is

(a) 30 (b) 31

(c) 32 (d) 33

35. The minimum number of flowers that could be offered to each deity is

(a) 0

(b) 15

(c) 16

(d) Cannot be determined

36. The minimum number of flowers with which Roopa leaves home is

(a) 16

(b) 15

(c) 0

(d) Cannot be determined

Passage-III

Directions for Questions 7 & 8:

Let S be the set of all pairs (i, j) where $\mathbf{1 \le i < j \le n}$, **and** $\mathbf{n \ge 4}$. Any two distinct members of S are called "friends" if they have one constituent of the pairs in common and "enemies" otherwise. For example, if n = 4, then S = { (1, 2), (1, 3), (1, 4), (2, 3), (2, 4), (3, 4)}.Here, (1, 2) and (1, 3) are friends, (1, 2) and (2, 3) are also friends, but (1, 4) and (2, 3) are enemies.

37. For general n, how many enemies will each member of S have?

(a) $\frac{1}{2}(n^2 - 3n - 2)$ (b) $2n - 7$

(c) $1/2(n^2 - 5n + 6)$ (d) $\frac{1}{2}(n^2 - 7n + 14)$

38. For general n, consider any two members of S that are friends. How many other members of S will be common friends of both these members?

(a) $2n - 6$ (b) $1/2n(n - 3)$

(c) $n - 2$ (d) $\frac{1}{2}(n^2 - 7n + 16)$

Passage-IV

Directions for questions 9 to 10:

Let $A_1, A_2, \ldots$ An be the n points on the straight-line $y = px + q$. The coordinates of A_k is (x_k, y_k), where k = 1, 2,... n such that $x_1, x_2, \ldots xn$ are in arithmetic progression. The coordinates of A_2 is (2, -2) and A_{24} is (68, 31).

39. The y-ordinates of A_8 is

(a) 13 (b) 10

(c) 7 (d) 5.5

40. The number of point(s) satisfying the above mentioned characteristics and not in the first quadrant is/are

(a) 1 (b) 2
(c) 3 (d) 7

Concept Eliminator (CE)

1. Ashish is given Rs.158 in one rupee denomination. He has been asked to allocate them into a number of bags such that any amount required between ₹1 and ₹158 can be given by handing out a certain number of bags without opening them. What is the minimum number of bags required?

(a) 11 (b) 12
(c) 13 (d) None of these

2. On 1st March, Timon arrived in a new city and was looking for a place to stay. He met a landlady who offered to rent her apartment at a reasonable price hut wanted him to pay the rent on a daily basis. Timon had a silver bar of 31 inches, and an inch of the silver bar was exactly equal to a day's rent. He agreed to pay an inch of the silver bar towards the daily rent. Timon warned to make the minimum number of pieces of the silver bar but did not want to pay any advance rent. How many pieces did he make ?

(a) 5 (b) 8
(c) 16 (d) 20

Direction for question number 3 & 4

From a group 545 contenders, a party has to select a leader. Even after holding a series of meetings, the politicians and the general body failed to reach a consensus. It was then proposed that all 545 contenders be given a number from 1 to 545. Then they will be asked to stand on a podium in a circular arrangement, and counting would start from the contender numbered 1. The counting would be done in a clockwise fashion. The rule is that every alternate contender would be asked to step down as the counting continued, with the circle getting smaller and smaller, till only one person remains standing. Therefore the first person to be eliminated would be the contender numbered 2.

3. Which position should a contender choose if he has to be the leader?

(a) 3 (b) 67
(c) 195 (d) 323

4. One of the contending politicians, Mr. Chanaya, was quite proficient in calculations and could correctly figure out the exact position. He was the last person remaining in the circle. Sensing foul play the politicians decided to repeat the game. However, this time, instead of removing every alternate person, they agreed on removing every 300th person from the circle. All other rules were kept intact. Mr. Chanaya did some quick calculations and found that for a group of 542 people the right position to become a leader would he 437. What is the right position for the whole group of 545 as per the modified rule?

(a) 3 (b) 194
(c) 249 (d) 437

5. There are two piles of pebbles, pile (*A*) and pile (*B*). When 100 pebbles are moved from (*A*) to (*B*), then the number of pebbles in (*B*) is double of that in (*A*). However, if some are moved from (*B*) to (*A*), then the number of pebbles in (*A*) is five times more than that in (*B*). What is the minimum possible number of pebbles in (*A*), and the number of pebbles in (*B*) in that case.

6. Given that n is a positive integer, and the equation $2x + 2y + z = n$ has a total of 28 positive integer solutions for (x, y, z). Then the value of n is

(a) 14 or 15 (b) 15 or 16
(c) 16 or 17 (d) 17 or 18

7. Consider triples (x, y, z) of three non- negative integers satisfying the system of equations

$$5x + 7y + 5z = 37$$

$$6x - y - 10z = 3.$$

Find the unit digit of sum of all the possible values of x

(a) 4 (b) 3

(c) 8 (d) None of these

8. Chocolates are available in three sizes big one \$ 5, medium one for \$3 and each small one for \$ 1/3. Someone has \$100 to buy 100 chocolates, how many Big, medium and small chocolates can a man purchase out of a total cost of \$100?

9. Given that a pile of 100 small weights have a total weight of 500 g, and the weight of a small weight is 1g, 10g, or 50g. Find the number of each kind of weights in the pile.

10. In a book with page numbers from 1 to 100, some pages are torn off. The sum of the numbers on the remaining pages is 4949. How many pages are torn off?

11. Positive integers are written on all the faces of a cube, one on each. At each corner (vertex) of the cube, the product of the numbers on the faces that meet at the corner is written. The sum of the numbers written at all the corners is 2076. If S denotes the sum of the numbers on all the faces, find the sum of all the possible values of S.

[Based on RMO 2004]

12. Find the smallest whole number such that its first digit is 4, and the value of the number obtained by moving this 4 to the last place is $\frac{1}{4}$ of the original value.

13. If a, b are positive integers such that 11 divides $a + 13b$ and 13 divides $a + 11b$. Then find the minimum value of a+b

[Based on RMO 2006]

14. A positive integer is called a " a good number "if it is equal to four times of the sum of its digits. Find the sum of all good numbers.

15. Find the smallest natural number n whose last digit is 6, and if the last digit is erased and placed in front of the remaining digits, the resulting number is four times as large as the original number. **[Based on IMO 1962]**

Solutions

Concept Applicator (CA)-1

1. If we eliminate 'y' then multiply eq(i) by 5 and eq (ii) by 4 and subtract it, we get $x = 3$ and substitute this value of x in 2nd equation, we get $y = 4$.

2. Given equations are

$3x-2y = 11$ ...(i)

$2x-3y = 4$ ... (ii)

Multiply equation (i) by 3 and equation (ii) by 2

$9x-6y = 33$ ...(iii)

$4x - 6y = 8$ ...(iv)

Now subtract equation (iv) from equation (iii) we get $5x = 25$ or $x= 5$ substitute this value of x in equation (i) and we get $y = 2$

3. Here ratio of co-efficients of 'x' is 2/6 =1/3 and ratio of co-efficients of 'y' is 3/9 =1/3 ratio of constant term is 7/21 =1/3, since these ratios are equal so in this case we have infinite roots.

4. Here co-efficients of x and y are in the same ratio so it has no solution.

5. Here in this case $\frac{6}{39} = \frac{8}{52}$, so in this case we have no roots.

6. Multiply equation (i) by 2 and add in equation (ii) we get $\frac{100}{3x+4y} = 20$

Or $x + 3y = 5$ ...(iii)

Multiply equation (i) by 12 and then subtract equation (ii), we get $\frac{200}{3x + 4y} = 20$

Or $3x + 4y = 10$ ...(iv)

On solving equations (iii) and (iv) we get $x = 2$, $y = 1$

7. Given equations are

$3x + 2y + 4z = 16$ ... (i)

$5x + 3y + z = 12$ ... (ii)

$2x + y + 3z = 11$ ...(iii)

From equations (i) and (ii) eliminating z, we get $17x + 10y = 32$ ... (iv)

From equations (ii) and (iii) eliminating z, we get $13x + 8y = 25$ ...(v)

Now solving equations (iv) and (v) we get $x = 1$ and $y = 1.5$

Put this value of x and y in equation (i), we get $z = 2.5$

8. Here given equations are

$6x + 7y + 8z = 10$ ...(i)

$27x + 27y + 36z = 18$ ....(ii)

We can find the value of y only when co-efficient of x and z should be in the same ratio. Since $\frac{6}{27} = \frac{8}{36} = \frac{2}{9}$. So multiply equation (i) by 9 and equation (ii) by 2 and then subtract, we get $y = 6$

9. Given equations are if $3x + 4y + 6z = 6$

$12x + 14y + 21z = 24$

Consider ratio of co-efficient of $x = \frac{3}{12} = \frac{1}{4}$

ratio of co-efficient of $y = \frac{4}{14} = \frac{2}{7}$

ratio of co-efficient of $z = \frac{6}{21} = \frac{2}{7}$

Since ratio of co efficient of y and z are same so we can eliminate y and z together and can find the value of x.

Multiply equation (i) with 7 and (ii) with 2 and subtract, we get $x = 2$

10. Given equations are $3x + 2y + 4z = 16$ .... (i)

$5x + 3y + 7z = 27$ ... (ii)

Multiply equation (i) by P and equation (ii) by Q, then comparing coefficient of x and y, we get

$3P + 5Q = 1$ ... (iii)

$2P + 3Q = 1$ ... (iv)

Solving equations (iii) and (iv), we get $P = 2$ and $Q = -1$

The value of P and Q also satisfy the condition of z so multiply equation (i) by 2 and equation (ii) by -1 and then add the two result, we get $x + y + z = 5$

Concept Applicator (CA)-2

1. Given in-equality is $3x - 5 < 2x + 1$
   or $3x - 2x < 5 + 1$ or $x < 6$
2. We can add two inequalities of the same type. So by adding above two inequalities we get
   ∴ The range of $x + y$ is given by, $2 < x + y < 38$
3. Given inequality $\frac{x}{2x+1} \geq \frac{1}{4}$

   $\frac{x}{2x+1} - \frac{1}{4} \geq 0$

   $\frac{4x-(2x+1)}{4(2x+1)} \geq 0$

   $\frac{2x-1}{4(2x+1)} \geq 0$

   Make $2x - 1 = 0$ So, $x = \frac{1}{2}$

   $2x + 1 = 0$ So, $x = -\frac{1}{2}$

   We can include $x = \frac{1}{2}$ but can't include $x = -\frac{1}{2}$

   Thus, solution set, $x \in \left(-\infty, -\frac{1}{2}\right) U \left[\frac{1}{2}, \infty\right)$
4. Given in equality can be written as $\frac{x+2}{x+3} - 1 \geq 0$

   so, $\frac{(x+2)-(x+3)}{(x+3)} \geq 0$

   $-\frac{1}{(x+3)} \geq 0$

   Multiply with –1 on both sides $\frac{1}{(x+3)} \leq 0$

   or $x + 3 < 0$

   Thus, solution set is $(-\infty, -3)$
5. Since LCM of 3 and 5 is 15 so multiply both sides by 15.

   $3 \cdot 3(x - 2) \geq 5 \cdot 5(2 - x)$

   $\Rightarrow 9x - 18 \geq 50 - 25x$

   or $9x + 25x \geq 50 + 18$

   $\Rightarrow 34x \geq 68 \qquad \Rightarrow x \geq 2$
6. Given in equation can be written as

   $\frac{6x}{4x-1} - \frac{1}{2} < 0$

   $\frac{12x-4x+1}{2(4x-1)} < 0$

   $\frac{(8x+1)}{2(4x-1)} < 0$

   When $8x+1 = 0$ so $x = -\frac{1}{8}$

   And $4x - 1 = 0$ so $x = \frac{1}{4}$

   Thus, solution set is $x \in \left(-\frac{1}{8}, \frac{1}{4}\right)$
7. Given in-equations are

   $-17 \leq 3x + 10 \leq -2$

   ∴ $-27 \leq 3x \leq -12$

   ∴ $-9 \leq x \leq -4$ ...(i)

   $-22 \leq 5x + 13 \leq 3$

   ∴ $-35 \leq 5x \leq -10$

   ∴ $-7 \leq x \leq -2$ ...(ii)

   $-19 \leq 2x - 9 \leq -3$

   ∴ $-10 \leq 2x \leq 6$

   ∴ $-5 \leq x \leq 3$ ...(iii)

   The common range of values that satisfies all the equations is $[-5, -4]$
8. Let Ankur got 'x' marks in fifth subject.

   So, average marks = $\frac{90+70+75+65+x}{5}$

   Given $70 \leq \frac{90+70+75+65+x}{5} \leq 75$

   $350 \leq 300 + x \leq 375$

   $50 \leq x \leq 75$
9. Let one integer = x

   Another consecutive even integer = $x + 2$

   Given $x > 3$ ... (i)

   Sum = $x + x + 2 = 2x + 2$

   $2x + 2 < 11$ so $2x < 9$

   $x < \frac{9}{2}$

   From (i) $3 < x < 4.5$ and x is an integer

   So, $x = 4$

   $x + 2 = 6$

   Thus pair of even integers (4, 6)

10. Profit = Revenue – Cost

Revenue – cost > 0 for some profit

$(5x + 20) - (2x + 80) > 0$

$3x - 60 > 0$

$x > 20$

Thus, the manufacturer have to sale more than 20 items to earn some profit.

Concept Applicator (CA)-3

1. (b) Given 3x + 7y = 37

When $y = 1$, $x = 10$ again when $y = 4$, $x = 3$ (since $x, y \in z^+$) only two combinations possible

2. (c) Given 5x + 16y = 100

Clearly $5x$ and 100 are multiples of 5

So y must be a multiple of 5

3. (c) Let x, y be the number of pens and copies purchased respectively, then 10x + 11y = 130

Since both 10x and 130 are divisible by 10, 11y should also be divisible by 10

$\Rightarrow y = 10, 20, 30, \ldots$ but for $y \geq 20$, $11y > 130$

$\therefore y = 10 \Rightarrow x = 2$

4. (c) Let m, n be the two parts such that m = 9p and n = 17q, where p, q ∈ N.

Then, 9p + 17q = 284. Dividing the equation throughout by 9, we get Rem $\left(\frac{9p+17q}{9}\right)$ = Rem $\left(\frac{284}{9}\right)$ = Rem $\left(\frac{8q}{9}\right)$ = 5 From observation, q = 4 satisfies the above equation, and the other values of q can be obtained by adding 9 successively.

$\therefore$ q = 4, 13, 22, ... but for q ≥ 22, 17q ≥ 284

So, possible values of q are 4, 13 and the corresponding values of p are 24, 7.

Hence, the required number of ways of dividing the number 284 is 2.

5. (a) Let x, y and z be the number of 10 marks, 5 marks and 2 marks questions respectively.

Then, $x + y + z = 35$ ----- (i)

and 10x + 5y + 2z = 100 ------ (ii)

(ii) – 2 • (i), we get, 8x + 3y = 30

Since, 3y and 30 are multiples of 3, 8x must be a multiple of 3 $\Rightarrow x = 3, 6, 9, 12, \ldots\ldots$ But for $x \geq 6$, $8x \geq 30$. So $x = 3$; $y = 2$ and $z = 30$. Hence, the number of 10 marks questions in the paper is 3.

6. The given equation can be written as $a = 14\left(\frac{x}{15} - 10\right)$, when a is a positive integer, then

$15 \mid x$ and $x > 150$, therefore $x_{min} = 165$ and $a_{min} = 14$.

7. (d) Given equation can be written as $x = \frac{3k+12}{k} = 3 + \frac{12}{k}$

So k may be 1, 2, 3, 4, 6, 12,

8. The given equality yields $\frac{3A + 11B}{33} = \frac{17}{33}$

So 12A + 11B = 17.

From observation we get (2,1) is the only positive integral solution.

9. Since $13x = 7y$ has integer solution (0, 0), so the general solution is $x = 7t$, $y = 13t$, $t \in \mathbb{Z}$.

Then $x + y = 20t$, so $80 < x + y < 120 \bullet 4 < t < 6$, i.e. $t = 5$.

Thus, $x = 35$, $y = 65$.

Is the unique desired solution .

10. (c) The given equation can be written as $xy = 2004\,(x + y)$.

Then $xy - 2004x - 2004y + 2004^2 = 2004^2$, so

$(x - 2004)\,(y - 2004) = 2004^2$

Since $2004^2 = (2^2 \times 3 \times 167)^2 = 2^4 \times 3^2 \times 167^2$,

The number of positive divisors of 2004^2 is $(4 + 1)\,(2 + 1)\,(2 + 1) = 45$, so there are 45 such solutions.

Exercis-A

Concept Builder (CB)

1. (d) Let the number to be multiplied be k. Then from the given data we can say

   53k – 35k = 540.

   Solving we get k = 30.

   Hence required product = 30 × 53 = 1590.

2. (c) Let us take two cases-

   Case (1) : He rented the care for 5 hrs and hence pay for it 60 × 5 = ₹ 300, now we have to check that whether the distance given 30km is possible or not.

   Charge for 30 km = 30 × 12 = 360, hence this is not possible as he has to pay ₹ 360 in this case.

   Case (2) : He rented car for more than 5 hrs, then the time = 300/50 = 6hrs

   Amount charged for distance = 7.50 × 30 = 225, hence this is a possible case and he rented the car for 6 hrs.

3. (d) According to merchant, (where a & b be the unequal no of golds)

   $\therefore a + b$ = total no of gold coins

   $48(a - b) = a^2 - b^2 \Rightarrow (a + b) = 48$.

4. (c) Let x be the length of the middle sized piece.

   According to question, $3x + x + 3x - 23 = 40$, $x = 9$,

   $\therefore$ length of shorter piece is 4

5. (b) Given equation is $2x + y = 40$, or $y = 40 - 2x$

   Given that both x and y are positive integers, minimum value of $x = 1$, then $y = 38$

   Next value is $x = 2, y = 36$

   $x = 3, y = 34$

   and so on … when $x = 13$, then $y = 14$

   When $x = 14$ then $y = 12$, but it is given that $x \leq y$ hence x can assume values from $x = 1$ to $x = 13$ so total 13 solutions of this equation.

6. (b) Linear equation relationship $y = ax + b$ [b = constant, x = over time, y = no. of accident]

   As per the given information => $8 = 1000x + b$ and $5 => 400x + b$

   We get $a = 1/200$ and $b = 3$

   Now, no over time means $x = 0$, then $y = b = 3$

7. (d) Let the portfolio 1, 2, 3 be x, y, z respectively.

   Common stocks- $12 = 2x + 4y + 2z$

   $\Rightarrow x + 2y + z = 6$

   Municipal stocks-$x + 2y + 2z = 6$ and preferred stock $= 3y + 3z = 6 => y + z = 2$

   We get $x = 2, y = 2, z = 0$.

   That means, zero portfolio III is offered.

8. (a) By solving equations (ii) and (iii) we will get $x + 5y = 11$ and solving eqs. (i) and (ii) we will get $5x + 7y = 19$

   Solving these two we will get ....

   $x = 1, y = 2, z = 3$.

9. (b) Let us assume that the software fails x, y and z times in a single stage, in two stages and in all stages respectively.

   Hence, $y + 3z = 6 + 7 + 4 = 17$

   Since $z = 4$, hence, $y = 5$,

   hence $x + 2y + 3z = 15 + 12 + 8$ or $x = 13$

10. (c) Total taxi fare : first 2 km

    = ₹ 20 + 8 × 5 + 4 × 8 = 20 + 40 + 32 = 92

    Distance traveled by Bullock Cart

    = (102 – 92) / 2 = 10 / 2 = 5 km

    Total distance = 14 + 5 = 19 km.

Concept Cracker (CC)

1. (d) Here, let eq (i) is multiplied by (p) and eq. (ii) is multiplied by (q) and then added to get the final result.

   $2px + 3py + 4pz = 10p$ …(i)

   $3qx + 5qy + 7qz = 5q$

   Adding there two equations and comparing with $3x + 4y + 6z$, we will get

   $2p + 3q = 3$ …(i)

   $3p + 5q = 4$ …(ii)

   $4p + 6q = 7$ …(iii)

Solving equations (i and (ii), we will get p = 3 and q = –1but when we put the values of p and q in the eq. (iii) we found that it dosenot satisfy the equation (iii) so we can not find the value of 3x + 4y + 6z.

2. (c) Here, let eq (i) is multiplied by (p) and eq. (ii) is multiplied by (q) and then added to get the final result.

$$px + 2py + 3pz = 5p$$
$$3qx + 5qy + kqz = 10q$$

Then,

$$p + 3q = 3 \quad ...(i)$$
$$2p + 5q = 4 \quad ...(ii)$$
$$3p + kq = 3 \quad ...(iii)$$

Solving equations (i) and (ii), we get $p = -3$ and $q = 2$.

Putting there values of p and q in equation (iii) we get $-9 + 2k = 3$ or $k = 6$.

3. (b) Value of y can be calculated only if x and z are of same ratio.

i.e $2/3 = 4/k$

$k = 6$

4. (c) Multiply 1st equation by p and 2nd by q and add then we will get

$$(9p + 8 q)x + (12p + 6q)y + (24p + 20q)z = 20 p + 30q$$

From options if p = 1 and q = –1, then we can calculate x + 6y + 4z

5. (a) Given equation is $4x - 17y = 1$ and $x \leq 1000$.

We can write down the equation as 17y + 1 = 4x maximum value of x = 1000, then maximum value of y is given by 17y + 1 = 4000 or y = 3999/17 = 235.23 hence maximum integral value of y = 235.

But from the given equation x = (17y + 1)/4 = 4y + (y + 1)/4, to have integral value of x, y must be 3, 7, 11…235. That is in the form of AP and we know as the 1st term as 3, last term as 235, common difference as 4, then the number of terms I given by, 3 + (n–1)4 = 235, or n = 59

6. (c) The important point to note is that all the three boxes are mislabeled.

If we select the box labeled Red, we know that it cannot be the box consisting red balls. When we pick *a* ball from this and it is red, we can then conclude that the box consists of Red & White balls (because we know this cannot be the box of red balls). However, if we get *a* white ball, then we cannot conclude whether the box consists of only White balls or both Red & White balls. Thus picking *a* ball from the box marked Red or from the box marked White may not help us conclude what is there in all the boxes.

So, we have to pick *a* ball from the box labeled Red & White. We know this Cannot consist of both red and white balls. When we pick a ball from this box, it can be a red ball or *a* white ball. Let us consider both the cases.

When the ball picked is a white ball

That means that this box is the WHITE box (because it cannot be the box with RED&WHITE balls). Then, the box labeled RED cannot be RED and it should be WHITE&RED and the third box marked WHITE should be the box which has both RED & WHITE balls. Thus we can identify all the three boxes.

When the ball picked is a red ball

That means that this box is the RED box (because it cannot be the box with RED & WHITE balls). Then, the box labeled WHITE cannot be WHITE and it should be labeled WHITE & RED and the third box marked RED should be the box which has the WHITE balls. Thus we can identify all the three boxes.

Thus, by picking *a* ball from the box labeled RED & WHITE, we can identify ALL the three boxes.

7. (b) Here in this type of problem since $x > 2$ hence it is always + ve but for $y > -1$ it can be –ve or it can be + ve, and it is the weak link, so we will take some values of x and y and eliminate options one by one,

When $x = 4$, $y = -1/2$, options (a) (c) and (d) are eliminated, if we don't get the correct option by once then we have to put another value to eliminate other incorrect options.

8. (d) We will solve this question by elimination method.

Option (a) – let $x = 6$ and $y = -10$ then $x + 4y < 1$ hence it is False.

Option (b) Let $x = 6$ and $y = -10$ then $x < -4y$, hence it is False.

Option (c) Let $x = 6$ and $y = -10$ then $-4x > 5y$, hence it is False.

9. (d) This is an example that don't ever directly jump on solving the question by writing an equation, just pause for a while, think for ***a*** moment, you may get a better way out in this question since Sita takes 1/3 of the total mints kept there..... which means total number of mints in the bowl should be *a* multiple of 3.

Let x be the initial number of mints in the bowl.

Sita took 1/3 of the mints but returned 4, that means, the number of mints let in the bowl are :

$(x - 1/3x + 4) = \left(\frac{2x}{3} + 4\right)$

Then Fatima took 1/4 and returned 3. So the number of mints left are : $(3/4\ (x - 1/3x + 4)) + 3$

Then Eswari took half of the remaining and returned 2 so the number of mints left are: $1/2((3/4\ (x - 1/3x + 4)) + 3) + 2 = 17$,

Solving for x, we will have $x = 48$.

10. (c) First we will find out the total number of pages in the book. Thus we need to find the sum of the first few natural numbers, such that we reach to *a* number just short of 1000. Let total number of pages n then $n(n + 1)/2 < 1000$. We will get $n = 44$, then sum of 1st 44 numbers $= 44 \times 45/2 = 22 \times 45 = 990$, but summation given is 1000 that means the no 10 was added twice.

11. (a) Let price of one burger, one shake and one fries is B, S and F respectively,

The from the given information

$3B + 7S + 1F = 120$ ...(i)

and $4B + 10\,S + 1F = 164.5$ ...(ii)

from (ii) – (i), we get $1B + 3S = 44.5$

Now from (i), we get $3B + 7S + 1F = 120$

$\Rightarrow\quad 1B + 1S + 1F + 2B + 6S = 120$

$\Rightarrow\quad 1B + 1S + 1F + 2(1B + 3S) = 120$

$\Rightarrow\quad 1B + 1S + 1F + 2 \times 44.5 = 120$

$\Rightarrow\quad 1B + 1S + 1\,F = 120 - 89$

$\Rightarrow\quad 1B + 1S + 1F =$ ₹ 31.

12. (a) The average is 35 7/17, = 602/17 we know that average of 1st n consecutive number is $(n + 1)/2$ but average is 35 7/17 hence value of n should be close to $2 \times 35 = 70$. If there are n numbers then after erasing we will have $n - 1$ numbers then summation of the $n-1$ numbers $= (n-1)\ x$ Avg

$= 602(n-1)/17$, hence $n-1$ should be divisible by 17, but n is close to 70 hence $n = 69$, so that $n-1 = 68$ is divisible by 17. Hence $n = 69$.

Sum of 1st 69 natural numbers

$= 69 \times 70/2 = 2415$

Sum of 68 number after erasing one number $= 602/17 \times 68 = 2408$

Hence the number that was erased

$= 2415 - 2408 = 7$

13. (b) Given, there are coins of denomination of ₹1, ₹ 2 and ₹5 and total number of coins 300. Total amount = ₹960. Let the number of coins of ₹1, ₹2, ₹5 be X, Y, and Z respectively.

Then $\quad X + Y + Z = 300$ ...(1)

$X + 2Y + 5Z = 960$ ...(2)

$Y + 2X + 5Z = 920$ ...(3)

Solving equations (1), (2) and (3), we get $X = 60$, $Y = 100$, $Z = 140$.

14. (b) Since for every D we need to buy 2 C, hence total cost = 220 for 3 items.

Since for every A we need to buy one B, hence total cost = 180 for 2 items

Since for every 2 E we need to buy one C,

hence total cost = 160 for 3 items

To get maximum points we have to maximize the expenditure, that is possible when we will buy 3 sets of (1D + 2B) = 3 × 3 = 9 items) along with 1 set of (1A + 1C) = 2 × 1 = 2 items) and 1 set of (2E + C) = 3 × 1 = 3, hence total items = 9 + 2 + 3 = 14.

15. (d) Let she added up to n numbers then sum of all the numbers = n(n + 1)/2, but it is given that she missed one number hence actual sum must be more than the sum that she got, 575.

Hence n(n + 1)/2 > 575, or n(n + 1) > 1150 the minimum value of n that satisfy this is 34,

Hence when n = 34 then sum is 34 × 35/2 = 595

The difference is 595–575 = 20 so she missed the number 20.

16. (b) Total no of loaves is 8. For equal distribution they have 8/3 part each.

Traveler one has to give traveler = 5–8/3 loaves = 7/3

Traveler 2 has to give traveler = 3–8/3 loaves = 1/3

∴ money has to be paid in the ratio 7:1.

17. (b) From the given information- M for Mayank, Mi for Mirza, L for little and J for Jaspal.

$M + Mi + L + J = 60$ ...(i)

$M = ½ (Mi + L + J)$, $Mi = 1/3 (M + L + J)$, $L = ¼ (M + Mi + J)$

By solving these, we get

$3M = 4Mi = 5L, 13/20\ M = J$

From (i) we get, $M + 3/4\ M + 3/5M + 13/20M = 60$

∴ $M = 20, J = 13$

18. (b) Let x be the number of diamond stolen by the thief.

∴ According to question, $x = (½\, x + 2) + \{ ½ (½\, x - 2) + 2\} + \{ ½ (x/4 - 3) + 2\} + 1$

∴ (solving the equation we get), $x = 36$.

19. (c) Let the no. of correct answers, wrong attempts and unanswered question be C, W and U respectively.

So, according to the question:

$C + W + U = 50$ ...(1)

$C - W/3 - U/6 = 32$

$\Rightarrow 6C - 2W - U = 192$ ...(2)

Adding 1 and 2 we get

$7C - W = 242$

$\Rightarrow W = 7C - 242$

So we see, the min. no. of wrong attempts by a student is 3 for C = 35;

20. (d) As per the given conditions:

Family	Possible no. of Adults	Possible no. of Children
1	0, 1, 2	3, 4, 5, ...
II	0, 1, 2	3, 4, 5, ...
III	0, 1, 2	3, 4, 5, ...

Also given:

Adults (A) > Boys (B) > Girls (G) > Families (F)

Going by the options, when we take the min. no. of families as 2, then the max. possible no. ofadults can be 4.

Since in any family there should be at least 2 boys and *a* girl, which means 4 boys in 2 families, and violates our condition of Adults > Boys.

When we take min. no of families as 3, then all the conditions are satisfied.

21. (c) Let 4 consecutive odd numbers be 2n–3, 2n–1, 2n + 1, 2n + 3 and given that some sum of all these 4 integers when divided by 10 it becomes a perfect square.

Hence 8n/10 is a perfect square, n must be *a* multiple of 5

If n = 5 then 8n/10 = 4 *a* perfect square. And the numbers are 7, 9, 11, 13

Next value of n = 10, and 15 that will not give us 8n/10 as perfect square.

n = 20 then 8n/10 = 16 a perfect square. And the numbers are 37, 39, 41, 43 and 41 is the given option hence 41 is one of the number.

22. (b) Let a two digit number 10a + b, when reversed it becomes 10b + a, and the difference is $(10b + a) – (10a + b) = 9(b-a) = 18$ (given)

Hence $b – a = 2$

So the values of a and b is (1, 3), (2, 4), (3, 5), (4, 6), (5, 7), (6, 8) and (7, 9)

Hence Other than 13, there are 6 more such numbers.

23. (a) Given that S = (2, 3, ….2n + 1), hence total numbers of element in S is 2n.

Odd elements are (3, 5, 7, ….2n + 1) total n elements

And even elements are (2, 4, ….2n) total n elements

Hence $X = [3 + 5 + 7 + ….(2n + 1)]/n$

And $Y = [2 + 4 + 6 + ….. + 2n]/2$

So $X – Y = [(3–2) + (5–4) + ….(2n + 1–2n)]/n = [1 + 1 + 1….n \text{ times}]/n = n/n = 1$

Hence $X–Y = 1$ (Point to note that it is independent from the value of n)

24. (d) I have seen people used to get confuse in understanding the concept of year, to get rid of this let present year is 2010.

Given that sum of ages of 8 members ten years ago (i.e in 2000) = 231.

After 3 year (i.e in 2003) their total age would be 231 + 8 × 3 = 255

But it is given that one member died at the age of 60 and one was born so total age on 2003 would be 255 – 60 = 195

After another 3 year (i.e in 2006) total age would be 195 + 8 × 3 = 219

It is given that one more member died at the age of 60 and one was born so total age on 2006 would be 219 – 60 = 159

The current total age (i.e in 2010 that means 4 years after 2006) = 159 + 8 × 4 = 191

And hence current average = 191/8 = 23.875 approx 24

25. (c) Let the amount that was written on the cheque is ₹ a and b Paisa and she got ₹ b and a Paisa,

From the given information we can write equation as

$$3 \times (100a + b) = 100b + a – 50$$

Or $299a = 97b – 50$

Or $b = (299a + 50)/97$

$= (291a + 8a + 50)/97$

$= 3a + (8a – 50)/97$

Now from the given options at $a = 18$ we will get integral value of b that value is $b = 56$.

26. (b) Let the initial quantity of rice is x kg.

It is given that the 1st customer buys half of the total quantity and half a kg more

So rice purchased by 1st customer

$= x/2 + 1/2 = (x + 1)/2$

Remaining quantity of rice in the store

$= x-(x + 1)/2 = (x–1)/2$

Now the second customer buys half of the remaining quantity and half a kg more.

So rice purchased by 2nd customer

$= (x – 1)/4 + 1/2 = (x + 1)/4$

Remaining quantity of rice in the shop

$= (x-1)/2 –(x + 1)/4 = (x-3)/4$

Now then the 3rd and last customer buys half of the remaining and half a kg more.

Rice purchased by 3rd customer $= (x – 3)/8 + 1/2 = (x + 1)/8$

Remaining quantity of rice $= (x-3)/4 – (x + 1)/8 = (x–7)/8$

But it is given that after last purchased there is no rice left in the store, hence $(x–7)/8 = 0$

Or $x = 7$kg

27. (c) After erasing the two numbers a and b and replacing it with $(a + b-1)$, we are actually reducing the sum of all the numbers on the board by 1, since this operation was performed 39 times hence total summation will be reduced by 39. After 39 operation we will have only one number left on the board as each operation reduces the number of numbers by 1, hence after 39 such operation 40 – 39 only one number will be left.

Initial sum of the numbers = 1 + 2 + 3.... + 40 = 40 × 41/2 = 820

So last number on the board = 820 – 39 = 781

28. (d) Clear morning = 9 and clear evening = 12

Maximum 9 days rain had in the evening which is followed by clear morning.

So, rainless day = 12 – 9 = 3.

29. (c) $2x + 3y + 4z = 16$ ...(i)

$4x + 4y + 5z = 26$ ...(ii)

$ax + by + cz = r$

$\Rightarrow \quad x + by + cz = 5$ ...(iii)

Infinite solution is possible, when any two or more lines lie on each other.

Subtracting equation (i) –(ii); $2x + y + z = 10$ and comparing with equation (iii)

$$2/1 = 1/b = 1/c = 10/5$$

we get, $b = 1/2$ and $c = 1/2$.

30. (d) Let the number of correct answer be x

Since total number of question = 30, hence number of incorrect answers = $30 - x$

$\therefore$ Marks of Ankur = $x - (30\text{-}x)/4 = 13.75$

$\Rightarrow \quad 4x - (30 - x) = 55$

$\Rightarrow \quad x = 17$

$\therefore$ Incorrect answers = 30 – 17 = 13

31. (b) Let a, b, c, d and e be the weight of five children in increasing order such that $a < b < c < d < e$. Adding all the weights, we get, $4 \times (a + b + c + d + e) = 408$

$\therefore (a + b + c + d + e) = 102$

The highest value must be given by d & e and the lowest must be given by $a + b$

Hence $a + b = 35$ and $d + e = 47$

$\therefore a + b + d + e = 82$ gives $c = 20$

Now the 2nd lowest is given by a & c

Hence $a + c = 36$ which gives $a = 16$ kg is the weight of lightest child.

32. (a) Let number of kaju barfi, Gulab jamun and Sandesh are x, y and z respectively. Then from the given information $x + y + z = 100$

And $10x + 3y + 0.5z = 100$ from these two equation we will get $19x + 5z = 100$

or $x = (100–5y)/19$ since x is a positive integer hence $y = 1$, then $x = 5$, and corresponding value of $z = 94$.

Hence he bought 1 Gulab jamun.

33. (d) Let cost of a Roti, a plate of Tadka and a cup of tea be ₹.x, y and z respectively. From the given conditions we can frame following equation-

$10x + 4y + z = 80,$ ... (i)

and $\quad 7x + 3y + z = 60$ ... (ii)

Now, by doing operation 3(ii) – 2 (i)

We will get $x + y + z = 20$

Hence $5x + 5y + 5z = 100$

34. (b) Let four different weights as a, b, c and d kgs then three taken at a time hence

We can assume $a + b + c = 13$ ...(i)

$a + b + d = 14$ ...(ii)

$b + c + d = 17$ ...(iii)

$a + c + d = 16$ ...(iv)

If we add all the equations then we will get

$3(a + b + c + d) = 60$

or $\quad a + b + c + d = 20$

now by subtracting the above equations we will get the values of a, b, c and d as 3,4,6 and 7 kgs respectively.

35. (d) Let he has initially \$x with him from that he pay \$10 for entry fee in the bar so rest amount with him is x–10, in casino he got his amount doubled so now after that amount left with him 2(x-10). Then after paying \$10 he came out from the casino and again pay \$10 for entry in the bar and got his amount double in the casino so now amount left with him 2[2(x–10)–20] and that amount must be equal to \$10 that he pay as exit charge in bar and left with no money so 2[2(x–10)–20] = 10 or x = 22.5 so he entered the bar with \$12.5.

Alternately : We can solve this type of question mentally without much calculation by approaching the question from back. Since after coming out from bar he left with no amount so he must had \$10 before leaving the bar, this \$10 he got after playing the casino where he doubled his amount so before playing the casino he must had \$5 only. This amount is left after paying \$10 for entry in the bar so before entering the bar he must had \$15 with him. Since while coming back from bar he has pay \$10 so before coming out from the bar he must had \$25 and that amount he got after playing the casino so before playing the casino he must had \$12.5, so when he entered the bar he had \$12.5.

36. (c) Let A had *a* rupee and *b* paisa, and B had *c* rupee and d paisa

Initial amount with A = 100a + *b* paisa and that with B = 100c + d paisa

After interchanging amount with A = 100b + *a* paisa and that with B = 100d + *c* paisa.

Increase in amount of A = 100b + *a*-100a-*b* = 99(*b*–*a*)

Decrease in amount of B = 100c + d–100d–*c* = 99(*c*–d) these two has to be equal hence

99(*b*–*a*) = 99(*c*–d), so *b* + d = *a* + *c*, the maximum value that *a*, *b*, *c*, and d can take is 99 as there is no paisa more than 99. So maximum value with A and B is 99.99 but in that case B is giving zero amount to A and that is not the case here so for maximum sum A had 98.99 and B had 99.98 and their sum is 198.97.

37. (b) Let initial amount with A and E as *a* and e respectively and amount with D after getting money from C is d then D gave half of his amount that is d/2 to E so total amount with E is e + d/2, amount left with D is d/2 and that with A after giving half of his amount to B is *a*/2 so from given condition *a*/2 + d/2 = e + d/2 hence *a*/e = 2/1

38. (c) Let son's present age of son = x that of mother = $x + y$ and that of father $x + y + z$. son will attain his mother's present age after y years then age of son = $x + y$, mother = $x + 2y$ and that of father = $x + 2y + z$. Son will attain father's present age after $y + z$ years, then age of son = $x + y + z$, mother = $x + 2y + z$ and that of father = $x + 2y + 2z$.

As per the given condition $(x + 2y + z)/(x + 2y + 2z) = 11/12$ from this we will get $x + 2y = 10z$. we need to find the ratio of

$(x + 2y)/(x + 2y + z) = 10z/11z = 10/11$

39. (b) Since 1st student and 10th student divided their total chocolate in 9 equal parts so the number of chocolates with them must be *a* multiple of 9. Hence let 1st student had initially 9y chocolates, and that with 10th student is 9z. And total number of chocolates with all the students from 2nd student to 9th student is x.

Then from 1st condition 10th student gave away his 9z chocolates to others. So all of them got z chocolates each and total number of chocolates with all the students from 2nd student to 9th student is $x + 8z = 40$ (since average for these 8 students is 5)

Or $\quad x/8 + z = 5 \quad$ ...(i)

Similarly from 2nd condition

$x + 8y = 32$

or $\quad x/8 + y = 4 \quad$ ...(ii)

Since y and z are integers hence x has to be a multiple of 8 so let $x = 8a$

Total number of chocolates = 61,

so $\quad x + 9y + 9z = 61$

or $\quad 8a + 9(x + z) = 61$

Let $x + z = b$ then $8a + 9b = 61$ and only integral solution of this equation is $a = 2$, and $b = 5$ hence $x = 16$, then $y = 2$ so total number of chocolates with 1st student initially was 18.

40. (c) Let daughter's age is d, that of son is s then that of mother is d + s while that of father is d + s + x. Now as per the 1st condition $[d + s + 2(d + s) + (d + s + x)]/4 = [s + (d + s)]/2$

$(3d + 3s + x)/4 = (d + 2s)/2$ so x = s-d hence father's age is d + s + s–d = 2s.

Now from the 2nd condition (s + 2s)/2 = 33 or s = 22

41. (a) Let number of steps behind A is x, in between them it is y and ahead of B it is z.

From 1st condition if both of them approaches towards each other then number of steps behind A is $x + 2$ and that in between them is $y - 4$ hence

$$x + 2 = y - 4$$

or $$x = y - 6 \quad \ldots(i)$$

Similarly from 2nd condition

$$z - 2 = y + 4$$

or $$z = y + 6 \quad \ldots(ii)$$

It is also given that total number of steps are 23 hence $x + y + z = 21$, since two steps are occupied by A and B. putting the value of x and z in above equation we will get $3y = 21$ or $y = 7$.

42. (b) Let on 1st day n^2 students are there and they form *a* square of n rows and n columns. Next day minimum number of students joined such that again they can form *a* square hence next day total number of students = $(n + 1)^2$, so number of students joined on 2nd day = $(n + 1)^2 - n^2 = 2n + 1$, similarly that on 3rd day is $2(n + 1) + 1 = 2n + 3$

It is given that $(2n + 1)/(2n + 3) = 25/27$ hence $n = 12$ and total number of students on 1st day $= 12^2 = 144$

43. (d) Let amount with A = 48a. With B = 48b, with C = 48c and with D = 0. Since A gave his half amount equally to B, C and D that means he gave away 24a and each B, C and D got 8a. So after this B had 48b + 8a, C had 48c + 8a and that with D is 8a. Now B gave away half of his amount that is 24b + 4a equally to C and D, so C and D received 12b + 2a each, so after this amount with A = 24a that with B = 4a + 24b that with C = 10a + 12b + 48c and with D = 10a + 12b . At the end C gave away half of his amount to D so C gave 5a + 6b + 24c to D. So at the end A = 24a, B = 4a + 24b, C = 5a + 6b + 24c and that with D = 15a + 18b + 24c. Since at end amount with A, B and C are equal so 24a = 4a + 24b or 5a = 6b similarly 7a = 12c. putting these value we will get amount with D at end is 44a and that with A, C and D is 24a so D has 83.33% more than A.

44. (c) Let initially A, B and C had a, b, and c gold coins with them, and A gave x coins to C then he gave 2x coins to B then after all these transactions A left with $a - 3x$ while B and C with $b + 2x$ and $c + x$. Since at end all of them had equal number of gold coins so $a - 3x = b + 2x$ or $x = (a - b)/5$ similarly $a - 3x = c + x$, hence $x = (a - c)/4$ equating these two values of x we will get

$$(a - b)/5 = (a - c)/4$$

or $$a + 4b = 5c \quad \ldots(i)$$

It is given that

$$a + b + c = 24 \quad \ldots(ii)$$

Since number of coins can't be in fraction so integral solution of above equations are $a = 14$, $b = 4$, $c = 6$ or $a = 11$ $b = 6$, and $c = 7$ and $a = 17$, $b = 2$and $c = 5$ so maximum possible value of $a = 17$.

Now here point is that how to solve when only two equations are given with three variables. In the above equations put $a = 5c - 4b$ in equation (ii) we will get $6c - 3b = 24$ or $2c = 8 + b$. here one more point to note that $a>b$ and $a>c$

Now if $b = 2$ then $c = 5$ and from eq(i) $a = 17$ and $a + b + c = 24$

If $b = 4$ then $c = 6$ and from eq(i) $a = 14$ and it satisfy $a + b + c = 24$

If $b = 6$ then $c = 7$ and from eq(i) $a = 11$ and it satisfy $a + b + c = 24$

If $b = 8$ then $c = 8$ and from Eq(i) $a = 8$ and it satisfy $a + b + c = 24$ but it does not satisfy $a>b$, $a>c$ so this solution set is ruled out. Hence after 8 all values of b will have the same result it will not satisfy $a > b$ and $a > c$.

45. (c) Let son's age is x, mother's is $x + y$ and that of father is $x + y + z$. After y years son will attain mother's present age then son's age $= x + y$, mother's $= x + 2y$ and that of father's

$= x + 2y + z$. and their average is 160/3 or their sum is 160 hence

$(x + y) + (x + 2y) + (x + 2y + z) = 160$

or $\quad 3x + 5y + z = 160 \quad$ ...(i)

Similarly after $(y + z)$ years son will attain father's present age and then average is 172/3 hence

$(x + y + z) + (x + 2y + z) + (x + 2y + 2z) = 172$

or $\quad 3x + 5y + 4z = 172 \quad$ ...(ii)

From equations (i) and (ii) we will get $3z = 12$ or $z = 4$.

46. (b) y years before mother was as old as son is now that is x and age of father $= x + z = x + 4$

it is given that $x/(x + 4) = 3/4$ hence $x = 12$.

47. (a) Let the contributions of A, B and C is a, b and c respectively.

So $a + b + c = 80$

From the 1st condition B's contribution is now 4/5b and that of C is 5/4c hence $a + b + c = a + 4b/5 + 5c/4$ or $4b = 5c$.

Similarly from the 2nd condition $a = 7c/4$, or $c = 4a/7$ put all these values in Eq(i) we will get $c = 20$.

48. (d) Let 1st group had a,b,c and d number of coins so $a + b + c + d = 48$, and 2nd group had e,f,g and h number of coins hence $e + f + g + h = 60$, so $a + b + c + d + e + f + g + h = 108$.

Now after interchange total number of coins with group A is $(a + b + c + d + e + f + g + h)/2 = 54$.

49. (c) Let total number of chairs $= x$ then number of students $= x + 4$.now as per the given condition $(x + 4)/2 = 2/3x$ or $x = 12$.

50. (b) Let cost of a pen is x and that of copy is y. According to the condition $40x > 15y$ and $16y > 40x$. Since $40x > 15y$ hence

$8x > 3y$ or $8x >= 3y + 1 \quad$ ...(i)

$16y > 40x$ or $2y > 5x$ or $5x <= 2y - 1 \quad$ ...(ii)

Multiply eq (i) with 5 and eq(ii) with 8 and then compare the result we will get

$16y - 8 >= 40x >= 15y + 5$

or $\quad 16y - 8 >= 15y + 5$

hence $y >= 13$ so minimum possible value of y is 13. Now $8x > 3y$ or $x > 3y/8$ or $x > 39/8$ so minimum possible value of y is 5 so minimum possible value of $x + y = 13 + 5 = 18$.

Exercis-B

Concept Deviator (CD)

(Single Correct MCQs)

1. (a) There are n terms on the left hand side of the given equation. Each of the terms will be either +1 or – 1 (because each term is formed by the product of +1s and/or –1s)

Since the sum of all the terms is zero, out of the n products that are being added, half should be +1 and the other half should be –1 so n must be an even number.

2. (b) Let the weights of the five boxes in decreasing order be a, b, c, d, e i.e (a>b>c>d>e) Then, we have

$a + b = 121 \quad$ ...(1)

$a + c = 120 \quad$ ...(2)

$d + e = 110 \quad$ ...(3)

$e + c = 112 \quad$ ...(4)

Out of 5 boxes 2 at a time was selected (a,b), (a,c), (a,d), (a,e), (b,c), (b,d), (b,e), (c,d), (c,e), (d,e), here we can make a note that each of a, b, c, d, and e comes four times hence summation of all 10 measurements must be given by

$4(a + b + c + d + e) = 1156$

OR

$a + b + c + d + e = 289 \quad$ ...(5)

Adding equations (1) and (3), we get

$a + b + d + e = 231 \quad$ ...(6)

From (5) and (6), we get $c = 58$

From equation (2), we get $a = 62$

3. (c) This is a typical Pigeon-hole principle problem. Here we have 128 boxes that can have 120, 121..... 144 oranges.

Now if we start filling the boxes with 120, 121, 122..... 144 oranges, we will be able to fill 25 boxes with distinct number of oranges. The remaining boxes will have a non-distinct number of oranges. Hence, we can fill 25 × 5 = 125 boxes with the number of oranges from 120 to 144 (both inclusive), i.e. 5 boxes having the same number of oranges. There are however 3 more boxes. If we start from 120 oranges again, we will have 6 boxes containing 120, 121 and 122 oranges each.

∴ the minimum number of boxes is 6.

Alternately

We can solve this question by Different number of oranges that we can fill = 144-120+1 (since both are inclusive) =25, let minimum number of boxes required =n then n> 128/25 $= 5\frac{3}{25}$, the lowest integral value of n is 6

4. (c) In one day there were two slots one for morning and one for evening in the morning they can go for yoga while in the evening for tenis.

They sat lazy for 24+14 = 38 slots, of which they did something (yoga or played) for 22days. ∴ 38-22 = 16 days total no of days be 16+14 = 30 days.

5. (c)

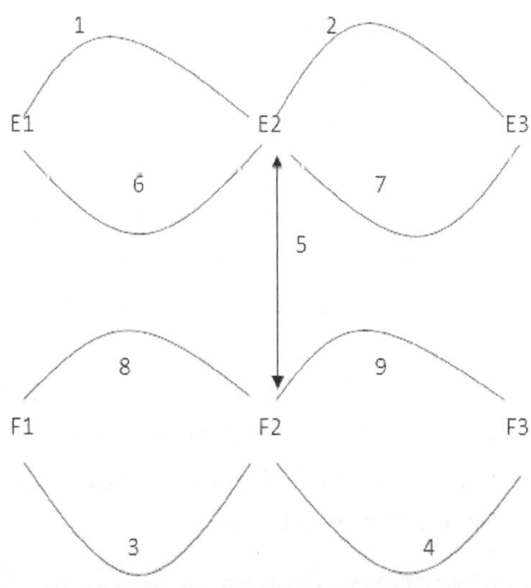

Let us assume that the three English men are E_1, E_2, E_3 and the Frenchmen are F_1, F_2, F_3 respectively.

Let us assume that E_2 is the only English man who knows French. Since our aim is to use the minimum number of calls, and there is only one possible communicator (i.e, E_2) between the two groups, all the information to be exchanged between the two groups can be done with a single call from F_2 to E_2. Before this E_2 should get the secrets from E_1 and E_3 and F_2 should get the secrets from F_1 and F_3.

Hence four calls are used till now, two in each group. (Shown in the diagram, from 1-4)

After this, E_2 calls up F_2 and both exchange the secrets of their groups.

Now E_2 and F_2 knows all the secrets o. So now E_2 uses two calls (to E_1 and E_3 calls 6 and 7 shown in the diagram) to convey the French secrets. Same is the case with F_2

Total minimum number of phone calls = (2 + 2) + 1 + (2 + 2) = 9 (as shown in the diagram)

6. (b) Let us assume that the room can accommodate 'a' tiles along the length and 'b' tiles along the breadth. Let the dimension of each tile be 1 × 1.

The total number of tiles in the room = $a \times b$

The total number of red tiles that is in the interior of the rectangle = $(a - 2)(b - 2)$

The number of white tiles along edge = $(ab) - (a - 2)(b - 2)$

Equating the number of red and white tiles we get

$$2(a - 2)(b - 2) = ab$$

$$2ab - 4a - 4b + 8 = ab$$

or $\quad ab - 4b = 4a - 8$

or $\quad b = (4a - 8)/(a - 4)$

Since a and b are number of tiles, they must be integral values. Now substituting from the Option, we find that only for n = 12, we get an integral value for b i.e., 5.

7. (d) Total number of teams = n

   Number of players in each team = k

   Number of players common to 2 teams = Number of teams = n

   Hence total number of students participating in the tournament = $nk - n = n(k-1)$

8. (b) Let the number of currency of 1 miso- x, 10 miso-y and 50 miso-z

   Then $x + 10y + 50z = 107$

   Possible value of z is o, 1 or 2

   Case (1) when $z = 0$, then $x + 10y = 107$, then possible solution of (y, x) is (0, 107), (1, 97), (2, 87)….(10, 7) hence total 11 value.

   Case (2) when $z = 1$ then $x + 10y = 57$, then possible values of (y, x) is (0, 57), (1, 47) … (5, 7) hence total 6 values.

   Case (3) when $z = 2$ then $x + 10y = 7$ then possible value of (y, x) is (0, 7) hence 1 value

   So total no of ways = 11 + 6 + 1 = 18

9. (d) Given equation is 1/m +4/n =1/12 or 1/m = 1/12- 4/n or m =12n/(n-48) since m is positive hence n>48 but n<60 hence possible values of n is 49, 51, 53, 55, 57 and 59, out of which only 49, 51 and 57 gives integral values of m.

10. (d) The easiest way to solve this question is that take some values of n such that seed(n) =9 these are 18, 27, 36 etc, that means multiples of 9, so what actually we need to find is the number of multiples of 9 less than 500,

    So 500/9 and quotient is 55 hence we will have 55 such numbers.

11. (d) Here if you are thinking that since ordinates are changing with a multiple of 2 hence you can use GP series formula then you are on the wrong track.

    The given points are A(7, 4), B(-10, 0), C(-10, 3), D(0, 10) and E(7, 7).

    It is also given that every second, these points halve their abscissas and double their ordinates.

    After 500 years, abscissas for all these points will be very much close to '0' hence doesn't play very important role in finding the distance. Points will be closest to each other after 500 years which are initially close to each other with respect to ordinates only. Here points A and C are closest to each other by 1 ordinate.

    Hence Points A and C will be closest.

12. (b) He can add 30 numbers in each group

    1000 – 1099, 1100 – 1199, 1200 – 1299, 1300 – 1399, 1400 – 1499, i.e. 150 numbers.

    He can also add

    1099 + 1100, 1199 + 1200

    1299 + 1300, 1399 + 1400

    1499 + 1500 & 1999 + 2000 i.e. 6 more numbers

    Total numbers = 156.

13. (b) Let the potter has given x pots to his sons.

    Hence, total amount made by his sons is ₹ x^2 after selling these pots.

    As the price of one banana wafer packet is less than ₹10 hence, x^2 will not be a multiple of 100.

    Assume that they bought n packets of potato wafers.

    Hence, total number of wafers packet = $n + 1$

    Hence, each son gets $(n + 1)/2$ packets. Hence, n is odd.

    Let b be the price of banana wafers. Hence, they have ₹ $(n \times 10 + b)$

    As n is odd, tens place of $(n \times 10 + b)$ is odd.

    Now, each brother can have equal money if total amount earned by them is even.

    Hence, from this we can conclude that b must be even.

    Then $x^2 = 10 \times n + b$ such that b is even and n is odd. So, x is an even number.

    Now, if unit's digit is of x is 2 or 8, then tens place of x^2 will be even.

It contradicts our proposition that tens digit of x^2 is odd and hence it is not possible.

So, unit's place digit of x is 4 or 6. In either case, unit digit of x^2 is 6. Hence, $b = 6$.

Hence, the son having banana wafers owes ₹ $((n-1) \times 10/2 + 6)$ and the other son owes $((n+1) \times 10/2)$ Hence, one of the son has ₹ $((n+1) \times 10/2) - ((n-1) \times 10/2 + 6) =$ ₹ 4 more than the other.

Hence, he must give Rs. 2 to the other to have financially equitable division.

14. (d) Given that $30 < 5n + 3 < 40$ it follows that

$\frac{27}{5} < n < \frac{37}{5}$, i.e. $5 < n < 8$, so $n = 6$ or 7.

When $n = 6$, then $3m = 5n + 1 = 31$ implies m is not an integer,

When $n = 7$, then $3m = 5n + 1 = 36$,

i.e. $m = 12$. Thus, $m + n = 19$.

15. (a) The given inequality $0 \le ax + 5 \le 4$ which can be written as $-5 \le ax \le -1$. Since the integer solutions for x are positive, so $a < 0$. Therefore,

$$-\frac{1}{a} \le x \le -\frac{5}{a}.$$

Since $0 < -\frac{1}{a} \le 1$ and $4 \le -\frac{5}{a}$,

so $a < -1$ and $a \ge -\frac{5}{4}$.

Thus the range of a is $-\frac{5}{4} \le a \le -1$.

16. (a) Since $x > y > z > 663$

So let $z = 663 + a$, $y = 663 + a + b$

and $z = 663 + a + b + c$

Then $x + y + z = 1998$

or

$(663 + a) + (663 + a + b) + (663 + a + b + c) = 1998$

or $\quad 3a + 2b + c = 9 \quad \ldots(i)$

And $\quad 2x + 3y + 4z = 4(663 + a) + 3(663 + a + b) + 2(663 + a + b + c) = 5992$

or $\quad 9a + 5b + 2c = 25 \quad \ldots(ii)$

From equation (i) and (ii) we get $3a + b = 7$

Integral solutions are (1, 4) and (2, 1)

But (1, 4) will lead negative value of c hence ruled out. So $(a, b, c) = (2, 1, 1)$

Or $\quad (x, y, z) = (665, 666, 667)$

17. (b) Let $x = 100a + 10b + c$,

then $a + b + c = 21$, and

$100a + 10b + c + 495 = 100c + 10b + a$

Therefore,

$$99(c-a) = 495,$$
$$c - a = 5,$$
$$\therefore \quad c = a + 5,$$

Hence the possible values of a are 1, 2, 3, 4. Then the condition $a + b + c = 21$.

So $\quad 2a + b = 16$,

so $\quad b = 2(8 - a) \le 9$, $a = 4$.

Hence, $c = 9$, $b = 8$, i.e. $x = 489$.

18. (d) Let the original four digit number be

$$abcd = 1000a + 100b + 10c + d$$

Then from the given condition

$(1000c + 100b + 10a + d) - (1000a + 100b + 10c + d) = 5940$

$$990(c - a) = 5940,$$
$$\therefore \quad c - a = 6.$$

When number is minimum, then a = 1, c = 7, so a + c = 8. Since a + b + c + b has a remainder 8 when it is divided by 9, so 2 b is divisible by 9, i.e.b = 9 or 0. Since n is odd, so b = 9. Thus, the four digit number is 1979

Sum of digits is 26

19. (b) Consider $xz + yz = z(x + y) = 23$

23 is a prime, so $z = 1$, $x + y = 23$. By substituting $y = 23x$ into the first equation, it follows that

$$(23 - x)(x + 1) = 63,$$
$$x^2 - 22x + 40 = 0,$$
$$(x - 2)(x - 20) = 0,$$

So $x_1 = 2, x_2 = 20$. Then $y_1 = 21, y_2 = 3$. Thus the solutions are (2, 21, 1) and (20, 3, 1),

20. (c) The given equation can be written as

$$xy + 42 = 9y$$

then $$y = \frac{42}{9 - x}.$$

y is a positive integer implies that $9 - x$ is a positive divisor of 42, so $9 - x = 1, 2, 3, 6, 7$, i.e. $x = 8, 7, 6, 3, 2$.

Correspondingly, $y = 42, 21, 14, 7, 6$. By checking the five solutions satisfy the original equation, so the answer is 5.

21. (b) The given equations can be written as $8y - 12x = xy$, so $xy + 12x - 8y - 96 = -96$, i.e. $(x - 8)(y + 12) = -96$.

Since $y + 12 \geq 13$ and $-7 \leq x - 8 < 0$. There are five possible cases to be considered:

$$\begin{cases} x - 8 = -1, \\ y + 12 = 96, \end{cases} \quad \begin{cases} x - 8 = -2, \\ y + 12 = 48, \end{cases}$$

$$\begin{cases} x - 8 = -3, \\ y + 12 = 32. \end{cases} \quad \begin{cases} x - 8 = -4, \\ y + 12 = 24, \end{cases}$$

$$\begin{cases} x - 8 = -6, \\ y + 12 = 16, \end{cases}$$

From them five positive integer solutions are obtained easily:

(7, 84), (6, 36), (5, 20), (4, 12), (2, 4)

Thus the answer is 5.

22. (b) Consider the second equation $xy + yz = y(x + z) = 31 = 1 \times 31$

It leads to $y = 1, x + z = 31$. By substituting them into the first equation, it follows that

$$x(1 + 31 - x) = 255,$$
$$x^2 - 32x + 255 = 0,$$
$$(x - 15)(x - 17) = 0,$$
$$\therefore \quad x_1 = 15, x_2 = 17.$$

Thus, the solutions are $x = 15, y = 1, z = 16$ and $x = 17, y = 1, z = 14$.

23. (a) The inequalities $b < \frac{7}{3} < 2a$ implies $\frac{2x+1}{3} < \frac{7}{3} < \frac{x+3}{2}$, so $2x + 1 < 7$ and $14 < 3x + 9$, hence $\frac{5}{3} < x < 3$. Thus the range of x is $\frac{5}{3} < x < 3$.

24. (b) From $\frac{2m + x}{3} \leq \frac{4mx - 1}{2}$ it follows that

$$4m + 2x \leq 12mx - 3,$$
$$2(6m - 1)x \geq 4m + 3.$$

The solution set is $x \geq \frac{3}{4}$ implies that

$$6m - 1 > 0 \text{ and } \frac{4m+3}{2(6m-1)} = \frac{3}{4},$$

$$\therefore \quad 2(4m + 3) = 3(6m - 1)$$

i.e. $(18 - 8)m = 6 + 3$,

$$\therefore \quad m = \frac{9}{10}.$$

25. (a) The solution set of the system in the set of the integers x satisfying $\frac{a}{9} \leq x < \frac{b}{8}$. Since each of 1, 2, 3 satisfies the two inequalities,

$$0 < \frac{a}{9} \leq 1 \Rightarrow 0 \leq a\ 9,$$

$$3 < \frac{b}{8} \leq 4 \Rightarrow 24 < b \leq 32.$$

So a has 9 choices and b has 8 choices, i.e. there are $9 \times 8 = 72$ required ordered pairs (a, b).

26. (b) We know that

$$a = \frac{1}{2}[(a - b) + (a + b)]$$

$$-2b = (a-b)-(a+b),$$

so $a-2b = -\frac{1}{2}(a+b)+\frac{3}{2}(a-b).a-2b$

will take its maximum value if

$$a+b=1,\ a-b=1,\ \text{i.e. } a=1,\ b=0.$$

Thus, $8a + 2016b = 8a = 8$.

27. (b) Given equations are

$$3x + 2y - z = 4, \quad \ldots(i)$$

$$2x - y + 2z = 6 \quad \ldots(ii)$$

$$x + y + z < 7. \quad \ldots(iii)$$

By $(i) + 2 \times (ii)$,

$$7x + 3z = 16,$$

$$\therefore \quad x = 1, z = 3.$$

From (i), $y = 2x + 2z - 6 = 2$. Since $x = 1$, $y = 2$, $z = 3$ satisfy the inequality (iii), the solution is $x = 1, y = 2, z = 3$.

28. (c) From the given condition

$$x + 2y + 5z - 100.$$

Here, x, y and z are non-negative integer

It is clear that $0 \le z \le 20$. For any possible value of z, $x + 2y = 100 - 5z$. Let $u = 100 - 5z > 0$. Then for solving the equation $x + 2y = u$, $(-u, u)$ is a special solution. So the general solution for (x, y) is $x = -u + 2t, y = u - t, t \in \mathbb{Z}$.

If $u = 2k$, then $k = \frac{u}{2} < t < u = 2k$,

i.e. there are $k + 1 = \frac{u}{2} + 1$ solutions for (x, y).

If $u = 2k + 1$,

then $k + 1\frac{u+1}{2} \le t \le u = 2k + 1$,

i.e. there are $k + 1 = \frac{u+1}{2}$ solutions for (x, y).

Thus, we have the following table :

u = 100 95 90 85 80 75 70 65 60 … 15 10 5 0

$k + 1$ = 51 48 46 43 41 38 36 33 31 … 8 6 3 1

The total number of solutions is given by

$4(10 + 20 + 30 + 40) + 5(8 + 6 + 3 + 1) + 51 = 541$.

Thus, there are 541 ways to get the 10 dollars.

29. (a) Let the number has (m+1) digit

$n = 10x + y$, where y is the last digit, and x is a m digit number.

Then $\quad 5(10x + y) = 10^m y + x,$

$$49x = (10^m - 5)y.$$

Since $1 \le y \le 9 < 49$, so $7 \mid (10^m - 5)$. The minimum value of m such that $7 \mid (10^m - 5)$ is 5, so

$$49x = 99995y$$

$$\Rightarrow \quad 7x = 14258y$$

$\because x|14258, \because y = 7, x = 14258.$

Thus, the number is 142587.

30. (a) As the maximum marks is 30. Assume that 19 questions asked correctly and one question is incorrect.

Case1: one mark question asked is wrong, 30 -1 -0.25 = 28.75

Case 2: two marks question asked is wrong, 30 -2 -0.33 = 27.66

Case3: one mark question asked is not attempted, 30 -1 -0.5 = 28.5

Case 4: two marks question asked is attempted, 30 -2 -0.5 = 27.5

Hence ranked 5.

Passage Type

31. (d) Given that the coins that the four people got were 1, 2, 3 and 4. Also, Q > P and S < R Q = 2P. This gives us two cases:

Case (1)- Q gets 4 and P gets 2

Case (2)- Q gets 2 and P gets 1

Since S < R, in the first case S = 1 & R = 3 and in second case S = 3 & R = 4

We can find out that in both cases S is odd.

32. (b) Here given information is R gets at least 2 more coins than S, that leaves us a possibility of

From case (1) R =3, and S =1

From case (2) r=4, S =3, but then difference is 1 hence this case is ruled out

So result is S =1, P =2, R =3, and Q =4.

33. (a) Given that Q < R. Possible cases are (Note that Q cannot be equal to 1 because P < Q)

Q = 2, R = 4, P = 1, S = 3

Q = 3, R = 4, P = 1, S = 2

Q = 3, R = 4, S = 1, P = 2

In case of the first case, we find that P + Q = 3 which is less than 4. And it negates the option (a)

34. (c) Initial number of flowers –x, After 1st dip in river flowers left =2x

Number of flowers offered to deity =y,

Remaining number of flowers = 2x-y,

remaining after 2nd dip 2(2x-y)

Remaining number of flowers after visit to 2nd temple = 2(2x-y) – y = 4x-3y,

Remaining after 3rd dip 2(4x-3y)

Remaining number of flowers after visit to 3rd temple = 2(4x-3y) – y = 8x-7y,

Remaining after 3rd dip 2(8x-7y)

Remaining number of flowers after visit to 4th temple = 2(8x-7y) – y = 16x-15y,

Remaining after 3rd dip 2(16x-15y)

It is given that finally she left with no flowers hence 2(16x-15y) =0, or x/y= 15/16 since x and y are integers hence they must be a multiple of 15 and 16.

given that if x=30 then y=32,

35. (c) Y is minimum when x is minimum i.e. x = 15 (x has to be multiple of 15). So, Y = 16.

36. (b) The minimum number with which Roopa leaves home is x= 15 (as discussed above, the least value of n is 15).

37. (c) let us assume a member of the set is (a,b) and another member (c, d) here a,b, c, and d all are distinct and less than n, then (a, b) is an enemy of (c, d)

Now question is in how many ways we can select such pair out of n members, if we select 2 out of n numbers (say a & b) then remaining members are n-2 and we have to select two more (say c & d) from these remaining n-2.

Condition is same as selecting 2 out of n – 2

Hence number of ways are ${}^{n-2}C_2$

So number of enemies ${}^{n-2}C_2$ = (n-2)(n-3)/2 =1/2(n^2 -5n +6)

38. (c) Let us consider two friends (a, b) and (a, c) here all a, b and c are distinct from each other,

One of their common friend is (b, c)

There are many more common friends in the form of (a, d) since an element a is common in all the sets.

Now question is in how many ways we can select d?

After taking out a, b and c from n elements we have n-3 elements and out of these we have to select d in n-3 ways or n-3 such element in the form of (a, d)

So total number of common friends = (n – 3) +1 = n – 2

39. (c) It is given that $x_1, x_2, x_3, \ldots\ldots x_n$ are in arithmetic progression Let the first term and the common difference of the AP be a and d respectively.

In this AP It is given $T_2 = a + d = 2$ and $T_{24} = a + 23d = 68$

Solving the 2 equations, we get a = –1 and d = 3

Therefore $x_8 = a + 7d = -1 + 7(3) = 20$

Now $y_2 = px_2 + q$, so we need to find the value of p and q

$-2 = p(2) + q$

Again $y_{24} = px_{24} + q$ or, $31 = p(68) + q$

Solving above two equations we get p = 1/2 and q = –3

Therefore y-coordinate of A_8, $y_8 = px_8 + q$

$= 1/2\,(20) - 3 = 7.$

Hence Y-coordinate is 7.

40. (c) For the points to lie in the first quadrant, the coordinates of both x_k and y_k must be positive. Now $A_1 = (x_1, y_1)$ where $x_1 = -1$ and $y_1 = 1/2\,(-1)-3 = -3.5$.

$A_1 = (-1, -3.5)$ (In 3rd quadrant)

$A_2 = (x_2, y_2)$ where $x_2 = 2$ and $y_2 = -2$. Or $A_2 = (2, -2)$ 4th quadrant.

$A_3 = (x_3, y_3)$ where $x_3 = -1 + 2(3)$ and $y_3 = \frac{1}{2}(5)-3$, $A_3 = (5, 1/2)$

$A_4 = (x_4, y_4)$ where $x_4 = -1 + 3(3)$ and $y_3 = \frac{1}{2}(8)-3$, and $A_4 = (8, 1)$ 1st quadrant.

Since d is positive and y is a linear relation in x, the corresponding coordinates of A_k i.e., A_5 onwards will be increasing and thus none of them will be negative. Thus, only for A_1, A_2 and A_3, we do not have both the coordinates as positive. Therefore only three points do not lie in the first quadrant.

For all the other points, both *x* and *y* coordinates are positive.

Hence only A1, A2 and A3 do not lie in the first quadrant.

Concept Eliminator

1. (d) We shall need to break up the 158 Re. 1 coins as: 1, 2, 4, 8, 16, 32, 64 (totalling to 127)

Now, we are left with 31 coins so next bag will have 31 coins. Total number of bags = 8

2. (a) The parts should be made of lengths: 1, 2, 4, 8, 16

On day 1 he will give first piece

On day 2 he will take first piece back and will give second piece of length 2.

Similar pattern continues

Thus minimum parts = 5.

3. (b) Here we will try to find the pattern

1st round - 2x (All the numbers in the form of 2x eliminated)

2nd round = 1 + 4x (All the numbers in the form of 1+ 4x eliminated) similarly

3rd round = 3 + 8x

4th round = 7 + 8x

5th round = 15 + 16x

6th round = 31 + 32x

7th round = 63 + 64x

8th round = 127 + 128x

9th round = 255 + 256x

Now we will eliminate options one by one,

(a) The number 3 is in the form of 3+8x (when x=0) hence Eliminates in 3rd round

(c) Eliminates in 3rd round

(d) Eliminates in 3rd round

(e) Eliminates in 9th round

Option (b) is correct.

4. (c) Similar to above solution Option c is correct.

5. Let *x* and *y* be the numbers of pebbles in the piles (*A*) and (*B*) respectively and *z* pebbles are moved from (*B*) to (*A*) then

$$2(x - 100) = y + 100 \qquad ...(i)$$

$$x + z = 6(y - z). \qquad ...(ii)$$

[5 times more means 6 times of original value]

(*i*) gives $y = 2x - 300$, so from (ii) gives $11x - 7z = 1800$

$11x = 1800 + 7z = 1793 + 7(z + 1)$

$$\Rightarrow x = 163 + \frac{7(z+1)}{11}$$

minimum value of $z = 10$

and $x_{min} = 170$

and for this $y = 40$

So there are 40 pebbles in (B).

6. (d) Let $u = x + y$, then $2u + z = n$

Here x and y are positive integer so $u \geq 2$

We will have $(u-1)$ solution for the equation $x + y = u$,

Now we have two cases

Case (i) When n is even

Hence $\sum_{u=2}^{\frac{n}{2}} (u-1) = 28$

$$(2-1) + (3-1) + \ldots + \left(\frac{n}{2} - 1\right) = 28$$

On solving we get n = 18

Case (ii) When n is odd

Hence $\sum_{u=2}^{\frac{n-1}{2}} (u-1) = 28$

$$1 + 2 + \ldots . + \left(\frac{n-1}{2} - 1\right) = 28$$

On solving we get n = 17

7. (a) Here we have three unknown and only two equations so 1st we eliminate one variable. By eliminating variable 'z' we get

$$16x + 13y = 77$$

By observation we get $x = 4$, $y = 1$ is a solution. No other positive integral value of x and y satisfy this equation. So $x = 4$, $y = 1$ and $z = 2$ is the only set of values.

8. Let the numbers of Big, medium and small chocolates that the buyer bought be x, y, z respectively, then

$$5x + 3y + \frac{1}{3}z = 100$$

$$x + y + z = 100.$$

or $\quad 15x + 9y + z = 300, \qquad \ldots(i)$

$\quad x + y + z = 100 \qquad \ldots(ii)$

(i) – (ii) yields $14x + 8y = 200$ or,

$\quad 7x + 4y = 100, \qquad \ldots(iii)$

One solution of this equation is (0, 25)

2nd solution is (0 + 4, 25 – 7) = (4, 18)

3rd solution is (4 + 4, 18 – 7) = (8, 11)

4th solution is (8 + 4, 11 – 7) = (12, 4)

Remaining solutions are negative hence ruled out.

Thus, the solutions for (x, y, z) are

(0, 25, 75), (4, 18, 78), (8, 11, 81), (12, 4, 84)

9. Let the numbers of weights of 1g, 10g, 50g be x, y, z respectively. Then

$$\begin{cases} x + y + z = 100, \\ x + 10y + 50z = 500. \end{cases}$$

By eliminating x, we get $9y + 49z = 400$,

From observation we get ;

$z = 1$, $y = 39$, $x = 60$.

Thus, in the 100 weights there are 60 of 1g, 39 of 2g and 1 of 50g, respectively.

10. Suppose r pages of the book are torn off. Note that the page numbers on both the sides of a page are of the form $2k - 1$ and $2k$, and their sum is $4k - 1$. The sum of the numbers on the torn pages must be of the form

$$4k_1 - 1 + 4k_2 - 1 + \ldots + 4k_r - 1$$

$$= 4(k_1 + k_2 + \ldots + k_r) - r.$$

The sum of the numbers of all the pages in the untorn book is

$$1 + 2 + 3 + \ldots + 100 = 5050.$$

Hence the sum of the numbers in the torn pages is

$$5050 - 4949 = 101.$$

We therefore have

$$4(k_1 + k_2 + \ldots + k_r) - r = 101.$$

This shows that $r \equiv 3 \pmod 4$. Thus $r \equiv 4l + 3$ for some $l \geq 0$.

Suppose $r \geq 7$, and suppose $k_1 < k_2 < \ldots < k_r$. Then we see that

$$4(k_1 + k_2 + \ldots + k_r) - r \geq 4(k_1 + k_2 + \ldots + k_7) - 7$$

$$\geq 4(1 + 2 + \ldots + 7) - 7$$

$$= 4 \times 28 - 7 = 105 > 101.$$

Hence $r = 3$. This leads to $k_1 + k_2 + k_3 = 26$ and one can choose distinct positive integers k_1, k_2, k_3 in several ways.

11. Let ABCDPQRS be a cube, and the numbers a, b, c, d, e, f be written on the faces ABCD, BQRS, PQRS, APSD, ABQP, CRSD respectively. Then the products written at the corners A, B, C, D, P, Q, R, S are respectively ade, abe, adf, cde, bce, bef, bcf, cdf. The sum of these 8 numbers is :

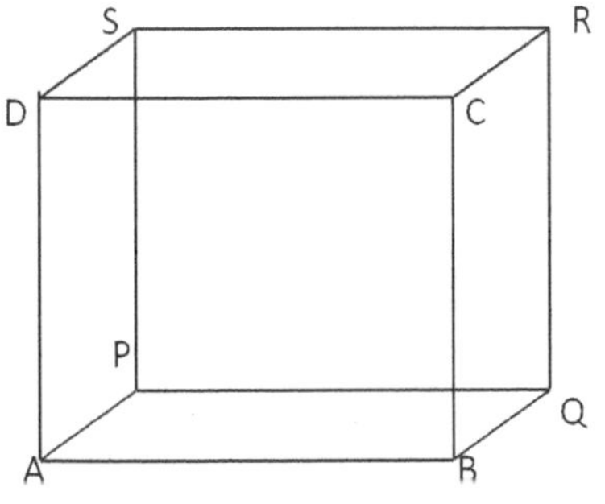

$= (e + f)(ab + bc + cd + ad) = (e + f)(a + c)(b + d)$.

This is given to be equal to $2076 = 2^2.3.173$. Since none of the factors $a + c$, $b + d$, $e + f$ is equal to 1.

Thus $(a + c)(b + d)(e + f) = 2076$

But $2076 = 4 \times 3 \times 173 = 2 \times 3 \times 346 = 2 \times 2 \times 519 = 2 \times 6 \times 173$

So possible values of summation

$$(a + c) + (b + d) + (e + f) = 4 + 3 + 173 = 180,$$
$$2 + 3 + 346 = 351,$$
$$2 + 2 + 519 = 523,$$
$$2 + 6 + 173 = 181.$$

Sum of all the possible values of S is $180+351+523+181 = 1235$

12. Let the desired whole number N has $n + 1$ digits, and it has 4 as first digit then $N = 4.10^n + x$, where x is an n-digit number.

From the given condition

$$4(10x + 4) = 4.10^n + x$$

i.e., $$39x = 4(10^n - 4) = 4\left(\underbrace{99\ldots96}_{n-1}\right)$$

$$\therefore \quad 13x = 4\left(\underbrace{33\ldots32}_{n-1}\right)$$

So, $13 \mid \underbrace{33\ldots32}_{n-1}$.

By checking the cases $n = 1, 2, \ldots$, one by one, it is easy to see that the minimal value of n is 5:

$$33332 \div 13 = 2564.$$

$\therefore \quad x = 4 \times 2564 = 10256$

and $\quad N = 410256.$

13. Since 13 divides $a + 11b$,

So 13 also divides $a + 11b - 13b = a - 2b$

Or 13 also divides $13a - (a - 2b) = 12a + 2b = 2(6a + b)$

Or 13 divides $(6a + b)$

Similarly given that

$11 \mid (a + 13b)$

$\Rightarrow 11 \mid (a + 2b)$

$\Rightarrow 11 \mid (6a + 12b)$

$\Rightarrow 11 \mid (6a + b)$

Hence we can say that both 11 and 13 divides $(6a + b)$ or $11 \times 13 = 143$ divides $(6a + b)$

Thus we may write $6a + b = 143k$ for some natural number k.

or

$6a + 6b = 143k + 5b = 144k + 6b - (k + b)$.

or 6 divides $k + b$ and hence $k + b \geq 6$.

So $6(a + b) = 143k + 5b = 138k + 5(k + b) \geq 138 + 5 \times 6 = 168$.

So we can conclude that $a + b > 28$.

From some observation we get $a = 23$ and $b = 5$, satisfy the condition. Thus the minimum value of $a + b$ is 28.

14. Consider following cases

Case (i) If number is of single digit:- then $a = 4a$, i.e., $a = 0$, so no single digit good number exists.

Case (ii) If number is of two digit:- Let $ab = 10a + b$ be a two digit good number, then $10a + b = 4(a + b)$ implies $2a = b$, so there are four good numbers 12, 24, 36, 48, and their sum is 120.

Case (iii) If number is of three digits, then $100a + 10b + c = 4(a + b + c)$. i.e., $96a + 6b - 3c = 0$. Since $96a + 6b - 3c \geq 96 + 0 - 27 > 0$, always, so no solution for (a, b, c)so no three digit good number exists.

Similarly no good number with four digits with $n(n \geq 4)$ exist.

Thus, the sum of all good numbers is 120.

15. From question it is clear that n can not be a one – digit number

Let $n = 10x + 6$, where x is a natural number of m digits. Then

$$6.10^m + x = 4(10x + 6)$$

$$\Rightarrow \quad 39x = 6(10)^m - 24$$

$$\Rightarrow \quad 13x = 2(10)^m - 8$$

So, $13 \mid (2(10)^m - 8)$ for some m, i.e. the remainder of $2(10)^m$ is 8

when divided by 13. By observation we find that $m = 5$.

$$\text{So} \quad x = \frac{2(10)^m - 8}{13} = \frac{199992}{13} = 15384,$$

$$n = 153846$$

CHAPTER

2

LOGARITHM

PART-A

2.1 LOGARITHM INTRODUCTION

2.1.1 Definition

Every positive real number N can be expressed in exponential form as

$N = a^x$ e.g. $64 = 4^3$

Where 'a' is a positive real number different than unity and is called the base and 'x' is called exponent.

We can write the above relation in logarithmic form as

$$\log_a N = x$$

Hence the two relations

$$\left.\begin{array}{c} a^x = N \\ \text{and} \\ \log_a N = x \end{array}\right] \text{ are identical where } N > 0, a > 0, a \neq 1$$

Hence we can define 'logarithm' as

"The logarithm of any number (say a positive number N) to a given base (say any positive number other than 1) is the index or the power (say x) to which the base must be raised in order to equal the given number"

Hence from definition If $a^x = N \quad \Rightarrow x = \log_a N$

Consider an equation $a^x = N$, here we are expressing N in terms of a and x.

Another way to write the same relation is $a = N^{1/x}$.

Here we are expressing a in terms of N and x.

If we have to express x in terms of the other two parameters a and N then we have to use logarithms.

As per definition of logarithm $x = \log_a N$

Point to note in the above equation is that, N is a **positive Number** and 'a' is a **positive Number other than 1**.

Logarithms can be expressed to any base (from definition positive number other than 1.) Logarithms from one base can be converted to logarithms to any other base. However, there are two types of logarithms that are commonly used, on the basis of bases.

(i) **Common Logarithms:** These are logarithms expressed to the base 10. For most of the problems under logarithms, it is common logarithms that we deal with. If logarithms are given without mentioning any base, it can normally be taken to be logarithms to the base 10.

(ii) **Natural logarithms or Napierian Logarithms:** These are logarithms expressed to the base of a number called "e".

Important Points

1. Unity cannot be the base of the logarithm as in this case

   $\log_1 N$ will not be possible and if $N = 1$

   Then $\log_1 1$ will have infinitely many solutions and will not be unique which is necessary in the functional notation.

   Another way to explain the same is that, for any value of x, $1^x = 1$

2. $a^{\log_a N} = N$ is an identify for all $N > 0$ and $a > 0$; $a \neq 1$ e.g. $2^{\log_2 7} = 7$

3. From the basic definition of log, we have 3 important deductions:

(i) $\log_N N = 1$ — i.e. logarithm of a number to the same base is 1.

(ii) $\log_{\frac{1}{N}} N = -1$ — i.e. logarithm of a number to its reciprocal is – 1.

(iii) $\log_N 1 = 0$ — i.e. logarithm of unity to any base is zero.

2.1.2 Logarithmic Formulae

In all these formulae, $\log_a N$ is defined when both a and N are positive number and $a \neq 1$

$$\log_b 1 = 0$$

$$\log_a a = 1$$

$$\log_a b \times \log_b a = 1$$

$$\log_b (m \times n) = \log_b m + \log_b n$$

$$\log_b \left(\frac{m}{n}\right) = \log_b m - \log_b n$$

$$\log_b m^n = n \log_b m$$

$$\log_b m = \frac{\log_a m}{\log_a b} = \log_a m \times \log_b a$$

$$b^{\log_b n} = n$$

If $\log_a m = \log_b n$ and if $m = n$, then a will be equal to b

If $\log_a m = \log_b n$ and if $a = b$, then m will be equal to n.

Illustration 1. Find value of $\log_{243} 27$

Solution: We know from definition that logarithm of a number to some base is the exponent by which the base must be raised in order to get that number, hence $x = \log_{243} 27$ can be written as $27 = 243^x$

$\Rightarrow \; 3^3 = 3^{5x}$

Or $x = \frac{3}{5}$

Alternately:-

Using formula $\log_{p^a} q^b = \frac{b}{a} \log_p q$

$$\log_{3^5} 3^3 = \frac{3}{5} \log_3 3 = \frac{3}{5}$$

Illustration 2. Find value of $\log_{1/3} 27\sqrt{3}$

Solution: We know from definition that logarithm of a number to some base is the exponent by which the base must be raised in order to get that number,

Let $x = \log_{\frac{1}{3}} 27\sqrt{3}$ can be written as $27\sqrt{3} = \frac{1^x}{3}$

$\Rightarrow \; 3^{\frac{7}{2}} = 3^{-x}$

Or $x = -\frac{7}{2}$

Alternately:-

Using formula $\log_{p^a} q^b = \frac{b}{a} \log_p q$

$$\log_{\frac{1}{3}} 27\sqrt{3} = \log_{3^{-1}} 3^{\frac{7}{2}} = -\frac{7}{2} \log_3 3 = -\frac{7}{2}$$

2.1.3 Proof of Basic Properties

If m, n are arbitrary positive real numbers and $a > 0$; $a \neq 1$

Property 1: $\log_a m + \log_a n = \log_a mn$
$(m > 0, n > 0)$

Proof: Let $x_1 = \log_a m$; $m = a^{x_1}$

$x_2 = \log_a n$; $n = a^{x_2}$

Now $mn = a^{x_1} a^{x_2} = a^{x_1 + x_2}$

$x_1 + x_2 = \log_a m + \log_a n = \log_a mn$

$\log_a m + \log_a n = \log_a mn$

Property 2: $\log_a \frac{m}{n} = \log_a m - \log_a n$

Proof: Let $x_1 = \log_a m$; $m = a^{x_1}$

$x_2 = \log_a n$; $n = a^{x_2}$

Now $\frac{m}{n} = \frac{a^x}{a^{x_2}} = a^{x_1 - x_2}$

$x_1 - x_2 = \log_a m - \log_a n = \log_a \frac{m}{n}$

$\log_a \frac{m}{n} = \log_a m - \log_a n$

Property 3: $a^{\log_b c} = c^{\log_b a}$

Proof: $a^{\log_b c} = a^{(\log_b c)(\log_a c)(\log_c a)}$

$= a^{\log_a c\,(\log_b c \log_c a)} = c^{\log_b a}$

Hence $a^{\log_b c} = c^{\log_b a}$

Illustration 3. Find the value of $\log_4 64$

Solution: Given $\log_4 64 = \log_4 4^3 = 3 \log_4 4 = 3$

Illustration 4. If log 2 = 0.3010 then find the value of log 5.

Solution: Since log 10 = 1 = log (2 × 5) = log 2 + log 5 = 0.3010 + log 5 = 1 or log 5 = 0.699

Illustration 5. Find the value of log 800 if log 2 = 0.3010

Solution: Since log 800 = log (8 × 100)

= log 8 + log 100 = $\log 2^3$ + 2 = 2 + 3 log 2

= 2 + 3 × 0.3010 = 2.9030.

Illustration 6. Find the value of (log 40) (log 60) (log 80) if it is given that log 2 = 0.3010 and log 3 = 0.4771

Solution: Consider log 40 = log (4 × 10) = log 4 + log 10 = 2 log 2 + 1 = 2 × 0.3010 +1 = 1.6020

Then log 60 = log (2 × 3 × 10) = log 2 + log 3 + log 10 = 0.3010 + 0.4771 + 1 = 1.7781

And finally log 80 = log (8 × 10) = log 8 + log 10 = 3 log 2 + 1 = 1.9030

Hence (log 40) (log 60) (log 80) = 1.6020 × 1.7781 × 1.9030 = 5.42

Illustration 7. Find the value of $(\log_b a) . (\log_c b) . (\log_a c)$

Solution:

$$\frac{\log a}{\log b} \times \frac{\log b}{\log c} \cdot \frac{\log c}{\log a} = \frac{\log a}{\log b} = 1$$

2.1.4 Characteristic & Mantissa

Since logarithm of a number is a number (Not necessary an integer), it will have an "integral" part and a "decimal" part. The integral part of the logarithm of a number is called the **Characteristic** and the decimal part of the logarithm is called the **Mantissa**.

To understand this property, have a look at these logarithms

Logarithms	Characteristic	Mantissa	
Log2 = 0.3010	0	0.0310	These are log of **1 digit** number &**Characteristic is 0**
Log3 = 0.4771	0	0.4771	
Log5 = 0.699	0	0.699	
Log10 = 1.0	1	0	These are log of **2 digit** numbers &**Characteristic is 1**
Log20 = 1.3010	1	0.3010	
Log40 = 1.6020	1	0.6020	
Log80 = 1.9030	1	0.9030	

Log100 = 2.0	2	0	These are log of **3 digit** numbers &**Characteristic is 2**
Log200 = 2.3010	2	0.3010	
Log400 = 2.6020	2	0.6020	
Log800 = 2.9030	2	0.9030	
Log1000 =3.0	3	0	These are log of **4 digit** numbers & **Characteristic is 3**
Log1200 = 3.0791	3	0.0791	
Log1500 = 3.1760	3	0.1760	

Hence we can conclude that "**the characteristic of the common logarithm of a number greater than unity is positive and is less by one than the number of digits in its integral part**".

Consider following logarithms:-

Logarithms	Characteristic	Mantissa	
Log(0.1) = –1.0	–1	0	These are log of numbers with no 0's after decimal & **Characteristic is –1**
Log(0.2) = –0.6989	–1	0.3010	
Log(0.3) = –0.5228	–1	0.4771	
Log(0.01) = –2	–2	0	These are log of numbers with one 0's after decimal & **Characteristic is –2**
Log(0.02) = –1.6989	–2	0.3010	
Log(0.03) = –1.5228	–2	0.4771	
Log(0.001)= –3	–3	0	These are log of numbers with two 0's after decimal & **Characteristic is –3**
Log(0.002) = –2.6989	–3	0.3010	
Log(0.003) = –2.5228	–3	0.4771	

In a common logarithms of a number less than 1, magnitude of characteristic is one more than the number of zeros after the decimal point. (Here we have assumed that mantissa is positive).

Illustration 8. Find the number of digits in 2^{100} if log 2 = 0.3010

Solution: Let $N = 2^{100}$ hence $\log N = \log (2^{100}) = 100 \log 2 = 100 \times 0.3010 = 30.10$ here characteristics of log is 30 hence number of digits must be 31.

Illustration 9. Find the number of digits in 6^{100} if log 2 = 0.3010, log3 = 0.4771.

Solution: Let $N = 6^{100}$ taking log we will get $\log N = 100 \log 6 = 100\{\log 2 + \log 3\} = 100 \{0.3010 + 0.4771) = 77.81$ here characteristics of log is 77 hence number of digits must be 78.

Illustration 10. Find the number of zeros between the decimal point and 1st non zero digit $\left(\frac{4}{9}\right)^{100}$, given that log 2 = 0.3010, log 3 = 0.4771

Solution: Let $N = \left(\frac{4}{9}\right)^{100}$

Taking logarithm $\log N = 100 [\log 4 - \log 9] = 100 [2 \log 2 - 2 \log 3] = 200 [\log 2 - \log 3]$

$= 200 [0.3010 - 0.4771] = -35.22$

Hence required number of zeros is 35.

Concept Applicator (CA)-1

1. Find the value of $\frac{1}{8^{\log_3 2}}$

2. Find the value of $\frac{1}{\log_3 2}+\frac{2}{\log_9 4}-\frac{3}{\log_{27} 8}$

3. Find the value of

$$\log 2+16\log\frac{16}{15}+12\log\frac{25}{24}+7\log\frac{81}{80}$$

4. Find the value of $(\log_2 8)(\log_8 16)-(\log_2 16)(\log_{16} 32)$

5. If $\log_{30} 3 = c$, $\log_{30} 5 = d$ then the value of $\log_{30} 8$

6. Find the value of $\log_5 \sqrt{5\sqrt{5\sqrt{5}}} \ldots$ °

7. Find the value of $\frac{2^{\sqrt{\log_2 3}}}{3^{\sqrt{\log_3 2}}}$

8. Find the number of digits in 6^{50}

9. Find the number of digits in 5^{25}

10. Find the number of zeros after decimal before a significant figure (a non zero digit) start in 3^{-50}

2.2 LOGARITHMIC EQUATION & INEQUATION

2.2.1 Logarithmic Equation

While solving Logarithmic equation we need to take care of the following things

(i) The function $\log x$ is defined only for positive x.

(ii) $a^{\log_a x} = x$ here both x and a are positive.

(iii) If $\log_a x = \log_a k$ then $x = k$

(iv) If $\log_a x = 0$ then $x = 1$

(v) If $\log_a x = 1$ then $x = a$

(vi) If $\log_a x = -ve$ then either $x \in (0, 1)$ & $a \in (1, \infty)$ or $x \in (1, \infty)$ & $a \in (0, 1)$

(vii) If $\log_a x = +ve$ then both x & $a \in (0, 1)$ or x & $a \in (1, \infty)$

Illustration 1. Solve $7^{\log_7 x} + 3x + 16 = 0$

Solution: Since we know that $7^{\log_7 x} = x$ hence given equation can be written as

$$4x + 16 = 0$$

$\Rightarrow$ $x = -4$ but at this value of x log is not defined

$$x = \phi$$

Illustration 2. $7^{\log_7 (x-3)} + 3(x-4) - 13 = 0$

Solution: Since $a^{\log_a x} = x$, so $7^{\log_7 (x-3)} = (x-3)$

$$x - 3 + 3x - 12 - 13 = 0$$

$$4x = 28 \quad \Rightarrow \quad x = 7$$

Illustration 3. Find the value of x such that

$x^2 + 11^{\log_{11} x} - 2 = 0$

Solution. We know that $a^{\log_a x} = x \; (>0)$

$$\Rightarrow x^2 + x - 2 = 0$$

$$= (x+2)(x-1) = 0$$

$= x = -2$ or 1 but x can not be negative.

$\therefore \; x = 1$

Illustration 4. Solve : $(x+3)^{\log_7(x+3)} = 49\,(x+3)$

Solution: By definition $(x+3) > 0$

$\therefore$ Taking log with base 7 both side we get

$= \log(x+3)^{\log_7(x+3)} = \log\{49\,(x+3)\}$

$= \log_7(x+3)\,.\,\log_7(x+3) = \log_7 49 + \log_7(x+3)$

$=$ let $\log_7(x+3) = y$ then equation becomes

$= y^2 - y - 2 = 0$

$= (y-2)(y+1) = 0$

$= y = 2$ or -1

When $y = 2$ the $\log_7(x+3) = 2$

$\therefore \; x + 3 = 7^2 = 49$

$x = 46$

When $y = -1$ then or $\log_7(x+3) = -1$

or $x + 3 = 7^{-1} = \dfrac{1}{7}$

or $x = -3 + \dfrac{1}{7} = -\dfrac{20}{7}$

Illustration 5. Solve the equation $3^{(\log_3 x)^2} + x^{\log_3 x} = 162$.

Solution: $(3^{\log_3 x})^{\log_3 x} + x^{\log_3 x} = 162$

$$\Rightarrow x^{\log_3 x} + x^{\log_3 x} = 162 \qquad \left[a^{\log_a x} = x\right]$$

$$\Rightarrow x^{\log_3 x} = 81$$

Taking log both side with base 3

$\Rightarrow (\log_3 x)^2 = \log_3 81 = 4 \quad \Rightarrow \log_3 x = \pm 2$

$\Rightarrow x = 9$ or $\dfrac{1}{9}$

2.2.2 Logarithmic Inequation

In order to solve the in equation involving logarithmic function we need to understand the nature of graph of $y = \log x$

In order to understand the graph of $y = \log x$ take different values of x and then try to understand the nature of graph here we have to take care that log function is defined only for $x > 0$

$x =$	$y = \log x$
1	0
2	0.3010
10	1
100	2
1000	3
0.1	– 1
0.01	– 2

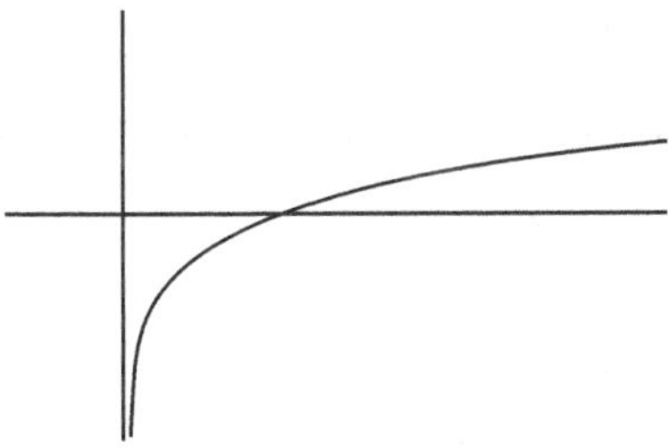

Features of the graph-

(1) Domain of the function (o, ∞)

(2) Range of the function $(-\infty, \infty)$

(3) Function is strictly increasing

(4) $y =$ is negative if $x \in (0, 1)$

(5) $y =$ is zero if $x = 1$

(6) $y =$ is positive if $x \in (1, \infty)$

(7) if $\log a > \log b$ then $a > b$

Logarithm Inequality

(i) (a) When $y > 1$ and $\log_y x > z \Rightarrow x > y^z$

(b) When $y > 1$ and $\log_y x < z \Rightarrow 0 < x < y^z$

(ii) (a) When $0 < y < 1$ and $\log_y x > z \Rightarrow 0 < x < y^z$

(b) When $0 < y < 1$ and $\log_y x < z \Rightarrow x > y^z$

(iii) Let a is a real number such that $a > 0, \neq 1$, then

(a) $\log_a f(x) \geq 0 \Rightarrow \begin{cases} f(x) \geq 1. & a > 1 \text{or} \\ 0 < f(x) \leq 1, & 0 < a < 1 \end{cases}$

(b) $\log_a f(x) \leq 0 \Rightarrow \begin{cases} f(x) \geq 1. & 0 < a < 1 \text{ or} \\ 0 < f(x) \leq 1, & a > 1 \end{cases}$

(iv) Let a is a real number, then

(a) $\log_a f(x) > \log_a g(x)$, if $a > 1, f(x) < g(x)$

(b) $\log_a f(x) \geq \log_a g(x)$

$$\Rightarrow \begin{cases} f(x) \geq g(x) > 0, & a > 1 \\ 0 < f(x) \leq g(x), & 0 < a < 1 \end{cases}$$

(c) $\log_a f(x) < \log_a g(x)$, if $0 < a < 1$, $f(x) > g(x)$

(d) $\log_a f(x) \leq \log_a g(x)$

$$\Rightarrow \begin{cases} 0 < f(x) \leq g(x), & a > 1 \\ f(x) \geq g(x) > 0, & 0 < a < 1 \end{cases}$$

(v) Let a is a real number such that $a > 0, \neq 1$, then

(a) $\log_a f(x) > m \Rightarrow f(x) > a^m$, $a > 1$ and

$\log_a f(x) > \text{m} \Rightarrow > 0 < f(x) < a^m, 0 < a < 1$

(b) $\log_a f(x) \geq m \Rightarrow \begin{cases} f(x) \geq a^m, & a > 1 \\ f(x) \leq a^m, & 0 < a < 1 \end{cases}$

(c) $0 < \log_a f(x) < m \Rightarrow 0 < f(x) < a^m$, $a > 1$

(d) $\log_a f(x) \leq m \Rightarrow \begin{cases} f(x) \leq a^m, & a > 1 \\ f(x) \geq a^m, & 0 < a < 1 \end{cases}$

Illustration 6. Find x if $x > 2$.

Solution: Given that $\log x > 2$ i.e $\log_{10} x > 2$

Or $x > 10^2 = 100$

So required answer is $x > 100$

Illustration 7. Find x if $\log (x + 1) < \log (x + 2)$

Solution: Since $\log x$ is an increasing function so $\log(x + 1) < \log(x + 2)$ for all the values where log functions are defined.

Now the given equation is defined if $x + 1 > 0$ or $x > -1$

Or $(x + 2) > 0$ or $x > -2$.

Combining all above result we get $x \in (-1\ \infty)$

Illustration 8. Find x if $\log(x^2) > \log(x^3)$

Solution: Since is $\log x$ an increasing function so $\log(x^2) > \log(x^3)$ for all the values where log functions are defined and $x^2 > x^3$ which is possible when $x \in (0, 1)$

<table>
<tr><th colspan="2">Graph of $y = \log_a x$</th></tr>
<tr><td>

(1) $y = \log_a x$ when $a > 1$

The graph of $y = \log_a x$ when $a > 1$ is similar to log x as $y = \log x$ can also be expressed as $y = \log_{10} x$

Features of the graph-

(1) Domain of the function $(0, \infty)$

(2) Range of the function $(-\infty, \infty)$

(3) Function is strictly increasing

(4) $y = \log_a x$ is negative if $x \in (0, 1)$

(5) $y = \log_a x$ is zero if $x = 1$

(6) $y = \log_a x$ is positive if $x \in (1, \infty)$

(7) if $\log_a p > \log_a q$ then $p > q$

</td><td>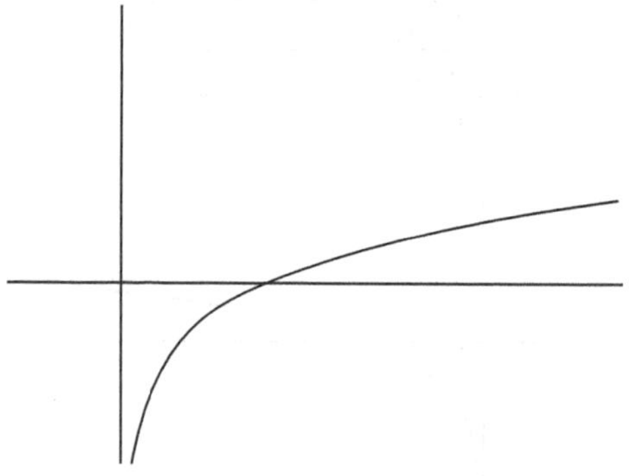</td></tr>
<tr><td>

(2) $y = \log_a x$ when $0 < a < 1$

The graph of $y = \log_a x$ when $a < 1$ is as shown

Features of the graph-

(1) Domain of the function (o, ∞)

(2) Range of the function $(-\infty, \infty)$

(3) Function is strictly decreasing

(4) $y = \log_a x$ is negative if $x \in (1,\infty)$

(5) $y = \log_a x$ is zero if x = 1

(6) $y = \log_a x$ is positive if $x \in (0, 1)$

(7) if $\log_a p > \log_a q$ then $p < q$

</td><td>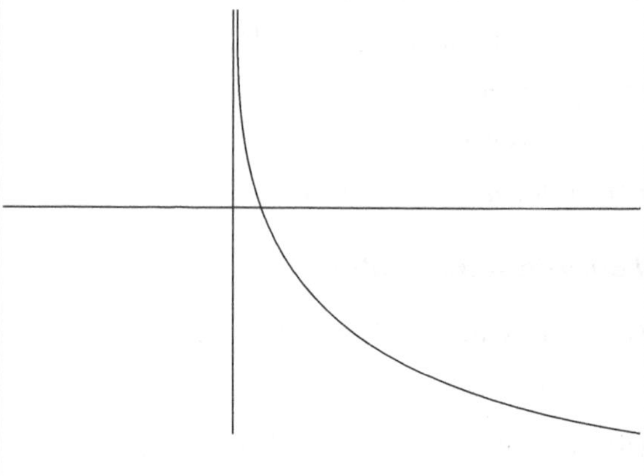</td></tr>
</table>

Illustration 9. Find x if $\log_{\frac{1}{2}}(x+3) < \log_{\frac{1}{2}}(x+1)$

Solution: Since $\log_{\frac{1}{2}} x$ is a decreasing function so $\log_{\frac{1}{2}}(x+3) < \log_{\frac{1}{2}}(x+1)$ if $(x + 1) < (x + 3)$ which is true for all the values of x. So all the values where log functions are defined and $x + 3 > 0$ or $x > -3$ and $x + 1 > 0$ or $x > -1$ which is possible when $x \in (-1,\infty)$

Illustration 10. Find x if $\log_{\frac{1}{2}}(x^2) < \log_{\frac{1}{2}}(x^3)$

Solution: Since $\log_{\frac{1}{2}} x$ is a decreasing function so $\log_{\frac{1}{2}}(x^2) > \log_{\frac{1}{2}}(x^3)$ if $x^2 < x^3$ which is possible when $x \in (1,\infty)$

Concept Applicator (CA)-2

1. Find the value of x if $(\log_2 3)(\log_3 4)(\log_2 5) \ldots \log_x (x+1) = 10$.

2. Solve the equation $\log_{10}(2^x + 2x - 8) = x(1 - \log_{10} 5)$

3. A circle has a radius of $\log_{10}(a^3)$ and a circumference of $\log_{10}(b^6)$. Then find the value of $\log_b a$

4. If $\log_2 \log_3 \log_4 \log_5 \text{A} = x$, then find the value of A

(a) 120^x (b) $2^{60}x$ (c) $2^{3^{4^{5^x}}}$ (d) $5^{4^{3^{2^x}}}$

5. Find the value of x if

$$3^{\log_3 \log\sqrt{x}} - \log x + (\log x)^2 - 3 = 0$$

6. Solve the equation:

$$2(\log_x \sqrt{5})^2 - 3\log_x \sqrt{5} + 1 = 0$$

7. Solve the equation $(\log x)^2 - \log(x^3) = \log(x^2) - 4$

8. Solve the equation $x^{\frac{\log x + 7}{4}} = 10^{(\log x)+1}$

9. Solve the value of x :

$$(\log 100x)^2 + (\log 10x)^2 = 14 + \log\frac{1}{x}$$

10. If $(a^{\log_b x})^2 - 5\,x^{\log_b a} + 6 = 0$, where $a > 0$, $b > 0$ & $ab \neq 1$, then which of the following could be the value of x.

(a) $2^{\log_b a}$ (b) $3^{\log_a b}$

(c) $b^{\log_a 2}$ (d) $b^{\log_a 2}$

Exercise A

Concept Builder (CB)

1. If $(0\,.\,pqr)^p = (p\,.\,qr)^q = (pq\,.\,r)^r = pqr$, then which of the following is true?

(a) $p - 1 = q + r$ (b) $p + q + r = 1$

(c) $\frac{1}{p} + \frac{1}{q} = \frac{1}{r}$ (d) $\frac{1}{p} + 1 = \frac{1}{q} + \frac{1}{r}$

2. Simplify: $\log\frac{21}{8} + 2\log\frac{6}{7} + \log\frac{7}{27}$

(a) $\log\frac{1}{2}$ (b) $\log 2$

(c) $\log\frac{1}{3}$ (d) $\log\frac{1}{7}$

3. $\log_3 \log_3 \log_3 k = p$ then which one of the following is a correct relation between p and k

(a) $K = 3^{3^p}$ (b) $k = 3^{3^{3^p}}$

(c) $p = 3^{3^k}$ (d) $p = 3^{3^{3^k}}$

4. Find the value of $\log 32^{\frac{1}{4}}\, 64^{\frac{1}{5}}\, 128^{\frac{1}{6}}$ in base $4^1\, 8^{\frac{1}{2}}\, 16^{\frac{1}{3}}$

(a) $\frac{219}{290}$ (b) $\frac{290}{219}$

(c) $\frac{290}{217}$ (d) $\frac{217}{290}$

5. Find x if $\log x + \log x^2 + \log x^3 + \ldots \log x^{10} = 11 \log 32$

(a) 2 (b) 4

(c) 8 (d) $\sqrt{2}$

6. If $a = \log_2 4$ $b = \log_4 16$ $c = \log_8 64$ then find $\frac{abc}{a+b+c}$

(a) $\frac{1}{2}$ (b) $\frac{2}{3}$ (c) $\frac{4}{3}$ (d) $\frac{8}{3}$

7. If $\log(x^3 + y^3) - \log(x^3 - y^3) = \log\frac{35}{19}$ then find k if $\log\left\{\frac{x^2 + y^2}{x^2 + y^2}\right\} = \log k$

(a) $\frac{27}{8}$ (b) $\frac{29}{37}$

(c) $\frac{18}{7}$ (d) $\frac{13}{5}$

8. If $\log_p 9 = q$, $\log_p 8 = r$ then find $\log_{\sqrt{8}} 3$

(a) $\frac{q}{r}$ (b) $\frac{r}{q}$

(c) pq (d) $\frac{pq}{p+q}$

9. Find the minimum value of expression $3\log_{10} x - \log_x 0.001$ givne that $x > 1$.

(a) 3 (b) –3

(c) 0 (d) None of these

10. If $K = 1/(\log_2\pi) + 1/(\log_4\pi) + 1/(\log_8\pi)$

(a) $2 < k < 3$ (b) $1 < k < 2$

(c) $3 < k < 4$ (d) $4 < k < 5$

11. Find the number of digits before decimal start in $(2.5)^{200}$

(a) 79 (b) 80

(c) 81 (d) None of these

12. Find the number of roots of the equation

$x^{\frac{\log x+5}{3}} = 10^{5+\log x}$

(a) 2 (b) 0

(c) 1 (d) None of these

13. Simplify $(\log_3 4 + \log_2 9)^2 - (\log_3 4 - \log_2 9)^2$

(a) 2 (b) 1

(c) 4 (d) None of these

14. Simplify $10^{3\log 2 - 2\log(\log 10^3) + \log((\log 10^6)^2)}$

(a) 23 (b) 32

(c) 41 (d) None of these

15. Simplify $(0.8)(1 + 9^{\log_3 8})^{\log_{65} 5}$

(a) 2 (b) 32

(c) 4 (d) None of these

16. Find the sum of the roots of the equation $\log_{\frac{1}{3}} x - 3\sqrt{\log_{\frac{1}{3}} x} + 2 = 0$

(a) $\frac{28}{81}$ (b) $\frac{28}{27}$

(c) $\frac{26}{81}$ (d) None of these

17. Simplify $\frac{\log_3 135}{\log_{15} 3} - \frac{\log_3 5}{\log_{405} 3}$

(a) 7 (b) 6

(c) 5 (d) None of these

18. Simplify $\log_2 \log_9 \sqrt{6 + \sqrt{6 + \sqrt{6 + \ldots\infty}}}$

(a) 7 (b) 6

(c) 1 (d) None of these

19. Simplify $2^{(\log_2 3.\, \log_3 4.\, \log_4 5 \ldots \log_{99} 100)}$

(a) 16 (b) 100

(c) 1 (d) None of these

20. Simplify $\frac{1}{\log_5 3} + \frac{1}{\log_6 3} - \frac{1}{\log_{10} 3}$

(a) 16 (b) 100

(c) 1 (d) None of these

21. Find the number of zeros after decimal before a significant figure (Non zero digit) start in $\left(\frac{9}{8}\right)^{-100}$

(a) 7 (b) 6

(c) 5 (d) None of these

22. If $x = \dfrac{\sqrt{10}+\sqrt{2}}{2}$ and $y = \dfrac{\sqrt{10}-\sqrt{2}}{2}$, then the value of $\log_2(x^2 - xy + y^2)$, is equal to

(a) 0 (b) 2

(c) 3 (d) 4

23. Find the number of zeros after decimal before a significant figure (non zero digit) start in $(0.35)^{12}$ given $\log 7 = 0.8451$

(a) 7 (b) 6

(c) 4 (d) None of these

24. Find the value of k if $2^{2\log_2\left(3^{3\log_3 4}\right)} = 2^k$

(a) 12 (b) 16

(c) 24 (d) None of these

25. The number $N = 5\log_{10} 2 + 2\log_{10} 31$, lies between two successive integers whose sum is equal to

(a) 5 (b) 7

(c) 9 (d) None of these

26. Find the sum of all the rational roots of the equation $2^{\log_2 e^{ln 5^{\log_5 7^{\log_7 10^{\log_{10}(8x-3)}}}}}$

(a) 4 (b) 2

(c) 18/5 (d) None of these

Concept Cracker (CC)

1. If $a = \sqrt[2]{b} = \sqrt[3]{c} = \sqrt[4]{d} = \sqrt[5]{5}$. If $N = (a \times b \times c \times d)^{100}$ then find the number of digits in N. Given that $\log 2 = 0.3010$.

(a) 210 (b) 140

(c) 240 (d) 180

2. Find the sum of the series–

$$\frac{1}{\log_2 4} + \frac{1}{\log_4 4} + \frac{1}{\log_8 4} \ldots\ldots \frac{1}{\log_{(2^{20})} 4}$$

(a) 81 (b) 51log2

(c) 105 (d) 144

3. If $a = b^2 = c^3 = d^3 = \ldots\ldots z^{26}$ then find the value of $\log_a(abcdef..xyz)$

(a) $\sum_{n=1}^{n=26} n$ (b) $\frac{1}{26}\sum_{n=1}^{n=26} n$

(c) $\sum_{n=1}^{n=26} \frac{1}{n}$ (d) $\frac{1}{26}\sum_{n=1}^{n=26} \frac{1}{n}$

4. If $N = \dfrac{1}{\log_2 a} + \dfrac{1}{\log_2 b} + \dfrac{1}{\log_2 c} \ldots\ldots\ldots\ldots \dfrac{1}{\log_2 z}$

If $N = \log_a(2^k)$. Find the value of k.

(a) 351 (b) 261

(c) 2^{26} (d) 3389

5. If $81^x = y$ and $9y^2 = 3$ then which of the following is a correct relation between x and y.

(a) $\dfrac{1}{x^2} + \dfrac{1}{y^2} = 69$ (b) $\dfrac{1}{x^2} + \dfrac{1}{y^2} = \dfrac{1}{69}$

(c) $\dfrac{1}{x^2} + \dfrac{1}{y^2} = 67$ (d) $\dfrac{1}{x^2} + \dfrac{1}{y^2} = \dfrac{1}{67}$

6. Find the value of x if

$$\frac{x}{\log 3} + \frac{x}{\log 9} + \frac{x}{\log 81} + \ldots\infty = \log 81$$

Assume $\log 3 = k$

(a) $4k^2$ (b) $3k^2$

(c) $2k^2$ (d) k^2

7. Find the value of x if

$$3^{\left\{\frac{1}{\log_x 9}\right\}} \times 3^{\left\{\frac{1}{\log_x 81}\right\}} \times 3^{\left\{\frac{1}{\log_x 6561}\right\}} \times \ldots = 81$$

(a) 3 (b) 9

(c) $\sqrt{3}$ (d) 81

8. If $\log(x + y) - \log 2 = \log_k k$ and $\log(x + y) - \log xy = \log\frac{5}{16}$

Find the value of $\log_x y + \log_y x$

(a) 5/3 (b) 6/3
(c) 20/3 (d) None of these

9. If $\log_{(x+y)}(x^2 - y^2) = k$, then $\log_{(x-y)}(x^2 - y^2)$ is equal to

(a) $\frac{k}{k+1}$ (b) $\frac{k+1}{k-1}$
(c) $\frac{k-1}{k+1}$ (d) None of these

10. $\log_{\sqrt{a}} b^2 \times \log_{\sqrt{b}} c^2 \times \log_{\sqrt{c}} d^2 \times$ ......$\log_{\sqrt{y}} z^2 \times \log_{\sqrt{z}} a^2$

is equal to

(a) 2^{27} (b) 2^{52}
(c) 2^{54} (d) 2^{26}

11. Find the sum of all the values of

x : $\log_5\left(5^{1/x} + 125\right) = \log_5(6) + 1 + \frac{1}{2x}$

(a) ¾ (b) ½
(c) 2/3 (d) None of these

12. Find the product xyz where x, y and z satisfy

$\log_3 x = \log 7 - \log 11$

$\log_7 y = \log 11 - \log 3$

$\log_{11} z = \log 3 - \log 7$

(a) 0 (b) 1
(c) 241 (d) None of these

13. If the value of $\left(\frac{1}{\log_2 \pi} + \frac{1}{\log_6 \pi}\right)$ lies between two consecutive integer then find the sum of those two integers.

(a) 7 (b) 6
(c) 5 (d) None of these

14. Let $a > 1$ be a real number then find the number of roots of the equation, for a unique value of 'a'. $a^{2\log_2 x} = 5 + 4x^{\log_2 a}$

(a) 0 (b) 1
(c) 2 (d) None of these

15. Find the number of non-integral roots of the equation

$$\log_2^2(x^2 - x) - 4\log_2(x-1)\log_2 x = 1,$$

(a) 0 (b) 1
(c) 2 (d) None of these

16. Find the sum of all the rational roots of the equation, $x^{\left(\frac{3}{4}\right)(\log_2 x)^2 + \log_2 x - \left(\frac{5}{4}\right)} = \sqrt{2}$

(a) 9/4 (b) 9/2
(c) 8 (d) None of these

17. Find the number of integral roots of the equation

$\log_4\left(2\log_3\left(1 + \log_2\left(1 + 3\log_3 x\right)\right)\right) = \frac{1}{2}$

(a) 2 (b) 1
(c) 0 (d) None of these

18. If $\log \sin x + \log \cos x = -2$ and $\log(\sin x + \cos x) = \frac{(\log k) - 2}{2}$ then the value of 'K' is

(a) 101 (b) 51
(c) 120 (d) 102

19. Find the sum of all the value of x satisfying $\log_{10}(2^x + 3x - 51) = x(1 - \log_{10} 5)$

(a) 51 (b) 17
(c) 21 (d) None of these

20. Find the number of positive integers which have the characteristic 4, when the base of the logarithm is 5 ?

(a) 2500 (b) 625
(c) 2499 (d) None of these

PART-B

2.3 ADVANCE MATERIAL FOR JEE ADVANCED, KVPY AND OLYMPIAD

2.3.1 Logarithmic & Exponential Functions

Learning in this Chapter:

- How to plot the graph of log function
- How to plot graph of a exponential function
- How to plot graph of $e^{[x]}$
- How to plot graph of $e^{\{x\}}$
- How to plot graph of $e^{\sin x}$, $e^{\cos x}$, $e^{\tan x}$, etc…
- How to plot graph of log $[x]$
- How to plot graph of log $\{x\}$
- How to plot graph of log(sin x), log(cos x) etc…

Graph 1: $y = 2^x$

In order to understand the graph of $y = 2^x$ take different values of x and then try to understand the nature of graph:

x	$y = 2^x$
0	1
1	2
2	4
3	8
-1	½
-2	¼
-3	1/8

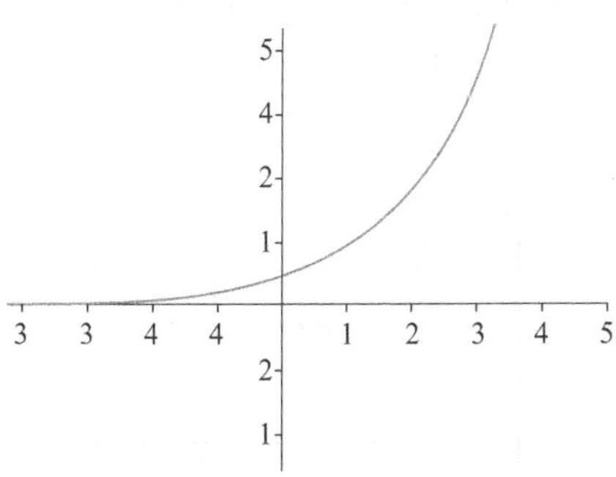

Features of the graph:

(1) Domain of the function $(-\infty, \infty)$

(2) Range of the function $(0, \infty)$

(3) Function is strictly increasing

Question That can be asked:

1. Find the number of roots of the equation $x^2 = 2^x$

(a) 0 (b) 1

(c) 2 (d) None of these

1. From the graph there exist two roots of the equation.

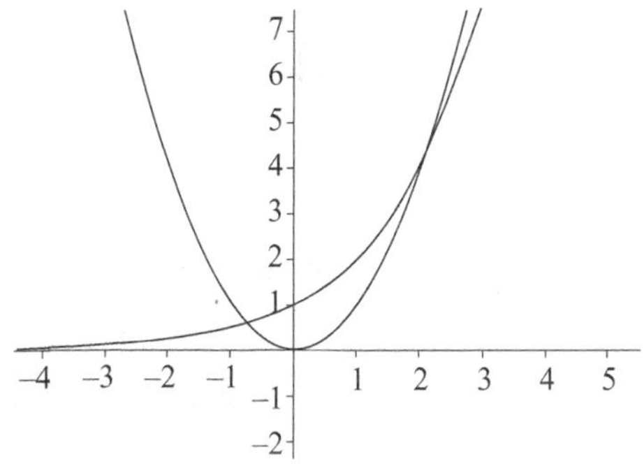

Option (c)

Graph 2: $y = 2^x$, 3^x, & 4^x

In order to understand the graph of $y = 2^x$, 3^x, 4^x take different values of x and then try to understand the nature of graph-

x	$y = 2^x$	$y = 3^x$	$y = 4^x$
0	1	1	1
1	2	3	4
2	4	9	16
3	8	27	64
– 1	½	1/3	¼
– 2	¼	1/9	1/16
– 3	1/8	1/27	1/64

Features of the graph:

(1) $2^x < 3^x < 4^x$ for $x > 0$

(2) $2^x > 3^x > 4^x$ for $x < 0$

Question That can be asked:

2. Find the number of roots of the equation $4^x = 2^x$

(a) 0 (b) 1

(c) 2 (d) None of these

2. From the graph there exist one root of the equation.

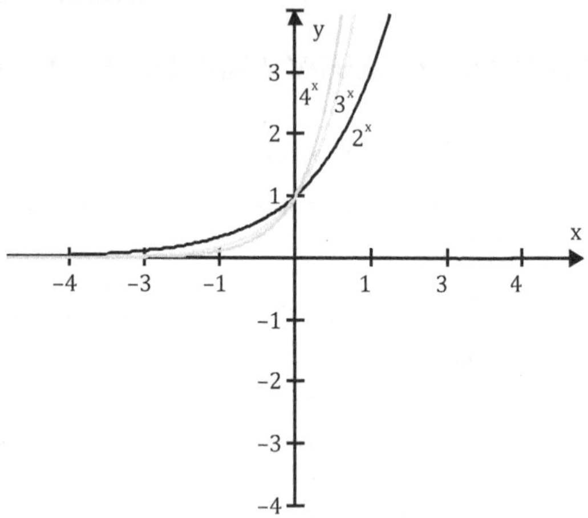

Option (b)

Graph 3 : $y = a^x$ when $0 < a < 1$

When $a > 1$ then graph of $y = a^x$ is similar to 2^x but when $0 < a < 1$ then graph is rotated with 180° as shown in the graph

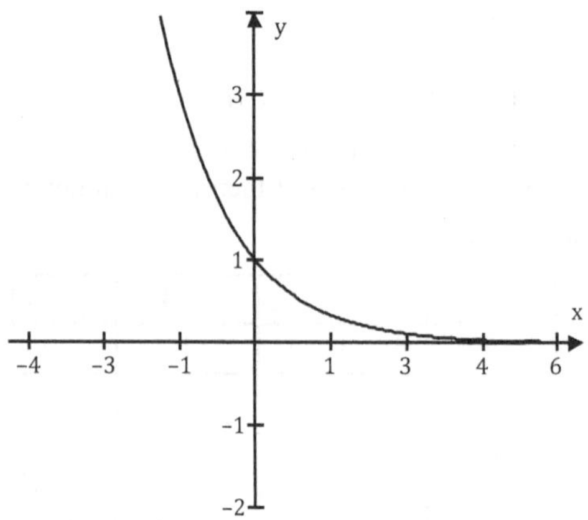

Features of the graph:

(1) Domain of the function $(-\infty, \infty)$

(2) Range of the function $(0, \infty)$

(3) Function is strictly decreasing

Graph 4 : $y = max(2^x, 2^{-x})$

In order to plot the graph of $y = \max(2^x, 2^{-x})$ 1st plot both the graphs $y = 2^x$ & 2^{-x}
when $x < 0$ then $2^x < 2^{-x}$ so $y = \max(2^x, 2^{-x}) = 2^{-x}$
when $x > 0$ then $2^x > 2^{-x}$ so $y = \max(2^x, 2^{-x}) = 2^x$

The Graph is as shown below

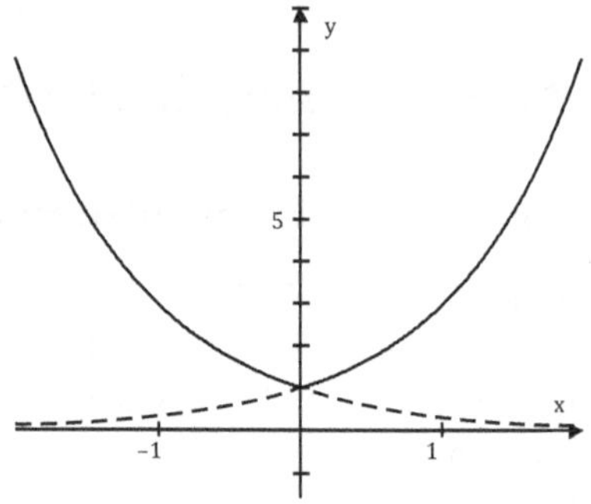

Question That can be asked:

3. Find the number of roots of the equation $x^2 = \min(2^x, 2^{-x})$

(a) 0 (b) 1
(c) 2 (d) None of these

3. From the graph there exist two roots of the equation.

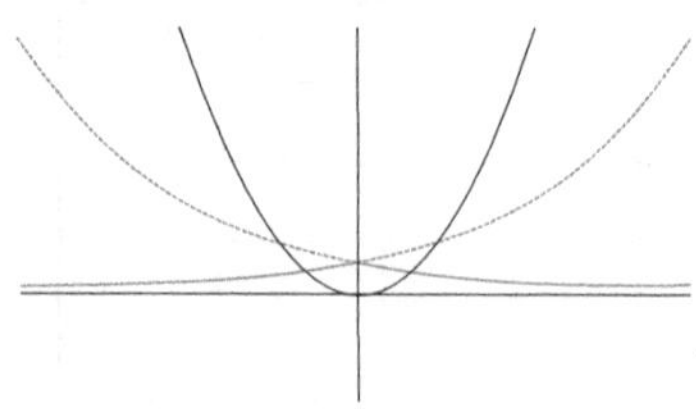

Option (c)

Graph 5 : $y = e^{(x^2)}$

In order to understand the graph of $y = e^{x^2}$ take different values of x and then try to understand the nature of graph:

x	$y = e^{x^2}$
0	1
1	e
2	e^4
3	e^9
– 1	e
– 2	e^4
– 3	e^9

Features of the graph:

(1) Domain of the function $(-\infty, \infty)$

(2) Range of the function $[1, \infty)$

(3) Function is strictly decreasing when $x \in (-\infty, 0]$ and strictly increasing in $[0, \infty)$

Question That can be asked:

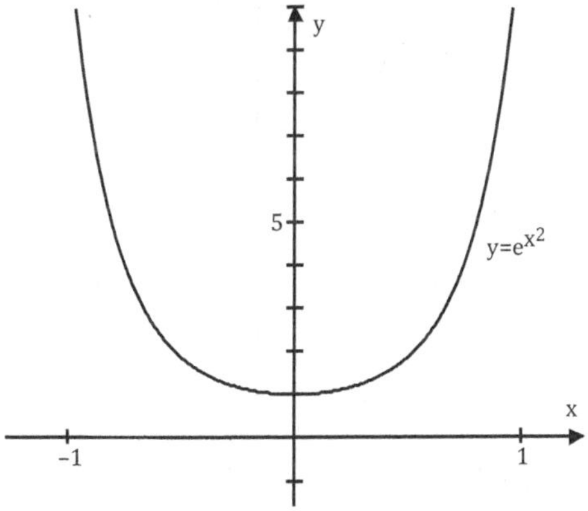

4. Find the minimum value of $y = e^{x^2}$

(a) 0 (b) 1

(c) e (d) None of these

4. From the graph minimum value is 1 when $x = 0$

Option (b)

Graph 6 : $y = \log x$

In order to understand the graph of $y = \log x$ take different values of x and then try to understand the nature of graph here we have to take care that log function is defined only for $x > 0$

x	$y = \log x$
1	0
2	0.3010
10	1
100	2
1000	3
0.1	–1
0.01	–2

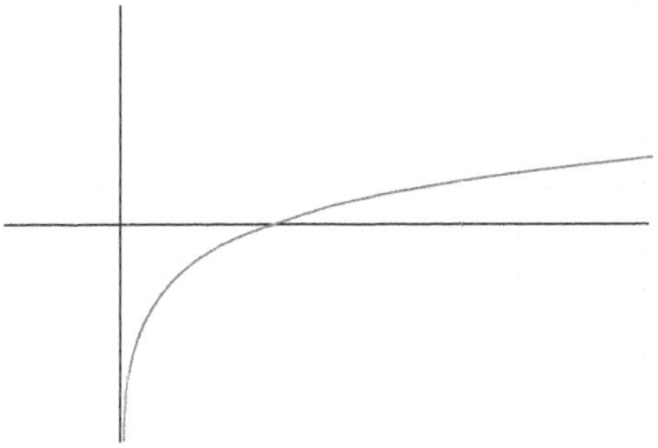

Features of the graph:

(1) Domain of the function (o, ∞)

(2) Range of the function $(-\infty, \infty)$

(3) Function is strictly increasing

Question That can be asked:

5. Find the number of roots of the equation $\log x = \sin x$

(a) 0 (b) 1

(c) 2 (d) None of these

5. At $x = 5\pi/2$ $\sin x = 1$, but since $5\pi/2 < 10$ so $\log x < 1$

From the graph number of roots is 3

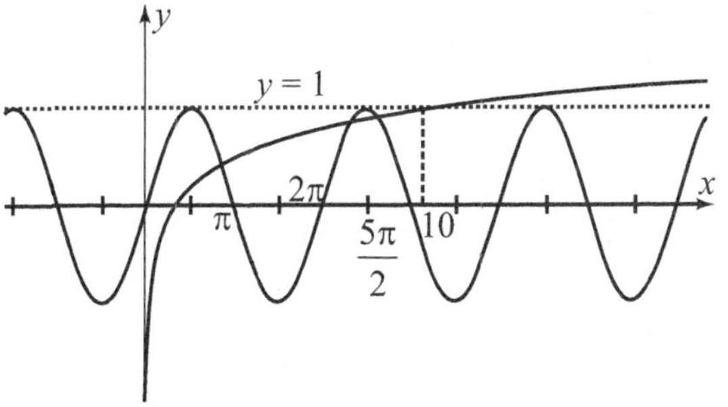

Option (d)

Graph 7 : $y = \log_a x$

(1) $y = \log_a x$ when $a > 1$

The graph of $y = \log_a x$ when $a > 1$ is similar to $\log x$ as $y = \log x$ can also be expressed as $y = \log_{10} x$

In fact $y = \log_a x$ is reflection with respect to $y = x$ line of the function $y = a^x$ (i.e they are inverse functions)

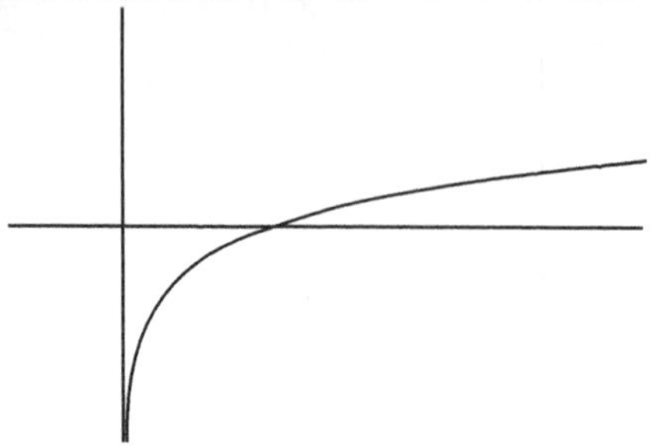

(2) $y = \log_a x$ when $0 < a < 1$

The graph of $y = \log_a x$ when $a < 1$ is as shown

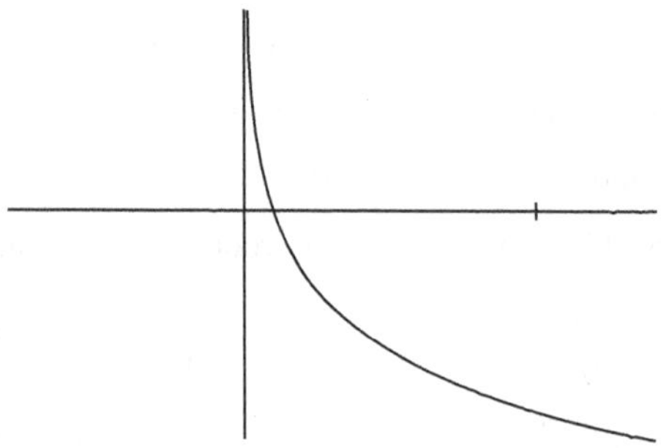

In fact $y = \log_a x$ is reflection with respect to $y = x$ line of the function $y = a^x$ (i.e they are inverse functions) when $0 < a < 1$

Advance Graph

Graph 8 : $y = e^x$ Family

1. $y = e^x$

Features of the graph:

(1) Domain of the function $(-\infty, \infty)$

(2) Range of the function $(0, \infty)$

(3) Function is strictly increasing

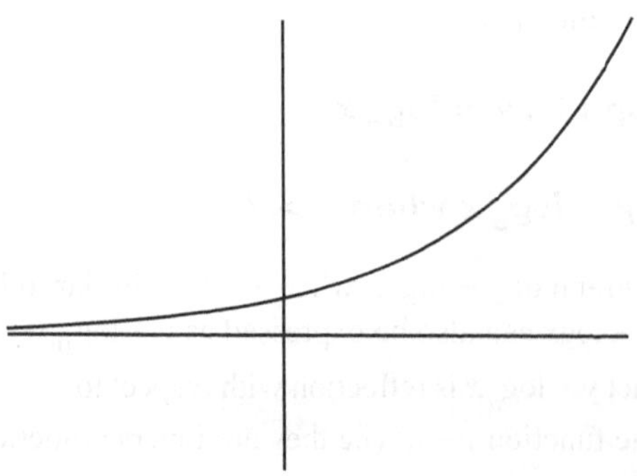

2. $y = e^{-x}$

Features of the graph-

(1) Domain of the function $(-\infty, \infty)$

(2) Range of the function $(0, \infty)$

(3) Function is strictly decreasing

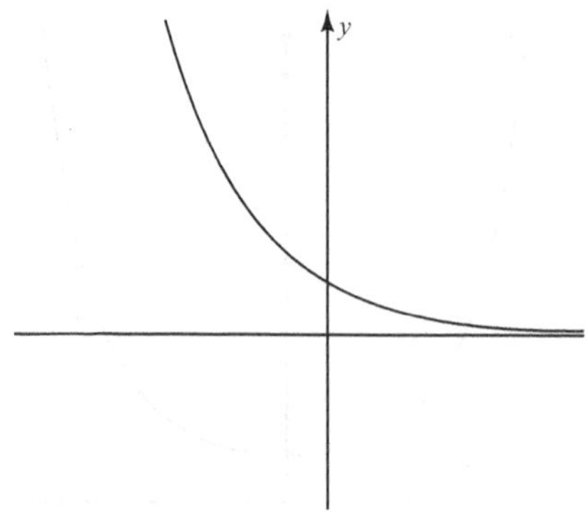

3. $y = e^{|x|}$

The function $y = e^{|x|}$ is defined as:

$$y = e^{|x|} = \begin{cases} e^{-x} \text{ when } x < 0 \\ e^{x} \text{ when } x > 0 \end{cases}$$

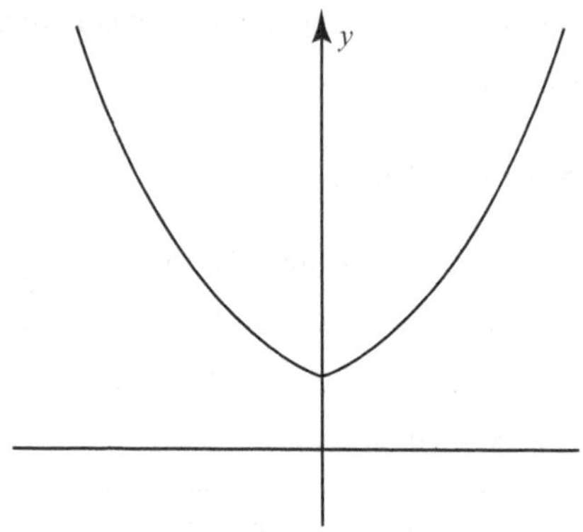

Features of the graph:

(1) Domain of the function $(-\infty, \infty)$

(2) Range of the function $[1, \infty)$

(3) Function is continuous

4. $y = e^{[x]}$

In order to plot $y = e^{[x]}$ 1st plot $y = e^x$ and then convert this graph of $y = f(x)$ to $y = f([x])$

The function $y = e^{[x]}$ is defined as:

$y = e^{[x]} = 1$ when $x \in [0, 1)$

$y = e^{[x]} = e$ when $x \in [1, 2)$ etc

Features of the graph-

(1) Domain of the function $(-\infty, \infty)$

(2) Function is discontinuous

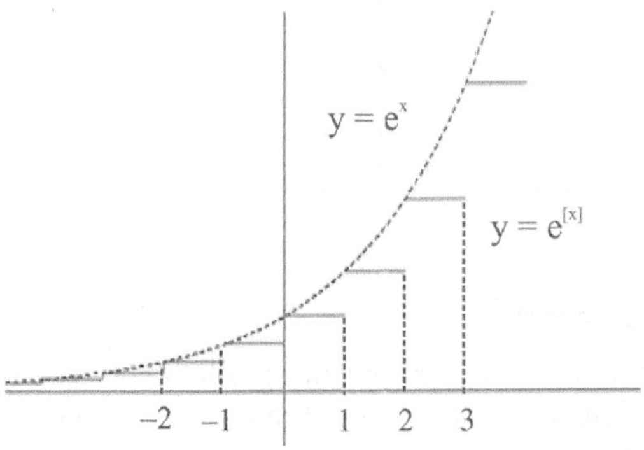

5. $y = [e^x]$

In order to plot $y = [e^x]$ 1st plot $y = e^x$ and then convert this graph of $y = f(x)$ to $y = [f(x)]$

The function $y = [e^x]$ is defined as-

$y = [e^x] = 0$ when $x \in (-\infty, 0)$

$y = [e^x] = 1$ when $x \in [0, \ln 2)$

$y = [e^x] = 2$ when $x \in [\ln 2, \ln 3)$ etc

Features of the graph:

(1) Domain of the function $(-\infty, \infty)$

(2) Function is discontinuous

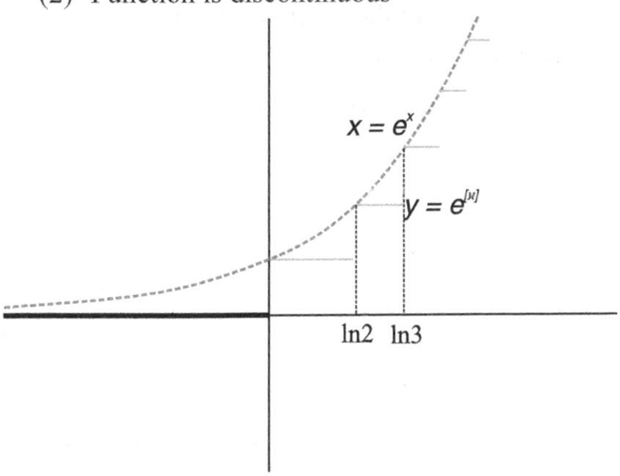

6. $y = e^{\{x\}}$

In order to plot $y = e^{\{x\}}$ we should define the function piecewise and it is periodic with period 1

Minimum value of the function is $e^0 = 1$ and maximum value is e

(1) Domain of the function $(-\infty, \infty)$

(2) Function is discontinuous at integral values of x

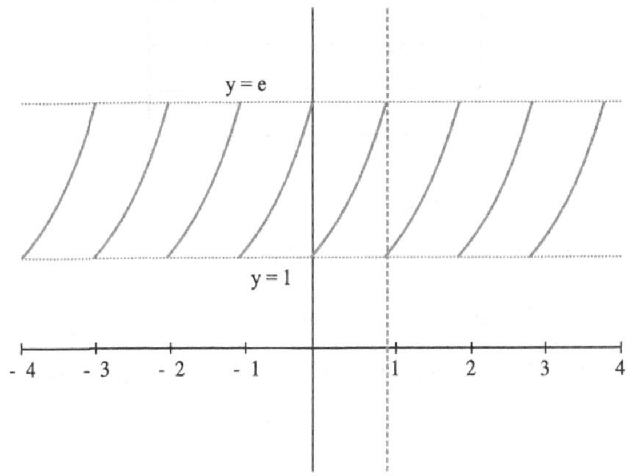

7. $y = \{e^x\}$

In order to plot $y = \{e^x\}$ we should define the function piecewise

When $x \in (-\infty, 0)$ then $y = \{e^x\} = e^x$

When $x \in [0, \ln 2)$ then $y = \{e^x\} = e^x - 1$

When $x \in [\ln 2, \ln 3)$ then $y = \{e^x\} = e^x - 2$ and so on

(1) Domain of the function $(-\infty, \infty)$

(2) Function is discontinuous

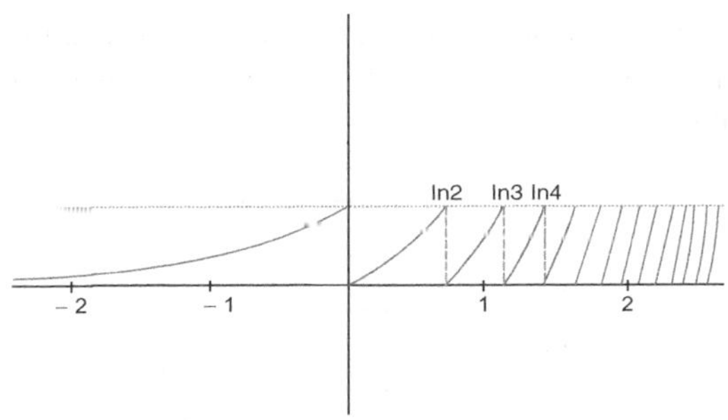

8. $y = e^{x^2}$

In order to understand the graph of $y = e^{x^2}$ take different values of x and then try to understand the nature of graph:

x	$y = e^{x^2}$
0	1
1	e
2	e^4
3	e^9
− 1	e
− 2	e^4
− 3	e^9

Features of the graph:

(1) Domain of the function $(-\infty, \infty)$

(2) Range of the function $[1, \infty)$

(3) Function is strictly decreasing when $x \in (-\infty, 0]$ and strictly increasing in $[0, \infty)$

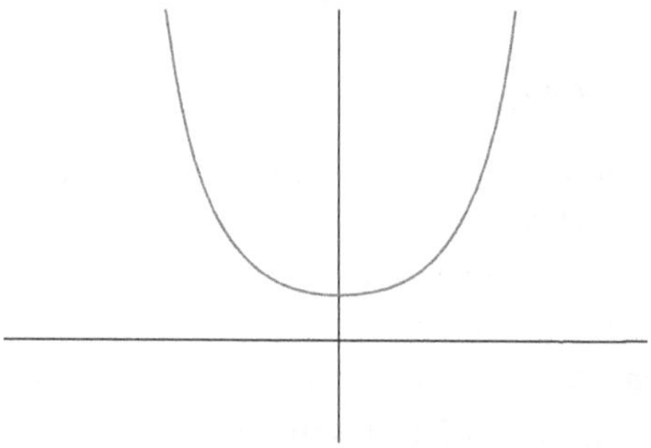

9. $y = e^{x^3}$

In order to understand the graph of $y = e^{x^3}$ take different values of x and then try to understand the nature of graph:

x	$y = e^{x^3}$
0	1
1	e
2	e^9
3	e^{27}
− 1	$1/e$
− 2	$\frac{1}{e^8}$
− 3	$\frac{1}{e^{27}}$

Features of the graph:

(1) Domain of the function $(-\infty, \infty)$

(2) Range of the function $(0, \infty)$

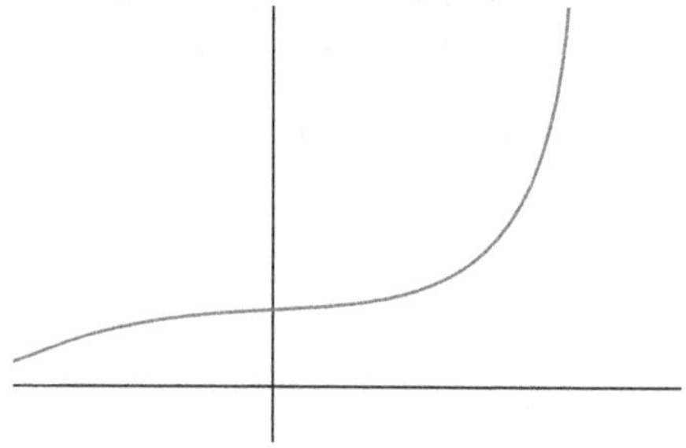

10. $y = e^{\sqrt{x}}$

In order to understand the graph of $y = e^{\sqrt{x}}$ take different values of x and then try to understand the nature of graph:

x	$y = e^{\sqrt{x}}$
0	1
1	e
4	e^2
9	e^3

Features of the graph:

(1) Domain of the function $[0, \infty)$

(2) Range of the function $[1, \infty)$

(3) Function is strictly increasing in $[0, \infty)$

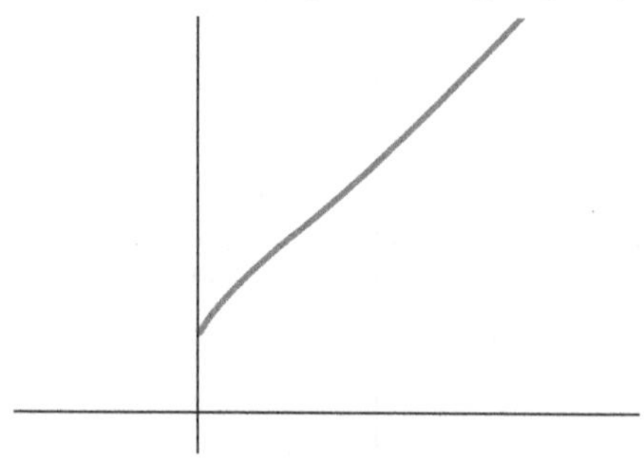

11. $y = e^{\sin x}$

In order to understand the graph of $y = e^{\sin x}$ take different values of x and then try to understand the nature of graph:

x	Sin x	$y = e^{\sin x}$
0	0	1
$\frac{\pi}{2}$	1	e
π	0	1
$\frac{3\pi}{2}$	– 1	$\frac{1}{e}$

Features of the graph:

(1) Domain of the function $(-\infty, \infty)$

(2) Range of the function $[\frac{1}{e}, \infty)$

(3) Function is periodic with period 2π

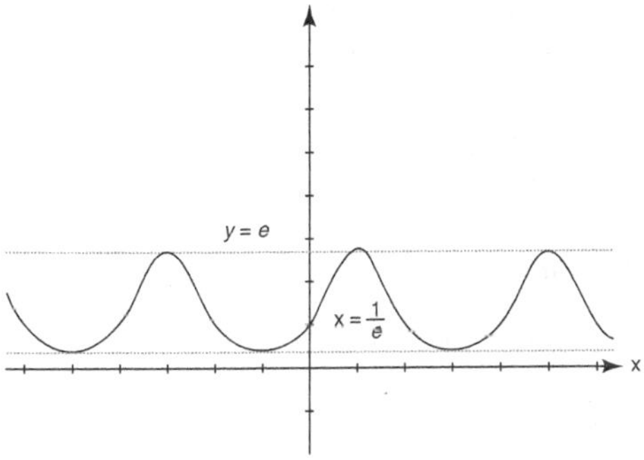

12. $y = e^{\cos x}$

In order to understand the graph of $y = e^{\cos x}$ take different values of x and then try to understand the nature of graph:

x	cos x	$y = e^{\cos x}$
0	1	e
$\frac{\pi}{2}$	0	1
π	– 1	$\frac{1}{e}$
$\frac{3\pi}{2}$	0	1

Features of the graph:

(1) Domain of the function $(-\infty, \infty)$

(2) Range of the function $[\frac{1}{e}, e]$

(3) Function is periodic with period 2π

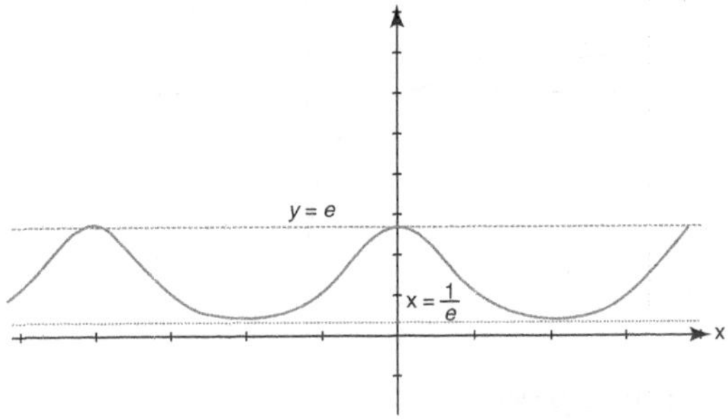

To understand some more graph like $e^{\tan x}$, $e^{\text{cosec } x}$, $e^{\sec x}$ etc, follow "Magical Graphs & Key Revision Notes (Disha publication)" by Gajendra Kumar.

$y = \log_a x$ Family

1. y = log x

Features of the graph:

(1) Domain of the function (o, ∞)

(2) Range of the function $(-\infty, \infty)$

(3) Function is strictly increasing

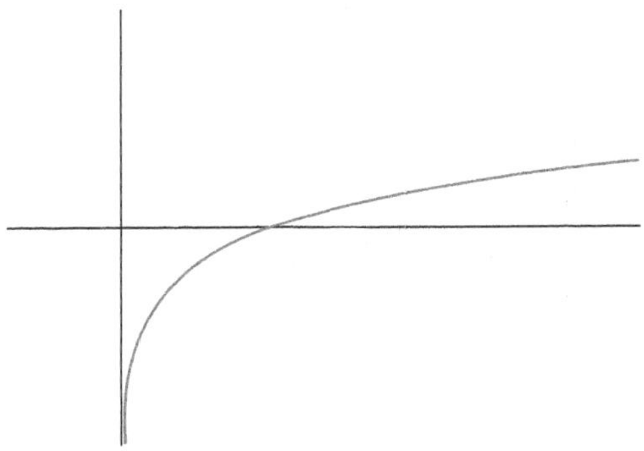

2. y = log (– x)

Graph of y = log (– x) can be obtained when graph of is rotated by 180°

Features of the graph:

(1) Domain of the function $(-\infty, 0)$

(2) Range of the function $(-\infty, \infty)$

(3) Function is strictly decreasing

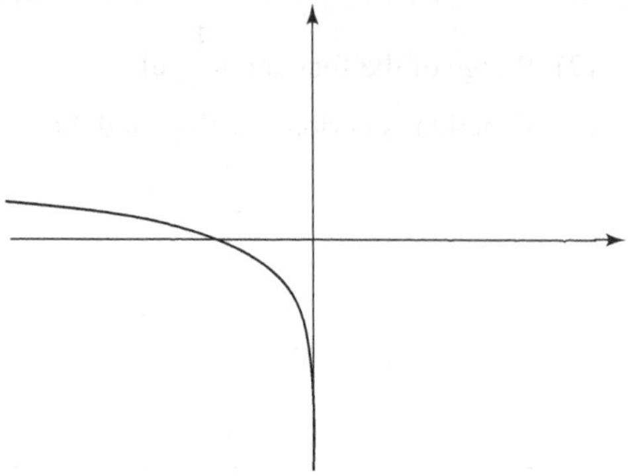

3. $y = \log|x|$

In order to understand the graph of $y = \log|x|$ take different values of x and then try to understand the nature of graph

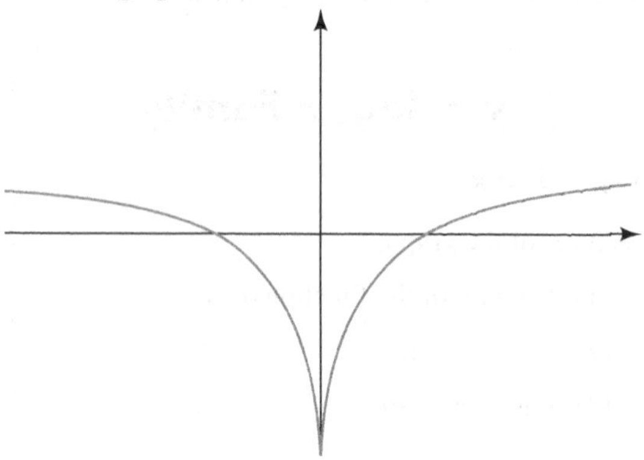

x	$\|x\|$	$y = \log\|x\|$
1	1	0
– 1	1	0
10	10	1
– 10	10	1
100	100	2
– 100	100	2

Features of the graph:

(1) Domain of the function $(-\infty, \infty) - \{0\}$
(2) Range of the function $(-\infty, \infty)$
(3) Function is discontinuous

4. $y = \log (x^2)$

In order to understand the graph of $y = \log (x^2)$ take different values of x and then try to understand the nature of graph

x	x^2	$y = \log x^2$
1	1	0
– 1	1	0
$\sqrt{10}$	10	1
$-\sqrt{10}$	10	1
10	100	2
– 10	100	2

Features of the graph:

(1) Domain of the function $(-\infty, \infty)-\{0\}$
(2) Range of the function $(-\infty, \infty)$
(3) Function is discontinuous

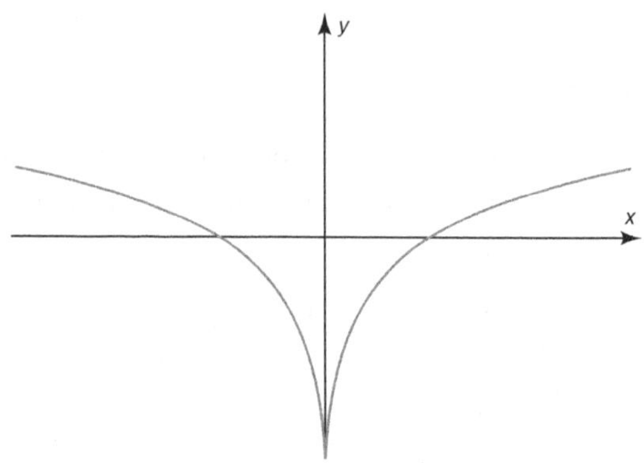

5. $y = \log x^3$

Graph of $y = \log x^3$ is similar to $y = \log x$ but

$\log x^3 > \log x$ when $x > 1$ and

$\log x^3 < \log x$ when $x < 1$ and

Features of the graph:

(1) Domain of the function $(0, \infty)$

(2) Range of the function $(-\infty, \infty)$

(3) Function is continuous

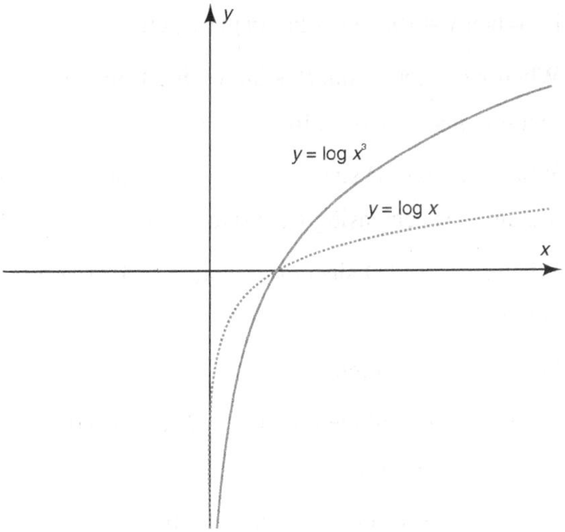

6. y = |log x|

Graph of $y = |\log x|$ is reflection of $y = \log x$ with respect to X-axis as shown in the figure.

Features of the graph:

(1) Domain of the function $(0, \infty)$

(2) Range of the function $[0, \infty)$

(3) Function is continuous

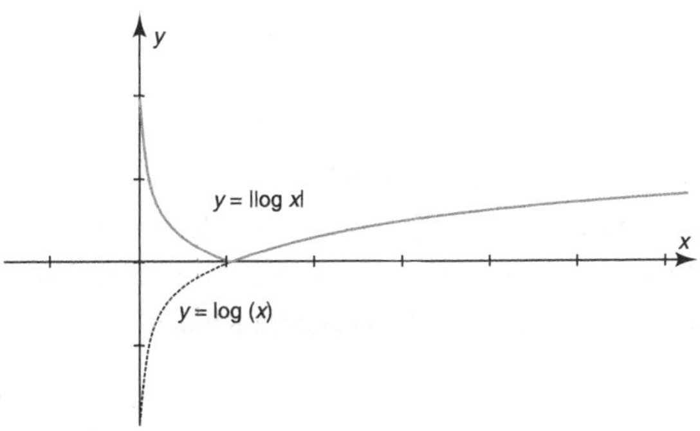

7. y = [log x]

In order to plot the graph of $y = [\log x]$ 1st plot the graph of $y = \log x$ and then transform the graph of $y = f(x)$ to $y = [f(x)]$ as shown in the figure.

Features of the graph:

(1) Domain of the function $(0, \infty)$

(2) Range of the function is whole numbers

(3) Function is discontinuous

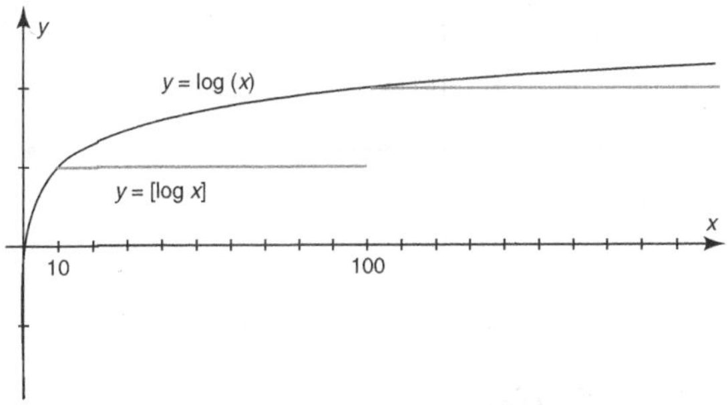

8. y = {log x}

In order to plot the graph of $y = \{\log x\}$ 1st plot the graph of $y = \log x$ and then transform the graph of $y = f(x)$ to $y = \{f(x)\}$ as shown in the figure.

Features of the graph:

(1) Domain of the function $(0, \infty)$

(2) Range of the function is $[0, 1)$

(3) Function is discontinuous

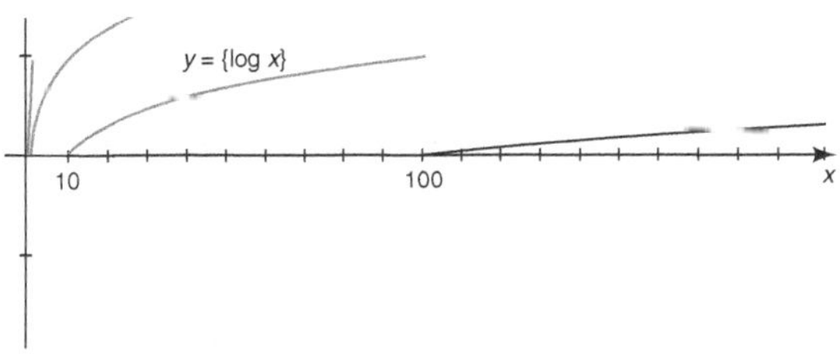

9. y = log[x]

In order to plot the graph of $y = \log[x]$ 1st define the function piece wise

When $0 < x < 1$ function is not defined

When $1 \leq x < 2$ then $\log [x] = 0$

When $2 \leq x < 3$ then $\log [x] = \log 2$

Features of the graph:

(1) Domain of the function $[1, \infty)$

(2) Function is discontinuous

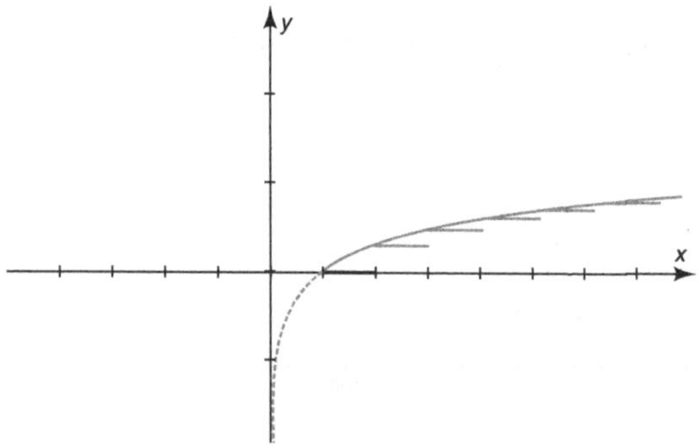

10. y = log{x}

In order to plot the graph of $y = \log\{x\}$ 1st define the function piece wise since range of $y = \{x\}$ is $[0, 1)$ so range of graph $\log \{x\}$ is $(-\infty, 0)$

Features of the graph:

(1) Domain of the function $(-\infty, \infty)$

(2) Range of the function $(-\infty, 0)$

(3) Function is discontinuous

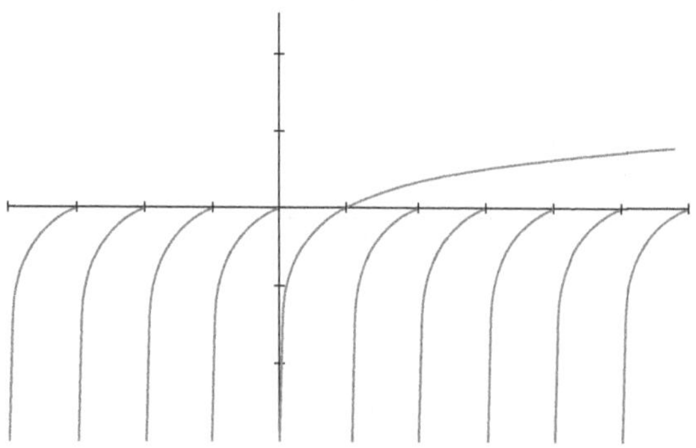

11. y = log{sin x}

In order to plot the graph of $y = \log(\sin x)$ 1st understand the graph of $y = \sin(x)$, since log function is defined only when for positive numbers so function $y = \log(\sin x)$ is defined only when $\sin(x)$ is positive ie. When $x \in (0, \pi) \cup (2\pi, 3\pi) \cup \ldots$ etc.

When $x \in (0, \pi/2)$ $\sin(x)$ is increasing from 0 to 1 so $\log(\sin x)$ is increasing from $-\infty$ to 0.

When $x \in (\pi/2, \pi)$ $\sin(x)$ is decreasing from 1 to 0 so $\log(\sin x)$ is decreasing from 0 to $-\infty$

When $x \in (\pi, 2\pi)$ $\sin(x)$ is negative so $\log(\sin x)$ is not defined.

Features of the graph:

(1) Domain of the function $(-2\pi, -\pi) \cup (0, \pi) \cup (2\pi, 3\pi)$ etc

(2) Range of the function $(-\infty, 0)$

(3) Function is discontinuous

(4) Function is periodic with period 2π

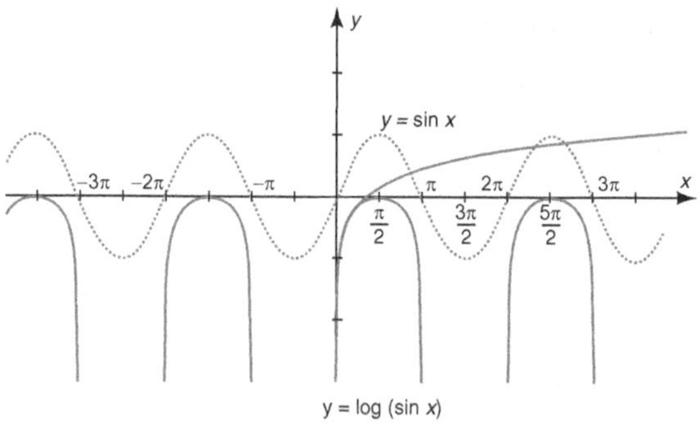

To learn some more graphs like $y = \log\{\cos x\}$, $y = \log\{\tan x\}$, $y = \log \{\text{cosec } x\}$ $y = \log(x^2 - 6x + 5)$, follow "Magical Graphs & Key Revision Notes (Disha publication)" by Gajendra Kumar.

Exercise B

Concept Deviator (CD)

(Single Correct MCQs)

1. Find the number of digits in $6 \times 8^8 \times 10^{10} \times 12^{12}$ given that log2 = 0.3010 and log 3 = 0.4771
   (a) 23 (b) 32
   (c) 34 (d) None of these

2. If $\log_7 \log_5 \left(\sqrt{x+5}+\sqrt{x}\right) = 0$, what is the value of x?
   (a) 2 (b) 3
   (c) 4 (d) 5

3. Consider an equation $[\log_{10}1] + [\log_{10}2] + \ldots [\log_{10}x] = x$ here $[x]$ is greatest integer function. Find the value of x.
   (a) 107 (b) 108
   (c) 109 (d) 110

4. Consider an equation $[\log_{20}1] + [\log_{20}2] + \ldots [\log_{20}x] = x$ here $[x]$ is greatest integer function. Find the value of x.
   (a) 417 (b) 418
   (c) 209 (d) 420

5. Solve for x, $8^{\log_6 x} - 36^{\log_x 6} = 58$
   (a) 6 (b) 8
   (c) 36 (d) 64

6. Find the sum of the series S = $1/\log_3 9 + 1/\log_9 9 + 1/\log_{27} 9 + \ldots + 1/\log_{(3^n)} 9$
   (a) $n(n+1)/2$ (b) $2/n(n+1)$
   (c) $n(n+1)(2n+1)/12$
   (d) $n(n+1)/4$

7. Find the value of x where x is given by the following relationship-
   $$27^{\log_3 48} \times 3^{\log_{27}(48)^2} \times 27^{\log_3 (48)^3} \times 3^{\log_{27}(48)^4} \times \ldots \times 27^{\log_3 (48)^{99}} \times 3^{\log_{27}(48)^{96}}$$
   (a) 8350 (b) 8250
   (c) 8100 (d) None of These

8. Find the smallest integer satisfying the inequality $\log_{(x^2+1)}(x+2) < 1$
   (a) 1 (b) 2
   (c) –1 (d) 3

9. Find x if $\log_2 x \log_{(x/64)} 2 = \log_{(x/16)} 2$
   (a) 2 (b) 4
   (c) 16 (d) 12

10. If $p = \log_{12} 18$, $q = \log_{24} 54$ then the value of $pq + 5(p-q)$ is
    (a) 0 (b) 4
    (c) 1 (d) 9

11. If $x \in R$ satisfies $(\log_{10}(100)x))^2 + (\log_{10} 10x)^2 + \log_{10} x \le 14$ then find the range of x
    (a) $10^{-9/2} \le x \le 10$ (b) $10^{-7/2} \le x \le 10$
    (c) $10^{-9/2} \le x \le 12$ (d) $10^{-9/2} \le x \le 20$

12. Find the sum of all the rational roots of the equation $2\log_3(x-2) + \log_3(x-4)^2 = 0$
    (a) 0 (b) 4
    (c) 1 (d) 3

13. Positive numbers x, y and z satisfying $xyz = 10000$ and $(\log_{10} x)(\log_{10} y) + (\log_{10} x)(\log_{10} z) + \log_{10} y)(\log_{10} z) = 6$. Find the value of $(\log_{10} x)^2 + (\log_{10} y)^2 + (\log_{10} z)^2$
    (a) 4 (b) 6
    (c) 5 (d) None of these

14. Find the sum of all the possible values of x which satisfy $|x-2|^{(10x^2-1)} = |x-2|^{3x}$
    (a) 4.3 (b) 3.4
    (c) 5.2 (d) None of these

15. Find the number of roots of the equation $\log_{x^2} 16 + \log_{2x} 64 = 3$
    (a) 4 (b) 3
    (c) 2 (d) None of these

16. Find xyz if
    $$\log_2 x + \log_4 y + \log_4 z = 2$$
    $$\log_9 x + \log_3 y + \log_9 z = 2$$
    And
    $$\log_{16} x + \log_{16} y + \log_4 z = 2$$
    (a) 24 (b) 32
    (c) 12 (d) None of these

17. Find the sum of all the values of x which satisfy the equation $|x-3|^{(3x^2-10x+3)} = 1$

(a) $\frac{16}{3}$ (b) $\frac{17}{3}$

(c) $\frac{20}{3}$ (d) None of these

18. Find the number of solutions of the equations $\log_4(x^2-1) - \log_4(x-1)^2 = \log_4\sqrt{(4-x)^2}$

(a) 4 (b) 3

(c) 2 (d) None of these

19. For $x, y \in N$, if $3^{2x-y+1} = 3^{y-2x+1} - 8$ and $\log_6 |2x^2y - xy^2| = 1 + \log_{36}(xy)$, then point (x, y) lies in which quadrant?

(a) lies on axis (b) 1st quadrant

(b) 4th quadrant (d) None of these

20. If 'x' and 'y' are real numbers such that, $2\log(2y-3x) = \log x + \log y$, find $\frac{x}{y}$.

(a) 9/4 (b) 4/9

(c) 2/9 (d) None of these

(Passage Type)

Passage-I

Consider the system of equations

$$\log_{10}(2000xy) - \log_{10}x \cdot \log_{10}y = 4$$

$$\log_{10}(2yz) - \log_{10}y \cdot \log_{10}z = 1$$

And $\log_{10}(zx) - \log_{10}z \cdot \log_{10}x = 0$

21. Find the number of triplets (x, y, z) which satisfy the above system of equation

(a) 0 (b) 1

(c) 2 (d) None of these

22. Find the sum of all the possible values of x that satisfy the above system of equation

(a) 110 (b) 100

(c) 201 (d) None of these

Passage-II

If (x_1, y_1) and (x_2, y_2) are the solution of the system of equation

$$\log_{25}(x) + \log_{64}(y) = 4$$

$$\log_x(25) - \log_y(64) = 1$$

Then

23. Find the value of x_1x_2

(a) 5^{12} (b) 2^{12}

(c) 10^{12} (d) None of these

24. Find the value of y_1y_2

(a) 5^{12} (b) 2^{12}

(c) 10^{12} (d) None of these

Passage-III

If x satisfy the equation,

$$\sqrt{(\log_3\sqrt[3]{3x} + \log_x\sqrt[3]{3x}\cdot\log_3 x^3} + \sqrt{(\log_3\sqrt[3]{3/x} + \log_x\sqrt[3]{3/x}\cdot\log_3 x^3} = 2$$

25. Number of integral values of 'x' is given by

(a) 0 (b) 2

(c) 1 (d) None of these

26. Which one of the following is incorrect about 'x'

(a) 'x' must be less than 3

(b) There exist some integral value of 'x'

(c) At least one value of $x > 10$ exist

(d) None of these

(Integer Type)

27. Find the value of $16x$ if $2^{(\log_2 3)^x} = 3^{(\log_3 2)^x}$.

28. If a, b are prime numbers and satisfying $(2+\sqrt{3})^{\frac{1}{\log_a(2-\sqrt{3})} + \frac{1}{\log_b\left(\frac{\sqrt{3}-1}{\sqrt{3}+1}\right)}} = \frac{1}{35}$, then find $|a-b|$ is equal to

29. Find the number of integral solution of the equation $\log_{\sqrt{x}}(x + |x-4|) = \log_x(7x - 12 + 7|x-4|)$.

30. If $\log_{3x} 45 = \log_{4x} 40\sqrt{3}$ then find the characteristic of x^3 to the base 3 .

IIT JEE corner

31. The least value of the expression $2\log_{10}x - \log_x(0.01)$. for $x > 1$ is :
(a) 10 (b) 2
(c) – 0. 01 (d) None of thjese

32. Solve for x the following equation :
$$\log_{(2x+3)}(6x^2+23x+21) = 4 - \log_{(3x+7)}(4x^2+12x+9)$$

33. The equation $x^{\frac{3}{4}(\log_2 x)^2 + \log_2 x - \frac{5}{4}} = \sqrt{2}$ has :
(a) at least one real solution
(b) exactly three real solution
(c) exactly one irrational
(d) Complex roots

34. The number of solution of $\log_4(x-1) = \log_2(x-3)$ is :
(a) 3 (b) 1
(c) 2 (d) 0

35. Let (x_0, y_0) be the solution of the following equations
$(2x)^{\ln 2} = (3y)^{\ln 3}$
$3^{\ln x} = 2^{\ln y}$
Then x_0 is
(a) $\frac{1}{6}$ (b) $\frac{1}{3}$
(c) $\frac{1}{2}$ (d) 6

36. The value of
$$6 + \log_{\frac{3}{2}}\left(\frac{1}{3\sqrt{2}}\sqrt{4 - \frac{1}{3\sqrt{2}}\sqrt{4 - \frac{1}{3\sqrt{2}}\sqrt{4 - \frac{1}{3\sqrt{2}}\cdots}}}\right)$$
is

Solutions

Concept Applicator (CA)-1

1. $\frac{1}{8^{\log_3 2}} = 8^{\log_2 3} = 2^{3\log_2 3} = 2^{\log_2 27} = 27$

2. $\frac{1}{\log_3 2} + \frac{2}{\log_9 4} - \frac{3}{\log_{27} 8}$

$= \log_2 3 + 2 \log_4 9 - 3 \log_8 27$

$= 3 \log_2 3 - 3 \log_2 3 = 0$

3. $\log 2 + 16 \log \frac{16}{15} + 12 \log \frac{25}{24} + 7 \log \frac{81}{80}$

$= \log 2 + 16 \log 16 - 16 \log 15 + 12 \log 25 - 12 \log 24 + 7 \log 81 - 7 \log 80$

$= \log 2 + 64 \log 2 - 16 \log 5 - 16 \log 3 + 24 \log 5 - 12 \times 3 \log 2 - 12 \log 3 + 28 \log 3 - 7 \log 5 - 28 \log 2$

$= \log 2 + \log 5 = \log 10 = 1$

4. The given expression is

$(\log_2 8)(\log_8 16) - (\log_2 16)(\log_{16} 32)$

$=(\log_2 2^3)(\log_{(2^3)}(2^4)) - (\log_2(2^4))(\log_{(2^4)}(2^5))$

$= (3)\left(\frac{4}{3}\right) - (4)\left(\frac{5}{4}\right) = 4 - 5 = -1$

5. $\log_{30} 8 = 3 \log_{30} 2 = 3 \log_{30}(30/15) = 3(\log_{30} 30 - \log_{30} 3 - \log_{30} 5) = 3(1 - c - d)$

6. Let $\sqrt{5\sqrt{5\sqrt{5 \ldots \infty}}} = x$

$\Rightarrow \sqrt{5x} = x$

$\Rightarrow x^2 = 5x$

$\Rightarrow x = 5$

$\Rightarrow \log_5 5 = 1$

7. Consider

$2^{\sqrt{\log_2 3}} = 2^{(\log_2 3)\cdot\frac{1}{\sqrt{\log_2 3}}} = (2^{\log_2 3})^{\frac{1}{\sqrt{\log_2 3}}}$

$= 3^{\frac{1}{\sqrt{\log_2 3}}} = 3^{\sqrt{\log_3 2}}$

So $2^{\sqrt{\log_2 3}} = 3^{\sqrt{\log_3 2}}$ and $\frac{2^{\sqrt{\log_2 3}}}{3^{\sqrt{\log_3 2}}} = 1$

8. Let $N = 6^{50}$

Taking log both side with base 10 we get

$\log N = 50 [\log_{10} 6]$

$= 50 [\log_{10} 2 + \log_{10} 3]$

$= 50 [0.3010 + 0.4771]$

$= 50 [0.7781] = 38.9050$

Since Characteristic is 38 so number of digits = 38 + 1 = 39

9. Let $N = 5^{25}$

Taking log both side with base 10 we get

$\log N = 25 \log_{10} 5 = 25 [1 - \log_{10} 2]$

$= 25 [1 - 0.3010] = 17.4750$

Since Characteristic is 17 so number of digits 17 + 1 = 18

10. Let $N = 3^{-50}$

Taking log both side with base 10

$\log_{10} 3^{-50} = -50 \log_{10} 3 = -50 (0.4771)$

$= -23.8550$

Since Characteristic is –24 so number of zero after decimal before a significant figure start = 23

Concept Applicator (CA)-2

1. Given that $(\log_2 3)(\log_3 4)(\log_4 5) \ldots \log_x (x + 1) = 10$.

Or $\frac{\log 3}{\log 2} \times \frac{\log 4}{\log 3} \times \frac{\log 5}{\log 4} \times \ldots \times \frac{\log(x+1)}{\log x}$

$= \log_2 (x + 1) = 10$

$x + 1 = 1024; x = 1023$

2. Given equation is,

$\log_{10}(2^x + 2x - 8) = x(1 - \log_{10} 5)$

$\Rightarrow \log_{10}(2^x + 2x - 8) = x \log_{10} 2 = \log_{10}(2^x)$

$\Rightarrow 2^x + 2x - 8 = 2^x \quad \Rightarrow x = 4.$

3. Given that circumference is $C = \log_{10}(b^6) = 6\log_{10} b$ and radius $r = \log_{10}(a^3) = 3\log_{10} a$

We known that circumference $C = 2\pi r$

$\therefore 6 \log_{10} b = 2\pi. 3 \log_{10} a$

$$\log_b a = \frac{\log_{10} a}{\log_{10} b} = \frac{1}{\pi}$$

4. Given that $\log_3\log_4\log_5 A = 2^x$

$\log_4\log_5 A = 3^{2^x}$

$\log_5 A = 4^{3^{2^x}}$

$A = 5^{4^{3^{2^x}}}$

Option (d)

5. Given equation is

$$3^{\log_3 \log\sqrt{x}} - \log x + (\log x)^2 - 3 = 0$$

which can be written as

$$\log\sqrt{x} - \log x + (\log x)^2 - 3 = 0$$

$\Rightarrow \frac{1}{2}\log_{10} x - \log_{10} x + (\log_{10} x)^2 - 3 = 0$

$\Rightarrow \log_{10}x - 2\log_{10} x + 2(\log_{10}x)^2 - 6 = 0$

$\Rightarrow 2(\log_{10} x)^2 - \log_{10} x - 6 = 0$

Let $\log_{10} x = t$ then equation becomes

$2t^2 - t - 6 = (2t + 3)(t - 2) = 0$ or $t = 2, -\frac{3}{2}$

$\Rightarrow \log_{10} x = 2$ or $\log_{10} x = -\frac{3}{2}$

$\Rightarrow x = 10^2 = 100$ or $x = 10^{-3/2}$

So values of x which satisfy the equation are $100, \& 10^{-3/2}$

6. Given equation is

$$2(\log_x \sqrt{5})^2 - 3\log_x \sqrt{5} + 1 = 0$$

$\Rightarrow \frac{2}{(\log_{\sqrt{5}} x)^2} - \frac{3}{\log_{\sqrt{5}} x} + 1 = 0$

$\Rightarrow (\log_{\sqrt{5}} x)^2 - 3\log_{\sqrt{5}} x + 2 = 0$

$\Rightarrow$ Let $\log_{\sqrt{5}} x = t$ then equation becomes

$t^2 - 3t + 2 = (t - 1)(t - 2) = 0$, so $t = 1$ or 2

$\Rightarrow \log_{\sqrt{5}} x = 1$ or $\log_{\sqrt{5}} x = 2$

$x = \sqrt{5}$ or $x = (\sqrt{5})^2 = 5$

7. Given equation is $(\log x)^2 - \log(x^3) = \log(x^2) - 4$ it can be re-written as

$\log^2 x - 3\log x - 2\log x + 4 = 0$

$\Rightarrow \log_{10}^2 x - 5\log_{10} x + 4 = 0$

$\Rightarrow (\log_{10} x - 4)(\log_{10} x - 1) = 0$

$\Rightarrow \log_{10} x = 4, x = 10^4$

$\log_{10} x = 1$ or $x = 10$

$\Rightarrow x = 10$ or 10^4

8. Taking log both the side with respect to base 10, we get

$\Rightarrow \frac{\log x + 7}{4} \log x = \log x + 1$

$\Rightarrow \log_{10}^2 x + 3\log_{10} x - 4 = 0$

$\Rightarrow (\log_{10} x + 4)(\log_{10} x - 1) = 0$

$\Rightarrow x = 10^{-4}$ and $x = 10^1$

9. Given equation can be written as

$= (\log 100 + \log x)^2 + (\log 10 + \log x)^2$

$= 14 - \log x$

$= (2 + \log x)^2 + (1 + \log x)^2 = 14 - \log x$

Now put $\log x = t$ we get,

$= (2 + t)^2 + (1 + t)^2 = 14 - t$

$= 2t^2 + 7t - 9 = 0$

$= (2t + 9)(t - 1) = 0$

$t = 1$ or $t = -\frac{9}{2}$

$= \log_{10} x = 1$ or $\log_{10} x = -\frac{9}{2}$

$x = 10$ or $x = +10^{\frac{-9}{2}} = \sqrt{10^{-9}}$

10. (b) Let $a^{\log_b x} = x^{\log_b a} = y$

$y^2 - 5y + 6 = 0 \Rightarrow y = 2, 3$

When $y = 2 = a^{\log_b x} \Rightarrow \log_b x = \log_a 2$

$\Rightarrow x = b^{\log_a 2}$

When $y = 3 = a^{\log_b x} \Rightarrow x = 3^{\log_a b}$

Exercis-A

Concept Builder (CB)

1. (c) Since $(0 . pqr)^p = (p . qr)^q = (pq . r)^r$ $= pqr$ taking log both the sides

$p(\log pqr - 3) = q(\log pqr - 2) = r(\log pqr - 1)$ $= \log pqr$.

$\log (pqr) = 3p/(p-1) = 2q/(q-1) = r/(r-1)$ $= k$ (say)

Since $3p/(p-1) = k$ hence $3p = k_p - k$ or $p = k/(k-3)$ or $1/p = (k-3)/k$ similarly $1/q = (k-2)/k$ and $1/r = (k-1)/k$ hence $1/p = 1/q + 1/r$

2. (a) Since $\log \frac{21}{8} = \log 21 - \log 8$

$= \log 3 + \log 7 - 3 \log 2$

$2 \log \frac{6}{7} = 2 \log 6 - 2 \log 7 = 2 \log 2 + 2 \log$

$3 - 2 \log 7$

$\log \frac{7}{27} = \log 7 - \log 27 = \log 7 - 3 \log 3$

Hence final value is

$(\log 3 + \log 7 - 3 \log 2) + (2 \log 2 + 2 \log$

$3 - 2 \log 7) + (\log 7 - 3 \log 3)$

$= -\log 2 = \log 2^{-1} = \log \frac{1}{2}$

3. (b) Since $\log_3 \log_3 \log_3 k = p$ or $\log_3 \log_3 k = 3^p$

Or $\log_3 k = 3^{3^p}$

Or $k = 3^{3^{3^p}}$

4. (d) The required value is

$$\frac{\log (32^{\frac{1}{4}} x\, 64^{\frac{1}{5}} x\, 128^{\frac{1}{6}})}{\log 4 + \log 8^{\frac{1}{2}} + \log 16^{\frac{1}{3}}}$$

$$= \frac{\frac{5}{4}\log 2 + \frac{6}{5}\log 2 + \frac{7}{6}\log 2}{2\log 2 + \frac{3}{2}\log 2 + \frac{4}{3}\log 2}$$

$$= \frac{\left[\frac{5}{4} + \frac{6}{5} + \frac{7}{6}\right]}{\left[2 + \frac{3}{2} + \frac{4}{3}\right]} = \left(\frac{75+72+70}{60}\right)\left(\frac{12+9+8}{6}\right)$$

$$= \frac{217}{60} \times \frac{6}{29} = \frac{217}{290}$$

5. (a) L.H.S $= \log(x.\, x^2.x^3 \,.....\, x^{10})$

$= \log(x^{1+2+3+...+10}) = \log x^{55} = 55 \log x$

$= 11\log x$^5 $= 11\log 32$

or x^5 $= 32$ or $x = 2$

6. (c) $a = \log_2 4 = 2\log_2 2 = 2$

$b = \log_4 16 = 2\log_4 4 = 2$

$c = \log_8 64 = 2\log_8 8 = 2$

Hence $\frac{abc}{a+b+c} = \frac{8}{6} = \frac{4}{3}$

7. (d) Since $\log(x^3 + y^3) - \log(x^3 - y^3)$

$= \log \frac{x^3 + y^3}{x^3 - y^3} = \log \frac{35}{19}$

$\Rightarrow \frac{x^3 + y^3}{x^3 - y^3} = \frac{35}{19}$ and $\frac{x}{y} = \frac{3}{2}$

Then $\frac{x^2 + y^2}{x^2 - y^2} = \frac{13}{5}$

Hence $k = \frac{13}{5}$

8. (a) Since $\frac{\log 9}{\log p} = q$ or $\log p = \frac{2\log 3}{q}$

Similarly $\log p = \frac{3\log 2}{r} = \frac{2\log 3}{q}$

$\frac{q}{r} = \frac{2\log 3}{3\log 2} = \frac{2}{3}\log_2 3 = \log_{\sqrt{8}} 3$

9. (d) Given $3\log_{10}x - \log_x 0.001 = 3\log_{10}x + 3 \log_x 10 = 3(\log_{10}x + 1/\log_{10}x)$

Since $x > 1$ hence $\log x > 0$ and we know that minimum value of $a + 1/a$ is 2 if $a > 0$ hence required minimum value is $3 \times 2 = 6$.

10. (c) Since K = 1/($\log_2\pi$) + 1/($\log_4\pi$) + 1/($\log_8\pi$) = (log2)/(logπ) + (2log2)/(logπ) + (3log2)/(logπ) = (log64)/(logπ)

Since π^3 < 64 <π^4 hence 3 < (log64)/(logπ) < 4 or 3 < k < 4

11. (b) Let N = $(2.5)^{200}$

Taking log both side with base 10

$\log_{10} N = 200 \log_{10}(2.5) = 200$

$\log_{10}\left(\frac{5}{2}\right) = 200[\log_{10} 5 - \log_{10} 2]$

= 200 [1 – 2 $\log_{10}$ 2] = 200[1 – 2 × 0.3010] = 200 [0.3980] = 79.60

Since Characteristic is 79 so number of digits = 79 + 1 = 80

12. (a) Given equation is $x^{\frac{\log x+5}{3}} = 10^{5+\log x}$ taking log both the sides we get

$\log x.\frac{\log x+5}{3} = \log_{10} 10^{5+\log x}$

$\Rightarrow \log x.\frac{\log x+5}{3} = 5+\log x$

$\Rightarrow \log^2 x + 5\log_{10} x = 15 + 3\log_{10} x$

$\Rightarrow \log^2 x + 2\log_{10} x - 15 = 0$

$\Rightarrow (\log_{10} x - 3)(\log_{10} x + 5) = 0$

$\Rightarrow \log_{10} x = 3$ or $x = 10^3$

Or $\log_{10} x = -5$ or $x = 10^{-5}$

So number of roots are 2

13. (d) Consider $(\log_3 4 + \log_2 9)^2 - (\log_3 4 - \log_2 9)^2$

We know that $(A + B)^2 - (A - B)^2 = 4AB$

$= 4\log_3 4 . \log_2 9 \Rightarrow 4.\frac{\log 4}{\log 3}\times\frac{\log 9}{\log 2} \Rightarrow 16$

14. (b) Consider the exponent $\log_{10} 8 - \log_{10} 9$

$+ 2\log_{10} 6 = \log\left(\frac{8\times 36}{9}\right) = \log_{10} 32$

$\therefore 10^{3\log 2 - 2\log(\log 10^3)+\log\left(\left(\log 10^6\right)^2\right)} = 10^{\log_{10} 32} = 32$

15. (c) Given expression is $(0.8)(1+9^{\log_3 8})^{\log_{65} 5}$

$\Rightarrow (0.8)(1 + 64)^{\log_{65}5}$

$\Rightarrow (0.8)(65)^{\log_{65}(5)}$

$\Rightarrow (0.8)(5) = 4$

16. (a) Given equation is $\log_{\frac{1}{3}} x - 3\sqrt{\log_{\frac{1}{3}} x} + 2$

= 0 this can be re-written as

$-\log_3 x - 3\sqrt{-\log_3 x} + 2 = 0$

$\Rightarrow (2-\log_3 x)^2 = \left(3\sqrt{-\log_3 x}\right)^2$

$\Rightarrow 4 + (\log_3 x)^2 - 4\log_3 x = -9\log_3 x$

$\Rightarrow (\log_3 x + 4)(\log_3 x + 1) = 0$

$\Rightarrow \log_3 x = -4$ $\quad\log_3 x = -1$

$\Rightarrow x = 3^{-4}$ $\quad x = 3^{-1}$

$\Rightarrow x = \frac{1}{81}, \frac{1}{3}$

So sum of the roots is $\frac{1}{81} + \frac{1}{3} = \frac{28}{81}$

17. (d) $\frac{\log_3 135}{\log_{15} 3} - \frac{\log_3 5}{\log_{405} 3} = \log_3(5\times 3^3)$.

. $\log_3 (5 \times 3) - \log_3 5 \log_3 (81 \times 5)$

Let $\log_3 5 = y$,

Then given expression $(3 + y)(1 + y) - y(4 + y) = 3 + 3y + y + y^2 - 4y - y^2 = 3$

18. (d) Let $x = \sqrt{6+\sqrt{6+\sqrt{6+\ldots\infty}}}$

$x = \sqrt{6+x}$

$\Rightarrow x^2 - x - 6 = 0$ $\quad\Rightarrow x = 3, -2$

$\therefore x = 3\ (t > 0)$

Given expression

$\log_2 \log_9 3 = \log_2 \log_{3^2} 3 = \log_2 \frac{1}{2} \Rightarrow -1$

19. (b) Let $x = 2^{(\log_2 3.\log_3 4.\log_4 5 \ldots \log_{99} 100)}$ taking log with base 2, both the sides we get,

$\log_2 x = \log_2 3 . \log_3 4 . \log_4 5. \log_5 6 \ldots. \log_{99} 100$

$= \frac{\log 3}{\log 2}\times\frac{\log 4}{\log 3}\times\frac{\log 5}{\log 4}\times\ldots\ldots\frac{\log 100}{\log 99} = \frac{\log 100}{\log 2}$

$= \log_2 100$

So $x = 100$

20. (c) $\frac{1}{\log_5 3}+\frac{1}{\log_6 3}-\frac{1}{\log_{10} 3}$:

$= \log_3 5+\log_3 6-\log_3 10$

$\Rightarrow \log_3\left(\frac{5\times 6}{10}\right)=\log_3 3=1$

21. (c) Let $N=\left(\frac{9}{8}\right)^{-100}$

Taking log both side with base 10

$\log_{10}\left(\frac{9}{8}\right)^{-100} = -100\log_{10} 9/8 = -100[\log_{10} 9$

$-\log_{10} 8] = -100[2\log_{10} 3 - 3\log_{10} 2]$

$= -100[2 \times 0.4771 - 3 \times 0.03010]$
$= -100[0.0512] = -5.12$
Since Characteristic is –6 so number of zero after decimal before a significant (non zero digit) figure start = 5

22. (b) Consider $\log_2(x^2 - xy + y^2)$
$= \log_2((x+y)^2 - 3xy)$

But $x+y=\sqrt{10};\ x-y=\sqrt{2};\ xy=\frac{10-2}{4}=2$

$\log_2(10-6)=\log_2 4 = 2$

23. (d) Let $N = (0.35)^{12}$
Taking log both side with base 10

$\log_{10}(0.35)^{12} = 12\log_{10}(0.35) = 12\log_{10}\frac{7}{20}$

$= 12[\log_{10} 7 - \log_{10} 20]$

$= 12\ [\log_{10} 7 - \log_{10} 2 - 1] = 12[0.8451 - 0.3010 - 1] = -5.4708$
Since Characteristic is –6 so number of zero after decimal before a significant (non zero digit) figure start = 5

24. (a) $2\log_2(3^{3\log_3 4}) = \left(3^{\log_3 64}\right)^2$
$= 64^{12} = 2^{12}$, So $k = 12$

25. (c) $N = \log_{10} 32 + \log_{10} 961 = \log_{10} 30752$
$\therefore 4 < N < 5$
$\Rightarrow$ Required sum is $4 + 5 = 9$

26. (b) Since

$a^{\log_a b} = b$ so $2^{\log_2 e^{ln 5^{\log_5 7^{\log_7 10^{\log_{10}(8x-3)}}}}}$

$= e^{ln 5^{\log_5 7^{\log_7 10^{\log_{10}(8x-3)}}}} = 5^{\log_5 7^{\log_7 10^{\log_{10}(8x-3)}}}$

$= 7^{\log_7 10^{\log_{10}(8x-3)}} = 10^{\log_{10}(8x-3)} = (8x-3) = 13$

$8x = 16$ or $x = 2$

Concept Cracker (CC)

1. (b) Given that $a = \sqrt[2]{b} = \sqrt[3]{c} = \sqrt[4]{d} = \sqrt[5]{5}$ hence $a = 5^{1/5}$ or $a^{100} = 5^{20}$. $b = 5^{2/5}$ hence $b^{100} = 5^{40}$, $c = 5^{3/5}$ hence $c^{100} = 5^{60}$, $d = 5^{4/5}$ hence $d^{100} = 5^{80}$ then $N = (a \times b \times c \times d)^{100} = 5^{20+40+60+80} = 5^{200}$. Then logN = 200log5 = 200(1 – log2) = 200(1 – 0.3010) = 200 × 0.699 = 139.8 Hence N has 140 digits.

2. (c) We can write the given expression as

$\frac{\log 2}{\log 4}+\frac{\log 4}{\log 4}+\frac{\log 8}{\log 4}+\ldots\frac{\log 2^n}{\log 4}$

$=\frac{1}{\log 4}[\log(2x4x8x\ldots 2^n)] = \frac{\{\log 2^{1+2+3+\ldots+n}\}}{\{\log 4\}}$

$= [\log 2^{\left\{\frac{n(n+1)}{2}\right\}}]/[\log 4] = n(n+1)/4$

when $n = 20$ then the value of summation is $20 \times 21/4 = 105$

3. (c) Taking log we will get :log$a$ = 2log$b$ = 3log$c$ = ……26log$z$

Then log$a$ + log$b$ + log$c$ +…..log$z$

$= \log a + \frac{\log a}{2} + \frac{\log a}{3} + \ldots \frac{\log a}{26}$

$= \log a\left[1+\frac{1}{2}+\frac{1}{3}+\frac{1}{4}+\ldots+\frac{1}{26}\right]$

Hence $\log_a(\text{abcdef} \ldots \text{yz}) = \frac{\log(\text{abcd}\ldots\text{yz})}{\log a}$

$= \frac{\log a + \log b + \log c + \ldots \log z}{\log a}$

$$= \frac{\left[1+\frac{1}{2}+\frac{1}{3}+.....+\frac{1}{26}\right]\log a}{\log a}$$

$$= 1+\frac{1}{2}+\frac{1}{3}+\frac{1}{4}+.....+\frac{1}{26} = \sum_{n=1}^{n=26}\frac{1}{n}$$

4. (a) $\frac{1}{\log_2 a} = \frac{\log 2}{\log a}, \frac{1}{\log_2 b} = \frac{\log 2}{\log b} = \frac{2\log 2}{\log a}$

$$\frac{1}{\log_2 c} = \frac{\log 2}{\log c} = \frac{3\log 2}{\log a}$$

$$\frac{1}{\log_2 a}+\frac{1}{\log_2 b}+\frac{1}{\log_2 c}+.....+\frac{1}{\log_2 z}$$

$$\frac{\log 2}{\log a}+\frac{2\log 2}{\log a}+\frac{3\log 2}{\log a}+....+\frac{26\log a}{\log a}$$

$$= \frac{\log 2}{\log a}[1+2+3+...+26] = \frac{\log 2}{\log a}\left[\frac{26\times 27}{2}\right]$$

$$= \frac{351\log 2}{\log a}$$

$$= \log_a(2^{351})$$

5. (c) Given that $y = 81^x$ and $9y^2 = 3$, hence $9.(81)^{2x} = 3$

Or $3^2 \times 3^{8x} = 3^{8x+2} = 3^1$

Or $8x + 2 = 1$ or $x = -\frac{1}{8}$

Substitute this value in 1st equation and we will get $y = 81^x = 3^{4x} = 3^{-\frac{1}{2}} = \frac{1}{\sqrt{3}}$

6. (c) L.H.S $= \frac{x}{\log 3}+\frac{x}{2\log 3}+\frac{x}{4\log 3}+...\infty$

$$\frac{x}{\log 3}\left[1+\frac{1}{2}+\frac{1}{4}+.....\infty\right]$$

$$= \frac{x}{\log 3}\left[\frac{1}{1-\frac{1}{2}}\right]$$

(Using formula to find sum of infinite term of a G.P]

$$= \frac{2x}{\log 3} = 4\log 3$$

$x = 2(\log 3)^2 = 2(0.30)^2 = .18 = 2k^2$

7. (d) Consider

$$3^{\left\{\frac{1}{\log_x 9}\right\}} = 3^{\log 9^x} = 3^{\log 3^{\sqrt{x}}}\ \sqrt{x} = x^{\frac{1}{2}}$$

Similarly $3^{\left\{\frac{1}{\log_x(81)}\right\}} = x^{\frac{1}{4}}$ and so on

Hence, L.H.S. $= x^{\frac{1}{2}}\,.\,x^{\frac{1}{4}}\,.\,x^{\frac{1}{8}}.......$

$$= x^{\frac{1}{2}+\frac{1}{4}+\frac{1}{8}}.......... = x = 81$$

8. (d) Since $\log(x+y)-\log 2 = \log_k k = 1 = \log 10$

Or $\log\frac{x+y}{2} = \log 10$

$x + y = 20$ ...(i)

From 2nd information

$$\log(x+y)-\log xy = \log\frac{5}{16}$$

$$\Rightarrow \log\frac{x+y}{xy} = \log\frac{5}{16}$$

Or $\frac{x+y}{xy} = \frac{5}{16}$ or $xy = 64$ ...(ii)

From eqs (i) and (ii) $x - y = 12$

Hence $x = 16$ and $y = 4$ (or vice-versa)

$$\log_x y+\log_y x = \frac{\log y}{\log x}+\frac{\log x}{\log y} = \frac{\log 4}{\log 16}+\frac{\log 16}{\log 4}$$

$$= \frac{2}{4}+\frac{4}{2} = \frac{1}{2}+2 = \frac{5}{2}$$

9. (d) Since $\log_{(x+y)}(x+y)(x-y) = \log_{(x+y)}(x+y) + \log_{(x+y)}(x-y)$

$= 1 + \log_{(x+y)}(x-y) = k$ or $\log_{(x+y)}(x-y) = k-1$

Now $\log_{(x-y)}(x^2-y^2) = \log_{(x-y)}(x+y)(x-y)$ $= \log_{(x-y)}(x+y) + \log_{(x-y)}(x-y)$

$$1+\frac{1}{k-1} = \frac{k}{k-1}$$

10. (b) Since $\log_{\sqrt{a}} b^2 = 4\log_a b = \frac{4\log b}{\log a}$

Similarly $\log_{\sqrt{b}} c^2 = \frac{4\log c}{\log b}$

Hence given expression is equal to

$$4^{26}\left[\frac{\log b}{lob\, a}\times\frac{\log c}{\log b}\times\frac{\log d}{\log c}\times.....\frac{\log z}{\log y}\times\frac{\log a}{\log z}\right]$$

$= 4^{26} = 2^{52}$

11. (a) $\log_5(5^{1/x}+125)=\log_5 6+\left(1+\frac{1}{2x}\right)$

$= \log_5\left[\frac{5^{1/x}+125}{6}\right]=\left(1+\frac{1}{2x}\right)$

$= 5^{1/x} + 125 = (5^1.5^{1/2x}).6$

$= 5^{1/x} + 125 = 5.6\ .5^{1/2x}$

Let $5^{\frac{1}{2x}} = y$

$= y^2 - 30y + 125 = 0$

$= (y - 25)\ (y - 5) = 0$

$= y = 5$ or 25

$= 5^{\frac{1}{2x}} = 5^1$ or 5^2

$x = \frac{1}{2}$ or $\frac{1}{4}$

So sum of all the possible value of x is 3/4

12. (b) Consider

$\log_3 x = \log 7 - \log 11 = \log\frac{7}{11}$ so $x = 3^{\log\frac{7}{11}}$

$\log_7 y = \log 11 - \log 3 = \log\frac{11}{3}$ so $y = 7^{\log\frac{11}{3}}$

$\log_{11} z = \log 3 - \log 7 = \log\frac{3}{7}$ so $z = 11^{\log\frac{3}{7}}$

$xyz = 3^{\log\frac{7}{11}}\ 7^{\log\frac{11}{3}}\ 11^{\log\frac{3}{7}}$

$\log(xyz)$

$=\log\left(3^{\log\frac{7}{11}}\ 7^{\log\frac{11}{3}}\ 11^{\log\frac{3}{7}}\right)=\log\left(3^{\log\frac{7}{11}}\right).$

$+\log\left(7^{\log\frac{11}{3}}\right)+\log\left(11^{\log\frac{3}{7}}\right)$

$\log(xyz) = \log 3\ (\log 7 - \log 11) + \log 7\ (\log 11 - \log 3) + \log 11\ (\log 3 - \log 7) = 0$

So xyz = 1

13. (c)

$\left(\frac{1}{\log_2\pi}+\frac{1}{\log_6\pi}\right)=(\log_\pi 2+\log_\pi 6)=\log_\pi 12$

Since $\pi^2 < 12 < \pi^3$

$\Rightarrow \log_\pi \pi^2 < \log_\pi 12 < \log_\pi \pi^3$

$\Rightarrow 2 < \log_\pi 12 < 3$

So sum of the two integers is $2 + 3 = 5$

14. (b) Given equation is

$\left(x^{\log_2 a}\right)^2-4\left(x^{\log_2 a}\right)-5=0;$ Let t $= x^{\log_2 a}$ then

$t^2 - 4t - 5 = 0$ on solving this we get, $t = 5$, -1, ignoring the negative roots we get,

$t = 5$ so $x^{\log_a 2} = 5$ $(\log_2 a).\ \log_5 x = 1$

Take log w.r.t. Base 5 .

$\log_5 x = \log_a 2$

$x = (5)^{\log_a 2}$

15. (a) Let then $\log_2(x - 1) = P$ & $\log_2 x = Q$ then

$\Rightarrow (P + Q)^2 - 4\ PQ = 1$

$\Rightarrow (P - Q)^2 = 1$

$\Rightarrow (P - Q) = \pm1$

$\therefore \log_2\left(\frac{x-1}{x}\right)=1$ or -1 $\therefore x = -1$ or 2 but x can not be negative so only $x = 2$ is solution but that is integer, so there exist zero non-integral roots

16. (a) Given equation is

$$x^{\left(\frac{3}{4}\right)(\log_2 x)^2+\log_2 x-\left(\frac{5}{4}\right)}=\sqrt{2}$$

Take log on both the sides with base 2 we get

$\left(\frac{3}{4}(\log_2 x)^2+\log_2 x-\frac{5}{4}\right)\log_2 x=\frac{1}{2}$

Now let us assume $\log_2 x = y$

$3y^3 + 4y^2 - 5y - 2 = 0$

$\Rightarrow 3y^2\ (y - 1) + 7y(y - 1) + 2(y - 1) = 0$

$\Rightarrow (y - 1)\ (3y^2 + 7y + 2) = 0$

$\Rightarrow (y - 1)\ (3y + 1)\ (y + 2) = 0$

$\Rightarrow y = 1$ or $y = -2$ or $y=\frac{-1}{3}$

$\therefore\ x = 2; \frac{1}{4}; \frac{1}{3\sqrt{2}}$

So sum of the rational roots is $2+\frac{1}{4}=\frac{9}{4}$

17. (b) Consider the given equation

$$\frac{1}{2}\log_2(2\log_3(1+\log_2(1+3\log_3 x))) = \frac{1}{2}$$

$\Rightarrow 2\log_3(1+\log_2(1+3\log_3 x)) = 4^{\frac{1}{2}} = 2$

$\Rightarrow (1+\log_2(1+3\log_3 x)) = 3^1 = 3$

$\Rightarrow \log_2(1+3\log_3 x) = 3-1 = 2$

$\Rightarrow 1+3\log_3 x = 2^2 = 4$

$\Rightarrow 3\log_3 x = 3$

$\Rightarrow \log_3 x = 1$

$\Rightarrow x = 3^1 = 3$

$\Rightarrow$ Final solution $x \in \{3\}$

So number of integral roots is 1.

18. (d) Given $\log\left(\frac{\sin 2x}{2}\right) = -2$

$\Rightarrow \frac{\sin 2x}{2} = \frac{1}{100} \Rightarrow \sin 2x = \frac{1}{50}$

Also $\log(\sin x + \cos x) = \frac{\log\left(\frac{k}{100}\right)}{2}$

$\Rightarrow \log(\sin x + \cos x)^2 = \log\left(\frac{k}{100}\right)$

$\Rightarrow 1+\sin 2x = \frac{k}{100}$

$\Rightarrow 1+\frac{1}{50} = \frac{k}{100}$

$\Rightarrow \frac{51}{50} = \frac{k}{100} \Rightarrow k = 102$

19. (b) We have,

$\log_{10}(2^x+3x-51) = x(1-\log_{10}5)$

$\Rightarrow \log_{10}(2^x+3x-51) = x\log_{10}2 = \log_{10}(2^x)$

$\Rightarrow 2^x + 3x - 51 = 2^x$

$\Rightarrow x = 17$

20. (a) If the number characteristic = 4, then

$4 \le \log_5 x < 5$

$5^4 \le x < 5^5$

$625 \le x < 3125$

Number of positive integers = 2500

Exercis-B

Concept Deviator (CD)

(Single Correct MCQs)

1. (d) Since Let $N = 6^6 \times 8^8 \times 10^{10} \times 12^{12} = 2^{64} \times 3^{18} \times 5^{10}$

Since log2 = 0.3010 hence log5 = 0.699

Taking log both the side, logN = 64log2 + 18log3 + 10log5 = (64 × 0.3010) + (18 × 0.4771) + (10 × 0.699) = 34.8418, hence number of digits in N is 34 + 1 = 35

2. (c) $\log_7\log_5(\sqrt{x+5}+\sqrt{x}) = 0$ using $\log_a b = 0$

$\Rightarrow b = a^0$

$\Rightarrow \log_5(\sqrt{x+5}+\sqrt{x}) = 7^0 = 1$

$\Rightarrow \sqrt{x+5}+\sqrt{x} = 5 \Rightarrow \sqrt{x+5} = 5-\sqrt{x}$

Now squaring both side we get, $x = 4$

3. (b) Consider the value of function $[\log_{10}x]$ in different intervals-

When x is a single digit number then $[\log_{10}x] = 0$

When x is a two digit number then $[\log_{10}x] = 1$

When x is a 3 digit number then $[\log_{10}x] = 2$ and so on.

Now consider the given equation $[\log_{10}1] + [\log_{10}2] + \ldots..[\log_{10}x] = x$

Since there are 9 single digit numbers and for those values $[\log_{10}x] = 0$

That 9 numbers will be compensated by 9 three digit numbers hence. Hence $x = 108$.

4. (b) Consider the value of function $[\log_{20}x]$in different intervals-

When $x < 20$ then $[\log_{20}x] = 0$

When $20 \le x < 400$ then $[\log_{20}x] = 1$

When $400 \le x < 8000$ then $[\log_{20}x] = 2$ and so on.

Now consider the given equation $[\log_{20}1] + [\log_{20}2] + \ldots..[\log_{20}x] = x$

Since there are 19 numbers for which $[\log_{20}x] = 0$

That 19 numbers will be compensated by 19 numbers from 400 onwards hence. Hence $x = 418$.

5. (c) This question should be solved by eliminating options, The easiest option to eliminate is $x = 6$ as it will give us $\log_6 x = 1$,

we will find that it will not satisfy the given condition, then next option is $x = 36$ and it will satisfy the equation hence $x = 36$.

6. (d) Consider $1/\log_3 9 = ½$, while $1/\log_9 9 = 1$, and $1/\log_{27} 9 = 3/2$, and $1/\log(_3 n)9 = n/2$

Hence required summation is $½ + 1 + 3/2 + n/2 = ½[1 + 2 + 3 + \ldots + n] = n(n + 1)/4$

7. (d) Consider the odd terms

$27^{\log3\ 48} X\ 27^{\log3\ (48)^\wedge 3} X \ldots X\ 27^{\log3\ (48)^\wedge 99}$

Let $\log_3 48 = k$ then it becomes $27^k X\ 27^{3k} X \ldots X\ 27^{99k} = 27^{k(1+3+5+\ldots 99)} = 27^{2500k} = 3^{7500k}$

Consider the even terms

$3^{\log27(48)^\wedge 2} X\ 3^{\log27(48)^\wedge 4} X \ldots X\ 3^{\log27\ (48)^\wedge 96}$

Let $\log_{27} 48 = p = k/3$ then it becomes $3^{2p} X\ 3^{4p} X \ldots X\ 3^{98p} = 3^{p(2+4+\ldots+96)} = 3^{2352p} = 3^{2352k/3} = 3^{784k}$

Hence given expression $= 3^{7500k} x\ 3^{784k} = 3^{8284k} = 3^{\log_3 48^\wedge 8284} = 48^\wedge 8284$

8. (c) $\log_{(x^2+1)}(x+2) < 1$ then $\log(x+2)/\log(x^2+1) < 1$ or $\log(x+2) < \log(x^2+1)$ or $(x+2) < (x^2+1)$ or $x^2 - x - 1 > 0$

Now from the options -1 is the smallest number

9. (b) Option (b) The value of x can be 4 or 8

10. (c) We have

$$p = \log_{12} 18 = \frac{\log_2 18}{\log_2 12} = \frac{1 + 2\log_2 3}{2 + \log_2 3} \text{ and}$$

$$q = \log_{24} 54 = \frac{\log_2 54}{\log_2 24} = \frac{1 + 3\log_2 3}{3 + \log_2 3}$$

Putting $x = \log_2 3$, we have

$pq + 5(p - q)$

$$= \frac{1+2x}{2+x} \cdot \frac{1+3x}{3+x} + 5\left\{\frac{1+2x}{2+x} - \frac{1+3x}{3+x}\right\}$$

$$= \frac{6x^2 + 5x + 1 + 5(-x^2 + 1)}{(x+2)(x+3)} = \frac{x^2 + 5x + 6}{(x+2)(x+3)} = 1$$

11. (a) Writing $u = \log_{10} x$, the given inequality can be written as

$(2 + u)^2 + (1 + u)^2 + u \le 14$

$\to\ 2u^2 + 7u - 9 \le 0$

$\to\ (2u + 9)(u - 1) \le 0 \to -9/2 \le u \le 1$

$\to\ -9/2 \le \log_{10} x \le 1 \to 10^{-9/2} \le x \le 10$

12. (d) Given equation is $2\log_3(x-2) + \log_3(x-4)^2 = 0$ and it is defined when $[x - 2 > 0, x \ne 4]$

$\Rightarrow\ 2\log_3(x-2) + 2\log_3|x-4| = 0$

$\Rightarrow\ 2\log_3(x-2)|x-4|) = 0$

$\Rightarrow\ (x-2)|x-4| = 1$

Now we have following cases

Case (*i*) : $x \ge 4$

$\Rightarrow\ (x-2)(x-4) = 1 \Rightarrow x^2 - 6x + 7 = 0$

$$\Rightarrow\ x = \frac{6 \pm \sqrt{8}}{2} \Rightarrow x = 3 \pm \sqrt{2}$$

$\Rightarrow\ x = 3 - \sqrt{2}$ equation is not defined at this value of x, hence ruled out.

$\Rightarrow\ x = 3 + \sqrt{2}$

Case (*ii*) : $x < 4$

$\Rightarrow\ (x-2)(4-x) = 1 \Rightarrow 4x - x^2 - 8 + 2x = 1$

$\Rightarrow\ x^2 - 6x + 9 = 0 \Rightarrow (x-3)^2 = 0$

$\Rightarrow\ x = 3$

From above results we can conclude that

$x = \{3, 3 + \sqrt{2}\}$

So sum of rational roots is 3

13. (a) Let $\log_{10} x = a$; $\log_{10} y = b$ and $\log_{10} z = c$

Here $xyz = 10000$

$\Rightarrow \log_{10} x + \log_{10} y + \log_{10} z = 4$

i.e. $a + b + c = 4$

Also, $ab + bc + ca = 6$

Now, $a^2 + b^2 + c^2 = (a + b + c)^2 - 2\,\Sigma\, ab = (4)^2 - (2)(6) = 4$

14. (a) Taking log both side w. r. t. base 10 we get,

$\log_{10}|x-2|^{10x^2-1} = \log|x-2|^{3x}$

$\Rightarrow (10x^2 - 1)\log|x-2| = 3x\log|x-2|$

$\Rightarrow \log|x-2|(10x^2 - 3x - 1) = 0$

$\Rightarrow \log|x-2| = 0$ or $10x^2 - 3x - 1 = 0$

$\Rightarrow |x-2| = 1$ or $(2x-1)(5x+1) = 0$

$$x = 3\ ;\ 1\ x = \frac{1}{x}, -\frac{1}{5}$$

Hence sum of all the possible values of x is x

$$= 1 + 3 + \frac{1}{2} - \frac{1}{5} = \frac{43}{10} = 4.3$$

15. (c) L. H. S.

$$= \frac{1}{\log_{2^4} x^2} + \frac{1}{\log_{2^6} 2x} = 3$$

$$\Rightarrow \frac{1}{\frac{1}{2}\log_2 x} + \frac{1}{\frac{1}{6}(1+\log_2 x)} = 3$$

Let $\log_2 x = t$

$$\Rightarrow \frac{2}{t} + \frac{6}{1+t} = 3 \Rightarrow 2(1+t) + 6t = 3t(1+t)$$

$$\Rightarrow (t-2)(3t+1) = 0$$

$$\Rightarrow t = 2 \text{ or } t = -1/3$$

$\log_2 x = 2 \Rightarrow x = 4$ and $\log_2 x = \frac{-1}{3}$

$\Rightarrow x = 2^{-1/3}$

16. (a) From $\log_2 x + \log_4 y + \log_4 z = 2$,

$\log_4 (x^2yz) = 2$

$\Rightarrow x^2yz = 4^2$ .... (i)

Similarly from we get $\log_9 x + \log_3 y + \log_9 z = 2$ we get

$y^2zx = 9^2$ .... (ii)

And from $\log_{16} x + \log_{16} y + \log_4 z = 2$ we get

$z^2xy = 16^2$ .... (iii)

Multiply above three results we get

$\Rightarrow x^4y^4z^4 = (2.3.4)^4$

$\Rightarrow xyz = 24$

17. (d) Taking log both side with respect to 10

$\Rightarrow \log|x-3|^{3x^2-10x+3} = \log 1$

$\Rightarrow (3x^2 - 10x + 3)\log_{10}|x-3|$

$= 0 \quad [x \neq 3]$

$\Rightarrow (3x-1)(x-3)\log_{10}|x-3| = 0$

$\Rightarrow x = \frac{1}{3}, |x-3| = 1$

$\therefore x = \frac{1}{3}, 2, 4$

So required sum is $\frac{1}{3} + 2 + 4 = \frac{19}{3}$

18. (b) Given equation is $\log_4(x^2-1) - \log_4(x-1)^2$

$= \log_4\sqrt{(4-x)^2}$

The given equation is defined if

$[x \neq [-1, 1] \cup \{4\}]$

$\log_4(x-1)(x+1) - \log_4(x-1)^2 = \log_4|4-x|$

$$\log_4\left[\frac{(x-1)(x+1)}{(x-1)^2}\right] = \log_4|4-x|$$

$$\Rightarrow \frac{x+1}{x-1} = |4-x|$$

$\Rightarrow (x+1) = (x-1)|x-4| \; (\therefore |4-x| = |x-4|)$

Case (i) : $x \geq 4$

$\Rightarrow (x+1) = (x-1)(x-4)$

$\Rightarrow x + 1 = x^2 - 5x + 4$

$\Rightarrow x^2 - 6x + 3 = 0$

$$\Rightarrow x = \frac{6 \pm \sqrt{24}}{2}$$

but x > 4, so $\frac{6+\sqrt{24}}{2} = 3 + \sqrt{6}$

Case (ii) : $x < 4$

$\Rightarrow (x+1) = (x-1)(4-x)$

$\Rightarrow x + 1 = 4x - x^2 - 4 + x$

$\Rightarrow x^2 - 4x + 5 = 0$, has no real roots.

19. (b) Consider the equation $3^{2x-y+1} = 3^{y-2x+1} - 8$

Let $3^{2x-y} = t$

$\Rightarrow 3t = \frac{3}{t} - 8 \Rightarrow 3t^2 + 8t - 3 = 0$

$\Rightarrow t = -3, \frac{1}{3}$

But $t = -3$ (rejected)

So, $3^{2x-y} = t = 3^{-1} \Rightarrow 2x - y = -1$ ... (i)

Now consider the 2nd equation, $\log_6|2x^2y - xy^2|$ $= 1 + \log_{36}(xy)$

$\Rightarrow \log_6|xy(2x-y)| = 1 + \log_6\sqrt{xy}$

$\Rightarrow \log_6|xy| = \log_6(6\sqrt{xy}) = xy = 36$ ... (ii)

$\therefore$ On solving (i) and (ii), we get $x = 4$ and $y = 9$

So, point (x, y) lies in 1st quadrant

20. (b) Given equation can be written as $\log(2y-3x)^2 = \log xy$

or $(2y-3x)^2 = xy$

or $4y^2 - 12xy + 9x^2 - xy$

Dividing the equation by y^2

$$9\left(\frac{x}{y}\right)^2 - 13\frac{x}{y} + 4 = 0$$

$$\left(\frac{x}{y} - 1\right)\left(\frac{9x}{y} - 4\right) = 0$$

$$\frac{x}{y} = 1, \frac{x}{y} = \frac{4}{9}$$

$x = y$ ruled out as for $x = y$, $2y - 3x$ is negative and log $(2y - 3x)$ is not defined

Hence $\frac{x}{y} = \frac{4}{9}$.

Passage Type

(For 21-22)

Given equations are:

$\log_{10}(2000\,xy) - \log_{10} x \cdot \log_{10} y = 4$ ... (i)

$\log_{10}(2yz) - \log_{10} y \cdot \log_{10} z = 1$ ... (ii)

And $\log_{10}(zx) - \log_{10} z \cdot \log_{10} x = 0$ ... (iii)

Equation (i) can be written as

$3 + \log_{10}(2\,xy) - \log_{10} x \cdot \log_{10} y = 4$

or $\log_{10}(xy) - \log_{10} x \cdot \log_{10} y = 1 - \log_{10}(2)$ ... (iv)

While equation (ii) can be written as

$\log_{10}(yz) - \log_{10} y \cdot \log_{10} z = 1 - \log_{10}(2)$ ... (v)

From (iv) and (v), we get

$\log x + \log y - \log x \cdot \log y = \log y + \log z - \log y \cdot \log z$

$\Rightarrow \log x\,(1 - \log y) = \log z\,(1 - \log y)$

$\Rightarrow (\log x - \log z)(1 - \log y) = 0$

$\therefore$ Ether, $\log x = \log z$

or $\log_{10} y = 1 \Rightarrow y = 10$

But y = 10 does not satisfy equation (*i*), hence rejected.

$\therefore$ $\log x = \log z$

From equation (iii) we get

$(\log_{10} x)^2 = 2(\log_{10} x) \Rightarrow \log_{10} x\,[\log_{10} x - 2] = 0$

$\therefore$ $x = 1$ or $x = 100$

if $x \equiv z \equiv 1$ then $y \equiv 5 \Rightarrow (x_1, y_1, z_1) \equiv (1, 5, 1)$

$X = z = 100$ then $y = 20 \Rightarrow (x_2, y_2, z_2) = (100, 20, 100)$

21. (c) So number of triplets is 2
22. (d) Sum of all the possible values of x is 101

(For 23-24)

$\log_{25} x + \log_{64} y = 4$

$\log_x 25 - \log_y 64 = 1$

Let $\log_{25} x = a$, $\log_{64} y = b$ then

$a + b = 4$

$\frac{1}{a} - \frac{1}{b} = 1$

$b - (4 - b) = (4 - b)\,b \Rightarrow 2b - 4 = 4b - b^2$

$b^2 - 2b - 4 = 0$, Let b_1, b_2 be the roots

$b_1 + b_2 = 2$

$\log_{64} y_1 + \log_{64} y_2 = 2$

$\Rightarrow \log_{64} y_1 y_2 = 2$, or $y_1 y_2 = 64^2 = 2^{12}$

Similarly $a_1 + a_2 = (4 - b_1) + (4 - b_2) = 8 - (b_1 + b_2) = 8 - 2$

$a_1 + a_2 = 6$

$\log_{25} x_1 + \log_{25} x_2 = 6$

$x_1 x_2 = 25^6 = 5^{12}$.

23. (a)
24. (b)

(For 25-26)

$$y = \sqrt{\left(\frac{1}{3}\log_3(3x) + \frac{1}{3}\log_x(3x)\right) + 3\log_x x}$$

$$+ \sqrt{\left(\frac{1}{3}\log_3\left(\frac{x}{3}\right) + \frac{1}{3}\log_x\left(\frac{3}{x}\right)\right)3\log_3 x}$$

$$= \sqrt{(1 + \log_3 x + 1 + \log_x 3)\log_3 x} +$$

$$+ \sqrt{(\log_3 x - 1 + \log_x 3 - 1)\log_3 x}$$

$$= \sqrt{\log\frac{2}{3}x + 2\log\frac{x}{3} + 1} + \sqrt{\log\frac{2}{3}x + 2\log\frac{x}{3} + 1}$$

$$= \sqrt{(\log_3 x + 1)^2} + \sqrt{(\log_3 x - 1)^2}$$

$$= |\log_3 x + 1| + |\log_3 x - 1| = 2$$

Case – I : $\log_3 x > 1 \Rightarrow x > 3$

$\log_3 x + \log_3 x = 2$

$x = 3$

Case – II : $-1 \le \log_3 x \le 1$

$\Rightarrow \frac{1}{3} \le x \le 3$

$\log_3 x + 1 + 1 - \log_3 x = 2$

$2 = 2$

$x \in \left[\frac{1}{3}, 3\right] - \{1\}$

Case – III : $\log_3 x < -1 \Rightarrow x < \frac{1}{3}$

$-2 \log_3 x = 2 ; x = \frac{1}{3}$.

Hence solution is $x \in \left[\frac{1}{3}, 3\right] - \{1\}$

Since base cannot be equal to 1.

Integral values of x are 2 and 3

25. (b)
26. (c)

Integer Type

27. (8) $2^{(\log_2 3)^x} = 3^{(\log_3 2)^x}$

Taking log to the base 2 on the both the sides, we get $(\log_2 3)^x . \log_2 2 = (\log_3 2)^x \log_2 3$

$(\log_2 3)^{x-1} = (\log_3 2)^2 \Rightarrow \frac{(\log_2 3)^{x-1}}{(\log_3 2)^x} = 1$

$(\log_2 3)^{2x-1} = \quad 1 = (\log_2 3)^0$

$\Rightarrow 2x - 1 = 0 \Rightarrow x = \frac{1}{2}$.

28. (2) Consider $\frac{1}{\log_a(2-\sqrt{3})} + \frac{1}{\log_b\left(\frac{\sqrt{3}-1}{\sqrt{3}+1}\right)}$

$= \log_{2-\sqrt{3}} a + \log_{\frac{\sqrt{3}-1}{\sqrt{3}+1}} b$

$= \log_{2-\sqrt{3}} a + \log_{2-\sqrt{3}} b = \log_{2-\sqrt{3}} (ab)$

Now, $(2+\sqrt{3})^{\log_{2-\sqrt{3}}(ab)}$

$= (2-\sqrt{3})^{\log_{2-\sqrt{3}}\left(\frac{1}{ab}\right)}$

$\Rightarrow \frac{1}{ab} = \frac{1}{35} \Rightarrow ab = 35$

As a, b are prime numbers, so either $a = 5$, $b = 7$ or $a = 7$, $b = 5$

Hence, $|a - b| = 2$.

29. (3) Let $x + |x - 4| = y$

$\therefore$ Equation beomes

$\log_x y^2 = \log_x (7y - 12) \Rightarrow y^2 = 7y - 12$
$\Rightarrow y^2 - 7y + 12 = 0 \Rightarrow y = 3$ or 4

If $y = 3$ then $x + |x - 4| = 3$ no value of x satisfy this condition since minimum value of $x + |x - 4|$ is 4

If $y = 4$ then $x + |x - 4| = 4$ then $x \in [0, 4]$

So number of integral solution is 2, 3, 4 as $\log_{\sqrt{x}} (x + |x - 4|)$ is not defined when x is 0

or 1

30. (3) Let $\log_{3x} 45 = \log_{4x} 40\sqrt{3} = k$

$45 = (3x)^k, 40\sqrt{3} = (4x)^k$

$\frac{45}{40\sqrt{3}} = \left(\frac{3x}{4x}\right)^k = \left(\frac{3}{4}\right)^k$

$\Rightarrow k = \frac{3}{2}$

log3x 45 $= \frac{3}{2}$

$45 = (3x)^{\frac{3}{2}}$

or $45 \times 45 = (3x)^3$

$x^3 = 75$

$\log_3 x^3 = \log_3 75 \Rightarrow \log_3 27 < \log_3 75 < \log_3 81$

$3 < \log_3 75 < 4$; Characteristic = 3.

IIT JEE Corner

31. (d) $2 \log_{10} x - \log_x (0.01)$

$\Rightarrow 2 \log_{10} x - \log_x (10^{-2})$

$\Rightarrow 2 \log_{10} x + 2 \log_x 10$

$\Rightarrow 2 (\log_{10} x + \log_x 10)$

$\Rightarrow \quad 2\left(\log_{10} x + \dfrac{1}{\log_{10} x}\right)$

$\Rightarrow$ 2 (≥ 2) ≥ 4. So least values is 4.

32. $\log_{(2x+3)}(6x^2 + 23x + 21) = 4 - \log_{(3x+7)}(4x^2 + 12x + 9)$

$\Rightarrow \log_{(2x+3)}[(2x+3)(3x+7)]$

$= 4 - \log_{(3x+7)}(2x+3)^2$

$\Rightarrow \log_{(2x+3)}(2x+3) + \log_{(2x+3)}(3x+7)$

$= 4 - 2\log_{(3x+7)}(2x+3)$

Let $\log_{(3x+7)}(2x+3) = y$

$\Rightarrow 1 + y = 4 - \dfrac{2}{y} \Rightarrow y^2 + y = 4y - 2$

$\Rightarrow y^2 - 3y + 2 = 0$

$\Rightarrow (y-2)(y-1) = 0$

or $y = 1$ or 2

$\therefore \log_{(2x+3)}(3x+7) = 1$ or 2

or $(3x+7) = (2x+3)$ or $(3x+7) = (2x+3)^2$

or $x = -4$ or $3x + 7 = 4x^2 + 9 + 12x$

or $4x^2 + 9x + 2 = 0$

or $4x(x+2) + 1(x+2) = 0$

or $x = -2$ or $-\dfrac{1}{4}$

$\therefore$ By definition of $\log_{(2x+3)}(3x+7)$,

$x = -2$ ruled out

$\therefore$ Only $x = -\dfrac{1}{4}$ is solution.

33. (b) $x^{\frac{3}{4}(\log_2 x)^2 + (\log_2 x) \frac{-5}{4}} = \sqrt{2}$

Take log both side with base 2.

$\Rightarrow \left(\dfrac{3}{4}(\log_2 x)^2 + (\log_2 x) - \dfrac{5}{4}\right)\log_2 x = \log_2 \sqrt{2}$

$\Rightarrow \dfrac{3}{4}(\log_2 x)^3 + (\log_2 x)^2 - \dfrac{5}{4}(\log_2 x) = +\dfrac{1}{2}$

$\Rightarrow 3(\log_2 x)^3 + 4(\log_2 x)^2 - 5(\log_2 x) - 2 = 0$

Let $(\log_2 x) = t$

$\Rightarrow 3t^3 + 4t^2 - 5t - 2 = 0$

$\Rightarrow (t-1)(t+2)(3t+1) = 0$

$\Rightarrow t = 1$ or -2 or $-\dfrac{1}{3}$ $\therefore x = 2, \dfrac{1}{4}, \dfrac{1}{\sqrt[3]{2}}$

34. (b) $\dfrac{1}{2}\log_2(x-1) = \log_2(x-3)$

$\therefore \sqrt{x-1} = (x-3)$

$\Rightarrow (x-1) = (x-3)^2 \Rightarrow x - 1 = x^2 + 9 - 6x$

$\Rightarrow x^2 - 7x + 10 = 0$

$x = 5$ or 2

$\therefore x = 5$ is solution. $(x > 3)$

35. (c) $(2x)^{\ln 2} = (3y)^{\ln 3}$ ... (1)

$3^{\ln x} = 2^{\ln y}$ ... (2)

Take log both side w.r.t. e

$\ln 2 \ln(2x) = \ln 3 \ln(3y)$ ... (3)

or $\ln x \ln 3 = \ln y \ln 2$ ... (4)

By equation (4) put the value of $\ln y$ in eq 3

$\ln 2[\ln 2 + \ln x] = \ln 3[\ln 3 + \ln y]$

$\Rightarrow \ln 2[\ln 2 + \ln x]$

$= \ln 3\left[\ln 3 + \dfrac{\ln x \ln 3}{\ln 2}\right]$

$\Rightarrow (\ln 2)^2[\ln 2 + \ln x] = (\ln 3)^2[\ln 2 + \ln x]$

$\Rightarrow (\ln 2 + \ln x)[(\ln 2)^2 - (\ln 3)^2] = 0$

$\ln x = -\ln 2$

$\ln x = \ln 2^{-1}$

or $x = \dfrac{1}{2}$ so

$y = \dfrac{1}{3}$ $\therefore x_0 = \dfrac{1}{2}$

36. Let $x = \dfrac{1}{3\sqrt{2}}\sqrt{4 - \dfrac{1}{3\sqrt{2}}\sqrt{4 - \dfrac{1}{3\sqrt{2}}\sqrt{4 - \dfrac{1}{3\sqrt{2}}\ldots}}}$

$x = \dfrac{1}{3\sqrt{2}}\sqrt{4-x} \Rightarrow 18x^2 = 4 - x$

$\Rightarrow 18x^2 + x - 4 = 0$

$x =$

$\dfrac{-1 \pm \sqrt{1 + 18 \times 4 \times 4}}{36} = \dfrac{-1 \pm 17}{36} = \dfrac{16}{36} = \dfrac{4}{9} = \left(\dfrac{2}{3}\right)^2$

Hence, $6 + \log_{\frac{3}{2}}\left(\dfrac{2}{3}\right)^2 = 6 - 2 = 4.$

CHAPTER

3

BASIC MATHS (Piecewise Defined Function)

PART-A

3.1 MODULUS FUNCTION

3.1.1 Introduction of Modulus Function

Introduction

Modulus function is defined as:

$$y = |x| = \begin{cases} -x & \text{if } x < 0 \\ 0 & \text{if } x = 0 \\ x & \text{if } x > 0 \end{cases}$$

It can also be defined as $y = |x| = \begin{cases} -x & \text{if } x < 0 \\ x & \text{if } x \geq 0 \end{cases}$

From above definition $|5| = 5$ and $|-5| = 5$

So if $|x| = 5$ means $x = \pm 5$

In other words $|x|$ is defined as distance of a point from the origin so if $|x| = 5$ is given that means distance of point (on x-axis) from the origin is 5 units and we know that there two such points exist +5 and –5.

Illustration 1. If $|x - 2| = 5$, then find x

Solution: Since, we know that if $|x| = 5$, then $x = \pm 5$ here instead of x, given quantity is $x - 2$, so $x - 2 = \pm 5$

So $x = 7$ or -3

Alternately:-

Since $|x|$ is defined as distance of a point from the origin similarly $|x - 2|$ is defined as distance of a point from the point $x = 2$. From point 2 distance is 5 units so values of x are –3 and 7.

3.1.2 Basic Inequality Related to Modulus Function

Consider a positive number 'a'

If $|x| > a$, then $x < -a$ or $x > a$

If $|x| < a$, then $-a < x < a$

Illustration 2. If $|x| > 8$, then find x

Solution: Since, we know that if $|x| > a$, then $x < -a$ or $x > a$

So if $|x| > 8$, then $x < -8$ or $x > 8$

or $x \in (-\infty, -8) \cup (8, \infty)$

Alternately:-

Since $|x|$ is defined as distance of a point from the origin so $|x| > 8$ is defined as distance of a point from the point x=0 or origin is more than 8 units so $x < -8$ or $x > 8$ or $x \in (-\infty, -8) \cup (8, \infty)$

Illustration 3. If $|x - 2| > 5$, then find x

Solution: Since, we know that if $|x| > 5$, then $x < -5$ or $x > 5$ here instead of x, given quantity is $x - 2$ so When $x - 2 < -5$, then $x < -3$

When $x - 2 > 5$, then $x > 7$

If we combine the above two results we will get

$x \in (-\infty, -3) \cup (7, \infty)$

Alternately:-

Since $|x|$ is defined as distance of a point from the origin, similarly $|x - 2| > 5$ is defined as distance of a point from the point $x = 2$ is more than 5 units.

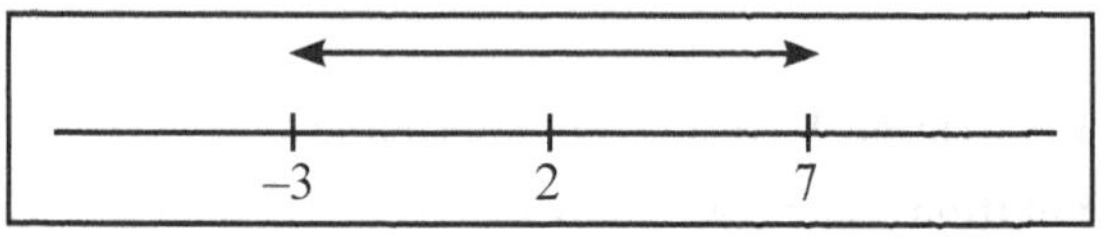

From point 2 distance is 5 units so values of x are –3 and 7.

3.1.3 Graphical Approach to Modulus Function

In order to plot the graph of $y = |x|$ use the standard definition of modulus. The definition has two parts:

Part 1: When $x < 0$ means in 2nd and 3rd quadrant plot $y = -x$

Part 2: When $x > 0$ means in 1st and 4th quadrant plot $y = x$.

Graph 1 : Y = |x|			
Part 1, when $x < 0$		Part 2 when $x > 0$	
X	Y	X	Y
–1	1	1	1
–2	2	2	2
–3	3	3	3

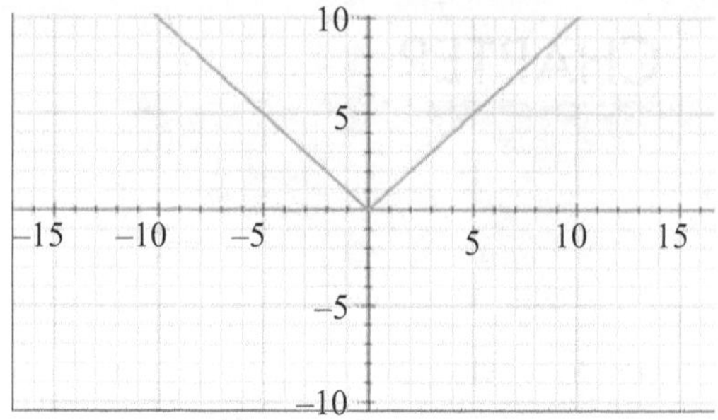

An alternate way of plotting y = |x|

Graph 2: $y = |x|$ **(Alternate Method)**

In order to plot $y = |x|$, 1st plot $y = x$ and then put modulus by doing that we are changing –1 to +1, –2 to +2 and so on, reflection about x-axis, as shown in the figure, The dotted line is converted to positive y (i.e got reflected with respect to x-axis).

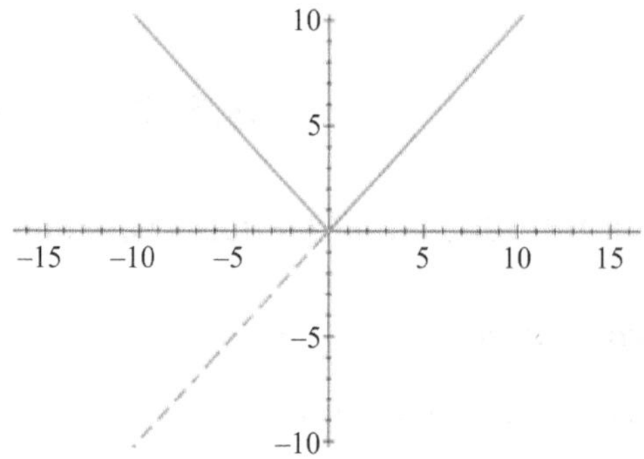

Transformation from *y = f(x) to y = |f(x)|*

If we know the graph of $y = f(x)$ then in order to plot $y = |f(x)|$ what we need to do is reflection of the graph $y = f(x)$ with respect to x-axis.

Consider a graph $y = f(x)$

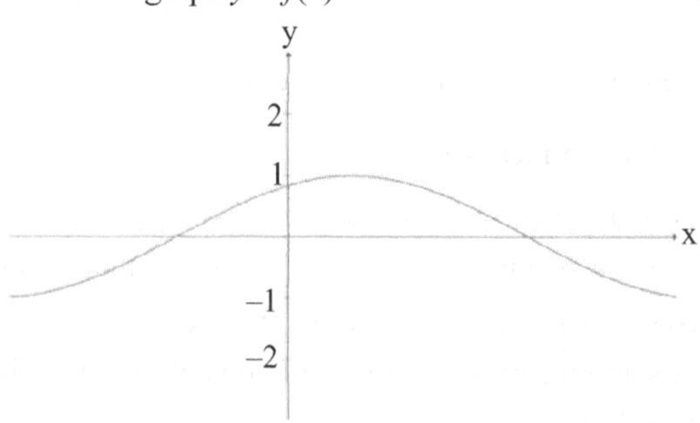

Now graph of $y = |f(x)|$ is reflection of the graph $y = f(x)$ with respect to x-axis.

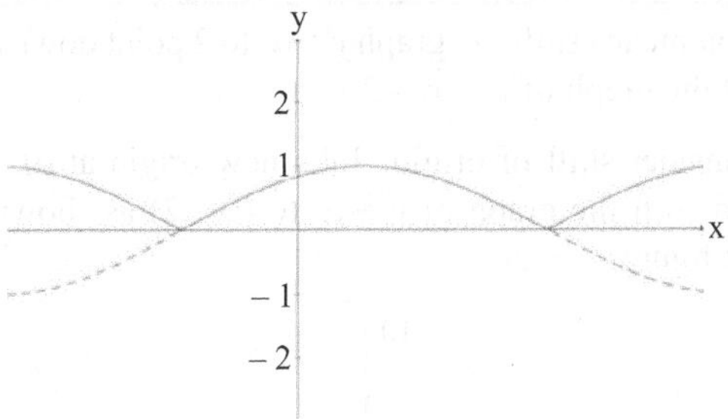

Transformation from *y = f(x)* to *y = f(|x|)*

If we know the graph of $y = f(x)$ then in order to plot $y = f(|x|)$ what we need to do is reflection of the graph $y = f(x)$ with respect to y-axis.

Consider a graph $y = f(x)$

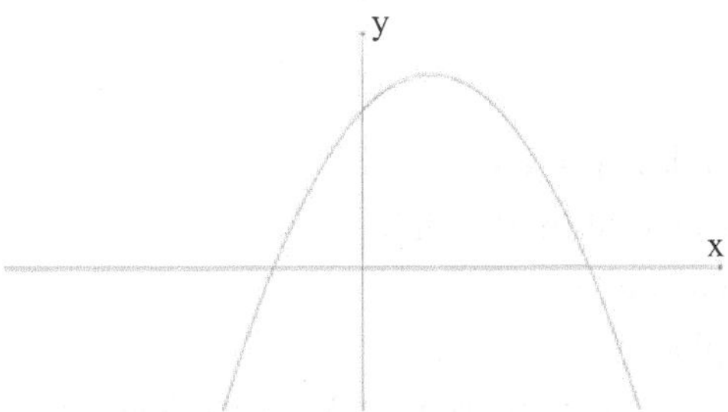

Now graph of $y = f(|x|)$ is reflection of the graph $y = f(x)$ with respect to y-axis.

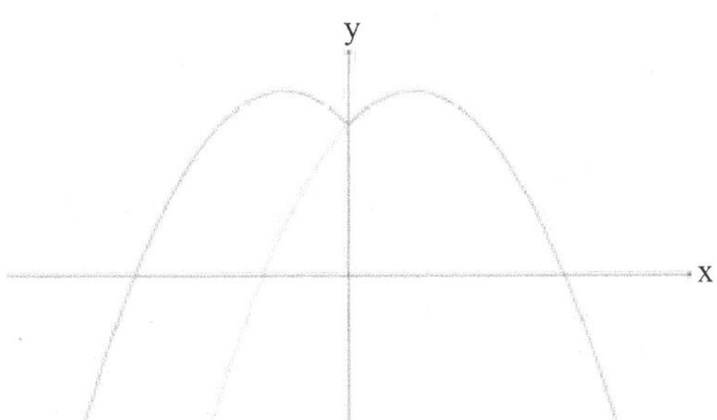

Graph 3: $y = |x - 2|$ **(Method-1)**

Method 1: From Definition

$$y = |x - 2| = \begin{cases} -(x-2) \text{ when } (x-2) < 0 \text{ or } x < 2 \\ (x-2) \text{ when } (x-2) \ge 0 \text{ or } x \ge 2 \end{cases}$$

So in total there are two straight lines:

Line 1: $y = -(x - 2)$ but it is only when $x < 2$ (remaining part is shown as dotted line)

Line 2: $y = (x - 2)$ and it is only when $x > 2$ (remaining part is shown as dotted line).

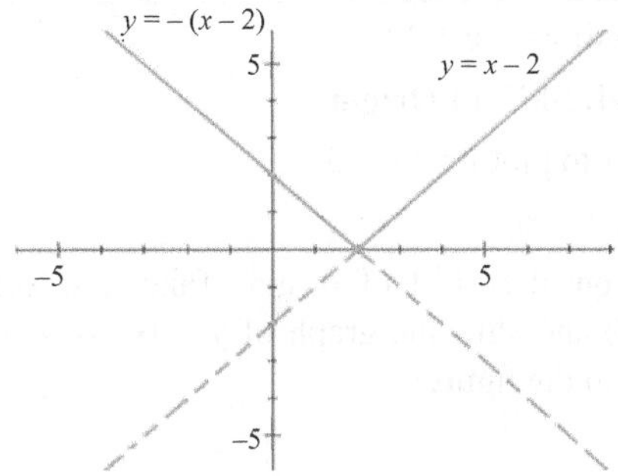

Graph 4: y = |x – 2| (Method-2)

Method 2: Reflection with respect to x-axis

In order to plot $y = |x - 2|$,

1st plot $y = x - 2$ and then put modulus, by doing that we are changing –1 to +1, –2 to +2 and so on, reflection about x-axis, as shown in the figure. The dotted line is converted to positive y (i.e got reflected with respect to x-axis).

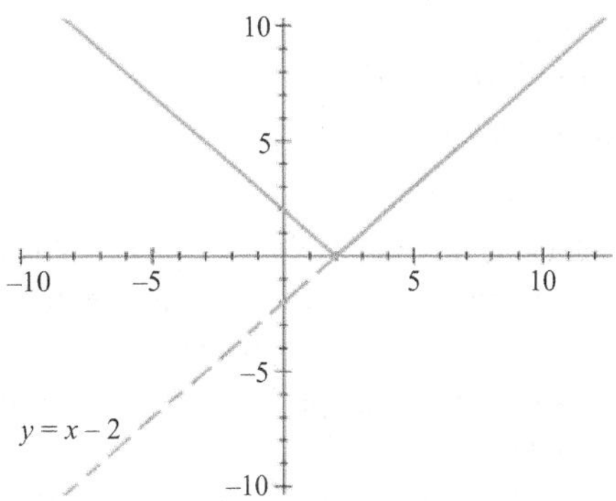

Graph 5: $y = |x - 2|$ **(Method-3)**

Method 3: Shift of Origin

In order to plot $y = |x - 2|$,

1st plot $y = |x|$

Now consider shift of origin. Take new origin at (– 2, 0) and shift the graph of $y = |x|$ to (– 2, 0) as shown in the figure.

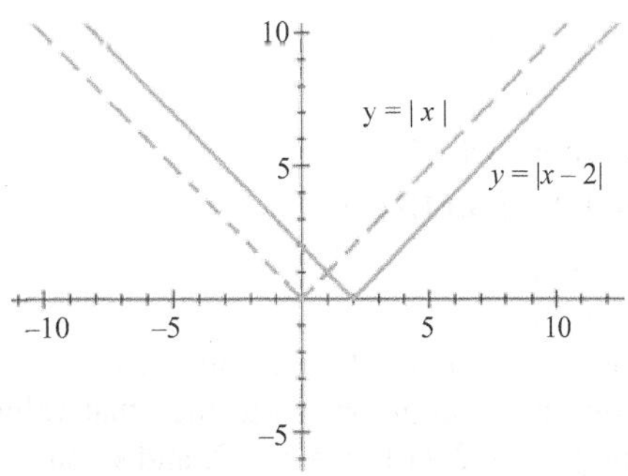

Graph 6: $y = |x + 3|$

Method: Shift of Origin

In order to plot $y = |x + 3|$

1st plot $y = |x|$

Now consider shift of origin. Take new origin at (– 3, 0) and shift the graph of $y = |x|$ to (– 3, 0) as shown in the figure.

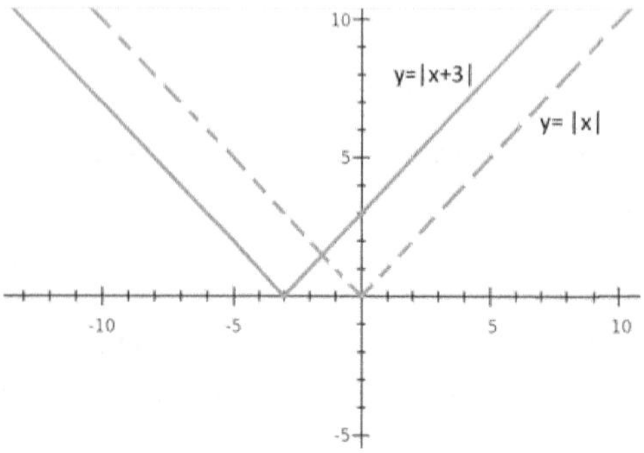

Graph 7: $y = |x| + 3$

Method: Shift of Origin

In order to plot $y = |x| + 3$

1st plot $y = |x|$

For each value of y add 3 to get $y = |x| + 3$, if at a point $y = 2$ in graph $y = |x|$ then that value will become $y = 2 + 3 = 5$ in $y = |x| + 3$, and so on,

That means shift the graph $y = |x|$ to 3 point up and get the graph of $y = |x| + 3$

Consider shift of origin. Take new origin at (0, 3) and shift the graph of $y = |x|$ to (0, 3) as shown in the figure.

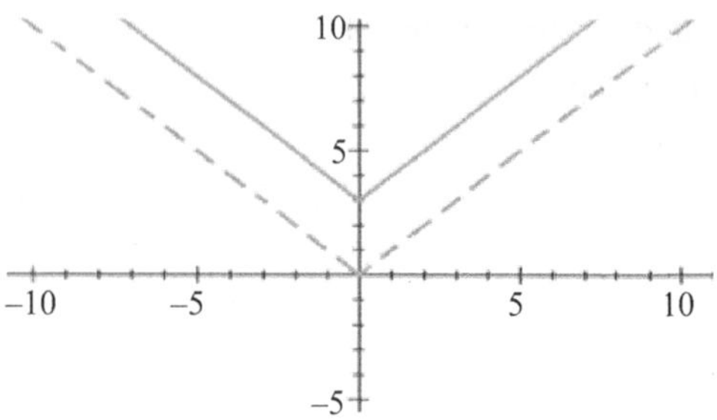

Graph 8: $y = |x| - 2$

Method: Shift of Origin

In order to plot $y = |x| - 2$

1st plot $y = |x|$

For each value of y subtract 2 to get $y = |x| - 2$, if at a point $y = 3$ in graph $y = |x|$ then that value will become $y = 3 - 2 = 1$ in $y = |x| - 2$, and so on,

That means shift the graph $y = |x|$ to 2 point down and get the graph of $y = |x| - 2$.

Consider shift of origin. Take new origin at (0, –2) and shift the graph of $y = |x|$ to (0, –2) as shown in the figure.

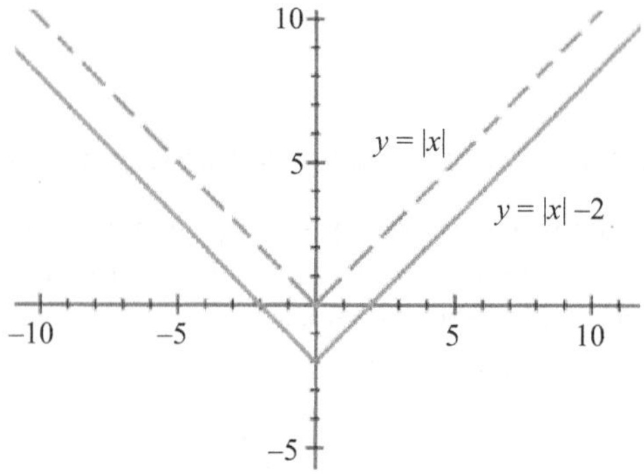

Graph 9: $y = |x + 2| - 3$

Method: Shift of Origin

In order to plot $y = |x + 2| - 3$

1st plot $y = |x|$

Consider shift of origin. Take new origin at (–2, –3) and shift the graph of $y = |x|$ to (–2, –3) as shown in the figure.

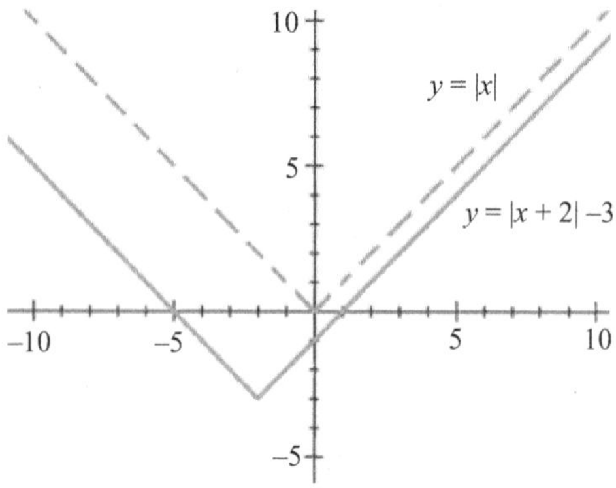

Illustration 4. Find the minimum value of

$$y = f(x) = |x + 2| - 3$$

(a) –3 (b) 0

(c) 3 (d) None of these

Solution. From the graph minimum value of function is at $x = -2$ and required minimum value is –3.

Option (a)

Graph 10: $y = 4|x - 2| + 2$

Step 1: Plot $y = 4|x|$

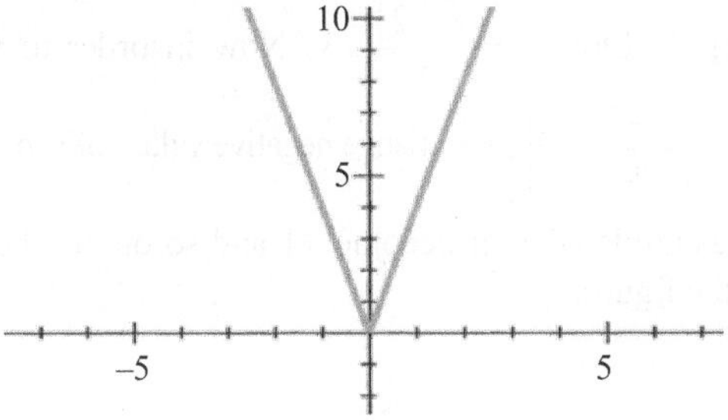

Step 2: Plot $y = 4|x - 2|$

Shift the graph of $y = 4|x|$ to new origin (2, 0) as shown in the figure.

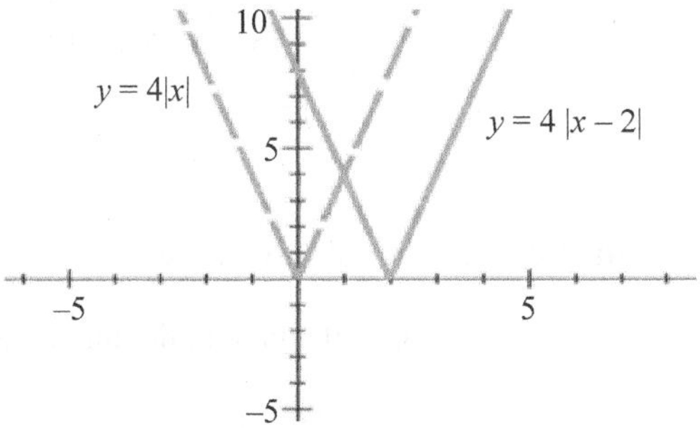

Step 3: Plot $y = 4|x - 2| + 2$

Shift the graph of $y = 4|x|$ to new origin (2, 2) as shown in the figure.

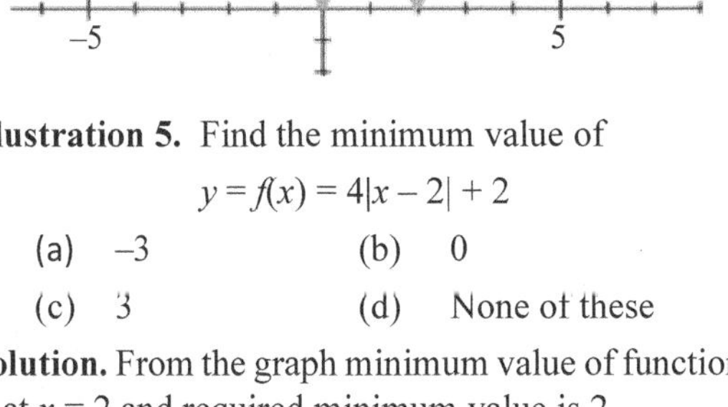

Illustration 5. Find the minimum value of

$$y = f(x) = 4|x - 2| + 2$$

(a) −3 (b) 0

(c) 3 (d) None of these

Solution. From the graph minimum value of function is at $x = 2$ and required minimum value is 2.

Option (d)

Graph 11: $y = \dfrac{|x+4|}{3} + 3$

Step 1: Plot $y = \dfrac{x}{3}$ as shown in the figure.

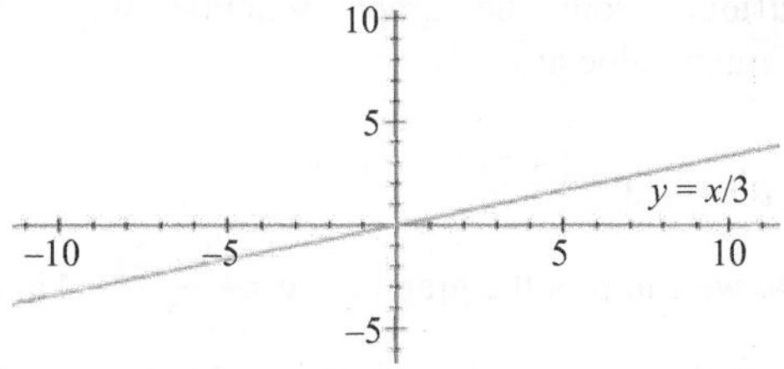

Step 2: Plot $y = \dfrac{|x|}{3}$ reflection of $y = \dfrac{x}{3}$ with respect to x-axis as shown in the figure.

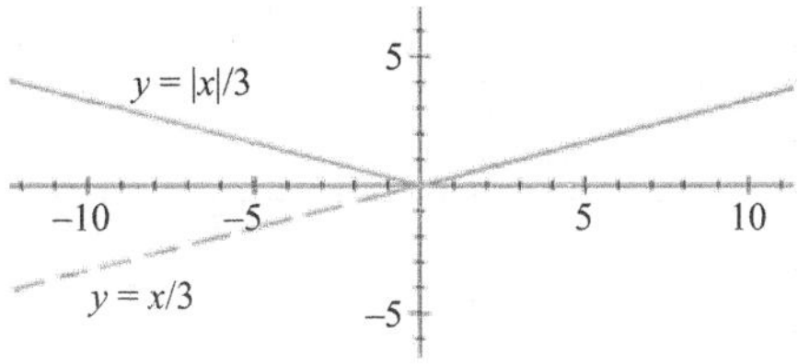

Step 3: Plot $y = \dfrac{|x+4|}{3}$ by shift of origin taking new origin at (– 4, 0) of graph $y = \dfrac{|x|}{3}$ as shown in the figure.

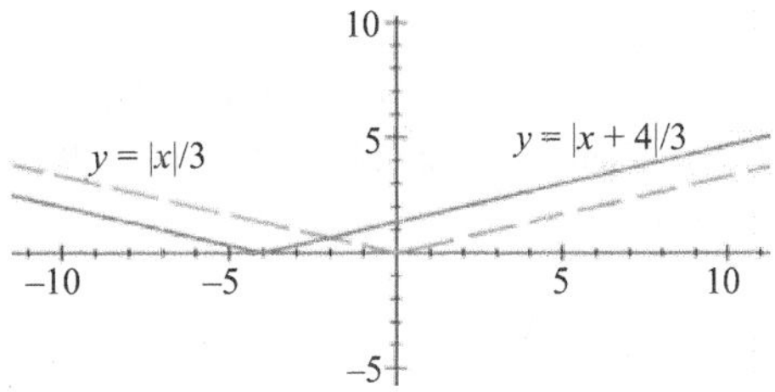

Step 4: Plot $y = \dfrac{|x+4|}{3} + 3$ by shift of origin taking new origin at (–4, 3) of graph $y = \dfrac{|x|}{3}$ as shown in the figure.

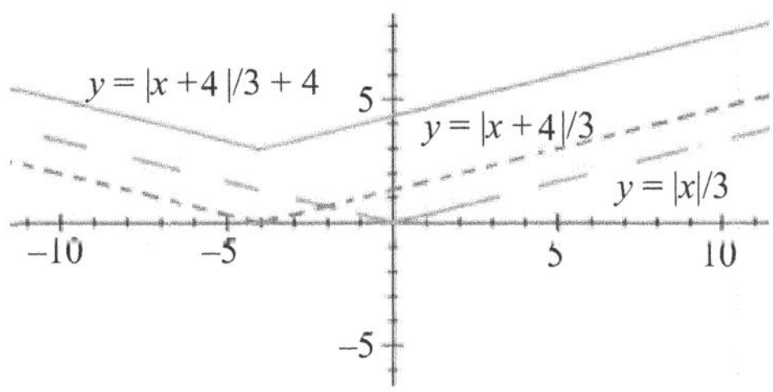

Illustration 6. Find the minimum value of

$$y = f(x) = \frac{|x+4|}{3} + 3$$

(a) −3 (b) 4

(c) 3 (d) None of these

Solution. From the graph function will attain minimum value at $x = -4$.

Graph 12: $y = \dfrac{|x-3|}{2} - 4$

Now, we can plot the graph of $y = \dfrac{|x-3|}{2} - 4$ in one step, just plot $y = \dfrac{|x|}{2}$ and shift it to a new origin (3, –4) as shown in the figure.

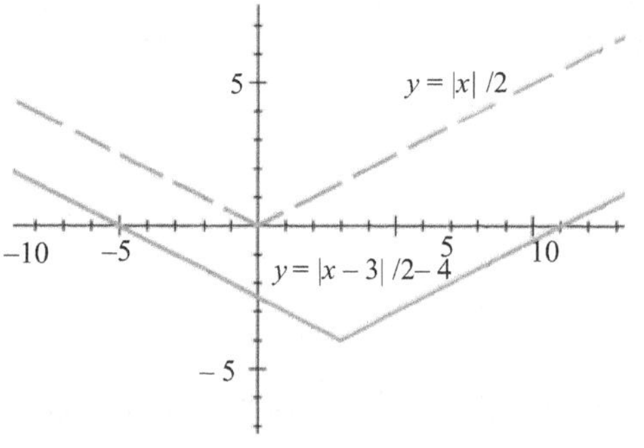

Illustration 7. Find the value of x where

$y = f(x) = \dfrac{|x-3|}{2} - 4$ will attain minimum value

(a) –3 (b) 4
(c) 3 (d) None of these

Solution. From the graph function will attain minimum value at $x = 3$.
Option (c)

Graph 13: $y = \left|\dfrac{|x-2|}{2} - 3\right|$

Step 1: Plot $y = \dfrac{|x-2|}{2} - 3$ as we did in previous graphs by 1st plotting $y = \dfrac{|x|}{2}$ and then shift of origin with new origin at (2, –3).

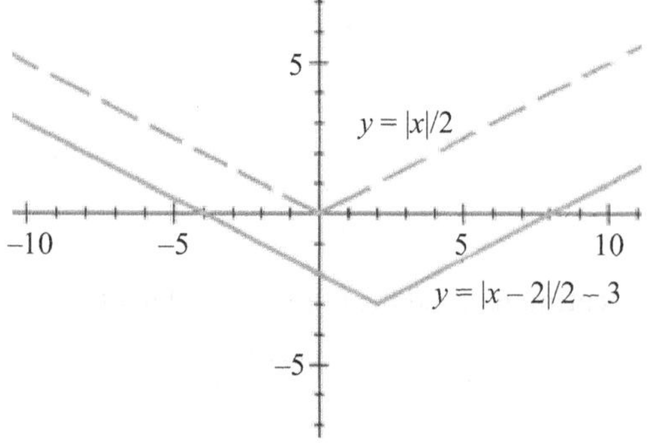

Step 1: Plot $y = \dfrac{|x-2|}{2} - 3$. Now in order to plot $y = \left|\dfrac{|x-2|}{2} - 3\right|$, we change negative values of y to +ve, as example –1 will become +1 and so on, as shown in the figure.

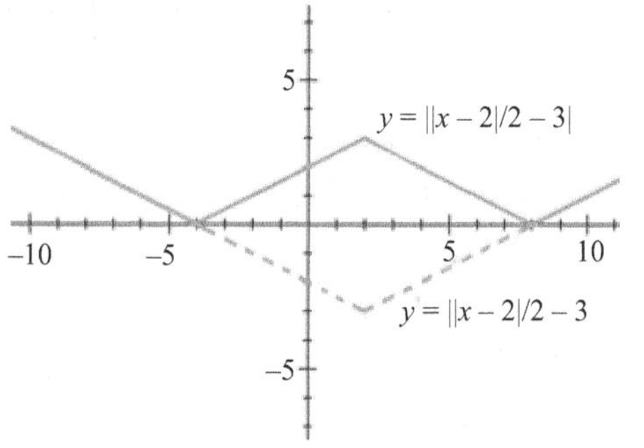

Illustration 8. Find the value of x where

$y = f(x) = \left|\dfrac{|x-2|}{2} - 3\right|$ will attain minimum value

(a) 6 (b) 4
(c) 3 (d) None of these

Solution. From the graph function will attain minimum value at $x = -4$ and 8.
Option (d)

Graph 14: $y = \left|\dfrac{|x-2|}{2} - 3\right| - 4$

Step 1: Plot $y = \dfrac{|x-2|}{2} - 3$ as we did in previous graphs by 1st plotting $y = \dfrac{|x|}{2}$ and then shift of origin with new origin at (2, –3).

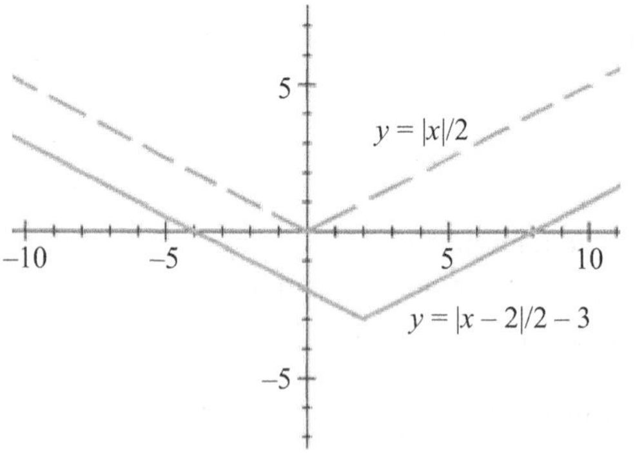

Step 2: Plot $y = \frac{|x-2|}{2} - 3$ now in order to plot $y = \left|\frac{|x-2|}{2} - 3\right|$, we change negative values of y to +ve, as example –1 will become +1 and so on, as shown in the figure.

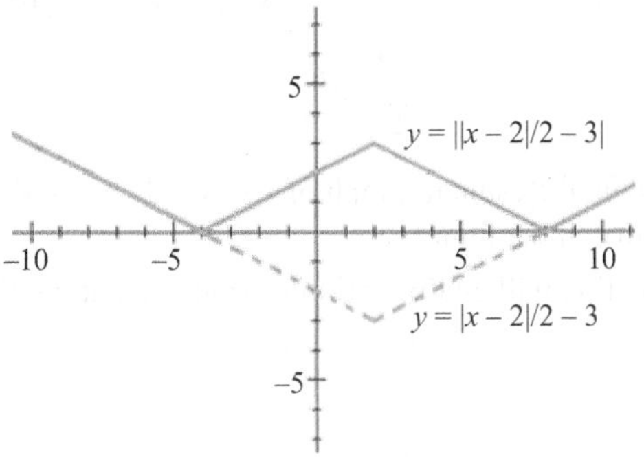

Step 3: In order to plot $y = \left|\frac{|x-2|}{2} - 3\right| - 4$ we will 1st plot $y = \left|\frac{|x-2|}{2} - 3\right|$ as we did in step 2.

If we pull this graph down by 4 unit then we get $y = \left|\frac{|x-2|}{2} - 3\right| - 4$, as shown in the figure.

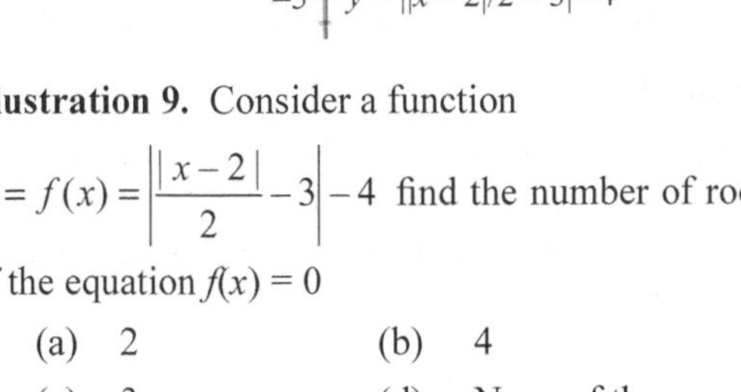

Illustration 9. Consider a function

$y = f(x) = \left|\frac{|x-2|}{2} - 3\right| - 4$ find the number of roots of the equation $f(x) = 0$

(a) 2 (b) 4

(c) 3 (d) None of these

Solution. The graph will intersect the x-axis twice so we will have 2 roots.

Option (a)

Graph 15: $\left|\left|\frac{|x-2}{2} - 3\right| - 4\right|$

Step 1: 1st plot $y = \left|\frac{|x-2|}{2} - 3\right| - 4$ as we did in previous graph.

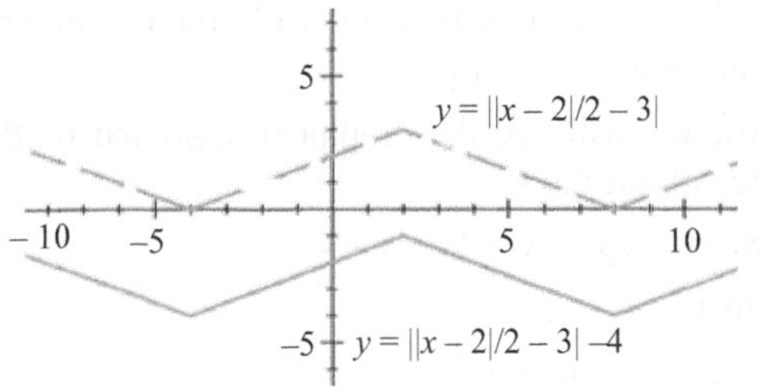

Step 2: After plotting $y = \left|\frac{|x-2|}{2} - 3\right| - 4$ if we put a modulus means we are changing each negative sign to positive sign as shown in the figure.

So the final graph of $y = \left|\left|\frac{|x-2|}{2} - 3\right| - 4\right|$

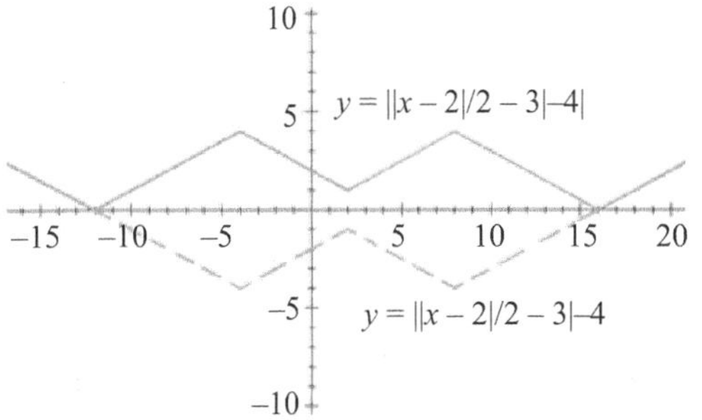

Illustration 10. Consider a function

$y = f(x) = \left|\left|\frac{|x-2|}{2} - 3\right| - 4\right|$ find the minimum value of the function

(a) 1 (b) 4

(c) 3 (d) None of these

Solution. From the graph function will attain minimum value at $x = -12$ and 16, the required minimum value is 0.

Option (d)

3.1.4 Summation of Two or More Modulus

In this section, we will discuss different types of questions where two or more modulus is added/ subtracted. General formulae for these types are discussed below.

Type 1: $f(x) = |x - a| + |x - b|$ if $a < b$, then function will attain a minimum value when $x \in [a, b]$

Type 2: $f(x) = |x-a| + |x-b| + |x-c|$ if $a < b < c$, then function will attain a minimum value at $x = b$.

Type 3: $f(x) = |x-a| + |x-b| + |x-c| + |x-d|$ if $a < b < c < d$ then function will attain a minimum value when $x \in [b, c]$.

Now, we will see the graphical approach of these types of questions:

Graph 16: $y = |x-2| + |x-4|$

Step 1:

When $x < 2$, then $|x-2| = -(x-2)$

And $|x-4| = -(x-4)$

So $y = |x-2| + |x-4| = -(x-2) - (x-4) = -2x + 6$

Plot $y = -2x + 6$ for $x < 2$ as shown in the figure

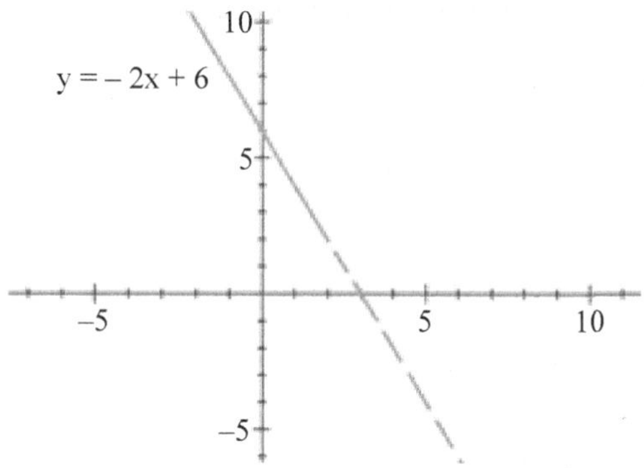

Step 2:

When $2 < x < 4$, then $|x-2| = (x-2)$

And $|x-4| = -(x-4)$

So $y = |x-2| + |x-4| = (x-2) - (x-4) = 2$

Plot $y = 2$ for $2 < x < 4$ as shown in the figure

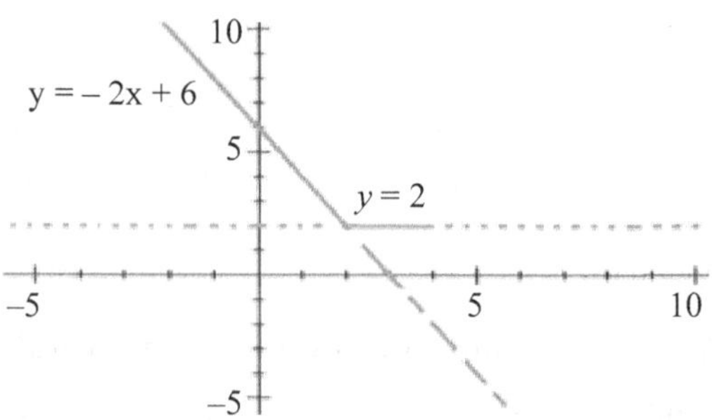

Step 3:

When $x > 4$, then $|x-2| = (x-2)$

And $|x-4| = (x-4)$

So $y = |x-2| + |x-4| = (x-2) + (x-4) = 2x - 6$

Plot $y = 2x - 6$ for $x > 4$ as shown in the figure

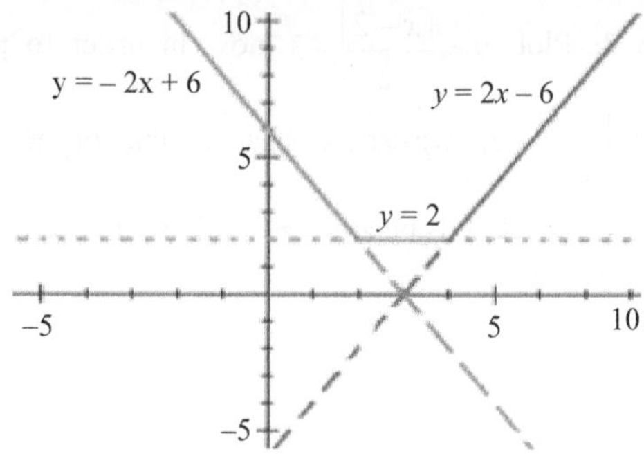

The final Resultant graph of $y = |x-2| + |x-4|$ is as shown in the figure.

Function will attain minimum value of 2 at $x = [2, 4]$

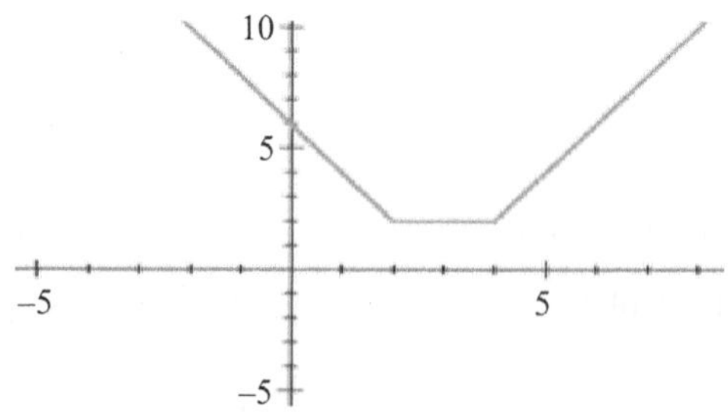

Illustration 11. Consider a function $y = f(x) = |x-2| + |x-4|$ find the minimum value of the function

(a) 1 (b) 4

(c) 3 (d) None of these

Solution. From the graph function will attain minimum value at $x = [2, 4]$ and the required minimum value is 2

Option (d)

Find the minimum value of $|x + 2| + |x - 2|$

Graph 17: $y = |x+2| + |x-2|$

Plot the graph of $y = |x+2| + |x-2|$ as we did in graph 16, we will find that value of y is constant between $x = -2$ to $x = 2$ and that value of $y (= 4)$ is the minimum value of y.

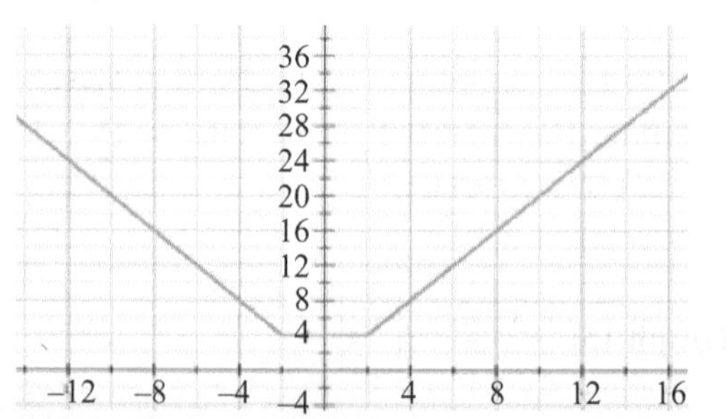

All the functions in the form of $y = |x - a| + |x - b|$ will have similar graph with minimum value of y at any value of x between $[a, b]$ here we have assumed that $a < b$.

Find the minimum value of $|x + 2| + |x - 2| + |x - 5|$

Graph 18: $y = |x + 2| + |x - 2| + |x - 5|$

Plot the graph of $y = |x + 2| + |x - 2| + |x - 5|$ as we did in graph 16, we will find that value of y is minimum at $x = 2$ and that minimum value of y is 7.

Alternately:- Find the critical points, these are –2, 2 and 5 arrange them in the increasing order on x-axis, middle one is 2 so the function will attain its minimum value at $x = 2$ as shown in the figure.

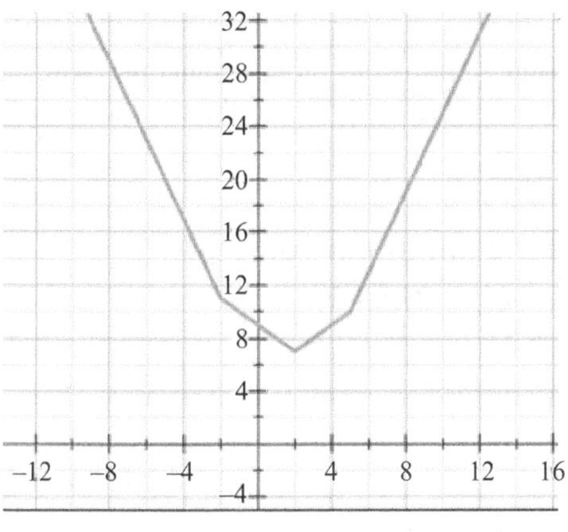

All the functions in the form of $y = |x - a| + |x - b| + |x - c|$ will have similar graph with minimum value of y at $x = b$ here we have assumed that $a < b < c$.

Illustration 12. Consider a function $y = f(x) = |x + 2| + |x - 2| + |x - 5|$ find the minimum value of the function

(a) 6 (b) 4

(c) 7 (d) None of these

Solution. From the graph function will attain minimum value at $x = 2$ and the required minimum value is 7

Option (c)

Find the minimum value of $|x + 2| + |x - 2| + |x - 5| + |x - 8|$

Graph 19: $y = |x + 2| + |x - 2| + |x - 5| + |x - 8|$

Plot the graph of $y = |x + 2| + |x - 2| + |x - 5| + |x - 8|$ as we did in graph 16, we will find that value of y is minimum at any value of x between 2 to 5.

Alternately:- Find the critical points, these are –2, 2, 5 and 8 arrange them in the increasing order on x-axis, since we have 4 critical points so we will get 2 middle values and these are 2 and 5 so the function will attain its minimum value at any value of x between 2 to 5.

Required minimum value is 13

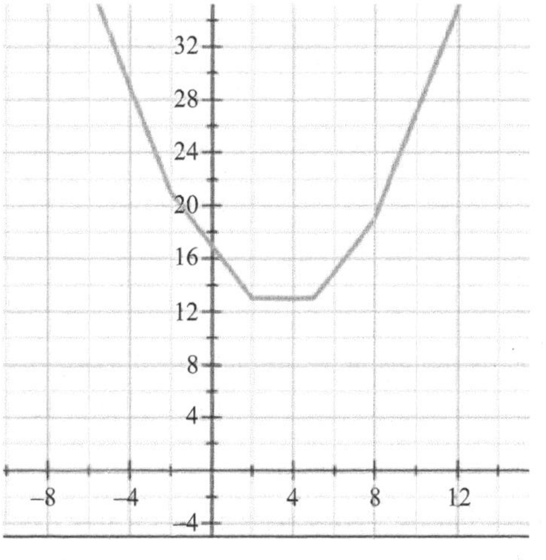

All the functions in the form of $y = |x - a| + |x - b| + |x - c| + |x - d|$ will have similar graph with minimum value of y at any value of x in $[b, c]$ here we have assumed that $a < b < c < d$.

Illustration 13. Consider a function $y = f(x) = |x + 2| + |x - 2| + |x - 5| + |x - 8|$ find the minimum value of the function

(a) 11 (b) 14

(c) 17 (d) None of these

Solution. From the graph function will attain minimum value at $x = [2, 5]$ and the required minimum value is 13

Option (d)

Find the minimum value of $|x + 2| + |x - 2| + |x - 5| + |x - 8| + |x - 10|$

Graph 20: $y = |x + 2| + |x - 2| + |x - 5| + |x - 8| + |x - 10|$

Plot the graph of $y = |x + 2| + |x - 2| + |x - 5| + |x - 8| + |x - 10|$ as we did in graph 16, we will find that value of y is minimum at $x = 5$.

Alternately:- Find the critical points, these are –2, 2, 5, 8 and 10 arrange them in the increasing order on x-axis, since we have 5 critical points so we will get only 1 middle value which is $x = 5$ so the function will attain its minimum value at $x = 5$.

Required minimum value is 18.

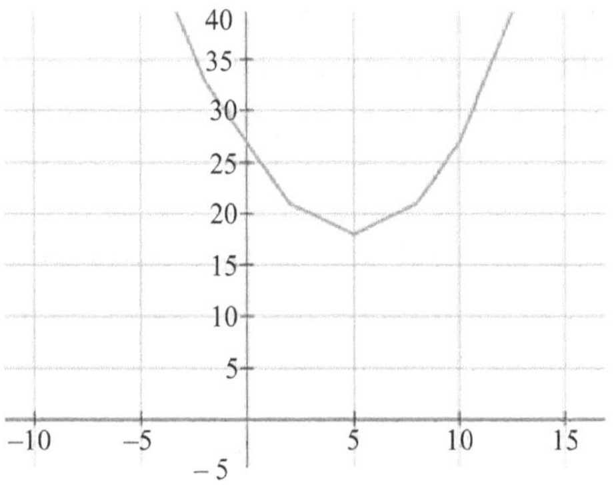

All the functions in the form of $y = |x - a| + |x - b| + |x - c| + |x - d| + |x - e|$ will have similar graph with minimum value of y at x = c here we have assumed that a < b < c < d < e.

Illustration 14. Consider a function $y = f(x) = |x + 2| + |x - 2| + |x - 5| + |x - 8| + |x - 10|$ find the minimum value of the function

(a) 11 (b) 14
(c) 18 (d) None of these

Solution. From the graph function will attain minimum value at $x = 5$ and the required minimum value is 18

Option (c)

Find the minimum value of $|x + 2| + 2|x - 2|$

Graph 21: $y = |x + 2| + 2|x - 2|$

1st question is how many critical points we have here 2 or 3? Answer is we have 3 critical points and these points are –2, 2, 2 (point 2 taken twice).

Out of these three points middle one is 2 so the function will attain its minimum value at $x = 2$

Required minimum value is 4.

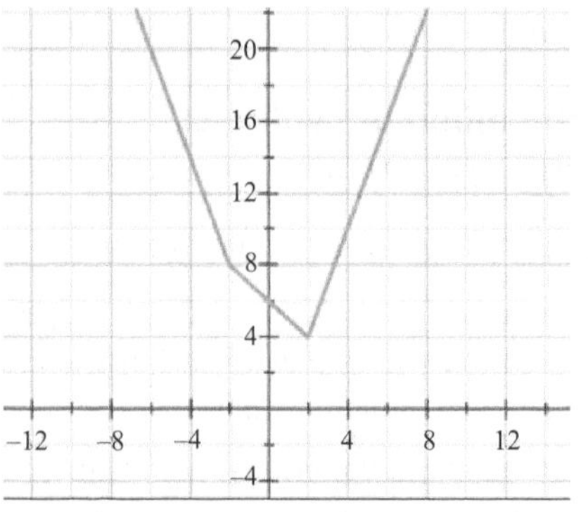

Illustration 15. Consider a function $y = f(x) = |x + 2| + 2|x - 2|$ find the minimum value of the function

(a) 1 (b) 4
(c) 3 (d) None of these

Solution. From the graph function will attain minimum value at $x = 2$ and the required minimum value is 4

Option (b)

Find the minimum value of $3|x + 2| + 2|x - 2|$

Graph 22: $y = 3|x + 2| + 2|x - 2|$

1st question is how many critical points we have here 2 or 5? Answer is we have 5 critical points and these points are –2, –2, –2, 2, 2 (point –2 is taken thrice and point 2 is taken twice)

Out of these 5 points middle one is –2 so the function will attain its minimum value at $x = -2$

Required minimum value is 8.

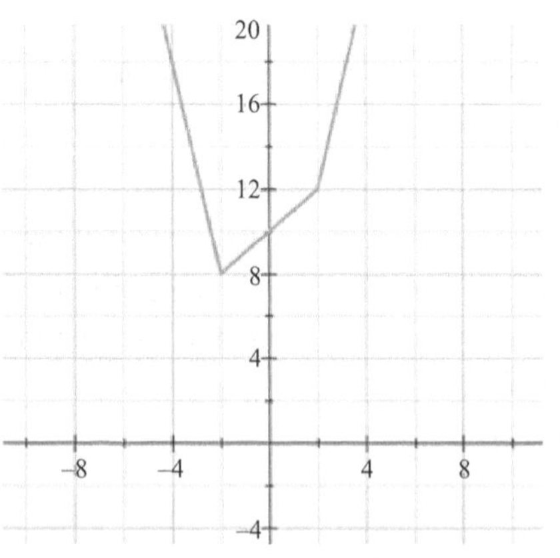

In general $a|x - p| + b|x - q|$ will attain its minimum value at $x = p$ if $a > b$ otherwise at $x = q$.

Illustration 16. Consider a function $y = f(x) = 2|x - 2| + 3|x + 2|$ find the minimum value of the function

(a) 7 (b) 4

(c) 8 (d) None of these

Solution. From the graph function will attain minimum value at $x = -2$ and the required minimum value is 8.

Option (c)

Find the minimum value of

$3|x + 2| + 2|x - 2| + |x - 4|$

Graph 23: $y = 3|x + 2| + 2|x - 2| + |x - 4|$

Here we have 6 critical points and these points are –2, –2, –2, 2, 2 and 4 (point –2 is taken thrice and point 2 is taken twice).

Here we have 6 critical points so we will have two middle values and these are 3rd and 4th one i.e $x = -2$ and $x = 2$, so the function will attain its minimum value at any value of x between –2 and 2, as shown in the graph.

Required minimum value is 14.

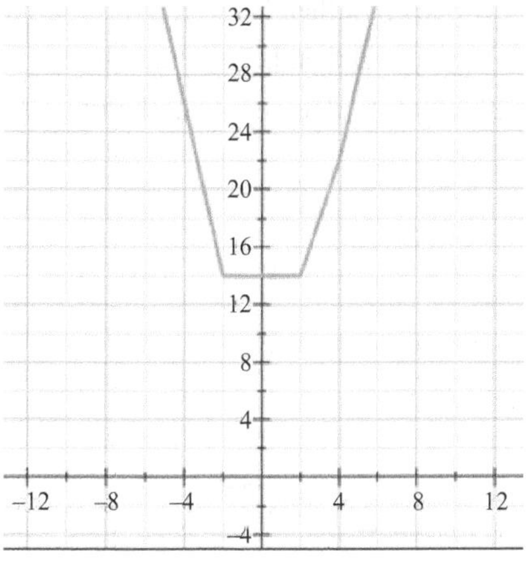

Concept Applicator (CA)-1

1. If $|x - \sqrt{2}| = \sqrt{5}$ then find x
2. If $|x| \leq 10$ then find x
3. Find the value of x where $y = f(x) = |x + 2| - 3$ will attain minimum value

   (a) –3 (b) 0

   (c) 3 (d) None of these
4. Find the value of x where $y = f(x) = 4|x - 2| + 2$ will attain minimum value

   (a) –3 (b) 0

   (c) 3 (d) None of these
5. Find the value of x where $y = f(x) = \frac{|x+4|}{3} + 3$ will attain minimum value

   (a) –3 (b) 4

   (c) 3 (d) None of these
6. Find the minimum value of

   $$y = f(x) = \left|\frac{|x-2|}{2} - 3\right|$$

   (a) 4 (b) 0

   (c) 3 (d) None of these
7. Consider a function $y = f(x) = \left|\frac{|x-2|}{2} - 4\right| - 4$ for what value of k the equation $f(x) = k$ has 4 roots.

   (a) (–5, 0) (b) (–5, –1)

   (c) (–1, 1) (d) None of these
8. Consider a function $y = f(x) = \left|\left|\frac{|x-2|}{3} - 3\right| - 4\right|$ for what value of k the equation $f(x) = k$ has maximum roots.

   (a) (1, 5) (b) (1, 4)

   (c) (3, 4) (d) None of these
9. Consider a function $y = f(x) = |x - 2| + |x - 4|$ find the number of roots of the equation $f(x) = 5$

   (a) 1 (b) 2

   (c) 3 (d) None of these
10. Consider a function $y = f(x) = 3|x + 2| + 2|x - 2| + |x - 4|$ find the minimum value of the function

    (a) 11 (b) 14

    (c) 13 (d) None of these

3.2 GREATEST INTEGER FUNCTION (GIF) AND FRACTIONAL PART

3.2.1 Greatest Integer Function (G.I.F)

Greatest Integer Function (G.I.F)

Greatest integer function (G.I.F) [.] is defined as greatest integer less than or equal to x.

When $x = 1.2$, then $[x] = [1.2] = 1$

When $x = 2.7$, then $[x] = [2.7] = 2$

When $x = 5.88$, then $[x] = [5.88] = 5$

When $x = 0.4$, then $[x] = [0.4] = 0$

When $x = -0.6$, then $[x] = [-0.6] = -1$

When $x = -1.2$, then $[x] = [-1.2] = -2$

> Consider a function $f(x) = [x] + [-x]$ it is defined as
> $$\begin{cases} [x] + [-x] = 0 \text{ if } x \text{ is an integer} \\ [x] + [-x] = -1 \text{ if } x \text{ is not an integer} \end{cases}$$

Illustration 1. Find the value of $[-0.6]$

Solution. Since $-1 < -0.6 < 0$, so $[-0.6] = -1$

Illustration 2. Find the value of $[0.1] + [1.1] + [2.1] + \ldots + [10.1]$

Solution. $[0.1] + [1.1] + [2.1] + \ldots + [10.1] = 1 + 2 + 3 + \ldots + 10 = 55$

Illustration 3. If $[x + 0.6] = 3$, then find the value of $[x - 0.6]$

Solution. Given that $[x + 0.6] = 3$, so $2.4 \leq x < 3.4$

And $1.8 \leq (x - 0.6) < 2.8$, so $[x - 0.6] = 1$ or 2

3.2.2 Graphical Approach

Graph 24: $y = [x]$

Greatest integer function [.] is defined as greatest integer less than or equal to x.

When $x = 1.2$, then $[x] = [1.2] = 1$

When $x = 2.7$, then $[x] = [2.7] = 2$

When $x = 5.88$, then $[x] = [5.88] = 5$

When $x = 0.4$, then $[x] = [0.4] = 0$

When $x = -0.6$, then $[x] = [-0.6] = -1$

When $x = -1.2$, then $[x] = [-1.2] = -2$

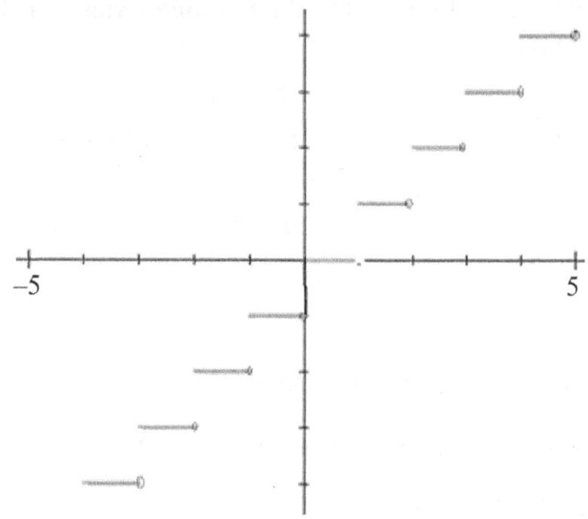

And so on, the graph of $y = [x]$ as shown in the figure.

The graph is discontinuous at integral values of x.

Graph 25: $y = [2x]$

When $x \in [0, \frac{1}{2})$ then $[x] = 0$

When $x \in [1/2, 1)$, then $[x] = 1$

When $x \in [1, 3/2)$, then $[x] = 2$

When $x \in [3/2, 2)$, then $[x] = 3$ and so on

When $x \in [-1/2, 0)$, then $[x] = -1$

When $x \in [-1, -\frac{1}{2})$ then $[x] = -2$ and so on

The graph of $y = [2x]$ as shown in the figure.

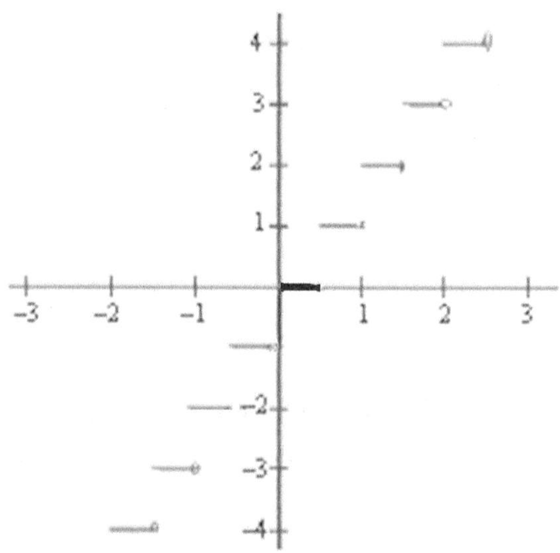

Graph 26: $y = [3x]$

When $x \in [0, 1/3)$, then $[x] = 0$

When $x \in [1/3, 2/3)$, then $[x] = 1$

When $x \in [2/3, 1)$, then $[x] = 2$

When $x \in [1, 4/3)$, then $[x] = 3$ and so on

When $x \in [-1/3, 0)$, then $[x] = -1$

When $x \in [-2/3, -1/3)$, then $[x] = -2$ and so on

The graph of $y = [3x]$ as shown in the figure.

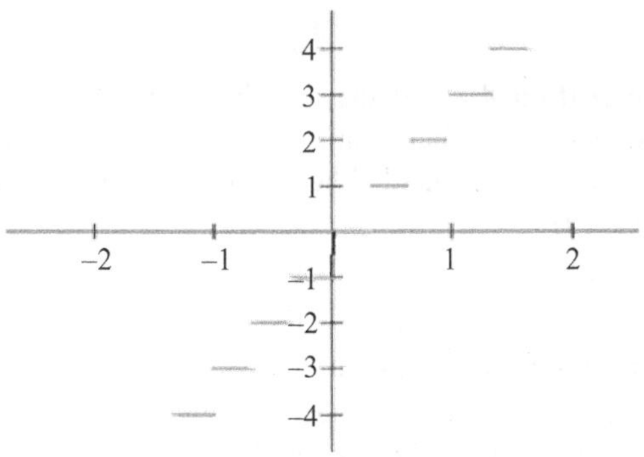

Graph 27: $\left[\frac{x}{2}\right]$

When $x \in [0, 2)$, then $[x] = 0$

When $x \in [2, 4)$, then $[x] = 1$

When $x \in [4, 6)$, then $[x] = 2$ and so on

When $x \in [-2, 0)$, then $[x] = -1$

When $x \in [-4, -2)$, then $[x] = -2$ and so on

The graph of $y = [x/2]$ as shown in the figure.

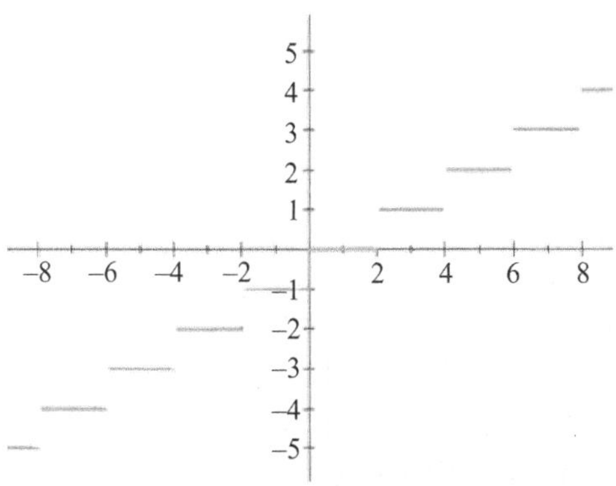

Graph 28: $\left[\frac{x}{3}\right]$

When $x \in [0, 3)$, then $[x] = 0$

When $x \in [3, 6)$, then $[x] = 1$

When $x \in [6, 9)$, then $[x] = 2$ and so on

When $x \in [-3, 0)$, then $[x] = -1$

When $x \in [-6, -3)$, then $[x] = -2$ and so on

The graph of $y = [x/3]$ as shown in the figure.

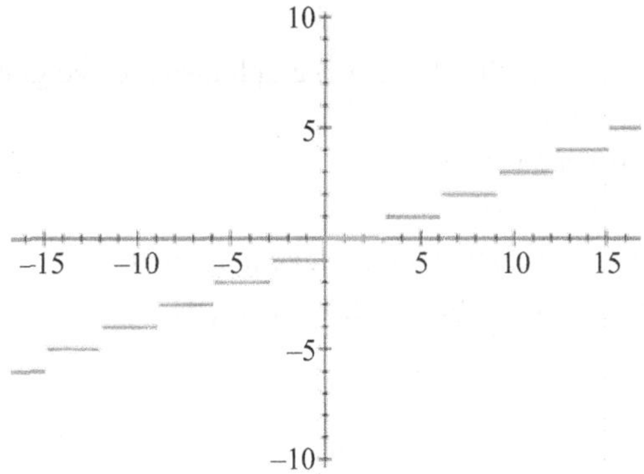

Graph 29: $y = -[x]$

Step 1: Plot $y = [x]$ as we did earlier

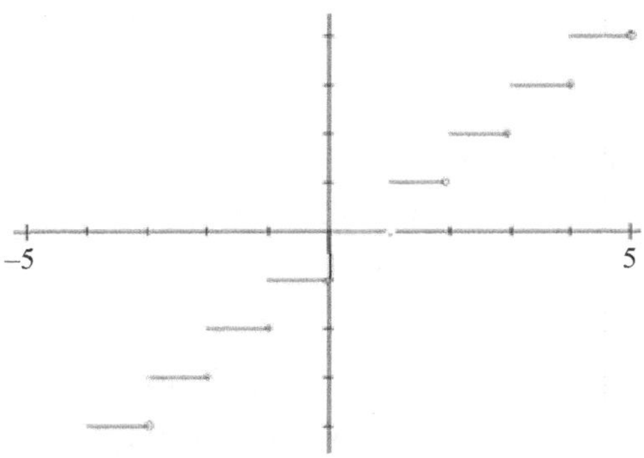

Step 2: Now plot $y = -[x]$ by taking reflection with respect to x-axis.

When $x \in [0, 1)$, then $[x] = 0$ so $[x] = 0$

When $x \in [1, 2)$, then $[x] = 1$ so $-[x] = -1$

When $x \in [2, 3)$, then $[x] = 2$ so $-[x] = -2$ and so on

When $x \in [-1, 0)$, then $[x] = -1$ so $-[x] = 1$

When $x \in [-2, -1)$, then $[x] = -2$ so $-[x] = 2$ and so on

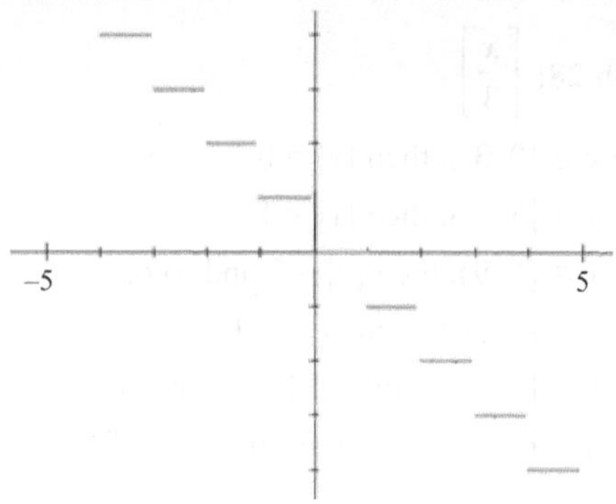

If we merge the above two graph then merged graph is as shown

Consider $f(x) = [x]$

And $g(x) = -[x]$

For $x \in [0, 1)$ both of the functions are 0.

$|f(x) - g(x)| = 4$ for x $(-2, -1) \cup (2, 3)$

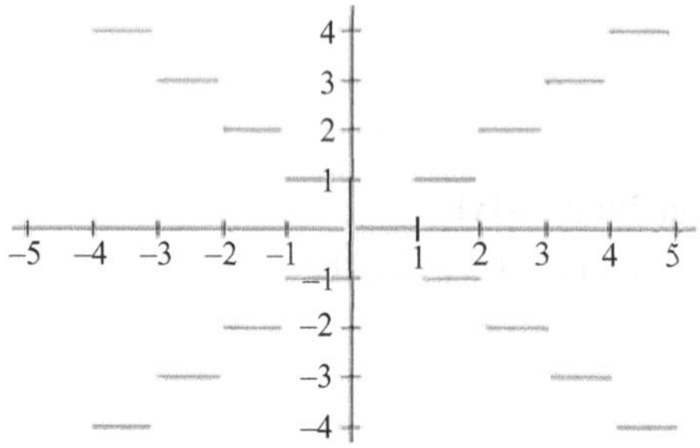

Graph 30: $y = [-x]$

Plot $y = [-x]$ by taking different values of x

When $x \in (0, 1]$, then $-x \in [-1, 0)$ so $[-x] = -1$

When $x \in (1, 2]$, then $-x \in [-2, -1)$ so $[-x] = -2$

When $x \in (2, 3]$, then $-x \in [-3, -2)$ so $[-x] = -3$

When $x \in (-1, 0]$, then $-x \in [0, 1)$ so $[-x] = 0$

When $x \in (-2, -1]$, then $-x \in [1, 2) = 0$ so $[-x] = 1$ so on

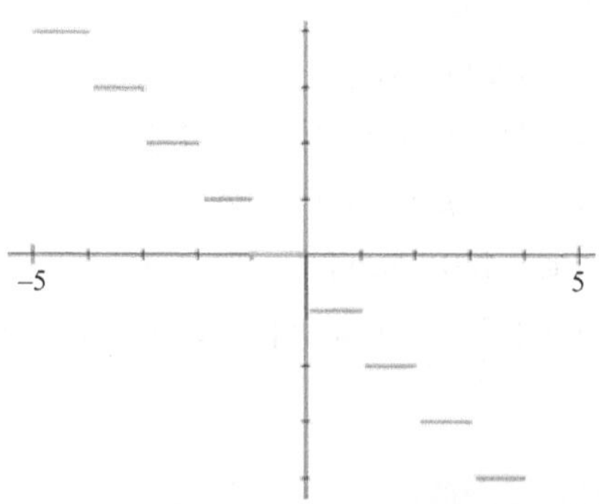

Plot and compare the graph of –

$y = -[x]$ and $y = [-x]$

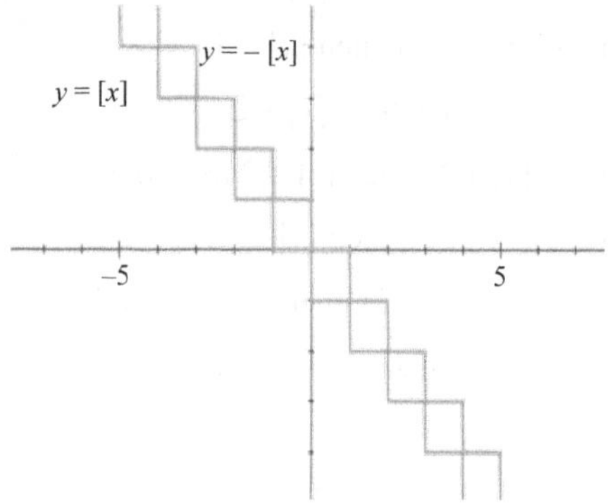

Illustration 4. Find the area between $F(x) = [-x]$ and $G(x) = -[x]$ between $x = -2$ to $x = 3$, (Although the area is not bounded hence take $x = t$ line for all integer t)

(a) 2 (b) 5

(c) 3 (d) None of these

Solution. From the graph:

For every unit of x there is a unit square so area bounded between the two graphs is 5 sq units.

Option (b)

Graph 31: $[|x|]$

When $x \in [0, 1)$, then $[x] = 0$

When $x \in [1, 2)$, then $[x] = 1$

When $x \in [2, 3)$, then $[x] = 2$ and so on

When $x \in [-1, 0)$, then $[x] = 0$

When $x \in [-2, -1)$, then $[x] = -1$ and so on

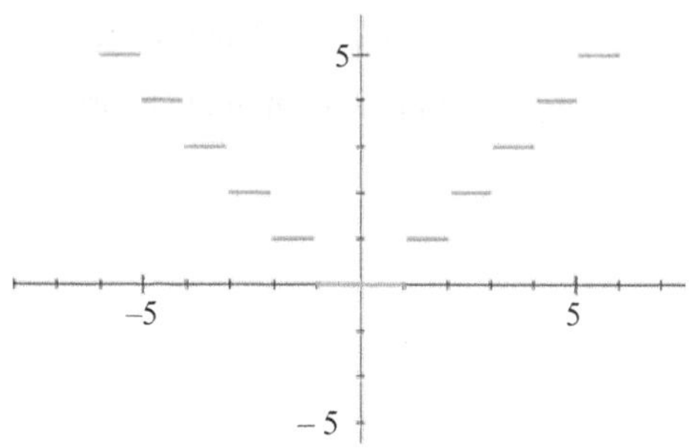

Graph 32: $|[x]|$

When $x \in [0, 1)$, then $[x] = 0$

When $x \in [1, 2)$, then $[x] = 1$

When $x \in [2, 3)$, then $[x] = 2$ and so on

When $x \in [-1, 0)$, then $[x] = -1$

When $x \in [-2, -1)$, then $[x] = -2$ and so on

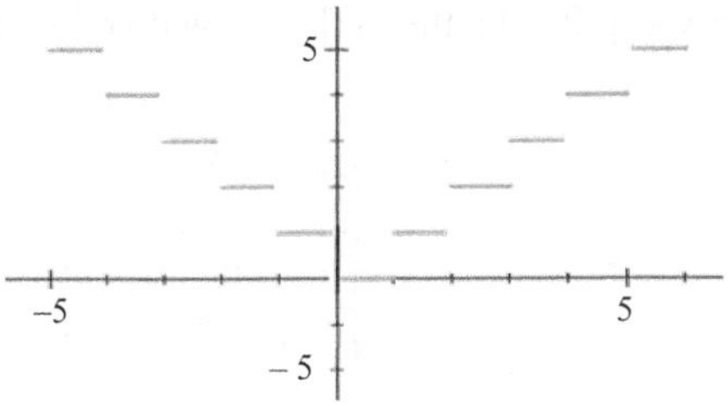

Illustration 5. If A_k is the area bounded between $y = [|x|]$, $y = |[x]|$, $x = k + 0.2$ and $x = k + 0.7$ where k is an integer, then find the value of $\sum_{-2}^{2} A_k$

(a) 2 (b) 5

(c) 3 (d) None of these

Solution. $\sum_{-2}^{2} A_k = A_{-2} + A_{-1} + A_0 + A_1 + A_2$

$A_{-2} = 1 \times 0.5 = 0.5$

$A_{-1} = 1 \times 0.5 = 0.5$

$A_0 = A_1 = A_2 = 0$

Hence $\sum_{-2}^{2} A_k = A_{-2} + A_{-1} + A_0 + A_1 + A_2$
$= 0.5 + 0.5 = 1$

Option (d)

Graph 33: $[x^2]$

When $x \in [0, 1)$, then $x^2 \in [0, 1)$ and $[x] = 0$
When $x \in [1, \sqrt{2})$, then $x^2 \in [1, 2)$ and $[x] = 1$
When $x \in [\sqrt{2}, \sqrt{3})$, then $x^2 \in [2, 3)$ and $[x] = 2$
When $x \in [\sqrt{3}, 2)$, then $x^2 \in [3, 4)$ and $[x] = 3$ and so on
When $x \in (-1, 0]$, then $x^2 \in (0, 1]$ and $[x] = 0$
When $x \in (-\sqrt{2}, -1]$ then $x^2 \in (1, 2]$ and $[x] = 1$ and so on

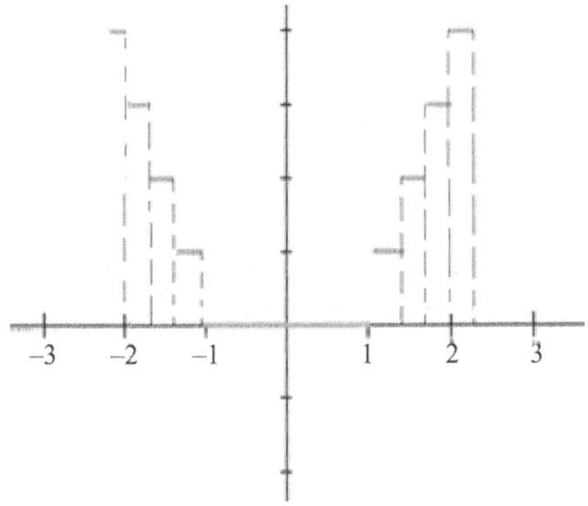

Illustration 6. Find the area between $[x^2]$, and $x = -1$ to $x = 2$

(a) $5 - \sqrt{3} - \sqrt{2}$ (b) $5 - \sqrt{3} - 2\sqrt{2}$
(c) $5 - 2\sqrt{3} - \sqrt{2}$ (d) none of these

Solution. When $x \in (-1, 1)$ the area is 0
When $x \in (1, \sqrt{2})$ the area is $(\sqrt{2} - 1)$
When $x \in (\sqrt{2}, \sqrt{3})$ the area is $2(\sqrt{3} - \sqrt{2})$
When $x \in (\sqrt{3}, 2)$ the area is $3(2 - \sqrt{3})$
So total area is $(\sqrt{2} - 1) + 2(\sqrt{3} - \sqrt{2}) + 3(2 - \sqrt{3})$
$= 5 - \sqrt{3} - \sqrt{2}$

Option (a)

Graph 34: $[x]^2$

When $x \in [0, 1)$, then $[x] = 0$ and $[x]^2 = 0$

When $x \in [1, 2)$, then $[x] = 1$ and $[x]^2 = 1$

When $x \in [2, 3)$, then $[x] = 2$ and $[x]^2 = 4$ and so on

When $x \in [-1, 0)$, then $[x] = -1$ and $[x]^2 = 1$

When $x \in [-2, -1)$ then $[x] = -2$ and $[x]^2 = 4$ and so on

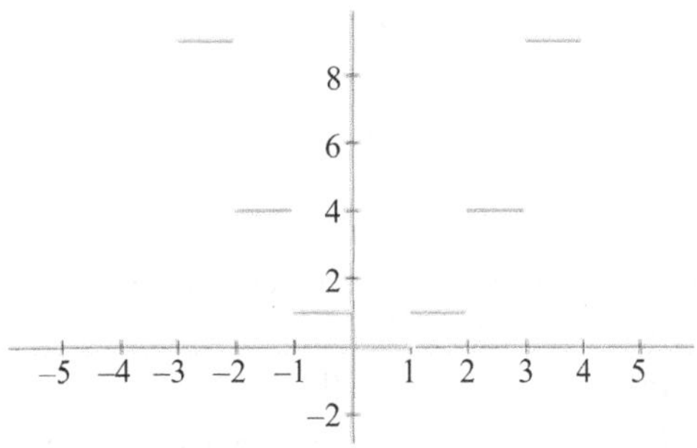

Illustration 7. Find x if $[x^2] = [x]^2$ when $x \in (0, 3)$

(a) $(0, \sqrt{2}) \cup [2, \sqrt{5})$
(b) $(0, \sqrt{2})$
(c) $(0, \sqrt{2}) \cup [2, \sqrt{6})$
(d) none of these

Solution. When $x \in (0, 1)$ all the values of x will satisfy.

When $x \in [1, \sqrt{2})$, $[x^2] = 1$ and $[x]^2 = 1$

When $x \in [\sqrt{2}, \sqrt{3})$, $[x^2] = 2$ and $[x]^2 = 1$

When $x \in [\sqrt{3}, 2)$, $[x^2] = 3$ and $[x]^2 = 1$

When $x \in [2, \sqrt{5})$, $[x^2] = 4$ and $[x]^2 = 4$ and so on

So the required value of x is $(0, \sqrt{2}) \cup [2, \sqrt{5})$
Option (a)

Graph 35: $[x^3]$

When $x \in [0, 1)$, then $x^3 \in [0, 1)$ and $[x^3] = 0$

When $x \in \left[1, 2^{\frac{1}{3}}\right)$, then $x^3 \in [1, 2)$ and $[x^3] = 1$

When $x \in \left[2^{\frac{1}{3}}, 3^{\frac{1}{3}}\right)$, then $x^3 \in [2, 3)$ and $[x^3] = 2$

When $x \in \left[3^{\frac{1}{3}}, 4^{\frac{1}{3}}\right)$, then $x^3 \in [3, 4)$ and $[x^3] = 3$ and so on

When $x \in (-1, 0]$ then $x^3 \in [-1, 0)$ and $= -1$

When $\left[-2^{\frac{1}{3}}, -1\right)$ then $x^3 \in [-2, -1)$ and $= -2$ and so on

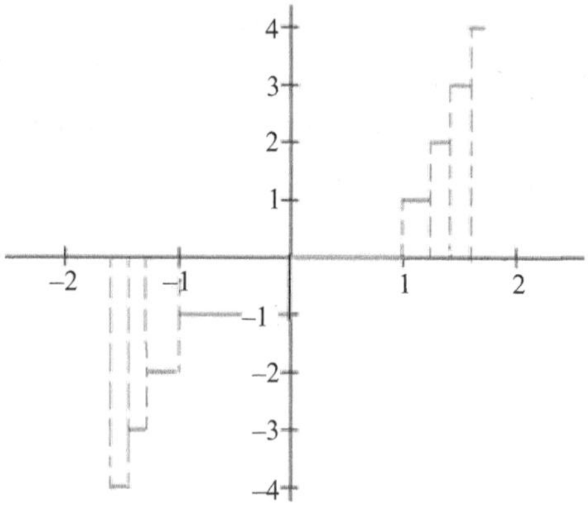

Illustration 8. How many times function y = $[x^3]$ is discontinuous in (–2, 2)?

(a) 12 (b) 14

(c) 15 (d) None of these

Solution. When $x > 0$ then function is discontinuous at 1, $2^{\frac{1}{3}}, 3^{\frac{1}{3}}, 4^{\frac{1}{3}}, 5^{\frac{1}{3}}, 6^{\frac{1}{3}}$ and $7^{\frac{1}{3}}$ total 7 points

Similarly, 7 points when $x < 0$ and one more point at origin, so total number of points is 15

Option (c)

Illustration 9. How many times function $y = [x^3]$ is discontinuous in (–10, 10)?

(a) 121 (b) 1999

(c) 1998 (d) None of these

Solution. Similar to the previous question when $x > 0$, then number of points is $10^3 - 1 = 999$ points, same number of points when $x < 0$, so total number of points is 999 + 999 + 1 = 1999

Option (b)

Graph 36: $[x]^3$

When $x \in [0, 1)$, then $[x] = 0$ and $[x]^3 = 0$

When $x \in [1, 2)$, then $[x] = 1$ and $[x]^3 = 1$

When $x \in [2, 3)$, then $[x] = 2$ and $[x]^3 = 8$ and so on

When $x \in [-1, 0)$, then $[x] = -1$ and $[x]^3 = -1$

When $x \in [-2, -1)$, then $[x] = -2$ and $[x]^3 = -8$ and so on

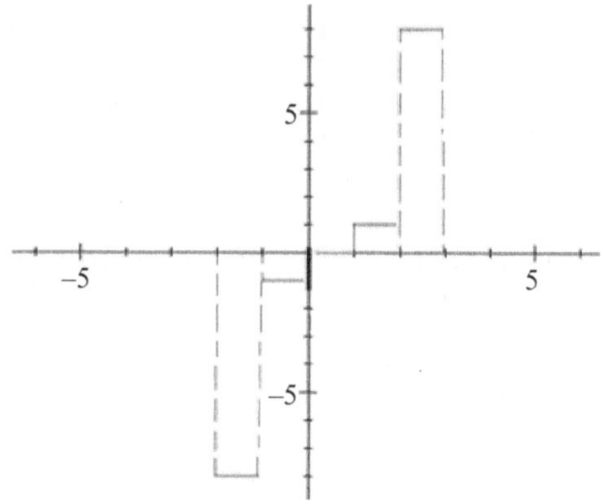

Illustration 10. How many times function $y = [x]^3$ is discontinuous in (–2, 2)?

(a) 2 (b) 4

(c) 5 (d) None of these

Solution. From the graph number of discontinuous points is 4

Option (b)

Graph 37: $|[x^3]|$

Step 1: Plot $[x]^3$ as we did in previous graph

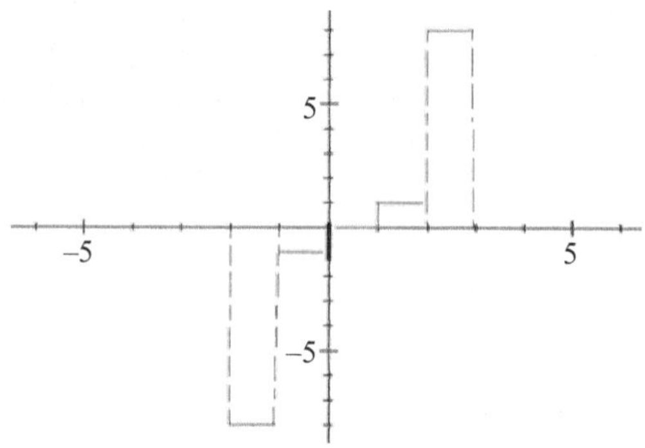

Step 2: Now plot $|[x]^3|$ by taking reflection of $[x]^3$ with respect to x-axis as shown in the figure.

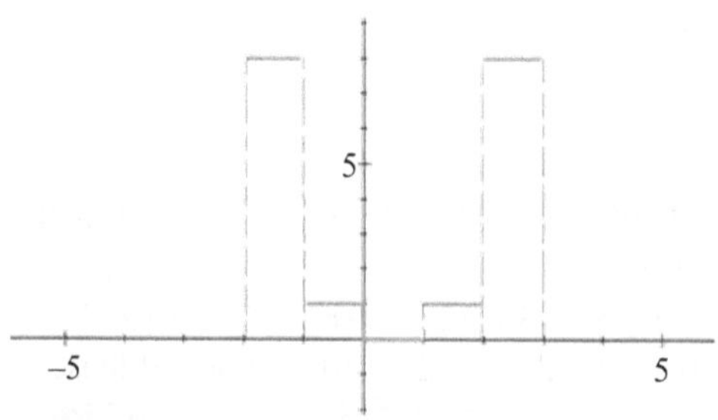

Graph 38: $|[x^3]|$

Step 1: Plot the graph of $y = [x^3]$ as we did in previous graph

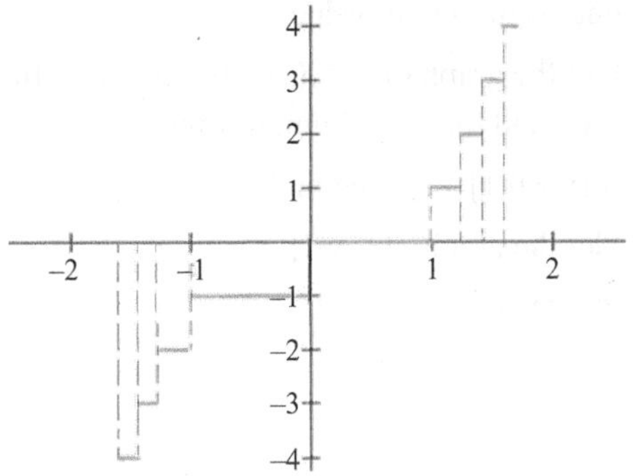

Step 2: Now do the reflection of $y = [x^3]$ with respect to x-axis to get $y = |[x^3]|$

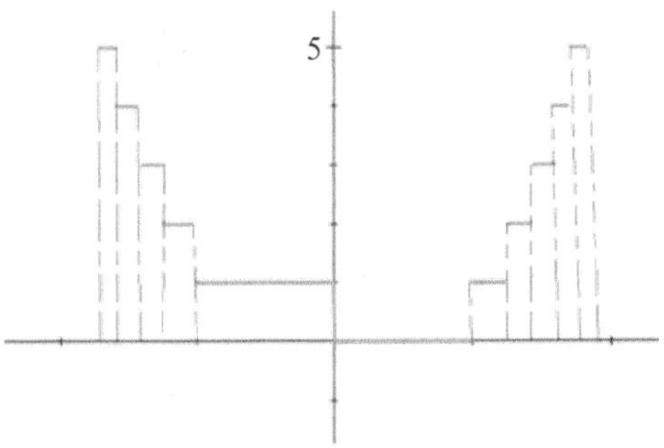

3.2.3 Fractional Part

Fractional Part {.}

Fractional part of x is defined as $\{x\} = x - [x]$

If $x = 0.3$, then $\{x\} = 0.3$

If $x = 1.6$, then $\{x\} = 0.6$

If $x = 3.8$, then $\{x\} = 0.8$

If $x = -0.3$, then $\{x\} = 0.7$

If $x = -3.9$, then $\{x\} = 0.1$ and so on

Graph 39: $\{x\}$

Fractional part of x is defined as $\{x\} = x - [x]$

When $x \in (-2, -1)$, then $y = x + 2$

When $x \in (-1, 0)$, then $y = x + 1$

When $x \in (0, 1)$, then $y = x$

When $x \in (1, 2)$, then $y = x - 1$

When $x \in (2, 3)$, then $y = x - 2$ and so on

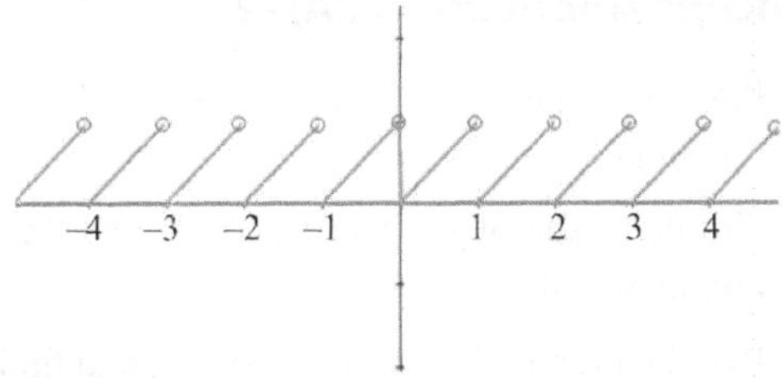

Illustration 11. Find the number of roots of the equation $x^2 = \{x\}$

(a) 1 (b) 2

(c) 3 (d) None of these

Solution. From the graph below number of roots is 2.

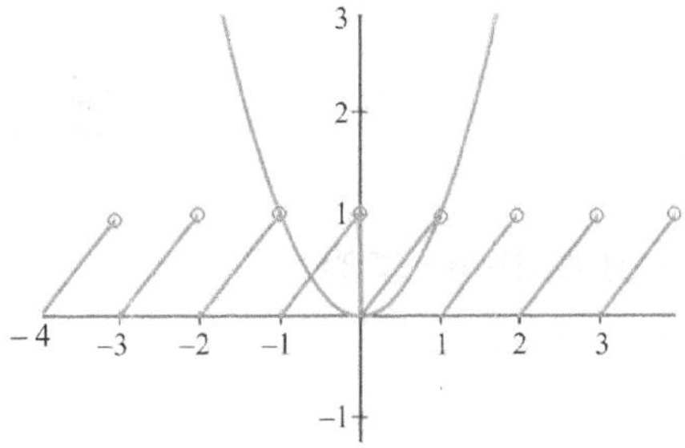

Option (b)

Graph 40: $\{x\}^2$

When $x \in [-2, -1)$, then $y = (x + 2)^2$

When $x \in [-1, 0)$, then $y = (x + 1)^2$

When $x \in [0, 1)$, then $y = x^2$

When $x \in [1, 2)$, then $y = (x - 1)^2$

When $x \in [2, 3)$, then $y = (x - 2)^2$ and so on

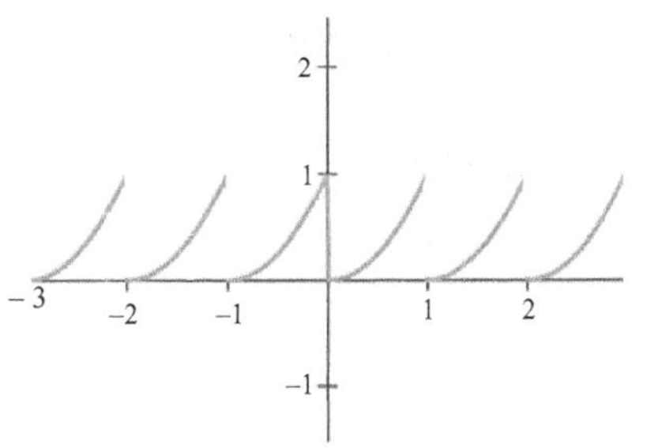

Illustration 12. Find the number of roots of the equation $x = \{x\}$

(a) 1 (b) 2

(c) 3 (d) None of these

Solution. From the graph above number of roots is 1 that is at $x = 0$

Option (a)

Concept Applicator (CA)-2

1. Plot the graph of $y = f(x) = |2x - 6|$ and find its minimum value.
2. Plot the graph of $y = f(x) = |3x - 8|$ and find its minimum value.
3. Plot the graph of $y = f(x) = -|3x - 9|$ and find its maximum value.
4. Plot the graph of $y = f(x) = 2|3x - 6| + 2$ and find its minimum value.
5. Plot the graph of $y = f(x) = 2|x + 2| + 2$ and find its minimum value.
6. Plot the graph of $y = f(x) = |x - 5| + |x - 10|$ and find its minimum value.
7. Plot the graph of $y = f(x) = |x - 5| + |x - 10| + |x - 12|$ and find its minimum value.
8. If $[x + 0.1] = 7$, then find x.
9. Plot the graph of $[-x/2]$
10. Plot the graph of $\{x/2\}$

Exercise A

Concept Builder (CB)

1. The solution of the inequation $1 + |x - 1| \geq 0$ is
   (a) $(-\infty, 0)$ (b) R
   (c) $(-2, 0)$ (d) $(0, 2)$
2. The set of real values of satisfying $|x - 1| - 1| \leq 1$ is
   (a) $[-1, 3]$ (b) $[0, 2]$
   (c) $[-1, 1]$ (d) None of these
3. Solution of $|x - 1| + |x - 2| + |x - 3| \geq 6$ is
   (a) $[0, 4]$
   (b) $(-\infty, -2) \cup [4, \infty)$
   (c) $(-\infty, 0] \cup [4, \infty)$
   (d) None of these
4. Value of x which satisfy $|3 - x| = x - 3$ is
   (a) $x < 3$ (b) $x > 3$
   (c) $x \geq 3$ (d) $x \leq 3$
5. Value of x which satisfy $2x - 1 = |x + 7|$ is
   (a) -2 (b) 8
   (c) $-2, 8$ (d) None of these
6. Value of x which satisfy $|x - 1| \geq |x - 3|$ is
   (a) $x \leq 2$ (b) $x \geq 2$
   (c) $[1, 3]$ (d) None of these
7. Value of x which satisfy $\left|\frac{1}{x} - 2\right| < 4$ is
   (a) $\left(-\infty, -\frac{1}{2}\right)$
   (b) $\left(\frac{1}{6}, \infty\right)$
   (c) $\left(-\frac{1}{2}, \frac{1}{6}\right)$
   (d) $\left(-\infty, -\frac{1}{2}\right) \cup \left(\frac{1}{6}, \infty\right)$
8. Value of x which satisfy $\left|1 + \frac{3}{x}\right| > 2$ is
   (a) $(0, 3]$ (b) $[-1, 0)$
   (c) $(-1, 0) \cup (0, 3)$ (d) None of these
9. The set of real values of x satisfying $|x - 1| \leq 3$ and $|x - 1| \geq 1$ is
   (a) $[2, 4]$
   (b) $(-\infty, 2] \cup [4, \infty)$
   (c) $[-2, 0] \cup [2, 4]$
   (d) None of these
10. Find the number of solutions of $|x|^2 - 3|x| + 2 = 0$
   (a) 4 (b) 1
   (c) 3 (d) 2

11. The sum of all real roots of the equation $|x-2|^2 + |x-2| - 2 = 0$ is
(a) 7 (b) 4
(c) 1 (d) 5

12. The solution of $||x| - 1| < |1 - x|, x \in R$ is
(a) $(-1, 1)$ (b) $(0, \infty)$
(c) $(-1, \infty)$ (d) None of these

13. Solution of $0 < |3x + 1| < \frac{1}{3}$ is
(a) $\left(-\frac{4}{9}, -\frac{2}{9}\right)$
(b) $\left[-\frac{4}{9}, -\frac{2}{9}\right]$
(c) $\left(-\frac{4}{9}, -\frac{2}{9}\right) - \left\{-\frac{1}{3}\right\}$
(d) $\left[-\frac{4}{9}, -\frac{2}{9}\right] - \left\{-\frac{1}{3}\right\}$

14. Solution of $|x^2 - 10| \le 6$ is
(a) $(2, 4)$ (b) $(-4, -2)$
(c) $(-4, -2) \cup (2, 4)$ (d) $[-4, -2] \cup [2, 4]$

Concept Cracker (CC)

1. Find the minimum value of $|x-2| + |x+2|$
(a) 2 (b) 4
(c) 6 (d) None of these

2. Find the maximum value of $5 - |x-2| - |x+2|$
(a) 2 (b) 4
(c) 6 (d) None of these

3. Find the minimum value of $|x-2| - |x+2| + |x-4|$
(a) 2 (b) 4
(c) –4 (d) None of these

4. Find the maximum value of (Local maxima)
$$y = f(x) = \left|\frac{|x-2|}{2} - 3\right|$$
(a) 4 (b) 0
(c) 3 (d) None of these

5. Find the number of integral values of x which satisfy, $-1 \le [2x^2 - 3] \le 1$ where [.] is GIF
(a) 0 (b) 1
(c) 2 (d) None of these

6. Find the value of x for which $|3x + 2| < 1$
(a) $\left[-1, -\frac{1}{3}\right]$ (b) $\left\{-\frac{1}{3}, -1\right\}$
(c) $\left(-1, -\frac{1}{3}\right)$ (d) None of these

7. Solution of $|2x - 3| < |x + 2|$ is
(a) $\left(-\infty, \frac{1}{3}\right)$
(b) $\left(\frac{1}{3}, 5\right)$
(c) $(5, \infty)$
(d) $\left(-\infty, \frac{1}{3}\right) \cup (5, \infty)$

8. If $|x-1| + |x| + |x+1| \ge 6$; then x lies in
(a) $(-\infty, 2]$ (b) $(-\infty, -2] \cup [2, \infty)$
(c) R (d) ϕ

9. Solution of $\left|x + \frac{1}{x}\right| > 2$ is
(a) R – {0} (b) R – {–1, 0, 1}
(c) R – {1} (d) R – {–1, 1}

10. Solution of $\left|x + \frac{1}{x}\right| < 4$ is
(a) $\left(2-\sqrt{3}, 2+\sqrt{3}\right) \cup \left(-2-\sqrt{3}, -2+\sqrt{3}\right)$
(b) $R - \left(2-\sqrt{3}, -2+\sqrt{3}\right)$
(c) $R - \left(-2-\sqrt{3}, 2+\sqrt{3}\right)$
(d) None of the above

11. The solution set of $\left|\frac{x+1}{x}\right| + |x+1| = \frac{(x+1)^2}{|x|}$ is
(a) $\{x \mid x \ge 0\}$ (b) $\{x \mid x > 0\} \cup \{-1\}$
(c) $\{-1, 1\}$ (d) $\{x \mid x \ge 1 \text{ or } x \le 1\}$

12. The product of all the solutions of the equation $(x-2)^2 - 3|x-2| + 2 = 0$, is
(a) 2 (b) – 4
(c) 0 (d) None of these

13. The set of all real x satisfy the inequality $\frac{3-|x|}{4-|x|} \geq 0$

(a) $[-3, 3] \cup (-\infty, -4) \cup (4, \infty)$

(b) $(-\infty, -4) \cup (4, \infty)$

(c) $(-\infty, -3) \cup (4, \infty)$

(d) $(-\infty, -3) \cup (3, \infty)$

14. The set of values of x satisfying $2 \leq |x-3| < 4$ is

(a) $(-1, 1] \cup [5, 7)$ (b) $-4 \leq x \leq 2$

(c) $-1 < x < 7$ or $x \geq 5$ (d) $x < 7$ or ≥ 5

15. Find x if $|x-1| - |x+2| < 1$

(a) $(-\infty, -1)$ (b) $(-1, \infty)$

(c) $(1, \infty)$ (d) None of these

16. Find the minimum value of $y = f(x)$ $= |x-2| + |x+3| - |x+1|$

(a) 2 (b) 3

(c) 5 (d) None of these

17. Find the minimum value of $y = f(x)$ $= |x-2| - |x+3| + |x+1|$

(a) 2 (b) -2

(c) 1 (d) None of these

18. Find the minimum value of $y = f(x)$ $= 2|x-2| - |x+1|$

(a) 2 (b) -2

(c) 1 (d) None of these

PART-B

3.3 ADVANCE MATERIAL FOR JEE ADVANCED, KVPY AND OLYMPIAD

3.3.1 Some Advance Graph

Plot the graph and find the area bounded by |x|+|y| =4

Graph 1:- |x|+ |y| =4

Here function has different definition in different quadrant so we will define it separately in each quadrant and then plot the graph in each quadrant.

- In 1st quadrant equation is $x + y = 4$
- In 2nd quadrant equation is $-x + y = 4$
- In 3rd quadrant $-x - y = 4$
- In 4th quadrant $x - y = 4$

These four lines will form a square and its area is $(8 \times 8)/2 = 32$sq unit

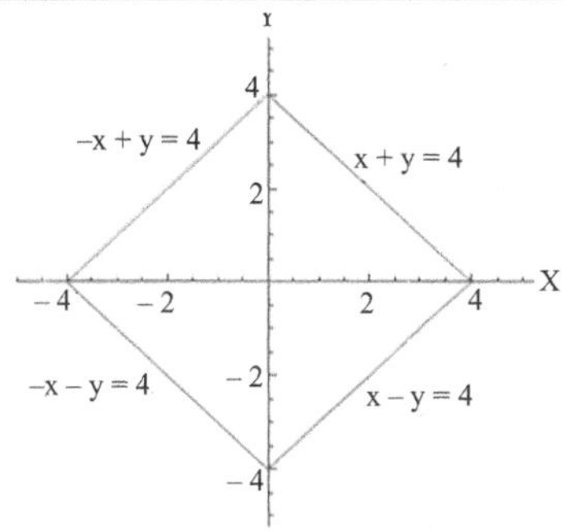

Area of the region bounded by $|x| + |y| = k$ is given by $2k^2$

Illustration 1 : Find the area bounded by $|x| + |y| = 4$

(a) 32 (b) 42 (c) 34 (d) None of these

Solution (a) : From the graph required area is 32.

Plot the graph and find the area bounded by 3|x|+2|y| = 12

Graph 2:- 3|x|+ 2|y| = 12

Here function has different definition in different quadrant so we will define it separately in each quadrant and then plot the graph in each quadrant.

- In 1st quadrant equation is $2y = 12 - 3x$
- In 2nd quadrant equation is $2y = 12 + 3x$
- In 3rd quadrant $-2y = 12 + 3x$
- In 4th quadrant $-2y = 12 - 3x$

These four lines will form a rhombus and its area is equal to 4 times the area of a triangle in one quadrant

Area of triangle in 1st quadrant $= \frac{1}{2}(4)(6) = 12$

so required area = 48 sq unit

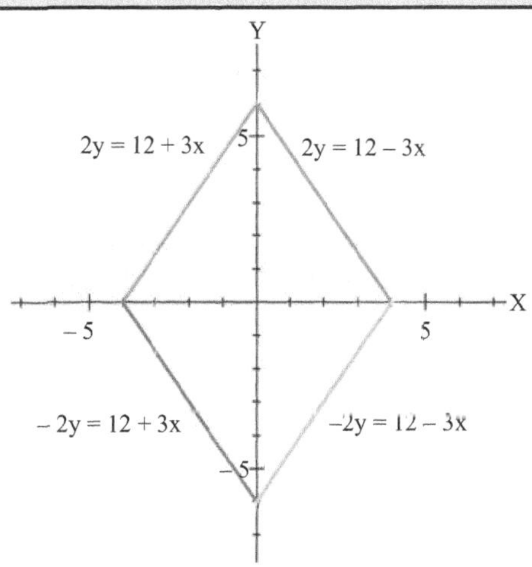

Area of the region bounded by a |x| + b|y| = k is given by $\frac{2(k^2)}{ab}$

Illustration 2. Find the area bounded by $3|x|+2|y|=12$

(a) 32 (b) 24 (c) 34 (d) None of these

Solution (d) : From the graph required area is 48

<table>
<tr><th colspan="2">Plot the graph and find the area bounded by |x – 2|+|y| = 10</th></tr>
<tr><th colspan="2">Graph 3:- |x – 2|+ |y| = 10</th></tr>
<tr><td>Step 1: 1st Plot $|x| + |y| = 10$ as we did in previous graph.</td><td>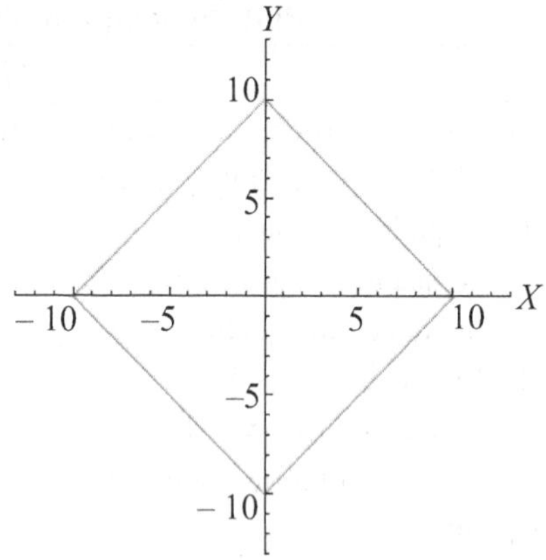</td></tr>
<tr><td>Step 2: Shift the graph of $|x| + |y| = 10$ to 2 unit right means to a new origin (2, 0) as shown in the graph.

So area of the graph $|x|+|y|=10$ and area of the graph $|x-2|+|y| = 10$ is same</td><td>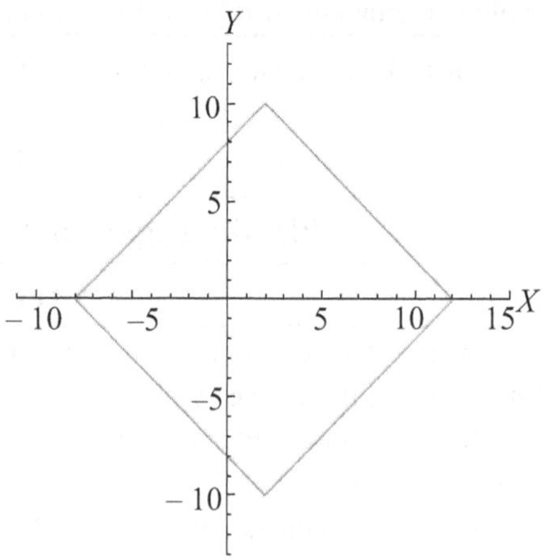</td></tr>
<tr><td colspan="2">Area of the region bounded by $|x-p|+|y-q|=k$ is given by $2k^2$ (Area is independent of p and q)</td></tr>
<tr><td colspan="2">Area of the region bounded by $a|x-p|+b|y-q|=k$ is given by $\frac{2(k^2)}{ab}$ (Area is independent of p and q)</td></tr>
</table>

Illustration 3. Find the area bounded by $|x-2|+|y|=10$

(a) 132 (b) 240 (c) 200 (d) None of these

Solution (c) : From the graph area is 200.

<table>
<tr><th colspan="2">Plot the graph and find the area bounded by |x|-|y| =4</th></tr>
<tr><th colspan="2">Graph 4:- |x|- |y| =4</th></tr>
<tr><td>Here function has different definition in different quadrant so we will define it separately in each quadrant and then plot the graph in each quadrant.
• In 1st quadrant equation is $x-y=4$
• In 2nd quadrant equation is $-x-y=4$
• In 3rd quadrant $-x+y=4$
• In 4th quadrant $x+y=4$
Here resultant figure is not a closed loop so we can not find the area bounded by the graph.</td><td>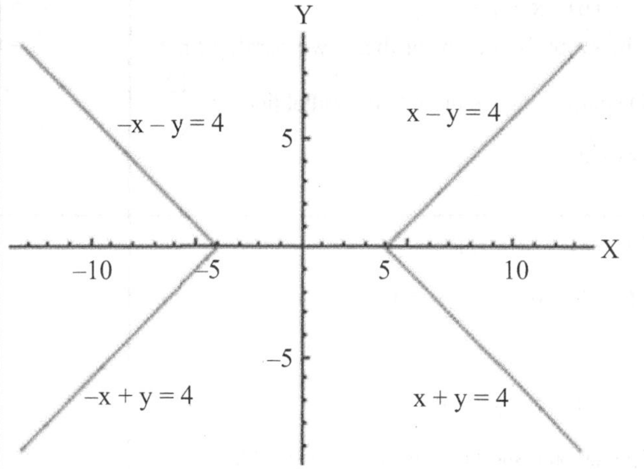</td></tr>
</table>

<table>
<tr><th colspan="2">Plot the graph and find the area bounded by |x+y| =6</th></tr>
<tr><th colspan="2">Graph 5:- |x+y| =6</th></tr>
<tr><td>Here function is defined as
(i) x+y=6
(ii) x+y = -6
The graph is shown for these two parallel lines.</td><td>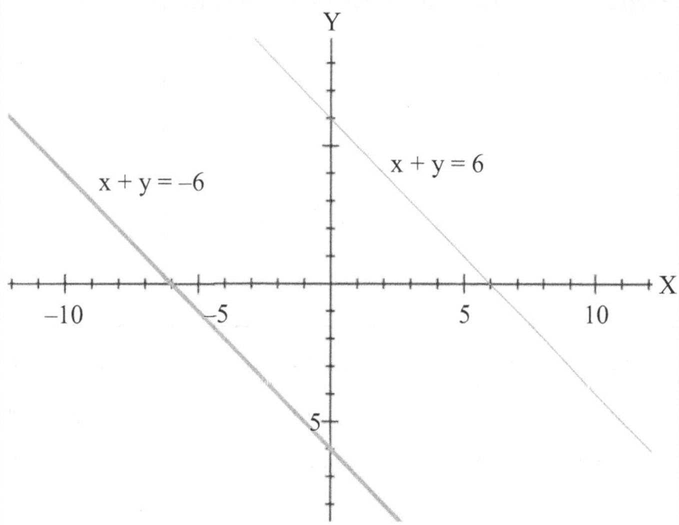</td></tr>
<tr><td colspan="2">These two lines will not form a closed loop so we cant find area bounded by these two lines</td></tr>
</table>

Plot the graph and find the area bounded by $|x+y| = 6$ and $|x - y| = 8$

Graph 6:- $|x + y| = 6$ & $|x - y| = 8$

Step 1:- Plot $|x + y| = 6$

Here function is defined as

(i) $x + y = 6$

(ii) $x + y = -6$

The graph is shown for these two parallel lines.

Distance between these two parallel lines is $\frac{12}{\sqrt{2}}$

$= 6\sqrt{2}$

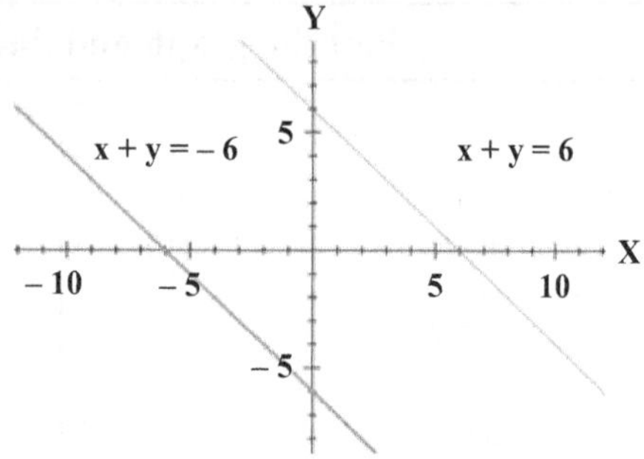

Step 2:- Plot $|x-y|=8$

Here function is defined as

(i) $x - y = 8$

(ii) $x - y = -8$

The graph is shown for these two parallel lines $\frac{16}{\sqrt{2}}$

Distance between these two parallel lines is

$= 8\sqrt{2}$

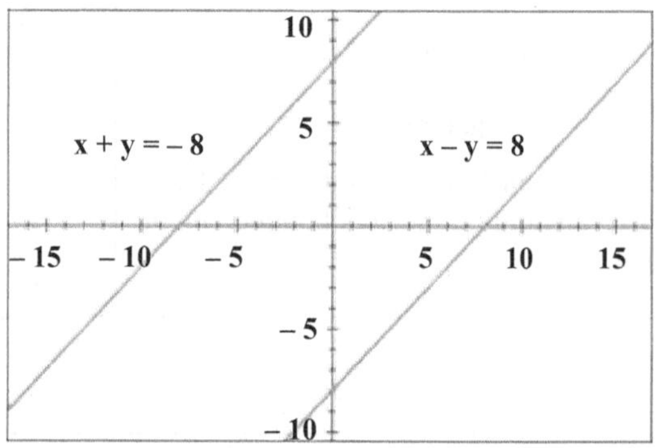

Step 3 : If we merge the above two graph then we will get a rectangle of side $6\sqrt{2}$ and $8\sqrt{2}$ so required area is $6\sqrt{2} \times 8\sqrt{2}$ = 96 sq unit

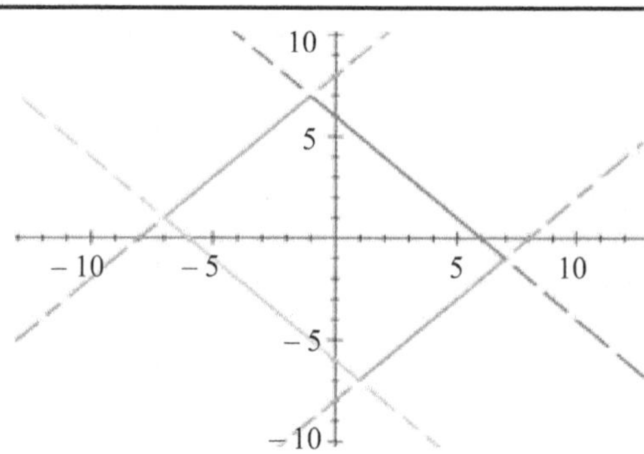

Illustation 4. Find the area bounded by $|x + y| = 6$ and $|x - y| = 8$

(a) 192 (b) 240 (c) 200 (d) None of these

Solution (d) : From the graph area is 96

Some more properties

Property 1:- $[x + y] \geq [x] + [y]$ for any real x , y

In general :- $[a + b + \ldots n] \geq [a] + [b] + \ldots .[n]$ for any real a, b,..n

Property 2:- $[xy] \geq [x]\,[y]$ for any real x , $y \geq 0$

In general : $[a\, b \ldots n] \geq [a]\,[b] \ \ldots .[n]$ for any real a, b,..$n \geq 0$

Property 3:- $\left[\frac{x}{n}\right] = \left[\frac{[x]}{n}\right]$ for $n \in N$ and $x \in R$

Property 4 (Hermite Identity):-

$$[x] + \left[x + \frac{1}{n}\right] + \left[x + \frac{2}{n}\right] + \ldots\ldots + \left[x + \frac{n-1}{n}\right] = [nx]$$

for $n \in N$ and $x \in R$

Property 5 :- For $x > 1$ be the inequality

$\left[\sqrt{\left[\sqrt{x}\right]}\right] = \left[\sqrt{\sqrt{x}}\right]$ must true

Proof :- By definition of integer part, it must be true that $\left[\sqrt{\sqrt{x}}\right] \leq \sqrt{\sqrt{x}} < \left[\sqrt{\sqrt{x}}\right] + 1$

Let $\left[\sqrt{\sqrt{x}}\right] = n$, then $n^4 \leq x < (n+1)^4$, so $n^2 \leq \left[\sqrt{x}\right]$ $< (n+1)^2$ which implies that $n^2 \leq \left[\sqrt{x}\right] < (n+1)^2$

So $n \leq \sqrt{\left[\sqrt{x}\right]} < n + 1$, and it implies that

$\left|\sqrt{\left|\sqrt{x}\right|}\right| = n = \left|\sqrt{\sqrt{x}}\right|$ i.e the given equality must be true.

Property 6 : If n is a positive integer, then .

$$\left[\sqrt{n} + \sqrt{n+1}\right] = \left[\sqrt{4n+2}\right]$$

Proof : Since

$$\left(\sqrt{n} + \sqrt{n+1}\right)^2$$

$$= 2n + 1 + 2\sqrt{n(n+1} > 2n + 1 + 2\sqrt{n^2} = 4n + 1$$

and

$$\left(\sqrt{n} + \sqrt{n+1}\right)^2 = 2n + 1 + 2\sqrt{n(n+1} < 2n + 1 + 2\sqrt{(n+1)^2}$$

$= 4n + 3$, letting $k = \left[\sqrt{4n+2}\right]$,

then $k \leq \left[\sqrt{4n+2}\right] < k + 1$, i.e. $k^2 \leq 4n + 2 < (k+1)^2$. Since $4n + 2$ and $4n + 3$ both are not perfect square numbers (a perfect square cannot have a remainder 2 or 3 when divided by 4), therefore $k^2 < 4n + 2 < 4n + 3 < (k+1)^2$, then $k^2 \leq 4n + 1 < \left(\sqrt{n} + \sqrt{n+1}\right)^2 4n + 3 \leq (k+1)^2$, i.e. $k < \sqrt{n} + \sqrt{n+1} < (k+1)$

Thus $\left[\sqrt{n} + \sqrt{n+1} = k = \left[\sqrt{4n+2}\right]\right]$

Property 7 : For any $m \geq 2$ the number of positive integers x such that $\left[\frac{x}{m-1}\right] = \left[\frac{x}{m+1}\right]$ is

$\frac{m^2 \quad 4}{4}$ if m is even and $\frac{m^2 - 5}{4}$ if m is odd

For example and proof see question number 17 of Concept Eliminator.

Exercise B

Concept Deviator (CD)

(Single Correct MCQs)

1. Consider a function $y = f(x) = 3\,|x + 2| + 2\,|x - 2| + |x - 4|$ find the minimum value of the function

   (a) 11 (b) 14

   (c) 13 (d) None of these

2. Find the area bounded by $|x| + |y| = 4$

   (a) 32 (b) 42

   (c) 34 (d) None of these

3. Find the area bounded by $3|x| + 2|y| = 12$

   (a) 32 (b) 24

   (c) 34 (d) None of these

4. Find the value of k such that the number of solutions of $y = |x|^2 - |x| - 6$ and $y = k$ is maximum

   (a) (– 6.25, – 6) (b) (– 6.25, –5)

   (c) (– 6, –5) (d) None of these

5. If A_k is the area bounded between $y = [|x|]$, $y = |[x]|$, $x = k + 0.2$ and $x = k + 0.7$ where k is an integer, then find the value of $\sum_{-2}^{2} A_k$
(a) 2 (b) 5
(c) 3 (d) None of these

6. Find the area between x-axis & $[x^2]$ and $x = -1$ to $x = 2$
(a) $5-\sqrt{3}-\sqrt{2}$ (b) $5-\sqrt{3}-2\sqrt{2}$
(c) $5-\sqrt{3}-2\sqrt{2}$ (d) none of these

7. How many times function $y = [x]^3$ is discontinuous in $(-2, 2)$?
(a) 2 (b) 4
(c) 5 (d) None of these

8. If x is a positive real number, then
$\left[\frac{x}{2}\right]+\left[\frac{x+1}{2}\right]=$
(a) $\left[x-\frac{1}{2}\right]$ (b) $[x]$
(c) $\left[x+\frac{1}{4}\right]$ (d) $\left[2x+\frac{1}{4}\right]$

9. The solution set of the equation, $4\{x\} = x + [x]$, where $\{x\}$ and $[x]$ denote the fractional and integral parts of a real number respectively, is
(a) $\{0\}$ (b) $\left\{0, \frac{5}{3}\right\}$
(c) $[0, \infty)$ (d) None of these

10. Let F (x) be a function defined by $F(x) = x - [x]$, $0 \neq x \in R$, where $[x]$ is the greatest integer less than or equal to x. Then the number of solutions of $F(x) + F\left(\frac{1}{x}\right) = 1$ is
(a) 0 (b) infinite
(c) 1 (d) 2

11. If $[x]^2 = [x+2]$, where $[x]$ = the greatest integer less than or equal to x then x must be such that
(a) $x = 2, -1$ (b) $x \in \{2, 3\}$
(c) $x \in [-1, 0) \cup [2, 3)$ (d) None of these

12. If $5\{x\} = x + [x]$ and $[x] - \{x\} = \frac{1}{2}$ when $\{x\}$ and $[x]$ are fractional and integral part of x then x is
(a) $\frac{1}{2}$ (b) $\frac{3}{2}$
(c) $\frac{5}{2}$ (d) $\frac{7}{2}$

13. The solution of $x - 1 = (x - [x])(x - \{x\})$ (where $[x]$ and $\{x\}$ are the integral and fractional part of x) is
(a) $x \in R$ (b) $x \in R \sim [1, 2)$
(c) $x \in [1, 2)$ (d) $x \in R \sim [1, 2]$

14. The solution set of $(x)^2 + (x+1)^2 = 25$, where (x) is the least integer greater than or equal to x, is
(a) $(2, 4)$ (b) $(-5, -4] \cup (2, 3]$
(c) $[-4, -3) \cup [3, 4)$ (d) None of these

15. For $x \in R$, consider a function $||x||$ which is defined as follows :
$$||x|| = \begin{cases} x+1, 0 \le x < 2 \\ |x-4|, x \ge 2 \end{cases}$$
Then the solution set of the equation
$||x||^2 + x = ||x|| + x^2$ is
(a) $\{-1, 1\}$ (b) $[2, \infty)$
(c) $[0, 2)$ (d) $\{0, 2\}$

16. The area bounded by $y \le 3-|3-x|$ and $y \ge |x-3|$
(a) 4 (b) 11/2
(c) 9/2 (d) 7/2

17. Find the sum of last three digits of the number $\left[\frac{10^{93}}{10^{31}+3}\right]$
(a) 7 (b) 8
(c) 6 (d) None of these

18. If $[x]$ denotes the greatest integer $\le x$ then value of $S = [\sqrt{1}] + [\sqrt{2}] + [\sqrt{3}] + \ldots + [\sqrt{2024}]$ is
(a) 59000 (b) 58750
(c) 59730 (d) 65138

19. Find the sum of all the roots of the equation $2[x] = x + 2\{x\}$.
(a) 1 (b) 2
(c) 0 (d) None of these

(Passage Type)

Passage-I

Consider two functions f(x) and g(x) the number of roots of the equation f(x) = g(x) is same as the number of point of intersection of these two graphs.

20. Consider a function $y=f(x)=\left|\frac{|x-2|}{2}-3\right|-4$ find the number of roots of the equation f(x) = 0

(a) 2 (b) 4
(c) 3 (d) None of these

21. Consider a function $y=f(x)=\left|\frac{|x-2|}{2}-3\right|-4$ for what value of k the equation f(x) = k has 4 roots.

(a) (– 5, 0) (b) (– 5, –1)
(c) (– 1, 1) (d) None of these

Passage-II

Consider a function of type f(x) = | x – a| + p|x – b| will attain – ∞ when x tends to – ∞ and will attain ∞ when x tends to ∞ so function will attain a minimum value.

22. Consider a function $y = f(x) = |x + 2| + |x - 2| + |x - 5| + |x - 8| + |x - 10|$ find the minimum value of the function

(a) 11 (b) 14
(c) 18 (d) None of these

23. Consider a function $y = f(x) = |x + 2| + 2|x - 2|$ find the minimum value of the function

(a) 1 (b) 4
(c) 3 (d) None of these

Passage-III

Consider two functions f(x) and g(x) the number of roots of the equation f(x) = g(x) is same as the number of point of intersection of these two graphs but if they intersect p times and touch each other q times then number of common roots is p+q

24. Find the value of k such that the number of roots of $y = ||x|^2 - 4|x| - 5|$ and $y = k$ is maximum

(a) (6, 9) (b) (5, 9)
(c) (0, 5) (d) None of these

25. Find the number of roots of the equation $y = ||x|^2 - 4|x| - 5| = 4$ is

(a) 6 (b) 2
(c) 4 (d) None of these

Passage-IV

A function is discontinuous at a point if value of a function is different on left hand and right hand of that point. One of the example of discontinuous function is $y = [x^3]$

26. How many times function $y = [x^3]$ is discontinuous in (–2, 2)?

(a) 12 (b) 14
(c) 15 (d) None of these

27. How many times function $y = [x^3]$ is discontinuous in (–10, 10)?

(a) 121 (b) 1999
(c) 1998 (d) None of these

(Integer Type)

28. Consider a function $y = f(x) = 2|x - 2| + 3|x + 2|$ find the minimum value of the function
29. The function $y = f(x) = |x - 2| + |x - 4| + |x - 6| + |x - 8| + |x - 10|$ will attain a minimum value at-
30. The function $y = f(x) = 2|x - 2| + 3|x - 4|$ will attain a minimum value at-
31. The function $y = f(x) = |x - 1| + 2|x - 2| + 3|x - 3| +\dots 10|x - 10|$ will attain a minimum value at-
32. Find the minimum value of $y = f(x) = ||x - 4| - 6| + 3$
33. If K is the area bounded by $|x - 2| + |y| = 10$ then find the difference between largest and smallest digit that appear in K
34. If K is the area bounded by $|x + y| = 6$ and $|x - y| = 8$ then find the difference between largest and smallest digit that appear in K
35. If K is the area enclosed by two functions f and g then find the sum of digits in K

$$f = \max(|x|, |y|) = 8$$
$$g = \min(|x|, |y|) = 3$$

36. If area bounded by |x + 2y| + |2x – y| = p is P then area bounded by |x + 3y| + |3x – y| = 2p is KP then the value of [K] is , where [K] is G.I.F

37. Find the number of distinct roots of the equation $[x]^2 = \{x\}x$.

(More Than One Correct MCQs)

38. The equation $|x + 1|\,|x - 1| = a^2 - 2a - 3$ can have real solution in x if a belongs to
(a) $[1-\sqrt{5}, -1]$ (b) $[-1, 3]$
(c) $[3, 1+\sqrt{5}]$ (d) $[1+\sqrt{3}, 1+\sqrt{5}]$

39. The equation $||x - 1| + a| = 4$ can have real solutions for x if a belongs to the interval
(a) $(-\infty, +\infty)$ (b) $(-\infty, -4]$
(c) $(4, +\infty)$ (d) $[-4, 4]$

40. The set of real values of k for which the equation $x^2 - 4|x| + 3 - |k - 1| = 0$ will have exactly four roots is
(a) $(-2, 4)$ (b) $(-4, 4)$
(c) $(-4, 2)$ (d) $(-1, 0)$

41. If the equation $x^2 - 4|x| + 3 - (k - 1) = 0$ has exactly four real roots then k cannot be equal to
(a) -2 (b) 0
(c) 2 (d) 4

42. If $[x + 0.19] + [x + 0.20] + [x + 0.21] + \ldots + [x + 0.91] = 542$ then
(a) $[10x] = 73$ (b) $[100x] = 739$
(c) $[10x + 0.4] = 74$ (d) $[100x + 0.1] = 740$

Concept Eliminator (CE)

1. Find the number of real solutions to the equation $4x^2 - 40[x] + 51 = 0$.
(a) 3 (b) 2
(c) 0 (d) None of these

2. Let x, y, z be three positive real numbers such that

$$x + [y] + \{z\} = 13.2$$
$$[x] + \{y\} + z = 14.3$$
$$\{x\} + y + [z] = 15.1$$

where [a] denotes the greatest integer $\leq$ a and {b} denotes the fractional part of b, then
(a) $xyz = 351.32$ (b) $x + y + z = 21.3$
(c) $x + y - z = 3.9$ (d) None of these

3. Determine the sum of all the real solutions of $\left[\frac{x}{2}\right] + \left[\frac{2x}{3}\right] = x$.
(a) 21 (b) 28
(c) 15 (d) None of these

4. Find the number of solutions of the equation $[x^3] + [x^2] + [x] = \{x\} - 1$
Here [.] is G.I.F and $\{x\} = x - [x]$
(a) 1 (b) 2
(c) 0 (d) None of these

5. Find the units digit of $\left[\frac{10^{20000}}{10^{100} + 3}\right]$
(a) 9 (b) 7
(c) 3 (d) None of these

6. Find the value of $S = \sum_{K=1}^{502}\left[\frac{305\,K}{503}\right]$.
(a) 76204 (b) 76404
(c) 76304 (d) one of these

7. If 'K' is the total number of integers n between 1 and 10000 (both inclusive) such that n is divisible by $[\sqrt{n}]$, then unit digit of 'K' is
Here [] is G.I.F
(a) 7 (b) 8
(c) 1 (d) None of these

8. Find the sum of all roots solve equation $[3x + 1] = 2x - \frac{1}{2}$
(a) -1 (b) 2
(c) 1 (d) None of these

9. Given that n is a natural number, and S_n is the number of roots of the equation $x^2 - [x^2] = (x - [x])$ in the interval $1 \leq x \leq n$ then which statement is correct.
(a) $S_{10} = 91$ (b) $S_{10} = 55$
(c) $S_{20} = 101$ (d) None of these

10. Find the number of non integral solutions to the equation $x^2 - 8[x] + 7 = 0$
(a) 0 (b) 2
(c) 1 (d) None of these

11. Find the number of distinct terms in the sequence $\left[\frac{1^2}{1980}\right], \left[\frac{2^2}{1980}\right], \left[\frac{3^2}{1980}\right], \ldots, \left[\frac{1980^2}{1980}\right]$?
(a) 990 (b) 496
(c) 1486 (d) None of these

12. Find the number of solutions of the equation $[x]\{x\} + x = 2\{x\} + 10$.

(a) 7 (b) 8
(c) 6 (d) None of these

13. How many of the first 1000 positive integers can be expressed in the form $[2x] + [4x] + [6x] + [8x]$ where x is a real number, and $[x]$ denotes the greatest integer less than or equal to x?

(a) 559 (b) 401
(c) 301 (d) None of these

14. For $x \in R$, let [x] denote the greatest integer $\leq$ x. Largest natural number for which
$$E = \left[\frac{\pi}{2}\right] + \left[\frac{1}{100} + \frac{\pi}{2}\right] + \left[\frac{2}{100} + \frac{\pi}{2}\right] \ldots + \left[\frac{n}{100} + \frac{\pi}{2}\right]$$
< 43 is

(a) 40 (b) 41
(c) 42 (d) None of these

15. Find the minimum natural number n, such that the equation $\left[\frac{10^n}{x}\right] = 1989$ has integer solution x.

(a) 7 (b) 8
(c) 6 (d) None of these

16. Find the number of integral solution of the equation $3x^2 - |xy| - 2y^2 + 7 = 0$

(a) 0 (b) 2
(c) 4 (d) None of these

17. If 'k' is the number of positive integers x which satisfy the condition $\left[\frac{x}{99}\right] = \left[\frac{x}{101}\right]$, then find the remainder when k divided by 9. Here [x] is greatest integer function.

(a) 0 (b) 3
(c) 6 (d) None of these

[Based on RMO]

18. If n is an integer greater than 7, then find the remainder when $\left({}^nC_7\right) - \left[\frac{n}{7}\right]$ is divided by 7.

(a) 0 (b) 3
(c) 4 (d) None of these

[Based on RMO]

Solutions

Concept Applicator (CA)-1

1. Since we know that if $|x| = \sqrt{5}$, then $x = \pm\sqrt{5}$ here instead of x, given quantity is $x - \sqrt{2}$, so $x - \sqrt{2} = \pm \sqrt{5}$

   So $x = \sqrt{5} + \sqrt{2}$ or $-\sqrt{5} + \sqrt{2}$

2. Since we know that if $|x| \leq a$ then $-a \leq x \leq a$

   So, if $|x| \leq 10$, then $-10 \leq x \leq 10$ or $x \in [-10, 10]$

   Alternately:- Since $|x|$ is defined as distance of a point from the origin so $|x| \leq 10$ is defined as distance of a point from the point $x = 0$ or origin is less than 10 units so $-10 \leq x \leq 10$ or $x \in [-10, 10]$

3. (d) From the graph (refer theory Graph no. 9) function will attain minimum value at $x = -2$.

4. (d) From the graph (refer theory Graph no. 10) function will attain minimum value at $x = 2$.

5. (d) From the graph (refer theory Graph no. 11) minimum value of function is at $x = -4$ and required minimum value is 3.

6. (b) From the graph (refer theory Graph no. 13) minimum value of function is at $x = -4$ and 8 and required minimum value is 0.

7. (d) From the graph (refer theory Graph no. 14), we can say that $f(x) = k$ will intersect the graph 4 times if k lies between local maxima and local minima which is -1 and -4.

8. (b) From the graph (refer theory Graph no. 15), we can say that $f(x) = k$ will intersect the graph maximum number of times if k lies between local maxima and local minima which is 4 and 1.

9. (b) From the graph (refer theory Graph no. 16) we can say that $f(x) = 5$ will intersect the graph twice.

10. (b) From the graph (refer theory Graph no. 16) function will attain minimum value at $x = [-2, 2]$ and the required minimum value is 14.

Concept Applicator (CA)-2

1. Function $y = f(x) = |2x - 6|$ is defined as

$$|2x-6| = \begin{cases} -2x+6 \text{ when } x < 3 \\ 2x-6 \text{ when } x \geq 3 \end{cases}$$

Its minimum values is 0 at $x = 3$.

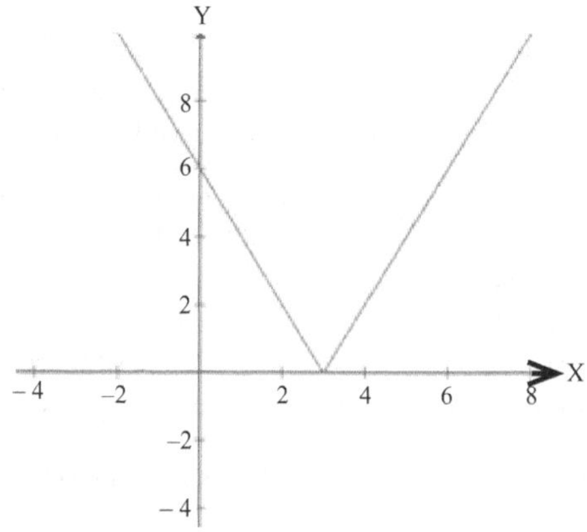

2. Function $y = f(x) = |3x + 8|$ is defined as

$$|3x+8| = \begin{cases} -3x-8 \text{ when } x < \dfrac{8}{3} \\ 3x+8 \text{ when } x \geq \dfrac{8}{3} \end{cases}$$

Its minimum values is 0.

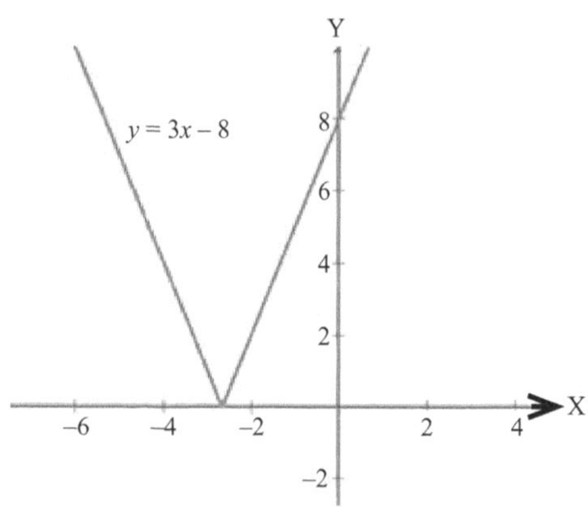

3. Function $y = f(x) = -|3x - 9|$ is defined as

$$-|3x-9| = \begin{cases} 3x-9 \text{ when } x < 3 \\ -3x+9 \text{ when } x \geq 3 \end{cases}$$

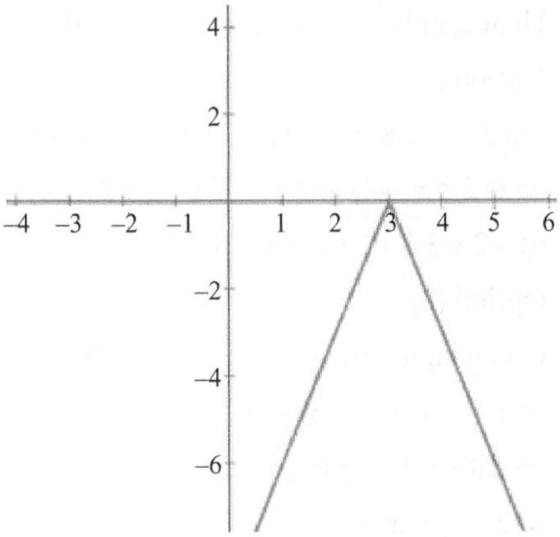

Its maximum values is 0

4. In order to plot the graph of $y = f(x) = 2|2x - 6| + 2$, 1st plot $y = 4x$, and shift the graph to a new origin (3, 2) as shown below

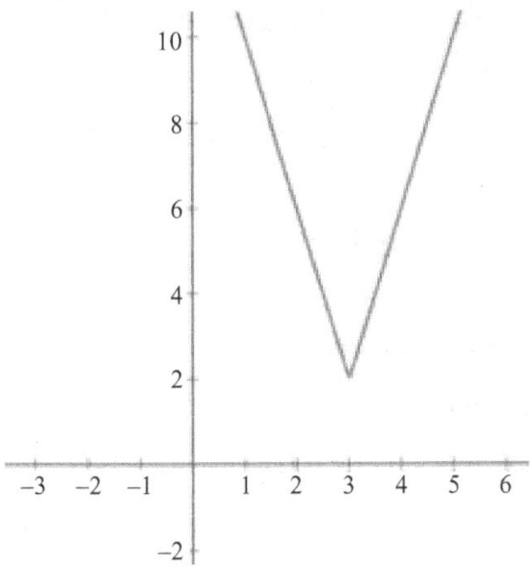

Its minimum value is 2.

5. In order to plot the graph of $y = f(x) = 2|x + 2| + 2$, 1st plot $y = 2x$, and shift the graph to a new origin (–2, 2) as shown below,

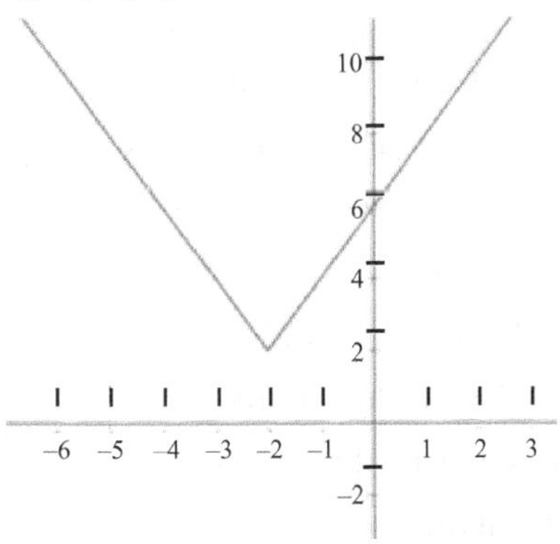

Minimum value is 2.

6. Graph of $y = f(x) = |x - 5| + |x - 10|$

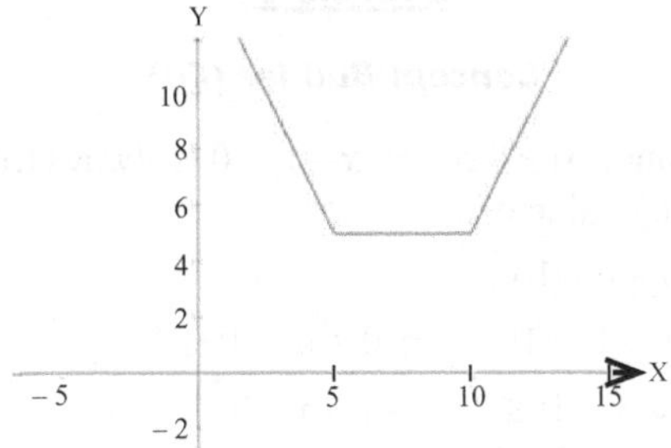

Function will attain minimum value at $x \in [5, 10]$ and required minimum value is 5.

7. The graph is as follows

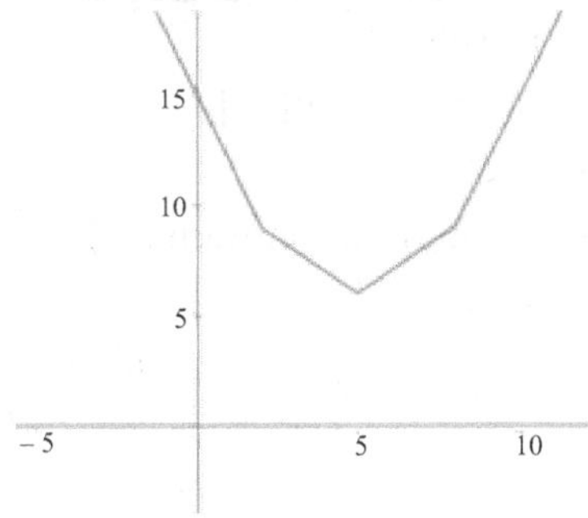

The function will attain minimum value at $x = 5$.

8. Given $[x + 0.1] = 7$

So $7 \leq x + 0.1 < 8$

$\Rightarrow 6.9 \leq x < 7.9$

9.

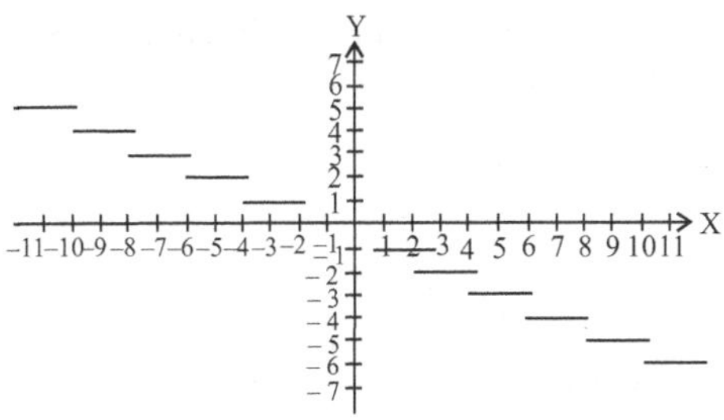

10.

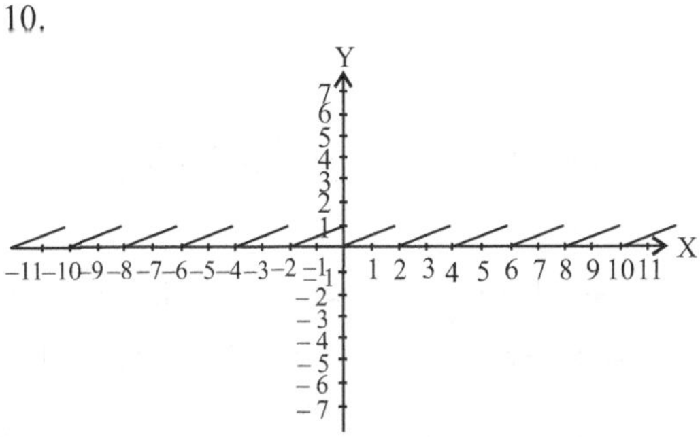

Exercis-A

Concept Builder (CB)

1. Since $|x| \geq 0$ so $1 + |x-1| \geq 0$ is always true, for any value of x

   Option (b)

2. $||x-1|-1| \leq 1 \Rightarrow 0 \leq |x-1| \leq 2$

   $\Rightarrow |x-1| \leq 2 \Rightarrow -1 \leq x \leq 3$

   Option (a)

3. For $x \leq 1$, the given inequation gives $x \leq 0$ and for $x \geq 3$, the given inequation gives $x \geq 4$

   For $1 < x \leq 2$ or $2 < x < 2$, the given inequation gives no solution.

   $\therefore$ Solution is $(-\infty, 0] \cup [4, \infty)$

   Option (c)

4. $|3-x| = x-3$ is true only when $x-3 \geq 0$ i.e., $x \geq 3$

   Option (c)

5. $2x - 1 = |x+7| = \begin{cases} x+7, & \text{if } x \geq -7 \\ -(x+7), & \text{if } x < -7 \end{cases}$

   $\therefore$ If $x \geq -7$, $2x - 1 = x + 7 \Rightarrow x = 8$
   If $x < -7$, $2x - 1 = -(x+7) \Rightarrow 3x = -6$
   $\Rightarrow x = -2$ not possible.
   Option (b)

6. $|x-1|$ is the distance of x from 1
   $|x-3|$ is the distance of x from 3
   The points $x = 2$ is equidistant from 1 and 3.
   Hence, the solution consists of all $x \geq 2$.
   Option (b)

7. $\left|\frac{1}{x} - 2\right| < 4 \Leftrightarrow -4 < \frac{1}{x} - 2 < 4$

   $\Leftrightarrow -2 < \frac{1}{x} < 6$

   Hence, $x \in \left(-\infty, -\frac{1}{2}\right) \cup \left(\frac{1}{6}, \infty\right)$

   Option (d)

8. $\left|1 + \frac{3}{x}\right| > 2$

   Case I. $1 + \frac{3}{x} > 2 \quad \Rightarrow \frac{3}{x} > 1$ (clearly $x > 0$)
   $\Rightarrow 3 > x$ or $x < 3$

   Case II. $1 + \frac{3}{x} < -2 \quad \Rightarrow \frac{3}{x} < -3$ (hence $x < 0$)
   $\Rightarrow 3 > -3x \qquad \Rightarrow -1 < x$ or $x > -1$

   Hence, either $0 < x < 3$ or $-1 < x < 0$

   Option (c)

9. $(-3 \leq x-1 \leq 3)$ and $(x-1 \leq -1$ or $x-1 \geq 1)$

   $\Rightarrow (-2 \leq x \leq 4)$ and $(x \leq 0$ or $x \geq 2)$

   $\Rightarrow -2 \leq x \leq 0$ or $2 \leq x \leq 4$

   Option (c)

10. Given equation is $|x|^2 - 3|x| + 2 = 0$

    $\Rightarrow (|x|-1)(|x|-2) = 0$

    $\Rightarrow$ Either $|x| = 1$ or $|x| = 2$

    $\Rightarrow x = \pm 1, x = \pm 2$

    Hence, the equation has four solutions.

    Option (a)

11. Given, $|x-2|^2 + |x-2| - 2 = 0$

    $\therefore y^2 + y - 2 = 0$ where $y = |x-2|$

    $\Rightarrow (y+2)(y-1) = 0$

    $\Rightarrow$ y = 1 [since y = $|x-2| = -2$ not possible]

    $\Rightarrow |x-2| = 1$

    $\Rightarrow x - 2 = \pm 1$

    $\Rightarrow x = 3, 1$

    $\therefore$ Sum of roots is 4.

    Option (b)

12. $||x|-1| < |1-x|$ ...(i)

    Case I. $x \geq 0$

    $\therefore$ Eq. (i) becomes $|x-1| < x-1$ or $|1-x| < 1-x$ which is not satisfied by any x, because

    $|a| \geq a, \forall a \in R$

    Case II. $-1 \leq x < 0$

    $\therefore$ Eq. (i) becomes $|-1-x| < 1-x$ or $|x+1| < 1-x$ or $x + 1 < 1 - x$ or $x < 0$

    Thus, (i) is satisfied for $-1 \leq x < 0$

    Case III. $x < -1$

    Eq. (i) becomes $|-1-x| < 1-x$

    $\Rightarrow |1+x| < 1-x \Rightarrow -(1+x) < 1-x$

    $\Rightarrow -2 < 0$, which is true

    So, solution set is $(-\infty, 0)$

    Option (d)

13. Consider $|3x + 1| < \frac{1}{3}$

$\Leftrightarrow -\frac{1}{3} < 3x + 1 < \frac{1}{3}$

$\Leftrightarrow -\frac{4}{3} < 3x < -\frac{2}{3}$

$\Leftrightarrow -\frac{4}{9} < x < -\frac{2}{9}$

Also, $0 < |3x + 1|$ is satisfied by each x except when $3x + 1 = 0$, i.e., $x = -\frac{1}{3}$

∴ Solution is

Option (c)

14. $|x^2 - 10| \le 6$

$\Leftrightarrow -6 < x^2 - 10 \le 6$

$\Leftrightarrow 4 \le x^2 \le 16$

$\Leftrightarrow 2 \le |x| \le 4$

∴ Either $2 \le x \le 4$ or $-4 \le x \le -2$

∴ $x \in [-4, -2] \cup [2, 4]$

Option (d)

Concept Cracker (CC)

1. The function will attain minimum value at any value of x between –2 to 2 and required minimum value is 4.

   Option (b)

2. Consider graph

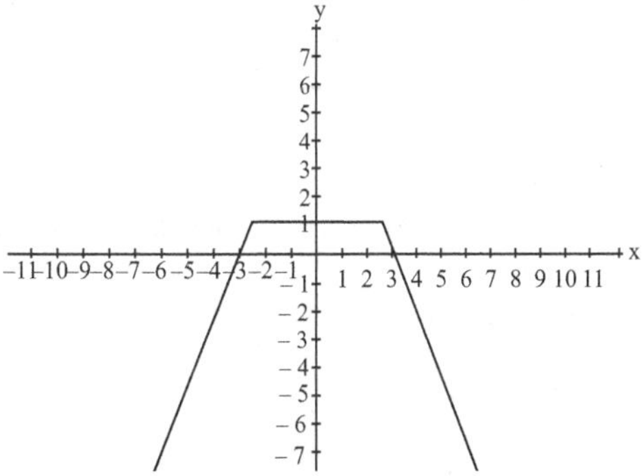

Its maximum value is 1.

Option (d)

3. The function $f(x) = |x - 2| + |x - 4| - |x + 2|$ will attain its minimum value at x = 4 as shown in the figure

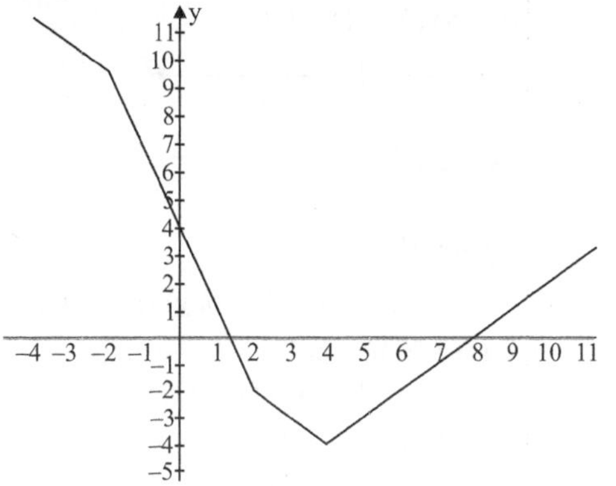

Required minimum value is – 4.

Option (c)

4. Function will attain local maxima at x =2 as shown in the graph

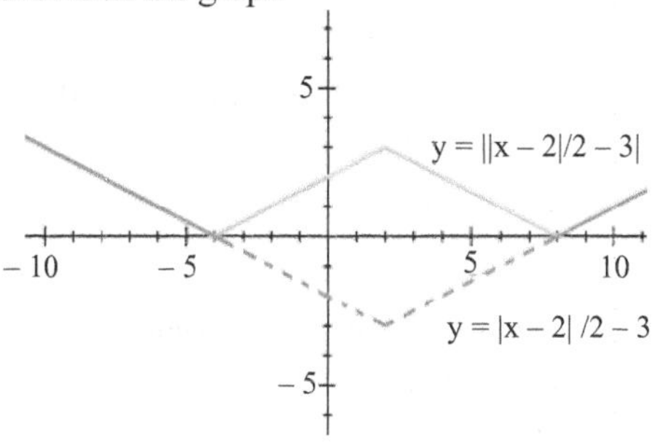

Required local maximum value is 3, but actual maximum value is ∞.

Option (c)

5. From the given condition $[2x^2 - 3] = -1, 0, 1$

   So $-1 \le 2x^2 - 3 < 2$

   or $1 \le x^2 < \frac{5}{2}$

   or $x \in \left(-\sqrt{\frac{5}{2}}, -1\right] \cup \left[1, \sqrt{\frac{5}{2}}\right)$ required integral values of x is – 1, and 1

   Option (c)

6. Given that $|3x + 2| < 1 \Leftrightarrow -1 < 3x + 2 < 1$
   $\Leftrightarrow -3 < 3x < -1$

$\Leftrightarrow -1 < x < -\frac{1}{3}$

Option (c)

7. $|2x-3| < |x+2|$

Squaring both sides, $(2x-3)^2 - (x+2)^2 < 0$

$(x-5)(3x-1) < 0$

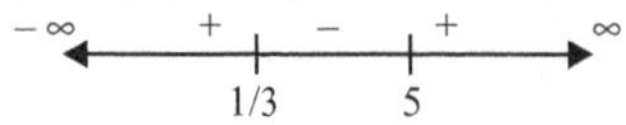

$\Rightarrow x \in (\frac{1}{3}, 5)$

Option (b)

8. $|x-1| + |x| + |x+1| \geq 6$, gives four cases

Case I. When $x < -1$ .... (i)

$-(x+1) - x - (x-1) \geq 6$

$\Rightarrow -x - 1 - x - x + 1 \geq 6$

$\Rightarrow -3x \geq 6$ ... (ii)

From Eqs. (i) and (ii), $x \leq -2$

Case II. When $-1 \leq x \leq 0$ .... (iii)

$\Rightarrow x + 1 - x - (x-1) \geq 6$

$\Rightarrow 1 - x + 1 \geq 6 \Rightarrow x \leq -4$ ..... (iv)

$\therefore$ no value of x [From (iii) and (iv)]

Case III. When $0 \leq x \leq 1$ ....(v)

$\Rightarrow x + 1 + x - (x-1) \geq 6$

$\Rightarrow x \geq 4$ ... (vi)

No solution, using Eqs. (v) and (vi)

Case IV. When $x > 1$ ... (vii)

$x + 1 + x + x - 1 \geq 6$

$\Rightarrow 3x \geq 6$ or $x \geq 2$ ....(viii)

From Eqs. (vii) and (viii), $x \geq 2$

Thus, from above four cases,

$x \leq -2$ or $x \geq 2$

$\Rightarrow x \in (-\infty, -2] \cup [2, \infty)$

Option (b)

9. $\left|x + \frac{1}{x}\right| > 2$ (clearly $x \neq 0$)

$\Rightarrow \left|\frac{x^2+1}{x}\right| > 2 \Rightarrow \frac{x^2+1}{|x|} > 2$ ($\because x^2+1 > 0$)

$\Rightarrow x^2 + 1 > 2|x|$

$\Rightarrow |x|^2 - 2|x| + 1 > 0 \Rightarrow (|x|-1)^2 > 0$

$\Rightarrow |x| \neq 1 \Rightarrow x \neq -1, 1$

$\therefore x \in R - \{-1, 0, 1\}$

Option (b)

10. $\left|x + \frac{1}{x}\right| < 4$ (clearly $x \neq 0$)

$\frac{x^2+1}{|x|} < 4 \Rightarrow x^2 + 1 < 4|x|$

$\Rightarrow |x|^2 - 4|x| + 1 < 0$

$\Rightarrow (|x|-2)^2 < 3 \Rightarrow (|x|-2)^2 < (\sqrt{3})^2$

$\Rightarrow ||x|-2)| < \sqrt{3}$

$\Rightarrow -\sqrt{3} < |x| - 2 < \sqrt{3}$

$\Rightarrow 2 - \sqrt{3} < |x| < 2 + \sqrt{3}$

$\therefore$ When $x > 0$, $2 - \sqrt{3} < x < 2 + \sqrt{3}$

When $x < 0$, $-2 - \sqrt{3} < x < -2 + \sqrt{3}$

Option (a)

11. $\frac{|x+1|}{|x|} + |x+1| = \frac{(x+1)^2}{|x|}$

Since, $|a| + |b| = |a+b|$ iff $ab \geq 0$

$\therefore \frac{(x+1)}{x}.(x+1) \geq 0$ or $\frac{(x+1)^2}{x} \geq 0$

$\Rightarrow x > 0$ or $\{-1\}$

Option (b)

12. $|x-2|^2 - 3|x-2| + 2 = 0$

$\Rightarrow (|x-2|-1)(|x-2|-2) = 0$

$\Rightarrow x = 3, 1, 4, 0$

$\therefore$ Product of roots = 0

Option (c)

13. Let $y = |x|$

$\therefore$ From given, we have

$\frac{3-y}{4-y} \geq 0$

$\Rightarrow y \leq 3$, or $y > 4$

$\Rightarrow |x| \leq 3$ or $|x| > 4$

$\Rightarrow x \in [-3, 3] \cup (-\infty, -4) \cup (4, \infty)$

Option (a)

14. Given, $2 \leq |x-3| < 4$

We have two cases, consider each case separately-

Case I. $2 \leq |x-3|$

$\Rightarrow x - 3 \leq -2$ or $x - 3 \geq 2$

$\Rightarrow x \leq 1$ or $x \geq 5$ ... (i)

Case II. $|x-3| < 4$

$\Rightarrow -4 < x-3 < 4$

$\Rightarrow -1 < x < 7$ .... (ii)

From Eqs. (i) and (ii), we get $x \in (-1, 1] \cup [5, 7)$.

Option (a)

15. Lets define the function piecewise

Case I. When $x < -2$, then $y = f(x) = 1 + 2 = 3$

Case II. When $-2 < x < 1$, then $y = f(x) = -2x - 1$ and $-2x - 1 < 1$ or $x < 1$

Case III. When $x > 1$ then $y = f(x) = -3$ all value of x will satisfy,

So $x > -1$

Option (b)

Graphical approach:-

Graph of the function is-

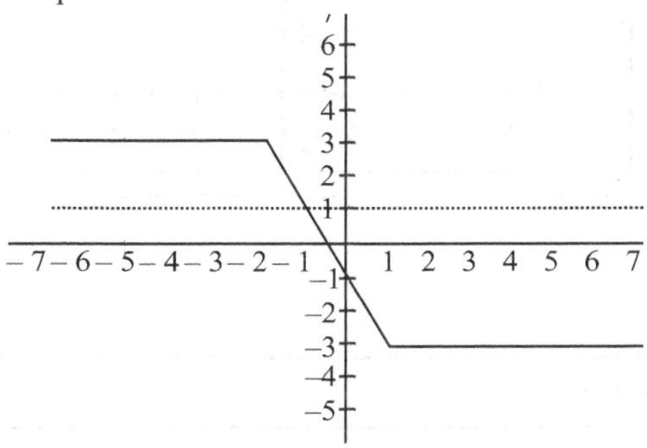

From the graph the function $y = f(x) < 1$ if $x > -1$

Option (b)

16. Maximum or minimum value of function will attain at $x = -3, -1$ or 2

At $x = -3$ the value of function is 3

At $x = -1$ the value of function is 5

At $x = 2$ the value of function is 2

So required minimum value of the function is 2

The graph is shown below

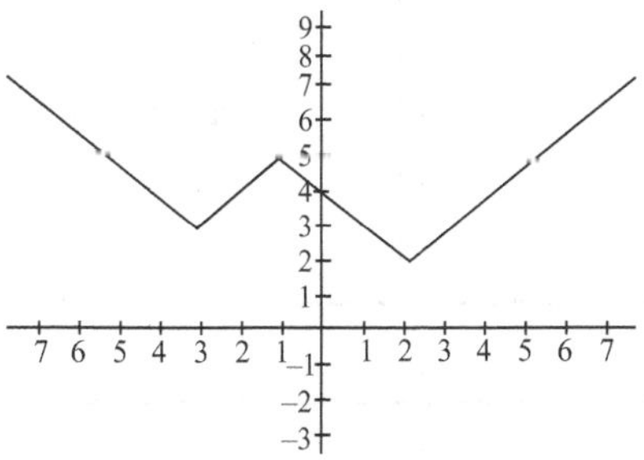

Option (a)

17. Maximum or minimum value of function will be attained at $x = -3, -1$ or 2

At $x = -3$ the value of function is 7

At $x = -1$ the value of function is 1

At $x = 2$ the value of function is -2

So required minimum value of the function is -2

The graph is shown below

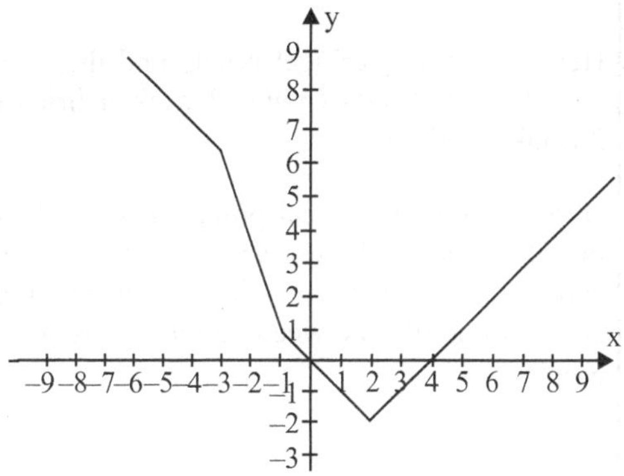

Option (b)

18. Maximum or minimum value of function will attain at $x = -1$ or 2

At $x = -1$ the value of function is 6

At $x = 2$ the value of function is -3

So required minimum value of the function is -3

The graph is shown below

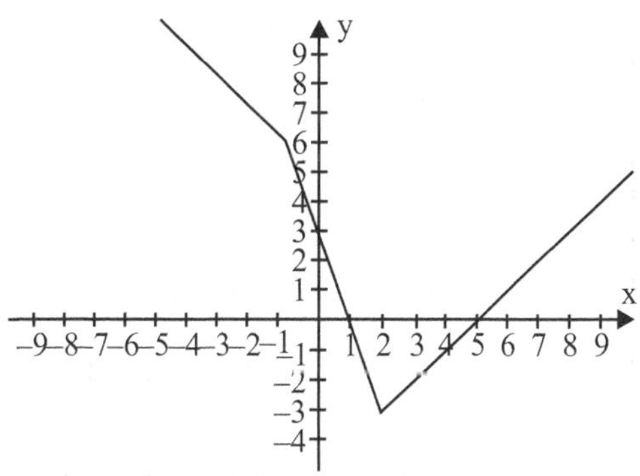

Option (d)

Exercis-B

Concept Deviator (CD)
(Single Correct MCQs)

1. (b) Lets see the graph of the function

Graph of y = 3|x+2| + 2|x-2|+|x-4|

Here we have 6 critical points and these points are –2, –2, –2, 2, 2 and 4 (point –2 is taken thrice and point 2 is taken twice)

Here we have 6 critical points so we will have two middle values and these are 3rd and 4th one i.e x= –2 and x = 2 so the function will attain its minimum value at any value of x between –2 and 2, as shown in the graph

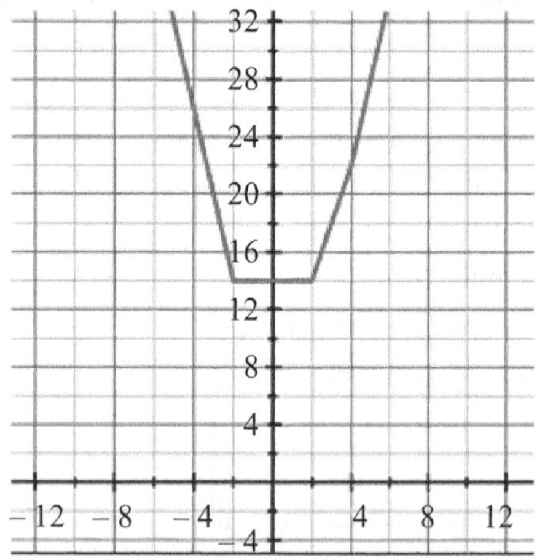

Required minimum value is 14

2. (a) Lets see the graph of the function

Graph of |x|+ |y| =4

Here function has different definition in different quadrant so we will define it separately in each quadrant and then plot the graph in each quadrant.

- In 1st quadrant equation is x + y = 4
- In 2nd quadrant equation is –x + y = 4
- In 3rd quadrant –x – y = 4
- In 4th quadrant x – y = 4

These four lines will form a square and its area is (8 × 8)/2 = 32 sq unit

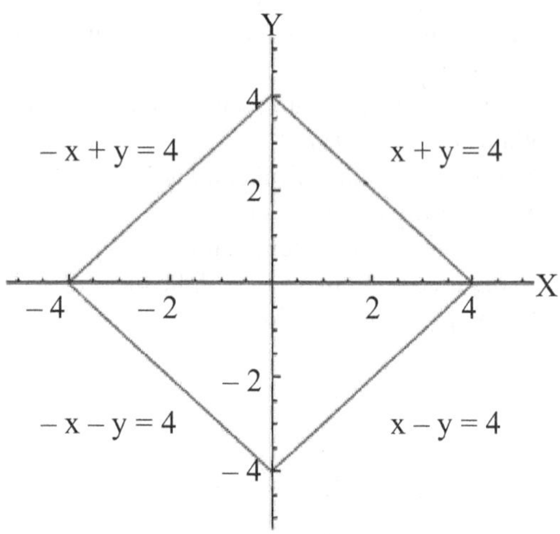

Area of the region bounded by |x| +|y| = k is given by $2k^2$

Required area is 32 sq unit

3. (d) Lets see the graph of the function

Graph of $3|x| + 2|y| = 12$

Here function has different definition in different quadrant so we will define it separately in each quadrant and then plot the graph in each quadrant.

- In 1st quadrant equation is $2y = 12 - 3x$
- In 2nd quadrant equation is $2y = 12 + 3x$
- In 3rd quadrant $-2y = 12 + 3x$
- In 4th quadrant $-2y = 12 - 3x$

These four lines will form a rhombus and its area is equal to 4 times the area of a triangle in one quadrant

Area of triangle in 1st quadrant = $\frac{1}{2}(4)(6) = 12$

So required area = 48 sq unit

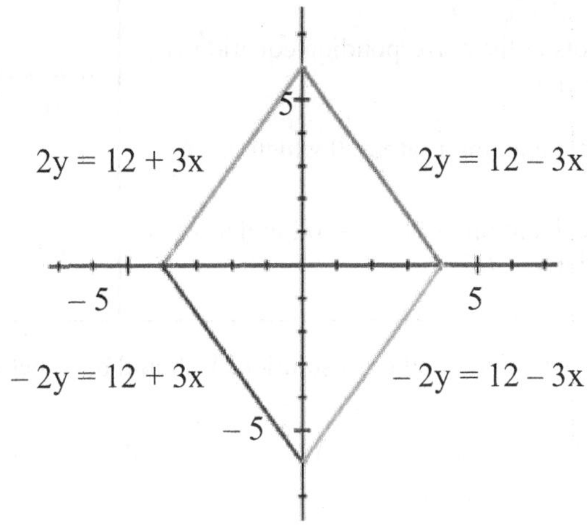

Area of the region bounded by $a|x| + b|y| = k$ is given by $\frac{2(k^2)}{ab}$

4. (a) Lets see the graph of the function

Plot the graph of $y = |x|^2 - |x| - 6$

Step 1:- When $x > 0$, then $y = x^2 - x - 6$
It has two roots – 2 and 3.

Minimum value of function is at $x = \frac{1}{2}$

$$y = f\left(\frac{1}{2}\right) = \left(\frac{1}{2}\right)^2 - \frac{1}{2} - 6 = -6.25$$

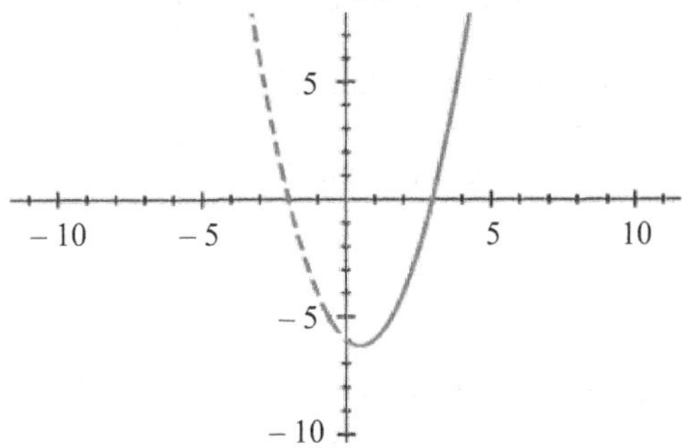

Step 2:- When $x < 0$ then $y = x^2 + x - 6$
It has two roots – 3 and 2.

Minimum value of function is at $x = -\frac{1}{2}$

$$y = f\left(\frac{1}{2}\right) = \left(-\frac{1}{2}\right)^2 - \frac{1}{2} - 6 = -6.25$$

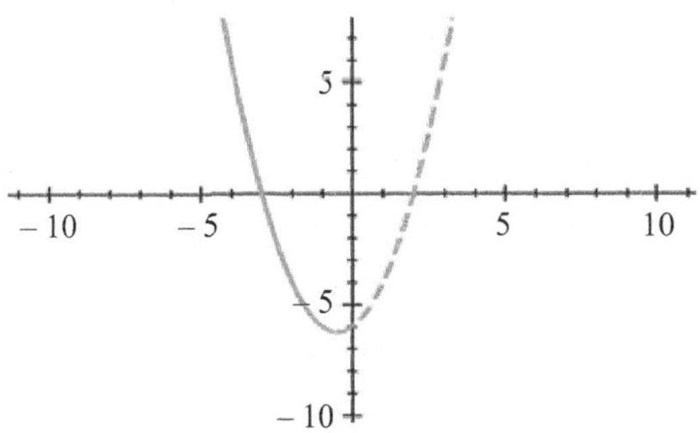

<table>
<tr><td>Step 3:- Combine the above two results and plot graph of the function
$y = |x|^2 - |x| - 6$
Roots of the corresponding equation is -3 and 3

Local maxima is at x = 0 which is -6

Local minima is at $x = -1/2$ and at $x = \frac{1}{2}$ which is -6.25</td><td>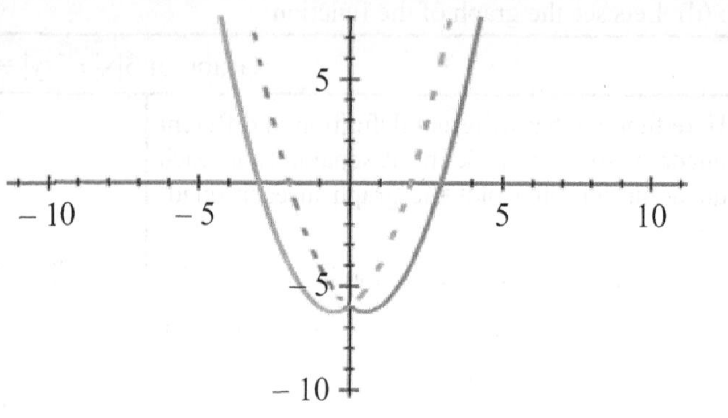</td></tr>
</table>

For maximum number of solutions k should be in between $(-6.25, -6)$

5. (d) $\sum_{-2}^{2} A_k = A_{-2} + A_{-1} + A_0 + A_1 + A_2$

$$A_{-2} = 1 \times 0.5 = 0.5$$
$$A_{-1} = 1 \times 0.5 = 0.5$$
$$A_0 = A_1 = A_2 = 0$$

Hence $\sum_{-2}^{2} A_k = A_{-2} + A_{-1} + A_0 + A_1 + A_2 = 0.5 + 0.5 = 1$

6. (a) In order to find the area, let us plot the graph.

<table>
<tr><th colspan="2">Graph of $[x^2]$</th></tr>
<tr><td>When $x \in [0, 1)$, then $x^2 \in [0, 1)$ and $[x^2] = 0$

When $x \in [1, \sqrt{2})$, then $x^2 \in [1, 2)$ and $[x^2] = 1$

When $x \in [\sqrt{2}, \sqrt{3})$, then $x^2 \in [2, 3)$ and $[x^2] = 2$

When $x \in [\sqrt{3}, 2)$, then $x^2 \in [3, 4)$ and $[x^2] = 3$ and so on

When $x \in (-1, 0]$, then $x^2 \in [0, 1)$ and $[x^2] = 0$

When $x \in (-\sqrt{2}, -1]$, then $x^2 \in [1,2)$ and $[x^2] = 1$ and so on</td><td>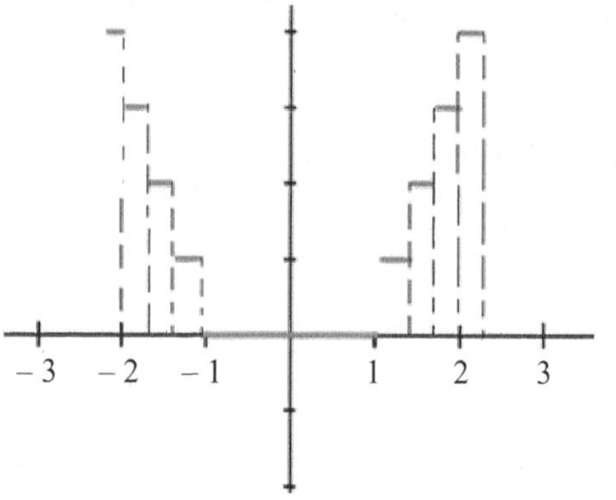</td></tr>
</table>

When $x \in (-1, 1)$ the area is 0

When $x \in (1, \sqrt{2})$ the area is $(\sqrt{2} - 1)$

When $x \in (\sqrt{2}, \sqrt{3})$ the area is $(\sqrt{3} - \sqrt{2})$

When $x \in (\sqrt{3}, 2)$ the area is $(2 - \sqrt{3})$

So total area is $(\sqrt{2} - 1) + 2(\sqrt{3} - \sqrt{2}) + 3(2 - \sqrt{3}) = 5 - \sqrt{3} - \sqrt{2}$

7. (d) Lets see the graph of the function

Graph of $[x^3]$	
When x ∈ [0, 1), then [x] = 0 and $[x]^3 = 0$ When x ∈ [1, 2), then [x] = 1 and $[x]^3 = 1$ When x ∈ [2, 3) then [x] = 2 and $[x]^3 = 8$ and so on When x ∈ [–1, 0), then [x] = –1 and $[x]^3 = -1$ When x ∈ [–2, –1) then [x] = –2 and = –8 and so on	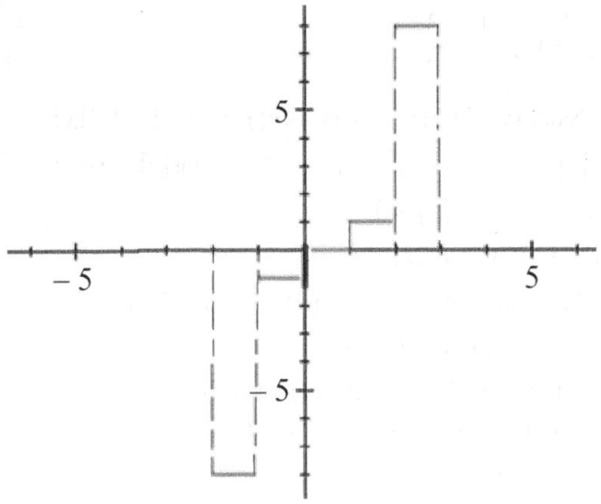

From the graph number of discontinuous points is 3

8. (b) $x = 2m + y$ where m is an integer and $0 \le y < 1$, then $[x] = 2m$ and

$$\left[\frac{x}{2}\right] = m, \left[\frac{x+1}{2}\right] = \left[\frac{2m+y+1}{2}\right] = m$$

$$\left(\text{since } \frac{1}{2} < \frac{1}{2} < 1\right)$$

$$\therefore \left[\frac{x}{2}\right] + \left[\frac{x+1}{2}\right] = [x]$$

Also if $x = (2m + 1) + y$, then

$$\left[\frac{x}{2}\right] = m, \left[\frac{x+1}{2}\right] = \left[\frac{2m+y+2}{2}\right] = m+1$$

But $[x] = 2m + 1 = \left[\frac{x}{2}\right] + \left[\frac{x+1}{2}\right]$

$\therefore \left[\frac{x}{2}\right] + \left[\frac{x+1}{2}\right] = [x]$ for every real number.

9. (b) Since $x = [x] + \{x\}$, the equation becomes $4\{x\} = [x] + \{x\} + [x] \Rightarrow 3\{x\} = 2[x]$

$$\Rightarrow \{x\} = \frac{2}{3}[x] \qquad .... (1)$$

We know that $0 \le \{x\} < 1 \Rightarrow 0 \le \frac{2}{3}[x] < 1$

$\Rightarrow 0 \le [x] < 1$ and $[x]$ is an integer

$\therefore$ $[x] = 0$ or 1, from (1), we get $\{x\} = 0$ if $[x] = 1$

and $\{x\} = \frac{2}{3}$ if $[x] = 1$

$\therefore x = 0 + 0$ or $1 + \frac{2}{3} \Rightarrow x = 0$ or $\frac{5}{3}$

The solutions set is $x \in \left\{0, \frac{5}{3}\right\}$

10. (a) Given $F(x) = x - [x]$ So, $F(x) + F\left(\frac{1}{x}\right) = 1$

$$\Rightarrow x - [x] + \frac{1}{x} - \left[\frac{1}{x}\right] = 1$$

$$\Rightarrow x + \frac{1}{x} = [x] + \left[\frac{1}{x}\right] + 1 \quad ... (1)$$

since RHS is an integer

$\therefore$ LHS must be integer, which is possible when x = 1, but at x = 1, LHS ≠ RHS

11. (c) Since $[x + n] = [x] + n$, if n ∈ ***I***, therefore

$[x]^2 = [x] + 2 \Rightarrow [x]^2 - [x] - 2 = 0$

$\Rightarrow ([x] + 1)([x] - 2) = 0 \Rightarrow [x] = -1$ or 2

If $[x] = -1$ then $-1 \le x < 0$

If $[x] = 2$, then $2 \le x < 3$

$\therefore x \in [-1, 0) \cup [2, 3)$

12. (b) $5\{x\} = x + [x]$ and $[x] - \{x\} = -$, since $x = [x] + \{x\}$

$$\Rightarrow 4\{x\} = 2[x] \text{ and } [x] - \{x\} = \frac{1}{2}.$$

After solving $[x]=1$ and $\{x\}=\frac{1}{2}$

$\therefore x = 1 + \frac{1}{2} = \frac{3}{2}$

13. (c) Since $(x-1)=(x-[x])(x-\{x\}) \Rightarrow x = 1+\{x\}[x]$
$\Rightarrow [x]+\{x\} = 1+\{x\}[x] \quad \Rightarrow (\{x\}-1)([x]-1) = 0$
$\{x\}-1 \neq 0 \therefore [x]-1=0$
$[x]=1 \Rightarrow x \in [1,2)$

14. (b) $(x)^2+(x+1)^2=25 \quad \Rightarrow (x)^2+\{(x)+1\}^2 = 25$
$\Rightarrow 2(x)^2+2(x)-24=0$
$\Rightarrow (x)^2+(x)-12=0 \Rightarrow (x)-4$ or 3
Now $(x) = -4 \Rightarrow -5 < x \leq -4$
and $(x) = 3$
$\Rightarrow 2 < x \leq 3$
Solution set is $(-5, -4] \cup (2, 3]$

15. (d) Case I : Let $0 \leq x < 2$, then $||x|| = x+1$ and the equation becomes
$(x+1)^2 + x = (x+1) + x^2 \quad \Rightarrow 2x = 0$
$\Rightarrow 2x = 0 \Rightarrow x = 0$
Case 2: Let $x \geq 2$, then $||x|| = |x-4|$ and the equation becomes
$|x-4|^2 + x = |x-4| + x^2$
$\Rightarrow x^2 - 8x + 16 + x = |x-4| + x^2$
$\Rightarrow |x-4| = 16-7x$
$\therefore x-4 = \pm(16-7x)$, provided $16-7x \geq 0$
$\therefore x = \frac{5}{2}$ or 2, but for $x = \frac{5}{2}$, $16-7x<0$, hence rejected
$\therefore x = 2$. The solution set is $\{0, 2\}$

16. (c) The area bounded is as shown

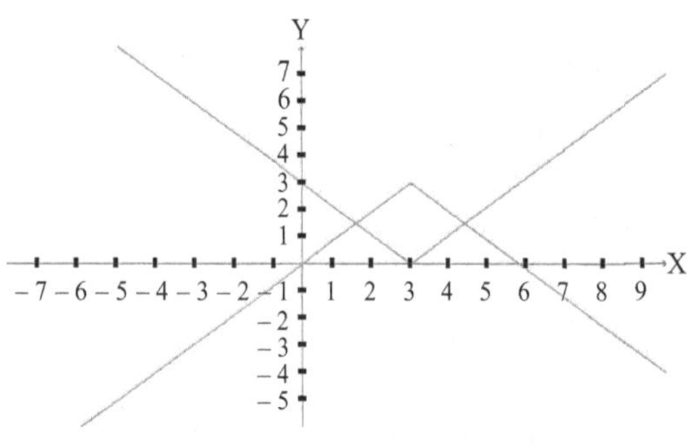

Required area is $\frac{3}{\sqrt{2}} \times \frac{3}{\sqrt{2}} = \frac{9}{2}$

17. (b) Let $10^{31} = t$, then

$$\left[\frac{10^{93}}{10^{31}+3}\right] = \left[\frac{t^3}{t+3}\right] = \left[\frac{t^3+3^3}{t+3} - \frac{3^3}{t+3}\right]$$

$$= t^2 - 3t + 3^2 + \left[-\frac{3^3}{t+3}\right] = t^2 - 3t + 3^2 - 1$$
$$= t(t-3) + 8 = 10^{31}(10^{31}-3) + 8,$$

thus the last three digits of $\left[\frac{10^{93}}{10^{31}+3}\right]$ is 008

18. (c) For any positive integer k and x, we have

$\sqrt{x} = k \Leftrightarrow k^2 \leq x < (k+1)^2 \Leftrightarrow x \in [k^2, k^2+2k]$

$\therefore \left[\sqrt{x}\right] = 1$ for $x = 1, 2, 3$, $\left[\sqrt{x}\right] = 2$ for

$x = 4,5,6,7,8$, $\left[\sqrt{x}\right] = 3$ for $x = 9, 10, 11, 12, 13, 14, 15$,

$\left[\sqrt{x}\right] = 44$ for $x = 1936,\ldots, 2024$

Thus, $S = (1)(3) + (2)(5) + 3(7) + \ldots + 44(89)$

$= \sum_{k=1}^{44}(k)(2k+1) = 2\sum_{k=1}^{44}(k)^2 + \sum_{k=1}^{44}(k)$

$= \frac{2}{6}(44)(45)(89) + \frac{1}{2}(44)(45) = 59730$

19. (d) To reduce the number of variables, $x = [x] + \{x\}$ yields the equation

$2[x] = [x] + 3\{x\}$,

$\therefore [x] = 3\{x\} < 3$.

When $[x] = 0, 1, 2$ respectively, then $\{x\} = 0\ \frac{1}{3}, \frac{2}{3}$ correspondingly, so the solutions are

$x = 0, \frac{4}{3}, \frac{8}{3}$.

Sum of all the roots is $0 + \frac{4}{3} + \frac{8}{3} = \frac{12}{3} = 4$

(Passage Type)

(For 20-21)

In order to solve this type of questions 1st plot the graph which is explained below-

Graph $y=\left|\frac{|x-2|}{2}-3\right|-4$

Step 1:- Plot $y=\frac{|x-2|}{2}-3$ as we did in previous graphs by 1st plotting $y=\frac{|x|}{2}$ and then shift of origin with new origin at (2, –3)

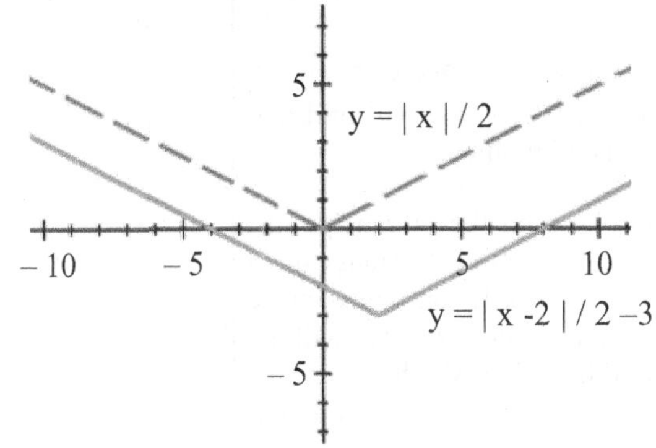

Step 2:- Plot $y=\frac{|x-2|}{2}-3$ now in order to plot $y=\left|\frac{|x-2|}{2}-3\right|$ we change negative values of y to +ve, as example –1 will become +1 and so on, as shown in the figure.

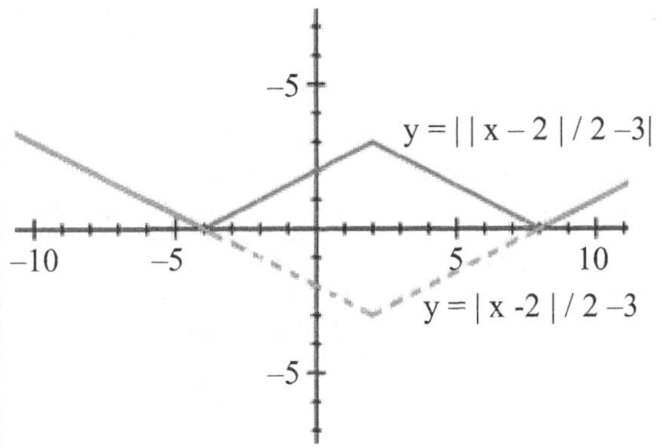

Step 3:- In order to plot $y=\left|\frac{|x-2|}{2}-3\right|-4$ we will 1st plot $y=\left|\frac{|x-2|}{2}-3\right|$ as we did in step 2, If we pull this graph down by 4 unit then we get $y=\left|\frac{|x-2|}{2}-3\right|-4$, as shown in the figure.

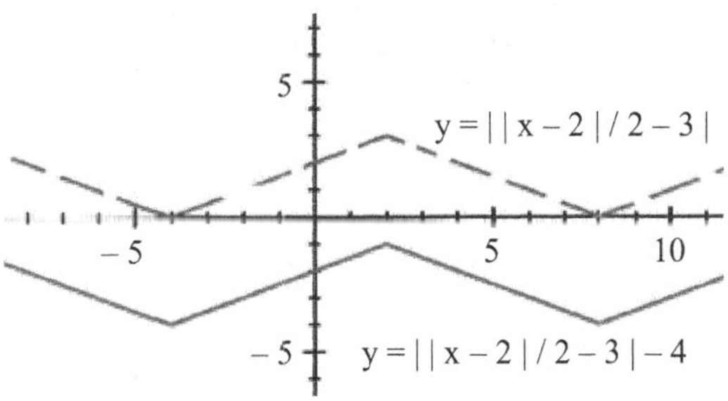

20. (a) The graph will intersect the x-axis twice so we will have 2 roots.
21. (d) From the graph we can say that f(x) = k will intersect the graph 4 times if k lies between local maxima and local minima which is –1 and –4

(For 22-23)

22. (c) Consider the graph of each function separately and then find the minimum value of the function-

Graph of $y = \|x + 2\| + \|x - 2\| + \|x - 5\| + \|x - 8\| + \|x - 10\|$	
Plot the graph of $y = \|x+2\| + \|x-2\| + \|x-5\| + \|x-8\| + \|x-10\|$, we will find that value of y is minimum at x = 5 **Alternately:-** Find the critical points, these are –2, 2, 5, 8 and 10 arrange them in the increasing order on x-axis, since we have 5 critical points so we will get only 1 middle value which is x = 5 so the function will attain its minimum value at x = 5. Required minimum value is 18	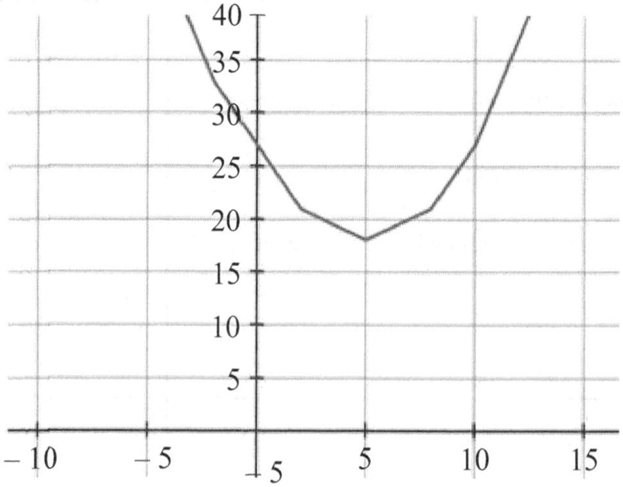
All the functions in the form of $y = \|x-a\| + \|x-b\| + \|x-c\| + \|x-d\| + \|x-e\|$ will have similar graph with minimum value of y at x = c here we have assumed that a < b < c < d < e.	

23. (b) Lets see the graph

Find the minimum value of $\|x + 2\| + 2\|x - 2\|$	
Graph of $y = \|x + 2\| + 2\|x - 2\|$	
1st question is how many critical points we have here 2 or 3? Answer is we have 3 critical points and these points are –2, 2, 2 (point 2 taken twice) Out of these three points middle one is 2 so the function will attain its minimum value at x = 2 Required minimum value is 4.	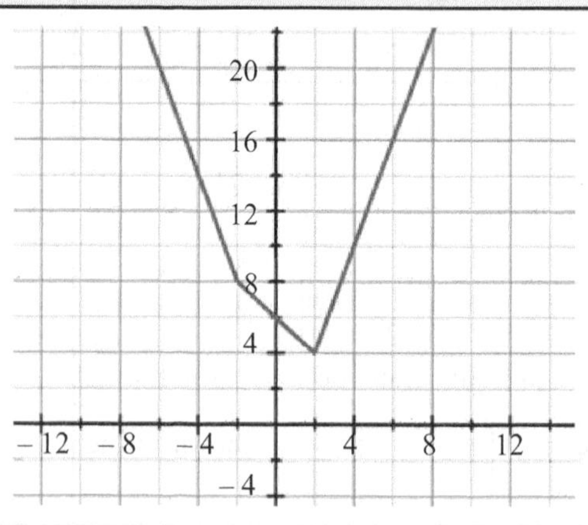

(For 24-25)

Plot the graph of $y = \|\|x\|^2 - 4\|x\| - 5\|$	
Step 1:- 1st plot $y = \|x\|^2 - 4\|x\| - 5$ As we did in previous graphs It has two roots –5 and 5. Local maxima is at x = 0 which is –5 Local minima is at x = –2 and at x= 2 which is –9	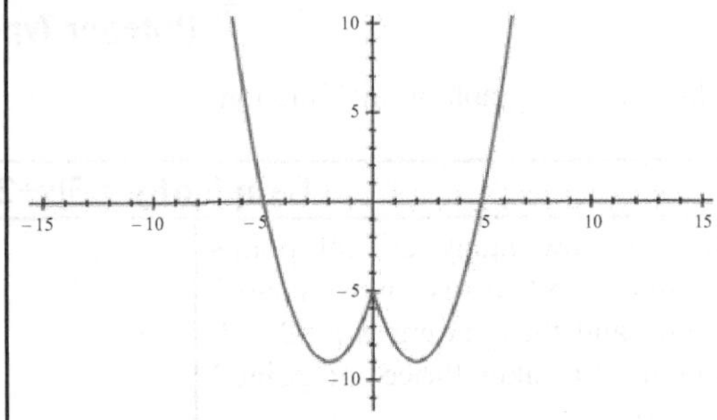
Step 2:- Now in order to plot $y = \|\|x\|^2 - 4\|x\| - 5$ do the reflection with respect to x-axis of the graph. $y = \|x\|^2 - 4\|x\| - 5$ as shown in the graph.	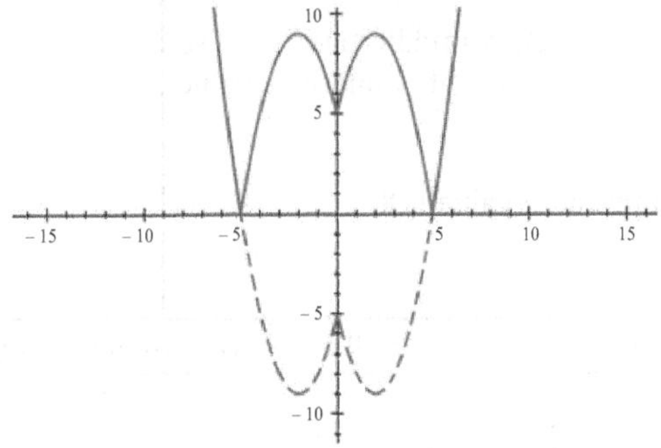

24. (a) For maximum number of solutions k should be in between (5, 9)
25. (c) The line y = 4 will intersect the graph 4 times.

Consider the graph of the function

Graph of $[x^3]$	
When $x \in [0, 1)$ then $x^3 \in [0, 1)$ and $[x^3] = 0$ When $x \in \left[1, 2^{\frac{1}{3}}\right)$ then $x^3 \in [1, 2)$ and $[x^3] = 1$ When $x \in \left[2^{\frac{1}{3}}, 3^{\frac{1}{3}}\right)$ then $x^3 \in [2, 3)$ and $[x^3] = 2$ When $x \in \left[3^{\frac{1}{3}}, 4^{\frac{1}{3}}\right)$ then $x^3 \in [3, 4)$ and $[x^3] = 3$ and so on When $x \in (-1, 0]$, then $x^3 \in [-1, 0)$ and $[x^3] = -1$ When $x \in \left[-2^{\frac{1}{3}}, -1\right)$, then $x^3 \in [-2, -1)$ and $[x^3]$ = –2 and so on	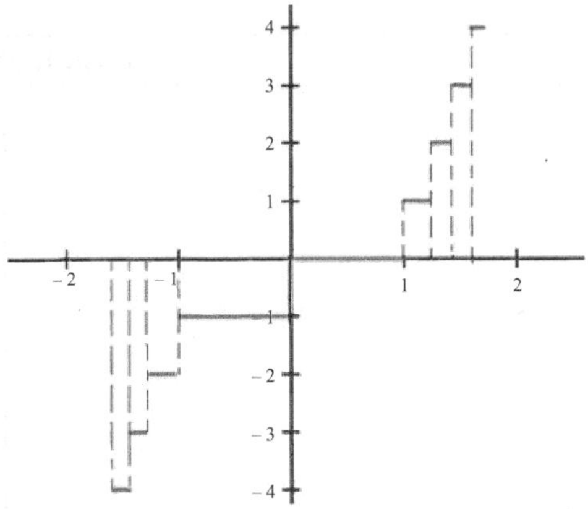

26. (c) When x > 0 then function is discontinuous at $1, 2^{\frac{1}{3}}, 3^{\frac{1}{3}}, 4^{\frac{1}{3}}, 5^{\frac{1}{3}}, 6^{\frac{1}{3}}$, and $7^{\frac{1}{3}}$, total 7 points.
Similarly 7 points when x < 0 and one more point at origin, so total number of points is 15

27. (b) Similar to the previous question when $x > 0$ then number of points is $10^3 - 1 = 999$ points, same number of points when $x < 0$ so total number of points is $999 + 999 + 1 = 1999$

(Integer Type)

28. (8) Lets see the graph of this function

<table>
<tr><th colspan="2">Graph of $y = 3|x+2| + 2|x-2|$</th></tr>
<tr><td>1st question is how many critical points we have here 2 or 5? Answer is we have 5 critical points and these points are –2, –2, –2, 2, 2 (point –2 is taken thrice and point 2 is taken twice)

Out of these 5 points middle one is –2 so the function will attain its minimum value at $x = -2$

Required minimum value is 8</td><td>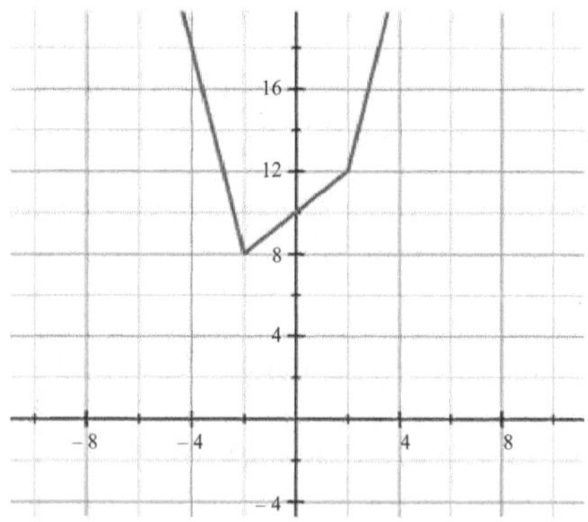
</td></tr>
<tr><td colspan="2">In general $a|x - p| + b|x - q|$ will attain its minimum value at $x = p$ if $a > b$ otherwise at $x = q$</td></tr>
</table>

Required minimum value is 8.

29. (6) The given function will attain a minimum value at $x = 6$
30. (4) The given function will attain a minimum value at $x = 4$
31. (7) Total number of critical points is $1 + 2 + 3 + ... + 10 = 55$ and middle one is 27 which is achieved at $x = 7$
32. (3) The function will attain a minimum value of 3 (at $x = 10$)
33. (2) Lets see the graph of the function

<table>
<tr><th colspan="2">Graph of $|x - 2| + |y| = 10$</th></tr>
<tr><td>Step 1:- 1st plot $|x| + |y| = 10$</td><td>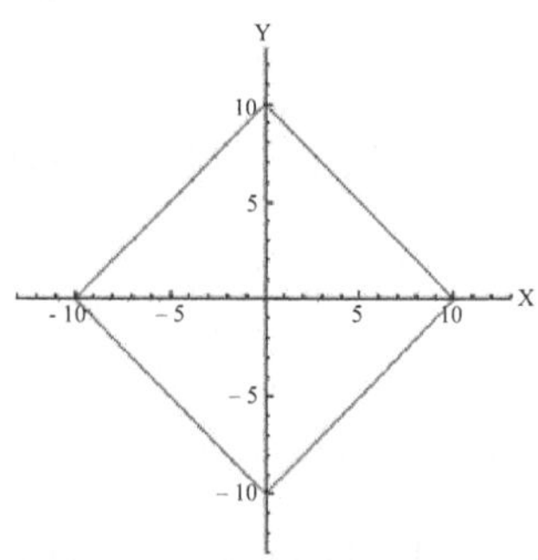
</td></tr>
</table>

Step 2:- Shift the graph of $|x| + |y| = 10$ to 2 unit right means to a new origin (2, 0) as shown in the graph.

So area of the graph $|x| + |y| = 10$ and area of the graph $|x - 2| + |y| = 10$ is same

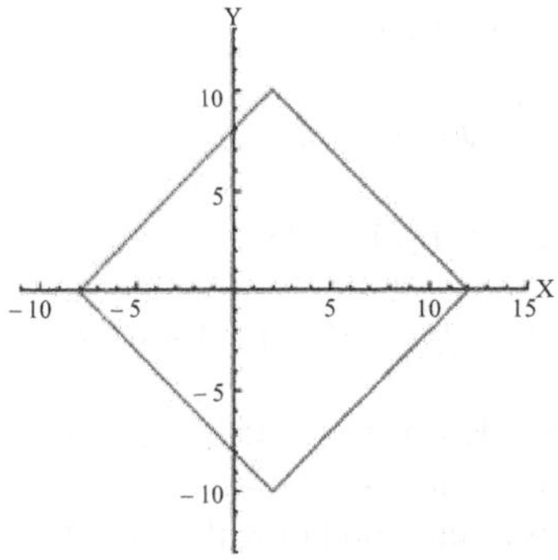

Area of the region bounded by $|x - p| + |y - q| = k$ is given by (Area is independent of p and q)

Area of the region bounded by $a|x - p| + b|y - q| = k$ is given by $\dfrac{2(k^2)}{ab}$ (Area is independent of p and q)

Required area = $4 \times (10 \times 10/2) = 200$

And required answer is $2 - 0 = 0$

34. (3) We already discussed this in theory part. Area is 96, so required answer is $9 - 6 = 3$.

35. (1) Consider the graph of these two functions

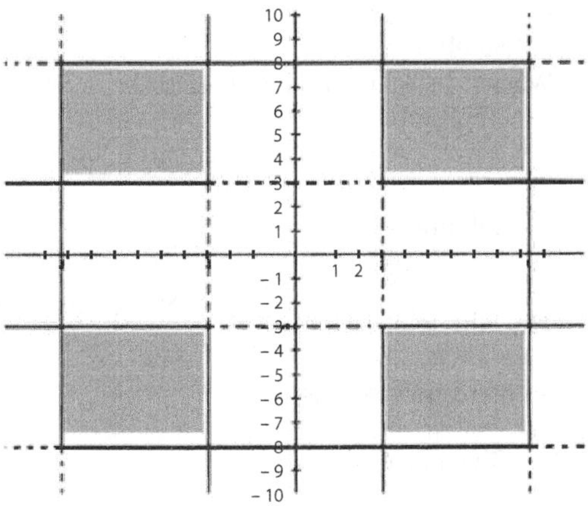

In every quadrant it will form a square, area of such a square is $5 \times 5 = 25$

Total area = $25 \times 4 = 100$

Sum of digits $1 + 0 + 0 = 1$

36. (4) Since line $x + 2y = 0$ and $2x - y = 0$ are perpendicular to each other similarly $x + 3y$ and $3x - y$ are perpendicular to each other so bounded area by $|x + 2y| + |2x - y| = p$ and by $|x + 3y| + |3x - y| = 2p$ are similar so their ratio of area is 1 : 4
So K = 4

37. (2) Let $[x] = n$, $t = \{x\}$, then $n^2 = t(n + t) \geq 0$. $0 \leq t < 1$ implies $n \geq 0$. Since $n^2 = t(n + t)$, so $n = 0$ or 1. If $n = 0$, then $t = 0$, i.e. $x = 0$. If $n = 1$, then

$1 = t(1 + t)$,

$t^2 + t - 1 = 0$,

$t = \dfrac{-1+\sqrt{5}}{2}$ $\therefore$ $x = n + t = \dfrac{1+\sqrt{5}}{2}$.

Thus, $x = 0$ or $x = \dfrac{1+\sqrt{5}}{2}$

(More Than One Correct MCQs)

38. (a, c) Equation has solution only if $a^2 - 2a - 3 \geq 0$

$\Rightarrow a \leq -1$ or $a \geq 3$

Then equation becomes $x^2 - 1 = \pm(a^2 - 2a - 3)$

$\Rightarrow x^2 = a^2 - 2a - 2$ or $x^2 = -a^2 + 2a + 4$

For real solution $a^2 - 2a - 2 \geq 0$

or $-a^2 + 2a + 4 \geq 0$

$\Rightarrow a \leq 1 - \sqrt{3}$ or $a \geq 1 + \sqrt{3}$

or $1 - \sqrt{5} \leq a \leq 1 + \sqrt{5}$

$\therefore a \in [1 - \sqrt{5}, -1] \cup [3, 1 + \sqrt{5}]$

39. (b, d) $||x - 1| + a| = 4 \quad \Rightarrow |x - 1| + a = \pm 4$
$\Rightarrow |x - 1| = -a \pm 4$
The above equation holds if $-a + 4 \geq 0$
or $-a - 4 \geq 0$
$\Rightarrow a \leq 4$ or $a \leq -4 \quad \Rightarrow a \in (-\infty, 4]$

40. (a, d) $x^2 - 4|x| + 3 = |k - 1|$ has exactly four distinct roots if $|k - 1| < 3 \Rightarrow -2 < k < 4$

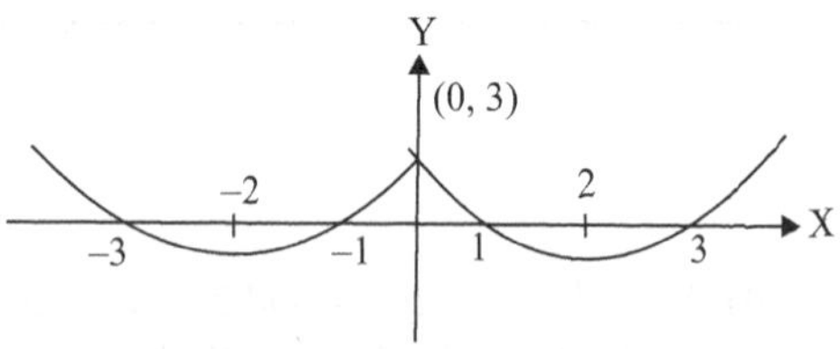

41. (a, b, d) $x^2 - 4|x| + 3 = k - 1$.
Clearly for four distinct roots $-1 < k - 1 < 3$
$\Rightarrow 0 < k < 4$

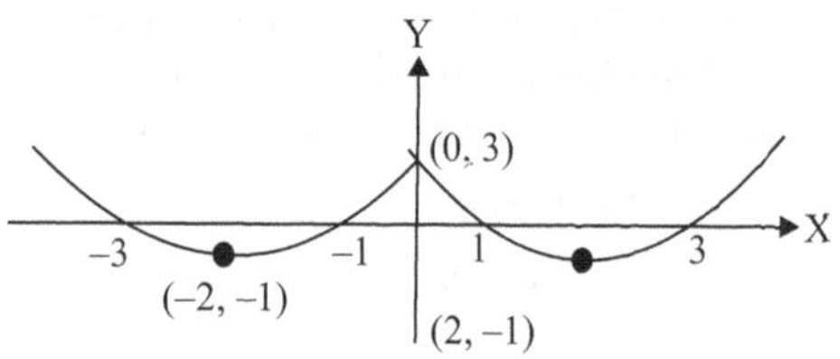

42. (a, b, c) Number of terms is 73
Since $542 = 7 \times 73 + 31$
$\Rightarrow [x + 0.60] = 7$ and $[x = 0.61] = 8$
so $x = 7.39$

Concept Eliminator (CE)

1. (d) To reduce the number of variables in the given equation, $x \geq [x]$ gives an inequality in x.
$0 = 4x^2 - 40[x] + 51 \geq 4x^2 - 40x + 51$
$= (2x - 3)(2x - 17)$, so $\frac{3}{2} \leq x \leq \frac{17}{2}$. It implies that [x] may
be 1, 2, …., 8.

When [x] = 1, the equation becomes $4x^2 + 11 = 0$, no real solution for x.
When [x] = 2, the equation becomes $4x^2 - 29 = 0$, $x = \frac{\sqrt{29}}{2}$, which has no integer part 2, hence it is a solution.

When [x] = 3, the equation becomes $4x^2 - 69 = 0$, $x = \frac{\sqrt{69}}{2}$, which has no integer part 3, a contradiction, hence no solution for x.
When [x] = 4, the equation becomes $4x^2 - 109 = 0$, $x = \frac{\sqrt{109}}{2} > 5$, a contradiction, hence no solution.
When [x] = 5, the equation becomes $4x^2 - 149 = 0$, $x = \frac{\sqrt{149}}{2} > 6$, hence no solution.
When [x] = 6, the equation becomes $4x^2 - 189 = 0$, $x = \frac{\sqrt{189}}{2}$, which has integer part 6, hence it is a solution.
When [x] = 7, the equation becomes $4x^2 - 229 = 0$, $x = \frac{\sqrt{229}}{2}$, which has integer part 7, hence it is a solution.
When [x] = 8, the equation becomes $4x^2 - 269 = 0$, $x = \frac{\sqrt{269}}{2}$, which has integer part 8, hence it is a solution.

Thus,the solutions are $\frac{\sqrt{29}}{2}, \frac{\sqrt{189}}{2}, \frac{\sqrt{229}}{2} \frac{\sqrt{269}}{2}$

So there are 4 solutions

2. (b) Given that

$x + [y] + \{z\} = 13.2$ …….(i)

$[x] + \{y\} + z = 14.3$ ……..(ii)

$\{x\} + y + [z] = 15.1$………(iii)

Adding the three equations we get $2(x + y + z) = 42.6$,

or $x + y + z = 21.3$ …………(iv)

Now given that $x + [y] + \{z\} = 13.2$

So $\{y\} + |z| = 8.1$, therefore $[z] = 8$ and $\{y\} = 0.1$.

Then from equation (ii) we get $[x] + z = 14.2$, so $\{z\} = 0.2$ and $z = 8.2$ which gives $[x] = 6$.

From equation (i) $x + [y] = 13$, so x is an integer, i.e $x = [x] = 6$.

And $y = 15.1 - 0 - 8 = 7.1$

So $xyz = (6)(7.1)(8.2) = 349.32$

And $x + y + z = 6 + 7.1 + 8.2 = 21.3$

And $x + y - z = 6 + 7.1 - 8.2 = 4.9$

3. (a) The given equation indicates that any solution x must be an integer. Let $x = 6q + r$, where $r = 0, 1, 2, 3, 4, 5$ and q is an integer. Then the given equation becomes

$q + \left[\frac{r}{2}\right] + \left[\frac{2r}{3}\right] = r$. Now we have following cases-

Case (i) $r = 0$ gives $q = 0$, so $x = 0$ is a solution.

Case (ii) $r = 1$ gives $q = 1$, so $x = 7$ is a solution.

Case (iii) $r = 2$ gives $q = 0$, so $x = 2$ is a solution.

Case (iv) $r = 3$ gives $q = 0$, so $x = 3$ is a solution.

Case (v) $r = 4$ gives $q = 0$, so $x = 4$ is a solution.

Case (vi) $r = 5$ gives $q = 0$, so $x = 5$ is a solution.

Thus, there are a total of 6 real solutions.

Sum of the solutions of the equation is $0+2+3+4+5+7= 21$

4. (a) Since the left hand side of the equation is an integer, therefore $\{x\} - 1$ is an integer. Then $0 \le \{x\} < 1$ implies $\{x\} = 0$, i.e $x = [x]$.

Thus, $x^3 + x^2 + x = -1$,

$(x^3 + x^2) + (x + 1) = 0$,

$(x + 1)(x^2 + 1) = 0$,

Therefore $x = -1$.

5. (c) Assume that $n = 10^{100}$. Then

$$\frac{10^{20000}}{10^{100}+3} = \frac{n^{200}}{n+3} = \frac{(n^2)^{100} - (3^2)^{100}}{n+3} + \frac{9^{100}}{n+3}$$

Since $9^{100} < n$ so $9^{100} < n + 3$

And $(n^2)^{100} - (3^2)^{100}$ is divisible by $n + 3$ so required quotient is given by $(n^2)^{100} - (3^2)^{100}$

Let $(n^2)^{100} - (3^2)^{100} = (n^2 - 3^2)^k$

$$\frac{(n^2)^{100} - (3^2)^{100}}{n+3} = \frac{n^{200} - 9^{100}}{n+3} = \frac{10^{20000} - 81^{50}}{10^{100}+3}$$

Since the units digit of $10^{20000} - 81^{50}$ is 9 and the units digit of $10^{100} + 3$ is 3, the units digit of the quotient must be 3.

6. (c) We know that If each of two positive real numbers x, y is not an integer but x + y is an integer, then $[x] + [y] = x + y - 1$.

Now $\frac{305k}{503} + \frac{305(503-k)}{503} = 305$ for $1 \le k \le 502$,

$$S = \left[\frac{305}{503}\right] + \left[\frac{305.2}{503}\right] + \ldots + \left[\frac{305.502}{503}\right]$$

$$= \left(\left[\frac{305.1}{503}\right] + \left[\frac{305.502}{503}\right]\right) + \ldots +$$

$$\left(\left[\frac{305.251}{503}\right] + \left[\frac{305.252}{503}\right]\right) = (304)(251) = 76304.$$

7. (b) Let $t = \left[\sqrt{n}\right]$ then $1 \le t \le 10^2$, and $t^2 \le n < (t + 1)^2$, or equivalently, $t^2 \le n \le t^2 + 2t$, i.e $n = t^2 + r$ where r is a non-negative integer with $0 \le r \le 2t$.

Now as per the given condition $(t^2 + r)$ is divisible by 't' which implies r can take 0, t and 2t for $1 \le t \le 99$ but when $t = 100$ then 'r' has only one value which is $r = 0$

Thus, total number of integers are $3 \times 99 + 1 = 298$

8. (d) Let $n = 2x - \frac{1}{2}$, then $n \in Z$ and $x = \frac{n}{2} + \frac{1}{4}$

$$3x + 1 = \frac{3n}{2} + \frac{3}{4} + 1 = \frac{3n}{2} + \frac{7}{4}$$

From the given condition $\left[\frac{3n}{2} + \frac{7}{4}\right] = n$

Clearly n can't be positive, by slight observation we find that $n = -3$ and -2, in remaining all the difference between LHS and RHS is increasing. Therefore $n = -3$ or $n = -2$. When $n = -3$,

then $x = -\frac{3}{2} + \frac{1}{4} = -\frac{5}{4}$;

when $n = -2$, then $x = -1 + \frac{1}{4} = -\frac{3}{4}$. Thus, the sum of roots is -2.

9. (a) It is obvious that $x = n$ is a required root. Let $1 \le x < n$ be another root of the given equation.

Let $I = [x]$, $F = \{x\}$, then $x = I + F$

the equation becomes $(I + F)^2 - [(I + F)^2] = F^2$,

$2IF - [2IF + F^2] = 0$,

therefore 2 IF is an non - negative integer, i.e., F = 0,

$\frac{1}{2I}, \frac{2}{2I}, \ldots, \frac{2I-1}{2I}$ for I = 1, 2, … , n–1.

Thus , the number of required roots is

$2[1 + 2 + 3 + \ldots + (n-1)] + 1$

$= (n-1)\, n + 1$

$= n^2 - n + 1.$

$S_{10} = 91$

10. (b) Let x be a solution with $[x] = I$.
Then the we can write the given equation as $x^2 + 7 = 8I$
Since LHS is positive so $I > 0$.
Now $I \le x < I + 1$ yields $I^2 + 7 \le x^2 + 7 = 8I < (I + 1)^2 + 7 = I^2 + 2I + 8$,
On simplification we get two relations
$\therefore I^2 - 8I + 7 \le 0$ and $I^2 - 6I + 8 > 0$.
So, $1 \le I < 2$ or $4 \le I < 7$, i.e possible values of I = 1, 5 , 6, 7.
Correspondingly, $x^2 + 7 = 8, 40, 48, 56$,
so $x = 1, \sqrt{33}, \sqrt{41}$ 7 since $x > 0$
So number of non integral solution of x is 2

11. (c) Since $(k + 1)^2 - k^2 = 2k + 1$ so

$\left[\frac{(k+1)^2}{1980}\right] > \left[\frac{k^2}{1980}\right]$ when

$2k + 1 > 1980$, i.e $k \ge 990$.

Therefore $\left[\frac{k^2}{1980}\right]$, , k = 990, 991, … , 1980

are 991 different values. For the sequence
Now consider the term ,

$\left[\frac{1^2}{1980}\right], \left[\frac{2^2}{1980}\right], \left[\frac{3^2}{1980}\right], \ldots, \left[\frac{990^2}{1980}\right],$

Since $\left[\frac{990^2}{1980}\right] = 495$

And any two consecutive terms have difference 0 or 1 only, the sequence must take

495 + 1 = 496 different values.

So total number of distinct values is

496 + (991 – 1) = 1486

12. (d) By using $x = [x] + \{x\}$, the given equation becomes $[x]\{x\} + [x] - \{x\} = 10$, so $([x] - 1)(\{x\} + 1) = 9$.

Since $[x] - 1$ is an integer, so $\{x\} + 1$ is a rational number. Let $\{x\} = p/q$ with $0 \le p < q$, then $([x] - 1)(p + q) = 9q$.

When p = 0, then $\{x\} = 0$, so $x = [x] = 10$.

When p > 0, then we have following two cases

Case (i) When p + q = 3, then p = 1, q = 2,

so $\{x\} = \frac{1}{2}$, $[x]=7$, $x = 7\frac{1}{2}$.

Case (ii) When p + q = 9, then there are three possible cases :

$p = 1, q = 8 \rightarrow \{x\} = \frac{1}{8}, [x] = 9, x = 9\frac{1}{8}$

$p = 2, q = 7 \rightarrow \{x\} = \frac{2}{7}, [x] = 8, x = 8\frac{2}{7}$

$p = 4, q = 5 \rightarrow \{x\} = \frac{4}{5}, [x] = 6, x = \frac{4}{5}$

Thus, there are five solutions for

$x : 10\ 7\frac{1}{2}, 9\frac{1}{8}, 8\frac{2}{7}, 6\frac{4}{5}$.

13. (d) Let the sum be N and $x = [x] + \{x\} = I + f$ where $I \in Z_0^+$ and $0 \le f < 1$. Then

$N = 20I + [2f] + [4f] + [6f]\ [8f] \le 1000$,

$\therefore 0 \le I \le 50$.

For any fixed $0 \le I \le 49$,

Consider [8f] its value will change when

$f = \left\{0, \frac{1}{8}, \frac{2}{8}, \frac{3}{8}, \frac{4}{8}, \frac{5}{8}, \frac{6}{8}, \frac{7}{8},\right\}$

$= \left\{0, \frac{1}{8}, \frac{1}{4}, \frac{3}{8}, \frac{1}{2}, \frac{5}{8}, \frac{3}{4}, \frac{7}{8},\right\}$

Consider [6f] its value will change when

$f = \left\{0, \frac{1}{6}, \frac{2}{6}, \frac{3}{6}, \frac{4}{6}, \frac{5}{6}\right\} = \left\{0, \frac{1}{6}, \frac{1}{3}, \frac{1}{2}, \frac{2}{3}, \frac{5}{6}\right\}$

Then [4f] its value will change when

$f = \left\{0, \frac{1}{4}, \frac{2}{4}, \frac{3}{4}\right\} = \left\{0, \frac{1}{4}, \frac{1}{2}, \frac{3}{4}\right\}$

And [2f] its value will change when $f = \left\{0, \frac{1}{2}\right\}$

Combining all the above values we get the universal set of

$$f=\left\{0,\frac{1}{8},\frac{1}{4},\frac{3}{8},\frac{1}{2},\frac{5}{8},\frac{3}{4},\frac{7}{8},\frac{1}{6},\frac{1}{3},\frac{2}{3},\frac{5}{6}\right\}$$

Therefore there are totally 12 different values of N for each I = 1, … , 49, but there are total of 11 different values of N when I = 0 (since in this case f cannot take value 0), and I = 50, t = 0 means N = 1000 is also a required value.

Therefore required number of values is $50 \times 12 = 600$ different values not greater than 1000.

14. (b) Since $3.14 < \pi < 3.142$, $1.57 < \frac{\pi}{2} < 1.571$

Thus, $\left[\frac{\pi}{2}+\frac{n}{100}\right] = 1$ for n = 0, 1, 2, ... , 42

Thus, largest natural number n for which E < 43 is 41.

15. (a) $\frac{10^n}{x}-1<\left[\frac{10^n}{x}\right]<\frac{10^n}{x}$

implies $\frac{10^n}{x}-1<1989\le\frac{10^n}{x}$

i.e., $\frac{10^n}{1990}<x\le\frac{10^n}{1989}$, or $(10^n)(0.0005025126)$

$<x\le(10^n)(0.0005027652)$

From observation we find that only for $n \ge 7$ the difference of the two decimals is greater than 1,

So when n = 7 then we find that

$5025 < x \le 5027$, and x = 5026 or 5027.

16. (c) We have following cases

Case (i) $xy > 0, 3x^2 - xy - 2y^2 = -7$

or $(3x + 2y)(x - y) = -7$

x and y being integers we can take

$3x + 2y = 7$ and $x - y = -1$ which give $x = 1, y = 2$

If x and y are changed to $-x$, $-y$, the equation remains the same so that $x = -1$

$y = -2$ is also a solution set

$x + y = 7$ do not give integral solutions

Case (ii) $xy = 0$ will not lead to integral solutions

Case (iii) $xy < 0$. The equations is

$3x^2 + xy - 2y^2 = -7$

or $(3x - 2y)(x + y) = -7$, $3x - 2y = -7$ and $x + y = 1$ leads to $x = -1, y = 2$

As before $x = 1$ $y = -2$ is also a solutions set

So total 4 roots exists

17. (c) We have following cases

If $\left[\frac{x}{99}\right]=\left[\frac{x}{101}\right]=0$ it is possible if and only if $x \in \{1, 2, 3, ..., 98\}$, and there are 98 such numbers.

If $\left[\frac{x}{99}\right]=\left[\frac{x}{101}\right]=1$, then x should lie in the set {101, 102, ..., 197}, there are 97 numbers.

In general , if $\left[\frac{x}{99}\right]=\left[\frac{x}{101}\right]=k$ where $k \ge 1$,

then x must be in the set $\{101k, 101k + 1, ..., 99(k + 1) - 1\}$, and there are $99 - 2k$ such numbers.

Now the condition for k is $99(k + 1) - 1 \ge 101\ k$ and this requirement is met only if $k \le 90$. Thus the total number of positive integers x for which $\left[\frac{x}{99}\right]=\left[\frac{x}{101}\right]$ is given by

$98 + \sum_{k=1}^{49}(99-2k) = 2499.$

Important Points:

For any $m \ge 2$ the number of positive integers x such that $\left[\frac{x}{m-1}\right]=\left[\frac{x}{m+1}\right]$ is $\frac{m^2-4}{4}$ if m is even

and $\frac{m^2-5}{4}$ if m is odd

18. (a) We know that have

$$({}^{n}_{7}c) = \frac{n(n-1)(n-2)...(n-6)}{7!}.$$

In numerator we have 7 consecutive integers so out of them one must be divisible by 7 and the other six factors leave the remainders 1, 2, 3, 4, 5, 6 is some order when divided by 7.

Hence the numerator may be written as $7k.(7k_1 + 1).(7k_2 + 2)...(7k_6 + 6)$

Also we conclude that $\left[\frac{n}{7}\right] = k$, as one factor is a multiple of 7.

$$({}^{n}_{r}c) - \left[\frac{n}{7}\right] = 7k.\frac{(7k_1+1).(7k_2+2)...(7k_6+6)}{7!} - k$$

$$= k\left[\frac{(7k_1+1)...(7k_6+6)-6!}{6!}\right]$$

$$= k\frac{[7t+6!-6!]}{6!}$$

$$= \frac{7tk}{6!} = 7\left(\frac{tk}{6!}\right)$$

Which is always divisible by 7.

CHAPTER 4

QUADRATIC EQUATION

PART-A

4.1 ROOTS AND NATURE OF ROOTS

4.1.1 Introduction

An equation of the form $ax^2 + bx + c = 0$ where $a \neq 0$, is called a quadratic equation. The numbers a, b, c are called the coefficients of the equation.

Roots of the quadratic equation: The value/ values of x that satisfy the given equation is called roots of the quadratic equation.

Example:

(i) $x^2 + 3x - 2 = 0$ is a quadratic equation, where coefficients a, b, and c are 1, 3 and –2 respectively.

(ii) $x^2 - x^{1/2} + 1 = 0$ is not a quadratic equation.

4.1.2 Quadratic Expression

An expression of the form $ax^2 + bx + c$, where a, b, c are some numbers and $a \neq 0$ is called a quadratic expression. The graph of quadratic function is parabolic in nature and it will cut the X-axis maximum 2 times.

4.1.3 Polynomial

An expression of the form $a_0x^n + a_1x^{n-1} + a_2x^{n-2} + \ldots a_{n-1}\, x + a_n$, where $a_0, a_1, a_2 \ldots a_n$ are constants $(a_0 \neq 0)$ is called a polynomial in x of degree n.

If $n = 2$ then function is called quadratic

The equation $a_0x^n + a_1x^{n-1} + a_2x^{n-2} + \ldots + a_n = 0$ is called a polynomial equation in of degree n.

Example:

(i) $3x^2 + 7x - 9$ is a polynomial in x of degree 2 (or quadratic function)

(ii) $2x^{10} + 2x^4 - 2$ is a polynomial in x of degree 10

(iii) $5x^0 = 5$ is a polynomial in x of degree 0 (this is a special case, a constant is also called a polynomial of degree zero)

(iv) $x + \frac{1}{x}$ is not a polynomial in x

A polynomial equation of degree has n and only n roots (equal roots being counted as many times as they occur). If number of roots is more than 'n' then polynomial is an identity.

4.1.4 Identity

An identity is defined as the statement of equality between two expressions in x that are equal for all values of x, for which expressions on the two sides are defined.

Alternatively, $f(x) = k$ is said to be an identity in x if it is satisfied by all values of x in the domain of $f(x)$.

Difference between Equation and Identity:

An identity is satisfied by all values of x for which expression is defined, whereas an equation in x is satisfied by some particular values of x.

Example:

(i) $(x-1)^2 = x^2 - 2x + 1$ is an identity in x because it is satisfied for all the values of x

A polynomial of nth degree cannot have more than n distinct roots. In the above relation the highest power of x is 2 but the relation is satisfied by more than two different values of $x(x = 0, x = 1$ and $x = -1$ satisfy the relation). Hence, it is an identity.

(ii) $2x^{90} - x^7 + 5$ is an equation in x because it is not satisfied by all the values of x e.g , $x = 0, 1$ etc

Illustration 1. If $(x^3 - 1)x^2 + (a + b - 1)x + (a + b + c) = 0$ has three roots then find a, b and c

Solution: If a quadratic equation has more than 2 roots then it must be identity

We know that if $ax^2 + bx + c = 0$ is an identity then $a = b = c = 0$

So $\quad a^3 - 1 = 0$ or $a = 1$,

Then $\quad a + b - 1 = 0$ so $b = 0$

And $\quad a + b + c = 0$ so $c = -1$

Important Points:

1. A polynomial of nth degree cannot have more than n distinct roots.
2. If a polynomial of nth degree has more than n distinct roots then polynomial is an identity.
3. The symbol '≡' is used to denote an identity where as '=' is used for an equation.
4. In an identity in x, the coefficients of similar powers of x on the two sides are equal. Thus, if $ax^3 + bx^2 + cx + d = 2x^3 - 6$ is an identity in x, then $a = 2$, $b = 0$, $c = 0$ and $d = -6$.

4.1.5 Identical Equations

Two equations are said to be identical if they have the same roots.

Example :

$x^2 - 5x + 6 = 0$ and $2x^2 - 10x + 12 = 0$ are identical equations because they have the same roots 2 and 3.

Some important points:

(i) Two equations are identical if and only if the coefficients of similar powers of x in the two equations are proportional. Thus if $ax^2 + bx + c = 0$ and $a_1x^2 + b_1x + c_1 = 0$ are identical equations, then $\frac{a}{a_1} = \frac{b}{b_1} = \frac{c}{c_1}$

(ii) Roots of an equation remains unchanged if it is multiplied or divided by a non-zero number. Thus, the equations $ax^2 + bx + c = 0$ and $m(ax^2 + bx + c) = 0$, where m is a number, are identical.

4.1.6 Rational Expression

An expression of the form $\frac{P(x)}{Q(x)}$, where $P(x)$ and $Q(x)$ are polynomials in x and $Q(x) \neq 0$ is called a rational expression in x. In the particular case when $Q(x)$ is a non-zero constant, $\frac{P(x)}{Q(x)}$, reduces to a polynomial in x. Thus every polynomial is a rational expression but the converse is not true.

Example :

(i) $\frac{x^2 - 5x + 4}{x^3 - 4x - 2}; x^3 - 4x - 2 \neq 0$

(ii) $x^2 - 5x + 4$

(iii) $\frac{8}{x-6}; x \neq 6$

(iv) $x + \frac{1}{x}$ *i.e.* $\frac{x^2 + 1}{x}; x \neq 0$

4.1.7 One Important Theorem

Every Quadratic equation has two and only two roots

First part: Prove that every quadratic equation has two roots.

Proof: Let us consider a quadratic equation $ax^2+bx+c=0$, where $a \neq 0$ ... (i)

Multiplying both sides of equation (i) by a, we get $a^2x^2 + abx + ac = 0$

or, $(ax)^2 + 2.ax.\frac{b}{2} + \left(\frac{b}{2}\right)^2 - \frac{b^2}{4} + ac = 0$

or $\left(ax + \frac{b}{2}\right)^2 = \frac{b^2}{4} - ac$

or, $\left(ax + \frac{b}{2}\right)^2 = \frac{b^2 - 4ac}{4}$ ... (ii)

Equations (i) and (ii) represents the same equation and will have the same roots,

From (ii), $ax + \frac{b}{2} = \pm\frac{\sqrt{b^2-4ac}}{2}$

or $ax = \frac{-b}{2} \pm \frac{\sqrt{b^2-4ac}}{2}$

$\therefore \quad x = \frac{-b \pm \sqrt{b^2-4ac}}{2a}$

Hence roots of equation $ax^2+bx+c=0$ are,

$\alpha = \frac{-b-\sqrt{b^2-4ac}}{2a}$ and $\beta = \frac{-b+\sqrt{b^2-4ac}}{2a}$

Hence proved that the quadratic equation $ax^2+bx+c=0$ has two roots α and β.

Second part: Prove that every quadratic equation has only two roots i.e. a quadratic equation cannot have more than two roots.

Proof: If possible let us suppose that the quadratic equation $ax^2+bx+c=0$ has three distinct roots α, β and γ.

Since α, β, γ are roots of the equation $ax^2+bx+c=0$

$\therefore \quad a\alpha^2 + b\alpha + c = 0$ ...(i)

$a\beta^2 + b\beta + c = 0$ ...(ii)

$a\gamma^2 + b\gamma + c = 0$ ...(iii)

Substracting (ii) from (i), we get

$a(\alpha^2-\beta^2)+b(\alpha-\beta)=0]$

or, $(\alpha-\beta)[a(\alpha+\beta)+b]=0$

But $\alpha \neq \beta \therefore a(\alpha+\beta)+b=0$ ... (iv)

Subtracting (iii) from (ii), we get

$a(\beta^2-\gamma^2)+b(\beta-\gamma)=0$

or, $(\beta-\gamma)[a(\beta+\gamma)+b]=0$

But $\beta \neq \gamma \therefore a(\beta+\gamma)+b=0$ ... (v)

Subtracting (v) from (iv), we get $a(\alpha-\gamma)=0$ or, $\alpha-\gamma=0$

$\therefore \alpha = \gamma$ which contradicts our supposition that α, β, γ are distinct. Hence our supposition is wrong, therefore, every quadratic equation has only two roots. Hence Proved.

4.1.8 Sum and Product of the Roots

If α and β be the roots of the equation $ax^2 + bx + c = 0$ then,

$$\alpha = \frac{-b-\sqrt{b^2-4ac}}{2a} \text{ and } \beta = \frac{-b+\sqrt{b^2-4ac}}{2a}$$

Sum of the roots,

$$\alpha+\beta = \frac{-b-\sqrt{b^2-4ac}}{2a} + \frac{-b+\sqrt{b^2-4ac}}{2a}$$

$$= \frac{-b-\sqrt{b^2-4ac}-b+\sqrt{b^2-4ac}}{2a} = \frac{-2b}{2a} = -\frac{b}{a}$$

Product of the roots,

$$\alpha\beta = \frac{\left(-b-\sqrt{b^2-4ac}\right)\left(-b+\sqrt{b^2-4ac}\right)}{4a^2}$$

$$= \frac{b^2-b^2+4ac}{4a^2} = \frac{4ac}{4a^2} = \frac{c}{a}$$

Thus sum of the roots,

$$\alpha+\beta = -\frac{\text{co-eff. of } x}{\text{co-eff of } x^2} = -\frac{b}{a}$$

And product of the roots, $\alpha\beta = \frac{\text{constant term}}{\text{coeff of } x^2} = \frac{c}{a}$.

Illustration 2. Find the sum and the product of the roots of the equation $x^2 - 7x + 30 = 0$ without finding its roots.

Solution: Sum of the roots 7

Product of the roots 30.

4.1.9 Nature of Roots of the Quadratic Equation

Let the quadratic equation be $ax^2 + bx + c = 0$ and $a \neq 0$ where *a*, *b*, *c* are real

The roots α and β of equation $ax^2 + bx + c = 0$ are

given by $\alpha = \dfrac{-b - \sqrt{b^2 - 4ac}}{2a}$

and $= \dfrac{-b + \sqrt{b^2 - 4ac}}{2a}$ ... (i)

Now there are four possibilities, Here $b^2 - 4ac = D$ is called discriminant.

Case I: When $D < 0$ *i.e,* $b^2 - 4ac < 0$

In this case $\sqrt{b^2 - 4ac}$ will be imaginary

Hence from (i), α and β will be both complex

Let $b^2 - 4ac = -k^2$ then roots of the equation will be $\alpha = a + ik$ and $b = a - ik$ or roots are conjugate to each other.

Case II: When $D = 0$ i.e., $b^2 - 4ac = 0$

In this case $\sqrt{b^2 - 4ac} = 0$

$\therefore$ From (i), $\alpha = -\dfrac{b}{2a}$ and $\beta = -\dfrac{b}{2a}$

hence both the roots α and β will be real and equal.

The expression $ax^2 + bx + c = a(x - \alpha)^2$ or expression is a perfect square.

Case III: When $D > 0$ i.e., $b^2 - 4ac > 0$ but not a perfect square

In this case $\sqrt{b^2 - 4ac}$ will be real.

Let $\sqrt{b^2 - 4ac} = \sqrt{D}$

Then from (i), $\alpha = \dfrac{-b - \sqrt{D}}{2a}$ and $\beta = \dfrac{-b + \sqrt{D}}{2a}$

Hence the roots α and β will be real and different (distinct).

In this case, if one root is $p + \sqrt{q}$ then other root is $p - \sqrt{q}$

If co-efficients are rational then irrational roots are conjugate of each other.

Case IV: When *D* i.e., $b^2 - 4ac$ is a perfect square (square of a rational number) and *a*, *b*, *c* are rationals

In this case $b^2 - 4ac$ = square of a rational number

$\therefore \sqrt{b^2 - 4ac} =$ a rational number

Let $\sqrt{b^2 - 4ac} = k$

Then from (i), $\alpha = \dfrac{-b - k}{2a}$ and $\beta = \dfrac{-b + k}{2a}$ where *a*, *b*, *c* are rationals.

Therefore, both the roots α and β will be rational.

If roots of equation $ax^2 + bx + c = 0$ be real and distinct and $\alpha < \beta$, then

$$\alpha = \frac{-b - \sqrt{b^2 - 4ac}}{2a} \text{ and } \beta = \frac{-b + \sqrt{b^2 - 4ac}}{2a}$$

To summarize the different cases:

Nature of the roots of the $\boldsymbol{ax^2 + bx + c = 0}$ means whether roots are real, real and rational, real and irrational, complex, or equal. Nature of the roots depends on the coefficients of the quadratic equation and its D (Discriminant)

Case	D	Nature of Roots: Here a, b & c are rational
Case (i)	$D < 0$	Complex roots
Case (ii)	$D = 0$	Real and equal roots
Case (iii)	$D > 0$	Real and distinct roots (Irrational roots if D is not a perfect square)
Case (iv)	$D > 0$	Real and distinct roots (rational roots if D is a perfect square)

4.1.10 Factorization Method to find the Roots of the Quadratic Equation

Given equation is $ax^2 + bx + c = 0$

Follow the following steps:

Step – 1: Make coefficient of x^2 as 1

Step – 2: Factorise the constant term.

Step – 3: Find the two factors whose summation is negative of coefficient of x.

Write $ax^2 + bx + c = a(x - \alpha)(x - \beta)$ then α and β are called roots of the equation.

Illustration 3. Find the roots of the equation $3x^2 - 15x + 18 = 0$

Solution: Since

$3x^2 - 15x + 18 = 3(x^2 - 5x + 6) = 3(x - 2)(x - 3)$

so the roots of the above equation is 2 and 3

Illustration 4. Find the roots of the equation $x^2 - 6x + 8 = 0$.

Solution: Discriminant of the equation and $D = \Delta$ $= b^2 - 4ac = 36 - 4(1)(8) = 4$

So the roots of the equation is given by

$$\frac{-(-6) \pm \sqrt{4}}{2} = \frac{6 \pm 2}{2} = 2 \text{ and } 4$$

Some more properties related to nature of roots:

1. If $a - 1$, b, $c \in I$ and the roots of the equation are rational numbers, then these roots must be integers.
2. If $a = c \Rightarrow$ Roots are reciprocal to each other.
3. If the equation is satisfied by more than two distinct complex numbers, then the equation becomes an identity, that is $a = b = c = 0$.
4. Conjugate Roots: The irrational and complex roots of a quadratic equation always occur in pairs. Therefore,
   (a) If one root be $\alpha + i\beta$, then other root will be $\alpha - i\beta$, Here a, b, c are real
   (b) If one root be $\alpha + \sqrt{\beta}$, then other root will be $\alpha - \sqrt{\beta}$, Here a, b, c are rational.
5. If D_1 and D_2 be the discriminants of two quadratic equations, then
   (a) If $D_1 + D_2 \geq 0$, then
   At least one of D_1 and $D_2 \geq 0$
   If $D_1 < 0$, then $D_2 > 0$
   (b) If $D_1 + D_2 < 0$, then
   At least one of D_1 and $D_2 < 0$
   If $D_1 > 0$, then $D_2 < 0$
6. For the quadratic equation $ax^2 + bx + c = 0$.
   (i) If $b = 0$ Roots are real/complex as $(c < 0/\, c > 0)$ and equal in magnitude but of opposite sign.
   (ii) If $c = 0$ One roots is zero, other is $-b/a$.
   (iii) If $b = c = 0$ Both roots are zero.
   (iv) If $\left.\begin{matrix} a > 0, & c < 0 \\ a < 0 & c > 0 \end{matrix}\right\} \Rightarrow$ Roots are of opposite sign.
   (v) If $\left.\begin{matrix} a > 0, & b > 0, & c > 0 \\ a < 0, & b < 0, & c < 0 \end{matrix}\right\} \Rightarrow$ Both roots are negative, provided $D \geq 0$
   (vi) If $\left.\begin{matrix} a > 0, & b < 0, & c > 0 \\ a < 0, & b > 0, & c < 0 \end{matrix}\right\} \Rightarrow$ Both roots are positive, provided $D \geq 0$
   (vii) If sign of a = sign of $b \neq$ sign of $c \Rightarrow$ Greater root in magnitude is negative.
   (viii) If sign of b = sign of $c \neq$ sign of $a \Rightarrow$ Greater root in magnitude is positive.
   (xi) If $a + b + c = 0 \Rightarrow$ One root is 1 and second root is c/a.
   (x) If $4a + 2b + c = 0$ then one root is 2
   (xi) If $4c + 2b + a = 0$ then one root is ½ etc.

Illustration 5. Find the roots of the equation $\sqrt{2}x^2 + \sqrt{3}x - 6\sqrt{2} = 0$

Solution: Discriminant of the equation and $D = \Delta$

$= b^2 - 4ac = 3 - 4(\sqrt{2})(-6\sqrt{2}) = 51$

So the roots of the equation is given by

$$\frac{-(\sqrt{3}) \pm \sqrt{51}}{2\sqrt{2}}$$

Illustration 6. Find the discriminant, roots and nature of roots of the following equations.

Sl. No	Equation	Discriminant	Roots	Nature of roots
(A)	$x^2 - 8x + 15 = 0$			
(B)	$x^2 + 12x + 32 = 0$			
(C)	$x^2 - 6x + 2 = 0$			
(D)	$x^2 - 6x - 2 = 0$			
(E)	$x^2 - 8x + 16 = 0$			
(F)	$x^2 - 6x + 8 = 0$			
(G)	$x^2 + 4 = 0$			

Solutions: Solution of these equations is as follows:

Sl. No	Equation	Discriminant	Roots	Nature of roots
(A)	$x^2 - 8x + 15 = 0$	4	3, 5	Real, rational, unequal,
(B)	$x^2 + 12 + 32 = 0$	16	–4, –8	Real, rational, unequal
(C)	$x^2 - 6x + 2 = 0$	28	$3 \pm \sqrt{7}$	Real, irrational, unequal
(D)	$x^2 - 6x - 2 = 0$	44	$3 \pm \sqrt{11}$	Real, irrational, unequal
(E)	$x^2 - 8x + 16 = 0$	0	4	Real and equal roots
(F)	$x^2 - 6x + 8 = 0$	4	2,4	Real, rational, unequal
(G)	$x^2 + 4 = 0$	–16	$\pm 2i$	Imaginary roots

Illustration 7. For the following equations find $\alpha + \beta$ and $\alpha\beta$ without finding the roots of the equation:

Sl. No	Equation	$\alpha + \beta$	$\alpha\beta$	$\alpha - \beta$
(A)	$x^2 - 8x + 15 = 0$			
(B)	$x^2 + 12x + 32 = 0$			
(C)	$x^2 - 6x + 2 = 0$			
(D)	$x^2 - 6x - 2 = 0$			
(E)	$x^2 - 8x + 16 = 0$			
(F)	$x^2 - 6x + 8 = 0$			
(G)	$x^2 + 4 = 0$			

Solutions: If an equation has two roots α and β then:

Sum of the roots $\alpha + \beta = -\dfrac{b}{a}$

Product of the roots $\alpha\beta = \dfrac{c}{a}$

Then find $\alpha - \beta$ by using the property

$$(\alpha - \beta)^2 = (\alpha + \beta)^2 - 4\alpha\beta$$

Then complete the table

Sl. No	Equation	$\alpha+\beta$	$\alpha\beta$	$\alpha-\beta$
(A)	$x^2 - 8x + 15 = 0$	8	15	± 2
(B)	$x^2 + 12x + 32 = 0$	–12	32	± 2
(C)	$x^2 - 6x + 2 = 0$	6	2	$\pm 2\sqrt{7}$
(D)	$x^2 - 6x - 2 = 0$	6	–2	$\pm 2\sqrt{11}$
(E)	$x^2 - 8x + 16 = 0$	8	16	0
(F)	$x^2 - 6x + 8 = 0$	6	8	± 2
(G)	$x^2 + 4 = 0$	0	4	$\pm 4i$

Illustration 8. Find the value of $\frac{\alpha}{\beta}+\frac{\beta}{\alpha}$ if α and β are the roots of the equation $x^2 - 6x + 8 = 0$.

Solution: If α and β are the roots of the equation $x^2 - 6x + 8 = 0$ then $\alpha + \beta = 6$ and $\alpha\beta = 8$.

Now

$$\frac{\alpha}{\beta}+\frac{\beta}{\alpha}=\frac{\alpha^2+\beta^2}{\alpha\beta}=\frac{\left\{(\alpha+\beta)^2-2\alpha\beta\right\}}{\alpha\beta}=\frac{36-16}{8}=\frac{20}{8}$$

$$=\frac{5}{2}$$

Illustration 9. Find the value of p for which the quadratic equation $x^2 - px + p + 3$ has reciprocal roots.

Solution: Let the roots of the given equation be α and $\frac{1}{\alpha}$ then, $\alpha\cdot\frac{1}{\alpha}=p+3 \Rightarrow p=-2$

Illustration 10. If $(a^2 - 1)x^2 + (a - 1)x + a^2 - 4a + 3 = 0$ is an identity in x, then the value of a is

(a) 1 (b) 3

(c) – 1 (d) – 3

Solution: For an identity both LHS and RHS are equal for any value of x

Equating the coefficients of similar powers of x, we get

$$\left.\begin{aligned} a\ \ -1=0 &\Rightarrow a=\pm 1\\ a-1=0 &\Rightarrow a=1\\ a\ \ -4a+3=0 &\Rightarrow a=1,3 \end{aligned}\right\}$$

$\therefore$ Common value of $a = 1$.

Illustration 11. If $p,q \in \{1,2,3,4\}$, the number of equations of the form $px^2 + qx + 1 = 0$ having real roots is

(a) 15 (b) 9

(c) 7 (d) 8

Solution: For real roots, we, must have $D \ge 0 \Rightarrow q^2 - 4p \ge 0 \Rightarrow q^2 \ge 4p$

Now we have following cases

If $p = 1$, then $q^2 \ge 4p \Rightarrow q^2 \ge 4 \Rightarrow q = 2,3,4$

If $p = 2$, then $q^2 \ge 4p \Rightarrow q^2 \ge 8 \Rightarrow q = 3,4$

If $p = 3$, then $q^2 \ge 4p \Rightarrow q^2 \ge 12 \Rightarrow q = 4$

If $p = 4$, then $q^2 \ge 4p \Rightarrow q^2 \ge 16 \Rightarrow q = 4$

Thus, we see that there are 7 cases. Hence, (c) is the correct answer.

Illustration 12. If p and q are the roots of the equation $x^2 + px + q = 0$, then

(a) $p = 1, q = -2$ (b) $p = 0, q = 1$

(c) $p = -2, q = 0$ (d) $p = -2, q = 1$

Solution: Since, p and q are roots of the equation $x^2 + px + q = 0, p + q = -p$ and $pq = q$.

$pq = q \Rightarrow q = 0$ or $p = 1$

If $q = 0$, then $p = 0$ and if $p = 1$, then $q = -2$

Hence, (a) is the correct answer.

Illustration 13. $a, b, c, d \in R$ are in G. P then find the nature of roots of the equation

$$(a^2 + b^2 + c^2)\,x^2 + 2\,(ab + bc + ca)\,x + b^2 + c^2 + d^2 = 0$$

Solution: Let r is the common ratio of the G. P. then put $b = ar$, $c = ar^2$, and $d = ar^3$ in the given equation

Then the given equation can be written as

$(x^2 + 2\,rx + r^2)(1 + r^2 + r^4) = 0$

$\Rightarrow \quad (x + r)^2 = 0 \Rightarrow x = -r, -r$

So the equation has real and repeated roots.

Illustration 14. Find the value of k such that the roots of $x^2 - (2k-1)x + k^2 + 2 = 0$ are in the ratio 1:2

Solution: Let the roots are α and 2α then

$$\alpha + 2\alpha = (2k - 1),$$

$$\alpha(2\alpha) = k^2 + 2$$

Eliminating we get

$$\Rightarrow \quad \frac{2k-1)^2}{9} = \frac{k^2+2}{2} \Rightarrow k = -4$$

Illustration 15. The set of values of $k \in R$ for which $kx^2 - (k + 1)x + 2k - 1 = 0$ has no real roots is

Solution: The given equation has no real roots so $D < 0$.

$(k + 1)^2 - 4k(2k - 1) < 0, \; k \neq 0$

$\Rightarrow \quad 7k^2 - 6k - 1 > 0$

$\Rightarrow \quad (7k + 1)(k - 1) > 0$

$\Rightarrow \quad k < -1/7$ or $k > 1$

Concept Applicator (CA)-1

1. Find the roots of the equation $x^2 - 10x + 21 = 0$
2. Find the difference between the roots of the equation $x^2 - 11x + 24 = 0$
3. The sum of the roots of the equation $x^2 + px + q = 0$ is equal to the sum of their squares, then

   (a) $p^2 - q^2 = 0$ (b) $p^2 + q^2 = 2q$

   (c) $p^2 + p = 2q$ (d) None of these
4. If $b < 0$, then the roots x_1 and x_2 of the equation $2x^2 + 6x + b = 0$, satisfy the condition $\left(\frac{x_1}{x_2}\right) + \left(\frac{x_2}{x_1}\right) < k$ then k is equal to

   (a) -3 (b) -5

   (c) -6 (d) -2
5. Sum of the real roots of the equation

   $x^2 + 9|x| + 6 = 0$

   (a) equals to 5 (b) equals to 10

   (c) equals to -5 (d) does not exist
6. A student was asked his age by his friend. The student said, "The number you get when you subtract 25 times my age from twice the square of my age will be thrice your age.". if the friend's age is 14, then age of the student is

   (a) 21 (b) 28

   (c) 14 (d) 25
7. The equation $\sqrt{x+1} - \sqrt{x-1} = \sqrt{4x-1}$ has

   (a) no solution

   (b) one solution

   (c) two solutions

   (d) more than two solutions
8. If α, β are the roots of the equation $8x^2 - 3x + 27 = 0$, then find the value of $\left(\frac{\alpha^2}{\beta}\right)^{1/3} + \left(\frac{\beta^2}{\alpha}\right)^{1/3}$

   (a) $\frac{1}{3}$ (b) $\frac{1}{4}$

   (c) $\frac{7}{2}$ (d) 4
9. If α, β are the roots of the equation $x^2 + x\sqrt{\alpha} + \beta = 0$, then the values of α and β are

   (a) $\alpha = 1$ and $\beta = -1$

   (b) $\alpha = 1$ and $\beta = -2$

   (c) $\alpha = 2$ and $\beta = 1$

   (d) $\alpha = 2$ and $\beta = -2$
10. Let $a > 0$, $b > 0$ and $c > 0$. Then both the roots of the equation $ax^2 + bx + c = 0$

    (a) are real and negative

    (b) have negative real parts

    (c) are rational numbers

    (d) None of these

4.2 COMMON/CONJUGATE ROOTS

4.2.1 Formation of Quadratic Equation whose roots are given

Let the quadratic equation whose roots are α and β be $ax^2+bx+c=0$

Then, $\alpha+\beta=\frac{-b}{a}$ and $\alpha\beta=\frac{c}{a}$

Now the quadratic equation whose roots are α and β is $ax^2+bx+c=0$ or, $a\left(x^2+\frac{b}{a}x+\frac{c}{a}\right)=0$

or, $x^2-\left(-\frac{b}{a}\right)x+\frac{c}{a}=0$

or, $x^2-(\alpha+\beta)x+\alpha\cdot\beta=0$

Thus, quadratic equation whose roots are α and β is $x^2-(\alpha+\beta)x+\alpha\beta=0$

or, x^2 – (sum of the roots)x + product of the roots = 0

Illustration 1. Find the quadratic equation whose roots are 4 and 5

Solution: Given roots are 4 and 5

Since quadratic equation whose roots are α and β is $x^2-(\alpha+\beta)x+\alpha\cdot\beta=0$

or $x^2-(4+5)x+4\times5=0$

or $x^2-9x+20=0$

Illustration 2. Find the quadratic equation whose roots are $2+\sqrt{3}$ and $2-\sqrt{3}$.

Solution: Since quadratic equation whose roots are α and β is $x^2-(\alpha+\beta)x+\alpha\cdot\beta=0$

Sum of the roots, $\alpha+\beta=(2+\sqrt{3})+(2-\sqrt{3})=4$

Product of the roots,

$\alpha.\beta=(2+\sqrt{3}).(2-\sqrt{3})=4-3=1$

So required equation is $x^2-4x+1=0$

Illustration 3. The coefficient of x in the quadratic equation ax^2+bx+c was wrongly taken as 17 in place of 13 and its roots were found to be – 2 and – 15, the actual roots of the equation are

(a) – 2 and 15 (b) – 3 and – 10
(c) – 4 and – 9 (d) – 5 and – 6

Solution: New roots are – 2, – 15

$\therefore$ Quadratic equation is $(x+2)(x+15)=0$

$\Rightarrow$ $x^2+17x+30=0$

Correct quadratic is $x^2+13x+30=0$

$\Rightarrow$ $(x+3)(x+10)=0$

$\Rightarrow$ $x=-10,-3$

Hence, actual roots are – 3, – 10.

Illustration 4. If $\frac{1}{4-3i}$ is a root of $ax^2+bx+1=0$, where a, b are real, then

(a) $a=25, b=-8$ (b) $a=25, b=8$
(c) $a=5, b=4$ (d) None of these

Solution: $\alpha=\frac{1}{4-3i},\beta\frac{1}{4+3i}\Rightarrow\alpha\beta=\frac{1}{a}$

or $a=25$

Also, $\alpha+\beta=\frac{-b}{a}\Rightarrow\frac{8}{25}=-\frac{b}{25}$ or $b=-8$

Option (a).

4.2.2 Transformation of Equations

In this section we will learn how to form or transform a given equation from the given condition, Let the given equation be

$$f(x)=a_0x^n+a_1x^{n-1}+\ldots+a_{n-1}x+a_n=0 \quad \ldots(1)$$

Case (i): To form an equation whose roots are $k\ (\neq 0)$ times roots of the equations in (1), replace x by x/k in equation (1).

Case (ii): To form an equation whose roots are the negatives of the roots in equation (1), replace x by $-x$ in equation (1).

Case (iii): To form an equation whose roots are k more than the roots of equation in (1), replace x by x-k in equation (1).

Case (iv): To form an equation whose roots are reciprocals of the roots in equation (1), replace x by $1/x$ in equation (1) and then multiply both the sides by x^n.

Case (v): To form an equation whose roots are square of the roots of the equation in (1) then we have to follow these steps

Step 1: Replace x by $\sqrt{x}$ in equation (1)

Step 2: Collect all the terms involving $\sqrt{x}$ on one side.

Step 3: Square both the sides and simplify.

Illustration 5. Form an equation whose roots are squares of the roots of

$$x^3 + 2x^2 - x + 2 = 0,$$

Solution:

Step 1: Replace x by $\sqrt{x}$ the result is $x\sqrt{x} + 2x - \sqrt{x} + 2 = 0$

Step 2: Collect all the terms involving $\sqrt{x}$ on one side the result is $\sqrt{x}\,(x - 1) = -2(x + 1)$

Step 3: Square both the sides and simplify to get $x(x - 1)^2 = 4(x + 1)^2$

$$\Rightarrow \quad x^3 - 6x^2 - 7x - 4 = 0$$

Case (vi): to form an equation whose roots are cubes of the roots of the equation then we have to follow these steps:

Step 1: Replace x by $x^{\frac{1}{3}}$

Step 2: Collect all the terms involving $x^{\frac{1}{3}}$ and $x^{\frac{2}{3}}$ on one side.

Step 3: Cube both the sides and simplify

4.2.3 Conjugate Roots

Part (i) Complex Roots

Complex roots of a quadratic equation with real coefficients always occur in conjugate pair i.e. if $\alpha + i\beta$ be one root, then other root will be $\alpha - i\beta$.

In other words for the quadratic equation $ax^2 + bx + c = 0$ where a, b, c are real numbers has one root $\alpha + i\beta$, then other root will be $\alpha - i\beta$.

Proof: Let $\alpha + i\beta$ be a root of the quadratic equation $ax^2 + bx + c = 0$ ... (i) where a, b, c are real numbers

It is to be proved that $\alpha - i\beta$ is the other root of equation (Or roots are conjugate to each other)

Since $\alpha + i\beta$ is a root of the equation (i)

$$\therefore \quad a(\alpha + i\beta)^2 + b(\alpha + i\beta) + c = 0$$

or, $$(a\alpha^2 - a\beta^2 + b\alpha + c) + (2.a.\alpha.\beta + b\beta)i = 0$$

Equating real and imaginary parts, we get

$$\left.\begin{array}{l} a\alpha^2 - a\beta^2 + b\alpha + c = 0 \\ \text{and} \quad 2a\alpha.\beta + b\beta = 0 \end{array}\right\} \quad \text{... (ii)}$$

Now put $x = \alpha - i\beta$ in the expression we get

$$a(\alpha - i\beta)^2 + b(\alpha - i\beta) + c = a(\alpha^2 + i^2\beta^2 - 2\alpha\beta i) + b\alpha - b\beta i + c$$

$$\Rightarrow \quad (a\alpha^2 + \alpha\beta^2 + b\alpha + c) - (2a\alpha\ \beta + b\beta)i$$

$$= 0 - 0i = 0 \text{ [From (ii)]}$$

Thus $\alpha - i\beta$ also satisfies equation (i)

Hence $\alpha - i\beta$ is also a root of equation (i)

Part (ii) Irrational roots

Irrational roots of a quadratic equation with rational coefficients always occur in conjugate pair i.e. if $m + \sqrt{n}$ be one root of the equation, $ax^2 + bx + c = 0$, where a, b, c are rationals then its other root will be $m - \sqrt{n}$.

Proof: Let $ax^2 + bx + c = 0$ be a quadratic equation, where a, b, c are rational... (i)

Let $m + \sqrt{n}$ be one root of equation (i)

It is to be proved that the other root of equation (i) is $m-\sqrt{n}$

Since $m+\sqrt{n}$ is a root of the equation (i)

$\therefore \quad a(m+\sqrt{n})^2+b(m+\sqrt{n})+c=0$

or, $\quad or, a(m^2+n+2m\sqrt{n})+(bm+b\sqrt{n})+c=0$

or, $\quad (am^2+an+bm+c)+(2am+b)\sqrt{n}=0$

Equating rational and irrationals parts, we get

$$\left.\begin{aligned} am^2+an+bm+c&=0 \\ \text{and} \quad 2am+b&=0 \end{aligned}\right\} \quad \ldots \text{(ii)}$$

Now put $x = m-\sqrt{n}$ in the expression we get

$$a(m-\sqrt{n})^2+b(m-\sqrt{n})+c=a(m^2+n-2m\sqrt{n}) + b\,(m-\sqrt{n}\,)+c$$

$$=(am^2+an+bm+c)-(2am+b)\sqrt{n}=0-0\sqrt{n}=0$$

[From (ii)]

Thus $m-\sqrt{n}$ also satisfies equation (i)

Hence $m-\sqrt{n}$ is the other root of equation (i) Hence Proved.

Illustration 6. Find the quadratic equation with real coefficient whose one root is $3 + 4i$

Solution: Since complex roots are conjugate so if one root is $3 + 4i$ then other root is $3 - 4i$

Sum of the roots $\quad \alpha+\beta=(3+4i)+(3-4i)=6$

Product of the roots

$\alpha.\beta=(3+4i)(3-4i)=9+16=25$

So required equation is $\; x^2-6x+25=0$

Illustration 7. Find the quadratic equation with real coefficient whose one root is $2+\sqrt{3}$

Solution: Since irrational roots are conjugate so if one root is $2+\sqrt{3}$ then other root is $2-\sqrt{3}$

Since quadratic equation whose roots are α and β is $x^2-(\alpha+\beta)x+\alpha\cdot\beta=0$

Sum of the roots $\alpha+\beta=\left(2+\sqrt{3}\right)+\left(2-\sqrt{3}\right)=4$

Product of the roots

$\alpha.\beta=\left(2+\sqrt{3}\right).\left(2-\sqrt{3}\right)=4-3=1$

So required equation is $\; x^2-4x+1=0$.

4.2.4 Condition for Common Root

Let us consider two quadratic equations

$$ax^2+bx+c=0 \quad \ldots \text{(i)}$$

$$\text{and} \quad a_1x^2+b_1x+c_1=0 \quad \ldots \text{(ii)}$$

Now assume that they have a common root,

Let $\quad c, c_1 \neq 0$ and $ab_1 - a_1b \neq 0$.

Let the common root be α, then

$$a\alpha^2+b\alpha+c=0 \quad \ldots \text{(iii)}$$

$$\text{and} \quad a_1\alpha^2+b_1\alpha+c_1=0 \quad \ldots \text{(iv)}$$

From equations (iii) and (iv)

$$\therefore \quad \frac{\alpha^2}{bc_1-b_1c}=\frac{\alpha}{ca_1-ac_1}=\frac{1}{ab_1-a_1b}$$

$$\alpha=\frac{bc_1-b_1c}{ca_1-ac_1}=\frac{ca_1-ac_1}{ab_1-a_1b} \quad \ldots \text{(v)}$$

or, $\left(bc_1-b_1c\right)\left(ab_1-a_1b\right)=\left(ca_1-ac_1\right)$

This is the required condition for the two equations to have a common root (at least one root common).

Special Case

Case (i): If $c = c_1 = 0$,

The equations $ax^2 + bx + c = 0$ and $a_1x^2 + b_1x + c_1 = 0$ will reduce to $ax^2 + bx = 0$ and $a_1x^2 + b_1x = 0$.

These two equations will have a common root 0. Other roots of the two equations will be $-\frac{b}{a}$ and $-\frac{b_1}{a_1}$

If $\frac{b}{a}=\frac{b_1}{a_1}$ or, $\frac{b}{b_1}=\frac{a}{a_1}$, then both roots will be common.

Case (ii): Condition for both roots common

Given equations are $ax^2+bx+c=0$ ... (i)

and $a_1x^2+b_1x+c_1=0$ ... (ii)

Let equations (i) and (ii) have both roots common. Since a quadratic equation has two and only two roots and equation (i) and (ii) have both roots common, therefore equations (i) and (ii) will be identical

Second Method

Let α and β be the common roots of equation (i) and (ii), then

$$\alpha+\beta=-\frac{b}{a} \quad ...\text{(iii)} \quad \alpha\beta=\frac{c}{a} \quad ...\text{(iv)}$$

$$\text{and,} \quad \alpha+\beta= \quad — ...\text{(v)} \quad \alpha\beta=\frac{c_1}{a_1} \quad ...\text{(vi)}$$

From (iii) and (v), $-\frac{b}{a}=-\frac{b_1}{a_1}$ or $\frac{a}{a_1}=\frac{b}{b_1}$ ... (vii)

From (iv) and (vi), $\frac{c}{a}=\frac{c_1}{a_1}$ or $\frac{a}{a_1}=\frac{c}{c_1}$ ... (viii)

From (vii) and (viii), $\frac{a}{a_1}=\frac{b}{b_1}=\frac{c}{c_1}$

This is the required condition.

Illustration 8. If $a, b, c \in R$ and equations $ax^2 + bx + c = 0$ and $x^2 + 2x + 9 = 0$ have a common root, then find $a : b : c =$

Solution: Given equations are

$x^2+2x+9=0$ ... (i)

and $ax^2+bx+c=0$ ... (ii)

Since, roots of equation(i) are complex and equation (i) and (ii) have a common root, therefore common root must be complex and hence both roots will be common.

Therefore, Eqs. (i) and (ii) are identical.

$$\therefore \frac{a}{1}=\frac{b}{2}=\frac{c}{9} \quad \therefore a:b:c=1:2:9$$

Illustration 9. If the equations $x^2 + 2x + 3\lambda = 0$ and $2x^2 + 3x + 5\lambda = 0$ have a non-zero common root, then λ is equal to

(a) 1 (b) – 1

(c) 3 (d) None of these

Solution: $\alpha^2+2\alpha+3\lambda=0$ and $2\alpha^2+3\alpha+5\lambda=0$

$$\Rightarrow \quad \frac{\alpha^2}{\lambda}=\frac{\alpha}{\lambda}=\frac{1}{-1}$$

$$\Rightarrow \quad \alpha = 1 \text{ and } \alpha = -\lambda$$

Option (b).

Illustration 10: If the equations $x^2 + ax + b = 0$ and $2x^2 + 3x + 5 = 0$ have a non-zero common root, then a – b is equal to (a and b are real).

(a) 1 (b) – 1

(c) 3 (d) None of these

Solution: Since the equation $2x^2 + 3x + 5 = 0$ has complex roots hence both the roots must be common so

$a = 3/2$, and $b = 5/2$ so $a - b = -1$

Option (b).

Concept Applicator (CA)-2

1. If $x^2 - ax + b = 0$ and $x^2 - px + q = 0$ have a root in common and the second equation has equal roots, then prove that $b+q=\frac{ap}{2}$.
2. If the roots of $a_1x^2 + b_1x + c_1 = 0$ are α_1, β_1 and those of $a_2x^2 + b_2x + c_2 = 0$ are α_2, β_2 such that $\alpha_1\alpha_2 = \beta_1\beta_2 = 1$ then

   (a) $\frac{a_1}{a_2}=\frac{b_1}{b_2}=\frac{c_1}{c_2}$ (b) $\frac{a_1}{c_2}=\frac{b_1}{b_2}=\frac{c_1}{a_2}$

   (c) $a_1a_2=b_1b_2=c_1c_2$ (d) None of these
3. If the equations $ax^2 + bx + c = 0$ and $x^2 + 2x + 3 = 0$ have a common root, then $a : b : c$ is equal to

   (a) 2 : 4 : 5 (b) 1 : 3 : 4

   (c) 1 : 2 : 3 (d) None of these
4. Find the nature of roots of the equation $(a + b - c - d)x^2 + (c - a)x + (d - b) = 0$, here a, b, c, and d are real and rational.

   (a) Complex (b) Real and rational

   (c) real and irrational (d) None of these

5. If the equations $x^2 - px + q = 0$ and $x^2 - ax + b = 0$ have a common root and the other root of the second equation is the reciprocal of the other root of the first, then find the value of $(q - b)^2$
   (a) $aq(p - b)^2$ (b) $bq(p - a)^2$
   (c) $bq(p - q)^2$ (d) None of these

6. If a and b are rational and b is not a perfect square, then the quadratic equation with rational coefficients whose one root is $\frac{1}{a+\sqrt{b}}$, is
   (a) $x^2 - 2ax + (a^2 - b) = 0$
   (b) $(a^2 - b)x^2 - 2ax + 1 = 0$
   (c) $(a^2 - b)x^2 - 2bx + 1 = 0$
   (d) None of these

7. The equation formed by decreasing each root of $ax^2 + bx + c = 0$ by 1 is $2x^2 + 8x + 2 = 0$, then
   (a) $a = -b$ (b) $b = -c$
   (c) $c = -a$ (d) $b = a + c$

8. If α. β are the roots of $9x^2 + 6x + 1 = 0$, then the equation with roots $\frac{1}{\alpha}, \frac{1}{\beta}$ is
   (a) $2x^2 + 3x + 18 = 0$ (b) $x^2 + 6x - 9 = 0$
   (c) $x^2 + 6x + 9 = 0$ (d) $x^2 - 6x + 9 = 0$

9. If α. β are the roots of $ax^2 + bx + c$, the equation whose roots are $2 + \alpha$, $2 + \beta$ is
   (a) $ax^2 + (4a - b)x + 4a - 2b + c = 0$
   (b) $ax^2 + (4a - b)x + 4a + 2b + c = 0$
   (c) $ax^2 + (b - 4a)x + 4a - 2b + c = 0$
   (d) $ax^2 + (b + 4a)x + 4a - 2b + c = 0$

10. The quadratic equations whose roots are cubes of the roots of $x^2 + bx + c = 0$ is
   (a) $x^2 + b(b^2 - 3c)x - c^3 = 0$
   (b) $x^2 + b(b^2 - 3c)x + c^3 = 0$
   (c) $x^2 - b(b^2 - 3c)x + c^3 = 0$
   (d) None of the above

4.3 QUADRATIC FUNCTION AND ITS GRAPH

4.3.1 Introduction of Quadratic Graph

In order to understand the graph of quadratic function lets start with the simplest form of quadratic function i.e $f(x) = y = x^2$. To plot this graph lets take some values of 'x' and find corresponding value of 'y'

Graph of $y = x^2$

x	$y = x^2$
0	0
1	1
–1	1
2	4
–2	4
3	9
–3	–9

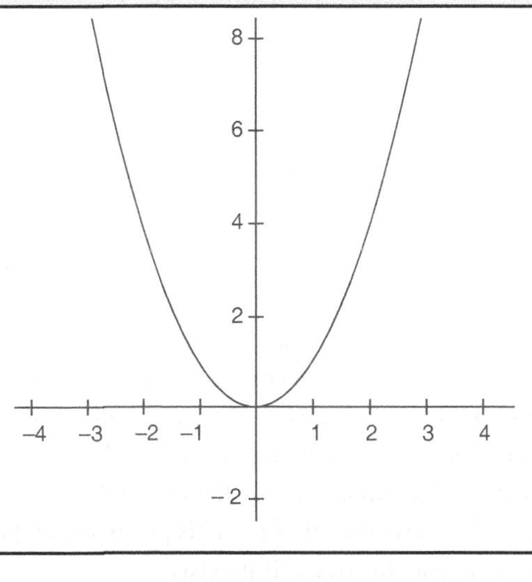

Graph is parabolic in nature. It touches x-axis at origin.

Now consider another graph $y = x^2 - 4$

Graph of $y = x^2 - 4$

x	$y = x^2 - 4$
0	– 4
1	–3
– 1	– 3
2	0
– 2	0
3	5
– 3	– 5

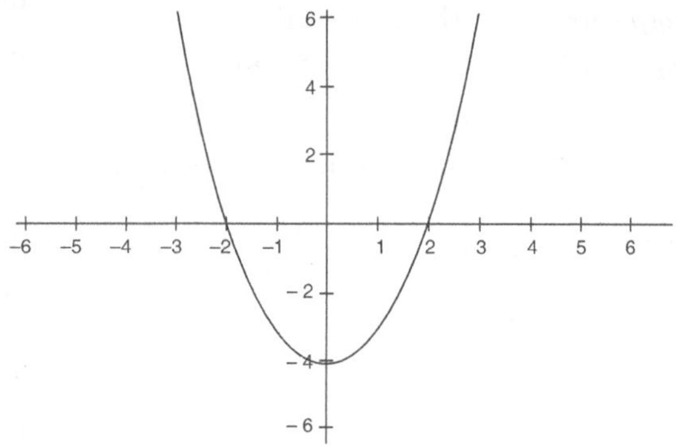

Features of this graph

(i) Graph is parabolic in nature

(ii) It cuts the X-axis twice (at $x = -2$ and at $x = 2$)

(iii) The graph attains minimum value at $x = 0$

4.3.2 Six Different Cases of Quadratic Graph

Consider a quadratic function is $f(x) = \boldsymbol{ax^2 + bx + c}$, If we plot the graph of a quadratic function then we have 6 cases and these 6 cases are:

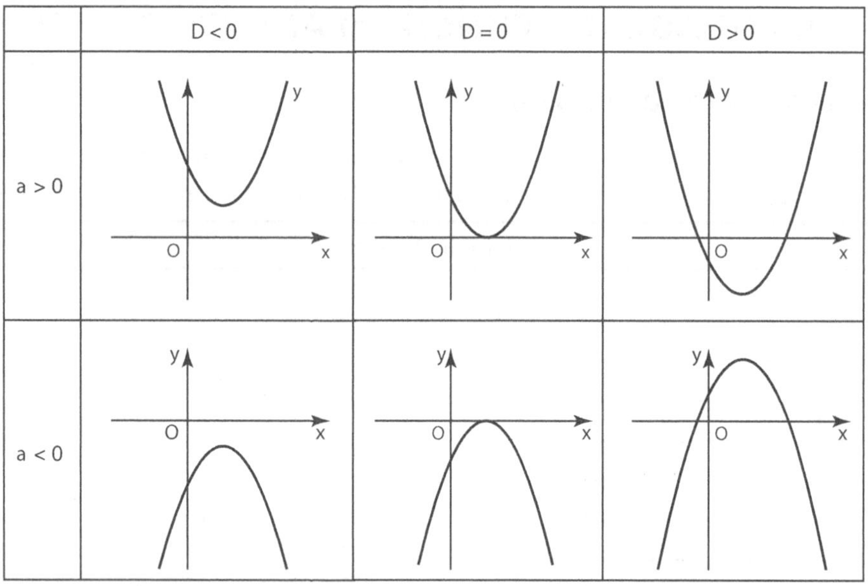

*Position of "Y" axis depends on nature of function.

Steps that should be followed to plot a quadratic graph

Step 1: Find the discriminant D of the quadratic function.

Step 2: If discriminant is –ve then no real root exist, otherwise it has real root/s

Step 3: If coefficient of x^2 i.e 'a' is positive then parabola is open upwards otherwise open downwards

Step 4: Locate the roots (if it exist)

Case	Coefficient of x^2 i.e 'a'	Discriminant D		No of roots
Case (i)		$D < 0$	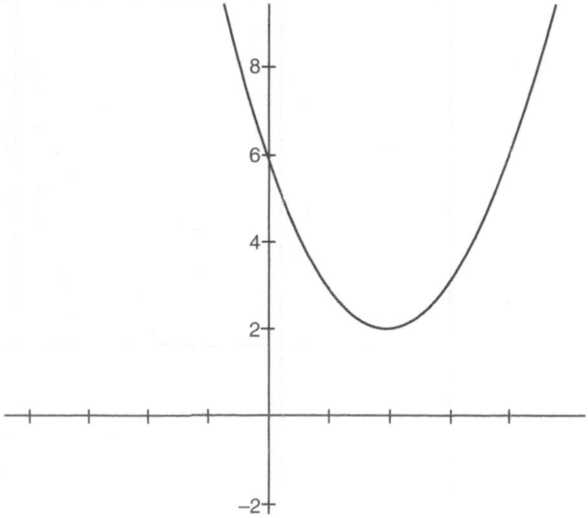	0
Case (ii)	$a > 0$	$D = 0$	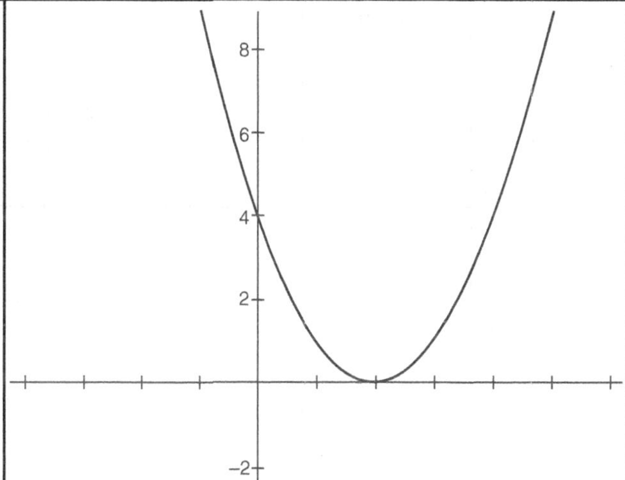	1
Case (iii)		$D > 0$	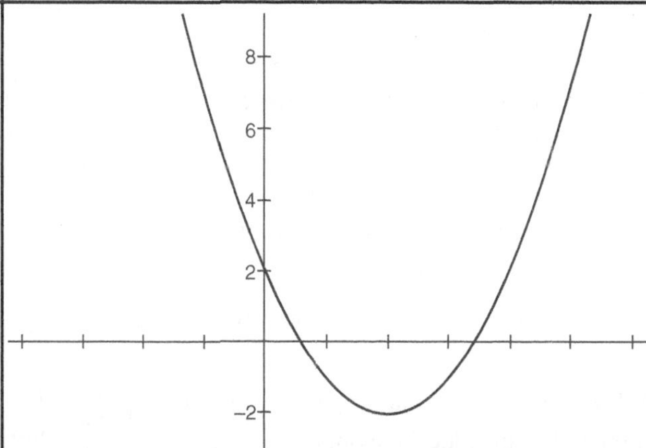	2

Case (iv)	a < 0	D < 0	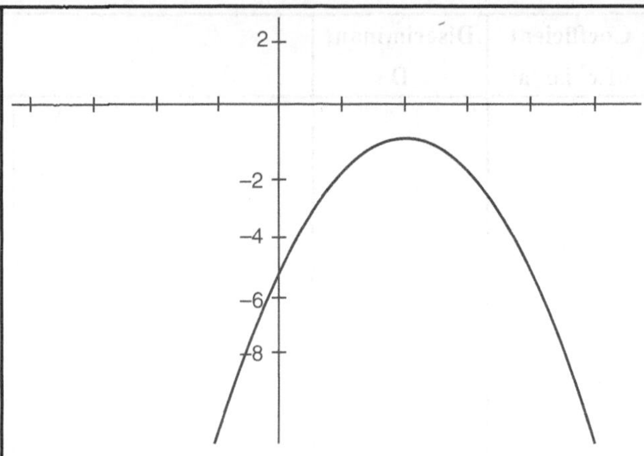	0
Case (v)		D = 0	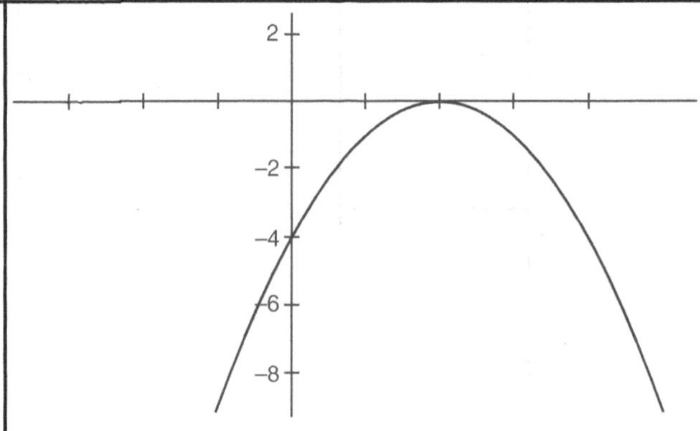	1
Case (vi)		D > 0	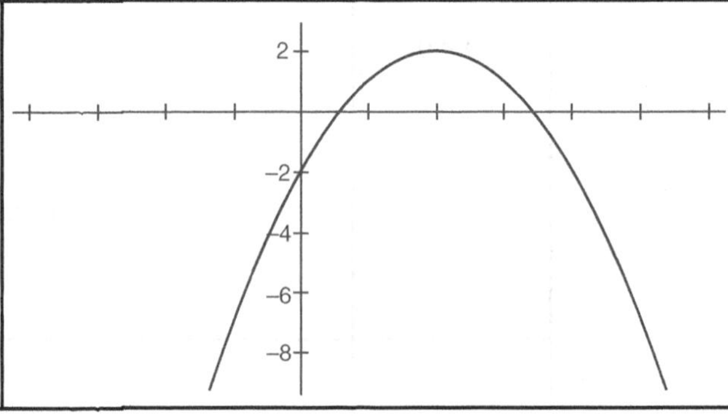	2

4.3.3 Maximum and Minimum Value of Quadratic Function

Consider a quadratic function $f(x) = y = ax^2 + bx + c$, we have seen the graph of a quadratic function from the graph we can conclude two cases:

Case (i): If $a > 0$ then parabola is open upwards and function will attain a local minima.

Case (ii): If $a < 0$ then parabola is open downwards and function will attain a local maxima.

Let us consider a quadratic equation

$$ax^2 + bx + c = 0, \text{ where } a \neq 0 \qquad \text{... (i)}$$

Divide both sides of equation (i) by a, we get

$$x^2 + \left(\frac{b}{a}\right)x + \left(\frac{c}{a}\right) = 0$$

or, $$(x)^2 + 2.x.\frac{b}{2a} + \left(\frac{b}{2a}\right)^2 - \frac{b^2}{4a^2} + \left(\frac{c}{a}\right) = 0$$

or, $$\left(x + \frac{b}{2a}\right)^2 - \frac{b^2}{4a^2} + \left(\frac{c}{a}\right) = \left(x + \frac{b}{2a}\right)^2 - \frac{b^2 - 4ac}{4a^2} = 0$$

The function will attain local minima or maxima at $x=-\left(\frac{b}{2a}\right)$

Case (i): If $a>0$ then parabola is open upwards and function will attain a local minima at $x=-\left(\frac{b}{2a}\right)$

And required minimum value is $-\frac{b^2-4ac}{4a^2}$

Case (ii): If $a<0$ then parabola is open downwards and function will attain a local maxima at $x=-\left(\frac{b}{2a}\right)$

And required maximum value is $-\frac{b^2-4ac}{4a^2}$

4.3.4 Quadratic Inequality

Consider a quadratic expression $f(x)=y=ax^2+bx+c$ as discussed graphically above we have 6 different cases:

Case (i): If $a>0$ and $D<0$ then $y=ax^2+bx+c>0\ \forall x\in R$

Case (ii): If $a>0$ and $D=0$ then $y=ax^2+bx+c\geq 0\ \forall x\in R$

Case (iii): If $a>0$ and $D>0$ then equation has two roots α & β then $y=ax^2+bx+c=a(x-\alpha)(x-\beta)<0\ \forall x\in(\alpha,\beta)$

Case (iv): If $a<0$ and $D<0$ then $y=ax^2+bx+c<0\ \forall x\in R$

Case (v): If $a<0$ and $D=0$ then $y=ax^2+bx+c\leq 0\ \forall x\in R$

Case (vi): If $a<0$ and $D>0$ then equation has two roots α & β then $y=ax^2+bx+c=a(x-\alpha)(x-\beta)>0\ \forall x\in(\alpha,\beta)$

Illustration 1: Discuss the graph of $f(x)=y=x^2+x-6$

Solution:

Graph 1: Plot the graph of x^2+x-6

Consider the $y=x^2+x-6$

Here $a=1$, $b=1$ and $c=-6$

(i) Coefficient of x^2 is +ve so graph should be open up wards.

(ii) Discriminant D = 25 > 0 so we have two distinct roots.

(iii) The two roots are –3 and 2

(iv) Plot the graph as shown.

(v) The function will attain a minimum value at $x=-1/2$

(vi) The required minimum value is $f\left(-\frac{1}{2}\right)=\left(-\frac{1}{2}\right)^2+\left(-\frac{1}{2}\right)-6=-\frac{25}{4}$

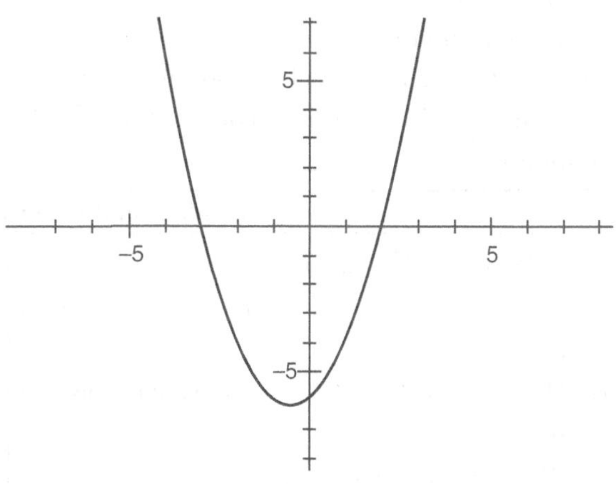

Illustration 2. Find the value of x such that $y=x^2+x-6<0$

(a) $(-\infty,-3)\cup(2,\infty)$

(b) $(-3, 2)$

(c) $(-3, 3)$

(d) None of these

Solution (b) : From the graph above the value of x is $(-3, 2)$.

Illustration 3. Discuss the graph of $f(x) = y = x^2 + 6x + 9$

Solution:

<table>
<tr><th colspan="2">Graph 2: Plot the graph of $x^2 + 6x + 9$</th></tr>
<tr><td>Consider the $y = x^2 + 6x + 9$
(i) Coefficient of x^2 is +ve so graph should be open upwards.
(ii) Discriminant D = 0 so we have one repeated root.
(iii) The root of the equation is – 3
(iii) Plot the graph as shown.
(iv) The function will attain a minimum value at $x = -3$
(v) The required minimum value is $f(-3) = (-3)^2 + 6(-3) + 9 = 0$</td><td>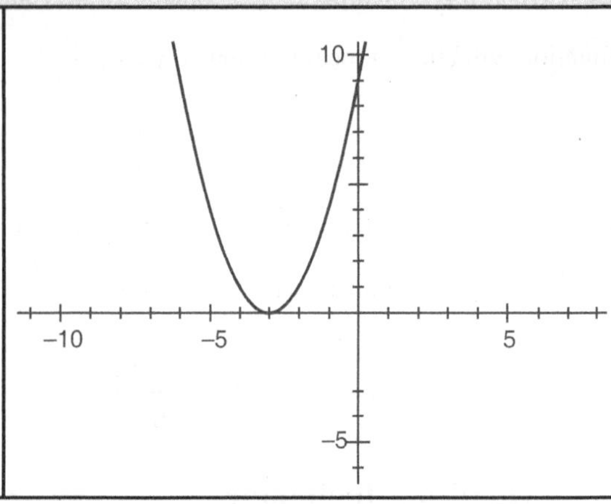</td></tr>
<tr><td colspan="2">If inequality is $y = x^2 + 6x + 9 > 0$ then from graph the value of x is $(-\infty, -3) \cup (-3, \infty)$ or R–{–3}</td></tr>
<tr><td colspan="2">If inequality is $y = x^2 + 6x + 9 < 0$ then from graph no such value of x exist.</td></tr>
</table>

Illustration 4. Find the value of x such that $y = x^2 + 6x + 9 < 0$

(a) $(-\infty, -3) \cup (2, \infty)$

(b) $(-3, 2)$

(c) $(-3, 3)$

(d) None of these

Solution (d) : From the graph, for no value of 'x' $y = x^2 + 6x + 9 \geq 0$

Illustration 5. Discuss the graph of $y = f(x) = x^2 - 4x + 6$.

Solution:

<table>
<tr><th colspan="2">Graph 3: Plot the graph of $y = x^2 - 4x + 6$</th></tr>
<tr><td>Consider the $y = x^2 - 4x + 6$
(i) Coefficient of x^2 is +ve so graph should be open upwards.
(ii) Discriminant D = – 8 < 0 so we have no real roots.
(iii) Plot the graph as shown.
(iv) The function will attain a minimum value at $x = 2$
(v) The required minimum value is $f(2) = (2)^2 - 4(2) + 6 = 2$</td><td>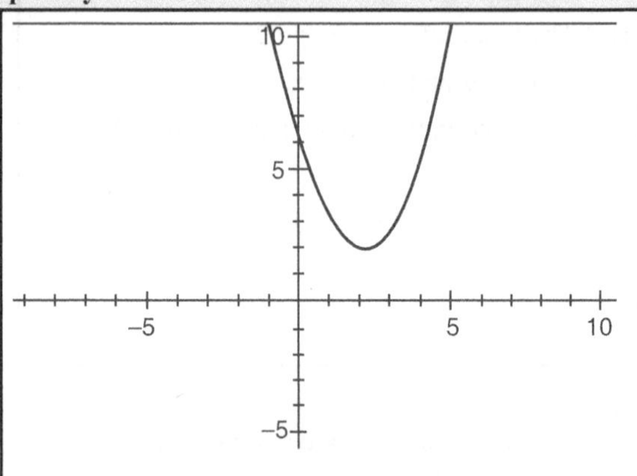</td></tr>
<tr><td colspan="2">If inequality is $y = x^2 - 4x + 6 > 0$ then from graph any value of x will satisfy this condition.</td></tr>
<tr><td colspan="2">If inequality is $y = x^2 - 4x + 6 < 0$ then from graph no such value of x exist.</td></tr>
</table>

Illustration 6. Find the value of k such that $y = x^2 - 4x + 6$ and $y = k$ has two distinct roots.

(a) $(2, \infty)$ (b) $(-3, 2)$

(c) $(-3, 2)$ (d) None of these

Solution (a) : From graph minimum value of function is 2 so $k > 2$

Illustration 7. Discuss the graph of $y = -x^2 - x + 6$.

Solution:

Graph 4: Plot the graph of $y = -x^2 - x + 6$

Consider the $y = -x^2 - x + 6$

(i) Coefficient of x^2 is –ve so graph should be open downwards.

(ii) Discriminant D = 25 > 0 so we have 2 real roots.

(iii) The two roots are –3 and 2

(iv) Plot the graph as shown.

(v) The function will attain a maximum value at $x = -1/2$

(vi) The required maximum value is

$$f\left(-\frac{1}{2}\right) = -\left(-\frac{1}{2}\right)^2 - \left(-\frac{1}{2}\right) + 6 = 25/4$$

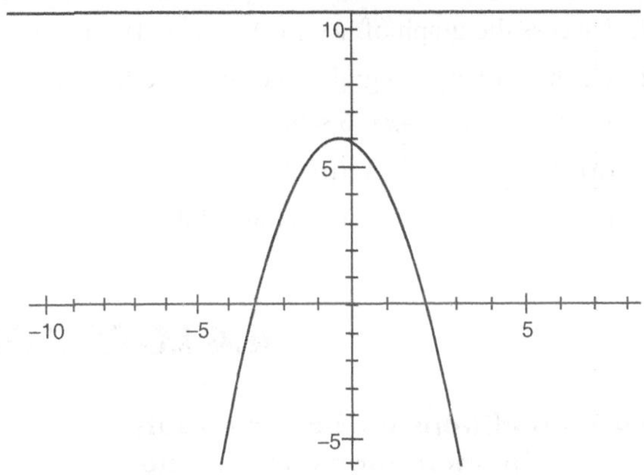

If inequality is $\boldsymbol{y = -x^2 - x + 6 > 0}$ then from graph value of x is $(-3, 2)$

If inequality is $\boldsymbol{y = -x^2 - x + 6 < 0}$ then from graph value of x is $(-\infty, -3) \cup (2, \infty)$

Illustration 8. For how many integral values of x the function $y = f(x) = x^2 - 4x + 6 > 0$.

(a) 2 (b) 3

(c) 4 (d) None of these

Solution (c) : From graph integral values of x is –2, –1, 0 and 1

Illustration 9. Discuss the graph of $y = f(x) = -x^2 + 4x - 4$

Graph 5: Plot the graph of $y = -x^2 + 4x - 4$

Consider the $\boldsymbol{y = -x^2 + 4x - 4}$

(i) Coefficient of x^2 is –ve so graph should be open downwards.

(ii) Discriminant D = 0 so we have only 1 real root.

(iii) Root of the equation is 2.

(iv) Plot the graph as shown.

(v) The function will attain a maximum value at $x = 2$.

(vi) The required minimum value is $f(2) = -(2)^2 + 4(2) - 4 = 0$

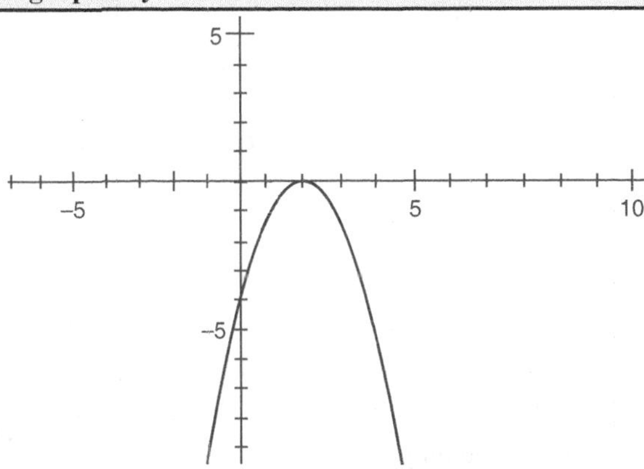

If inequality is $y = -x^2 + 4x - 4 > 0$ then from graph no value of x exist

If inequality is $y = -x^2 + 4x - 4 \le 0$ then from graph all values of x will satisfy the condition.

Illustration 10. For how many integral values of x the function $y = f(x) = -x^2 + 4x - 4 > 0$.

(a) 2 (b) 3

(c) 4 (d) None of these

Solution (d) : From graph no value of x will satisfy the given condition.

Concept Applicator (CA) - 3

1. Discuss the graph of $y = f(x) = -x^2 + 4x - 6$.
2. For how many integral values of x the function $y = f(x) = -x^2 + 4x - 6 < 0$.
   (a) 2 (b) 3
   (c) 4 (d) None of these
3. Discuss the graph of $y = |x^2 - 6x + 5|$
4. For what value of k the equation $y = f(x) = |x^2 - 6x + 5| = k$ has 4 distinct roots
   (a) (2,4)
   (b) (0,3)
   (c) (0,4)
   (d) None of these
5. Find x if $x^2 + x + 1 < 0$
6. Find x if $x^2 - 9 < 0$
7. Solve for $-x^2 + x - 1 < 0$
8. Solve simultaneously $x^2 - 9 > 0$ and $x^2 - 25 < 0$
9. Solve for $x^2 + 11x + 24 < 0$
10. Solve for $x^2 + 11x + 18 > 0$

4.4 LOCATION OF ROOTS

4.4.1 Conditions for both the roots to be more than a real number

Case (i) If a > 0

If $a > 0$, then the parabola $y = ax^2 + bx + c$ opens upwards and intersects the x-axis at α and β where α, β are two roots of the equation $ax^2 + bx + c = 0$

$$\alpha, \beta = \frac{-b \pm \sqrt{b^2 - 4ac}}{2a}$$

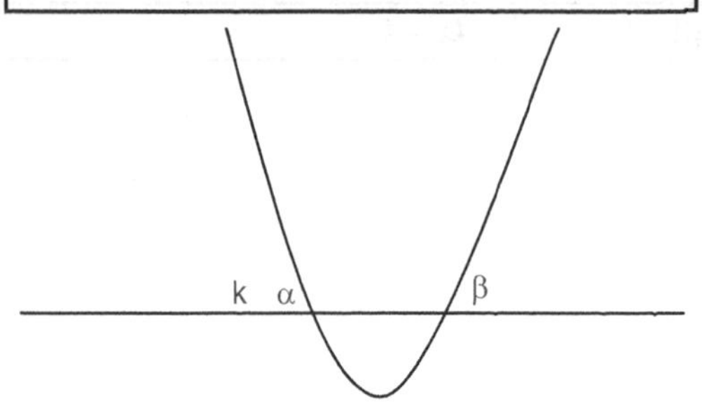

In this case both the roots **α and β** will be more than K if K lies to left of both **α and β**.

From the figure we note that both the roots are more than k if and only if

(i) $D > 0$ (ii) $k < \frac{-b}{2a}$

(iii) $f(k) > 0$

Case (ii) If a < 0

If $a < 0$, then the parabola $y = ax^2 + bx + c$ opens downwards and intersects the x-axis at α and β where α, β are two roots of the equation $ax^2 + bx + c = 0$

$$\alpha, \beta = \frac{-b \pm \sqrt{b^2 - 4ac}}{2a}$$

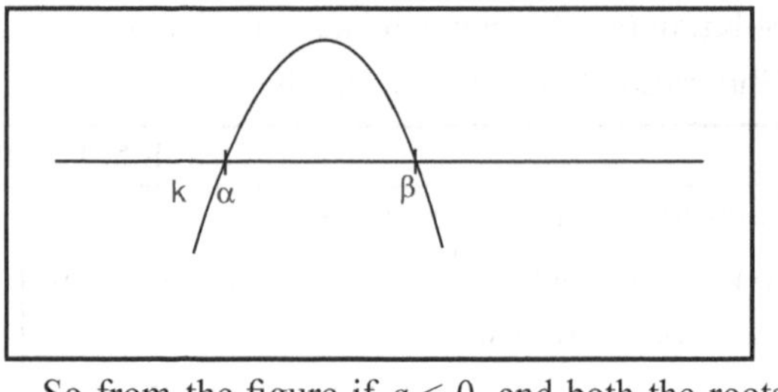

So from the figure if $a < 0$, and both the roots will be more than k then

(i) $D > 0$ (ii) $k < \frac{-b}{2a}$

(iii) $f(k) < 0$

Final Conclusion

Combining the above two sets of conditions, we get both the roots of $ax^2 + bx + c = 0$ are more than a real number k if only if

(i) $D > 0$ (ii) $k < \frac{-b}{2a}$

(iii) $a f(k) > 0$

Illustration 1. Let $x^2 - (k-3)x + k = 0$ $(k \in R)$ be a quadratic equation. Find the value of k for which both the roots are greater than 2.

Solution:

Both roots are greater than 2

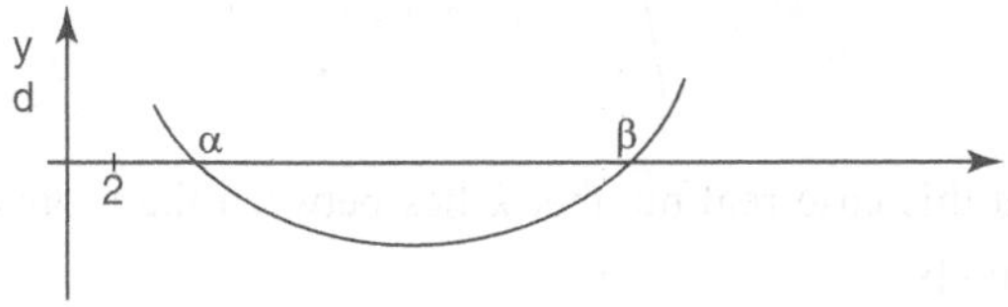

We have three conditions

$\Rightarrow$ (i) $f(2) > 0$, (ii) $D \geq 0$ (iii) $\dfrac{-b}{2a} > 2$

$\Rightarrow$ $f(2) = 4 - 2(k-3) + k > 0$ or $k < 10$ ... (i)

Since $D > 0, D = (k-3)^2 - 4k$
$= k^2 - 10k + 9 = (k-1)(k-9$

or $k \in (-\infty, 1) \cup (9, \infty)$ ... (ii)

Since $\dfrac{-b}{2a} > 2$ or $k - 3 > 4$ $\Rightarrow$ or $k > 7$ ... (iii)

Combining above three results we get $k \in (9, 10)$.
*Here we have annumed two distinct roots.

4.4.2 Conditions for both the roots to be less than a real number 'k'

Case (i) If a > 0

If $a > 0$, then the parabola $y = ax^2 + bx + c$ opens up wards and intersects the x-axis at α and β where α, β are two roots of the equation $ax^2 + bx + c = 0$

$$\alpha, \beta = \frac{-b \pm \sqrt{b^2 - 4ac}}{2a}$$

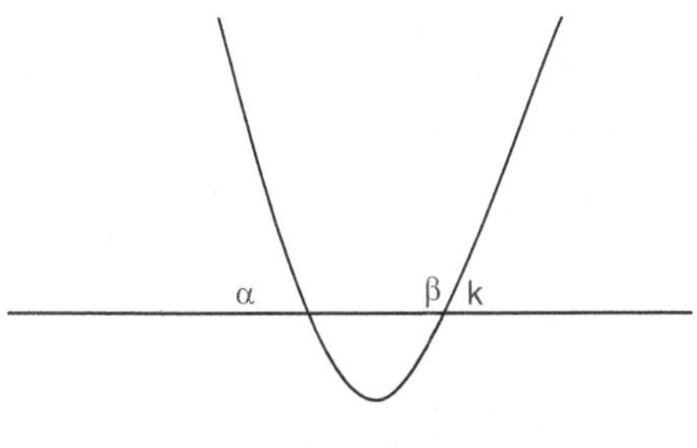

In this case both the roots **α and β** will be less than k if k lies to right of both **α and β**.

From the figure we note that both the roots are less than k if and only if

(i) $D > 0$ (ii) $k > \dfrac{-b}{2a}$

(iii) $f(k) > 0$

Case (ii) If a < 0

If $a < 0$, then the parabola $y = ax^2 + bx + c$ opens downwards and intersects the x-axis at **α and β** where α, β are two roots of the equation $ax^2 + bx + c = 0$

$$\alpha, \beta = \frac{-b \pm \sqrt{b^2 - 4ac}}{2a}$$

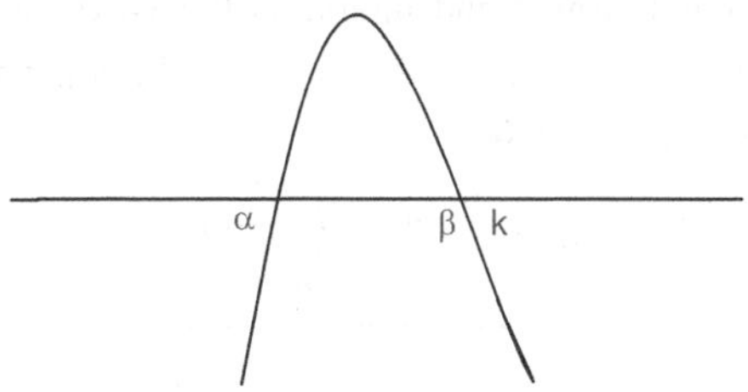

So from the figure if $a < 0$, and both the roots will be less than k then

(i) $D > 0$ (ii) $k > \dfrac{-b}{2a}$

(iii) $f(k) < 0$

Final Conclusion

Both the roots of $ax^2 + bx + c = 0$ are less than a real number k if and only if

(i) $D > 0$ (ii) $k > \dfrac{-b}{2a}$

(iii) $\alpha f(k) > 0$

Illustration 2. Let $x^2 - (k-3)x + k = 0(k \in R)$ be a quadratic equation. Find the value of k for which both the roots are less than 2.

Solution:

Both the roots are smaller than 2 then we have

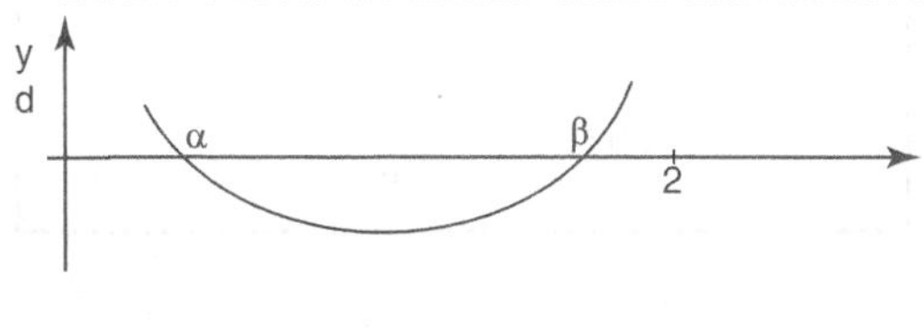

$\Rightarrow$ (i) $f(2) > 0$ (ii) $D \geq 0$ (iii) $\frac{-b}{2a} < 2$

$\Rightarrow$ $f(2) = 4 - 2(k-3) + k > 0$ or $k < 10$... (i)

Since $D > 0$, $D = (k-3)^2 - 4k$

$= k^2 - 10k + 9 = (k-1)(k-9)$

or $k \in (-\infty, 1) \cup (9, \infty)$ ... (ii)

Since $\frac{-b}{2a} < 2$ or $k - 3 < 4 \Rightarrow$ or $k < 7$ ... (iii)

Combining above three results we get $k \in (-\infty, 1)$.

4.4.3 Conditions for a real number to be in between both the roots

Case (i) If a > 0

If $a > 0$, then the parabola $y = ax^2 + bx + c$ opens upwards and intersects the x-axis at α **and** β where α, β are two roots of the equation $ax^2 + bx + c = 0$

$$\alpha, \beta = \frac{-b \pm \sqrt{b^2 - 4ac}}{2a}$$

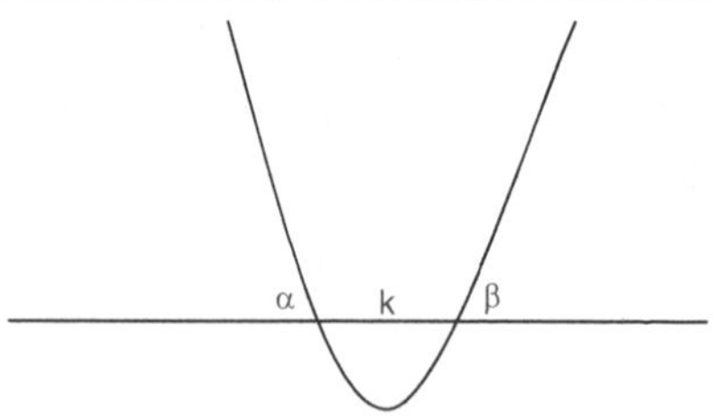

In this case real number k lies between the roots α **and** β.

From the figure we note that k lies between the roots α **and** β if and only if

(i) $D > 0$ (ii) $f(k) < 0$

Case (ii) If a < 0

If $a < 0$, then the parabola $y = ax^2 + bx + c$ opens downwards and intersects the x-axis at α and β where α, β are two roots of the equation $ax^2 + bx + c = 0$

$$\alpha, \beta = \frac{-b \pm \sqrt{b^2 - 4ac}}{2a}$$

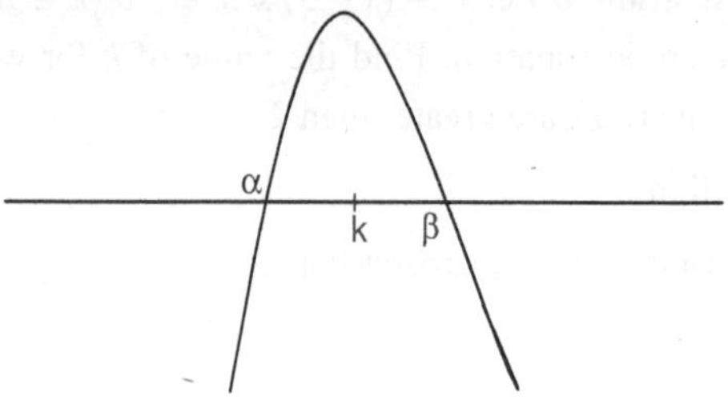

In this case real number k lies between the roots α and β.

From the figure we note that k lies between the roots α and β if and only if

(i) $D > 0$ (ii) $f(k) > 0$

Final Conclusion

The real number k lies between the roots of the quadratic equation $f(x) = ax^2 + bx + c = 0$ if and only if a and $f(k)$ are of opposite signs, that is, if and only if

(i) $a > 0$ (ii) $D > 0$ (iii) $f(k) < 0$

or

(i) $a < 0$ (ii) $D > 0$ (iii) $f(k) > 0$

Combining, we may say k lies between the roots of $f(x) = ax^2 + bx + c = 0$ if and only if

(i) $D > 0$ (ii) $af(k) < 0$

Illustration 3. Let $x^2 - (k-3)x + k = 0$ ($k \in R$) be a quadratic equation. Find the value of k for which one root is smaller than 2 and the other root is greater than 2

Solution:

One root is smaller than 2 and other root is greater than 2 $\Rightarrow$ 2 lies between the roots

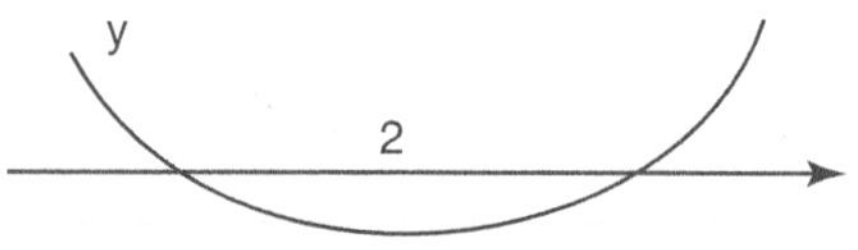

Since coefficient of x^2 is +ve so necessary and sufficient condition is

$f(2) < 0 \Rightarrow 4 - 2(k-3) + k < 0 \Rightarrow k > 10.$

4.4.4 Conditions for exactly one root of a quaratic equation to lie in the interval K_1, K_2 where $K_1 < K_2$

Case (i) If a > 0

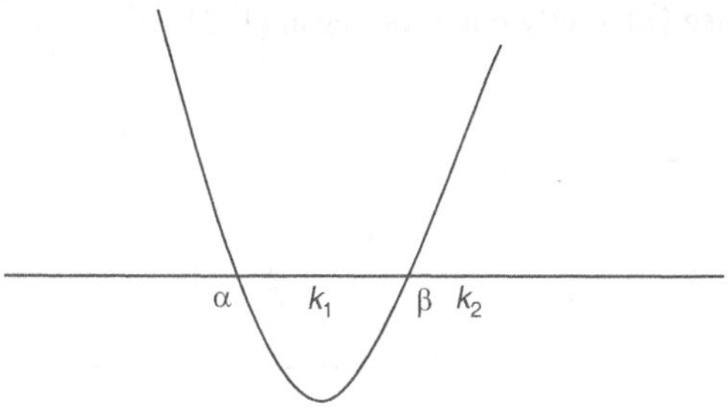

In this cae $f(k_1) < 0$ and $f(k_2) > 0$

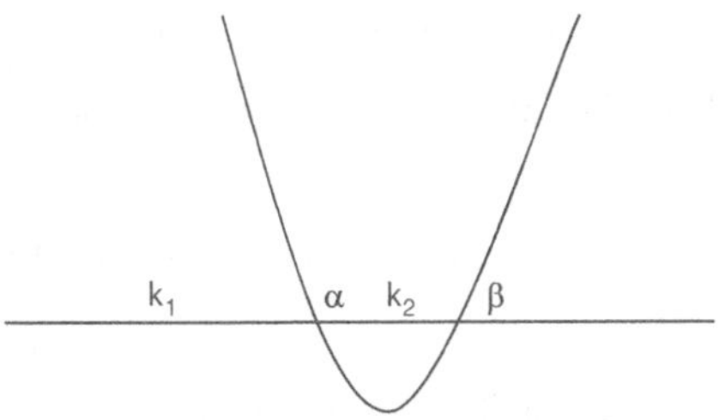

In this cae $f(k_1) > 0$ and $f(k_2) < 0$

If $a > 0$, then exactly one root of $\boldsymbol{f(x) = ax^2 + bx + c = 0}$ lies in the interval (k_1, k_2) if and only if $\boldsymbol{f(k_1) > 0}$ and $\boldsymbol{f(k_2) < 0}$. or $\boldsymbol{f(k_1) < 0}$ **and** $\boldsymbol{f(k_2) > 0}$

Final Conclusion: Thus, if $a > 0$, exactly one root of $\boldsymbol{f(x) = ax^2 + bx + c = 0}$ lies in the interval (k_1, k_2) if and only if $f(k_1)$ and $f(k_2)$ are of opposite sign i.e $f(k_1) f(k_2) < 0$.

Case (ii) If a < 0

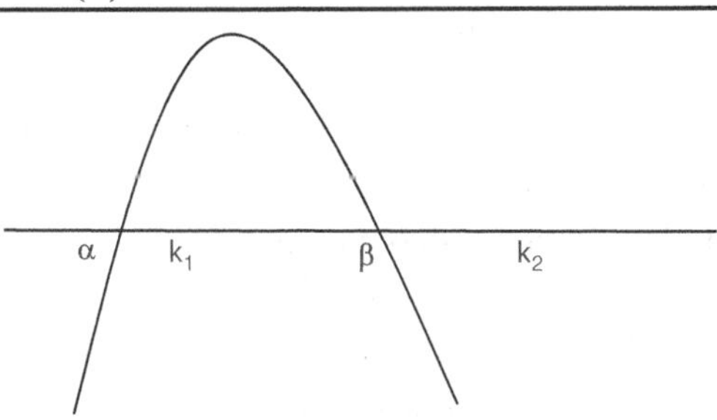

In this cae $f(k_1) > 0$ and $f(k_2) < 0$

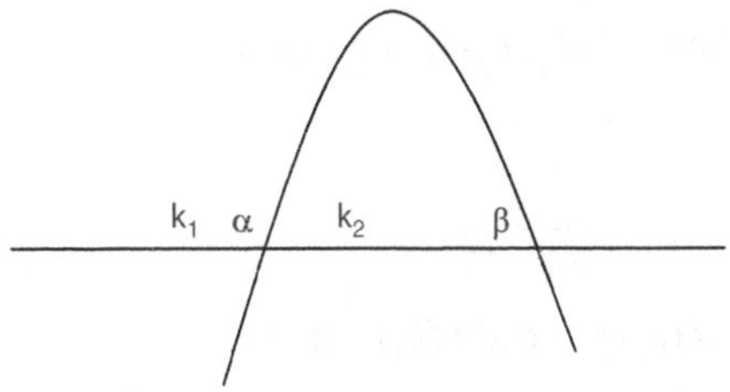

In this case $f(k_1) < 0$ and $f(k_2) > 0$

If $a < 0$, then exactly one root of $\boldsymbol{f(x) = ax^2 + bx + c = 0}$ lies in the interval (k_1, k_2) if and only if $\boldsymbol{f(k_1) > 0}$ **and** $\boldsymbol{f(k_2) < 0}$ Or $\boldsymbol{f(k_1) < 0}$ **and** $\boldsymbol{f(k_2) > 0}$

Final Conclusion: Thus, if $a < 0$, exactly one root of $\boldsymbol{f(x) = ax^2 + bx + c = 0}$ lies in the interval (k_1, k_2) if and only if $f(k_1) f(k_2) < 0$.

Combining above two cases if $f(k_1) f(k_2) < 0$ then there must be a root of quadratic equation in the interval (k_1, k_2).

Illustration 4. Let $x^2 - (k-3)x + k = 0 (k \in R)$ be a quadratic equation. Find the value of k for which exactly one root lies in the interval (1, 2).

Solution: We have two cases as shown below

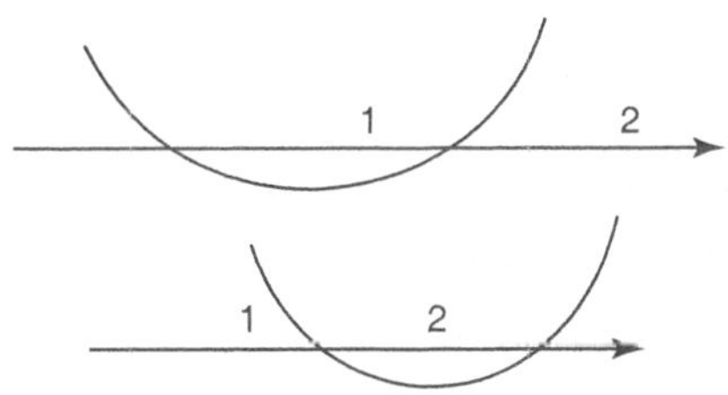

In each case $f(1) . f(2) < 0 \Rightarrow 4(10 - k) < 0$

$\Rightarrow k \in (10, \infty)$

4.4.5 Conditions for both the roots of a quadratic equation to lie in the interval k_1, k_2 where $k_1 < k_2$

Case (i) If a > 0

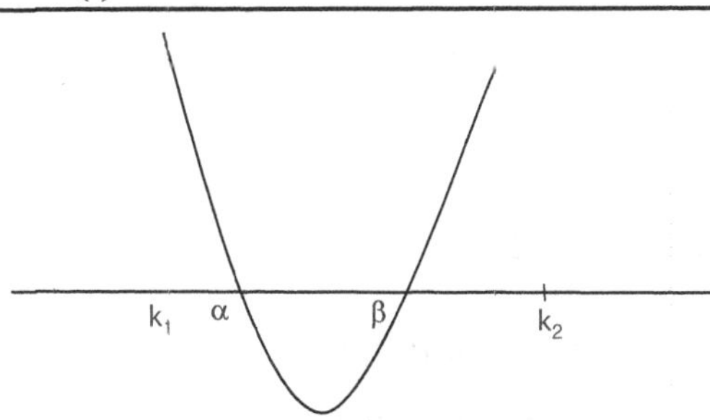

In this cae $f(k_1) > 0$ and $f(k_2) > 0$

If $a > 0$, both the roots of $f(x) = ax^2 + bx + c = 0$ lie in the interval (k_1, k_2) if and only if

(i) $D > 0$

(ii) $k_1 < -\dfrac{b}{2a} < k_2$

(iii) $f(k_1) > 0$ and $f(k_2) > 0$

Case (ii) If a < 0

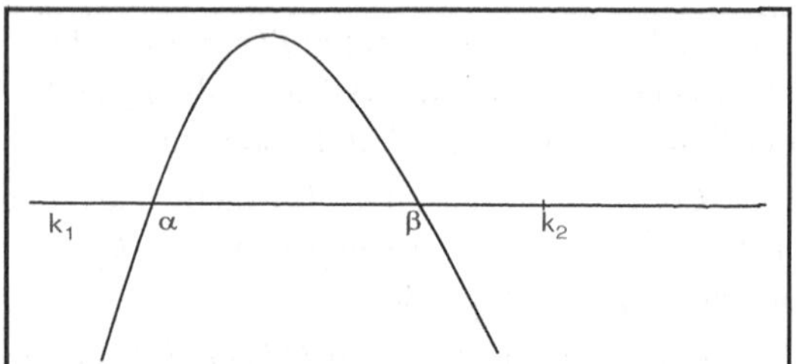

In this case f $(k_1) < 0$ and f $(k_2) < 0$

Final Conclusion:

In case $a < 0$, then the conditions are

(i) $D > 0$ (ii) $k_1 < -\dfrac{b}{2a} < k_2$

(iii) $f(k_1) < 0$ and $f(k_2) < 0$

Illustration 5. Let $x^2 - (k-3)x + k = 0 (k \in R)$ be a quadratic equation. Find the value of k for which both roots lie in the interval (1, 2).

Solution: We have two cases as shown below

Since Both roots lie in the interval (1, 2).

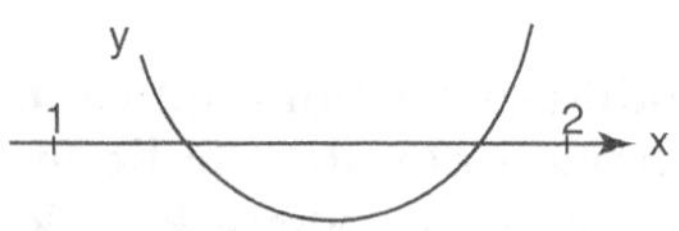

Then we have following conditions

(i) $f(1) > 0, 1 - k + 3 + k > 0 \Rightarrow 4 > 0$ which is true for all the values of k

(ii) $f(2) > 0, 4 - k + 6 > 0 \Rightarrow k < 10$

(iii) $1 < \dfrac{-b}{2a} < 2$ or $1 < \dfrac{k-3}{2} < 2 \Rightarrow 5 < k < 7$

(iv) $D \ge 0, k \in (-\infty, 1] \cup [9, \infty)$

No value of 'x' satisfy all the condition.

Illustration 6. Let $x^2 - (k-3)x + k = 0 (k \in R)$ be a quadratic equation. Find the value of k for which at least one root lie in the interval (1, 2)

Solution. We have two cases as shown below

Case I: Exactly one root lies in (1, 2)

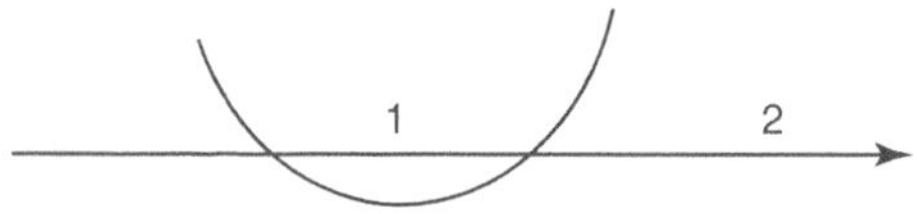

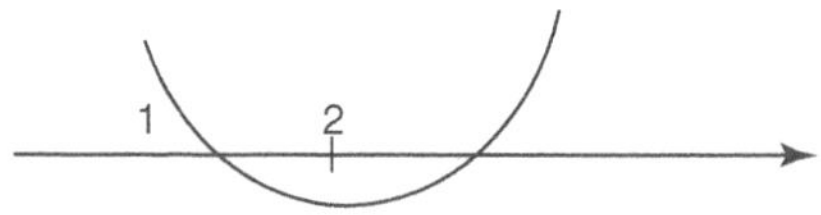

In all the cases we have

$$f(1) \cdot f(2) < 0 \Rightarrow k > 10$$

Case II: Both roots lie in (1, 2) From previous question no such value of 'k' exist

So for at least one root lies in (1, 2) $\Rightarrow k \in (10, \infty)$

Illustration 7. Let $x^2 - (k-3)x + k = 0 (k \in R)$ be a quadratic equation. Find the value of $\boldsymbol{k}$ for which one root is smaller than 1 and the other is greater than 2.

Solution: For one root greater than 2 and other is smaller than 1, conditions are

$f(1) < 0$ ... (i) and $f(2) < 0$ ... (ii)

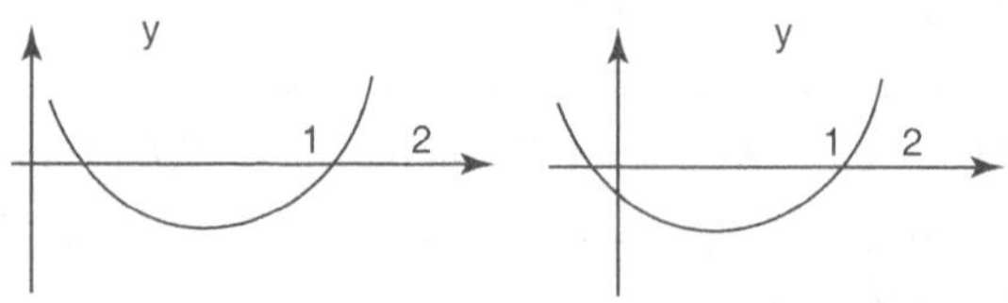

From (i), $f(1) < 0$, but $f(1) = 4$ which is not possible. So, no such 'k' exists.

Illustration 8. Let $x^2 - (k - 3)x + k = 0$ $(k \in R)$ be a quadratic equation. Find the value of k for which both the roots are positive.

Solution: Both the roots are positive $\Rightarrow D \ge 0$,

$D = (k-3)^2 - 4k = k^2 - 10k + 9 = (k-1)(k-9)$

or $k \in (-\infty, 1] \cup [9, \infty)$ ... (i)

Both sum and product of roots are positive

$\Rightarrow k - 3 > 0, k > 0$ and $k \in (-\infty, 1] \cup [9, \infty)$

$\Rightarrow k \in [9, \infty)$.

Illustration 9. Let $x^2-(k-3)x+k=0 (k\in R)$ be a quadratic equation. Find the value of k for which both the roots are equal but opposite in sign.

Solution: Since roots are equal in magnitude but opposite in sign

$\Rightarrow$ sum of roots is zero as well as $D \geq 0$

$\Rightarrow$ $k\in(-\infty,1]\cup[9,\infty)$ and $k-3=0$, i.e, $k=3$

$\Rightarrow$ no such k exists.

Illustration 10. If for real 'k' the roots of the equation $x^2-2kx+k^2+k-3=0$ are real and less than 3, then

(a) $k<2$ (b) $2\leq k\leq 3$
(c) $3<k\leq 4$ (d) $k>4$

Solution (a) : Consider the given function

$$f(x)=x^2-2kx+k^2+k-3=0$$

Since, $f(x)$ has real roots both less than 3. Therefore, $D>0$ and $f(3)>0$ and $-\dfrac{b}{2a}<3$

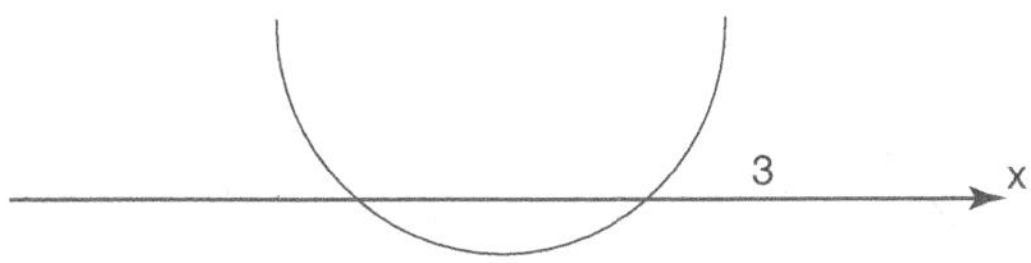

$\Rightarrow$ $k^2-(k^2+k-3)>0$ and $k^2-5k+6>$ and $k<3$

$\Rightarrow$ $k<3$ and $(k-2)(k-3)>0$

$\Rightarrow$ $k<3$ and $k<2$ or $k>3$

$\Rightarrow$ $k<2$

Illustration 11. If $a<b<c<d$ then roots of $(x-a)(x-c)+\pi(x-b)(x-d)=0$ are

(a) Real and distinct
(b) Complex
(c) Real and rational
(d) None of these

Solution (a) : Let $f(x)=(x-a)(x-c)+(x-b)(x-d)$

We know that a polynomial function is continuous on R and $f(a)>0$, $f(b)<0$, $f(c)<0$, $f(d)>0$

$\therefore$ $f(x)=0$ has a root in (a, b) and a root in (c, d)

Illustration 12. $a>0$, $a+b+c<0$ then roots of $ax^2+bx+c=0$ are

(a) Real and distinct
(b) Complex
(c) Real and rational
(d) None of these

Solution (a) : Since $a>0$, the parabola $y=ax^2+bx+c$ opens upwards.

Since $a+b+c<0$ so

$y(1)<0$, therefore we can say that $y=ax^2+bx+c$ meets the x-axis twice.

$\therefore$ $ax^2+bx+c=0$ has distinct real roots.

Concept Applicator (CA)-4

1. Let $x^2-(k-3)x+(k-3)=0 (k\in R)$ be a quadratic equation. Find the value of k for which the roots of the equation are real and distinct
   (a) $(-\infty, 3)\cup(7,\infty)$
   (b) $(-\infty, 0)\cup(9,\infty)$
   (c) $(-\infty, -9)\cup(9,\infty)$
   (d) None of these

2. Let $x^2-(k-3)x+k=0 (k\in R)$ be a quadratic equation. Find the value of k for which the roots of the equation are equal
   (a) $-9, -1$ (b) $-9, 1$
   (c) $3, 7$ (d) None of these

3. For what value of 'k' the equation $x^2-kx+8=0$ has exactly one root between 1 and 2
   (a) $(3, 9)$ (b) $(5, 9)$
   (c) $(6, 9)$ (d) None of these

4. The equation $x^2+bx+b^2=0$ has two roots each of which exceeds a number c, then which one must be true?
   (a) $a^2<4b^2$ (b) $c^2+ac+b^2>0$
   (c) $-\dfrac{a}{2}<c$ (d) None of these

5. If 'a' and 'b' be the roots of $4x^2 - 16x + k = 0$, $k \in R$ such that $1 < a < 2$ and $2 < b < 3$, then find the number of integral values of k
   (a) 5 (b) 6 (c) 2 (d) 3
6. If $x^2 - x + a - 3 < 0$ for at least one negative value of x, then complete set of values of 'a' is
   (a) $(-\infty, 4)$ (b) $(-\infty, 2)$
   (c) $(-\infty, 3)$ (d) $(-\infty, 1)$
7. If the roots of $x^2 + x + a = 0$ exceed a, then
   (a) $2 < a < 3$ (b) $a > 3$
   (c) $-3 < a < 3$ (d) $a < -2$
8. Find the necessary and sufficient condition for the equation $(1 - a^2) x^2 + 2ax - 1 = 0$ to have a roots lying in the interval (0, 1)
   (a) $a \neq 0$ (b) $a > 0$
   (c) $a < 2$ or $a > 2$ (d) None of these
9. Let (α, β) be the roots of the equation $(x - a)(x - b) = c$, $c \neq 0$. The roots of the quadratic $(x - a)(x - \beta) + c = 0$ are
   (a) a, c (b) b, c
   (c) a, b (d) $a + c, b + c$
10. If both the roots of the equation $x^2 - 2px + p^2 - = 0$ are less than 4 then find the range of p.
   (a) $p > -1$ (b) $p < 5$
   (c) $p < 3$ (d) None of these

4.5 WAVY CURVE AND INEQUALITY

4.5.1 Inequality with Linear Factors

Case (i): $f(x) = (x - a)$ function with one factor

$f(x) = (x - a) < 0$ if $x < a$

$f(x) = (x - a) > 0$ if $x > a$

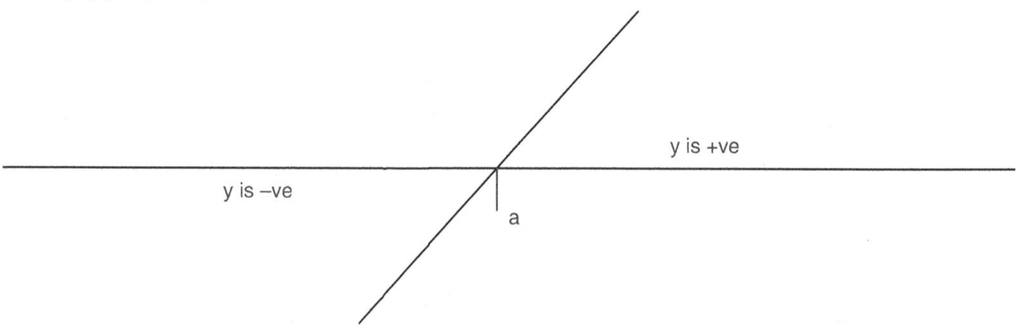

Case (ii): $f(x) = (x - a)(x - b)$

i.e function with two factors

$f(x) = (x - a)(x - b) < 0$ if $x \in (a, b)$

$f(x) = (x - a)(x - b) > 0$ if $x \in (-\infty, a) \cup (b, \infty)$

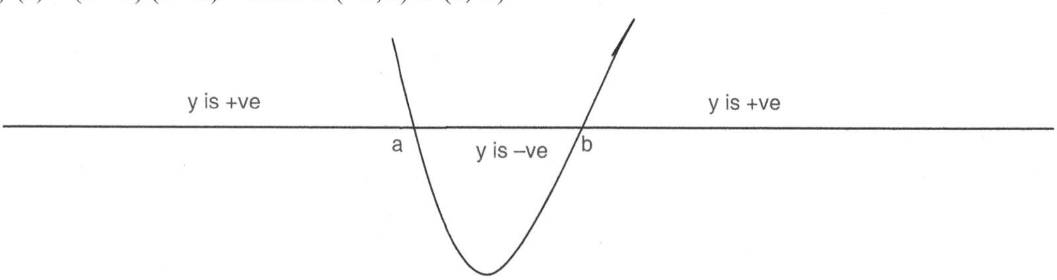

Now let us see the reason behind this.

Lets assume that $a < b$. Represent a and b on number line then we have three regions on the number line and these regions are $(-\infty, a)$, (a, b) and (b, ∞) let us see the sign of each factor and that of their product in each region.

Region	(x-a)	(x-b)	(x-a) (x-b)
$(-\infty, a)$	-ve	-ve	+ve
(a, b)	+ve	-ve	-ve
(b, ∞)	+ve	+ve	+ve

Case (iii): $f(x)\ (x-a)(x-b)(x-c)$ i.e function with three factors

In order to see the sign scheme of this function lets see sign of each factor.

Lets assume that $a < b < c$ represent a, b and c on number line then we have four regions on the number line and these regions are $(-\infty, a)$ (a, b), (b, c) and (c, ∞) let us see the sign of each factor and that of their product in each region.

Region	(x-a)	(a-b)	(x-c)	(x-a) (x-b) (x-c)
$(-\infty, a)$	–ve	–ve	–ve	–ve
(a, b)	+ve	–ve	–ve	+ve
(b, c)	+ve	+ve	–ve	–ve
(c, ∞)	+ve	+ve	+ve	+ve

Hence sign scheme is

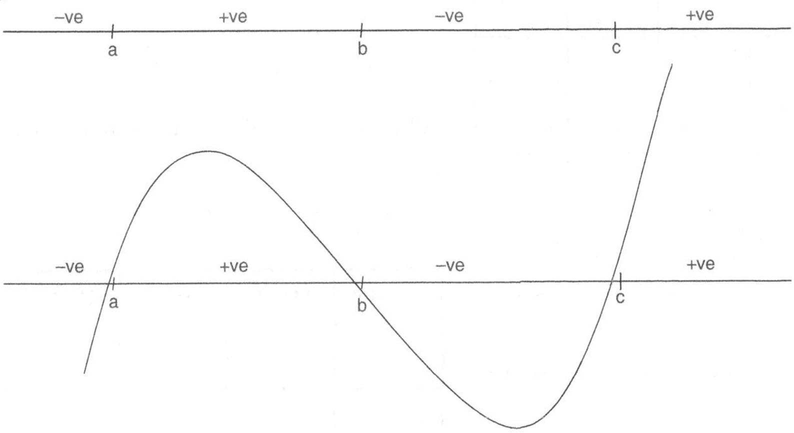

And this process will continue for 'n' number of factors

Important Points:

1. For function $f(x) = (x-a)(x-b)(x-c)\ ...(x-r)$
2. Mark the critical points i.e $a, b, c, ...r$ such that $a < b < c < .... < r$
3. Sign of $y = f(x)$ in the last region i.e (r, ∞) must be positive
4. Then at each of the critical points function will change sign from +ve to –ve or vice versa

Illustration 1. Solve $(x-1)(x-2)(x-3) \geq 0$

Solutions: First of all, equate all the factors to zero.

$$x - 1 = 0 \Rightarrow x = 1$$

$$x - 2 = 0 \Rightarrow x = 2$$

$$x - 3 = 0 \Rightarrow x = 3$$

Mark all these points on the number line

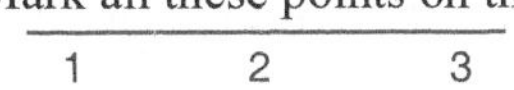

Region	$(x-1)$	$(x-2)$	$(x-3)$	$(x-1)(x-2)(x-3)$
$(-\infty, 1)$	–ve	–ve	–ve	–ve
$(1, 2)$	+ve	–ve	–ve	+ve
$(2, 3)$	+ve	+ve	–ve	–ve
$(3, \infty)$	+ve	+ve	+ve	+ve

Then draw the curve starting with positive side

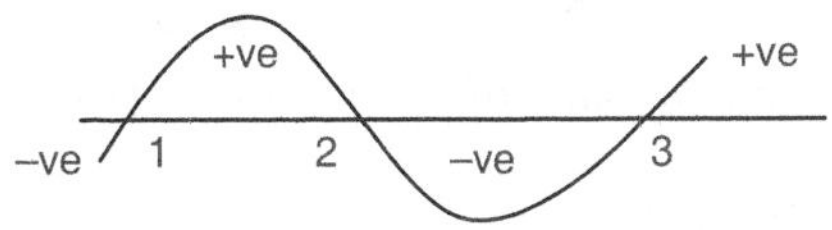

The required value of x is $[1, 2] \cup [3, \infty)$

Illustration 2. For what integral values of x,

$$(x-5)(x-10)(x-15)(x-20)<0$$

Solutions: First of all find the critical points, these are 5, 10, 15 and 20

Mark all these points on the number line

5 10 15 20

Region	(x–5)	(x–10)	(x–15)	(x–20)	(x–5)(x–10)(x–15)(x–20)
(–∞, 5)	–ve	–ve	–ve	–ve	+ve
(5, 10)	+ve	–ve	–ve	–ve	–ve
(10, 15)	+ve	+ve	–ve	–ve	+ve
(15, 20)	+ve	+ve	+ve	–ve	–ve
(20, ∞)	+ve	+ve	+ve	+ve	+ve

The sign scheme or wavy curve is as shown below:

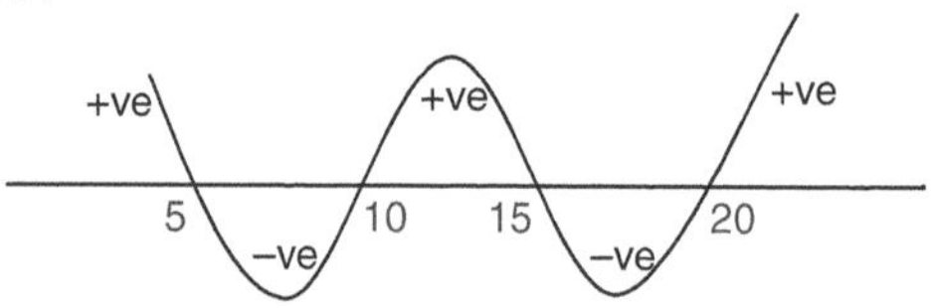

So integral values of x is {6, 7, 8, 9, 16, 17, 18, 19}
Total 8 integral values of x exist.

Illustration 3. For what values of x,

$$(x-5)(x-10)(x-15)(x-20)(x-25)>0$$

Solution: First of all find the critical points, these are 5, 10, 15, 20 and 25

Mark all these points on the number line

5 10 15 20 25

The sign scheme or wavy curve is as shown below:

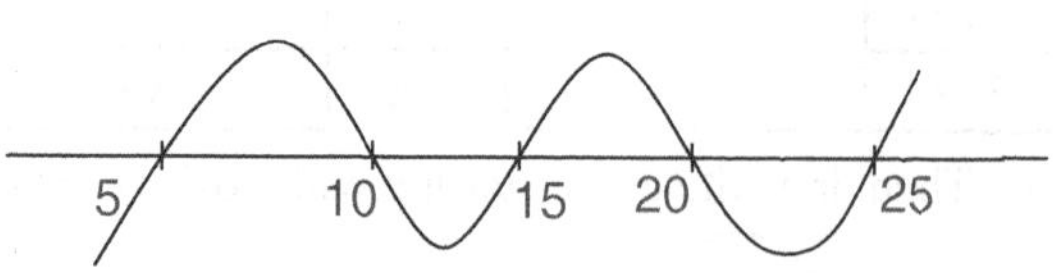

So required value of $x \in (5, 10) \cup (15, 20) \cup (25, \infty)$.

4.5.2 Inequality with Exponent of Linear Factors

We have already seen the wavy curve/ sign scheme of a function with linear factors like $f(x) = (x-a)(x-b)(x-c)$ now we will see when some of the factors will be raised to a certain power i.e $f(x) = (x-a)^p (x-b)^q (x-c)^r$. Nature of this curve depends on p, q, r whether they are even or odd.

If any of the exponent is odd then it will not change the process of sign scheme but if any exponent is even then at that factor function will not change its sign. Let us see this with an example.

Illustration 4. Solve $(x-1)^3(x-2)^5(x-3)^7 \geq 0$.

Solution: Mark all the critical points on the number line

1 2 3

Region	$(x-1)^3$	$(x-2)^5$	$(x-3)^7$	$(x-1)^3(x-2)^5(x-3)^7$
(-∞, 1)	-ve	-ve	-ve	-ve
(1, 2)	+ve	-ve	-ve	+ve
(2, 3)	+ve	+ve	-ve	-ve
(3, ∞)	+ve	+ve	+ve	+ve

Then draw the curve starting with positive side

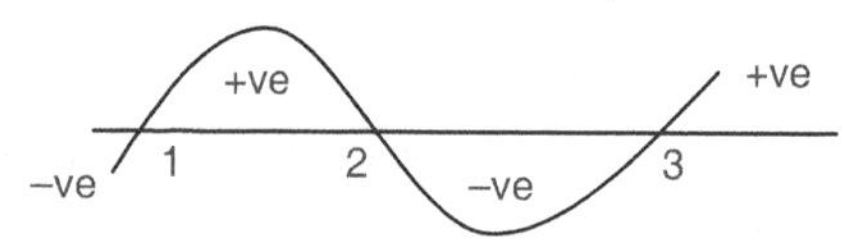

The required value of x is $[1, 2] \cup [3, \infty)$

If we observe then we will get that sign scheme of $f(x) = (x-1)(x-2)(x-3)$ as discussed in Illustration (1) and sign scheme of $(x-1)^3(x-2)^5(x-3)^7$ is same since all the exponents are odd.

Illustration 5. Solve $(x-1)^3(x-2)^2(x-3)^7 < 0$

Solution: Mark all the critical points on the number line

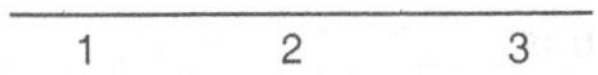

Region	$(x-1)^3$	$(x-2)^3$	$(x-3)^3$	$(x-1)^3$ $(x-2)^5$ $(x-3)^7$
$(-\infty, 1)$	–ve	+ve	–ve	+ve
(1, 2)	+ve	+ve	–ve	–ve
(2, 3)	+ve	+ve	–ve	–ve
$(3, \infty)$	+ve	+ve	+ve	+ve

Then draw the curve starting with positive side

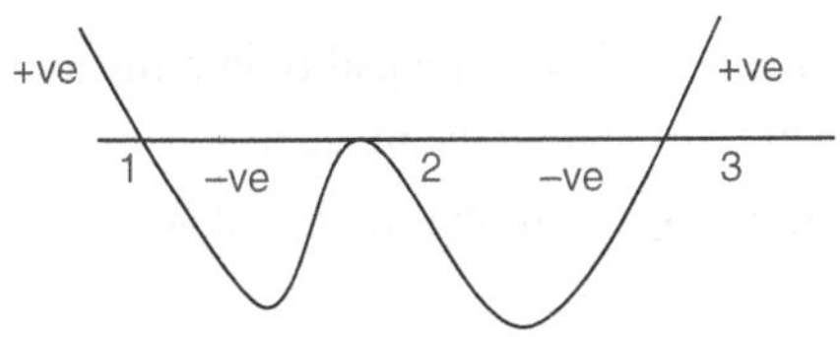

Here function will not change its sign at $x = 2$ because exponent of factor $(x-2)$ is even.

The required value of x is $(1, 2) \cup (2, 3)$

Illustration 6. Solve

$$(x-1)^3(x-2)^2(x-3)^7(x-4)^8 < 0$$

Solution: Mark all the critical points on the number line

Then draw the curve starting with positive side

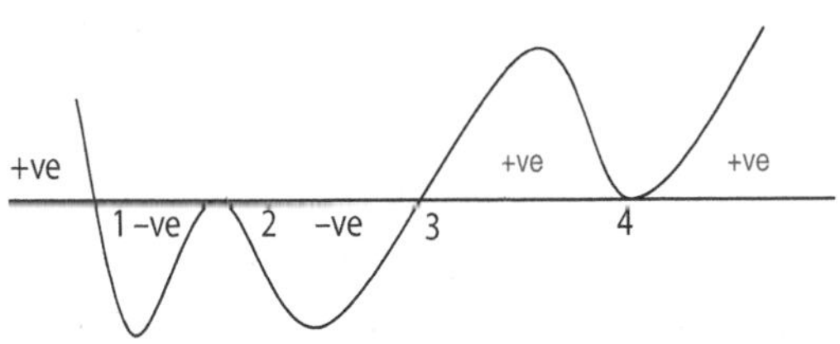

Here function will not change its sign at $x = 2$ and 4 because exponent of factor $(x-2)$ and $(x-4)$ is even.

The required value of x is $(1, 2) \cup (2, 3)$

4.5.3 Inequality with Rational Factors

Rational factor is defined as $f(x) = \dfrac{\{p(x)\}}{\{q(x)\}}$ provided $q(x) \neq 0$

Sign scheme of $\dfrac{\{p(x)\}}{\{q(x)\}}$ and that of $\{p(x)\}\{q(x)\}$ are same (only the sign scheme not the curve or wavy curve)

We have to take care of the values of x such that $q(x) \neq 0$, We will understand this by one example

Illustration 7. Find the value of 'x' such that $f(x) = \dfrac{\{x-2\}}{\{x-4\}} \geq 0$.

Solution: Sign scheme of $\dfrac{\{x-2\}}{\{x-4\}}$ & $(x-2)(x-4)$ is same (But not the curve)

We know that sign scheme of $(x-2)(x-4)$ is

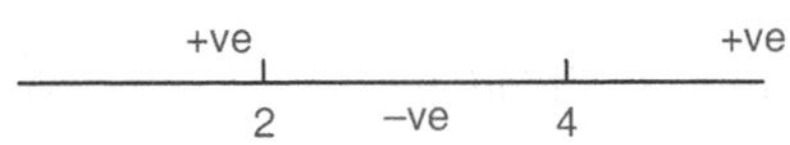

And this is only the sign scheme of $\dfrac{\{x-2\}}{\{x-4\}}$ only difference in that $x \neq 4$

If $(x-2)(x-4) \geq 0$ then $x \in (-\infty, 2] \cup [4, \infty)$, here 2 and 4 both are included

If $\dfrac{\{x-2\}}{\{x-4\}} \geq 0$ then $x \in (-\infty, 2] \cup (4, \infty)$ only 2 is included.

Illustration 8. Find the value of 'x' such that

$$f(x) = \frac{\{(x-2)(x-4)\}}{\{(x-3)(x-5)\}} \geq 0$$

Solution: Sign scheme of $\dfrac{\{(x-2)(x-4)\}}{\{(x-3)(x-5)\}}$ and $(x-2)(x-4)(x-3)(x-5)$ is same

We know that sign scheme of $(x\ \ 2)(x\ \ 4)$ $(x-3)(x-5)$ is

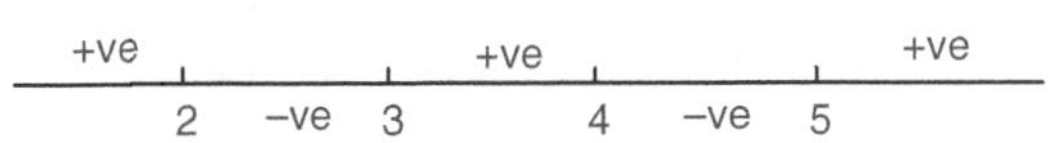

So $\dfrac{\{(x-2)(x-4)\}}{\{(x-3)(x-5)\}} \geq 0$ for

$x \in (-\infty, 2] \cup (3, 4] \cup (5, \infty)$

Note that point 3 and 5 can not be included

Illustration 9. Find the value of 'x' such that

$$f(x)=\frac{\left\{(x-2)^2(x-4)^3\right\}}{\left\{(x-3)^4(x-5)^5\right\}}\geq 0$$

Solution: Sign scheme of $\frac{\left\{(x-2)^2(x-4)^4\right\}}{\left\{(x-3)^3(x-5)^5\right\}}$ and $(x-2)^2(x-4)^4(x-3)^3(x-5)^5$ is same

We know that sign scheme of $(x-2)^2(x-4)^4(x-3)^3(x-5)^5$ is

+ve | 2 | +ve | 3 | +ve | 4 | −ve | 5 | +ve

So $\frac{\left\{(x-2)^2(x-4)^3\right\}}{\left\{(x-3)^4(x-5)^5\right\}}\geq 0$ for

$x \in (-\infty, 3) \cup (5, \infty) \cup (3, 4)$

Illustration 10. Find the value of 'x' such that

$$f(x)=\frac{\left(x^2+x+1\right)\left(x^2-x-6\right)}{\left(x^2-x+1\right)\left(x^2-9x+20\right)}\leq 0$$

Solution: Consider factors one by one

$$x^2+x+1 \geq 0 \text{ for all } x$$
$$x^2-x-6=(x+2)(x-3)$$
$$x^2-x+1 \geq 0 \text{ for all } x$$
$$x^2-9x+20=(x-4)(x-5)$$

So sign scheme of $f(x)=\frac{(x^2+x+1)(x^2-x-6)}{(x^2-x+1)(x^2-9x+20)}$

is same as that of $\frac{\{(x+2)(x-3)\}}{\{(x-4)(x-5)\}}$

Sign scheme of $\frac{\{(x+2)(x-3)\}}{\{(x-4)(x-5)\}}$ is

+ve | −2 | −ve | 3 | +ve | 4 | −ve | 5 | +ve

So required value of x is $[-2, 3] \cup (4, 5)$.

4.5.4 Range of a Rational Function

Consider a rational function $\frac{p(x)}{q(x)}$ here both $p(x)$ and $q(x)$ are quadratic function, then the rational function $\frac{p(x)}{q(x)}$ may not attain all real number and has a range for real x. We will take one example to understand this.

Let the rational function as $\frac{ax^2+bx+c}{dx^2+ex+f}$ we need to find its range for real x

Let $\frac{ax^2+bx+c}{dx^2+ex+f}=y$

or $ax^2+bx+c=y(dx^2+ex+f)$

or $(a-dy)x^2+(b-ey)x+(c-yf)=0$

Since x is a real number so $D \geq 0$

or $b^2+e^2y^2-2eby-4(c-ef)dy-4a(c-ef)\geq 0$

This is a quadratic in equality in terms of y, on solving this we can find the range of y.

Illustration 11. Find the range of the function $\frac{x^2-12}{2x-7}$.

Solution: Let $y=\frac{x^2-12}{2x-7}$

$\Rightarrow x^2-2yx+7y-12=0$

Since x is real so $4y^2-4(7y-12)\geq 0$

$\Rightarrow y^2-7y+12\geq 0 \Rightarrow (y-3)(y-4)\geq 0$

$\Rightarrow y \in (-\infty, 3] \cup [4, \infty)$.

Illustration 12. Find the range of the function $\frac{x^2-2x+3}{x^2-2x-8}$.

Solution: Let $y=\frac{x^2-2x+3}{x^2-2x-8}$

$\Rightarrow (x^2-2x-8)y=x^2-2x+3$

$\Rightarrow (y-1)x^2-2(y-1)x-(8y+3)=0$

As $x \in R$, $4(y-1)^2+4(y-1)(8y+3)\geq 0$.

$\Rightarrow (y-1)(9y+2)\geq 0$

$\Rightarrow y\leq -2/9$ or $y\geq 1$.

But if $y=1$ then

$\frac{x^2-2x+3}{x^2-2x-8}=1$ or $x^2-2x+3=x^2-2x-8$

$\Rightarrow \quad 3 = -8$ which is not possible so $y \neq 1$

Thus, $y \in (-\infty, -2/9] \cup (1, \infty)$.

Illustration 13. Find the range of the function $\dfrac{x^2-12x+100}{x^2+12x+100}$.

Solution. Let $y = \dfrac{x^2-12x+100}{x^2+12x+100}$

Then $(x^2 + 12x + 100)\, y = x^2 - 12x + 100$

$\Rightarrow \quad (y-1)\, x^2 + 12\,(y+1)\, x + 100\,(y-1) = 0$

As x is real $36\,(y+1)^2 - 100\,(y-1)^2 \geq 0$

$\Rightarrow \quad (y - 1/4)\ (y-4) \leq 0$

$\Rightarrow \quad \dfrac{1}{4} \leq y \leq 4.$

Illustration 14. Find the value of x where $\dfrac{x^2-12x+100}{x^2+12x+100}$ will attain maximum/minimum.

Solution: From the previous solution $1/4 \leq y \leq 4$

For $y = 4$, we get $x = -10$

For $y = \dfrac{1}{4}$, we get $x = 10$.

Concept Applicator (CA)-5

1. Find the number of integral values of x such that
$$(x-10)(x-20)(x-30)(x-40) < 0$$
2. Find the value of x such that
$$(x-10)(x-20)(x-30) < 0$$
3. Find the value of x such that
$$(x+1)(x-3)(x-2)^2 \geq 0$$
4. Find the value of x such that
$$x(x+6)(x+2)^2(x-3) > 0$$
5. Find the value of x such that
$$(x-1)^2(x+1)^3(x-4) < 0$$
6. Find the value of x such that $\dfrac{2x-3}{3x-7} \geq 0$
7. Find the value of 'x' such that
$$f(x) = \frac{\{(x-5)(x-15)\}}{\{(x-10)(x-20)\}} \geq 0$$
8. Find the value of 'x' such that
$$f(x) = \frac{\{(x-5)^2(x-15)^3\}}{\{(x-10)^4(x-20)^5\}} \geq 0$$
9. Find the value of 'x' such that
$$f(x) = \frac{(x^2+3x+10)(x^2-9x+20)}{(x^2-5x+15)(x^2-13x+42)} \leq 0$$
10. Find the value of 'x' such that $\dfrac{x^2-5x+12}{x^2-4x+5} > 3$.

4.6 CUBIC AND POLYNOMIAL EQUATION

4.6.1 Transformation of Equation

We already discussed about this in chapter 4.2.2, lets revise this once and then continue the same.

In this section we will learn how to form or transform a given equation from the given condition, Let the given equation be

$$f(x) = a_0x^n + a_1x^{n-1} + \ldots + a_{n-1}x + a_n = 0 \quad \ldots (1)$$

Case (i): To form an equation whose roots are $k\,(\neq 0)$ times roots of the equations in (1), replace x by x/k in equation (1).

Case (ii): To form an equation whose roots are the negatives of the roots in equation (1), replace x by $-x$ in equation (1).

Case (iii): To form an equation whose roots are k more than the roots of equation in (1), replace x by $x - k$ in equation (1).

Case (iv): To form an equation whose roots are reciprocals of the roots in equation (1), replace x by $1/x$ in equation (1) and then multiply both the sides by x^n.

Case (v): To form an equation whose roots are square of the roots of the equation in (1) then we have to follow these steps.

Step 1: Replace x by $\sqrt{x}$ in equation (1).

Step 2: Collect all the terms involving $\sqrt{x}$ on one side.

Step 3: Square both the sides and simplify.

Illustration 1. Form an equation whose roots are squares of the roots of $x^3 + 2x^2 - x + 2 = 0$.

Solution:

Step 1: Replace x by $\sqrt{x}$ the result is

$x\sqrt{x} + 2x - \sqrt{x} + 2 = 0$

Step 2: Collect all the terms involving $\sqrt{x}$ on one side the result is

$\sqrt{x}\ (x - 1) = -2\ (x + 1)$

Step 3: Square both the sides and simplify to get

$\mathbf{x(x - 1)^2 = 4(x + 1)^2}$

$\Rightarrow \quad \mathbf{x^3 - 6x^2 - 7x - 4 = 0}$

Case (vi): to form an equation whose roots are cubes of the roots of the equation then we have to follow these steps:

Step 1: Replace x by $x^{\frac{1}{3}}$

Step 2: Collect all the terms involving $x^{\frac{1}{3}}$ and $x^{\frac{2}{3}}$ on one side.

Step 3: Cube both the sides and simplify.

Important Note: Let α, β be roots of the equation $ax^2 + bx + c = 0$ then equation whose roots are

(i) $-\alpha, -\beta$ is $ax^2 - bx + c = 0$

(ii) $-1/\alpha \quad -1/\beta$

Replace x by $-1/x$ to obtain $a - bx + cx^2 = 0$

(iii) α^2, β^2

Replace x by $\sqrt{x}$ to obtain $ax + b\sqrt{x} + c = 0$

$\Rightarrow \quad a^2x^2 + 2\,acx + c^2 = b^2x$

$\Rightarrow \quad a^2x^2 + (2ac - b^2)\,x + c^2 = 0$

(iv) $1/\alpha^2, 1/\beta^2$

Change x to $1/x$ in the above equation (case iii) we get

$$\frac{a^2}{x^2} + \frac{2a\,c - b^2}{x} + c^2 = 0$$

or $\quad a^2 + (2ac - b^2)\,x + c^2x^2 = 0$

(v) $a\alpha + b, \alpha\beta + b$

Replace x to $\dfrac{x-b}{a}$ we get

$(x - b)^2 + b\,(x - b) + ac = 0$

(vi) $\alpha + \dfrac{b}{2a}, \beta + \dfrac{b}{2a}$

$4\,a^2x^2 - b^2 + 4\,ac = 0.$

Illustration 2. If α and β are the roots of the quadratic equation $x^2 + x + 1 = 0$ then find the quadratic equation whose roots are $\alpha + 2$ and $\beta + 2$.

Solution: We have to form an equation whose roots are $\alpha + 2$ and $\beta + 2$

Sum of the roots $= (\alpha + 2) + (\beta + 2) = \alpha + \beta + 4 = -1 + 4 = 3$

Product of the roots $= (\alpha + 2)(\beta + 2) = \alpha\beta + 2(\alpha + \beta) + 4 = 1 - 2 + 4 = 3$

So required quadratic equation is $x^2 - 3x + 3 = 0$

Alternately:

Since roots are $\alpha + 2$ and $\beta + 2$ so substitute $(x - 2)$ instead of 'x' in original equation then we get

Required equation is $x^2 - 3x + 3 = 0$.

Illustration 3. If α and β are the roots of the quadratic equation $x^2 - 5x + 6 = 0$ then find the quadratic equation whose roots are $\dfrac{1}{3\alpha+2}$ and $\dfrac{1}{3\beta+2}$.

Solution: Roots of $x^2 - 5x + 6 = 0$ are 2 and 3

So $\quad \dfrac{1}{3\alpha+2} = \dfrac{1}{6+2} = \dfrac{1}{8}$ and $\dfrac{1}{3\beta+2} = \dfrac{1}{9+2} = \dfrac{1}{11}$

Sum of the roots $= \dfrac{1}{8} + \dfrac{1}{11} = \dfrac{19}{88}$

Product of the roots = $\left(\frac{1}{8}\right)+\left(\frac{1}{11}\right)=\frac{1}{88}$

Required equation is

$$x^2-\left(\frac{19}{88}\right)x+\frac{1}{88}=88x^2-19x+1=0.$$

Illustration 4. Let α, β be roots of the equation $ax^2 + 2\,bx + c = 0$. Then the equation whose roots are $\pm\,\alpha^2.\pm\,\beta^2$ is

Solution: α^2, β^2 are roots of $ax + 2\,b\sqrt{x}\,+ c = 0$

$$(ax + c\,)^2 = 4\,b^2x$$

$\Rightarrow\quad a^2\,x^2\,+ 2\,(ac - 2\,b^2)\,x + c^2 = 0$

And, $-\,\alpha^2, -\,\beta^2$ are roots of

$a^2\,x^2\,- 2\,(ac - 2\,b^2)\,x + c^2 = 0$

$\Rightarrow$ Required equation is

$(a^2\,x^2 + c^2\,)^2 - 4\,(ac - 2\,b^2)x^2 = 0.$

Illustration 5. Let α, β be roots of the equation $ax^2\,+ 2\,bx + c = 0$. Then the equation whose roots are $\pm\,1/\alpha$, $\pm\,1/\beta$ is

Solution: $1/\alpha$, $1/\beta$ are roots of $a + 2\,bx + cx^2 = 0$ and

$-\,1/, -\,1/\beta$ are roots of $a - 2\,bx + cx^2 = 0$

$\therefore$ Required equation is

$(cx^2 + a)^2 - 4b^2\,x^2 = 0$

or, $c^2x^4 + 2\,(ac - 2\,b^2)\,x^2 + a^2 = 0.$

4.6.2 Cubic Equation

Cubic Equation:

A cubic equation is in the form of

$f(x) = ax^3 + bx^2 + cx + d = 0$ where $a \neq 0$

If α, β, γ are the roots of $f(x) = ax^3 + bx^2 + cx + d = 0$, then the equation can be written as

$ax^3 + bx^2 + cx + d = a(x - \alpha)\,(x - \beta)(x - \gamma)$

$= a\{x^3 - (\alpha + \beta + \gamma)x^2 + (\beta\gamma + \gamma\alpha + \alpha\beta)x - \alpha\beta\gamma\}$

Comparing the coefficients we get

Sum of the roots $\alpha+\beta+\gamma=-\frac{b}{a}$,

Sum of the roots taken two at a time

$\beta\gamma + \gamma\alpha + \alpha\beta = \frac{c}{a}$ and

Product of the roots $\alpha\beta\gamma = -\frac{d}{a}$

A cubic equation can have

(i) Three real roots

(ii) 1 real and one more real and repeated roots

(iii) 1 real and two complex roots

> Consider a cubic equation in the form of
>
> $x^3 + ax^2 + bx + c = 0$, it has
>
> (i) Three real roots if $(2a^3 - 9ab + 27c)^2 \leq 4(a^2 - 3b)^3$
>
> (ii) 1 real and one more real and repeated roots
> $(2a^3 - 9ab + 27c)^2 = 4(a^2 - 3b)^3$
>
> (iii) 1 real and two imaginary roots
> $(2a^3 - 9ab + 27c)^2 > 4(a^2 - 3b)^3$

Now we will analyze the graph of a cubic function $ax^3 + bx^2 + cx + d$

Case (i)	$a > 0$	Only one real root	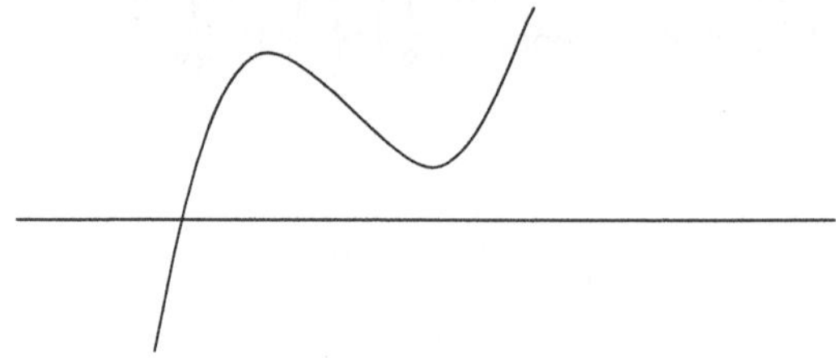
Case (ii)	$a > 0$	One real root and one repeated real root	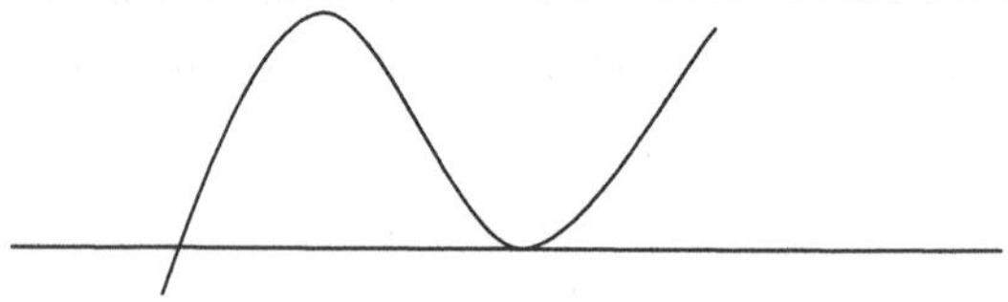
Case (iii)	$a > 0$	Three real roots	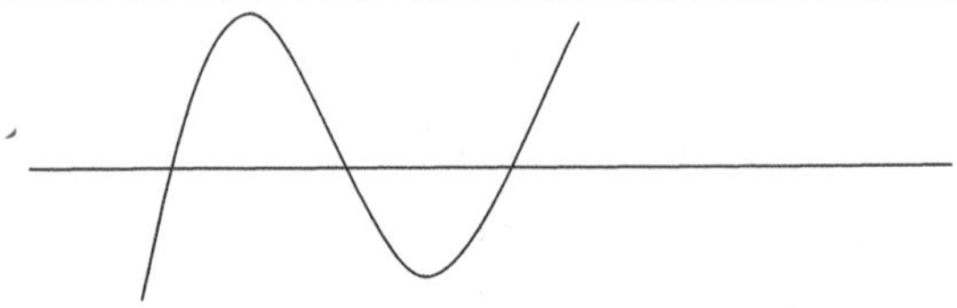
Case (iv)	$a < 0$	Only one real root	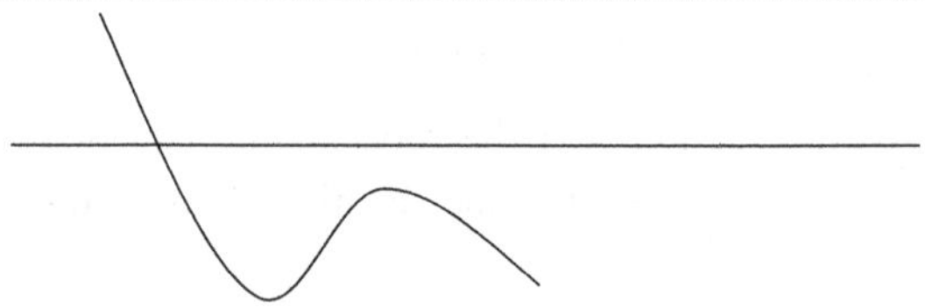
Case (v)	$a < 0$	One real root and one repeated real root	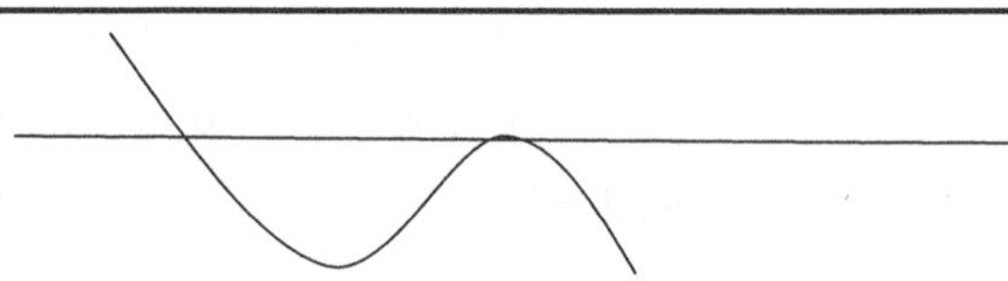
Case (vi)	$a < 0$	Three real roots	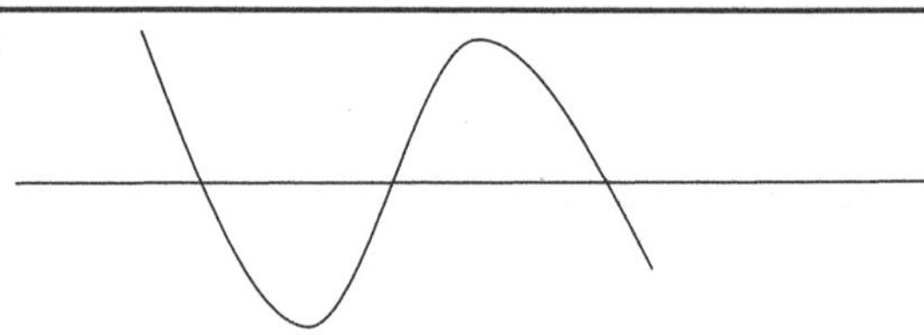

Illustration 6. Use factorization method to find the roots of $x^3 - 3x^2 - x + 3 = 0$

Solution: Consider $f(x) = x^3 - 3x^2 - x + 3$

Since $f(1) = 0$ so $x = 1$ is a root of the equation and $(x - 1)$ is a factor of the function

$x^3 - 3x^2 - x + 3 = (x - 1)(x^2 - 2x - 3) = (x - 1)(x + 1)(x - 3)$

So roots of the equation are – 1, 1 and 3.

Illustration 7. Plot the graph of $y = x^3 - 3x^2 - x + 3$

Solution: By factorizing we get

$$y = x^3 - 3x^2 - x + 3 = (x - 1)(x + 1)(x - 3)$$

So roots of the corresponding equation is – 1, 1 and 3

And graph is as shown below:

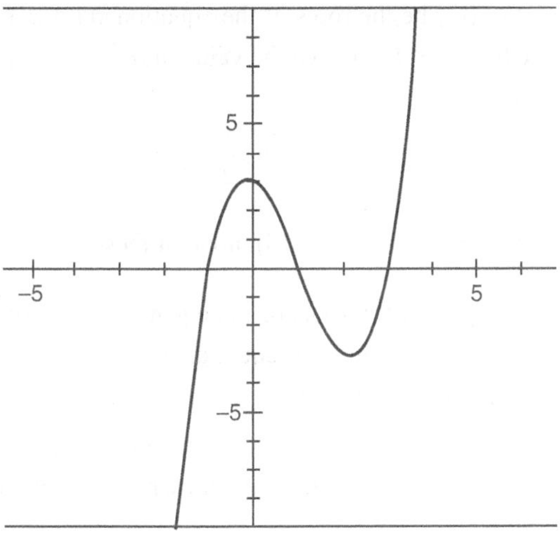

Illustration 8. Find the conditions on a, b, c and d such that equations $2ax^3 + bx^2 + cx + d = 0$ and $2ax^2 + 3bx + 4c = 0$ have a common root.

Solution: Let 'α' be a common root of the given equations, then $2a\alpha^3 + b\alpha^2 + c\alpha + d = 0$ ... (i)

and $2a\alpha^2 + 3b\alpha + 4c = 0$ ... (ii)

Now multiply Eq. (ii) by α and then subtract Eq. (i) from it, then we get

$$2b\alpha^2 + 3c\alpha - d = 0 \quad \text{... (iii)}$$

Now we have two quadratic equations (ii) and (iii) having a common root α, so

$$\frac{\alpha^2}{-3bd - 12c^2} = \frac{\alpha}{8bc + 2ad} = \frac{1}{6ac - 6b^2},$$

Eliminating α from these two equations, we get

$$(4bc + ad)^2 = \frac{9}{2}(bd + 4c^2)(b^2 - ac),$$

So the required condition is

$$(4bc + ad)^2 = \frac{9}{2}(bd + 4c^2)(b^2 - ac).$$

Illustration 9. If the roots of $x^3 - 12x^2 + 39x - 28 = 0$ are in AP, then their common difference is

(a) ± 1 (b) ± 2

(c) ± 3 (d) ± 4

Solution: If roots are $\alpha - k.\ \alpha.\ \alpha + k$ with common difference k, then $\alpha - k + \alpha + \alpha + k = 12$

$\Rightarrow$ $\alpha = 4$, and $(\alpha - k)\,\alpha\,(\alpha + k) = 28$

$\Rightarrow$ $(4^2 - k^2)4 = 28 \Rightarrow k^2 = 9 \Rightarrow k = \pm 3$

Option (c)

Illustration 10. Let α, β, γ be roots of $x^3 + px + q = 0$, then find an equation whose roots are $\alpha^3, \beta^3, \gamma^3$

Solution: Put $y = \alpha^3 \Rightarrow \alpha = y^{1/3}$

$\therefore$ $y + py^{1/3} + q = 0$

$\Rightarrow$ $(y + q)^3 = -p^3 y$

$\Rightarrow$ $y^3 + 3qy^2 + (3q^2 + p^3)y + q^3 = 0$

or required equation is

$$x^3 + 3qx^2 + (3q^2 + p^3)y + q^3 = 0$$

4.6.3 Bi-quadratic Equation

Consider a bi-quadratic function

$f(x) = ax^4 + bx^3 + cx^2 + dx + e$

If α, β, γ, δ are the roots of the equation $ax^4 + bx^3 + cx^2 + dx + e = 0$, then

$$\alpha + \beta + \gamma + \delta = -\frac{b}{a},$$

$$\alpha\beta + \alpha\gamma + \alpha\delta + \beta\gamma + \beta\delta + \gamma\delta = \frac{c}{a}$$

$$\alpha\beta\gamma + \alpha\beta\delta + \alpha\gamma\delta + \beta\gamma\delta = -\frac{d}{a},$$

$$\alpha\beta\gamma\delta = \frac{\mathbf{e}}{\mathbf{a}}$$

***Polynomial Equation*:**

The equation $f(x) = a_0x^n + a_1x^{n-1} + a_2x^{n-2} + \ldots + a_{n-1}x + a_n = 0$ where $a_0, a_1, \ldots, a_{n-1}, a_n \in C$, the set of complex numbers, and $a_0 \neq 0$, is said to be an equation of degree n. An equation of degree n has exactly n roots. Let $\alpha_1, \alpha_2, \ldots, \alpha_n \in C$ be the n roots of. Then

$$f(x) = \mathrm{a}_0(x - \alpha_1)(x - \alpha_2) \ldots (x - \alpha_n)$$

$$\alpha_1 + \alpha_2 + \ldots + \alpha_n = -\frac{a_1}{a_0}$$

$$\alpha_1\alpha_2\ldots\alpha_n = (-1)^n \frac{a_n}{a_0}$$

4.6.4 Factor Theorem

Consider a polynomial $f(x)$ by doing normal algebraic division we can find the quotient and remainder when this function is divided by a linear factor $(x - a)$. But according to remainder theorem when a polynomial $f(x)$ divided by a linear factor $(x - a)$ remainder is $f(a)$. We will try to understand this theorem by some examples.

Illustration 13. Find the quotient and remainder by doing normal algebraic division when $x^3 - 7x - 6$ divided by $(x - 4)$. Also use remainder theorem to find the remainder.

Solution: Consider polynomial $p(x) = x^3 - 7x - 6$, and let's divide by the linear factor $x - 4$

```
          x^2 - 4x^2 + 9
x - 4 ) x^3 + 0x^2 - 7x - 6
       -x^3 - 4x^2
       ----------------
             4x^2 - 7x^2 - 6
            -4x^2 - 16x
            ---------------
                   9x - 6
                  -9x - 36
                  --------
                       30
```

So we get a quotient of $q(x) = x^2 + 4x + 9$ and remainder of $r(x) = 30$.

Now by using remainder theorem we get required remainder as $f(4) = 4^3 - 7 \times 4 - 6$
$= 64 - 28 - 6 = 30$

Illustration 14. Find the remainder when $x^4 - 7x^2 + 6$ divided by $(x - 1)$.

Solution: Since $f(x) = x^4 - 7x^2 - 6$ so from remainder theorem required remainder is $f(1)$
$= 1 - 7 - 6 = -12$.

Concept Applicator (CA)-6

1. If α, β, γ be the roots of the equation $x(1 + x^2) + x^2(6 + x) + 2 = 0$, then the value of $\alpha^{-1} + \beta^{-1} + \gamma^{-1}$ is

   (a) -3 (b) $\frac{1}{2}$

   (c) $-\frac{1}{2}$ (d) none of these

2. If α, β, γ be the roots of the equation $x^3 - 13x + 11 = 0$, then find the equation whose roots are $\alpha + \beta$, $\beta + \gamma$ and $\alpha + \gamma$

3. If α, β, γ be the roots of the equation $x^3 + 3x + 2 = 0$, then find the equation whose roots are $\alpha\beta$, $\beta\gamma$ and $\alpha\gamma$

4. Using remainder theorem check that $x = 2$ is a root of the equation $x^3 + 3x + 2 = 0$ or not.

5. If $x = 3$ is a root of the equation $x^3 + 3x^2 - px + p = 0$ then find the value of p.

6. If $x^2 - 1$ is a factor of $x^4 + ax^3 + 3x - b$, then

   (a) $a = 3, b = -1$ (b) $a = -3, b = 1$

   (c) $a = 3, b = 1$ (d) None of these

7. Let α, β be roots of the equation $ax^2 + 2bx + c = 0$. Then the equation whose roots are $\pm 2\alpha$, $\pm 2\beta$ is

8. If $x^2 - 3x + 2$ is a factor of $x^4 - px^2 + q$, then p, q are

   (a) 2, 3 (b) 4, – 5

   (c) 5, 4 (d) 0, 0

9. If $f(x) = 2x^3 + mx^2 - 13x + n$ and 2, 3 are roots of the equation $f(x) = 0$, then find the value of m+n

   (a) 25 (b) – 35

   (c) 35 (d) None of these

10. Use remainder theorem to prove that 3 is a root of the equation $x^3 - 7x^2 + 36 = 0$.

Exercise A

Concept Builder (CB)

1. If $a > 1$, then find the nature of roots of the equation $(1-a)x^2 + 3ax - 1 = 0$
   (a) one positive and one negative
   (b) both negative
   (c) both positive
   (d) both non-real complex
2. The number of solutions of the equation $|x| = \cos x$ is
   (a) one (b) two
   (c) three (d) zero
3. The number of solutions of $|[x]-2x| = 4$, where $[x]$ is the greatest integer $\le x$, is
   (a) 2 (b) 4 (c) 1 (d) infinite
4. If a, b, c are three distinct positive real numbers, then the number of real roots of $ax^2 + 2b|x| - c = 0$ is
   (a) 4 (b) 2
   (c) 0 (d) None of these
5. If $a \in R$ and $a < 0, \ne -2$, then the equation $x^2 + a|x| + 1 = 0$
   (a) can not have any real root
   (b) must have exactly two real roots
   (c) must have either exactly two real roots or no real roots
   (d) must have either four real roots or no real roots
6. Let a, b, c be real numbers with $a \ne 0$ and let α, β be the roots of the equation $ax^2 + bx + c = 0$. Then $a^3x^2 + abcx + c^3 = 0$ has roots
   (a) $\alpha^2\beta\ \beta^2\alpha$ (b) α, β^2
   (c) $\alpha^2\beta, \beta\alpha$ (d) $\alpha^3\beta, \beta^3\alpha$
7. Let $P(x) = 0$ be the polynomial equation of least possible degree with rational coefficients having $\sqrt[3]{10} + \sqrt[3]{100}$ as a root. Then the product of all the roots of $P(x) = 0$ is
   (a) 30 (b) 42
   (c) 343 (d) 110
8. The number of real solutions of $x - \dfrac{1}{x^2-4} = 2 - \dfrac{1}{x^2-4}$ is
   (a) 0 (b) 1
   (c) 2 (d) infinite
9. The number of values of the pair (a, b) for which $a(x+1)^2 + b(x^2-3x-2) + x + 1 = 0$ is an identity in x is
   (a) 0 (b) 1
   (c) 2 (d) infinite
10. The polynomial $(ax^2 + bx + c)(ax^2 - dx - c)$, $ac \ne 0$ has
    (a) four real zeros
    (b) at least two real zeros
    (c) at most two real zeros
    (d) no real zeros
11. The equation $(a+2)x^2 + (a-3)x = 2a - 1$, $a \ne -2$ has roots rational for
    (a) all rational values of a except $a = -2$
    (b) all real values of a except $a = -2$
    (c) rational values of $a > \dfrac{1}{2}$
    (d) None of the above
12. The real roots of the equation $7^{\log_7(x^2-4x+5)} = x - 1$ are
    (a) 1 and 2 (b) 2 and 3
    (c) 3 and 4 (d) 4 and 5

13. If α, β are the roots of $ax^2 + bx + c = 0$ and $\alpha + k, \beta + k$ are the roots of $px^2 + qx + r = 0$, then find the value of $\frac{b^2 - 4ac}{q^2 - 4pr}$

(a) $\left(\frac{p}{a}\right)^2$ (b) 1

(c) $\left(\frac{a}{p}\right)^2$ (d) 0

14. If α, β are the roots of $ax^2 + bx + c = 0$ are in the ratio m : n, then

(a) $mnb^2 = ac(m+n)^2$

(b) $b^2(m+n) = mn$

(c) $m + n = b^2mn$

(d) $c^2(mn) = ab(m+n)^2$

15. The value of $\sqrt{6+\sqrt{6+\sqrt{6+\ldots\infty\infty}}}$ is

(a) 3 (b) 6

(c) – 2 (d) – 4

16. The roots of the equation $(q-r)x^2 + (r-p)x + (p-q) = 0$

(a) $\frac{r-p}{q-r}, 1$ (b) $\frac{p-q}{q-r}, 1$

(c) $\frac{q-r}{p-q}, 1$ (d) $\frac{r-p}{p-q}, 1$

17. If difference between the roots of the equation $x^2 + px + q = 0$ is 1, then

(a) $p^2 = 4q$ (b) $p^2 = 4q + 1$

(c) $p^2 = 4q - 1$ (d) None of these

18. If sum of the roots of the equation $\frac{x^2 - bx}{ax - c} = \frac{\lambda - 1}{\lambda + 1}$ is zero then the value of λ is

(a) $\frac{a-b}{a+b}$ (b) c

(c) $\frac{1}{c}$ (d) $\frac{a+b}{a-b}$

19. If the roots of the equation $(a^2 + b^2)y^2 - 2(ac + bd)y + c^2 + d^2 = 0$ are equal, then

(a) ab = dc (b) ac = bd

(c) ad + bc = 0 (d) $\frac{a}{b} = \frac{c}{d}$

20. If $x = \sqrt{7 + 4\sqrt{3}}$, then $x + \frac{1}{x}$ is equal to

(a) 4 (b) 6

(c) 3 (d) 2

21. The value of k for which the equation $x^2 + 2(k+1)x + k^2 = 0$ has equal roots is

(a) – 1 (b) $-\frac{1}{2}$

(c) 1 (d) None of these

22. The value of p for which the difference between the roots of the equation $x^2 + px + 8 = 0$ is 2 are

(a) ± 2 (b) ± 4

(c) ± 6 (d) ± 8

23. If the expression $a^2(b^2 - c^2)x^2 + b^2(c^2 - a^2)x + c^2(a^2 - b^2)$ is a perfect square, then

(a) a, b, c are in AP

(b) a^2, b^2, c^2 are in AP

(c) a^2, b^2, c^2 are in HP

(d) a^2, b^2, c^2 are in GP

24. If α is a root of $4x^2 + 2x - 1 = 0$, then the other root is

(a) $3\alpha^3 - 4\alpha$ (b) $4\alpha^3 - 3\alpha$

(c) $3\alpha^3 + 4\alpha$ (d) $4\alpha^3 + 3\alpha$

25. If the roots of the equation $(c^2 - ab)x^2 - 2(a^2 - bc)x + (b^2 - ac) = 0$, when a, b, c ≠ 0, are equal, then

(a) a = 0
(b) b = 0
(c) c = 0
(d) $a^3+b^3+c^3=3abc$

26. If the roots of the equation $ax^2+bx+c=0$ are in the ratio 1 : n, then
(a) $na^2=bc(1+n)^2$
(b) $nb^2=ca(1+n)^2$
(c) $na^2=ca(1-n)^2$
(d) None of these

27. The roots of the equation $(x-a)(x-b)=abx^2$ are always
(a) real
(b) imaginary
(c) cannot be determined
(d) None of these

28. A car travels 25km an hour faster than a bus for journey of 500 km. If the bus takes 10 h more than the car, then the speed of the bus and the car is
(a) 15 km/hr, 40 km/hr
(b) 35 km/hr, 60 km/hr
(c) 25 km/hr, 50 km/hr
(d) None of these

29. Find the harmonic mean of the roots of the equation
$(5+\sqrt{2})x^2-\left(4+\sqrt{5}\right)x+\left(8+2\sqrt{5}\right)=0$
(a) 2 (b) 4
(c) 7 (d) 8

30. Solve the equation $4^{x+2}=2^{x+3}+48$
(a) – 3/2 (b) – 2
(c) – 3 (d) 1

31. If the roots of the equation $x^2+a^2=8x+6a$ are real, then
(a) $a\in[2,8]$ (b) $a\in[-2,8]$
(c) $a\in(2,8)$ (d) $a\in(-2,8)$

32. The expression $x^2+2bx+c$ is always positive, if
(a) $b^2-4ac>0$ (b) $b^2-4ac<0$
(c) $c^2<b$ (d) $b^2<c$

33. Let a, b and c be real numbers such that $4a+2b+c=0$ and $ab>0$. Then the quadratic equation $ax^2+bx+c=0$ has
(a) real roots
(b) complex roots
(c) purely imaginary roots
(d) only one root

34. Let $D=a^2+b^2+c^2, a, b$ being consecutive integers and $c=ab$ then $\sqrt{D}$ is
(a) always an even integer
(b) can't be discussed
(c) always an odd integer
(d) always irrational

35. If α,β are the roots of the equation $ax^2+bx+c=0,$ then the equation whose roots are $\frac{1}{a\alpha+b}$ and $\frac{1}{a\beta+b}$ is
(a) $cax^2-bx+1=0$ (b) $cax^2+bx+1=0$
(c) $cax^2+bx-1=0$ (d) None of these

36. If the equations $ax^2+bx+c=0$ and $x^2+x+1=0$ have a common root here a, b, c are real, then
(a) a + b + c = 0
(b) a = b = c
(c) a = b or b = c or c = a
(d) a – b + c = 0

37. The number of values of a for which $(a^2-3a+2)x^2+(a^2-5a+6)x+a^2-4=0$ is an identity in x, is
(a) 0 (b) 2
(c) 1 (d) 3

38. If $a \in R, b \in R$ then the equation $x^2 - abx - a^2 = 0$ has
(a) one positive root and one negative root
(b) both positive roots
(c) both negative roots
(d) non-real roots

39. If l, m, n are real, $l \neq m$, then the roots of the equation are $(l-m)x^2 - 5(l+m)x - 2(l-m) = 0$, are
(a) real and equal (b) complex
(c) real and unequal (d) None of these

40. The condition that the equation $\frac{1}{x} + \frac{1}{x+b} = \frac{1}{m} + \frac{1}{m+b}$ has real roots that are equal in magnitude but opposite in sign is
(a) $b^2 = m^2$ (b) $b^2 = 2m^2$
(c) $2b^2 = m^2$ (d) None of these

41. If α and β be the roots of the equation $x^2 + px - \frac{1}{2p^2} = 0$, where $p \in R$. Then the minimum value of $\alpha^4 + \beta^4$ is
(a) 2 (b) $2+\sqrt{2}$
(c) $2\sqrt{2}$ (d) $2-\sqrt{2}$

42. If a, b, c are non-zero, unequal rational numbers then the roots of the equation $abc^2x^2 + (3a^2 + b^2)\ cx - 6a^2 - ab + 2b^2 = 0$ are
(a) rational (b) imaginary
(c) irrational (d) None of these

43. For the equation $3x^2 + px + 3 = 0, p > 0$, if one of the roots is the square of the other, then p is equal to
(a) $\frac{1}{3}$ (b) 1 (c) 3 (d) $\frac{2}{3}$

44. If a, b, c are positive real numbers, then the number of real roots of the equation $ax^2 + b|x| + c = 0$ is
(a) 1 (b) 4
(c) 0 (d) None of these

45. If the sum of the roots of the equation $ax^2 + bx + c = 0$ is equal to the sum of the reciprocals of their squares, then bc^2, ca^2, and ab^2 are in
(a) AP (b) GP
(c) HP (d) None of these

46. If $\sin\theta$ and $\cos\theta$ are the roots of the equation $ax^2 + bx + c$, then
(a) $(a-c)^2 = b^2 - c^2$
(b) $(a-c)^2 = b^2 + c^2$
(c) $(a+c)^2 = b^2 - c^2$
(d) $(a+c)^2 = b^2 + c^2$

47. If the roots of the equation $\frac{1}{x+p} + \frac{1}{x+q} = \frac{1}{r}$ are equal in magnitude and opposite in sign, then
(a) $p+q=r$
(b) $p+q=2r$
(c) product of roots $= \frac{1}{2}(p^2+q^2)$
(d) None of the above

48. Find the number of real roots of the equation (x + 1) (x + 2) (x + 3) (x + 4) = 24
(a) 2 (b) 0
(c) 4 (d) None of these

49. Find x such that $\sqrt{x+2\sqrt{x-1}} - \sqrt{x-2\sqrt{x-1}} = 2$

$$\sqrt{x+2\sqrt{x-1}} - \sqrt{x-2\sqrt{x-1}} = 2$$

(a) 2 (b) 0
(c) 4 (d) None of these

50. Given that the equation $(x-19)(x-97) = p$ has real roots r_1 and r_2. Find the minimum real root of quadratic equation, $(x-r_1)(x-r_2) = -p$.
(a) 97 (b) –97
(c) –19 (d) None of these

Concept Cracker (CC)

1. If $x^2 - 4x + \log_{1/2} a^2 = 0$ does not have two distinct real roots, then maximum value of a is
(a) ¼ (b) 1/16
(c) – ¼ (d) None of these

2. If the equations $ax^2 - 2bx + c = 0$, $bx^2 - 2cx + a = 0$ and $cx^2 - 2ax + b = 0$ all have only positive roots, then which one of the following is true?
(a) $a = b = c$ (b) $a \neq b \neq c$
(c) $a \neq b = c$ (d) $a \neq b \neq c$

3. Consider the equation $x^2 - 2x - n = 0$ where $n \in N$ and $n \in [100, 1000]$. Total number of different values of 'n' so that the given equation has integral roots, is
(a) 14 (b) 18
(c) 21 (d) 16

4. Find the values of 'a' for which equation $x^2 + (a-b)x + (1-a-b) = 0$, has two distinct real roots for all values of b, where $a, b \in R$
(a) $a < 1$ (b) $a > 1$
(c) $a \in R^+$ (d) None of these

5. Find the number of solutions of $\left(\sqrt{2+\sqrt{2}}\right)^x + \left(\sqrt{2-\sqrt{2}}\right)^x = 2 \cdot 2^{x/4}$ is
(a) 0 (b) 1
(c) 2 (d) None of these

6. If $c < a < b < d$, then roots of the equation $bx^2 + (1 - b(c+d))x + bcd - a = 0$
(a) Exactly one root lie between c and a
(b) Exactly one root lie between c and d
(c) Exactly one root lie between a and b
(d) complex roots

7. If p and q are prime numbers (more than 2), then which of the following statement about the roots of the equation $x^2 + 2px - 4q = 0$ is correct
(a) always integer
(b) always irrational
(c) always rational but never integer
(d) nothing can be said

8. Let x_1, x_2, x_3, x_4 are real numbers, then number of ordered tuples (x_1, x_2, x_3, x_4) satisfying the given equation $x_1 + x_2 = 4$ and $x_1x_2 - x_3^2 = 4 + x_4^2$ is
(a) 1 (b) 2
(c) infinite (d) 0

9. If x, y are rational numbers such that $x + y + (x - 2y)\sqrt{2} = 2x - y + (x - y - 1)\sqrt{6}$ then how many pairs of (x, y) exist?
(a) 1 (b) 2
(c) 0 (d) None of the above

10. If $x + \lambda y - 2$ and $x - \mu y + 1$ are factors of the expression $6x^2 - xy - y^2 - 6x + 8y - 12$, then
(a) $\lambda = \frac{1}{3}, \mu = \frac{1}{2}$ (b) $\lambda = 2, \mu = 3$
(c) $\lambda = \frac{1}{3}, \mu = -\frac{1}{2}$ (d) None of these

11. Find the number of the real solutions of $1 + |e^x - 1| = e^x(e^x - 2)$
(a) 0 (b) 1
(c) 2 (d) None of these

12. The equation $x^3 + 5x^2 + px + q = 0$ and $x^3 + 7x^2 + px + r = 0$ have two roots in common. If their third roots be γ_1 and γ_2 respectively, then find $\gamma_1 + \gamma_2$
(a) 12 (b) –12
(c) 2 (d) –2

13. If $x + \frac{1}{x} = 10$, then $\left(x^3 + \frac{1}{x^3}\right) - 5\left(x^2 + \frac{1}{x^2}\right) + \left(x + \frac{1}{x}\right)$ is equal to
(a) 470 (b) 490
(c) 510 (d) None of these

14. If p, q, r are positive and are in AP, the roots of the quadratic equation $px^2 + qx + r = 0$ are real for

(a) $\left|\frac{r}{p}-7\right|\geq 4\sqrt{3}$ (b) $\left|--7\right|<4\sqrt{3}$

(c) all p and r (d) no p and r

15. Find the sum of the roots of the equation $\frac{2x+1}{9}+\frac{x^2+7}{x^2-7}=\frac{2x+17}{9}$.

(a) 3 (b) 5

(c) 0 (d) None of these

16. The value of m for which the equation $(1+m)x^2-2(1+3m)x+(1+8m)=0$ has equal roots, is

(a) 0, 3 (b) 1

(c) 2 (d) 3

17. If the ratio of the roots of $lx^2+nx+n=0$ is p : q, then find the value of $\sqrt{\frac{p}{q}}+\sqrt{\frac{q}{p}}+\sqrt{\frac{n}{l}}$

(a) –1 (b) 0

(c) –1 (d) –2

18. If $(ax^2+bx+c)y+a'x^2+b'x+c'=0$ then the condition that x may be rational function of y is

(a) $(ac'-a'c)^2=(ab'-a'b)(bc'-b'c)$

(b) $(ab'-a'b)^2=(ac'-a'c)(bc'-b'c)$

(c) $(bc'-b'c)^2=(ab'-a'b)(ac'-a'c)$

(d) None of the above

19. In a $\Delta PQR, \angle R=\frac{\pi}{2}$. If $\tan\left(\frac{P}{2}\right)$ and $\tan\left(\frac{Q}{2}\right)$ are roots of the equation $ax^2+bx+c=0$ $(a\neq 0)$, then

(a) a + b = c (b) b + c = 0

(c) a + c = b (d) b = c

20. Find the sum of all the possible values of x that satisfy the condition $9^x-4\times 3^{x+2}+3^5=0$,

(a) 0 (b) 6

(c) 5 (d) None of these

21. The solution set of $\left|\frac{x+1}{x}\right|+|x+1|=\frac{(x+1)^2}{|x|}$ is

(a) $\{x\,|\,x\geq 0\}$

(b) $\{x\,|\,x\geq 0\}\cup\{-1\}$

(c) $\{-1, 1\}$

(d) $\{x\,|\,x\geq 1 \text{ or } x\leq -1\}$

22. If $5^x+(2\sqrt{3})^{2x}\geq 13^x$, then the solution set for x is

(a) $[2, +\infty)$ (b) $(2, +\infty)$

(c) $(4, +\infty)$ (d) None of these

23. For what value of p the expression $\cos^4 x+\sin^2 x-p, p\in R$ will be equal to zero for some real value of x.

(a) $p\leq 1$ (b) $\frac{3}{4}\leq p\leq 1$

(c) $p\geq\frac{3}{4}$ (d) None of these

24. Number of solutions of an equation $|x|^2-3|x|+2=0$ will be

(a) 4 (b) 1

(c) 3 (d) 2

25. If the roots of the equation $x^n - 1 = 0$ are $1, a_1, a_2, a_3, ..., a_{n-1}$ then the value of $(1-a_1)(1-a_2)(1-a_3)...(1-a_{n-1})$ will be

(a) n (b) n^2

(c) n^n (d) 0

26. If α, β are the roots of $x^2+px+q=0$ and also of $x^{2n}+p^nx^n+q^n=0$ and if $\frac{\alpha}{\beta}, \frac{\beta}{\alpha}$ are root of $x^n + 1 + (x + 1)^n = 0$, then n can be $i+p\neq 0$

(a) 13 (b) 14

(c) 15 (d) None of these

27. If cos θ, sin ϕ, sin θ are in GP, then roots of $x^2 + 2\cot\phi\cdot x + 1 = 0$ are always

(a) equal (b) real

(c) imaginary (d) greater than 1

28. If $p > 2$ and $p\in N$, then the equation $x^3 - px + 1 = 0$

(a) cannot have rational roots

(b) cannot have integral roots

(c) cannot have irrational roots

(d) None of these

29. Find the value of p for which the equations $x^2 = p(x + p)$ has its roots greater than p,

(a) $\left(-5,-\frac{1}{2}\right)$ (b) $\left(-\frac{1}{6},\frac{1}{4}\right)$

(c) null set ϕ (d) $(-\infty, 0)$

30. Find the minimum value of $f(x)=\dfrac{x^2-1}{x^2+1}$ for every real number x
    (a) does not exist because f is unbounded
    (b) is not attained even though f is bounded
    (c) is equal to 1
    (d) is equal to –1

31. If $a \le 0$ then the root of the equation $x^2 - 2a|x-a| - 3a^2 = 0$ is
    (a) $\left(1-\sqrt{2}\right)a, \left(\sqrt{6}-1\right)a$
    (b) $\left(-1+\sqrt{6}\right)a, \left(1+\sqrt{2}\right)a$
    (c) $\left(1+\sqrt{2}\right)a$
    (d) $-\left(1\pm\sqrt{6}\right)a, \left(1\pm\sqrt{2}\right)a$

32. If $(2x^2-3x+1)(2x^2+5x+1)=9x^2$, then equation has
    (a) four real roots
    (b) two real and two imaginary roots
    (c) all imaginary
    (d) none of the above.

33. If $a + c < b$ and the roots of $ax^2 + bx + c = 0$, where $a \neq 0$ are non-real Then which one of the following is true if a, b, c € R
    (a) $4a+c>2b$ (b) $4a+c<2b$
    (c) $4a+c=2b$ (d) None of these

34. For how many integral values of 'a' the quadratic equation $3x^2 + 2(a^2 + 1)x + (a^2 - 3a + 2) = 0$ possesses roots of opposite sign
    (a) 0 (b) 1
    (c) 2 (d) More than 2

35. If the expression $\dfrac{(ax-b)(dx-c)}{(bx-a)(cx-d)}$ will be capable of all values when x is real, then-
    (a) $(a^2-b^2)(c^2-d^2)>0$
    (b) $(a^2-b^2)(c^2-d^2)<0$
    (c) $(d^2-b^2)(c^2-a^2)>0$
    (d) None of these

36. If $y \neq 0$, then the number of values of the pair (x, y) such that $x + y + \dfrac{x}{y} = \dfrac{1}{2}$ and $(x+y)\,\dfrac{x}{y} = -\dfrac{1}{2}$ is
    (a) 1 (b) 2
    (c) 0 (d) None of these

37. If the equation $ax^2+bx+c=0$ has distinct real roots and $ax^2+b|x|+c=0$ also has two distinct real roots then
    (a) $a>0$ (b) $ac<0$
    (c) $ab<0$ (d) $bc<0$

38. If $P(x)=ax^2+bx+c$ and $(x)=-ax^2+dx+c$, $ac \neq 0$, then the equation $P(x)\,Q(x)=0$ has
    (a) At least two real roots
    (b) exactly two real roots
    (c) No real roots
    (d) None of these

39. For how many integral values of 'x' the inequality $\dfrac{x^2-4x+3}{x^2+x+1} \le 0$ holds true
    (a) 0 (b) 1
    (c) 3 (d) None of these

40. If $f(x)=x^2+2bx+2c^2$ and $g(x)=-x^2-2cx+b^2$ are such that $\min f(x) > \max g(x)$, then relation between b and c, is
    (a) $0>c>\dfrac{b}{2}$ (b) $0<c<\dfrac{b}{2}$
    (c) $|c|<\sqrt{2}\,|b|$ (d) $|c|>\sqrt{2}\,|b|$

41. If $c>0$ and $4a+c<2b$, then ax^2-bx+c has a root in the interval
    (a) (0, 2) (b) (2, 4)
    (c) (0, 1) (d) (–2, 0)

42. Find the number of integral values of x which satisfy $x^2-6x+5 \le 0$ and $x^2-2x>0$
    (a) 3 (b) 4
    (c) 2 (d) infinite

43. If $x \in R$, then $\frac{x^2 - x + 1}{x^2 + x + 1}$ takes values in the interval

(a) $\left(\frac{1}{3}, 3\right)$ (b) $\left[\frac{1}{3}, 3\right]$

(c) (0, 3) (d) None of these

44. Find the number of real roots of the equation $|x|^{6/5} - 26|x|^{3/5} - 27 = 0$

(a) 0 (b) 1

(c) 2 (d) None of these

45. Find the number of distinct real roots of $(x^2 + x - 2)(x^2 + x - 3) = 12$

(a) 1 (b) 0

(c) 2 (d) None of these

46. If both the roots of the equation $x^2 - 6ax + 2 - 2a + 9a^2 = 0$ exceed 3, then

(a) $a < 1$ (b) $a > 11/9$

(c) $a > 3/2$ (d) $a < 5/2$

IIT JEE Mains/AIEEE Corner

47. If $\alpha \neq \beta$ but $\alpha^2 = 5\alpha - 3$ and $\beta^2 = 5\beta - 3$ then the equation having α / β and β / α as its roots is [AIEEE 2002]

(a) $3x^2 - 19x + 3 = 0$

(b) $3x^2 + 19x - 3 = 0$

(c) $3x^2 - 19x - 3 = 0$

(d) $x^2 - 5x + 3 = 0$

48. Difference between the corresponding roots of $x^2 + ax + b = 0$ and $x^2 + bx + a = 0$ is same and $a \neq b$, then [AIEEE2002]

(a) $a + b + 4 = 0$ (b) $a + b - 4 = 0$

(c) $a - b - 4 = 0$ (d) $a - b + 4 = 0$

49. Product of real roots of the equation $t^2 x^2 + |x| + 9 = 0$ [AIEEE 2002]

(a) is always positive (b) is always negative

(c) does not exist (d) none of these.

50. If p and q are the roots of the equation $x^2 + px + q = 0$, then [AIEEE 2002]

(a) $p = 1, q = -2$ (b) $p = 0, q = 1$

(c) $p = -2, q = 0$ (d) $p = -2, q = 1$

51. If a, b, c are distinct + ve real numbers and $a^2 + b^2 + c^2 = 1$ then $ab + bc + ca$ is [AIEEE 2002]

(a) less than 1 (b) equal to 1

(c) greater than 1 (d) any real no.

52. If the sum of the roots of the quadratic equation $ax^2 + bx + c = 0$ is equal to the sum of the squares of their reciprocals, then $\frac{a}{c}, \frac{b}{a}$ and $\frac{c}{b}$ are in [AIEEE 2003]

(a) Arithmetic – Geometric Progression

(b) Arithmetic Progression

(c) Geometric Progression

(d) Harmonic Progression.

53. The value of 'a' for which one root of the quadratic equation $(a^2 - 5a + 3)x^2 + (3a - 1)x + 2 = 0$ is twice as large as the other is [AIEEE 2003]

(a) $-\frac{1}{3}$ (b) $\frac{2}{3}$

(c) $-\frac{2}{3}$ (d) $\frac{1}{3}$

54. The number of real solutions of the equation $x^2 - 3|x| + 2 = 0$ is [AIEEE 2003]

(a) 3 (b) 2

(c) 4 (d) 1

55. The real number x when added to its inverse gives the minimum value of the sum at x equal to [AIEEE 2003]

(a) – 2 (b) 2

(c) 1 (d) – 1

56. Let two numbers have arithmetic mean 9 and geometric mean 4. Then these numbers are the roots of the quadratic equation [AIEEE 2004]

(a) $x^2 - 18x - 16 = 0$

(b) $x^2 - 18x + 16 = 0$

(c) $x^2 + 18x - 16 = 0$

(d) $x^2 + 18x + 16 = 0$

57. If $(1 - p)$ is a root of quadratic equation $x^2 + px + (1 - p) = 0$ then its root are [AIEEE 2004]

(a) – 1, 2 (b) – 1, 1

(c) 0, – 1 (d) 0, 1

58. If one root of the equation $x^2 + px + 12 = 0$ is 4, while the equation $x^2 + px + q = 0$ has equal roots, then the value of 'q' is [AIEEE 2004]
(a) 4 (b) 12
(c) 3 (d) $\frac{49}{4}$

59. In a triangle PQR, $\angle R = \frac{\pi}{2}$. If $\tan\left(\frac{P}{2}\right)$ and $\tan\left(\frac{Q}{2}\right)$ are the roots of $ax^2 + bx + c = 0$, $a \neq 0$ then [AIEEE 2005]
(a) $a = b + c$ (b) $c = a + b$
(c) $b = c$ (d) $b = a + c$

60. If both the roots of the quadratic equation $x^2 - 2kx + k^2 + k - 5 = 0$ are less than 5, then k lies in the interval [AIEEE 2005]
(a) $(5, 6]$ (b) $(6, \infty)$
(c) $(-\infty, 4)$ (d) $[4, 5]$

61. If the roots of the quadratic equation $x^2 + px + q = 0$ are tan 30° and tan 15°, respectively, then the value of $2 + q - p$ is [AIEEE 2006]
(a) 2 (b) 3
(c) 0 (d) 1

62. All the values of m for which both roots of the equation $x^2 - 2mx + m^2 - 1 = 0$ are greater than -2 but less than 4, lie in the interval [AIEEE 2006]
(a) $-2 < m < 0$ (b) $m > 3$
(c) $-1 < m < 3$ (d) $1 < m < 4$

63. If x is real, the maximum value of $\frac{3x^2+9x+17}{3x^2+9x+7}$ is [AIEEE 2006]
(a) $\frac{1}{4}$ (b) 41
(c) 1 (d) $\frac{17}{7}$

64. If the difference between the roots of the equation $x^2 + ax + 1 = 0$ is less than $\sqrt{5}$, then the set of possible values of a is [AIEEE 2007]
(a) $(3, \infty)$ (b) $(-\infty, -3)$
(c) $(-3, 3)$ (d) $(-3, \infty)$.

65. The quadratic equations $x^2 - 6x + a = 0$ and $x^2 - cx + 6 = 0$ have one root in common. The other roots of the first and second equations are integers in the ration 4 : 3. Then the common root is [AIEEE 2009]
(a) 1 (b) 4
(c) 3 (d) 2

66. If the roots of the equation $bx^2 + cx + a = 0$ be imaginary, then for all real values of x, the expression $2b^2x^2 + 6bcx + 2c^2$ is : [AIEEE 2009]
(a) less than 4 a b
(b) greater than – 4 a b
(c) less than – 4 a b
(d) greater than 4 a b

67. Sachin and Rahul attempted to solve a quadratic equation. Sachin made a mistake in writing down the constant term and ended up in roots (4, 3). Rahul made a mistake in writing down coefficient of x to get roots (3, 2). The correct roots of equation are : [AIEEE 2011]
(a) 6, 1 (b) 4, 3
(c) – 6, – 1 (d) – 4, – 3

68. Let for $a \neq a_1 \neq 0$, $f(x) = ax^2 + bx + c$ $g(x) = a_1x^2 + b_1x + c_1$ and $p(x) = f(x) - g(x)$. If $p(x) = 0$ only for $x = -1$ and $p(-2) = 2$, then the value of $p(2)$ is : [AIEEE 2011]
(a) 3 (b) 9
(c) 6 (d) 18

69. The equation $e^{\sin x} - e^{-\sin x} - 4 = 0$ has : [AIEEE 2012]
(a) infinite number of real roots
(b) no real roots
(c) exactly one real roots
(d) exactly four real roots.

70. The real number of k for which the equation, $2x^3 + 3x + k = 0$ has two distinct real roots in [0, 1] [JEE, M 2013]
(a) lies between 1 and 2
(b) lies between 2 and 3
(c) lies between – 1 and 0
(d) does not exist.

71. If the equations $x^2 + 2x + 3 = 0$ and $ax^2 + bx + c = 0$, a, b, c ∈ R, have a common root, then a : b : c is [JEE Mains 2013]
(a) 1 : 2 : 3 (b) 3: 2 : 1
(c) 1 : 3 : 2 (d) 3 : 1 : 2

DIRECTIONS: Q. No. 72 is Assertion – reason type questions . Each of these questions contains two statements : Statement- 1 (Assertion) and statement–2 (Reason). Each of these questions also has four alternative choices, only one of which is the correct answer. You have to select correct choice.

72. Statement – 1 : For every natural number n ≥ 2, $\frac{1}{\sqrt{1}}+\frac{1}{\sqrt{2}}+\ldots+\frac{1}{\sqrt{n}}>\sqrt{n}$.

Statement – 2 : For every natural number n ≥ 2, $\sqrt{n(n+1)}<n+1$. [2008]

(a) Statement – 1 is false, statement – 2 is true
(b) Statement – 1 is true, statement – 2 is true; statement – 2 is a correct explanation for statement – 1
(c) Statement – 1 is true, statement – 2 is true; statement – 2 is not a correct explanation for statement – 1.
(d) Statement – 1 is true, statement – 2 is false

73. If $a \in R$ and the equation $-3(x-[x])^2+2(x-[x])+a^2=0$ (where $[x]$ denotes the greatest integer $\le x$) has no integral solution, then all possible values of a lie in the interval: [JEE M 2014]

(a) $(-2,-1)$ (b) $(-\infty,-2)\cup(2,\infty)$
(c) $(-1,0)\cup(0,1)$ (d) $(1,2)$

74. Let α and β be the roots of equation $px^2+qx+r=0, p\neq 0$. If p, q, r are in A.P. and $\frac{1}{\alpha}+\frac{1}{\beta}=4$. then the value of $|\alpha-\beta|$ is: [JEE M 2014]

(a) $\frac{\sqrt{34}}{9}$ (b) $\frac{2\sqrt{13}}{9}$
(c) $\frac{\sqrt{61}}{9}$ (d) $\frac{2\sqrt{17}}{9}$

75. Let α and β be the roots of equation $x^2-6x-2=0$. If $a_n=\alpha^n-\beta^n$, for $n\ge 1$, then the value of $\frac{a_{10}-2a_8}{2a_9}$ is equal to: [JEE M 2015]

(a) 3 (b) – 3
(c) 6 (d) – 6

76. The sum of all real values of x satisfying the equation $(x^2-5x+5)^{x^2+4x-60}=1$ is [JEE M 2016]

(a) 6 (b) 5
(c) 3 (d) – 4

Board Level Questions

Part 1

1. If $x \in R$ and consider the equation $\sqrt{2x+1}-\sqrt{2x-1}=1$
If P is the number of +ve real roots and Q is the number of –ve real roots then find the value of $|P-Q|$
2. If I, m, n, are real" I + m ≠ 0, then find the nature of the roots of the equation $(I+m)x^2-3(I-m)x-2(I+m)=0$
3. If p and q are distinct primes and $x^2-px+q=0$ has distinct positive integral roots, then find p, q.
4. If b < 0, x_1 and x_2 are the roots of the equation $2x^2+6x+b=0$, then find the maximum value of $\frac{x_1}{x_2}+\frac{x_2}{x_1}$
5. If ax^2+bx+c, a, b, c ∈ R has no real zeros, and if c < 0, then prove that a is negative
6. Let a > 0, b > 0 and c > 0. Then find the nature of the roots of the equation $ax^2+bx+c=0$
7. Find the number of integral roots of the equation $\sqrt{x+3-4\sqrt{x-1}}+\sqrt{x+8-6\sqrt{x-1}}=$
8. If p, q, r > 0, then prove that the equation $x^3-px^2+qx-r=0$ cannot have negative roots.

Part 2

9. The number of values of the triplet (a, b, c) for which $a\cos 2x+b\sin^2 x+c=0$ is satisfied by all real x, is
10. Find the quadratic equation whose roots are the AM and HM of the roots of the equations $x^2+7x-1=0$
11. If the sum of the roots of the quadratic equation $ax^2+bx+c=0$, $(a, b, c\neq 0)$ is equal to sum of the squares of their reciprocals, then prove that $\frac{a}{c},\frac{b}{a},\frac{c}{b}$, are in HP

12. Find the values of 'a' for which $4^t - (a-4)2^t + \frac{9}{4}a < 0, \forall t \in (1, 2)$

13. Find the set of values of k such that $x^2 - 2(4k-1)x + 15k^2 - 2k - 7 > 0$ for each x

14. If tan A and tan B are the roots of the quadratic equation $x^2 - px + q = 0$, then find the value of $\sin^2(A+B)$

15. If a & b are the roots of the equation $x^2 - (a-2)x - a - 1 = 0$ then find the real value of a for which $\alpha^2 + \beta^2$ is minimum

16. Find the real values of a for which the quadratic equation $3x^2 + 2(a^2+1)x + (a^2 - 3a + 2) = 0$ possesses roots of opposite signs

17. If x is real then find the range of $\frac{x^2 - x + 1}{x^2 + x + 1}$

18. If $a \le 0$, then number of real roots of $x^2 + 2a|x - a| + 3a^2 = 0$ is (For a specific value of a)

19. Find the number of integral roots of the equation $x^4 + \sqrt{x^4 + 20} = 22$

20. Find the number of roots of the equation $\frac{1}{x} + \frac{1}{\sqrt{1-x^2}} = \frac{35}{12}$ is

21. If all the roots of $x^3 + px + q = 0$ $p, q \in R$. $q \neq 0$ are real, then prove that p is negative

22. Find the set of all real values of a for which the roots of the equation $x^2 - 2ax + a^2 - 1 = 0$. Lies between 2 and 8

23. Find the set of all values of a for which the roots of the equation $(1+a)x^2 - 3ax + 4a = 0$ is more than 1

24. Find the number of real solutions of the equation $\sqrt{x^2 + \sqrt{x^2+11}} + \sqrt{x^2 - \sqrt{x^2+11}} = 4$

25. Let a, b, c , d be four integers such that ad is odd and bc is even, then prove that equation $ax^3 + bx^2 + cx + d = 0$ has at least one irrational root

26. If $b^2 \ge 4ac$ for the equations $ax^4 + bx^2 + c = 0$, then find the condition that all the roots of the equation will be real

27. If α, β are the roots of the equation $ax^2 + 2bx + c = 0$ and $\alpha + h$, $\beta + h$ are the roots of the equations $Ax^2 + 2Bx + C = 0$, then find the value of $\frac{b^2 - ac}{B^2 - AC}$ in terms of a and A

28. Find the number of the real roots of the equation $(3-x)^5 + (x-5)^5 = -32$

29. Prove that there exist no polynomials with integral coefficients satisfying the conditions. $P(1) = 2$, $p(3) = 1$

30. If a and b are integers and roots of $x^2 + ax + b = 0$ are rational then prove that these roots must be integers.

31. Given the equation $(x-a)(x-a-b) = 1$, where a, b are two constants. Prove that the equation has two real roots , where one is greater than a, and the other is less than a.

32. Find the value of
$$\sqrt{2\sqrt{2\sqrt{2\sqrt{2}}}}\ldots - \sqrt{2+\sqrt{2+\sqrt{2+\sqrt{2+\ldots}}}}$$

33. Let a be the minimum root of the equation $x^2 - 3|x| - 2 = 0$, find the value of $-\frac{1}{a}$.

34. If a is a root of the equation $x^2 - 3x + 1 = 0$, then find the value of $\frac{2a^5 - 5a^4 + 2a^3 - 8a^2}{a^2 + 1}$

35. a, b, c are three distinct non – zero real numbers. Prove that the following three equations $ax^2 + 2bx + c = 0$, $bx^2 + 2cx + a = 0$, and $cx^2 + 2ax + b = 0$ cannot all have two equal real roots.

36. For what values of b do the equations: $1988x^2 + bx + 8891 = 0$ and $8891x^2 + bx + 1988 = 0$ have a common root?

37. Given that a is a root of the equation $x^2 - x - 3 = 0$ Evaluate $\frac{a^3 + 1}{a^5 - a^4 - a^3 + a^2}$.

38. If the equation $x^2 + 2(1+a)x + (3a^2 + 4ab + 4b^2 + 2) = 0$ has real roots find the values of a and b.

39. Find the value of k, such that the equation $x^2 - kx - 7 = 0$ and $x^2 - 6x - (k+1) = 0$ have a common root, and find the common root and different roots.

PART-B

4.7 ADVANCE MATERIAL FOR JEE ADVANCED, KVPY AND OLYMPIAD

4.7.1 Combined Equation with Two Variables

Type 1:

Consider a general equation of 2nd degree

$$ax^2+by^2+2gx+2fy+2hxy+c=0$$

This is can be resolved into two linear factors, for this we will solve the equation for x

$ax^2+2x(hy+g)+(by^2+2fy+c)=0$ this is a quadratic equation for 'x', and it can be resolved into two linear factors then its discriminant must be a perfect square.

$ax^2+2hxy+by^2+2gx+2fy+c$ can be expressed as product of two linear factors, if

$$abc+2fgh-af^2-bg^2-ch^2=0$$

We will try to understand this with an example

Illustration 1. Find the point of intersection of straight lines given by $x^2+xy-6y^2+5y-1=0$

Solution: the given equation can be re-written as $x+x(y)+(-6y^2+5y-1)=0$

From Sri-Dharacharya formula

$$x=\frac{\left\{-(y)\pm\sqrt{\left\{(y)^2-4\left(-6y^2+5y-1\right)\right\}}\right\}}{2}$$

$$x=\frac{\left\{-(y)\pm\sqrt{\left\{25y^2-20y+4\right\}}\right\}}{2}$$

$$=\frac{\left\{-(y)\pm\sqrt{\{5y-2\}^2}\right\}}{2}=\frac{\{-(y)\pm(5y-2)\}}{2}$$

$$x=\frac{4y-2}{2}\ or\ \frac{-6y+2}{2}$$

Or $x=2y-1$ or $x=-3y+1$

Solving above two equations we get $y=\frac{2}{5}$ & $x=-\frac{1}{5}$

Type II:

If it is given that $y-mx$ is a factor of $f(x, y)$ then put $y=mx$ in $f(x, y)$ and it will satisfy $f(x, y)=0$

We will get similar result if we substitute $x=\frac{y}{m}$ in $f(x, y)$ and it will satisfy $f(x, y)=0$

Here $f(x, y)$ is a homogeneous equation

Illustration 2. If $x-y$ and $y-2x$ are two factors of the expression $x^3-3x^2y+\lambda xy^2+\mu y^3$, then find λ & μ

Solution: Given $f(x, y)=x^3-\ x^2y+\lambda xy^2+\mu y^3$

$x-y$ is a factor of $f(x, y)$ putting $y=x$, $f(x, y)=0$ which gives

$$1-3+\lambda+\mu=0\ \lambda+\mu=2 \quad ...(i)$$

$y-2x$ is a factor of $f(x, y)$

⇒ Putting $y=2x$, $f(x, y)=0$ which gives

$$1-6+4\lambda+8\mu=0 \text{ or } 4\lambda+8\mu=5 \quad ...(ii)$$

On solving equation (i) and (ii) we get

⇒ $\lambda=\frac{11}{4}$ and $\mu=-\frac{3}{4}$

Illustration 3. If $x+y$ and $y+3x$ are two factors of the expression $\lambda x^3-\mu x^2y+xy^2+y^3$, then find the third factor

Solution: As it is a third degree homogeneous expression in x, y, we have,

$$y^3+y^2x-\mu yx^2+\lambda x^3=(y+x)(y+3x)(y+mx)$$

$$=y^3+(m+3+1)y^2x+(3+m+3m)yx^2+3mx^3$$

$$\Rightarrow 1=m+4, -\mu=3+4m, \lambda=3m$$

$$\Rightarrow m=-3$$

∴ $y-3x$ is third factor

Alternately:

Given $f(x, y)=\lambda x^3-\mu x^2y+xy^2+y^3$

$x+y$ is a factor of $f(x, y)$ putting $y=-x$, $f(x, y)=0$ which gives

$$\lambda+\mu=0 \quad ...(i)$$

$y+3x$ is a factor of $f(x, y)$

⇒ Putting $y=-3x, f(x,y)=0$ which gives

$$\lambda+3\mu=18 \quad ...(ii)$$

On solving equation (i) and (ii) we get

⇒ $\lambda=-9$ and $\mu=9$

Now lets assume $y-mx$ is 3rd factor then put $y=mx$ and we get $m=3$

So 3rd factor is $y-3x$

Illustration 4. For what values of K $\in$ R the expression $2x^2 + Kxy + 3y^2 - 5y - 2$ can be expressed as

$$(a_1x+b_1y+c_1)\cdot(a_2x+b_2y+c_2)$$

Solution: As, $ax^2+2hxy+by^2+2gx+2fy+c$ can be expressed as product of two linear factors, if

$$abc+2fgh-af^2-bg^2-ch^2=0$$

$\Rightarrow$ $2x^2+Kxy+3y^2-5y-2$ can be expressed as $(a_1x+b_1y+c_1)(a_2x+b_2y+c_2)$, if

$$2(3)(-2)+2\left(-\frac{5}{2}\right)(0)\left(\frac{K}{2}\right)-2\left(-\frac{5}{2}\right)^2$$

$$-3(0)^2-(-2)\left(\frac{K}{2}\right)^2=0$$

$\Rightarrow K^2 - 49 = 0$ or $K = 7, -7$

Illustration 5. If the expression $y^2 + 2xy + 2x + my - 3$ can be resolved into two rational factors, then find m

Solution: The given equation can be written as $y^2+y(2x+m)+(2x-3)=0$

Here, $D=(2x+m)^2-4(2x-3)$

$=4x^2+4xm-8x+m^2+12$

$=4x^2+4x(m-2)+(m^2+12)$

$=\left[4x^2+4x(m-2)+(m-2)^2\right]+(m^2+12)-(m-2)^2$

$=\left[2x+(m-2)\right]^2+\left[12+4m-4\right]$

For rational factors, D should be a perfect square.

$\therefore 8+4m=0$ *or* $m=-2$

Alternately:

As, $ax^2+2hxy+by^2+2gx+2fy+c$ can be expressed as product of two linear factors, if

$abc+2fgh-af^2-bg^2-ch^2=0$

From this we get $\Rightarrow m = -2$

4.7.2 Rolle's Theorem

According to this theorem, if a function is-

(i) Continuous in [a, b]

(ii) Differentiable in (a, b)

(iii) And f(a) = f(b)

Then there exist at least one root of the equation f'(x) = 0 in the interval (a, b)

Illustration 6. If a, b, c $\in$ R. and a + b + c = 0 , then the quadratic equation $3ax^2 + 2bx + c = 0$ has

(a) at least one root in [0, 1]

(b) at least one root in [1, 2]

(c) at least one root in [3 /2, 2]

(d) none of these.

Solution: Let $f(x) = ax^3 + bx^2 + cx$.

Since f(x) is continuous and derivable on R.

Also $f(0) = 0$ and $f(1) = a + b + c = 0$.

By the Rolle's theorem, there exists at least one $\alpha \in (0, 1)$ such that

$f'(\alpha) = 0 \Rightarrow 3a\alpha^2 + 2b\alpha + c = 0$.

Thus, $3ax^2 + 2bx + c = 0$ has at least one root in [0, 1]

Illustration 7. Let a, b, c be three real numbers such that $2a+3b+6c=0$. Then, the quadratic equation $ax^2+bx+c=0$ has

(a) imaginary roots

(b) at least one root in (0, 1)

(c) at least one root in (–1, 0)

(d) both roots in (1, 2)

Solution: Consider a function $f(x) = 2ax^3 + 3bx^2 + 6cx$

Since f(0) = 0 and f(1) = 2a + 3b + 6c = 0

By the Rolle's theorem, there exists at least one root of equation f′(x) = 0 in [0, 1]

$$f'(x)=6ax^2+6bx+6c=0$$

Or equation $ax^2 + bx + c = 0$ has one root lies between 0 and 1.

Option (b)

4.7.3 Eisenstein's Irreducibility Criterion

Consider a polynomial $f(x)=a_0x^n+a_1x^{n-1}+a_2x^{n-2}+\ldots+a_{n-1}x+a_n$ here $a_i \in I$ and $a_0 \neq 0$. f(x) is said to be irreducible over I if f(x) can not be written as product of two non constant polynomials of degree < n.

Lets take some example to understand this

Ex (i) $f(x)=x^3+x^2+x+1=(x^2+1)(x+1)$. Here f(x) is written as product of two polynomials, so it is not irreducible function.

Ex (ii) $f(x)=x^2+x+1$. Here f(x) can not be written as product of two polynomials so f(x) is an irreducible function.

Eisenstein's Irreducibility Criterion

If p is a prime such that p divides every coefficient a_i except a_0 but p^2 does not divide a_n, then f(x) is irreducible

Illustration 8. Prove that $x^7+48x-24$ is irreducible.

Solution: Let $p = 3$ then 48 and 24 both are divisible by 3 but 24 is not divisible by 9 so the given expression is irreducible.

Illustration 9. Let n be a prime and p (x) = $x^{n-1} + x^{n-2} + ... + x + 1$, then prove that p(x) is irreducible over I

Solution: p (x) = $\dfrac{x^n-1}{x-1}$

$\Rightarrow$ p (x + 1) = $\dfrac{(x+1)^n-1}{x}$

$= x^{n-1} + {}^nc_1\, x^{n-2} + ... + {}^nC_{n-2}\, x + {}^nC_{n-1}$

For $1 \le i \le n-1$.

${}^nC_i = \dfrac{n!}{(n-i)!i!}$

As n is prime so n does not divide (n-1)!

And n does not divide i! for $1 \le i \le n - 1$.

Thus, p divides ${}^pC_i$ for $1 \le i \le p-1$.

p^2 does not divide ${}^pC_{p-1}$, and hence p (x) is irreducible over I by the Eisenstein's irreducible over I.

Illustration 10. Let a, b $\in$ I. Then prove that polynomial p (x) = $(x - a)^2 (x - b)^2 + 1$ is irreducible.

Solution: Here p(x) is a polynomial of degree 4

Suppose p (x) = q (x) r (x) where q (x) and r (x) are non - constant polynomials of degrees < 4.

Since p(a) = p(b) = 1, so q (a) = r (a) = q (b) = r (b) = ±1

$\Rightarrow$ Both p (x) ± 1 and q (x) ± 1 are divisible by (x – a) (x – b).

But degree of q (x) + degree of r (x) = 4.

We have q (x) ± 1 = r (x) ± 1 = (x – a) (x – b)

$\therefore$ q (x) r (x) = $[(x-a)(x-b) \pm 1]^2$

$\Rightarrow$ p (x) = p (x) ± 2 (x – a) (x – b)

$\Rightarrow$ (x – a) (x – b) = 0.

A contradiction, so p(x) is irreducible.

4.7.4 Reciprocal Equation

Apolynomial $f(x)=a_0x^n+a_1x^{n-1}+a_2x^{n-2}+\ldots+a_{n-1}x+a_n=0$ here $a_0 \ne 0$ is called a reciprocal equation, if $a_i = \pm a_{n-i}$

Suppose $a_0x^n+a_1x^{n-1}+a_2x^{n-2}+\ldots+a_{n-1}x+a_n=0$ has roots $\alpha_1,\alpha_2\ldots\alpha_n$

Now transform $x \to \dfrac{1}{y}$ we get

$a_ny^n+a_{n-1}y^{n-1}+a_{n-2}y^{n-2}+\ldots+a_1y+a_0=0$ has roots $\dfrac{1}{\alpha_1},\dfrac{1}{\alpha_2},\ldots\dfrac{1}{\alpha_n}$

From the definition of reciprocal equation if α is a root of the equation then $\dfrac{1}{\alpha}$ is also root of the equation.

We know that two equations have the same roots then

$$\frac{a_0}{a_n}=\frac{a_1}{a_{n-1}}=\ldots\frac{a_{n-1}}{a_1}=\frac{a_n}{a_0}$$

Since $\dfrac{a_0}{a_n}=\dfrac{a_n}{a_0} \Rightarrow \left(\dfrac{a_0}{a_n}\right)^2=1 \Rightarrow \dfrac{a_0}{a_n}=\pm 1$

So, $a_i = \pm a_{n-i}$

Properties of Reciprocal Equation:

Property 1 : Reciprocal equation of 1st class.

If $a_n = a_0$, $a_1 = a_{n-1}$ etc then it is called reciprocal equation of 1st class.

Property 2 : Reciprocal equation of 2nd class.

If $a_n = -a_0$, $a_1 = -a_{n-1}$ etc then it is called reciprocal equation of 2nd class.

Property 3 : If 'α' is a root of the equation then $\dfrac{1}{\alpha}$ is also one root of the equation. So roots always come in pairs.

If reciprocal equation is of an odd degree then one of the roots must be either 1 or –1.

If equation is of 1st class then root is –1.

If equation is of 2nd class then root is 1.

Property 4 : A reciprocal equation of even degree and 1st class is known as standard form and can be solved by a substitution $x+\frac{1}{x}=y$.

Property 5 : A reciprocal equation of an odd degree must have a root either +1 or –1, so divide it by (x – 1) or (x + 1) it will reduce to reciprocal equation of even degree and 1st class i.e. in standard form.

Property 6 : If reciprocal equation is of even degree and second class, then $x^2 - 1$ will be a factor of function so it can be reduced in standard form.

Illustration 11. Solve the reciprocal equation

$x^4-3x^3+4x^2-3x+1=0$

Solution: Let $x+\frac{1}{x}=t$ then $x^2+\frac{1}{x^2}=t^2-2$

The given equation can be written as

$x^2+\frac{1}{x^2}-3x-\frac{3}{x}+4=0$

Or $\left(x^2+\frac{1}{x^2}\right)-3\left(x+\frac{1}{x}\right)+4=0$

$\Rightarrow\left(t^2-2\right)\ \ 3t+4-t^2-3t+2=0$

We get t = 1 and 2

When $x+\frac{1}{x}=1$ then $x^2-x+1=0$. It has no real solution.

When $x+\frac{1}{x}=2$,then $x^2-2x+1=0$ or x = 1

Illustration 12. Solve the equation $x^5+3x^4+2x^3-2x^2-3x-1=0$

Solution: Since it is an reciprocal equation of odd degree, so by observation we can find that x=1 is a solution of the equation.

The given equation can be written as

$$(x-1)\left(x^4+4x^3+6x^2+4x+1\right)=0$$

$$(x-1)(x+1)^4=0$$

So, roots of equation are x = 1 and -1

Illustration 13. Solve the equation $x^5=\frac{133x-78}{133-78x}$

Solution: Given equation can be written as $78x^6-133x^5+133x-78=0$

From observation we can say that 1 and -1 are the roots of the equation, so (x^2-1) is a factor of the given expression. Divide the given expression by (x^2-1) we get

$$\left(x^2-1\right)\left(78x^4-133x^3+78x^2-133x+78\right)=0$$

Now, consider the reciprocal equation

$78x^4-133x^3+78x^2-133x+78=0$

Let $x+\frac{1}{x}=t$ then $x^2+\frac{1}{x^2}=t^2-2$

The given equation can be written as

$$78x^2+\frac{78}{x^2}-133x-\frac{133}{x}+78=0$$

Or $78\left(x^2+\frac{1}{x^2}\right)-133\left(x+\frac{1}{x}\right)+78=0$

$\Rightarrow 78\left(t^2-2\right)-133t+78=78t^2-133t-78=0$

On solving we get t = $\frac{13}{6}$ & $-\frac{6}{13}$

When $x+\frac{1}{x}=\frac{13}{6}$ then $x=\frac{2}{3}$ & $\frac{3}{2}$

When $x+\frac{1}{x}=-\frac{6}{13}$ no real root

So, roots of the equation are 1, -1, 2/3 and 3/2

Illustration 14. Find the remainder when $p(x^5)$ is divided by $p(x)=x^4+x^3+x+1$

Solutions: Given that $p(x)=x^4+x^3+x+1$

Or $(x-1)p(x)=x^5-1$

Now $p\left(x^5\right)=\left(x^{20}-1\right)+\left(x^{15}-1\right)+\left(x^{10}-1\right)+\left(x^5-1\right)+5=\left(x^5-1\right)q(x)+5$

Here, $q(x)=\left(x^{13}+x^{10}+x^5+1\right)+\left(x^{10}+x^5+11\right)+\left(x^5+1\right)+1$

∴ the required remainder is 5.

Illustration 15. Let m, n ∈ *N* and p (x) = 1 + x +... + x ^{m}. p (x) will divide p (x^n), then find the HCF of (m + 1) and n

Solution: Given that

$$p(x)=1+x+x^2+\ldots+x^m=\frac{1-x^{m+1}}{1-x}$$

Given that $\frac{p\left(x^n\right)}{p(x)}=\frac{\left(1-x^{(m+1)n}\right)(1-x)}{\left(1-x^n\right)\left(1-x^{m+1}\right)}$ is a polynomial.

Since both $1 - x^n$ and $1 - x^{m+1}$ divide $1 - x^{(m+1)n}$ and since factors of $1 - x^{(m+1)n}$ are distinct, it is necessary and sufficient that $x^{m+1} - 1$ and $x^n - 1$ have no factor in common except $x - 1$, that is HCF $(m + 1, n) = 1$.

4.7.5 Special Equations

Type 1: If equation is in the form of

$$(x-a)(x-b)(x-c)(x-d) = k$$

Where $a < b < c < d$ and $b - a = d - c$

Put $y = \frac{\{(x-a)+(x-b)+(x-c)+(x-d)\}}{4}$

$$= x - \frac{a+b+c+d}{4}$$

Illustration 16. Solve $(x+4)(x+8)(x+12)(x+16) = 144$

Solution: Put $y = x + \frac{4+8+12+16}{4} = x+10$

Or $x = y + 10$

$$(x+4)(x+8)(x+12)(x+16) =$$
$$(y-6)(y-2)(y+2)(y+6) = 144$$

Or $(y^2-4)(y^2-36) = 144$ *let* $y^2 = t$

So $(t-4)(t-36) = 144$ *or* $t^2 - 40t + 144 = 144$ *or* $t^2 - 40t = 0$ *or* $t = 0, 40$

So corresponding value of y = 0 or $\pm\sqrt{40}$

And that of x = 10, $10 \pm \sqrt{40}$

Alternative:

$$(x+4)(x+8)(x+12)(x+16) = 144$$
$$(x^2+20x+64)(x^2+20x+96) = 144$$

Let $x^2 + 20x = t$ *then* $(t+64)(t+96) = 144$

Now we can solve this for t and then find the value of 'x'

Type 2: If equation is in the form of

$$(x-a)(x-b)(x-c)(x-d) = kx^2$$

Where ab= cd

Put $y = x + \frac{ab}{x}$

Illustration 17. Solve $(x+2)(x+3)(x+8)(x+12) = 4x^2$

Solution: In the given equation since (-2)(-12) = (-3)(-8) so we can write the equation as

$$(x+2)(x+3)(x+8)(x+12) = 4x^2$$
$$(x^2+14x+24)(x^2+11x+24) = 4x^2$$

Since x = 0 is not the root of the equation so we can divide both side of the equation by x^2 then we get

$$\left(x+\frac{24}{x}+14\right)\left(x+\frac{24}{x}+11\right) = 4$$

Now let $x + \frac{24}{x} = t$ then we get

$(t+14)(t+11) = 4$ *or* $t^2 + 25t + 150$

From this we get t = -15 or -10

So $x + \frac{24}{x} = -15$ *or* -10

We get two quadratic equations $x^2 + 15x + 24 = 0$ *or*

or $x^2 + 10x + 24 = 0$

On solving these, we get x = $\frac{-15 \pm \sqrt{129}}{2}, -6, -4$

Type 3: If equation is in the form of

$(x-a)^4 + (x-b)^4 = k$

Put $y = x - \frac{a+b}{2}$

Illustration 18. Solve the equation

$(x-2)^4 + (x-4)^4 = 16$

Solution: put $y = x - \frac{2+4}{2} = x - 3$

Then we get $(y+1)^4 + (y-1)^4 = 16$

Or $y^4 + 6y^2 - 7 = 0$ *or* $(y^2+7)(y^2-1) = 0$

Only real value is $y^2 = 1$ or y = ±1

And corresponding value of x is 2 and 4

4.7.6 Newton's Theorem

Consider a polynomial of degree 'n'.

$P(x) = a_n x^n + a_{n-1} x^{n-1} + \ldots + a_1 x + a_0$

Let P(x) = 0, have roots $x_1, x_2 \ldots x_n$.

Now define the following sums

$P_1 = x_1 + x_2 + \ldots + x_n$

$P_2 = x_1^2 + x_2^2 + \ldots + x_n^2$

⋮

$P_k = x_1^k + x_2^k + \ldots + x_n^k$

Then according to Newton's formula

$a_n P_1 + a_{n-1} = 0$

$a_n P_2 + a_{n-1} P_1 + 2a_{n-2} = 0$

$a_n P_3 + a_{n-1} P_2 + a_{n-2} P_1 + 3a_{n-3} = 0$

and so on...

Now let us define elementry symmetric sum for the roots of equation P(x) = 0

$S_1 = x_1 + x_2 + ... + x_n$

$S_2 = \Sigma x_i x_j$

$\vdots$

$S_n = x_1; x_2 ... x_n$

Then

$P_1 = S_1$

$P_2 = s_1 P_1 - 2s_2$

and so on.

Example. If x_1, x_2 and x_3 are roots of $x^5 + 5x^2 - 4x - 20 = 0$, then find $x_1^5 + x_2^5 + x_3^5$.

Sol. $P_1 + 5 = 0 \quad \Rightarrow \quad P_1 = -5$

$P_2 + 5P_1 + 2(-4) = 0 \quad \Rightarrow \quad P_2 = 33$

$P_3 + 5P_2 - 4P_1 + 3(-20) = 0 \quad \Rightarrow \quad P_3 = -125$

$P_4 + 5P_3 - 4P_2 - 20P_1 = 0 \quad \Rightarrow \quad P_4 = 657$

$P_5 + 5P_4 - 4P_3 - 20P_2 = 0 \quad \Rightarrow \quad P_5 = -3225$

4.7.7 Cubic Equation (Tschirnhaus – Vieta) Approach

Consider a cubic equation $ax^3 + bx^2 + cx + d = 0$

Lets define $p = \dfrac{3ac - b^2}{3a^2}$

And $q = \dfrac{2b^3 - 9abc + 27a^2 d}{27a^3}$

Now we have following cases-

Case (i): p< 0 , all three real roots

Case (ii): p> 0 , only one real root.

Case (iii): p= 0, one repeated real root.

Now lets define following terms

$$A = 2\sqrt{-\frac{p}{3}}$$

$$\phi = \cos^{-1}\left(\frac{3q}{Ap}\right)$$

$$B = -\frac{b}{3a}$$

Then three roots are given as

$$\alpha = A\cos\left(\frac{\phi}{3}\right) + B$$

$$\beta = A\cos\left(\frac{\phi + 2\pi}{3}\right) + B$$

$$\gamma = A\cos\left(\frac{\phi + 4\pi}{3}\right) + B$$

Illustration 19. find the roots of the equation $x^3 - 27x + 27 = 0$

Solution: Here $p = \dfrac{3ac - b^2}{3a^2} = \dfrac{-81 - 0}{3} = -27$

$$q = \frac{2b^3 - 9abc + 27a^2 d}{27a^3} = \frac{0 - 0 + 729}{27} = 27$$

Then

$$A = 2\sqrt{-\frac{p}{3}} = 2 \times 3 = 6$$

$$\phi = \cos^{-1}\left(\frac{3q}{Ap}\right) = \cos^{-1}\left(\frac{81}{-162}\right) = \frac{2\pi}{3}$$

$$B = -\frac{b}{3a} = 0$$

Then three roots are given as

$$\alpha = 6\cos\left(\frac{2\pi}{9}\right)$$

$$\beta = 6\cos\left(\frac{8\pi}{9}\right)$$

$$\gamma = 6\cos\left(\frac{4\pi}{9}\right)$$

Exercise B

Concept Deviator (CD)

(Single Correct MCQs)

1. If $x^2 - 2r \cdot p_r x + r = 0; r = 1, 2, 3$ are three quadratic equations of which each pair has exactly one root common, then the number of solutions of the triplet (p_1, p_2, p_3) is

(a) 2 (b) 1 (c) 9 (d) 27

2. If *a*, *b*, *c* are all distinct, then $a\frac{(x-b)(x-c)}{(a-b)(a-c)} + b\frac{(x-c)(x-a)}{(b-c)(b-a)} + c\frac{(x-a)(x-b)}{(c-a)(c-b)} - x$ is equal to

(a) 0 (b) abc
(c) a + b + c (d) None of these

3. If 'x' 'a' & 'b' are real, then find the maximum value of $y = 2\ (a-x)\left(x+\sqrt{x^2+b^2}\right)$

(a) a^2+b^2 (b) a^2
(c) $\sqrt{a^2+b^2}$ (d) $a\sqrt{a^2+b^2}$

4. The number of real solutions of the equation $\left(\frac{7}{9}\right)^x = -3 + x - x^2$ is

(a) 0 (b) 1
(c) 2 (d) None of these

5. Roots of the equation $ax^3 + bx^2 + cx + d = 0$ remain unchanged by increasing each coefficient by one unit, then

(a) $a = b = c = d \neq 0$ (b) $a = b \neq c = d \neq 0$
(c) $a \neq b = c = d \neq 0$ (d) $a = b = c \neq d \neq 0$

6. Complete set of values of 'a' such that $\frac{x^2-x}{1-ax}$ attains all real values, is

(a) $[1, \infty)$ (b) $(0, 4]$
(c) $(0, 1]$ (d) None of these

7. If the equation $a x^2 + b x + c = 0$ $(a > 0)$ has two roots α and β such that $\alpha < -2$, and $\beta > 2$. Then how many of the following statements are incorrect?

(i) $b^2 - 4ac > 0$
(ii) $c < 0$
(iii) $a + |b| + c < 0$
(iv) $4a + 2|b| + c < 0$.

(a) 0 (b) 2 (c) 3 (d) All

8. If $x^2 - (a-3)x + a = 0$ has at least one positive root, then

(a) $a \in (-\infty, 0) \cup [3, 9]$
(b) $a \in (-\infty, 1) \cup [9, \infty)$
(c) $a \in (-\infty, 0] \cup [9, \infty]$
(d) None of these

9. If $\log_2(ax^2 + x + a) \geq 1\ \forall\ x \in R$, then exhaustive set of values of 'a 'is

(a) $\left(0, 1+\frac{\sqrt{5}}{2}\right)$ (b) $\left(1-\frac{\sqrt{5}}{2}, 1+\frac{\sqrt{5}}{2}\right)$
(c) $\left(-1, 1-\frac{\sqrt{5}}{2}\right)$ (d) $\left(1+\frac{\sqrt{5}}{2}, \infty\right)$

10. If $x_1, x_2, x_3, \ldots, x_n$ are the roots of $x^n + ax + b = 0$, then the value of $(x_1 - x_2)(x_1 - x_3)(x_1 - x_4) \ldots (x_1 - x_n)$ is equal to

(a) $nx_1 + b$ (b) $n(x_1)^{n-1}$
(c) $n(x_1)^{n-1} + a$ (d) $n(x_1)^{n-1} + b$

11. The values of x satisfying $\log_{|\sin x|}|\cos x| + \log_{|\cos x|}|\sin x| = 2$, when $x \in \left(0, \frac{\pi}{2}\right)$, is

(a) $\left\{\frac{\pi}{4}\right\}$ (b) $\left\{\frac{\pi}{3}\right\}$
(c) $\left(0, \frac{\pi}{4}\right)$ (d) $\left(\frac{\pi}{4}, \frac{\pi}{2}\right)$

12. If $a^3\frac{(x-b)(x-c)(x-d)}{(a-b)(a-c)(a-d)} + b^3\frac{(x-c)(x-d)(x-a)}{(b-c)(b-d)(b-a)} + c^3\frac{(x-d)(x-a)(x-b)}{(c-d)(c-a)(c-b)} + d^3\frac{(x-a)(x-b)(x-c)}{(d-a)(d-b)(d-c)} = x^3$, then x has

(a) no solution
(b) on real and two imaginary roots
(c) three real roots
(d) infinitely many roots

13. If $a,b,c \in I$ and $ax^2+bx+c=0$ has irrational root and $\lambda \in Q = \frac{p}{q}$ where $f(x)=ax^2+bx+c$, then find the minimum value of $|f(\lambda)|$

(a) $\frac{1}{q^2}$ (b) $\frac{p^2}{q^2}$

(c) $\frac{p}{q^2}$ (d) None of these

14. If an unknown polynomial is divided by $(x-1)$ and $(x-2)$, we obtain the remainder 2 and 1 respectively, then the remainder resulting from the division of this polynomial by $(x-1)(x-2)$ is

(a) 3 (b) -3 (c) $3-x$ (d) $3-2x$

15. Find the number of integers, which satisfy the inequality $\frac{(16)^{1/x}}{(2^{x+3})} > 1$

(a) infinite (b) 0 (c) 1 (d) 4

16. The equation $(x-n)+(x-n^2)^m+(x-n^3)^m+\ldots+(x-n^m)^m=0$, (m is odd positive integer), has

(a) all real roots

(b) one real and $(n-1)$ imaginary roots

(c) one real and $(m-1)$ imaginary roots

(d) no real root

17. If $x^2-(a+b+c)x+(ab+bc+ca)=0$ has imaginary roots, where $a,b,c \in R^+$, then $\sqrt{a},\sqrt{b},\sqrt{c}$

(a) can be sides of Δ (b) cant be side of Δ

(c) Are in GP (d) None of these

18. Number of solutions for $x^2-2-2[x]=0$, (where [.] denotes greatest integer function) is

(a) zero (b) one

(c) two (d) infinite

19. If $x,y \in [0,10]$, then number of solutions (x, y) of the inequation $3^{\sec^2 x-1}\sqrt{9y^2-6y+2} \le 1$ is

(a) 2 (b) 1

(c) 4 (d) infinite

20. If α, β are roots of $375x^2-25x-2=0$ and $s_n=\alpha^n+\beta^n$ then $\lim_{x\to\infty}\sum_{r=1}^{n} s_r$ is

(a) 7/116 (b) 1/12

(c) 29/348 (d) None of these

21. If $x=2+2^{1/3}+2^{2/3}$, then the value of x^3-6x^2+6x is

(a) 3 (b) 2

(c) 1 (d) None of these

22. How many values of x satisfy $x(x-3)\ge 0$ and $(3|x|-3)^2=|x|+7$ simultaneously.

(a) 0 (b) 1

(c) 3 (d) None of these

23. Let f(x) be a polynomial with integral coefficients. Let a, b, c be three distinct integers such that f(a) = f(b) = f(c) = –1. Find the number of integral roots of f(x) = 0.

(a) 1 (b) 3

(c) 2 (d) None of these

24. If the equation $x^2+ix+a=0, x^2-2x+ia=0$, $a \ne 0$ have a common root, then

(a) a is real (b) $a=\frac{1}{2}+i$

(c) $a=\frac{1}{2}-i$

(d) the other root is also common

25. If the roots of the equation $x^2-2ax+a^2+a-3=0$ are real and less than 3, then

(a) $a<2$ (b) $2\le a\le 3$

(c) $3<a\le 4$ (d) $a>4$

26. Find the values of 'a' for which exactly one root of the equation $e^a x^2-e^{2a}x+e^a-1=0$ lies between 1 and 2.

(a) $\log\frac{(5-\sqrt{17})}{4}<a<\log\frac{5+\sqrt{17}}{4}$

(b) $0<a<100$

(c) $\log\frac{5}{4}<a<\log\frac{10}{3}$

(d) None of the above

27. If $2a+3b+6c=0, a,b,c \in R$, then the quadratic equation $ax^2+bx+c=0$ has

(a) at least one root in [0, 1]

(b) at least one root in [2, 3]

(c) at least one root in (3, 4]

(d) None of the above

28. The constant term of the quadratic expression $\sum_{k=2}^{n}\left(x-\frac{1}{k-1}\right)\left(x-\frac{1}{k}\right)$, as $n\to\infty$ is

(a) – 1 (b) 0
(c) 1 (d) None of these

29. The system $y^{\left(x^2+7x+12\right)}=1$ and $x+y=6, y>0$ has

(a) no solution
(b) one solution
(c) two solutions
(d) more than two solutions

30. If $p\in[-1, 1]$, then $4x^3-3x-p=0$ has a root in –

(a) [0, 1] (b) $\left[\frac{1}{2}, 1\right]$
(c) $\left[0, \frac{1}{2}\right]$ (d) [– 1, 1]

31. Find the least integral value of m for which every solution of the inequality $1\le x\le 2$ is a solution of the inequality $x^2 - mx + 1 < 0$.

(a) 7 (b) 3
(c) 2 (d) None of these

32. If $f(x) = ax^2 + bx + c$, such that $c<0$ and $a - 2b + 4c > 0$, then f(x) has

(a) One root in the interval (0, 1/2)
(b) One root in the interval (-1/2, 0)
(c) Both roots are positive
(d) None of these

33. The number of negative integers belonging to the solution set of the equation $2^{|x+2|} - |2^{x+1} - 1| = 2^{x+1} + 1$ is equals to

(a) 0 (b) 2
(c) 3 (d) None of these

34. Find the sum of all the real roots of the equation $P(x) = 2x^{98} + 3x^{97} + 2x^{96} + ... + 2x + 3 = 0$.

(a) 7 (b) 3
(c) 2 (d) None of these

35. If the reciprocal of every root of $x^3 + x^2 + ax + b = 0$ is also a root then how many ordered pairs of (a,b) exist?

(a) 4 (b) 6
(c) 9 (d) None of these

36. The values of x satisfying the inequality $|x^3 - 1| \ge 1 - x$ belong to

(a) $(-\infty, -1]$ $[1,\infty)$ (b) $(-\infty, -1]$ $[0,\infty)$
(c) $(-\infty, 0]$ $[1,\infty)$ (d) None of these

37. The integral value(s) of a for which the equation $(x^2 + x + 2)^2 - (a - 3)(x^2 + x + 2)(x^2 + x + 1) + (a-4)(x^2 + x + 1)^2 = 0$ have at least one real root is /are

(a) 5 (b) 6
(c) 4 (d) None of these

38. If the expression $ax^2+8x-(20+4a)<0\,\forall x\in$ [0, 2] then a lies in-

(a) (0, ∞) (b) [0, ∞)
(c) [1, ∞) (d) None of these

39. Find the sum of all the integral values of 'k' here 'k' is such that there is exactly one root of the equation $x^2 - (k - 1)x + (k + 4) = 0$ lies between the roots of the equation $x^2 - (k + 3)x + (k + 2) = 0$

(a) 3 (b) 5
(c) 7 (d) None of these

40. Let $f(x) = ax^2+bx+c$, a, b, c ∈ R and a ≠ 0. If f(x) > 0 $\forall x\in R$ then how many of the following statements are true-

(i) $4a - 2b + c > 0$
(ii) $2a - b + c > 0$
(iii) $10a + 3b + c > 0$
(iv) $2a + b + c > 0$

(a) 0 (b) 2
(c) 3 (d) None of these

41. If a, b, and c are odd positive integers, then the roots of $ax^2+bx+c=0$, if real, are

(a) Integers
(b) Rational numbers but not integers
(c) Irrational
(d) Equal

42. For how many integral values of x inequality, $\frac{3x^2-7x+8}{x^2+1}\le 2$ holds true?

(a) 0 (b) 3
(c) 5 (d) None of these

43. Solve the inequality, $\frac{1}{x-2}-\frac{1}{x}\le\frac{2}{x+2}$

(a) $\left(-2,\frac{(3-\sqrt{17})}{2}\right]\cup(0,2)\cup\left[\frac{(3-\sqrt{17})}{2},+\infty\right)$

(b) $\left(-2,\frac{(3-\sqrt{17})}{2}\right]\cup\left[\frac{(3-\sqrt{17})}{2},+\infty\right)$

(c) $\left(-2,\frac{(3-\sqrt{17})}{4}\right]\cup\left[\frac{(3-\sqrt{17})}{4},+\infty\right)$

(d) None of these

44. If α, β are the roots of the equation $x^2-2\cos\theta x+1=0$, then equation whose roots are α^n, β^n are.

(a) $x^2-(2\cos n\theta)x+1=0$

(b) $2x^2-(2\cos n\theta)x-1=0$

(c) $x^2+(2\cos n\theta)x+1-0$

(d) $x^2-(2\cos n\theta)x-1=0$

45. If $(x-1)^3$ is a factor of $x^4+ax^3+bx^2+cx-1$, then the other factor is

(a) $x-3$ (b) $x+1$

(c) $x+2$ (d) None of these

46. If n and r are positive integers such that $0<r<n$, then the roots of the quadratic equation ${}^nC_{r-1}x^2+2\,{}^nC_rx+{}^nC_{r+1}-0$ arc

(a) real and distinct

(b) rational

(c) rational but not an integer

(d) imaginary

47. The least +ve integral value of 'a 'for which the equations $3x^4+4x^3-12x^2+5a=0$ has no real root is

(a) 2 (b) 4

(c) 6 (d) none of these

48. Let $f(x)=ax^2+bx+c$, a, b, c $\in$ R and a $\neq$ 0. suppose $f(x)>0$ for all x $\in$ R. and $g(x)=f(x)+f'(x)+f'(x)$. Then

(a) $g(x)>0\ \forall x\in R$.

(b) $g(x)<0\ \forall x\in R$.

(c) g (x) = 0 two distinct real roots.

(d) g (x) = 0 has real roots.

49. The equation $3^{x-1}+5^{x-1}=34$ has

(a) no solution

(b) one solution

(c) two solution

(d) more than two solution

50. In a triangle PQR, $\angle R=\pi/4$. If tan (P/3) and tan(Q/3) are the roots of the equation $ax^2+bx+c=0$, then

(a) a + b = c (b) b + c = 0

(c) a + c = b (d) b = c

51. Let (a_1, a_2, a_3, a_4, a_5) denote a rearrangement of (8, – 10, 6, 5, – 9), then the equation $a_1x^4+a_2x^3+a_3x^2+a_4x+a_5=0$ has

(a) at least two real roots

(b) all four real roots

(c) only imaginary roots

(d) none of these.

52. If x + y , z $\in$ R. x + y + z = 4 and $x^2+y^2+z^2=6$, find the maximum possible value of z.

(a) $\sqrt{5}$-1 (b) $\sqrt{3}$

(c) 2 (d) None of these

53. If a, b $\in$ R. and the quadratic equation $x^2+(a-b)x-a-b+1=0$ has unequal real roots for all b $\in$ R. then

(a) a > 0 (b) a > 1

(c) a < 1 (d) a can take any value.

54. Let a, b, c , p , q be five different non-zero real numbers and x, y, z be three numbers satisfying the system of equations

$$\frac{x}{a}+\frac{y}{a-p}+\frac{z}{a-q}=1$$

$$\frac{x}{b}+\frac{y}{b-p}+\frac{z}{b-q}=1$$

$$\frac{x}{c}+\frac{y}{c-p}+\frac{z}{c-q}=1$$

Then how many of the following statements are correct?

(i) $x + y + z = a + b + c - p - q$

(ii) $x = \frac{abc}{pq}$

(iii) $y = \frac{(a+p)(b+p)(c+p)}{p(p+q)}$

(iv) $z = \frac{(a+q)(b+p)(c+q)}{q(q+p)}$

(a) 2 (b) 3
(c) 4 (d) None of these

55. If α, β, γ are the roots of $x^3 + 64 = 0$, then the equation whose roots are $\left(\frac{\alpha}{\beta}\right)^2$ and $\left(\frac{\alpha}{\gamma}\right)^2$ is

(a) $x^2 - 14x + 6 = 0$
(b) $x + x + 1 = 0$
(c) $x^2 + 4x + 16 = 0$
(d) $x^2 - x + 1 = 0$

IIT CORNER

56. If l, m, n are real, $l \neq m$, then the roots of the equation : $(l - m)x^2 - 5(l + m)x - 2(l - m) = 0$ are (IIT JEE 1979)

(a) Real and equal
(b) Complex
(c) Real and unequal
(d) None of these.

57. If x, y and z are real and different $u = x^2 + 4y^2 + 9z^2 - 6yz - 3zx - 2xy$, then u is always, (IIT JEE 1979)

(a) non-negative (b) zero
(c) non-positive (d) none of these

58. Let $a > 0$, $b > 0$ and $c > 0$. Then the roots of the equation $ax^2 + bx + c = 0$ (IIT JEE 1979)

(a) are real and negative
(b) have negative real parts
(c) both (a) and (b)
(d) none of these

59. Both the roots of the equation $(x - b)(x - c) + (x - a)(x - c) + (x - a)(x - b) = 0$ are always (IIT JEE 1980)

(a) positive (b) real
(c) negative (d) none of these

60. If $(x^2 + px + 1)$ is a factor of $(ax^3 + bx + c)$, then (IIT JEE 1980)

(a) $a^2 + c^2 = -ab$ (b) $a^2 - c^2 = -ab$
(c) $a^2 - c^2 = ab$ (d) none of these

61. The largest interval for which $x^{12} - x^9 + x^4 - x + 1 > 0$ is (IIT JEE 1982)

(a) $-4 < x \leq 0$ (b) $0 < x < 1$
(c) $-100 < x < 100$ (d) $\infty < x < \infty$

62. The equation $x - \frac{2}{x-1} = 1 - \frac{2}{x-1}$ has (IIT JEE 1984)

(a) no root
(b) one root
(c) two equal roots
(d) infinitely many roots.

63. If $a^2 + b^2 + c^2 = 1$, then $ab + bc + ca$ lies in the interval (IIT JEE 1984)

(a) $[\frac{1}{2}, 2]$ (b) $[-1, 2]$
(c) $[-\frac{1}{2}, 1]$ (d) $[-1, \frac{1}{2}]$

64. If α and β are the roots of $x^2 + px + q = 0$ and α^4, β^4 are the roots of $x^2 - rx + s = 0$, then the equation $x^2 - 4qx + 2q^2 - r = 0$ has always (IIT JEE 1989)

(a) two real roots (b) two positive roots
(c) two negative roots
(d) one positive and one negative root.

65. Let a, b, c be real numbers, $a \neq 0$. If α is a root of $a^2x^2 + bx + c = 0$. β is the root of $a^2x^2 - bx - c = 0$ and $0 < \alpha < \beta$, then the equation $a^2x^2 + 2bx = 2c = 0$ has a root γ that always satisfies (IIT JEE 1989)

(a) $\gamma = \frac{\alpha + \beta}{2}$ (b) $\gamma = \alpha + \frac{\beta}{2}$
(c) $\gamma = \alpha$ (d) $\alpha < \gamma < \beta$

66. The number of solution of the equation $\sin(e^x) = 5^x + 5^{-x}$ is (IIT JEE 1990)

(a) 0 (b) 1
(c) 2 (d) Infinitely many

67. If p, q, r are + ve and are in A. P., the roots of quadratic equation $px^2 + qx + r = 0$ are all real for (IIT JEE 1994)

(a) $\left|\frac{r}{p} - 7\right| \geq 4\sqrt{3}$ (b) $\left|\frac{p}{r} - 7\right| \geq 4\sqrt{3}$

(c) all p and r (d) no p and r

68. Let $p, q \in \{1, 2, 3, 4\}$. The number of equations of the form $p x^2 + q x + 1 = 0$ having real roots is (IIT JEE 1994)

(a) 15 (b) 9

(c) 7 (d) 8

69. The equation $\sqrt{x+1} - \sqrt{x-1} = \sqrt{4x-1}$ has (IIT JEE 1997)

(a) no solution (b) one solution

(c) two solution

(d) more than two solutions

70. If the roots of the equation $x^2 - 2ax + a^2 + a - 3 = 0$ are real and less than 3, then (IIT JEE 1999)

(a) $a < 2$ (b) $2 \leq a \leq 3$

(c) $3 < a \leq 4$ (d) $a > 4$

71. If α and β (α < β) are the roots of the equation $x^2 + bx + c = 0$, where $c < 0 < b$, then (IIT JEE 2000)

(a) $0 < \alpha < \beta$ (b) $\alpha < 0 < \beta < |\alpha|$

(c) $\alpha < \beta < 0$ (d) $\alpha < 0 < |\alpha| < \beta$

72. If b > a, then the equation $(x - a)(x - b) - 1 = 0$ has (IIT JEE 2000)

(a) both roots in (a, b)

(b) both roots in $(-\infty, a)$

(c) both roots in $(b, +\infty)$

(d) one roots in $(-\infty, a)$ and the other in $(b + \infty)$

73. The set of all real numbers x for which $x^2 - |x + 2| + x > 0$, is (IIT JEE 2002)

(a) $(-\infty, -2)$

(b) $(-\infty, -\sqrt{2}) \cup (\sqrt{2}, \infty)$

(c) $(-\infty, -1) \cup (1, \infty)$

(d) $(\sqrt{2}, \infty)$

74. For all 'x', $x^2 + 2ax + 10 - 3a > 0$, then the interval in which ' a ' lies is (IIT JEE 2004)

(a) $a < -5$ (b) $-5 < a < 2$

(c) $a > 5$ (d) $2 < a < 5$

75. If one root is square of the other root of the equation $x^2 + px + q = 0$, then the relation between p and q is (IIT JEE 2004)

(a) $p^3 - q(3p - 1) + q^2 = 0$

(b) $p^3 - q(3p + 1) + q^2 = 0$

(c) $p^3 + q(3p - 1) + q^2 = 0$

(d) $p^3 + q(3p + 1) + q^2 = 0$

76. Let a, b, c be the sides of a triangle where $a \neq b \neq c$ and $\lambda \in R$. If the roots of the equation $x^2 + 2(a + b + c)x + 3\lambda(ab + bc + ca) = 0$ are real, then (IIT JEE 2006)

(a) $\lambda < \frac{4}{3}$ (b) $\lambda > \frac{5}{3}$

(c) $\lambda \in \left(\frac{1}{3}, \frac{5}{3}\right)$ (d) $\lambda \in \left(\frac{4}{3}, \frac{5}{3}\right)$

77. Let α, β be the roots of the equation $x^2 - px + r = 0$ and $\frac{\alpha}{2}$, 2 β be the roots of the equation $x^2 - qx + r = 0$. Then the value of r is (IIT JEE 2007)

(a) $\frac{2}{9}(p - q)(2q - p)$

(b) $\frac{2}{9}(q - p)(2p - q)$

(c) $\frac{2}{9}(q - 2p)(2q - p)$

(d) $\frac{2}{9}(2p - q)(2q - p)$

78. Let p and q be real numbers such that $p \neq 0$, $p^3 \neq q$ and $p^3 \neq -q$. If α and β are non zero complex numbers satisfying $\alpha + \beta = -p$ and $\alpha^3 + \beta^3 = q$, then a quadratic equation having $\frac{\alpha}{\beta}$ and $\frac{\beta}{\alpha}$ as its roots is (IIT JEE 2010)

(a) $(p^3 + q)x^2 - (p^3 + 2q)x + (p^3 + q) = 0$

(b) $(p^3 + q)x^2 - (p^3 - 2q)x + (p^3 + q) = 0$

(c) $(p^3 - q)x^2 - (5p^3 - 2q)x + (p^3 - q) = 0$

(d) $(p^3 - q)x^2 - (5p^3 + 2q)x + (p^3 - q) = 0$

79. Let α and β be the roots of $x^2 - 6x - 2 = 0$, with $\alpha > \beta$. If $a_n = \alpha^n - \beta^n$ for $n \geq 1$, then the value of $\frac{a_{10} - 2a_8}{2a_9}$ is (IIT JEE 2011)

(a) 1 (b) 2

(c) 3 (d) 4

80. A value of b for which the equations

$x^2 + bx - 1 = 0$

$x^2 + x + b = 0$

Have one root in common. (IIT JEE 2011)

(a) $-\sqrt{2}$ (b) $-i\sqrt{3}$

(c) $i\sqrt{5}$ (d) $\sqrt{2}$

81. The quadratic equation $p(x) = 0$ with real coefficients has purely imaginary roots. Then the equation $p(p(x)) = 0$ has (JEE Adv. 2014)

(a) one purely imaginary root
(b) all real roots
(c) two real and two purely imaginary roots
(d) neither real nor purely imaginary roots

82. Let $-\frac{\pi}{6} < \theta < -\frac{\pi}{12}$. Suppose α_1 and β_1 are the roots of the equation $x^2 - 2x \sec \alpha + 1 = 0$ and α_2 and β_2 are the roots of the equation $x^2 + 2x \tan \theta - 1 = 0$. If $\alpha_1 > \beta_1$ and $\alpha_2 > \beta_2$, then $\alpha_1 + \beta_2$ equals (JEE Adv. 2016)

(a) $2(\sec \theta - \tan \theta)$ (b) $2 \sec \theta$

(c) $-2 \tan \theta$ (d) 0

(Passage Type)

Passage-I

If roots of the equation $x^4 - 12x^3 + bx^2 + cx + 81 = 0$ are positive, given that b and c are rational then

83. Value of b-c is

(a) 162 (b) -54

(c) 54 (d) -162

84. The roots of equation $bx^2 + bx + c = 0$ are

(a) Complex roots
(b) Real and rational roots
(c) Real and irrational roots
(d) None of these

Passage-II

If real roots of $x^2 + b_1x + c_1 = 0$, is 'k' more than the real roots of $x^2 + b_2x + c_2 = 0$. If least value of $x^2 + b_1x + c_1$ be – ¼. And least value of $x^2 + b_1x + c_1$ occurs at $x = \frac{7}{2}$.

85. The least value of $x^2 + b_2x + c_2$ is

(a) -1 (b) $-1/2$

(c) $-1/4$ (d) $-1/3$

86. The value of b_1 is

(a) 6 (b) -7 (c) 8 (d) 0

Passage-III

Consider a quadratic function $f(x) = ax^2 + bx + c$ and the roots of $f(x) = x$ be α and β if α β λ and δ are roots of $f(f(x)) = x$. then-

87. If α and β and are real and unequal then

(a) λ and δ are imaginary
(b) λ and δ are real
(c) λ and δ may be real or imaginary
(d) cannot say

88. If α and β are real and equal then

(a) λ and δ are imaginary
(b) λ and δ are real
(c) λ and δ may be real or imaginary
(d) cannot say

Passage-IV

Let x_1, x_2, x_3, x_4 be the root (real or complex) of the equation $x^4 + ax^3 + bx^2 + cx + d = 0$ then

Sum of the roots $x_1 + x_2 + x_3 + x_4 = -a$

$$\sum x_i x_j = b$$
$$\sum x_i x_j x_k = -c$$
$$x_1 x_2 x_3 x_4 = d$$

If $x_1 + x_2 = x_3 + x_4$ and $a, b, c, d \in R$,

89. If $a = 2$, then the value of $b - c$ is

(a) -1 (b) 1

(c) -2 (d) 2

90. If $b + c = 1$ and $a \neq -2$, then for real values of 'c' $\in$

(a) $\left(-\infty, \frac{1}{4}\right)$ (b) $(-\infty, 3)$

(c) $(-\infty, 1)$ (d) $(-\infty, 4)$

Passage V

Let $f(x) = ax^3 + bx^2 + cx + d$ and $a > 0$, then $f'(x) = 3ax^3 + 2bx + c$ Now we have following cases

CASE (1): The equation $f'(x) = 0$ has no real roots. Then f(x) always increases as x increases and the equation $f(x) = 0$ has one real root i.e. graph of f(x) will cut the X-axis only once as shown in the figure.

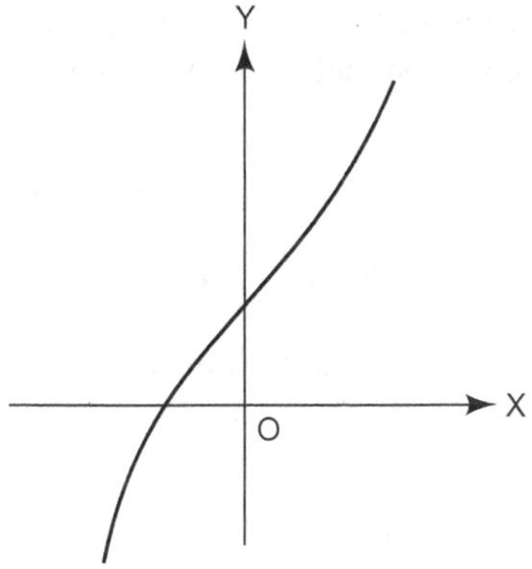

CASE (2): The equation $f'(x) = 0$ has two distinct roots α and β $(\alpha < \beta)$. If $f(\alpha)f(\beta) > 0$, then the equation $f(x) = 0$ has one real root.

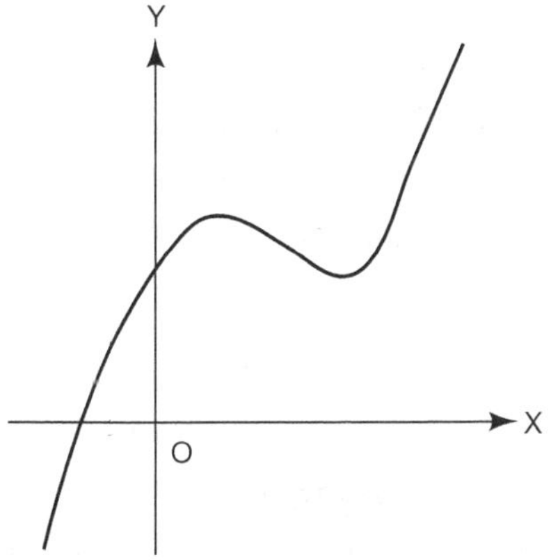

CASE (3) : The equation $f'(x) = 0$ has two distinct roots α and β and $f(\alpha)f(\beta) < 0$, then the equation $f(x) = 0$ has three real roots.

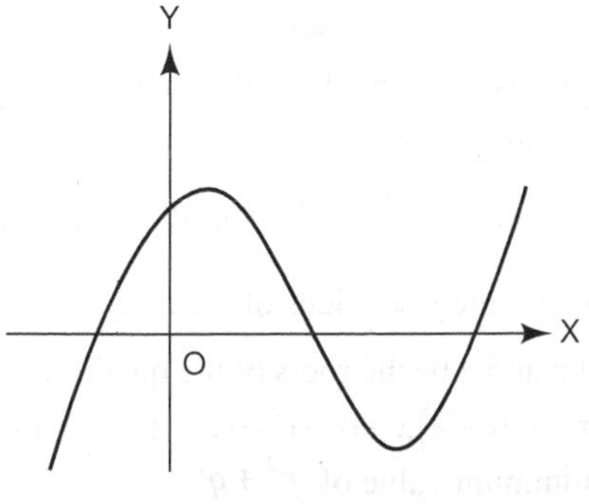

CASE (4) : The equation $f'(x) = 0$ has two distinct roots α and β but $f(\beta) = 0$ also. In this case $f(x) = 0$ has repeated root and one other root.

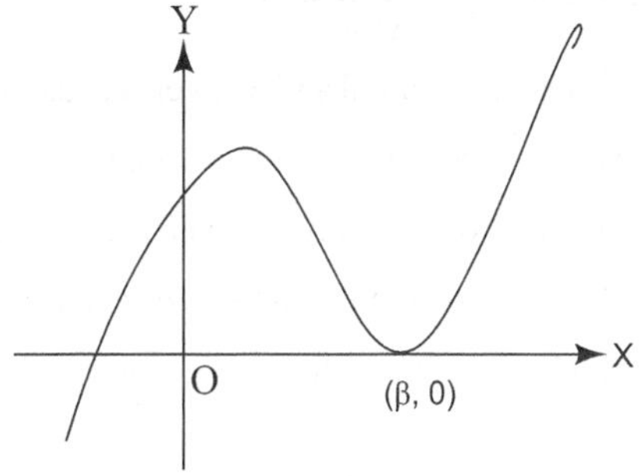

In general, the equation $g(x) = 0$ has a repeated root $x = \alpha$, if $g(\alpha) = g'(\alpha) = 0$.

Now answer the following questions:

91. The equation $8x^3 - 20x^2 + 6x + 9 = 0$ has
    (a) One repeated root
    (b) All distinct real roots
    (c) All roots repeated
    (d) Only one real root which is not repeated

92. If the roots of $x^3 - 3x + 1 = 0$ be α, β, γ then the value of $[\alpha] + [\beta] + [\gamma]$ is
    (a) 0 (b) 1
    (c) −1
    (d) Not defined as all roots are not real

Passage VI

Consider a function $f(x) = \dfrac{p(x)}{q(x)}$

Now investigate the nature of p(x) and q(x) and try to factorize it if possible. There may be limitations of discontinuity of f(x), whenever $q(x)$ vanishes. If y tends to $+\infty$ and $-\infty$ while crossing a finite values of x, say $x = a$ if $(x - a)$ is a factor of q(x) then the line $x = a$ is a vertical asymptote of the curve $y = f(x)$. Similarly the line $y = b$ is a horizontal asymptote if x tends to $+\infty$ and $-\infty$ while crossing $y = b$.

Try to understand with some example, consider a function $f(x) = \dfrac{(x-2)}{(x+4)}$ this function has an asymptote $x = -4$ and $y = 1$

Now based on above concept answer the following questions:

93. The curve $y = \dfrac{x-5}{x^2+4x}$ has

(a) One horizontal and one vertical asymptote.

(b) No horizontal but one vertical asymptote.

(c) One horizontal but no vertical asymptote.

(d) One horizontal and two vertical asymptote

94. Which one of the following is correct about the curve $y = \dfrac{x+5}{x^2+4x+5}$

(a) It has one horizontal and one vertical asymptote.

(b) It has no horizontal but one vertical asymptote.

(c) It has one horizontal but no vertical asymptote.

(d) It has one horizontal and two vertical asymptote.

Passage VII

Let us consider two quadratic equation

$$ax^2+bx+c = 0 \quad ...(i)$$

and $$a_1x^2+b_1x+c_1 = 0 \quad ...(ii)$$

Now assume that they have a common root,

Let $c, c_1 \neq 0$ and $ab_1 - a_1b \neq 0$.

Let the common root be α, then $a\alpha^2+b\alpha+c=0$ ... (iii) and

$$a_1\alpha^2+b_1\alpha+c_1 = 0 \quad ... (iv)$$

From equation (iii) and (iv)

$$\therefore \quad \frac{\alpha^2}{bc_1-b_1c} = \frac{\alpha}{ca_1-ac_1} = \frac{1}{ab_1-a_1b}$$

$$\alpha = \frac{bc_1-b_1c}{ca_1-ac_1} = \frac{ca_1-ac_1}{ab_1-a_1b} \quad ...(v)$$

Or, $(bc_1-b_1c)(ab_1-a_1b) = (ca_1-ac_1)^2$

Now consider three quadratic equations

$x^2 + 2rp_rx + r = 0$; r = 1, 2, 3.

Given that each pair has exactly one root common.

95. The sum of all the three common roots is equals to

(a) 0 or $\sqrt{6}$ (b) $\dfrac{11}{\sqrt{6}}$ or $-\dfrac{11}{\sqrt{6}}$

(c) $\dfrac{11}{\sqrt{6}}$ or $-\sqrt{11}$ (d) 0 or $\pm\dfrac{11}{\sqrt{\ }}$

96. What is the degree of lowest degree equation with rational co-efficient whose roots are all the common roots of the equation?

(a) 4 (b) 3

(c) 6 (d) None of these

Passage VIII

Let α, β, γ be three numbers such that $\alpha+\beta+\gamma=2$,

$$\frac{1}{\alpha}+\frac{1}{\beta}+\frac{1}{\gamma}=\frac{1}{2},$$

$$\frac{1}{\alpha^2}+\frac{1}{\beta^2}+\frac{1}{\gamma^2}=\frac{9}{4},$$

97. Find the cubic equation whose roots are α, β and γ

(a) $x^3-2x^2-x+2=0$

(b) $x^3-2x^2+x+2=0$

(c) $x^3-2x^2-x-2=0$

(d) None of these

98. Find the value of $\alpha^3+\beta^3+\gamma^3$

(a) –3 (b) 8

(c) 7 (d) None of these

(Integer Type)

99. The equation $x^3-6x^2+9x+\lambda=0$ have exactly one root in (1, 3) then find the number of integral values of $[\lambda+1]$ (where [.] denotes the greatest integer function)

100. If $x^2+\lambda x-8=0$ and $(b-c)x^2+(c-a)x+\lambda(a-b)=0$ have both the roots common, then find the value of λ.

101. If $-3 \le \dfrac{x^2-\lambda x-2}{x^2+x+1} \le 2$ for all $x \in R$, then how many integral values of λ exist?

102. If p and q be the roots of the quadratic equation $x^2-(\alpha-2)x-\alpha-1=0$, then find the minimum value of p^2+q^2.

103. If α and β are the roots of $2x^2+7x+k=0$ and $|\alpha^2-\beta^2|=\frac{7}{4}$, then find the value of 'k'

104. Find the number of solutions of equation $x^6-x^5+x^4-x^3+x^2-x+\frac{3}{4}=0$

105. Find the number of integral values of for a which the quadratic equation $2x^2-(a^3+8a-1)x+a^2-4a=0$ possesses roots of opposite signs.

106. If the equation $x^2+2(a+1)x+9a-5=0$ has only negative roots, then find the minimum value of a

107. If one root is $\sqrt{3}-\sqrt{2}$ then the equation of the lowest degree with rational coefficient is a degree of-

108. If the quadratic equation $3x^2+2(a^2+1)x+(a^2-3a+2)=0$ possesses roots of opposite sign then [a], where [a] represents integral part of a can be equal to

109. If the roots of the equation $x^4-8x^3+bx^2-cx+16=0$ are all positive then one of the roots is-

110. If the quadratic polynomial, $y=(\cot\alpha)x^2+2(\sqrt{\sin\alpha})x+\frac{1}{2}\tan\alpha$, $\alpha\in[0, 2\pi]$ can take negative values for all $x\in R$, then the value of $a\in\left(\frac{5\pi}{\lambda},\pi\right)$, then find the value of λ

111. Find the number of quadratic equations with real roots which remain unchanged even after squaring their roots.

112. Find the number of solutions for the equation $\log_4(2x^2+x+1)-\log_2(2x-1)=1$

113. If α,β be the roots $x^2+px-q=0$ and γ,δ be the roots of $x^2+px+r=0, q+r\neq 0$, then $\frac{(\alpha-\gamma)(\alpha-\delta)}{(\beta-\gamma)(\beta-\delta)}$ is

114. The least positive integer x, which satisfies the inequality $\log_{\log_2\frac{x}{2}}(x^2-10x+22)>0$ is equal to

115. Find the sum of all the integral value of a for which the equation (2 – x) (x + 1) = a has non – negative roots.

116. Let x_1, x_2, x_3 satisfying equation $x^3-x^2+\beta x+\gamma=0$ are in G.P. where (x_1, x_2, $x_3>0$), then find the maximum value of [β] +[γ] +2 , where [.] denotes the greatest integer function.

117. If the range of the expression $\frac{x^2+x+d}{x^2+2x+d}$ is $\left[\frac{5}{6},\frac{3}{2}\right]$ for all real x for which it is defined then the value of d is

118. Find the sum of absolute values of the of integral values of a for which the inequality $3-|x-a|>x^2$ is satisfied by at least one negative x

119. Find the number of integral values of x such that $x^2-2x+\cos^{10}\alpha=0$,

120. If α,β,γ are such that $\alpha+\beta+\gamma=2, \alpha^2+\beta^2+\gamma^2=6, \alpha^3+\beta^3+\gamma^3=8$, then find the value of $\frac{\alpha^4+\beta^4+\gamma^4}{3}$

121. Find the number of integral values of λ such that $(\lambda^2+\lambda-2)x^2+(\lambda+2)x<1$ for all $x\in R$,

122. The least integral value of k for which $(k-2)x^2+8x+k+4>0$ for all $x\in R$ is

123. For how many integral values of 'x' the inequality $1<\frac{3x^2-7x+8}{x^2+1}\leq 2$ satisfy.

124. Find the numerical value of sum of all the real roots of $(x+3)^4+(x+5)^4=16$

125. Find the number of distinct real roots of the equation $x^2+|x-1|=1$.

IIT CORNER

126. The smallest value of k, for which both the roots of the equation

$$x^2-8kx+16(k^2-k+1)=0$$

are real, distinct and have values at least 4, is

127. The number of distinct real roots of

$x^4-4x^3+12x^2+x-1=0$ is

(Assertion and Reason)

Directions : In the following questions, each question contains Statement I (Assertion) and Statement II (Reason). Each question has 4 choices (a), (b), (c) and (d) out of which only one is correct. The choices are

(a) Statement I is True, Statement II is true; Statement II is a correct explanation for Statement I

(b) Statement I is True, Statement II is True; Statement II is NOT a correct explanation for Statement I.

(c) Statement I is True, Statement II is False.

(d) Statement I is False, Statement II is True.

128. Statement I: The quadratic equation $(p-q)x^2 + (q-r)x + (q-r) = 0$ have real roots

Statement II: If sum of the coefficients in a quadratic equation vanishes then its one root is $x = 1$

129. **Statement I:** The equation $f(x) = 1 + \frac{x}{1} + \frac{x^2}{2!} + \frac{x^3}{3!} + \frac{x^4}{4!} \ldots + \frac{x^{100}}{100!} = 0$ has a pairs of repeated roots.

Statement II: Polynomial equation $f(x) = 0$ has a repeated root a if $f(\alpha) = 0$. and $f'(\alpha) = 0$

130. **Statement I:** If one roots is $\sqrt{5}-\sqrt{2}$, then the equation of lowest degree with rational coefficient is $x^4 - 14x^2 + 9 = 0$.

Statement II: For a polynomial equation with rational coefficient irrational roots occurs in pairs.

131. **Statement I:** If a, b, c $\in$ R. – [0]; then at least one $ax^2 + bx + c = 0$, $bx^2 + cx + a = 0$ and $cx^2 + ax + b = 0$ has imaginary roots.

Statement II: If a, b, c $\in$ R, a $\neq$ 0, then imaginary roots of the equation $ax^2 + bx + c = 0$ occur in conjugate pair.

132. **Statement I:** If roots of the equation $x^2 + bx + c = 0$, are two consecutive integers, then $b^2 - 4c = 1$

Statement II: If a, b, c are odd integer, then the roots of the equation $4abc\, x^2 + (b^2 - 4ac)x - b = 0$ are real and distinct.

133. **Statement I:** Let $ac \neq bc$ and let x_1, x_2 be the roots of $x^2 + ax + bc = 0$ and x_2, x_3 be the roots of $x^2 + bx + ca = 0$, then x_1, x_3 are the roots of $x^2 + cx + ab = 0$.

Statement II: If a, b $\in$ R. then $|ab| < \sqrt{a^2+b^2}$.

134. **Statement I:** The equation $(x-p)(x-r) + k(x-q)(x-s) = 0, p < q < r < s,$ has real roots, if $k > 0$

Statement II: The equation $ax^2 + bx + c = 0; a,b,c \in R,$ has non-real roots, if $b^2 - 4ac < 0$

135. **Statement I:** There exist only one value of 'a' for which $(a^2-3a+2)x^2 + (a^2-5a+6)x + a^2 - 4 = 0$, is an identity.

Statement II: If $a = b = c = 0$, then $ax^2 + bx + c = 0$ is an identity in x.

136. **Statement I:** The remainder obtained when the polynomial $1 + x + x^3 + x^9 + x^{27} + x^{81} + x^{243}$ is divided by $(x-1)$ is 7.

Statement II: If f(a) = p then f(x) gives remainder p when divided by (x-a).

137. **Statement I** : If $f(x) = ax^3 + bx^2 + cx + d$ and $f'(x) = 0$ has a repeated root $x = \alpha$(say). Then the curve $y = f(x)$ has an inflexion with a tangent parallel to the x-axis at $x = \alpha$.

Statement II: $f'(x) = 0$ has a repeated root $x = \alpha$, so the equation $f(x) = 0$ has all three equal roots, each equal to α and hence $f(\alpha) = f'(\alpha) = f''(\alpha) = 0$.

138. **Statement I:** Roots of the equation $(x^2-4x+3) + \lambda(x^2-6x+8) = 0$ are real and distinct for all $\lambda \in R$

Statement II: If $P(x) = 0$ and $Q(x) = 0$ are two different quadratic equations, each having real and distinct roots then $P(x) + \lambda Q(x) = 0$ has real and distinct roots for all $\lambda \in R$

139. Let $a \neq -2$ and $(a+2)x^2 + 2(a+1)x + a = 0$ has integral roots.

Statement I: a can take four distinct integral values

Statement II: $\frac{a}{a+2}$ must be integer

140. **Statement I:** Let $f(x) = ax^2 + bx + c$; a, b, c $\in$ R. If $f(x)$ assumes real values for real values of x and non – real values for non real value of x, then a = 0.

Statement II: If a, b, c are complex numbers, $a \neq 0$. Then $\alpha + i\beta$, $\beta \neq 0$ is a root of $ax^2 + bx + c = 0$ if and only if $\alpha - i\beta$ is a root of $ax^2 + bx + c = 0$

IIT CORNER

141. Let , b, c, p, q be real numbers. Suppose α, β are the roots of the equation $x^2 + 2px + q = 0$ and $\alpha, \frac{1}{\beta}$ are the roots of the equation $ax^2 + 2bx + c = 0$, where $\beta^2 \notin \{-1, 0, 1\}$

Statement 1: $(p^2 - q)(b^2 - ac) \geq 0$ and

Statement 2: $b \neq pa$ or $c \neq qa$

(IIT JEE 2008)

(a) Statement 1 is True, Statement 2 is True; Statement 2 is a Correct explanation for Statement 1

(b) Statement 1 is True, Statement 2 is True; Statement 2 is NOT a correct explanation for Statement 1

(c) Statement 1 is True, Statement 2 is False

(d) Statement 1 is False, Statement 2 is True.

(Matrix Match Type)

142. Match the columns.

	Column I		Column II
(A)	The graph of the curve $x^2 = 3x - y - 2$ is strictly below the line y equals to 'a' then 'a' is given by	(p)	$a > 6$
(B)	The value of a for which the equation $x^2 - 4x - 2 = -a$ has no roots.	(q)	$0 \leq a \leq 1$
(C)	The value of a for which the inequality $x^2 \quad 2(4a-1)x + 15a^2 - 2a$ $-7 > 0$ is valid for all $x \in R$, is	(r)	$a = 1/4$
(D)	If $x \in R$, then $\frac{x^2 + 2x + a}{x^2 + 4x + 3a}$ can take all real values, if	(s)	$2 < a < 4$

143. Match the columns

	Column I		Column II
(A)	If the equation $x^2 + 2(k+1)x + (9k-5) = 0$ has only negative roots, then	(p)	$k < 4$
(B)	If the equation $x^2 - 4x + k = 0$ has two real roots then	(q)	$k \geq 6$
(C)	If $x^2 - 2(k-1)x + (2k+1) = 0$ has both roots positive, then	(r)	$-8 < k < 8$
(D)	If $x^2 - kx + 16 > 0, \forall x \in R$ then	(s)	$k \geq 4$

144. Match the columns

	Column I		Column II
(A)	If the absolute value of the difference of roots of the equation $x^2 + kx + 1 = 0$ exceeds $\sqrt{3k}$, then	(p)	$2 < k < 4$
(B)	If the inequality $x^2 - 2(4k-1)x + 15k^2 - 2k - 7 > 0$ is valid for all x, then	(q)	$k > 4$
(C)	Find 'k' such that $x^2 + kx + k < 0$ for all x.	(r)	$k < -1$ or $k > 0$
(D)	If $2x^2 - 2(2k+1)x + k(k+1) = 0$ have one root less than k and other roots greater than k, then	(s)	$-0 < k < 4$

145. Match the columns.

	Column I		Column II
(A)	a, b, c, d are four distinct real numbers and they are in A. P. If $2(a-b)+x(b-c)^2+(c-a)^3=2(a-d)+(b-d)^2+(c-d)^3$ then the value of x can be	(p)	A rational number
(B)	If the roots of equation $ax^2+bx+c=0$ are α and β and the roots of the equation $a^5x^2+ba^2c^2x+c^5=0$ are 4 and 8 then $\lvert\alpha\beta\rvert$ is	(q)	An irrational number
(C)	If $\log_3(\log_5 x)+\log_{1/3}(\log_{1/5} y)=1$ and $x^2y=1$ $(x, y \in R)$ then the value of $5x+y$ is	(r)	2
(D)	Let a, b, c are rational numbers $a \neq -1$ and x_1, x_2 and x_1x_2 are the real roots of the equation $x^3-ax^2+bx-c=0$ then x_1x_2 is	(s)	6

146. Let α, β and γ are three real numbers such that $\alpha+\beta+\gamma=2$, $\alpha^2+\beta^2+\gamma^2=6$ and $\alpha^3+\beta^3+\gamma^3=8$. Then:

	Column I		Column II
(A)	The value of $\alpha^4+\beta^4+\gamma^4$ is equal to	(p)	20
(B)	$(1-\alpha)(1-\beta)(1-\gamma)$ is equal to	(q)	18
(C)	If $-1<x<1$, then $(x-\alpha)(x-\beta)(x-\gamma)$ is	(r)	a positive quantity
(D)	$(1+\alpha^2)(1+\beta^2)(1+\gamma^2)$ is equal to	(s)	a negative quantity
		(t)	zero

147. If α, β be the roots of $ax^2+2bx+c=0$ and γ, δ be the roots of $px^2+2qx+r=0$, then

	Column 1		Column 2
(A)	If $\alpha,\beta,\gamma,\delta$ are in A.P then	(p)	$\frac{b^2}{ac}=\frac{q^2}{pr}$
(B)	If $\alpha,\beta,\gamma,\delta$ are in G.P then	(q)	$\frac{b^2-ac}{c^2}=\frac{q^2-pr}{r^2}$
(C)	If $\alpha,\beta,\gamma,\delta$ are in H.P then	(r)	$\frac{b}{a}=\frac{q}{p}$
(D)	If $\alpha-\gamma=\delta-\beta$ then	(s)	$\frac{b^2-ac}{a^2}=\frac{q^2-pr}{p^2}$

148. If $a(p+q)^2+2bpq+c=0$ and $a(p+r)^2+2bpr+c=0$, then

	Column 1		Column 2
(A)	$-\frac{2(a+b)p}{a}$	(p)	$\frac{1}{q}+\frac{1}{r}$
(B)	$p^2+\frac{c}{a}$	(q)	qr
(C)	$\frac{2}{\lvert a\rvert}\sqrt{(2a+b)bp^2-ac}$	(r)	q + r
(D)	$-\frac{2(a+b)p}{ap^2+c}$	(s)	$\lvert q-r\rvert$

149. Let $f(x) = ax^2 + bx + c$, where a, b, c ∈ R. Suppose $|f(x)| \le 1 \ \forall \ x \in [0, 1]$, then

	Column 1		Column 2
(A)	Maximum value of $|a|$ is	(p)	1
(B)	Maximum value of $|c|$ is	(q)	8
(C)	Maximum value of $|a|+|b|+|c|$ is	(r)	4
(D)	Maximum value of $|a + 2b + 4c|$	(s)	17

150. Let x_0, x_1 and x_2 be three distinct real numbers. Define three polynomials $f_0(x)$, $f_1(x)$ $f_2(x)$ and f(x) as follows.

$$f_0(x) = \frac{(x-x_1)(x-x_2)}{(x_0-x_1)(x_0-x_2)}$$

$$f_1(x) = \frac{(x-x_0)(x-x_2)}{(x_1-x_0)(x_1-x_2)}$$

And $$f_2 = \frac{(x-x_0)(x-x_1)}{(x_2-x_0)(x_2-x_1)}$$

$$f(x) = (x - x_0)(x - x_1)(x - x_2)$$

then

	Column I		Column II
(A)	$f_0(x) + f_1(x) + f_2(x)$ equals	(p)	$x_0 + x_1 + x_2 - x \ \forall \ x$
(B)	$(x_1 + x_2)f_0(x) + (x_2 + x_0)f_1(x) + (x_0 + x_1)f_2(x)$ equal	(q)	0
(C)	$\sum_{k=0}^{2}(x)\, x^2$ equals	(r)	1
(D)	$\frac{f(x)}{(x-x_0)f'(x_0)} + \frac{f(x)}{(x-x_1)f'(x_1)} + \frac{f(x)}{(x-x_2)f'(x_2)} - 1$ equals	(s)	x^2

151. Match the Column I with Column II

If a ∈ and $(x - a)(x - 12) + 2 = (x + b)(x + c)$ For b, c ∈ I, then

	Column I		Column II
(A)	a =	(p)	9
(B)	*b*=	(q)	15
(C)	c=	(r)	–10
		(s)	–13
		(t)	–11
		(u)	–14

(More Than One Correct MCQs)

152. If the equations $x^2 + bx + c = 0$ and $bx^2 + cx + 1 = 0$ have a common root then

(a) $b + c + 1 = 0$

(b) $b^2 + c^2 + 1 = bc$

(c) $(b - c)^2 + (b - 1)^2 + (c - 1)^2 = 0$

(d) $b + c + 1 = 2bc$

153. Let a, b, c be three distinct real numbers such that each of the expression

$ax^2 + bx + c$, $bx^2 + cx + a$, *and* $cx^2 + ax + b$

is positive for each x ∈ R then which of the followings can be a possible value of $\frac{bc+ca+ab}{a^2+b^2+c^2}$

(a) $\frac{1}{2}$ (b) $\frac{3}{4}$ (c) $\frac{4}{5}$ (d) $\frac{1}{5}$

154. Equation $\frac{\pi^e}{x-e}+\frac{e^\pi}{x-\pi}+\frac{\pi^\pi+e^e}{x-\pi-e}=0$ has

(a) One real root in (e, π) and other in $(\pi - e, e)$

(b) One real root in (e, π) and other in $(\pi, \pi + e)$

(c) Two real roots in $(\pi - e, \pi + e)$

(d) Both the real roots are positive

155. The solutions of equations $x^2 + y^2 - 4x - 4y = 60$ and $xy + 2x + 2y = 20$ satisfy the following equation(s).

(a) $x + y = 10$ (b) $x + y = 20$

(c) $x - y = 10$ (d) $x + y = -10$

156. If the equation $x^4 + ax^3 - 13x^2 + bx - 4 = 0$ has one repeated root and one more root being $2 + \sqrt{5}$, then

(a) Repeated root is 2

(b) a = 10, b = - 20

(c) a = 0, b = - 20

(d) Repeated root is -2

157. If the equation $ax^3 + (a+b)x^2 + (b+c)x + c = 0$ and $2x^3 + x^2 + 2x - 5 = 0$ have a common root, then

(a) a+b+c = 0 (b) $\frac{a+b}{c}=1$

(c) $\frac{b+c}{a}=4$ (d) $\frac{a}{c}=\frac{2}{5}$

158. If the equation $x^4 - 4x^3 + ax^2 + bx + 1 = 0$ have four positive roots, then

(a) Roots are necessarily integers

(b) at least 2 roots are equal

(c) Roots are in HP

(d) Roots are in AP

159. The polynomial equation $x^6 + 2x^3 + 5 + ax^3 + a = 0$ has

(a) No real root if $|a| < 4$

(b) Some real roots if $|a| > 4$

(c) At most two real roots for all a ∈ R

(d) At least two non-real roots for all a ∈ R

160. The equation $|x + 1|\,|x - 1| = a^2 - 2a - 3$ can have real solution in x if a belongs to

(a) $[1-\sqrt{5}, -1]$

(b) $[-1, 3]$

(c) $[3, 1+\sqrt{5}]$

(d) $[1+\sqrt{3}, 1+\sqrt{5}]$

161. If $b^2 > 4ac$, then $a(x^2 + 4x + 4)^2 + b(x^2 + 4x + 4) + c = 0$ have all roots real and distinct if

(a) $b < a < 0 < c$ (b) $a < b < 0 < c$

(c) $0 < a < c < b$ (d) $a < c < 0 < b$

162. The question $x^2 + a^2x + b^2 = 0$ has two roots each of which exceeds a number c, then

(a) $a^4 > 4b^2$

(b) $c^2 + a^2c + b^2 > 0$

(c) $-\frac{a^2}{2} < c$

(d) $c^2 + a^2c + b^2 < 0$

163. Let $\frac{a}{2}<\frac{b}{3}<\frac{c}{4}$ then the equation $(2x - a)(3x - b) + (3x - b)(4x - c) + (4x - c)(2x - a) = 0$ has

(a) Both roots real

(b) Roots are real and equal

(c) One root lies between $\left(\frac{a}{2},\frac{b}{3}\right)$

(d) Other root lies between $\left(\frac{b}{3},\frac{c}{4}\right)$

164. If one root of the equation $ax^2 + bx + c = 0$ $(a < 0)$ lies in $(-\infty, -3)$ and other lies in $(3, \infty)$ then

(a) $9a + 3|b| + c > 0$

(b) $c > 0$

(c) $4a + 2|b| + c > 0$

(d) $a + |b| + c < 0$

165. If $2a - b - c < 0$ and the equation $cx^2 + bx - 2a = 0$ has no real roots then which of the following is/are correct?

(a) $ac < 0$

(b) $a > 0$

(c) $\frac{c-a}{2} > a$

(d) $\frac{c+2b}{8} > a$

166. If $4ac > b^2$ and $a + c > b$ for real number a, b and c, then which of the following is true?
   (a) $a > 0$
   (b) $a + 2b > 4c$
   (c) $a + b + c > 0$
   (d) $4a + c > 2b$

167. If $\frac{x^2+ax+3}{x^2+x+a}$, takes all real values for possible real values of x, then
   (a) $4a^3 + 39 \geq 0$
   (b) $4a^3 + 39 < 0$
   (c) $a < \frac{1}{4}$
   (d) $a \geq \frac{1}{4}$

168. If α is a real root of the equation $ax^2+bx+c=0$, and β is a real root of the equation $-ax^2+bx+c=0$, then the equation $ax^2+2bx+2c=0$ has
   (a) Real roots
   (b) None- real roots
   (c) A root lying between α and β
   (d) A root equal to α or β

169. If $a < 0$, then the root of the equation $x^2-2a|x-a|-3a^2=0$ is
   (a) $(-1-\sqrt{6})a$
   (b) $(1\ \ \sqrt{2})a$
   (c) $(-1+\sqrt{6})a$
   (d) $(1+\sqrt{2})a$

170. If $a < b < c < d$, then for any real non- zero λ, the quadratic equation $(x - a)(x - c) + \lambda(x - b)(x - d) = 0$ has
   (a) Non- real roots
   (b) One real root between a and c
   (c) One real root between b and d
   (d) Irrational roots

171. If a, b, c are positive rational numbers such that a > b > c and the quadratic equation $(a+b-2c)x^2+(b+c-2a)x+(c+a-2b)=0$ has a root in the interval (-1, 0), then
   (a) $c + a < 2b$
   (b) both roots of the given equation are rational
   (c) The equation $ax^2+2bx+c=0$ has both negative real roots
   (d) The equation $cx^2+2ax+b=0$ has both negative real roots

172. Which all of the following is correct about the values of 'k 'for which $x^2+kx+\sin^{-1}(x^2-4x+5)+\cos^{-1}(x^2-4x+5)=0$, has at least one solution,
   (a) Only one value of k exist
   (b) 'k' can not assume any integral value
   (c) $k=-2-\frac{\pi}{4}$
   (d) $k=-2+\pi$

173. Which one of the following is true about the number of real roots of the equation $1+2x+3x^2+\ldots+(n+1)x^n=0$, where n is even.
   (a) It is equal to 0
   (b) It is less than 'n' but independent of n
   (c) It is less than 'n' but dependent of n
   (d) It is equal to n/2

174. If the equation $x^2+9y^2-4x+3=0$ is satisfied for value of x and y, then
   (a) $1 \le x \le 3$ (b) $2 \le x \le 3$
   (c) $-\frac{1}{3} \le y \le \frac{1}{3}$ (d) $0 < y < \frac{2}{3}$

175. If $ax + by = 1$, $cx^2 + dy^2 = 1$, have only one solution , then
   (a) $\frac{a^2}{c}+\frac{b^2}{d}=1$ (b) $x=\frac{a}{c}$
   (c) $y=-\frac{b}{d}$ (d) $xy=\frac{ab}{cd}$

IIT CORNER

176. For real x, the function $\frac{(x-a)(x-b)}{x-c}$ will assume all real values provided**(IIT JEE 1984)**
   (a) $a > b > c$ (b) $a < b < c$
   (c) $a > c > b$ (d) $a < c < b$

177. If s is the set of all real x such that $\frac{2 \quad 1}{2x^2+3x^2+x}$ is positive $\frac{2x-1}{2x^2+3x^2+x}$, then S contains **(IIT JEE 1986)**

(a) $(-\infty, -\frac{3}{2})$ (b) $(-\frac{3}{2}, -\frac{1}{4})$

(c) $(-\frac{1}{4}, \frac{1}{2})$ (d) $(\frac{1}{2}, 3)$

(e) None of these.

178. If a,. b and c are distinct positive numbers, then the expression (b + c – a) (c + a – b) (a + b – c) - a b c is **(IIT JEE 1986)**

(a) positive (b) negative

(c) non – positive (d) non – negative

(e) none of these.

179. If a, b, c, d and p are distinct real numbers such that $(a^2 + bg^2 + c^2)p^2 - 2(ab + bc + cd)p + (b^2 + c^2 + d^2) \le 0$ then a, b, c, d **(IIT JEE 1987)**

(a) are in A. P. (b) are in G. P.

(c) are in H. P. (d) satisfy a b = c d

(e) satisfy none of these.

180. Number of divisor of the form $4n + 2$ ($n \ge 0$) of the integer 240 is **(IIT JEE 1998)**

(a) 4 (b) 8 (c) 10 (d) 3

181. Let *S* be the set of all non-zero real numbers α such that the quadratic equation $\alpha x^2 - x + \alpha = 0$ has two distinct real roots x_1 and x_2 satisfying the inequality $|x_1 - x_2| < 1$. Which of the following intervals is(are) *a* subset(s) of *S*? ***(JEE Adv. 2015)***

(a) $\left(-\frac{1}{2}, -\frac{1}{5}\right)$ (b) $\left(-\frac{1}{\sqrt{5}}, 0\right)$

(c) $\left(0, \frac{1}{\sqrt{5}}\right)$ (d) $\left(\frac{1}{\sqrt{5}}, \frac{1}{2}\right)$

(Subjective IIT JEE Questions)

182. Solve for x : $4^x - 3^{x-\frac{1}{2}} = 3^{x+\frac{1}{2}} - 2^{2x-1}$

(IIT JEE 1978)

183. If (m, n) = $\frac{(1-x^m)(1-x^{m-1}).....(1-x^{m-n+1})}{(1-x)(1-x^2)......(1-x^n)}$ where m and n are positive integers ($n \le m$), show that (m, n + 1) = (m – 1, n + 1) + x^{m-n-1} (m – 1, n). (IIT JEE 1978)

184. Solve for x : $\sqrt{x+1} - \sqrt{x-1} = 1$.

(IIT JEE 1978)

185. Solve the following equation for x : $2\log_x a + \log_{ax} a + 3\log_{a^2x} a = 0$, $a > 0$ [IIT JEE 1978]

186. Show that the square of $\frac{\sqrt{26-15\sqrt{3}}}{5\sqrt{2}-\sqrt{38+5\sqrt{3}}}$ is a rational number.

(IIT JEE 1978)

187. Find all integers x for which $(5x - 1) < (x + 1)^2 < (7x - 3)$. (IIT JEE 1978)

188. If α, β are the roots of $x^2 + px + q = 0$ and γ, δ are the roots of $x^2 + rx + s = 0$, evaluate $(\alpha - \gamma)(\alpha - \delta)(\beta - \gamma)(\beta - \delta)$ in terms of p, q, r and s. Deduce the condition that the equations have a common root. (IIT JEE 1979)

189. Show that for any triangle with sides a, b, and c, $3(ab + bc + ca) < (a + b + c)^2 < 4(bc + ca + ab)$ when are the first two expressions equal? **(IIT JEE 1979)**

190. Given $n^4 < 10^n$ for a fixed positive integer $n \ge 2$, prove that $(n + 1)^4 < 10^{n+1}$.

(IIT JEE 1980)

191. Let y = $\sqrt{\frac{(x+1)(x-3)}{(x-2)}}$ find all the real values of x for which y takes real values. **(1980)**

192. For what values of m, does the system of equations

$$3x + my = m$$
$$2x - 5y = 20$$

has solution satisfying the conditions $x > 0$, $y > 0$. **(1980)**

193. Show that the equation $e^{\sin x} - e^{-\sin x} - 4 = 0$ has no real solution. **(IIT JEE 1982)**

194. m n squares of equal size are arranged to from a rectangle of dimension m by n , where m

and n are natural numbers. Two squares will be called 'neighbors' if they have exactly one common side. A natural number is written in each square such that the number written in any square is the arithmetic mean of the numbers written in its neighboring squares. Show that this is possible only if all the numbers used are equal. **(IIT JEE 1982)**

195. If one root of the quadratic $a x^2 + b x + c = 0$ is equal to the n- th power of the other , then show that $\left(ac^n\right)^{\frac{1}{n+1}} + \left(a^n c\right)^{\frac{1}{n+1}} + b = 0$

(IIT JEE 1983)

196. Find all real values of x which satisfy $x^2 - 3x + 2 > 0$ and $x^2 - 3x - 4 \le 0$ **(IIT JEE 1983)**

197. Solve for x; $\left(5+2\sqrt{6}\right)^{x^2-3} + \left(5-2\sqrt{6}\right)^{x^2-3} = 10.$

(IIT JEE 1986)

198. For $a \le 0$, determine all real roots of the equation $x^2 - 2a|x - a| - 3a^2 = 0$

(IIT JEE 1986)

199. Find the set of all x for which

$$\frac{2x}{\left(2x^2+5x+2\right)} > \frac{1}{(x+1)}$$ **(IIT JEE 1987)**

200. Let α_1 , α_2 and β_1, β_2 be the roots of $a x^2 + b x + c = 0$ and $p x^2 + q x + r = 0$ respectively. If the system of equations $\alpha_1 y + \alpha_2 z = 0$ and $\beta_1 y + \beta_2 z = 0$ has a nontrivial solution, then prove that $\frac{b^2}{q^2} = \frac{ac}{pr}$. **(IIT JEE 1987)**

201. Solve $|x^2 + 4x + 3| + 2x + 5 = 0$

(IIT JEE 1988)

202. Let a , b, c be real. If $a x^2 + b x + c = 0$ has two real roots α and β, where $\alpha < -1$ and $\beta > 1$, then show that $1 + \frac{c}{a} + \left|\frac{b}{a}\right| < 0$

(IIT JEE 1995)

203. Find the set of all solutions of the equation

$$2^{|y|} - |2^{y-1} - 1| = 2^{y-1} + 1$$

(IIT JEE 1997)

204. Let S be a square of unit area. Consider any quadrilateral which has one vertex on each side of S. If a, b, c and d denote that the length of the sides of the quadrilateral. Prove that

$$2 \le a^2 + b^2 + c^2 + d^2 \le 4.$$

(IIT JEE 1997)

205. If α, β are the roots of $a x^2 + b x + c = 0$, $(a \ne 0)$ and $\alpha + \delta$, $\beta + \delta$ are the roots of

$$A x^2 + B x + C = 0 \ (A \ne 0)$$

for some constant δ , then prove that

$$\frac{b^2 - 4ac}{a^2} = \frac{B^2 - 4AC}{A^2}.$$ **(IIT JEE 2000)**

206. Let a, b, c be real numbers with $a \ne 0$ and let α, β be the roots of the equation $a x^2 + b x + c = 0$. Express the roots of $a^3 x^2 + a b c x + c^3 = 0$ in terms of α, β. **(IIT JEE 2001)**

207. If $x^2 + (a - b) x + (1 - a - b) = 0$ where a, b, $\in$ R then find all values of a for which equation has unequal real roots for all values of b.

(IIT JEE 2003)

208. Let a and b be the roots of the equation $x^2 - 10 c x - 11 d = 0$ and those of $x^2 - 10 a x - 11 b = 0$ are c, d then the value of $a + b + c + d$, when $a \ne b \ne c \ne d$, is **(IIT JEE 2006)**

Concept Eliminator (CE)

1. Find the sum of the eighth powers of all roots of the equation $x^3 - x + 1 = 0$

[Based on Vietnamese – Mathematical – Olympiad 1975]

2. Given that the equation $x^2 - ax + 3 - b = 0$ has two distinct real roots, $x^2 + (6-a)x + 6 - b = 0$ has two equal real roots, and $x^2 + (4-a)x + 5 - b = 0$ has no real roots. Then the ranges of a and b are

(a) $2 < a < 4, 2 < b < 5,$
(b) $1 < a < 4, 2 < b < 5,$
(c) $2 < a < 4, 1 < b < 5,$
(d) $1 < a < 4, 1 < b < 5.$

[Based on China – Mathematical – Competition 2005]

3. Find all real numbers a for which the equation $x^2 + (a-2)x + 1 = 3|x|$ has exactly three distinct real solutions in x. **[RMO – 2003]**

4. The equation $x^3+ax^2+bx+c=0$ has three (not necessarily distinct) real roots t, u, v. For what relation between a, b and c numbers t^3, u^3, v^3 are roots of the equation $x^3+a^3x^2+b^3x+c=0$

[Based on Vietnamese – Mathematical – Olympiad 1979]

5. Find the polynomial of the lowest degree with integer coefficients one of whose roots is $\sqrt{2}+\sqrt[3]{3}$.

[Based on Vietnamese – Mathematical – Olympiad 1984]

6. Solve the equation
$$\sqrt{1+\sqrt{1-x^2}}\left(\sqrt{(1+x)^3}-\sqrt{(1-x)^3}\right)=2+\sqrt{1-x^2}.$$

[Based on Vietnamese – Mathematical – Olympiad 1984]

7. Find all pairs (a, b) of real numbers such that whenever α is a root of $x^2+ax+b=0, \alpha^2-2$ is also a root of the equation. **RMO – 2007**

8. Find sum of all the possible values of m for which the equation $16x^4-mx^3+(2m+17)x^2-mx+16=0$ has four distinct roots forming a geometric progression.

[Based on Vietnamese – Mathematical – Olympiad 1985]

9. (1) Find all polynomial P (x) of the lowest degree with rational coefficients that satisfy
$$P\left(\sqrt[3]{3}+\sqrt[3]{9}\right)=3+3\sqrt{3}$$
(2) Does there exist a polynomial P(x) with integer coefficients that satisfies
$$P\left(\sqrt[3]{3}+\sqrt[3]{9}\right)=3+\sqrt[3]{3}?$$

[Based on Vietnam Maths Olympiad 1997]

10. Let $P_1(x)=ax^2-bx-c, P_2(x)=bx^2-cx-a,$ $P_3(x)=cx^2-ax-b$ be three quadratic polynomials where *a*, *b*, *c* are non-zero real numbers. Suppose there exists a real number a such that $P_1(\alpha)=P_2(\alpha)=P_3(\alpha)$. Prove that $a=b=c$. **RMO – 2010**

11. Find the value of $\dfrac{x^4-6x^3-2x^2+18x+23}{x^2-8x+15}$ when $x=\sqrt{19-8\sqrt{3}}$.

[Based on SSSMO (J) 2007]

12. Solve the equation $(m-2)x^2-(m+3)x-2m-1=0$ for real value of *m*, find the roots which are independent of m.

(Based on China Maths Competition 2005)

13. If the equation $(m^2-1)x^2-2(m+2)x+1=0$ has at least one real root , find the range of m.

(Based on China Maths Competition 2004)

14. a, b, c are real numbers with $a^2+b^2+c^2>0$ Then the equation $x^2+(a+b+c)x+(a^2+b^2+c^2)=0$ has
(a) 2 negative real roots,
(b) 2 positive real roots,
(c) 2 real roots with opposite signs,
(d) no real roots.

(Based on China Maths Competition 1997)

15. MR. Fat is going to pick three non-zero real numbers and Mr. Taf is going to arrange the three numbers as the coefficient of a quadratic equation
$$\_x^2+\_x+\_=0.$$
Mr. Fat wins the game if and only if the resulting equation has two distinct rational solutions. Who has a winning strategy?

(Based on ASUMO 1990)

16. *a*, *b* are two different positive integers, and the two quadratic equations
$(a-1)x^2-(a^2+2)x+(a^2+2a)=0$ and
$(b-1)x^2-(b^2+2)x+(b^2+2a)=0$
Have one common root. Find the value of $\dfrac{a^b+b^a}{a^{-b}+b^{-a}}$ **(Based on CHNMOL 2003)**

17. *m* is a real number. Solve the equation $|x^2-1|+|x^2-4|=mx.$

(Based on Canada Maths Competition)

18. If p, q_1 and q_2 are real numbers with $p=q_1+q_2+1$, prove that at least one of the following two equations $x^2+x+q_1=0, x^2+px+q_2=0$ has two distinct real roots.

(Based on China Maths Competition 1988)

19. x_1 and x_2 are roots of the equation $x^2+x-3=0$. Find the value of $x_1^3-4x_2^2+19$.

(Based on China Maths Competition 1996)

20. Given that the quadratic equation $x^2 - px + q = 0$ has two real roots $\alpha \& \beta$.
    (i) Find the quadratic equation that takes α^3, β^3 as its two roots:
    (ii) If the new equation is still $x^2 - px + q = 0$. Find all the possible pairs (p, q).

(Based on China Maths Competition 1996)

21. Given that m is a real number not less than – 1, such that the equation in x
$x^2 + 2(m-2)x + m^2 - 3m + 3 = 0$ has two distinct real roots x_1 and x_2.
    (i) If $x_1^2 + x_2^2 = 6$, find the value of m.
    (ii) Find the maximum value of $\frac{mx_1^2}{1-x_1} + \frac{mx_2^2}{1-x_2}$.

(Based on CHINMOL 2000)

22. Given that the two roots of the equation $x^2 + px + q = 0$ are 1 greater than the two roots of the equation $x^2 + 2qx + \frac{1}{2}p = 0$ respectively, and the difference of the two roots of $x^2 + px + q = 0$ is equal to that of $x^2 + 2qx + \frac{1}{2}p = 0$. Find the solutions to each of the two equations.

(Based on China Maths Competition 1993)

23. Given p + q – 198, find the integer solutions of the equation $x^2 + px + q = 0$.

(Based on RUSMO 1989)

24. Given that are the real roots of $x^2 + 19x - 97 = 0$, (α and β) and $\frac{1+\alpha}{1-\alpha} + \frac{1+\beta}{1-\beta} = -\frac{m}{n}$, where m and n are two relatively prime natural numbers.

Find the value of m + n.

(Based on China Maths Competition 1997)

25. Prove that if α and β are the roots of the equation $x^2 + px + 1 = 0$. and if γ and δ are the roots of the equation $x^2 + qx + 1 = 0$, then $(\alpha - \gamma)(\beta - \gamma)(\alpha + \delta)(\beta + \delta) = q^2 - p^2$.

(Based on USSR)

26. Given that α, β are the roots of the equation $x^2 - 7x + 8 = 0$. where α > β. Find the value of $\frac{2}{\alpha} + 3\beta^2$ without solving the equation.

(Based on China Maths Competition 1998)

27. Solve the equation $\frac{13x - x^2}{x+1}\left(x + \frac{13-x}{x+1}\right) = 42.$

(Based on CHINA Maths Competition 1999)

28. Let n be a positive integer such that $n^2 + 19n + 48$ is a perfect square. Find the value of n.

(Based on SSMO(J) Maths Competition 2008)

29. Find the value of the smallest positive integer m such that the equation $x^2 + 2(m+5)x + (100m+9) = 0$ has only integer solutions.

(Based on SSMO(J) Maths Competition 2009)

30. p, q are two integers roots of the equation $x^2 - \frac{p^2+11}{9}x + \frac{15}{4}(p+q) + 16 = 0$ Find the values of p and q.

(Based on CHNMOL Maths Competition 2005)

31. Determine all integral solution of
$a^2 + b^2 + c^2 = a^2b^2$

(Based on USAMO Maths Competition 1975)

32. Given that $\frac{1260}{a^2 + a - 6}$ is a positive integer, where a is a positive integer. Find the value of a.

(Based on CHINA Maths Competition 2003)

33. Let p be a positive prime number such that the equation $x^2 - px - 580p = 0$ has two integer solutions. Find the value of p.

(Based on SSMO Maths Competition 2003)

34. Let P be an integer such that both roots of the equation $5x^2 - 5px + (66p - 1) = 0$ are positive integers. Find the value of P.

(Based on SSMO Maths Competition 2006)

35. Find all the natural numbers p, q such that the equation $x^2 + pqx + p + q = 0$ has two integer roots.

(Based on RUSMO Maths Competition 1991)

36. If a ∈R and the equation
$(a-2)(x-[x])^2 + 2(x-[x]) + a^2 = 0$ (where

[x] denotes the greatest integer ≤ x) has no integral solution and has exactly one solution in (2, 3), then 'a' lies in the interval

(a) (–1, 2) (b) (0, 1) (c) (–1, 0)(d) (2, 3)

37. If $a > 0, a \neq 1$, then the equation $2\log_x a + \log_{ax} a + 3\log_{a^2x} a = 0$ has

(a) exactly one real root

(b) two real roots

(c) no real roots

(d) infinite number of real roots

38. If the quadratic equation $ax^2 + bx + 6 = 0$ does not have real roots and $b \in R^+$, then

(a) $a > \max\left\{\frac{b^2}{24}, b-6\right\}$

(b) $a < \max\left\{\frac{b^2}{24}, b-6\right\}$

(c) $a > \min\left\{\frac{b^2}{24}, b-6\right\}$

(d) $a < \min\left\{\frac{b^2}{24}, b-6\right\}$

39. Theequation $x^2 \cdot 2^{|x-3|+4} + 2^{x-1} = x^2 \cdot 2^{x+1} + 2^{|x-3|+2}$ has

(a) no solution

(b) two solutions

(c) $x = \pm\frac{1}{2}$ and $x \geq 3$

(d) $x \geq 3$

40. If $x_1 > x_2 > x_3$ and x_1, x_2, x_3 are roots of $\frac{x-a}{b} + \frac{x-b}{a} = \frac{b}{x-a} + \frac{a}{x-b}; (a, b > 0)$ and $x_1 - x_2 - x_3 = c$, then a, c, b are in

(a) AP (b) GP

(c) HP (d) none of these

41. Solve for x, $(x^2 + 3x + 1)(x^2 + 3x - 3) \geq 5$

(a) $(-\infty, -4]U[-2, -1]U[1, \infty)$

(b) $(-\infty, -4]U[1, \infty)$

(c) $(-\infty, -4]U[4, \infty)$

(d) None of these

42. If a, b, c, d are four consecutive terms of an increasing AP, then the roots of the equation $(x-a)(x-c) + 2(x-b)(x-d) = 0$ are

(a) real and distinct

(b) non-real complex

(c) real and equal

(d) integers

43. Consider a cubic equation $x^3 - x^2 + \beta x + \gamma = 0$, where β and γ are real. If the roots of the equation are in A.P. then β and γ lies in the interval

(a) (– ∞, 1/3) & (–1/27, ∞)

(b) (–1/27, ∞) & (– ∞, 1/3)

(c) (1/3, ∞) & (– ∞ , –1/27)

(d) (0,∞) & (-∞,0)

44. Let a and b be the roots of the equation $x^2 - 10cx - 11b = 0$ and those of $x^2 - 10ax - 11b = 0$ are c, d, then find the value of a + b + c + d, when a ≠ b ≠ c ≠ d ≠ 0.

(a) 1210 (b) 121

(c) 1331 (d) None of these

45. If $(x-2)^6 + (x-4)^6 = 64$, then equation has

(a) real and irrational roots

(b) real and rational roots

(c) two real and two irrational roots

(d) None of the above

46. Number of integral points (x, y) in the 1st quadrant that satisfy the equation $y^4 + 6xy^2 - 8x = 0$ is equal to

(a) 1 (b) 2

(c) infinity (d) None of these

47. If $f(x) = x^4 + 9x^3 + 35x^2 - x + 4$, then find the value of $f(-5 + 2\sqrt{-4})$

(a) – 160 (b) 160

(c) 0 (d) None of these

Solutions

Concept Applicator (CA)-1

1. Given equation is

   $x^2-10x+21=(x-3)(x-7)=0$ so $x = 3$ or 7.

2. Given equation is $x^2 - 11x + 24 = (x - 3)(x - 8) = 0$ so $x = 3$ or 8 so required difference is $8 - 3 = 5$.

3. $\alpha+\beta=-p, \alpha\beta=q$

   Given, $\alpha+\beta=\alpha^2+\beta^2$

   $=(\alpha+\beta)^2-2\alpha\beta$

   $\Rightarrow \quad -p = p^2 - 2q$

   $\Rightarrow \quad p^2 + p = 2q$

4. (d) The discriminant of the quadratic equation $2x^2 + 6x + b$ is given by $D = 36 - 8b > 0$. Therefore, the given equation has real roots. We have,

   $$\frac{x_1}{x_2}+\frac{x_2}{x_1}=\frac{x_1^2+x_2^2}{x_1x_2}=\frac{(x_1+x_2)^2-2x_1x_2}{x_1x_2}$$

   $$=\frac{(-3)^2-\frac{2b}{2}}{b/2}=\frac{18}{b}-2<-2 \quad [\because b<0]$$

   Hence, (d) is the correct answer.

5. (d) Since, x^2, $9|x|$ and 6 are positive so $x^2 + 9|x| + 6 = 0$ does not have any real root.

   Therefore, sum does not exist.

6. (c) $2x^2 - 25x = 3(14)$

   $\Rightarrow \quad 2x^2 - 25x - 42 = 0$

   $\Rightarrow \quad (2x + 3)(x - 14) = 0$

   $\Rightarrow \quad x = x=-\frac{3}{2}, 14$ but x cannot be negativ

   Hence, $x = 14$

7. (a) Given equation is $\sqrt{x+1}-\sqrt{x-1}=\sqrt{4x-1}$

   On squaring, we get $x = 5/4$ which does not satisfy the given equation. Hence, no solution is possible.

8. α, β are roots of $8x^2 - 3x + 27 = 0$

   $\Rightarrow \alpha+\beta = \frac{3}{8}, \alpha\beta=\frac{27}{8}$

   $$\therefore \left(\frac{\alpha^2}{\beta}\right)^{\frac{1}{3}}+\left(\frac{\beta^2}{\alpha}\right)^{\frac{1}{3}}=\frac{(\alpha^3)^{\frac{1}{3}}+(\beta^3)^{\frac{1}{3}}}{(\alpha\beta)^{\frac{1}{3}}}$$

   $$\frac{\alpha+\beta}{(\alpha\beta)^{1/3}}=\frac{3/8}{\left(\frac{27}{8}\right)^{1/3}}=\frac{3/8}{3/2}=\frac{1}{4}$$

9. (b) Since, α, β are roots of equation $x^2+x\sqrt{\alpha}+\beta=0$

   $\therefore \alpha+\beta = -\sqrt{\alpha}, \alpha\beta=\beta \Rightarrow \alpha=1$ [since $\beta=0$ does not satisfy the equation]

10. (b) We have, $D = b^2 - 4ac$ if $D \geq 0$, then the roots of the equation are given by $x=\frac{-b\pm\sqrt{D}}{2a}$

    As, $D = b^2 - 4ac < b^2$ ($\because a > 0, c > 0$), it follows that the roots of the quadratic equation are negative.

    In case $D < 0$, then the roots of the equation are given by $x=\frac{-b\pm i\sqrt{-D}}{2a}$ which have negative real parts.

Concept Applicator (CA)-2

1. Given equations are $x^2 - ax + b = 0$ ... (i)

   and $x^2 - px + q = 0$ ... (ii)

   Let α be the common root. Then roots of Eq. (ii) will be α and α. Let β be the other root of Eq. (i). So roots of Eq. (i) are α, β and those of Eq.(ii) are α, α.

   Now, $\alpha + \beta = a$ ... (iii) $\quad \alpha\beta = b$ ...(iv)

   $2\alpha = p$ ... (v) $\quad \alpha^2 = q$ ... (vi)

   Consider LHS $= b + q = \alpha\beta + \alpha^2 = \alpha(\alpha + \beta)$

   Consider RHS $=\frac{ap}{2}=\frac{(\alpha+\beta)2\alpha}{2}=\alpha(\alpha+\beta)$

   So RHS = LHS proved.

2. (b) Since, $\alpha_2 = \frac{1}{\alpha_1}$ and $\beta_2 = \frac{1}{\beta_1}$

Roots of the second equation are reciprocal of those of the first.

$\therefore$ Replacing x by $\frac{1}{x}$ in $a_1x^2 + b_1x + c_1 = 0$ we get, $c_1x^2 + b_1x + a_1 = 0$

$\therefore$ $c_1x^2 + b_1x + a_1 = 0$ and $a_2x^2 + b_2x + c_2 = 0$ have both roots common.

$\Rightarrow \frac{c_1}{a_2} = \frac{b_1}{b_2} = \frac{a_1}{c_2}$

3. (c) For the equation $x^2 + 2x + 3 = 0$,
Discriminant $= (2)^2 - 4\cdot1\cdot3 < 0$

$\therefore$ Roots of $x^2 + 2x + 3 = 0$ are complex. Since, the equations $x^2 + 2x + 3 = 0$ and $ax^2 + bx + c = 0$ are given to have a common root, therefore both roots will be common.

So, both the equations are identical.

$\Rightarrow a : b : c = 1 : 2 : 3$

4. (b) Since sum of the co-efficient is 0 so $x = 1$ is one root of the equation, since one root is real and rational so other root must be real and rational.

5. (b) Let α and β be the roots of $x^2 - px + q = 0$ and α and $\frac{1}{\beta}$ be the roots of $x^2 - ax + b = 0$

Then, $\alpha + \beta = p$ and $\alpha\beta = q$

Also, $\alpha + \frac{1}{\beta} = a$ and $\frac{\alpha}{\beta} = b$

Now, $(q-b)^2 = \left(\alpha\beta - \frac{\alpha}{\beta}\right)^2 = \alpha^2\left(\beta - \frac{1}{\beta}\right)^2 = \frac{\alpha}{\beta}\cdot$

$$\beta\alpha\left[(\alpha+\beta) - \left(a + \frac{1}{\beta}\right)\right]^2 = bq(p-a)^2$$

6. (b) We know that rational roots always occurs in pairs, ie, if one root of given equation is $\frac{1}{a+\sqrt{b}}$, then other root must be $\frac{1}{a-\sqrt{b}}$.

Thus, sum of roots $= \frac{2a}{a^2-b}$ and product of roots $= \frac{1}{a^2-b}$.

Hence, the quadratic equation

$$x^2 - \left(\frac{2a}{a^2-b}\right)x + \left(\frac{1}{a^2-b}\right) = 0$$

$$\Rightarrow \quad (a^2-b)x^2 - 2ax + 1 = 0$$

7. (b) Substituting $x + 1$ instead of x in 1st equation

$\Rightarrow ax^2 + (b+2a)x + a + b + c = 0$ but equation given is $2x^2 + 8x + 2 = 0$

$\Rightarrow \frac{a}{2} = \frac{b+2a}{8} = \frac{a+b+c}{2}$

or a : b : c = 1 : 2 : –2

8. (c) Since both the roots are reciprocal so changing x to $\frac{1}{x}$ in given equation, $\Rightarrow x^2 + 6x + 9 = 0$

9. (c) Replace x, by $(x - 2)$ to get equation whose roots are $\alpha + 2$, $\beta + 2$

$ax^2 - (4a - b)x + 4a + c - 2b = 0$

or $ax^2 + (b - 4a)x + 4a - 2b + c = 0$

10. (b) Let α, β be the roots of $x^2 + bx + c = 0$

Then, $\alpha + \beta = -b, \alpha\beta = c$

Roots of the required equation are α^3, β^3, so the equation is $x^2 - (\alpha^2 + \beta^3)x + \alpha^3\beta^3 = 0$

$\Rightarrow x^2 - \{(\alpha+\beta)^3 - 3\alpha\beta(\alpha+\beta)\}x + \alpha^3\beta^3 = 0$

$\Rightarrow x^2 - (-b^3 + 3cb)x + c^3 = 0$

$\Rightarrow x^2 + b(b^2 - 3c)x + c^3 = 0$.

Alternately:

Substitute $x \to (x)^{\frac{1}{3}}$ then equation will be $x^{\frac{2}{3}} + bx^{\frac{1}{3}} = -c$, cubing both the sides, we get

$[x^{\frac{1}{3}}(x^{\frac{1}{3}} + b)]^3 = -c^3$

$\Rightarrow x[x + b^3 + 3b(x^{\frac{2}{3}} + bx^{\frac{1}{3}}] = -c^3$

$\Rightarrow x[x + b^3 - 3bc] = -c^3 \Rightarrow x^2 + b(b^2 - 3c)x + c^3 = 0$

Concept Applicator (CA)-3

1.

Plot the graph of $y = -x^2 + 4x - 6$

Consider the $y = -x^2 + 4x - 6$

(i) Coefficient of x^2 is –ve so graph should be open downwards.

(ii) Discriminant D < 0 so we have no real root.

(iii) Plot the graph as shown.

(iv) The function will attain a maximum value at $x = 2$

(v) The required maximum value is

$f(2) = -(2)^2 + 4(2) - 6 = -2$

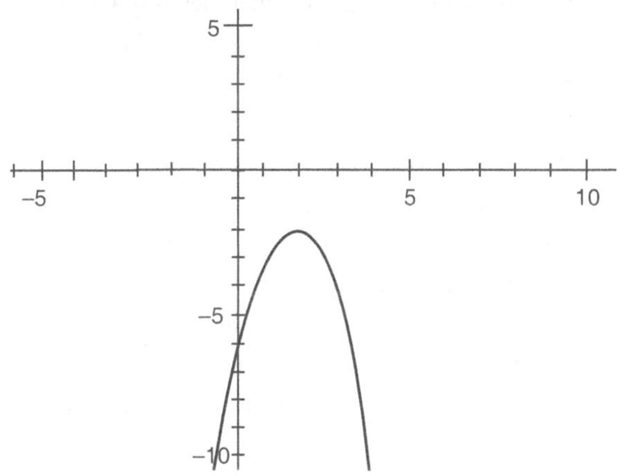

If inequality is $y = -x^2 + 4x - 6 > 0$ then from graph no value of x exist

If inequality is $y = -x^2 + 4x - 6 \leq 0$ then from graph all values of x will satisfy the condition.

2. (d) From graph all values of x will satisfy the given condition.

3.

Plot the graph of $y = |x^2 + 6x + 5|$

In order to plot $y = |x^2 - 6x + 5|$ 1st plot $y = x^2 - 6x + 5$ and then put modulas by doing that we are changing negative values of y to positive values i.e reflection w.r.t x-axis as shown in the figure.

In this case y will attain its local maxima at $x = 3$ which is 4

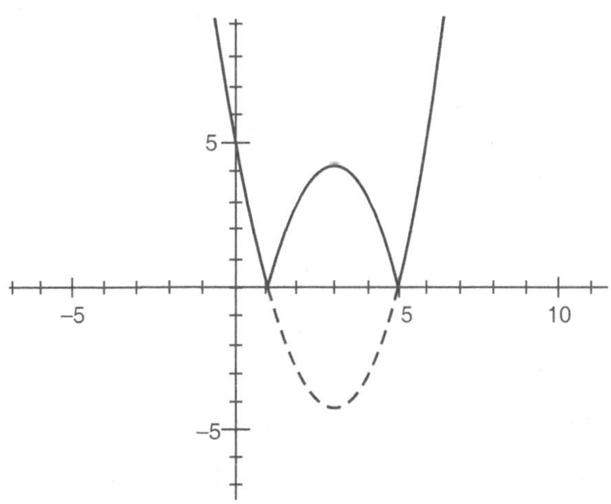

4. (c) From graph value of k should be in (0, 4)

5. Consider $x^2 + x + 1 < 0$

Discriminant D = 1 – 4 = – 3 < 0, so the equation has no real roots

Co-efficient of x^2 is 1 so parabola is open upwards with no real roots that means graph is similar to Case (i) we have discussed above.

So $x^2 + x + 1 > 0 \ \forall x \in R$

So no such 'x' exists.

6. $x^2 - 9 = (x - 3)(x + 3) < 0$ so $x \in (-3, 3)$

7. Consider $-x^2 + x - 1 < 0$

Descriminant D = 1 – 4 = – 3 < 0 , so the equation has no real roots

Co-efficient of x^2 is –1 so parabola is open downwards with no real roots that means graph is similar to Case (vi) we have discussed above.

So $-x^2 + x - 1 < 0 \ \forall x \in R$

So all values of 'x' will satisfy the above condition.

8. $x^2 - 9 = (x - 3)(x + 3) > 0$ so $x \in (-\infty, -3) \cup (3, \infty)$ ... (i)

$x^2 - 25 = (x - 5)(x + 5) < 0$ so $x \in (-5, 5)$ ... (ii)

From above two equations we get (–5, –3) U (3, 5)

9. $x^2 - 11x + 24 = (x + 3)(x + 8) < 0$ so $x \in (-8, -3)$

10. $x^2 + 11x + 18 = (x + 2)(x + 9) > 0$

so $x \in (-\infty, -9) \cup (-2, \infty)$

Concept Applicator (CA)-4

1. (a) The given equation is

$$f(x) = x^2 - (k-3)x + k - 3 = 0$$

Consider D of the equation,

$$D = (k-3)^2 - 4k + 12$$

$$= k^2 - 10k + 21 = (k-3)(k-7)$$

Both the roots are real and distinct $\Rightarrow D > 0$

$\Rightarrow (k-3)(k-7) > 0 \Rightarrow k \in (-\infty, 3) \cup (7, \infty)$

2. (c) Both the roots are equal $\Rightarrow D = 0$

$\Rightarrow$ D $= (k-3)^2 - 4k + 12 = (k-3)(k-7) = 0$

k = 3 & 7

3. (c) We know that if $f(x) = 0$ has exactly one root between (a, b) then $f(a).f(b) < 0$

or $(9 - k)(12 - 2k) < 0$ or $k \in (6, 9)$

4. $f(c) > \Rightarrow c^2 + ac + b^2 > 0$

5. (d) $f(1) > 0, f(2) < (0, f(3) > 0$ and $D > 0$

$\Rightarrow -12 + k > 0, k < 16, k > 12$ and $16(16 - k) > 0$

$\Rightarrow 12 < k < 16$

$\therefore k \in \{13, 14, 15\}$

$\Rightarrow$ Three integral values.

6. (c) The equation $x^2 - x + a - 3 = 0$ must have at least one negative root.

For real roots, $D \geq 0 \Rightarrow 1 - 4(a-3) \geq 0$

$$\Rightarrow a \leq \frac{13}{4} \Rightarrow a \in \left(-\infty, \frac{13}{4}\right] \quad \text{... (i)}$$

Both root will be non-negative, if $D \geq 0, a - 3 \geq 0,$

$$\Rightarrow a = \frac{13}{4}, a \geq 3$$

$$\Rightarrow a \in \left[3, \frac{13}{4}\right] \quad \text{... (ii)}$$

Thus, equation will at least one negative root, if

$a \in \left(-\infty, \frac{13}{4}\right] \cup \left[3, \frac{13}{4}\right]$ [Using Eqs. (i) and (ii)]

$\Rightarrow a \in (-\infty, 3)$

7. (d) Since, both roots of $f(x) = 0$ exceed a. therefore,

(i) Discriminant > 0 so $1 - 4a > 0$ or a < 1/4

(ii) $f(a) > 0$

so $a^2 + 2a > 0$ or $a(a-2) > 0$

so $a \in (-\infty, -2) \cup (0, \infty)$

(iii) $-\frac{1}{2} > a$

Solving Eqs. (i), (ii) and (iii) $a < -2$

8. (d) We know that if $f(x) = 0$ has exactly one root between (a, b) then $f(a).f(b) < 0$

$(-1)(1 - a^2 + 2a - 1) < 0$

$a^2 - 2a < 0 \Rightarrow a \in (0, 2)$

9. (c) α, β are roots of equation $(x - a)(x - b) - c = 0$

$\therefore (x-a)(x-b) - c = (x-\alpha)(x-\beta)$

$\Rightarrow (x-\alpha)(x-\beta) + c = (x-a)(x-b)$

$\Rightarrow$ a, b are roots of equation

$$(x-a)(x-b)+c=0$$

10. (c) $x^2 - 2px + p^2 - 1 = 0 \Rightarrow (x-p)^2 = 1$

$\therefore$ $x = p \pm 1$

Both the roots are less than 4 if $p+1<4$ and $p-1 < 4 \Rightarrow p < 3$

Concept Applicator (CA)-5

1. Sign scheme of $(x-10)(x-20)(x-30)(x-40)$ is as shown below

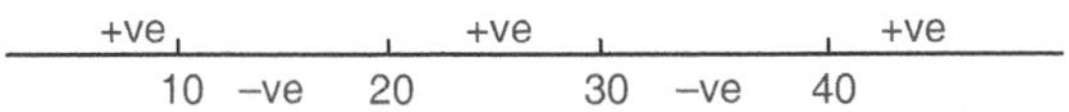

So required value of x is $(10, 20) \cup (30, 40)$ total number of integral values is 18.

2. Sign scheme of $(x-10)(x-20)(x-30)$ is as shown below

So required value of x is $(-\infty, 10) \cup (20, 30)$.

3. In case of $(x+1)(x-3)(x-2)^2$, here, the power is even for the factor $x-2$, so there will not be any change in the sign of the function at $x = 2$.

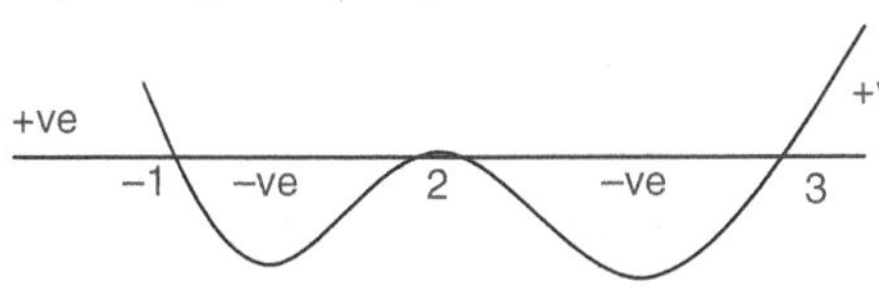

So required answer is $(-\infty, -1] \cup [3, \infty) \cup \{2\}$.

4. Find the critical points, locate it on the number line, The sign scheme(wavy curve) is shown below

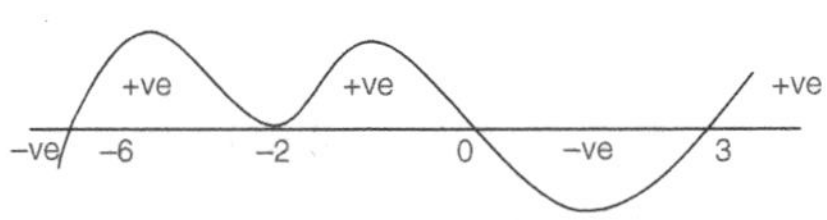

$$(-6, -2) \cup (-2, 0) \cup (3, \infty)$$

5. Here critical points are –1, 1, and 4, exponent of factor $(x-1)$ is even so function will not change its sign at $x = 1$.

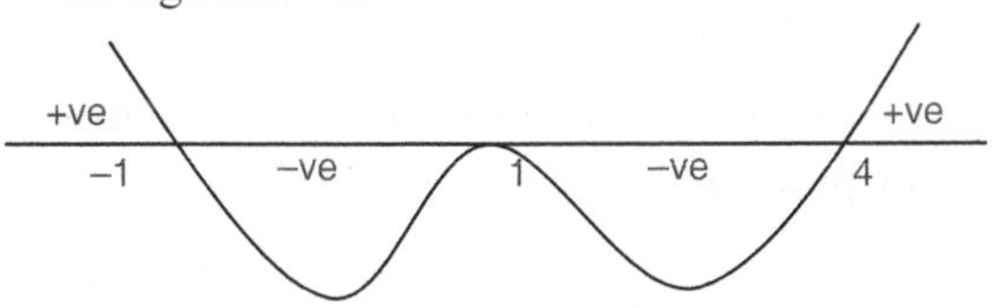

$$x \in (-1, 1) \cup (1, 4).$$

6. Sign scheme of $\dfrac{2x-3}{3x-7}$ and $(2x-3)(3x-7)$ is same, only difference is $x \neq 7/3$.

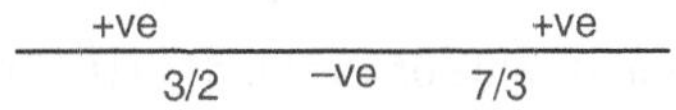

$$\left(-\infty, \frac{3}{2}\right] \cup \left(\frac{7}{3}, \infty\right).$$

7. Sign scheme of $\dfrac{\{(x-5)(x-15)\}}{\{(x-10)(x-20)\}}$ and $(x-5((x-10)(x-15)(x-20)$ is same

We know that sign scheme of

$(x-5)(x-10)(x-15)(x-20)$ is

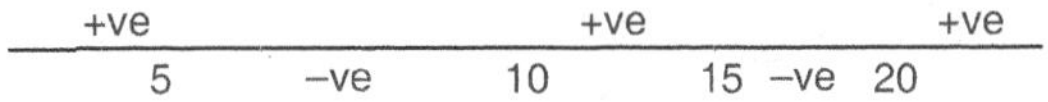

So $\dfrac{\{(x-5)(x-15)\}}{\{(x-10)(x-20)\}} \geq 0$ for

$x \,\varepsilon\, (-\infty, 5] \cup (10, 15] \cup (20, \infty)$.

8. Sign scheme of $\dfrac{\{(x-5)^2(x-15)^3\}}{\{(x-10)^4(x-20)^5\}}$ and $(x-5)^2(x-15)^3(x-10)^4(x-20)^5$ is same. We know that sign scheme of

$(x-5)^2(x-15)^3(x-10)^4(x-20)^5$ is

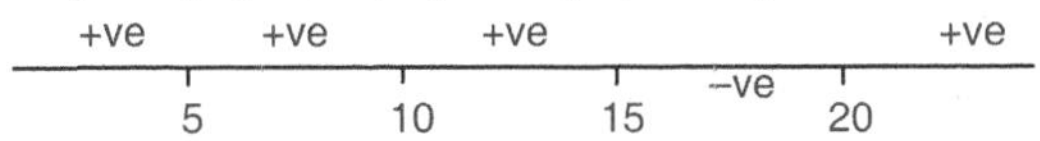

So $\dfrac{\{(x-5)^2(x-15)^3\}}{\{(x-10)^4(x-20)^5\}} \geq 0$ for

$x \in (-\infty, 10) \cup (10, 15) \cup (20, \infty)$.

9. Consider factors one by one

$$x^2 + 3x + 10 \geq 0 \text{ for all } x$$

$$x^2 - 9x + 20 = (x-4)(x-5)$$

$$x^2 - 5x + 15 \geq 0 \text{ for all } x$$

$$x^2 - 13x + 42 = (x-6)(x-7)$$

So sign scheme of $f(x) =$

$\dfrac{(x^2+3x+10)(x^2-9x+20)}{(x^2-5x+15)(x^2-13x+42)}$ is same as that of

$$\frac{\{(x-4)(x-5)\}}{\{(x-6)(x-7)\}}$$

Sign scheme of $\frac{\{(x-4)(x-5)\}}{\{(x-6)(x-7)\}}$ is

+ve 4 −ve 5 +ve 6 −ve 7 +ve

So required value of x is $[4, 5] \cup (6, 7)$

10. Given that $\frac{x^2-5x+12}{x^2-4x+5} - 3 > 0$

$$\frac{x^2-5x+12-3x^2+12x-15}{(x^2-4x+5)} > 0$$

$$\frac{-2x^2+7x-3}{(x^2-4x+5)} > 0$$

$$\frac{2x^2-7x+3}{(x^2-4x+5)} < 0$$

$$\frac{2x^2-6x-x+3}{(x^2-4x+5)} < 0$$

$$\Rightarrow \frac{2x(x-3)-1(x-3)}{(x^2-4x+5)} < 0$$

$$\Rightarrow \frac{(2x-1)(x-3)}{(x^2-4x+5)} < 0$$

Consider $y = x^2 - 4x + 5$ here discriminant of this expression is $(-4)^2 - 4 - 5 < 0$

So, $x^2 - 4x + 5 > 0$

Sign scheme of $\frac{(2x-1)(x-3)}{(x^2-4x+5)}$ and $(2x - 1)(x - 3)$ is same

+ve ½ −ve 3 +ve

So, $x \in \left(\frac{1}{2}, 3\right)$.

Concept Applicator (CA)-6

1. The given equation is $x(1 + x^2) + x^2(6 + x) + 2 = 2x^3 + 6x^2 + x + 2 = 0$ has roots α, β, γ.

So, $2x^3 + x^2 + 6x + 2 = 0$ has roots $\alpha^{-1}, \beta^{-1}, \gamma^{-1}$ (By substituting $1/x$ instead of x)

Hence, Sum of the roots $= -\frac{(\text{coefficient of } x^2)}{(\text{coefficient of } x^3)}$

$$\therefore \ \alpha^{-1} + \beta^{-1} + \gamma^{-1} = -\frac{1}{2}.$$

Option (c)

2. Since α, β, γ be the roots of the equation $x^3 - 13x + 11 = 0$ so

$$\alpha + \beta + \gamma = 0 \quad \dots \text{(i)}$$

Now we need to frame an equation whose roots are $\alpha + \beta$, $\beta + \gamma$ and $\alpha + \gamma$

But from (i) $\alpha + \beta = -\gamma$, $\beta + \gamma = -\alpha$ and $\alpha + \gamma = -\beta$

It means we have to form an equation whose roots are $-\alpha$, $-\beta$, and $-\gamma$

So to get required equation just substitute $-x$ instead of x

Required equation is $-x^3 + 13x + 11 = 0$.

3. Since α, β, γ be the roots of the equation $x^3 + 3x + 2 = 0$ so

$$\alpha + \beta + \gamma = 0 \quad \dots \text{(i)}$$

$$\alpha\beta + \beta\gamma + \alpha\gamma = 3 \quad \dots \text{(ii)}$$

$$\alpha\beta\gamma = -2 \quad \dots \text{(iii)}$$

Now we need to frame an equation whose roots are $\alpha\beta$, $\beta\gamma$ and $\alpha\gamma$

Sum of the roots $\alpha\beta + \beta\gamma + \alpha\gamma = 3$

Sum of the roots taken two at a time $= \alpha^2\beta\gamma + \beta^2\alpha\gamma + \gamma^2\alpha\beta = \alpha\beta\gamma(\alpha + \beta + \gamma) = (-2)(0) = 0$

Product of the roots $(\alpha\beta)(\beta\gamma)(\alpha\gamma) = (\alpha\beta\gamma)^2 = 4$

So required equation is $x^3 - 3x^2 - 4 = 0$

Alternate Solution

Since α, β, γ be the roots of the equation $x^3 + 3x + 2 = 0$ so

And $\alpha\beta\gamma = -2$

And we need to find equation whose roots are $\alpha\beta$, $\beta\gamma$ and $\alpha\gamma$

But $\alpha\beta = -\frac{2}{\gamma}$, $\beta\gamma = -\frac{2}{\alpha}$ and $\alpha\gamma = -\frac{2}{\beta}$

So we need to find equation whose roots are $-\frac{2}{\gamma}, -\frac{2}{\alpha}$, and $-\frac{2}{\beta}$

From transformation of equation we get the equation whose roots are $-\frac{1}{\gamma}, -\frac{1}{\alpha}$, and $-\frac{1}{\beta}$

Is $\left(-\frac{1}{x}\right)^3 + 3\left(-\frac{1}{x}\right) + 2 = 0$ or $2x^3 - 3x^2 - 1 = 0$

And equation whose roots are $-\frac{2}{\gamma}, -\frac{2}{\alpha}$, and $-\frac{2}{\beta}$

Is $2\left(\frac{x}{2}\right)^3 - 3\left(\frac{x}{2}\right)^2 - 1 = 0$ or $x^3 - 3x^2 - 4 = 0$

4. Let the function be $f(x) = x^3 + 3x + 2$ remainder when $f(x)$ is divided by $(x - 2)$ is given by $f(2) = 8 + 6 + 2 = 16$

   Since remainder is not zero so 2 is not a root of the equation.

5. Let $f(x) = x^3 + 3x^2 - px + p$ if $x = 3$ is a root of the equation then $(x - 3)$ must be a factor of $f(x)$ or $f(3) = 0$

   So $f(3) = 27 + 27 - 3p + p = 0$ or $p = 27$

6. Let $f(x) = x^4 + ax^3 + 3x - b$

   If $x^2 - 1$ is a factor then $(x - 1)$ and $(x + 1)$ is also a factor

   Then, $f(1) = 0$

   $\Rightarrow \quad a - b + 4 = 0 \quad$ ... (i)

   and $f(-1) = 0$

   $\Rightarrow \quad a + b + 2 = 0 \quad$ (ii)

   On solving eqs.(i) and (ii), we get $a = -3$, $b = 1$

7. 2α, 2β are roots of $ax^2 + 4bx + 4c = 0$ and -2α, -2β are roots of $ax^2 - 4bx + 4c = 0$

   $\therefore$ Required equation is

   $(ax^2 + 4c)^2 - 16b^2x^2 = 0$

   or, $a^2x^4 + 8(ac - 2b^2)x^2 + 16c^2 = 0$

8. Solution: Since, $x^2 - 3x + 2$ ie, $(x - 2)(x - 1)$ is factor of $x^4 - px^2 + q$

   $\therefore \quad x = 2, x = 1$ are roots of $x^4 - px^2 + q = 0$

   $\Rightarrow \quad 16 - 4p + q = 0$ and $1 - p + q = 0$

   On solving these equations, we get $p = 5$, $q = 4$

9. Since, 2 and 3 are roots of $f(x) = 0$

   $\therefore \quad f(2) = 0$ and $f(3) = 0$

   $\Rightarrow \quad 4m + n = 10$ and $9m + n + 15 = 0$

   Solving the two equations, we get $m = -5$, $n = 30$.

   So $m + n = 25$

   Option (a)

10. Since $f(x) = x^3 - 7x^2 - 36$ so from remainder theorem $f(3) = 27 - 63 - 36 = 0$ so $x = 3$ is a root of the equation.

Exercis-A

Concept Builder (CB)

1. (c) $\alpha + \beta = \frac{3a}{a-1}, \alpha\beta = \frac{1}{a-1}$. As $a > 1, \alpha + \beta > 0$ and $\alpha\beta > 0$ and $D = 9a^2 + 4(1 - a)$

   $= 9\left(a^2 - \frac{4}{9}a + \frac{4}{9}\right) = 9\left\{a\left(a - \frac{4}{9}\right) + \frac{4}{9}\right\} > 0$, as a > 1

   $\therefore$ Real and positive roots.

2. (b) Here, $|x| = \cos x$. Thus, to find value of x for which both curves have point of intersections.

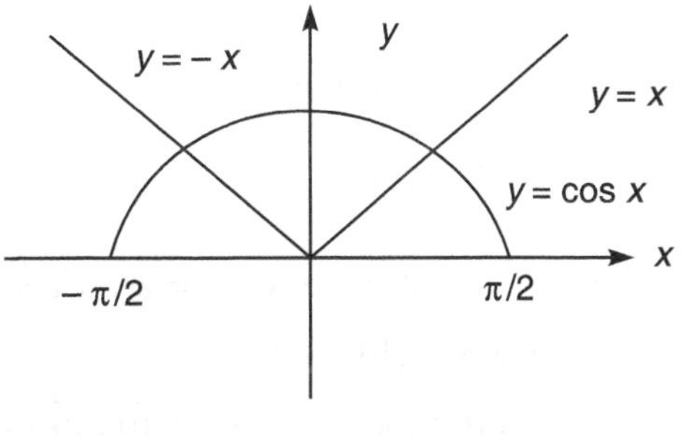

Clearly, there are two points of intersection of $y = |x|$ and $y = \cos x$. hence, there are two real solutions.

3. (b) If $x = n \in I, |n - 2n| = 4 \quad \therefore n = \pm 4$

If $x = n + k, n \in I, 0 < k < 1$ then

$$|n - 2(n+k)| = 4$$

$\therefore |-n - 2k| = 4$. It is possible, if k $= \frac{1}{2}$.

Then,

$|-n-1| = 4$, ie, $n + 1 = \pm 4 \quad \therefore n = 3, -5$

4. (b) Given equation is $a|x|^2 + 2b|x| - c = 0$

$\therefore |x| = \frac{-2b \pm \sqrt{4b^2 + 4ac}}{2} = -b \pm \sqrt{b^2 + 4ac}$

Since, a, b, c are positive. So $|x| = -b + \sqrt{b^2 + ac}$. We will get two real values of x.

5. (d) $x^2 + a|x| + 1 = 0$

$\Rightarrow |x|^2 + a|x| + 1 = 0$

$|x| = \frac{-a \pm \sqrt{a^2 - 4}}{2}$

Since, $a < 0$ and $\neq -2$

$\Rightarrow \frac{-a + \sqrt{a^2 - 4}}{2}$ and $\frac{-a - \sqrt{a^2 - 4}}{2}$ both are positive.

Thus, there are four roots if $a < -2$ else no real root.

6. (a) Dividing the equation, $a^3x^2 + abcx + c^3 = 0$ by c^2, we get $a\left(\frac{ax}{c}\right)^2 + b\left(\frac{ax}{c}\right) + c = 0$

$\Rightarrow \frac{ax}{c} = \alpha, \beta$ are roots

$\Rightarrow x = \frac{c}{a}\alpha$ and $x = \frac{c}{a}\beta$

$\Rightarrow$ So roots of above equation are $\alpha^2\beta$ and $\alpha\beta^2$

7. (d) Let $x = \sqrt[3]{10} + \sqrt[3]{100}$

$\Rightarrow x^3 = 10 + 100 + 3 \cdot \sqrt[3]{10} \cdot \sqrt[3]{100}\left(\sqrt[3]{10} + \sqrt[3]{100}\right)$

$\Rightarrow x^3 = 110 + 30x$

$\Rightarrow x^3 - 30x - 110 = 0$; where $P(x)$ is least possible degree 3.

Product of all roots = 110

8. (a) $x = 2$, But $x^2 - 4 \neq 0$

$\therefore \quad x \neq \pm 2$

$\therefore$ No solution

9. (a) $(a + b)x^2 + x(2a - 3b + 1)x + (a - 2b + 1) = 0$ This is an identity if

$\Rightarrow a + b = 0, 2a - 3b + 1 = 0, a - 2b + 1 = 0$

$\Rightarrow$ No value of (a, b) is possible

10. (b) $D = b^2 - 4ac, D' = d^2 + 4ac$

$\Rightarrow D + D' = b^2 + d^2 > 0$

$\therefore$ At least one of D, D' is positive.

11. (a) The sum of the coefficients

$= (a + 2) + (a - 3) - (2a - 1) = 0$

So, $x = 1$ is a root (If one root is rational then other must be rational)

$\therefore$ Other root $= -\frac{2a-1}{a+2}$ = rational for all rational values $a, a \neq -2$.

12. (b) We know that : $a^{\log_a x} = x$, so given equation can be reduced to

$x^2 - 4x + 5 = x - 1$

$\Rightarrow x^2 - 5x + 6 = 0$

$\Rightarrow x = 2, 3$

13. (c) From 1st equation $\alpha + \beta = -\frac{b}{a}, \ \alpha\beta = \frac{c}{a}$

$\Rightarrow [\{(\alpha + k) + (\beta + k)\}^2 - 4(\alpha + k)(\beta + k)]$

$= (\alpha + \beta)^2 - 4\alpha\beta$

$\Rightarrow \frac{q^2}{p^2} - 4\frac{r}{p} = \frac{b^2}{a^2} - \frac{4c}{a}$

$\Rightarrow \frac{q^2 - 4pr}{p^2} = \frac{b^2 - 4ac}{a^2}$

$\Rightarrow \frac{b^2 - 4ac}{q^2 - 4pr} = \left(\frac{a}{p}\right)^2$

14. (d) Since, $\frac{\alpha}{\beta}=\frac{m}{n}$

$\Rightarrow \frac{\alpha+\beta}{\alpha-\beta}=\frac{m+n}{m-n}$

$\Rightarrow \frac{(\alpha+\beta)^2}{(\alpha+\beta)^2-4\alpha\beta}=\frac{(m+n)^2}{(m+n)^2-4mn}$

$\Rightarrow \frac{\{(\alpha+\beta)^2-4\alpha\beta\}}{\{(\alpha+\beta)^2\}}=\frac{\{(m+n)^2-4mn\}}{\{(m+n)^2\}}$

$\Rightarrow \frac{4\alpha\beta}{(\alpha+\beta)^2}=\frac{4mn}{(m+n)^2}$

$\Rightarrow \frac{c/a}{b^2/a^2}=\frac{mn}{(m+n)^2}$

$\Rightarrow ac\,(m+n)^2 = mn\,b^2$

> ***Smart Cracker:-***
> Above approach was for board, the smart way to solve is
>
> Let m = 1 and n = 2 or roots are ½ and 1 then equation is $x^2-\left(\frac{3}{2}\right)x+\frac{1}{2}=0$ or $2x^2-3x+1=0$ so a = 2, b = –3 and c = 1. From options only
>
> $\Rightarrow ac\,(m+n)^2 = mn\,b^2$ satisfies the condition.

15. (a) Let $x=\sqrt{6+\sqrt{6+\sqrt{6+\ldots.\text{ to }\infty}}}=\sqrt{6+x}$

Squaring both sides, we get $x^2 = 6 + x$

$\Rightarrow x = -2, 3$ but x is positive. Hence, x = 3

16. (b) We know that if sum of the co-efficient is equal to zero then one root is 1

Here, $q-r+r-p+p-q=0$

$\therefore$ Its roots are 1, $\frac{p-q}{q-r}$

17. (b) Let α, $\alpha + 1$ be the roots of the given equation, then $\alpha + \alpha + 1 = -p$

$\Rightarrow 2\alpha = -p-1$

$\Rightarrow \alpha = -\frac{p+1}{2}$, Also $\alpha(\alpha+1) = q$

$\Rightarrow -1 + p^2 = 4q$

$\Rightarrow p^2 = 4q + 1$

18. (a) Given, $\frac{x^2-bx}{ax-c}=\frac{\lambda-1}{\lambda+1}$

$\Rightarrow (\lambda+1)(x^2-bx) = (\lambda-1)(ax-c)$

$\Rightarrow (\lambda+1)x^2 - [\lambda b + b + \lambda a - a]x + (\lambda-1)c = 0$

Since, $\alpha + \beta = 0$

$\Rightarrow \frac{\lambda b+b+\lambda a-a}{\lambda+1}=0$

$\Rightarrow \lambda\,(a+b) = a-b$

$\Rightarrow \lambda=\frac{a-b}{a+b}$

19. (d) Since roots of the equation are equal so, Discriminant = 0

$\Rightarrow 4(ac+bd)^2 - 4(a^2+b^2)(c^2+d^2) = 0$

$\Rightarrow 2abcd - a^2d^2 - b^2c^2 = 0$

$\Rightarrow (ad-bc)^2 = 0$

$\Rightarrow ad = bc$

$\Rightarrow \frac{a}{b}=\frac{c}{d}$

20. (a) Given, $x=\sqrt{7+4\sqrt{3}}$

$\Rightarrow \frac{1}{x}=\frac{1}{\sqrt{7+4\sqrt{3}}}=\sqrt{7-4\sqrt{3}}$

$\Rightarrow x+\frac{1}{x}=\sqrt{7+4\sqrt{3}}+\sqrt{7-4\sqrt{3}}$

$=\left(\sqrt{3}+2\right)+\left(2-\sqrt{3}\right)=4$

21. (b) If roots are equal then discriminant must equal to zero. $[2(k+1)]^2 - 4k^2 = 0$

$\Rightarrow 4(k+1)^2 - 4k^2 = 0$

$\Rightarrow 8k + 4 = 0 \quad \Rightarrow k = -\frac{1}{2}$

22. (c) $\alpha + \beta = -p, \alpha\beta = 8$

Given, $\alpha - \beta = 2 \quad \Rightarrow (\alpha - \beta)^2 = 4$

$\Rightarrow (\alpha + \beta)^2 - 4\alpha\beta = 4$

$\Rightarrow p^2 - 32 = 4$

$\Rightarrow p^2 = 36 \quad \Rightarrow p = \pm 6$

23. (c) If a quadratic function is a perfect square then the equation has two equal roots.

In the given expression sum of the coefficient is zero so, $f(1) = 0$

$\therefore \alpha = \beta = 1$

$\Rightarrow \alpha\beta = 1$ or $c^2 (a^2 - b^2) = a^2 (b^2 - c^2)$

$\Rightarrow b^2 (a^2 + c^2) = 2a^2c^2$ or $b^2 = \dfrac{2a^2c^2}{a^2 + c^2}$

$\therefore a^2, b^2, c^2$ are in HP

24. (b) Let the other root be β, then

$\alpha + \beta = -\dfrac{2}{4} = -\dfrac{1}{2} \Rightarrow \beta = -\dfrac{1}{2} - \alpha$ ... (i) and

$4\alpha^2 + 2\alpha - 1 = 0$

Now, $4\alpha^3 - 3\alpha = \alpha(4\alpha^2 - 3) = \alpha(1 - 2\alpha - 3)$

[since $4\alpha^2 + 2\alpha - 1 = 0$]

$= -2\alpha^2 - 2\alpha = -\dfrac{1}{2}(4\alpha^2) - 2\alpha$

$= -\dfrac{1}{2}(1 - 2\alpha) - 2\alpha$ [because $4\alpha^2 = 1 - 2\alpha$]

$= -\dfrac{1}{2} - \alpha = \beta$

Hence, $4\alpha^3 - 3\alpha$ is the other root.

25. (d) If roots of the equation are equal then discriminant is equal to zero.

$\Rightarrow 4(a^2 - bc)^2 - 4(c^2 - ab)(b^2 - ac) = 0$

$\Rightarrow [a^4 - 2a^2 bc + b^2 c^2]$
$\qquad - [b^2c^2 - ab^3 - ac^3 + a^2bc] = 0$

$\Rightarrow a(a^3 + b^3 + c^3 - 3abc) = 0$

$\therefore$ Either $a = 0$ or $a^3 + b^3 + c^3 - 3abc = 0$, but given that $a, b, c \neq 0$

26. (b) Let α be a root. then, the root is $n\alpha$

$\therefore \alpha + n\alpha = \dfrac{-b}{a}$ .... (i)

and $\alpha \cdot n\alpha = \dfrac{c}{a}$ .... (ii)

From Eq. (i), $\alpha = \dfrac{-b}{a(1+n)}$

Putting this value in Eq. (ii), we get

$n \cdot \left\{\dfrac{-b}{a(1+n)}\right\}^2 = \dfrac{c}{a}$; or $nb^2 = \dfrac{ca^2(1+n)^2}{a}$ or

$nb^2 = ca(1 + n)^2$

27. The given equation is $(ab - 1)x^2 + (a + b)x - ab = 0$

here, $D = (a + b)^2 + 4ab(ab - 1)$
$= (a + b)^2 - 4ab + 4(ab)^2$
$= (a - b)^2 + (2ab)^2 \geq 0$ for all a, b

$\therefore$ Roots are always real.

28. (c) Let the speed of the bus = x km/h, then the speed of the car = (x + 25) km/h

Time taken by the bus to cover 500 km $= \dfrac{500}{x}$ h and

time taken by the car to cover same distance $= \dfrac{500}{x+25}$ h

From the given condition,

$\dfrac{500}{x} = \dfrac{500}{x+25} + 10 \Rightarrow \dfrac{50}{x} = \dfrac{50}{x+25} + 1$

$\Rightarrow \dfrac{50}{x} - \dfrac{50}{x+25} = 1 \Rightarrow \dfrac{50(x + 25 - x)}{x(x+25)} = 1$

$\Rightarrow 1250 = x^2 + 25x$

$\Rightarrow (x + 50)(x - 25) = 0$ since $x \neq -50$,

$\therefore \quad x = 25$

Hence, the speed of the bus = 25 km/h and speed of the car = (25 + 25) = 50 km/h

29. (b) Sum of the roots $\alpha + \beta = \dfrac{4 + \sqrt{5}}{5 + \sqrt{2}}$ and

product $\alpha\beta = \dfrac{8 + 2\sqrt{5}}{5 + \sqrt{2}}$

Harmonic mean is defined as HM

$= \dfrac{2\alpha\beta}{\alpha + \beta} = \dfrac{16 + 4\sqrt{5}}{4 + \sqrt{5}} = 4$

30. (d) Put $2^x = y$, the equation becomes $16y^2 = 8y + 48$

$\Rightarrow (y-2)(2y+3) = 0$

$\Rightarrow y = 2, -\frac{3}{2}$. But 2^x cannot be negative

So, $2^x = 2 \Rightarrow x = 1$

31. (b) Since, its roots are real so $D \geq 0$ or

$\therefore 64 - 4(a^2 - 6a) \geq 0$

$4a^2 - 24a - 64 \leq 0$

$a^2 - 6a - 16 = (a+2)(a-8) \leq 0$

$\Rightarrow -2 \leq a \leq 8 \quad \Rightarrow a \in [-2, 8]$

32. (d) Given expression $x^2 + 2bx + c$ is always positive if

$\Rightarrow D < 0 \quad \Rightarrow 4b^2 - 4c < 0 \quad \Rightarrow b^2 < c$

33. (a) Clearly, $ax^2 + bx + c = 0$ is satisfied by $x = 2$ [since $4a + 2b + c = 0$]

$\Rightarrow$ one real root is 2

$\therefore$ other root is also real.

34. (c) $D = a^2 + b^2 + c^2 = a^2 + (a+1)^2 + a^2(a+1)^2$

$= a^4 + 2a^3 + 3a^2 + 2a + 1$

$= (a^2 + a + 1)^2$

$\therefore \sqrt{D} = a(a+1) + 1 =$ always an odd integer.

35. (a) Since, α, β are roots of the equation $ax^2 + bx + c = 0$

$\therefore \alpha + \beta = -\frac{b}{a}$ and $\alpha\beta = \frac{c}{a}$

We know that equation with given roots is $x^2 -$ (Sum of the roots) $x +$ Product of the roots

The required equation is

$$x^2 - \left(\frac{1}{a\alpha + b} + \frac{1}{a\beta + b}\right)x + \frac{1}{a\alpha + b} \cdot \frac{1}{a\beta + b} = 0$$

$\Rightarrow \{a^2\alpha\beta + ab(\alpha + \beta) + b^2\}x^2$

$-\{a(\alpha + \beta) + 2b\}x + 1 = 0$

$\Rightarrow (ca - b^2 + b^2)x^2 - (2b - b)x + 1 = 0$

$\Rightarrow cax^2 - bx + 1 = 0$

36. (b) $x^2 + x + 1 = 0$ has complex roots, so roots are conjugate and hence both the roots are common.

$\Rightarrow \frac{a}{1} = \frac{b}{1} = \frac{c}{1}$

37. (c) A polynomial is an identity if coefficients are 0

So $a^2 - 3a + 2 = 0$, $a^2 - 5a + 6 = 0$, $a^2 - 4 = 0$

$\Rightarrow a = 1, 2$ and $a = 2, 3$ and $a = 2, -2$. $\therefore$ Equation is identity, if $a = 2 \Rightarrow$ only one solution.

38. (a) Let α, β be the roots of $x^2 - abx - a^2 = 0$, Then product of the roots $\alpha\beta = -a^2$ which shows product of roots < 0; ie, one root must be negative and other must be positive.

39. (c) Discriminant of the given equation is $D = 25(l+m)^2 + 8(l-m)^2 > 0$

Hence, roots are real and unequal.

40. (b) From observation, $x = m$ is a root of the equation. Therefore, the other root must be $-m$. That is,

$$\frac{1}{-m} + \frac{1}{-m+b} = \frac{1}{m} + \frac{1}{m+b}$$

$$\Rightarrow \frac{1}{b-m} - \frac{1}{b+m} = \frac{2}{m}$$

$$\Rightarrow \frac{b+m-b+m}{b^2 - m^2} = \frac{2}{m}$$

$\Rightarrow 2m^2 = 2b^2 - 2m^2$ or $2m^2 = b^2$

41. (b) Here, $\alpha^4 + \beta^4 = (\alpha^2 + \beta^2)^2 - 2\alpha^2\beta^2$

$= \left\{(\alpha+\beta)^2 - 2\alpha\beta\right\}^2 - 2(\alpha\beta)^2$

$= \left(p^2 + \frac{1}{p^2}\right)^2 - \frac{1}{2p^4} = p^4 + \frac{1}{2p^4} + 2$

$= \left(p^2 - \frac{1}{\sqrt{2}p^2}\right)^2 + 2 + \sqrt{2} \geq 2 + \sqrt{2}$

Thus, minimum value of $\alpha^4 + \beta^4$ is $2 + \sqrt{2}$.

42. (a) $D = c^2(3a^2 + b^2)^2 - 4abc^2(-6a^2 - ab + 2b^2)$
$= c^2(3a^2 - b^2 + 4ab)^2$
= a perfect square so roots are rational

43. (c) Let the roots of the equation be α & α^2

then $\alpha + \alpha^2 = -\frac{p}{3}$

and $\alpha \cdot \alpha^2 = \frac{3}{3}$ ie, $\alpha^3 = 1$

$\Rightarrow \alpha = 1, \omega, \omega^2$

If $\alpha = 1 \Rightarrow p = -6$
If $\alpha = \omega \Rightarrow p = 3$
but given that $p > 0 \therefore p \neq -6$ Hence, $p = 3$

44. (c) Since, a, b, c are positive therefore, $ax^2 + b|x| + c > 0$.

Hence, the equation $ax^2 + b|x| + c = 0$ has no real root.

45. (a) Let α and β be the roots of the given equation.

Then, $\alpha + \beta = -\frac{b}{a}$ and $\alpha\beta = \frac{c}{a}$

Given,

$$\alpha + \beta = \frac{1}{\alpha^2} + \frac{1}{\beta^2} = \frac{\alpha^2 + \beta^2}{(\alpha\beta)^2} = \frac{(\alpha+\beta)^2 - 2\alpha\beta}{(\alpha\beta)^2}$$

$$\Rightarrow -\frac{b}{a} = \frac{\frac{b^2}{a^2} - \frac{2c}{a}}{\frac{c^2}{a^2}} = \frac{b^2 - 2ca}{c^2}$$

$\Rightarrow 2ca^2 = bc^2 + ab^2$

Hence, bc^2, ca^2 and ab^2 are in AP.

46. (d) Since, $\sin\theta$ and $\cos\theta$ are the roots of the equation

$ax^2 + bx + c = 0$

$\therefore \sin\theta + \cos\theta = -\frac{b}{a}$ and $\sin\theta\cos\theta = \frac{c}{a}$

Now, $(\sin\theta + \cos\theta)^2 = 1 + 2\sin\theta\cos\theta$

$$\therefore \frac{b^2}{a^2} = 1 + \frac{2c}{a} = \frac{a + 2c}{a}$$

$\Rightarrow b^2 = a(a + 2c) = a^2 + 2ac$

$\Rightarrow b^2 + c^2 = a^2 + 2ac + c^2 = (a + c)^2$

Hence, $(a + c)^2 = b^2 + c^2$

47. (b) We have, $\frac{1}{x+p} + \frac{1}{x+q} = \frac{1}{r}$

$\Rightarrow r(x + q + x + p) = (x + p)(x + q)$

or $x^2 + x(p + q - 2r) + pq - r(p + q) = 0$

Let its roots be α and $-\alpha$

Then, $\alpha + (-\alpha) = -(p + q - 2r)$ or $p + q = 2r$

And, product $= \alpha(-\alpha) = pq - r(p + q)$

$$= pq - \frac{(p+q)^2}{2} = -\frac{1}{2}(p^2 + q^2)$$

48. (a) The given equation can be written as $(x^2 + 5x + 4)(x^2 + 5x + 6) = 24$

Now substitute $x^2 + 5x = t$ we get $t^2 + 10t = 0$
$\Rightarrow t = 0, t = -10$.

For $t = 0$, $x^2 + 5x = 0 \Rightarrow x = 0, x = -5$

For $t = -10$, $x^2 + 5x = -10$ does not have real roots.

49. (d) Put $x - 1 = t^2$, we get

$\sqrt{(t^2 + 1 + 2t) - (t^2 + 1 - 2t)} = 2$

$\sqrt{(1+t)^2 - (1-t)^2}$

$|t + 1| - |t - 1| = 2$ which is satisfied for each $t \geq 1$ so $x \geq 2$

50. (d) The problem can be solved by manipulating the left hand side of the equation.

$(x - 19)(x - 97) = p$ has real roots r_1 and r_2
$\Rightarrow (x - 19)(x - 97) - p = 0$ has real roots r_1 and r_2
$\Rightarrow (x - 19)(x - 97) - p = (x - r_1)(x - r_2)$ for any x
$\Rightarrow (x - r_1)(x - r_2) + p = (x - 19)(x - 97)$ are any x
$\Rightarrow$ the roots of $(x - r_1)(x - r_2) + p = 0$ are 19 and 97
$\Rightarrow$ the roots of $(x - r_1)(x - r_2) = -p$ is 19.
Thus the minimum root of $(x - r_1)(x - r_2) = -p$ is 19.

Concept Cracker (CC)

1. (a) Since the equation does not have two distinct real roots, discriminant ≤ 0

$\Rightarrow 16-4\log_{1/2} a^2 \le 0 \quad \Rightarrow \log_{1/2} a^2 \ge 4$

$\Rightarrow a^2 \le \frac{1}{16}$

2. (a) From the given information

$\frac{b}{a}>0, \frac{c}{a}>0, \frac{c}{b}>0, \frac{a}{b}>0, \frac{a}{c}>0, \frac{b}{c}>0.$

Thus, a, b, c, all have same signs. Also, $b^2 \ge ac, c^2 \ge ab, a^2 \ge bc$

It is possible if and only if $a = b = c$.

3. (c) $x^2 - 2x - n = 0 \Rightarrow (x-1)^2 = n+1$

$\Rightarrow x = 1 \pm \sqrt{n+1}$

So, $n + 1$ should be a perfect square.

But, $n \in [100, 1000] \Rightarrow n+1 \in [101, 1001]$

Number of perfect square between 101 and 1001 is 21.

4. (b) For two distinct real roots, $D > 0$

$\Rightarrow (a-b)^2 - 4(1-a-b) > 0$

$\Rightarrow b^2 + b(4-2a) + a^2 + 4a - 4 > 0$ for the above quadratic expression to be true, $\forall\, b \in R$, the D of its corresponding equation should be less than zero.

ie, $(4-2a)^2 - 4(a^2+4a-4) < 0$

$\Rightarrow -32a + 32 < 0$

$\Rightarrow a > 1$

5. (b) As we know, AM $\ge$ GM and equality holds only when values are equal

$$\frac{\left(\sqrt{2+\sqrt{2}}\right)^x + \left(\sqrt{2-\sqrt{2}}\right)^x}{2} \ge (2^{x/2})^{1/2}$$

$$\Rightarrow \left(\sqrt{2+\sqrt{2}}\right)^x + \left(\sqrt{2-\sqrt{2}}\right)^x \ge 2 \cdot 2^{\frac{x}{4}}$$

This equality holds only, if

$$\left(\sqrt{2+\sqrt{2}}\right)^x = \left(\sqrt{2-\sqrt{2}}\right)^x \text{ ie, } x = 0$$

Only 1 real root exists.

6. (b) Here, $f(x) = bx^2 + (1 - b(c+d))x + bcd - a$ or $f(x) = b(x-c)(x-d) + (x-a)$ then,

$f(c) = (c-a) = -\text{ve}$

$f(d) = (d-a) = +\text{ve}$

So, one root lies between c and d.

7. (b) We know that nature of roots depends on the discriminant

Here given equation is, $x^2 + 2px - 4q = 0$

$\Rightarrow D = 4p^2 + 16q > 0$; as $p, q \in$ odd prime numbers.

$\Rightarrow p^2 + 4q = d^2 \Rightarrow 4q = d^2 - p^2$ ....(i)

But $d^2 - p^2 = 8k$; $k \in$ integer ...(ii)

From Eqs. (i) and (ii), $q = 2k$ which is not possible as q is prime number more than 2

$\therefore$ D is not a perfect square, but positive so roots are always irrational.

8. (a) Here, $x_1 + x_2 = 4 \Rightarrow x_2 = 4 - x_1$ and $x_1x_2 - x_3^2 = 4 + x_4^2$

$\Rightarrow x_1(4-x_1) - x_3^2 = 4 + x_4^2$

$\Rightarrow x_1^2 - 4x_1 + 4 + x_3^2 + x_4^2 = 0$

$\Rightarrow (x_1 - 2)^2 + x_3^2 + x_4^2 = 0$, which is possible only when $x_1 = 2, x_3 = 0, x_4 = 0$ and $x_2 = 2$.

So, there exist only one solution.

9. (a) Equating the rational and irrational part, we get

$x + y = 2x - y,$

$x - 2y = 0, x - y - 1 = 0$

$\Rightarrow x = 2$ and $y = 1$

10. (a) Factorise the given expression

$6x^2 - xy - y^2 - 6x + 8y - 12 = 6(x + \lambda y - 2)(x - \mu y + 1)$

$\Rightarrow \lambda = \frac{1}{3}, \mu = \frac{1}{2}$

11. (b) $2 + |e^x - 1| = (e^x)^2 - 2e^x + 1 = |e^x - 1|^2$

$|e^x - 1|^2 - |e^x - 1| - 2 = 0$ or $|e^x - 1| = 2, -1$

$\Rightarrow |e^x - 1| = 2$

$\Rightarrow e^x - 1 = 2, -2 \Rightarrow e^x = 3, -1 \Rightarrow e^x = 3$ and it has only one real solution.

12. (b) Since, two common roots must satisfy $(x^3 + 5x^2 + px + q) - (x^3 + 7x^2 + px + r) = 0$
i.e., $-2x^2 + q - r = 0$ sum of its roots is zero
So sum of common roots = 0
Let $x^3 + 5x^2 + px + q = 0$ and $x^3 + 7x^2 + px + r = 0$ has roots α, β, γ_1 and α, β, γ_2 respectively.
$\therefore \alpha + \beta + \gamma_1 = -5$ and $\alpha + \beta + \gamma_2 = -7$ where $\alpha + \beta = 0$
$\therefore \gamma_1 = -5$ or $\gamma_2 = -7$ so $\gamma_1 + \gamma_2 = -12$

13. (b) $\left(x^3 + \frac{1}{x^3}\right) - 5\left(x^2 + \frac{1}{x^2}\right) + \left(x + \frac{1}{x}\right)$

$$\Rightarrow \left(x + \frac{1}{x}\right)^3 - 3\left(x + \frac{1}{x}\right) - 5\left[\left(x + \frac{1}{x}\right)^2 - 2\right] + \left(x + \frac{1}{x}\right)$$

$$\Rightarrow \left(x + \frac{1}{x}\right)^3 - 5\left(x + \frac{1}{x}\right)^2 - 2\left(x + \frac{1}{x}\right) + 10$$

$\Rightarrow 1000 - 500 - 20 + 10 = 490$

14. (a) Since, p, q, r are in AP $\Rightarrow 2q = p + r$
The roots of $px^2 + qx + r = 0$ are real, if

$$q^2 - 4pr \geq 0 \Rightarrow \left(\frac{p+r}{2}\right)^2 - 4pr \geq 0$$

$$\therefore \left(\frac{r}{p}\right)^2 - 14\left(\frac{r}{p}\right) + 1 \geq 0 \Rightarrow \left|\frac{r}{p} - 7\right| \geq 4\sqrt{3}$$

15. (c) Given, $\frac{2x+1}{9} + \frac{x^2+7}{x^2-7} = \frac{2x+17}{9}$, on simplifying we get
$\Rightarrow 7x^2 = 175$
$\Rightarrow x^2 = 25 \Rightarrow x = \pm 5$ so sum of the roots is 0

16. (a) Given equation is, $(1 + m)x^2 - 2(1 + 3m)x + (1 + 8m) = 0$ .... (i)
Given that roots of the equation are equal so D = 0
$4(1 + 3m)^2 - 4(1 + m)(1 + 8m) = 0$
$\Rightarrow m^2 - 3m = 0 \quad \Rightarrow m(m - 3) = 0$
$\therefore m = 0, 3$

17. (b) Let the roots be α and β
Then, $\alpha + \beta = \frac{-n}{l}; \alpha\beta = \frac{n}{l}$; and given that $\frac{\alpha}{\beta} = \frac{p}{q}$
Now, consider

$$\sqrt{\frac{p}{q}} + \sqrt{\frac{q}{p}} + \sqrt{\frac{n}{l}} = \sqrt{\frac{\alpha}{\beta}} + \sqrt{\frac{\beta}{\alpha}} + \sqrt{\alpha\beta}$$

$$= \frac{\alpha + \beta + \alpha\beta}{\sqrt{\alpha\beta}} = \frac{-\frac{n}{l} + \frac{n}{l}}{\sqrt{\frac{n}{l}}} = 0$$

18. (a) The given equation can be written as $(ay + a')x^2 + (by + b')x + (cy + c') = 0$
Since x is rational function of, $(by + b')^2 - 4(ay + a')(cy + c')$ is a perfect square
that is, $(b^2 - 4ac)y^2 + (2bb' - 4ac' - 4a'c)y + b'^2 - 4a'c'$ is a perfect square $\Rightarrow$ D = 0
So $4(bb' - 2ac' - 2a'c)^2 - 4(b^2 - 4ac)(b'^2 - 4a'c') = 0$
Or $(ac' + a'c)^2 - 4aa'cc' = abb'c + a'bb'c - a'bb'c - a'c'b^2 - acb'^2$
Or $(ac' - a'c)^2 = (ab' - a'b)(bc' - b'c)$

19. (a) Since $\angle R = \frac{\pi}{2}$ so $\frac{P}{2} + \frac{Q}{2} = \frac{\pi}{4}$

$$\therefore \tan\left(\frac{P}{2} + \frac{Q}{2}\right) = \tan\frac{\pi}{4}$$

$$\Rightarrow \frac{\tan\frac{P}{2} + \tan\frac{Q}{2}}{1 - \tan\frac{P}{2}\tan\frac{Q}{2}} = 1$$

$$\Rightarrow \frac{-b/a}{1 - \left(\frac{c}{a}\right)} = 1 \quad \Rightarrow a + b = c$$

20. (c) $(3^x)^2 - 36 \cdot 3^x + 243 = 0$
Put $3^x = t$, the equation becomes $t^2 - 36t + 243 = 0$
$\Rightarrow (t - 9)(t - 27) = 0 \quad \Rightarrow t = 9, 27$
$\Rightarrow 3^x = 9$ or $3^x = 27 \Rightarrow x = 2$ or $x = 3$

∴ Solution pair is (2, 3) so sum of all the possible values of x is 5

21. (b) Solution: $\frac{|x+1|}{|x|}+|x+1|=\frac{|x+1|^2}{|x|}$

$$\Rightarrow \quad |x+1|\left\{\frac{1}{|x|}+1-\frac{|x+1|}{|x|}\right\}=0$$

$\therefore |x+1|=0$

$\therefore x-1$

or $1+|x|-|x+1|=0$
$\Rightarrow |x+1|-|x|=1$

$x \geq 0$

22. (d) The given expression can be written as $\left(\frac{5}{13}\right)^x+\left(\frac{12}{13}\right)^x \geq 1$

Assume, $\cos\alpha=\frac{5}{13}$ and $\sin\alpha=\frac{12}{13}$

$\therefore \cos^x\alpha+\sin^x\alpha \geq 1$,

Equality holds for $x = 2$. If $x < 2$, both $\cos\alpha$ and $\sin\alpha$ increase (being positive fractions). So,

$\cos^x\alpha+\sin^x\alpha > 1$ if $x < 2$. Thus, $x \leq 2$

23. (b) $\cos^4 x-\cos^2 x+1-p=0$; $0 \leq \cos^2 x \leq 1$, the roots of $y^2-y+1-p=0$ lie between 0 and 1.

$D \geq 0 \Rightarrow 1-4(1-P) \geq 0$

$\Rightarrow P \geq \frac{3}{4}$

$f(0)$ and $f(1) \geq 0 \Rightarrow 1-P \geq 0 \Rightarrow P \leq 1$

$\Rightarrow \frac{3}{4} \leq p \leq 1$

Alternately :-

Consider $y=f(x)=\cos^4 x+\sin^2 x$

$=\cos^4 x-\cos^2 x+1$

Let $t=\cos^2 x$

Then $y=t^2-t+1$, here $-1 \leq t \leq 1$

Graph of $y=t^2-t+1$ is

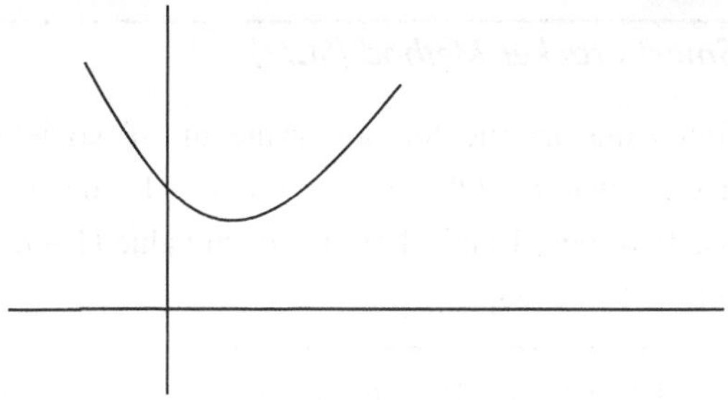

It will attain minimum value when t = ½

Required minimum value is ¾

So range of this is [3/4, 1]

or $\frac{3}{4} \leq p \leq 1$

24. (a) Given equation is $|x|^2-3|x|+2=0$

$\Rightarrow (|x|-1)(|x|-2)=0$

$\Rightarrow$ Either $|x|=1$ or $|x|=2$

$\Rightarrow x=\pm 1, x=\pm 2$

Graphically-

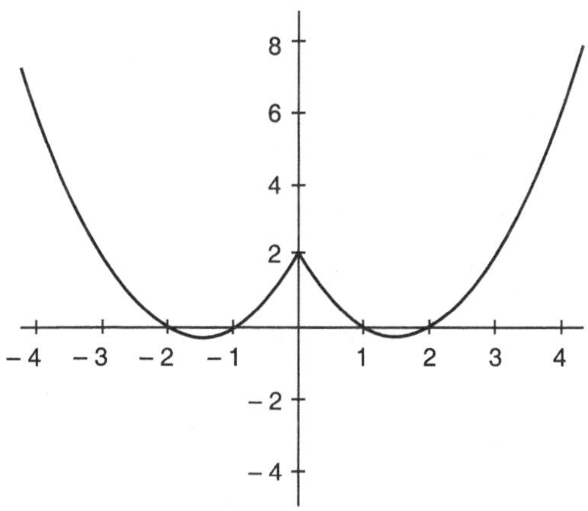

Hence, the given equation has four solutions.

25. (a) Given that $1, a_1, a_2, ..., a_{n-1}$ are roots of $x^n-1=0$

$$\Rightarrow \frac{x^n-1}{x-1}=(x-a_1)(x-a_2)...(x-a_{n-1})$$

$\Rightarrow x^{n-1}+x^{n-2}+x^{n-3}+...+x+1$
$=(x-a_1)(x-a_2)...(x-a_{n-1})$

Put $x = 1$, on both the sides, we get $1+1+1+...+1=(1-a_1)(1-a_2)...(1-a_{n-1})$

$\Rightarrow (1-a_1)(1-a_2)...(1-a_{n-1})=n$

Smart Cracker Method (SCM)

Since this is true for any value of 'n' so let's assume that $n = 2$ then equation is $x^2 - 1 = 0$ and it has two roots 1 and –1 so the given value $(1 - a_1) = (1 + 1) = 2 = n$

26. (b) Here, $\alpha + \beta = -p, \alpha\beta = q$ Let $t = x^n \Rightarrow$ when $x = \alpha, t = \alpha^n$ when $x = \beta, t = \beta^n$

$\therefore t^2 + p^n t + q^n = 0 \quad \therefore \alpha^n + \beta^n = -p^n, \alpha^n \beta^n = q^n$

Again $\frac{\alpha}{\beta}$ and $\frac{\beta}{\alpha}$ are roots of $x^n + 1 + (x + 1)^n = 0$

$$\Rightarrow \left(\frac{\alpha}{\beta}\right)^n + 1 + \left(\frac{\alpha}{\beta} + 1\right)^n = 0$$

$\Rightarrow \alpha^n + \beta^n + (\alpha + \beta)^n = 0 \Rightarrow -p^n + (-p)^n = 0$

which is true only if n is even integer.

27. (b) As $\cos\theta, \sin\phi, \sin\theta$ are in GP,

$\Rightarrow \sin^2\phi = \cos\theta \sin\theta$

$\Rightarrow \cos 2\phi = 1 - \sin 2\theta$

Now consider, discriminant of the equation,

$$D = 4\cot^2\phi - 4 = \frac{4\cos 2\phi}{\sin^2\phi} = \frac{4(1 - \sin 2\theta)}{\sin^2\phi} > 0;$$

as $\sin 2\theta < 1$ and $\sin^2\phi > 0$ as

$\Rightarrow$ Roots are real.

28. (a) Let $\frac{a}{b}$ (in reduced form) be a rational root of $x^3 - px + 1 = 0$.

Then, $\frac{a^3}{b^3} - \frac{pa}{b} + 1 = 0$ or $a^3 - pab^2 + b^3 = 0$

$\Rightarrow a(pb^2 - a^2) = b^3 \Rightarrow \frac{a}{b^3} = \frac{1}{pb^2 - a^2}$

As $\gcd(a, b) = 1$, Therefore, $a = 1$. Thus,

$1 - pb^2 + b^3 = 0$

$\Rightarrow b^2(b - p) = 1 \Rightarrow b^2 = 1 \Rightarrow b = \pm 1$

Therefore, $\frac{a}{b} = \pm 1 \Rightarrow 1$ or -1 is a root of $x^3 - px + 1 \Rightarrow 1 - p + 1 = 0$ or $-1 + p + 1 = 0$

$\Rightarrow p = 2$ or $p = 0$.

But it is given that $p > 2$ $\therefore$ our assumption that $x^3 - px + 1$ has a rational root must be wrong.

29. (c) The given equation is $x^2 - px - p^2 = 0$ has roots $> p$

$\Rightarrow \alpha - p > 0, \beta - p > 0$ and $D \geq 0$

$\Rightarrow \alpha + \beta - 2p > 0, \alpha\beta - p(\alpha + \beta) + p^2 > 0$ and $p^2 + 4p^2 \geq 0$ and

$\therefore p - 2p > 0, -p^2 - p \cdot p + p^2 > 0 \Rightarrow p < 0$, $p^2 < 0$ (not possible for any real value)

30. (d) $f(x) = \frac{x^2 + 1 - 2}{x^2 + 1} = 1 - \frac{2}{x^2 + 1}$

$f(x)$ is minimum when $\frac{2}{x^2 + 1}$ is maximum or $(x^2 + 1)$ is minimum.

ie, $x^2 + 1 \geq 1$ for all x.

Hence, $\left(\frac{2}{x^2 + 1}\right)_{\max} = 2$ or $f(x) = 1 - \frac{2}{x^2 + 1}$ is minimum, when $\frac{2}{x^2 + 1}$ is maximum

or $f(x) = 1 - 2 = -1$.

31. (a) We have two cases

Case (i) If $x - a \geq 0, |x - a| = x - a$

$\therefore$ The equation becomes $x^2 - 2a(x - a) - 3a^2 = 0$

$\Rightarrow x^2 - 2ax - a^2 = 0$

$\Rightarrow x = \left(1 + \sqrt{2}\right)a, \left(1 - \sqrt{2}\right)a$

Since $x \geq a$ and $a \leq 0 \quad \therefore x = \left(1 - \sqrt{2}\right)a$

Case (ii) If $x - a < 0, |x - a| = -(x - a)$

$\therefore x^2 + 2a(x - a) - 3a^2 = 0$

$\Rightarrow x^2 + 2ax - 5a^2 = 0$

$\Rightarrow$ Roots are given by

$$x = -\left(1 + \sqrt{6}\right)a, \left(-1 + \sqrt{6}\right)a$$

since $x < a \leq 0$ so $x = \left(-1 + \sqrt{6}\right)a$

32. (a) Clearly, $x = 0$ does not satisfy $(2x^2 - 3x + 1)(2x^2 + 5x + 1) = 9x^2$. Therefore, we can write the above equation as follows

$$\left(2x + \frac{1}{x} - 3\right)\left(2x + \frac{1}{x} + 5\right) = 9 \quad \text{... (i)}$$

Put $2x + \frac{1}{x} = y$. So that the Eq. (i) becomes

$(y - 3)(y + 5) = 9$ or $y^2 + 2y - 24 = 0$

$\Rightarrow (y + 6)(y - 4) = 0 \Rightarrow y = -6, 4$. When $y = -6$, we get $2x + \frac{1}{x} = -6$

$\Rightarrow 2x^2 + 6x + 1 = 0$

$$\Rightarrow x = \frac{-6 \pm \sqrt{36 - 8}}{4} = \frac{-3 \pm \sqrt{7}}{2}.$$

When $y = 4$, we get $2x + \frac{1}{x} = 4$

$$\Rightarrow 2x^2 - 4x + 1 = 0 \Rightarrow x = \frac{4 \pm \sqrt{16 - 8}}{4} = \frac{2 \pm \sqrt{2}}{2}$$

Thus, $x = \frac{-3 \pm \sqrt{7}}{4}$ and $\frac{2 \pm \sqrt{2}}{2}$

33. (b) Since, $ax^2 + bx + c = 0$ has non-real roots.

Thus, either $f(x) = ax^2 + bx + c$

$\Rightarrow f(x) > 0$ or $f(x) < 0$ for all x.

Since $a + c < b \Rightarrow f(-1) < 0$

$\therefore f(x) < 0$ for all or at $x = -2$; $4a - 2b + c < 0$

$\Rightarrow 4a + c < 2b$.

34. (a) The quadratic equation $3x^2 + 2(a^2 + 1)x + a^2 - 3a + 2 = 0$ will have two roots of opposite sign, if it has real roots and the product of the roots is negative, that is, if $D > 0$ and $\alpha\beta < 0$

$\Rightarrow 4(a^2 + 1)^2 - 12(a^2 - 3a + 2) > 0$ and $\frac{a^2 - 3a + 2}{3} < 0$. Both of these conditions are satisfied, if $a^2 - 3a + 2 < 0$ ie, if $(a - 1)(a - 2) < 0$ or $1 < a < 2$.

Hence there exist no integral values of a

35. (a) Here, $(a^2 - b^2)(c^2 - d^2) > 0$ (note that $ad \neq bc$) $y = \frac{adx^2 - (ac + bd)x + bc}{bcx^2 - (bd + ac)x + ad}$

$\Rightarrow ybcx^2 - (bd + ac)xy + y\,ad$

$= adx^2 - (ac + bd)x + bc$

$(ybc - ad)x^2 - [y(bd + ac) - (ac + bd)]x + (y\,ad - bc) = 0$

as $x \in R$ $[y(bd + ac) - (ac + bd)]^2 \geq 4(y\,bc - ad)(y\,ad - bc)$

$\Rightarrow (bd + ac)^2 \cdot y^2 + (ac + bd)^2 - 2(bd + ac)(ac + bd)y \geq 4[(abcd)y^2 - (b^2c^2 + a^2d^2)y + abcd]$

$\Rightarrow (ac - bd)^2 \cdot y^2 - 2[(ac + bd)^2 - 2(b^2c^2 + a^2d^2)]y + (ac - bd)^2 \geq 0$

$\Rightarrow (ac - bd)^2 \cdot y^2 - 2[(ac - bd)^2 - 2(bc - ad)^2]y + (ac - bd)^2 \geq 0$

$\Rightarrow 4[(ac - bd)^2 - 2(bc - ad)^2]^2 - 4[(ac - bd)^2]^2 \leq 0$

$\Rightarrow [(ac - bd)^2 - 2(bc - ad)^2] + (ac - bd)^2]\,[(ac - bd)^2 - 2(bc - ad)^2 - (ac - bd)^2] \leq 0$

$\Rightarrow 2\{(ac - bd)^2 - (bc - ad)^2\} \cdot 2\{-(bc - ad)^2\} \leq 0$

$\Rightarrow \{(ac - bd)^2 - (bc - ad)^2\} > 0$

$\Rightarrow \{(ac - bd)^2 - (bc - ad)^2\} > 0$

$\Rightarrow \{(ac - bd) - (bc - ad)\}\{(ac - bd) + (bc - ad)\} > 0$

$\{(a - b)c + (a - b)d\}$

$\{(a + b)c - (a + b)d\} > 0$

$\Rightarrow (a - b)(c + d)(a + b)(c - d) > 0$

$\Rightarrow (a^2 - b^2)(c^2 - d^2) > 0$

36. (b) $xy + y^2 + x = \frac{y}{2}, x^2 + xy = -\frac{y}{2}$

Adding, $(x + y)^2 + x = 0$, and subtracting $y^2 - x^2 = y - x$

$\Rightarrow y - x = 0$, $x + y = 1$ and $(x + y)^2 + x = 0$, $y \neq 0$

Solving these equations, $x = -1$, $y = 2$ and

$x = -\frac{1}{4} = y$

37. (b) The equation $ax^2 + bx + c = 0$ must have the opposite sign $\Rightarrow$ a and c are of opposite sign.

38. (a) $P(x) = 0$ has real roots, if $b^2 - 4ac \geq 0$ and $Q(x) = 0$ has real roots, if $d^2 + 4ac \geq 0$.

Since, $ac \neq 0$. If $ac < 0$, $b^2 - 4ac \geq 0$ and hence, $P(x) = 0$ has real roots or if $ac > 0$ then $d^2 + 4ac \geq 0$ and $Q(x) = 0$ has real roots. Hence, at least two roots of $P(x)\, Q(x) = 0$ are real.

39. (c) Let $P(x) = x^2 - 4x + 3$ and $Q(x) = x^2 + x + 1$. Consider $Q(x)$ the coefficient of x^2 is positive and $D < 0 \Rightarrow Q(x)$ is positive for all. Hence,

$$\frac{p(x)}{q(x)} \leq 0 \Rightarrow p(x) \leq 0$$

$\Rightarrow x^2 - 4x + 3 \leq 0 \Rightarrow (x-1)(x-3) \leq 0$

$\Rightarrow 1 \leq x \leq 3$

Values of x is $\{1, 2, 3\}$

40. (d) We have $f(x) = (x + b)^2 + 2c^2 - b^2$

$\Rightarrow \min f(x) = 2c^2 - b^2$ and $g(x) = b^2 + c^2 - (x + c)^2$

$\Rightarrow \max g(x) = b^2 + c^2$.

Thus, $\min f(x) > \max g(x)$

$\Rightarrow 2c^2 - b^2 > b + c^2 \Rightarrow c^2 > 2b^2$

$\Rightarrow |c| > \sqrt{2}|b|$

41. (a) Let $f(x) = ax^2 - bx + c$

$f(0) = c > 0$ and $f(2) = 4a - 2b + c < 0$ so that $f(x) = 0$ has a root in the interval $(0, 2)$.

42. (a) $x^2 - 6x + 5 \leq 0$ and $x^2 - 2x > 0$

$\Rightarrow 1 \leq x \leq 5$ and $x < 0$ or $x > 2$

$\Rightarrow 2 < x \leq 5 \Rightarrow x \in \{3, 4, 5\}$

43. (b) Let $\dfrac{x^2 - x + 1}{x^2 + x + 1} = k$

$\Rightarrow x^2 - x + 1 - k(x^2 + x + 1) = 0$

$(1 - k)x^2 - (1 + k)x + (1 - k) = 0$

Discriminant ≥ 0

$\Rightarrow (1 + k)^2 - 4(1 - k)^2 \geq 0$

$\Rightarrow (k-3)\left(k - \dfrac{1}{3}\right) \leq 0$

$\Rightarrow k \in \left[\dfrac{1}{3}, 3\right]$

44. (c) Put $|x|^{3/5} = y$, then

$y^2 - 26y - 27 = 0$

$\Rightarrow (y + 1)(y - 27) = 0 \Rightarrow y = -1$ or 27

As, $y = |x|^{3/5} \geq 0 \quad \forall\, x$, we get $y = 27$

$\Rightarrow |x|^{3/5} = 3^3 \Rightarrow |x| = 3^5 \Rightarrow x = \pm 3^5$

So the equation has 2 real roots

45. (c) Put $x^2 + x = y$, then the equation becomes

$(y - 2)\,(y - 3) = 12$

$\Rightarrow y^2 - 5y - 6 = 0$

$\Rightarrow (y - 6)(y + 1) = 0 \Rightarrow y = 6, -1$.

When $y = 6$, we get $x^2 + x - 6 = 0$

$\Rightarrow (x + 3)(x - 2) = 0$ or $x = -3, 2$

When $y = -1$, we get $x^2 + x + 1 = 0$ and it has no real roots

46. (b) The given equation can be re-written as $(x - 3a)^2 = 2(a - 1)$

So $a \geq 1$ and

Let $x = 3a \pm \sqrt{2a - 2}$

As per the given condition both the roots will exceed 3

$3a - \sqrt{2a - 2} > 3$

$\Rightarrow 3(a - 1) > \sqrt{2}\sqrt{a - 1}$

$\therefore\ a > 1$ and $\sqrt{a-1} > \sqrt{2}/3$

$\Rightarrow a > 1 + \dfrac{2}{9} = \dfrac{11}{9}$

Which is necessary and sufficient condition to have both the roots greater than 3.

IIT JEE mains/AIEEE corner

47. (a) Given that $\alpha^2 = 5\alpha - 3$ and $\beta^2 = 5\beta - 3$;

$\Rightarrow \alpha$ & β are roots of equation, $x^2 = 5x - 3$ or $x^2 - 5x + 3 = 0$

Thus, the equation having α/β and β/α as its roots is

$$x^2 - x\left(\frac{\alpha}{\beta}+\frac{\beta}{\alpha}\right)+\frac{\alpha\beta}{\alpha\beta}=0$$

$$\Rightarrow x^2 - x\left(\frac{\alpha^2+\beta^2}{\alpha\beta}\right)+1=0$$

or $3x^2 - 19x + 3 = 0$

48. (a) Let us assume that α, β and γ, δ be the roots of the equations
$x^2 + ax + b = 0$ and $x^2 + bx + a = 0$ respectively.
$\Rightarrow \alpha + \beta = -a, \alpha\beta = b$ and $\gamma + \delta = -b, \gamma\delta = a$
Given $|\alpha - \beta| = |\gamma - \delta| \Rightarrow (\alpha - \beta)^2 = (\gamma - \delta)^2$
$\Rightarrow (\alpha + \beta)^2 - 4\alpha\beta = (\gamma + \delta)^2 - 4\gamma\delta$
$\Rightarrow a^2 - 4b = b^2 - 4a \Rightarrow (a^2 - b^2) + 4(a - b) = 0$
$\Rightarrow a + b + 4 = 0 \quad (\because a \neq b)$.

49. (a) Product of real roots $= \dfrac{9}{t^2} > 0, \ \forall t \in R$
$\therefore$ product of real roots is always positive.

50. (a) Since p and q are the roots of the equation $x^2 + px + q = 0$, so
$p + q = -p$ and $pq = q \Rightarrow q(p-1) = 0$
$\Rightarrow q = 0$ or $p = 1$.
If $q = 0$, then $p = 0$. i.e., $p = q$
$\Rightarrow p = 1$ and $q = -2$.

51. (a) Since $(a - b)^2 + (b - c)^2 + (c - a)^2 > 0$
$\Rightarrow 2(a^2 + b^2 c^2 - ab - bc - ca) > 0$
$\Rightarrow 2 > 2(ab + bc + ca) \Rightarrow ab + bc + ca < 1$

52. (d) $ax^2 + bx + c = 0$, $\alpha + \beta = \dfrac{-b}{a}, \alpha\beta = \dfrac{c}{a}$

as for given condition, $\alpha + \beta = \dfrac{1}{\alpha^2} + \dfrac{1}{\beta^2}$

$$\alpha + \beta = \frac{\alpha^2+\beta^2}{\alpha^2\beta^2} - \frac{b}{a} = \frac{\frac{b^2}{a^2} - \frac{2c}{a}}{\frac{c^2}{a^2}}$$

on simplification $2a^2 c = ab^2 + bc^2$

$$\Rightarrow \frac{2a}{b} = \frac{c}{a} + \frac{b}{c} \Rightarrow \frac{c}{a}, \frac{a}{b}, \frac{b}{c}, \text{ are in A. P.}$$

$$\therefore \frac{a}{c}, \frac{b}{a}, \& \frac{c}{a} \text{ are in H. P.}$$

53. (b) Let the roots of given equation be α and 2α then

Sum of the roots $\alpha + 2\alpha = 3\alpha = \dfrac{1-3a}{a^2-5a+3}$

Product of the roots, $\alpha . 2\alpha = 2\alpha^2 = \dfrac{2}{a^2-5a+3}$

$$\therefore 2\left[\frac{1}{9}\frac{(1-3a)^2}{(a^2-5a+3)^2}\right] = \frac{2}{a^2-5a+3}$$

$$\frac{(1-3a)^2}{(a^2-5a+3)} = 9 \text{ or } 9a^2 - 6a + 1$$
$$= 9a^2 - 45a + 27$$

Or $39a = 26$ or $a = \dfrac{2}{3}$

54. (c) $x^2 - 3|x| + 2 = 0 \Rightarrow |x|^2 - 3|x| + 2 = 0$
$(|x| - 2)(|x| - 1) = 0$
$|x| = 1, 2$ or $x = \pm 1, \pm 2$
$\therefore$ No. Of solution = 4

55. (c) As per the given condition $y = x + \dfrac{1}{x}$
$= \left(\sqrt{x} - \dfrac{1}{\sqrt{x}}\right)^2 + 2$

Which has minimum value 2 at $x = 1$

56. (b) Let two numbers be a and b then from the given condition $\dfrac{a+b}{9} = 9$ and $\sqrt{ab} = 4$
$\therefore$ Equation with roots a and b is $x^2 - (a + b)x + ab = 0 \Rightarrow x^2 - 18x + 16 = 0$

57. (c) Let the second root be α.
Then $\alpha + (1 - p) = -p \Rightarrow \alpha = -1$
Also $\alpha(1 - p) = 1 - p$
$\Rightarrow (\alpha - 1)(1 - p) = 0 \Rightarrow p = 1 \quad [\because \alpha = -1]$
$\Rightarrow$ Roots are $\alpha = -1$ and $1 - p = 0$

58. (d) Since 4 is a root of $x^2 + px + 12 = 0$
$\Rightarrow 16 + 4p + 12 = 0 \Rightarrow p = -7$
Now, the equation $x^2 + px + q = 0$ has equal roots.
$\therefore p^2 - 4q = 0 \Rightarrow q = \dfrac{p^2}{4} = \dfrac{49}{4}$.

59. (b) Given that $\tan\left(\frac{P}{2}\right)$, $\tan\left(\frac{Q}{2}\right)$ are the roots of $a x^2 + b x + c = 0$

$$\tan\left(\frac{P}{2}\right) + \tan\left(\frac{Q}{2}\right) = -\frac{b}{a},$$

$$\tan\left(\frac{P}{2}\right) . \tan\left(\frac{Q}{2}\right) = \frac{c}{a}$$

$$\frac{\tan\left(\frac{P}{2}\right) + \tan\left(\frac{Q}{2}\right)}{1 - \tan\left(\frac{P}{2}\right)\tan\left(\frac{Q}{2}\right)} = \tan\left(\frac{P}{2} + \frac{Q}{2}\right) = 1$$

$$\Rightarrow \frac{-\frac{b}{a}}{1 - \frac{c}{a}} = 1 \Rightarrow -\frac{b}{a}, \frac{a}{c} = \frac{c}{a}$$

$\Rightarrow - b = a - c$ or $c = a + b$.

60. (c) Given that both roots are less than 5

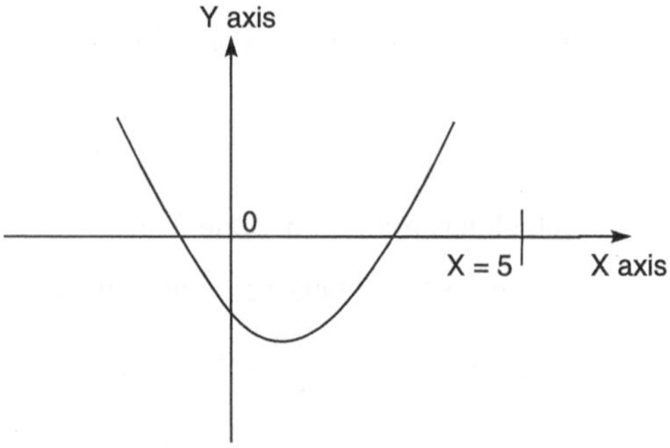

Then (i) Discriminant ≥ 0

(ii) $p(5) > 0$

(iii) $\frac{\text{sum of roots}}{2} < 5$

From (i) $4 k^2 - 4 (k^2 + k - 5) \geq 0$

$4 k^2 - 4 k^2 - 4 k + 20 \geq 0$

$4 k \leq 20 \Rightarrow k \leq 5$

From (ii) $f(5) > 0$; $25 - 10 k + k^2 - 5 > 0$

Or $k^2 - 9 k + 20 > 0$

Or $k (k - 4) - 59 k - 4) > 0$

Or $(k - 5) (k - 4) > 0$

$\Rightarrow k \in (-\infty, 4) \cup (-\infty, 5)$

From (ii) $\frac{\text{sum of roots}}{2} = -\frac{b}{2a} = \frac{2k}{2} < 5$

From the information (i), (ii) & (iii) we get $k \in (-\infty, 4)$.

61. (b) Given equation is $X^2 + p x = q = 0$

Sum of roots = $\tan 30° + \tan 15° = -p$

Product of roots = $\tan 30° . \tan 15° = q$

$$\tan 45° = \frac{\tan 30° + \tan 15°}{1 - \tan 30° . \tan 15°} = \frac{-p}{1-q} = 1$$

$\Rightarrow -p = 1 - q \Rightarrow q - p = 1 \quad \therefore 2 + q - p = 3.$

62. (c) Equation $x^2 - 2 m x + m^2 - 1 = 0$

$(x - m)^2 - 1 = 0$ or $(x - m + 1)(x - m - 1) = 0$

$x = m - 1, m + 1$

$m - 1 > -2$ and $m + 1 < 4$

$\Rightarrow m > -1$ and $m < 3$ or, $-1 < m < 3$.

63. (b) Let $Y = \frac{3x^2 + 9x + 17}{3x^2 + 9x + 7}$

$3 x^2 (y - 1) + 9 x (y - 1) + 7 y - 17 = 0$

$D \geq 0$ $\quad$ x is real

$81 (y - 1) 2 - 4 \times 3 (y - 1)(7 y - 17) \geq 0$

$\Rightarrow (y - 1)(y - 41) \leq 0 \quad 1 \leq y \leq 41$

$\therefore$ Max value of y is 41

ALTERNATE SOLUTION:

Consider $f(x) = \frac{3x^2 + 9x + 17}{3x^2 + 9x + 7}$

$\Rightarrow f(x) = 1 + \frac{10}{3x^2 + 9x + 7}$

Clearly $f(x)$ is maximum when

$g(x) = 3 x^2 + 9 x + 7$ is min.

Here $g(x) = 3\left(x^2 + 3x + \frac{9}{4}\right) + 7 - \frac{27}{4}$

$$= 3\left(x + \frac{3}{2}\right)^2 + \frac{1}{4}$$

Which is minimum when $x = \frac{-3}{2}$

$$\Rightarrow \quad f_{max} = 1 + \frac{10}{3\times\frac{9}{4} - 9\times\frac{3}{2} + 7}$$

$$= 1 + \frac{10\times 4}{27 - 54 + 28} = 41.$$

64. (c) Let α and β are roots of the equation

$x^2 + ax + 1 = 0$

$\alpha + \beta = -a$ and $\alpha\beta = 1$

given $|\alpha - \beta| < \sqrt{5} \Rightarrow \sqrt{(\alpha+\beta)^2 - 4\alpha\beta} < \sqrt{5}$

$(\because (\alpha - \beta)^2 = (\alpha + \beta)^2 - 4\alpha\beta)$

$\Rightarrow \sqrt{a^2 - 4} < \sqrt{5} \Rightarrow a^2 - 4 < 5$

$\Rightarrow a^2 - 9 < 0 \Rightarrow a^2 < 9 \Rightarrow -3 < a < 3$

$\Rightarrow a \in (-3, 3)$.

65. (d) Let the roots of equation $x^2 - 6x = a = 0$ be α and 4 β and that of the equation $x^2 - cx + 6 = 0$ be α and 3 β. Then

$\alpha + 4\beta = 6; \quad 4\alpha\beta = a$

and $\alpha + 3\beta = c; \quad 3\alpha\beta = 6$

$\Rightarrow a = 8$

$\Rightarrow$ The equation becomes $x^2 - 6x + 8 = 0$

$\Rightarrow (x - 2)(x - 4) = 0$

$\Rightarrow$ roots are 2 and 4

$\Rightarrow \alpha = 2, \beta = 1 \quad \therefore$ Common roots is 2.

66. (b) Given that roots of the equation

$bx^2 + cx + a = 0$ are imaginary

$\therefore c^2 - 4ab < 0 \quad \ldots$ (i)

Let $y = 3b^2x^2 + 6bcx + 2c^2$

$\Rightarrow 3b^2x^2 + 6bcx + 2c^2 - y = 0$

As x is real, $D \geq 0$

$\Rightarrow 36b^2c^2 - 12b^2(2c^2 - y) \geq 0$

$\Rightarrow 12b^2(3c^2 - 2c^2 + y) \geq 0$

$\Rightarrow c^2 + y \geq 0 \Rightarrow y \geq -c^2$

But from eqn. (i), $c^2 < 4ab$ or $-c^2 > -4ab$

$\Rightarrow$ we get $y \geq -c^2 > -4ab$

$\Rightarrow y > -4ab$

67. (a) Let the correct equation be $ax^2 + bx + c = 0$

Now sachin's equation $\Rightarrow ax^2 + bx + c' = 0$

Roots found by Sachin's are 4 and 3

Rahul's equation, $ax^2 + b'x + c = 0$

Roots found by Rahul 's are 3 and 2

$$-\frac{b}{a} = 7 \quad \ldots \text{(i)}$$

$$\frac{c}{a} = 6 \quad \ldots \text{(ii)}$$

From (i) and (ii), roots of the correct equation $x^2 - 7x + 6 = 0$ are 6 and 1.

68. (d) If P (x) = 0

$\Rightarrow p(x) = ax^2 + bx + c - (a_1x^2 + b_1x + c_1)$

$\Rightarrow (a - a_1)x^2 + (b - b_1)x + (c - c_1) = 0$.

It has only one solution, x = – 1

$\Rightarrow b - b_1 = a - a_1 + c - c_1 \quad \ldots (1)$

Vertex = (– 1, 0)

$$\Rightarrow \frac{b - b_1}{2(a - a_1)} = -1$$

$\Rightarrow b - b_1 = 2(a - a_1) \quad \ldots (2)$

Now P (– 2) = 2

$\Rightarrow f(-2) - g(-2) = 2$

$\Rightarrow 4a - 2b + c - 4a_1 + 2b_1 - c_1 = 2$

$\Rightarrow 4(a - a_1) - 2(b - b_1) + (c - c1) = 2 \quad \ldots (3)$

From equations, (1), (2) and (3)

$$a - a_1 = c - c_1 = \frac{1}{2}(b - b_1) = 2$$

Now , $P(2) = f(2) = g(2)$

$= 4(a - a_1) + 2(b - b_1) + 9(c - c_1)$

$= 8 + 8 + 2 = 18$

69. (b) Given equation is $e^{\sin x} - e^{-\sin x} - 4 = 0$

put $e^{\sin x} = t$ in the given equation, we get

$$t^2 - 4t - 1 = 0$$

$$\Rightarrow t = \frac{4 \pm \sqrt{16+4}}{2} = \frac{4 \pm \sqrt{20}}{2} = \frac{4 \pm 2\sqrt{5}}{2}$$

$= 2 \pm \sqrt{5}$

$\Rightarrow e^{\sin x} = 2 \pm \sqrt{5}$ $(\because t = e^{\sin x})$

$\Rightarrow e^{\sin x} = 2 - \sqrt{5}$ and $e^{\sin x} = 2 + \sqrt{5} > e$, so this option is rejected

$\Rightarrow e^{\sin x} = 2 - \sqrt{5} < 0$ and $\sin x = \ln(2 + \sqrt{5}) > 1$, so this option is also rejected

$\therefore$ The equation has no real roots.

70. (d) $f(x) = 2x^3 + 3x + k$

$f'(x) = 6x^2 + 3 > 0 \quad \forall x \in R\ (\because x^2 > 0)$

$\Rightarrow f(x)$ is strictly increasing function

$\Rightarrow f(x) = 0$ has only one real root, so two roots are not possible.

71. (a) Given equations are

$x^2 + 2x + 3 = 0$ . . . (i)

$ax^2 + bx + c = 0$ . . . (ii)

Roots of equation (i) are imaginary roots. According to the question (ii) will also have both roots same as (i). Thus

$\frac{a}{1} = \frac{b}{2} = \frac{c}{3} = \lambda$ (say) $\Rightarrow a = \lambda, b = 2\lambda, c = 3\lambda$.

Hence, required ratio is 1 : 2 : 3.

72. (b) Statement 2 is $\sqrt{n(n+1)} < n+1$, $n \geq 2$

$\Rightarrow \sqrt{n} < \sqrt{n+1}$, $n \geq 2$ which is true

$\Rightarrow \sqrt{2} < \sqrt{3} < \sqrt{4} < \sqrt{5} < ----\sqrt{n}$

Now $\sqrt{2} < \sqrt{n} \Rightarrow \frac{1}{\sqrt{2}} > \frac{1}{\sqrt{n}}$

$\sqrt{3} < \sqrt{n} \Rightarrow \frac{1}{\sqrt{3}} > \frac{1}{\sqrt{n}}$;

$\sqrt{n} \leq \sqrt{n} \Rightarrow \frac{1}{\sqrt{n}} \geq \frac{1}{\sqrt{n}}$

Also $\frac{1}{\sqrt{1}} > \frac{1}{\sqrt{n}}$ $\quad \therefore$ Adding all, we get

$$\frac{1}{\sqrt{1}} + \frac{1}{\sqrt{2}} + \frac{1}{\sqrt{3}} + \ldots\ldots + \frac{1}{n} > \frac{n}{\sqrt{n}} = \sqrt{n}$$

Hence both the statements are correct and statement 2 is a correct explanation of statement-1.

73. (c)Consider $-3(x - [x])^2 + 2[x - [x]) + a^2 = 0$

$\Rightarrow 3\{x\}^2 - 2\{x\} - a^2 = 0 \quad (\because x - [x] = \{x\})$

$\Rightarrow 3\left(\{x\}^2 - \frac{2}{3}\{x\}\right) = a^2, a \neq 0$

$\Rightarrow a^2 = 3\{x\}\left(\{x\} - \frac{2}{3}\right)$

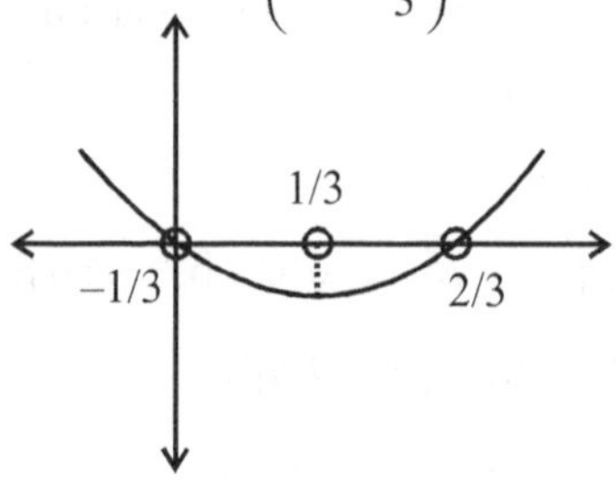

Now, $\{\ \} \in (0,1)$ and $\frac{-2}{3} \leq a^2 < 1$ (by graph)

Since , x is not an integer

$\therefore \ a \in (-1,1) - \{0\} \Rightarrow a \in (-1,0) \cup (0,1)$

74.(b) Let p, q, r are in AP

$\Rightarrow 2q = p + r$ ...(i)

Given $\frac{1}{\alpha} + \frac{1}{\beta} = 4 \Rightarrow \frac{\alpha+\beta}{\alpha\beta} = 4$

We have $\alpha + \beta = -q/p$ and $\alpha\beta = \frac{r}{p}$

$\Rightarrow \frac{-\frac{q}{p}}{\frac{r}{p}} = 4 \Rightarrow q = -4r$ ....(ii)

From (i), we have

$2(-4r) = p + r \Rightarrow p = -9r$

$q = -4r$

Now $|\alpha - \beta| = \sqrt{(\alpha+\beta)^2 - 4\alpha\beta}$

$$= \sqrt{\left(\frac{-q}{p}\right)^2 - \frac{4r}{p}} = \frac{\sqrt{q^2 - 4pr}}{|p|}$$

$$= \frac{\sqrt{16r^2 + 36r^2}}{|-9r|} = \frac{2\sqrt{13}}{9}$$

75.(a) $\alpha, \beta = \frac{6 \pm \sqrt{36+8}}{2} = 3 \pm \sqrt{11}$

$\alpha = 3+\sqrt{11}, \beta = 3-\sqrt{11}$

$\therefore a_n = \left(3+\sqrt{11}\right)^n - \left(3-\sqrt{11}\right)^n$

$\frac{a_{10} - 2a_8}{2a_9}$

$$= \frac{\left(3+\sqrt{11}\right)^{10} - \left(3-\sqrt{11}\right)^{10} - 2\left(3+\sqrt{11}\right)^8 + 2\left(3-\sqrt{11}\right)^8}{2\left[\left(3+\sqrt{11}\right)^9 - \left(3-\sqrt{11}\right)^9\right]}$$

$$= \frac{\left(3+\sqrt{11}\right)^8\left[\left(3+\sqrt{11}\right)^2 - 2\right] + \left(3-\sqrt{11}\right)^8\left[2 - \left(3-\sqrt{11}\right)^2\right]}{2\left[\left(3+\sqrt{11}\right)^9 - \left(3-\sqrt{11}\right)^9\right]}$$

$$= \frac{\left(3+\sqrt{11}\right)^8\left(9+11+6\sqrt{11}-2\right) + \left(3-\sqrt{11}\right)^8\left(2-9-11+6\sqrt{11}\right)}{2\left[\left(3+\sqrt{11}\right)^9 - \left(3-\sqrt{11}\right)^9\right]}$$

$$= \frac{6\left(3+\sqrt{11}\right)^9 - 6\left(3-\sqrt{11}\right)^9}{2\left[\left(3+\sqrt{11}\right)^9 - \left(3-\sqrt{11}\right)^9\right]} = \frac{6}{2} = 3$$

76.(c) $(x^2 - 5x + 5)^{x^2+4x-60} = 1$

Case I

$x^2 - 5x + 5 = 1$ and $x^2 + 4x - 60$ can be any real number

$\Rightarrow x = 1, 4$

Case II

$x^2 - 5x + 5 = -1$ and $x^2 + 4x - 60$ has to be an even number

$\Rightarrow x = 2, 3$

where 3 is rejected because for $x = 3$, $x^2 + 4x - 60$ is odd.

Case III

$x^2 - 5x + 5$ can be any real number and $x^2 + 4x - 60 = 0$

$\Rightarrow x = -10, 6$

$\Rightarrow$ Sum of all values of $x = -10 + 6 + 2 + 1 + 4 = 3$

(Board Level Questions)

Part 1

1. For the equation to be defined, we have

   $2x + 1 \geq 0$ and $2x - 1 \geq 0 \Rightarrow 2x \geq -1$ and $2x \geq 1 \Rightarrow x \geq 1/2$.

   We rewrite the equation as

   $\sqrt{2x+1} = 1 + \sqrt{2x-1}$

   $\Rightarrow 1 = 2\sqrt{2x-1} \Rightarrow 2x - 1 = 1/4 \Rightarrow x = 5/8$.

   Since $x = 5/8$ satisfies the given equation, the given equation has just one solution which is positive, so P = 1 and Q = 0

   So $|P - Q| = 1$

2. Discriminant of the given equation is

   $D = 9(I - m)^2 + 8(I + m)^2$

   As $I + m \neq 0$, $(I + m)^2 > 0$. Also, $(I - m)^2 \geq 0$.

   Thus, $D > 0$.

   Hence, roots of the given equation are real and unequal.

3. Let α and β be two roots of $x^2 - px + q = 0$ and assume $\alpha < \beta$. we have $\alpha + \beta = p$, $\alpha\beta = q$. As q is prime, we must have $\alpha = 1$, $\beta = q$.
   Thus, $p = 1 + q \Rightarrow p - q = 1$.
   As p and q are prime, this is possible if $p = 3$, $q = 2$.

4. The Discriminant of the quadratic equation $2x^2 + 6x + b = 0$ is given by
   $D = 36 - 8b > 0$. Therefore the given equation has real roots.

   Now consider $\frac{x_1}{x_2} + \frac{x_2}{x_1} = \frac{x_1^2 + x_2^2}{x_1 x_2}$

   $$= \frac{(x_1 + x_2)^2 - 2x_1x_2}{x_1x_2}$$

   $$= \frac{(-3)^2 - 2(b/2)}{b/2} = \frac{18}{b} - 2 < -2$$

   So required maximum value is –2

5. Let if $f(x) = ax^2 + bx + c$. Since $f(x)$ has no real zeros, either $f(x) > 0$, or $f(x) < 0$ for all $x \in R$. therefore, $a < 0$ as the parabola $y = f(x)$ must open downwards.

6. Consider $D = b^2 - 4ac$. If $D \geq 0$, Then the roots of the equation are given by $\frac{-b \pm \sqrt{D}}{2a}$

Since $D = b^2 - 4ac < b^2$ ($\because a > 0, c > 0$).

It follows that the roots of the quadratic equation are negative.

In case $D < 0$. Then the roots of the equation are given by

$= \frac{-b \pm \sqrt{D}}{2a}$ Which have negative real parts.

7. Put $\sqrt{x-1} = t$ or $x = t^2 + 1$ so that the given equation becomes

$\sqrt{t^2+4-4t} + \sqrt{t^2+9-6t} = 1$ or

$\sqrt{(t-2)^2} + \sqrt{(t-3)^2} = 1$

Or $|t-2| + |t-3| = 1$

$2 \leq t \leq 3$ or $5 \leq x \leq 10$

So number of integral roots is 6

8. Since $p, q, r > 0$

And when $x < 0$ then

$x^3 - px^2 + qx - r < 0$ $[\because p, q, r > 0]$

So equation $x^3 - px^2 + qx - r = 0$ cannot have negative roots.

Part 2

9. If an equation is satisfied by all the values of x then it is an identity

$a(1 - 2\sin^2 x) + b\sin^2 x + c = 0$, ie,

$(b - 2a)\sin^2 x + (a + c) = 0$.

It is an identity, if $b - 2a = 0$, $a + c = 0$. So, $\frac{a}{1} = \frac{b}{2} = \frac{c}{-1}$. thus, number of triplet (a, b, c) are infinite.

10. Let α & β are the roots of equation $x^2 + 7x - 1 = 0$ then $\alpha + \beta = -7$ and $\alpha\beta = -1$

$\therefore$ Quadratic whose roots are AM and HM of the roots of the equation $x^2 + 7x - 1 = 0$.

$\Rightarrow \alpha' = \frac{\alpha+\beta}{2}$ and $\beta' = \frac{2\alpha\beta}{\alpha+\beta}$ or $\alpha' = -\frac{7}{2}$

and $\beta' = \frac{-2}{-7} = \frac{2}{7}$

$\therefore$ Required equation is, $x^2 - (\alpha' + \beta')x + \alpha'\beta' = 0$

$x^2 - \left(-\frac{45}{14}x\right) - 1 = 0$

or $14x^2 + 45x - 14 = 0$

11. From the given condition

$$\alpha + \beta = \frac{1}{\alpha^2} + \frac{1}{\beta^2} = \frac{\alpha^2+\beta^2}{\alpha^2\beta^2} = \frac{(\alpha+\beta)^2 - 2\alpha\beta}{\alpha^2\beta^2}$$

$$\Rightarrow -\frac{b}{a} = \frac{b^2-2ac}{c^2} \Rightarrow \frac{2a}{c} = \frac{b^2}{c^2} + \frac{b}{a} = \frac{ab^2+bc^2}{ac^2}$$

$$\Rightarrow 2a^2c = ab^2 + bc^2 \Rightarrow \frac{2a}{b} = \frac{b}{c} + \frac{c}{a} \Rightarrow \frac{c}{a}, \frac{a}{b}, \frac{b}{c}$$

are in AP. $\Rightarrow \frac{a}{c}, \frac{b}{a}, \frac{c}{b}$ are in H.P.

12. Let $2^t = x$ and $f(x) = x^2 - (a-4)x + \frac{9}{4}a$. As per the given condition $f(x) < 0, \forall x \in (2,4)$. one of the roots of $f(x)$ should be less than 2 and the other must be greater than 4. ie, $f(2) < 0$ and $f(4) < 0$, $a < -48$ and $a > 128/7$, which is not possible. hence, no such 'a' exists.

13. Since coefficient of x^2 is positive so for the required condition $D < 0$

$(4k-1)^2 - (15k^2 - 2k - 7) < 0$

$\Rightarrow k^2 - 6k + 8 < 0 \Rightarrow (k-2)(k-4) < 0$

$\Rightarrow 2 < k < 4$

14. Since tan A and tan B are the roots of the quadratic equation $x^2 - px + q = 0$,

So, sum of the roots, $\tan A + \tan B = p$, and

Product of the roots, $\tan A \tan B = q$,

$$\therefore \tan(A+B) = \frac{\tan A + \tan B}{1 - \tan A \tan B} = \frac{p}{1-q}$$

Now, $\sin^2(A+B) = \frac{\tan^2(A+B)}{\sec^2(A+B)}$

$$= \frac{p^2/(1-q)^2}{1+p^2/(1-q)^2} = \frac{p^2}{(1-q)^2+p^2}$$

15. Discriminant of the equation is

$$(a-2)^2 + 4(a+1)$$
$$= a^2 - 4a + 4 + 4a + 4$$
$$= a^2 + 8 > 0 \text{ as } a \in R.$$

$\therefore$ Roots of the given equation are real.

Now $\alpha + \beta = a - 2,\ \alpha\beta = -(a+1)$.

We have

$$\alpha^2 + \beta^2 = (\alpha+\beta)^2 - 2\alpha\beta = (a-2)^2 + 2(a+1)$$
$$= a^2 - 4a + 4 + 2a + 2 = a^2 - 2a + 6$$
$$= (a-1)^2 + 5$$

Thus, the minimum value of $\alpha^2 + \beta^2$ is 5 when $a = 1$.

16. The quadratic $3x^2 + 2(a^2+1)x + a^2 - 3a + 2 = 0$ will have two roots of opposite signs if it has real roots and the product of the roots is negative, that is, if

$$4(a^2+1)^2 - 12(a^2 - 3a + 2) \geq 0$$

And $\dfrac{a^2 - 3a + 2}{3} < 0.$

Both of these conditions are met if

$a^2 - 3a + 2 < 0$ i. e. if $(a-1)(a-2) < 0$ or $1 < a < 2$.

17. Let $k = \dfrac{x^2 - x + 1}{x^2 + x + 1} \Rightarrow kx^2 + kx + k = x^2 - x + 1$

$\Rightarrow (k-1)x^2 + (k+1)x + k - 1 = 0$

As x is real, the Discriminant $D \geq 0$

$D = (k+1)^2 - 4(k-1)^2 \geq 0$

$\Rightarrow (3k-1)(-k+3) \geq 0$

$\Rightarrow (k - 1/3)(k-3) \leq 0 \Rightarrow 1/3 \leq k \leq 3$

So the required range is $\left[\dfrac{1}{3}, 3\right]$

18. Put $x - a = t$ then the given equation can be written as-

$$(a+t)^2 + 2a|t| + 3a^2 = 0$$

$\Rightarrow t^2 + 2a(t + |t|) + 4a^2 = 0$

If $t < 0$, then $t^2 + 4a^2 = 0$ which has no real roots

Now consider when $t \geq 0$.

$$t^2 + 4at + 4a^2 = 0$$

$\Rightarrow t = -2a$

$\Rightarrow x = -a =$ positive number

So for a specific value of a there exist only one root.

19. Put $x^4 = y$ then we can write the given equation as $y + 20 = (22 - y)^2$

$\Rightarrow y = 16,\ 29.$

But, $y = 29$ is not possible for integer value of x

and $x^4 = y = 16$

$\Rightarrow x = \pm 2$ as x is an integer.

Number of integral roots is 2

20. Let $x = 1/u$, and $\sqrt{1-x^2} = 1/v$

Then

$$u + v = 35/12,$$

and $u^2 + v^2 = u^2 v^2$

From this we get $uv = 25/12,\ -49/12$

When $uv = 25/12$, we get $x = 3/5, 4/5$

When $uv = -49/12$, we get $x = -(5+\sqrt{73})/14$.

So there are three roots.

21. If α, β, γ are the roots of $x^3 + px + q = 0$, then

$\alpha + \beta + \gamma = 0,\ \beta\gamma + \gamma\alpha + \alpha\beta = p,\ \alpha\beta\gamma = -q$

As, $q \neq 0$, none of α, β, γ is zero. Now, use

$2p = (\alpha+\beta+\gamma)^2 - (\alpha^2 + \beta^2 + \gamma^2) < 0$

Or $p < 0$ Proved

22. Consider the given equation

$x^2 - 2ax + a^2 - 1 = 0$, we can re-write it as

$(x-a)^2 = 1 \Rightarrow x = a \pm 1$

Given that roots lies between 2 and 8 so

Now, $2 < a \pm 1 < 8$

$\Rightarrow 1 < a < 7,\ 3 < a < 9$

$\therefore a \in (3, 7)$.

23. Clearly $a \neq 0, -1$.

Given that both the roots are more than 1 it is possible if and only if

(i) $D > 0$

$\Rightarrow 9a^2 - 4(4a)(1 + a) \geq 0$

$a(7a + 16) \leq 0$

$$a \varepsilon \left(-\infty, \frac{16}{7}\right] \cup [0, \infty)$$

(ii) $a \,.\, f(a) < 0$

$\Rightarrow (1 + a)[(1 + a) - 3a + 4a] > 0$

$\Rightarrow (1 + a)(2a - 1) > 0$

$$\Rightarrow (-\infty, -1) \cup \left(\frac{-1}{2}, \infty\right)$$

(iii) $\dfrac{3a}{2(1+a)} > 1$

$$\Rightarrow \frac{a-2}{1+a} > 0$$

$\Rightarrow a\varepsilon(-\infty, -1) \cup (2, \infty)$

Combining we get

$$-\frac{16}{7} \leq a < -1$$

24. Put $\sqrt{x^2 + 11}$ = t then given equation can be written as

$$\sqrt{t^2 + t - 11} + \sqrt{t^2 - t - 11} = 4 \quad .....(i)$$

Since $(t^2 + t - 11) - (t^2 - t - 11) = 2t$

From above two results we get

$$\sqrt{t^2 + t - 11} - \sqrt{t^2 - t - 11} = \frac{t}{2} \quad ....(ii)$$

Adding equation (i) and (ii), we get

$$2\left(\sqrt{t^2 + t - 11}\right) = 4 + \frac{t}{2}$$

$$\Rightarrow t^2 + t - 11 = 4 + t + \frac{t^2}{16}$$

Or $t^2 = 5$

$\Rightarrow x = \pm\sqrt{5}$

So number of roots is 2.

25. Given equation is $ax^3 + bx^2 + cx + d = 0$

Assume that it has all rational roots

Putting $x = \dfrac{y}{a}$ in the equation we get

$y^3 + by^2 + acy + a^2d = 0$

Here coefficient of y^3 is 1 so it must have all integral roots (It is known that it has rational roots)

If α, β, γ are roots of this equation then α, β, γ are divisors of a^2d and as such must be odd integers. Now,

$\alpha + \beta + \gamma = -b \Leftrightarrow b$ is odd

And $ac = \beta\gamma + \gamma\alpha + \alpha\beta \Leftrightarrow c$ is odd

$\Rightarrow bc$ is odd.

A contradiction

Hence proved that equation $ax^3 + bx^2 + cx + d = 0$ has at least one irrational root

26. Put $x^2 = y$. The given equation becomes $f(x) = ay^2 + by + c = 0$

The given equation will have four real roots if equation $ay^2 + by + c = 0$ has two non-negative roots. This can happen if

$$\frac{-b}{a} \geq 0,$$

$b^2 - 4ac \geq 0$

$\Rightarrow -ab \geq 0, ac \geq 0$ [$\because b^2 - 4ac \geq 0$ is given]

Thus, a and b must have opposite sign whereas a, – b and c must have the same sign.

So the required condition is

$a > 0, b < 0, c > 0$ or

$a < 0, b > 0, c < 0$.

27. We have $\alpha + \beta = \dfrac{-2b}{a}$

And $(\alpha + h) + (\beta + h) = \dfrac{-2B}{A}$

$$\Rightarrow 2h = -\frac{2B}{A} + \frac{2b}{a} \Rightarrow h = \frac{b}{a} - \frac{B}{A}$$

Also,

$(\alpha - \beta)^2 = (\alpha + \beta)^2 - 4\alpha\beta$

$$= \frac{4b^2}{a^2} - \frac{4c}{a} = \frac{4(b^2 - ac)}{a^2}$$

And $(\alpha - \beta)^2 = [(\alpha + h) - (\beta + h)]^2$

$$= \frac{4(B^2 - AC)}{A^2}$$

Thus, $\frac{4(b^2 - ac)}{a^2} = \frac{4(B^2 - AC)}{A^2}$

$\Rightarrow \frac{b^2 - ac}{B^2 - Ac} = \frac{a^2}{A^2}$

28. Put $4 - x = t$, so that (1) becomes

$(t - 1)^5 - (t + 1)^5 = -32$

$\Rightarrow 10t^4 + 20t^2 - 30 = 0$

$(t^2 + 3)(t^2 - 1) = 0$

When x is real, $t^2 + 3 > 0$

$\therefore t = \pm 1 \Rightarrow x = 3, 5.$

Hence number of real roots is 2

29. Let p (x) be one such polynomial, Since it has integral coefficient so both p(I) must be an integer here I is an integer.

Given that p(3) =1 that means p(x) –1 has a factor (x – 3), hence

P (x) – 1 = (x – 3) f (x) here f (x) is another polynomial with integral coefficient.

$\Rightarrow p(1) - 1 = -2f(1)$

$\Rightarrow 1 = -2f(1)$

Or $f(1) = -1/2$ which is not possible

Hence there exist no polynomial

30. Let $\frac{p}{q}$ where p, q ∈ I. here p and q are co-prime to each other i.e h c*f* (p, q) = 1 be a root of $x^2 + ax + b = 0$.

Then $\frac{p^2}{q^2} + \frac{ap}{q} + b = 0$

$\Rightarrow p^2 + apq - bq^2 = 0$

$\Rightarrow p^2 = -q(ap + bq)$

$\Rightarrow q \mid p^2 \Rightarrow q \mid p$ [$\therefore$ h c*f* (p, q) = 1]

As $q \mid p$ and h c*f* (p, q) = 1 so q = 1

Thus, $\frac{p}{q} = p$ is an integer

Hence proved that roots are integer.

31. The two roots should be compared with a, it is natural to use the substituting $y = x - a$. Under the substitution, the problem becomes to prove one root is positive and the other is negative for y.

Then the equation becomes $y^2 - by - 1 = 0$, and the conclusion is obvious : By Viete theorem, the product of its two roots is – 1, so they must have opposite signs.

32. Let $x = \sqrt{2\sqrt{2\sqrt{2\sqrt{2}}}\ldots}$,

$$y = \sqrt{2 + \sqrt{2 + \sqrt{2 + \sqrt{2 + \ldots}}}}$$

Then x satisfies the equation $x^2 = 2x$, and its solution is $x = 2$ (since $x > 0$).

Similarly, y satisfies the equation $y^2 = 2 + y$. Then $(y - 2)(y + 1) = 0$ and $y > 0$ yields the solution $y = 2$. Thus

$$\sqrt{2\sqrt{2\sqrt{2\sqrt{2}}}\ldots} - \sqrt{2 + \sqrt{2 + \sqrt{2 + \sqrt{2 + \ldots}}}}$$

$$= x - y = 0.$$

33. The given equation can be written as $|x|^2 - 3|x| - 2 = 0$

Or $|x| = \frac{3 \pm \sqrt{17}}{2}$

But $|x|$ cannot be negative so $|x| = \frac{3 + \sqrt{17}}{2}$

So minimum root is $a = -\frac{3 + \sqrt{17}}{2}$

Or $-\frac{1}{a} = \frac{2}{3 + \sqrt{17}} = \frac{2(\sqrt{17} - 3)}{8} = \frac{(\sqrt{17} - 3)}{4}$

34. Since a is a root of the equation so $x^2 - 3x + 1 = 0$ so

$a^2 - 3a + 1 = 0$ yiels $a^2 + 1 = 3a$ and $a \neq 0$. Therefore by division,

$$\frac{2a^5 - 5a^4 + 2a^3 - 8a^2}{a^2 + 1}$$

$$= \frac{(a^2 - 3a + 1)(2a^3 + a^2 + 3a) - 3a}{a^2 + 1}$$

$$= -\frac{3a}{3a} = -1.$$

35. Let Δ_i, $i = 1, 2, 3$ we denote the Discriminant of the ith equation.

For the given condition to be true it suffices to show $\Delta_1 + \Delta_2 + \Delta_3 > 0$.

$\Delta_1 + \Delta_2 + \Delta_3 = (4b^2 - 4ac) + (4c^2 - 4ab) + (4a^2 - 4bc)$
$= 2(2a^2 + 2b^2 + 2c^2 - 2ab - 2bc - 2ca)$
$= 2[(a - b)^2 + (b - c)^2 + (c - a)^2] > 0$

Hence all Δ_1, Δ_2 and Δ_3 can not be negative at the same time. Hence proved that the given equation cannot all have two equal real roots.

36. Let x_0 be the common root of the two given equations, then

$1988x_0^2 + bx_0 + 8891 = 8891x_0^2 + bx_0 + 1988,$
$(8891 - 1988)x_0^2 = (8891 - 1988),$
$\Rightarrow x_0 = \pm 1$

Substituting back such x_0 into the first equation,

$\pm b = -(8891 + 1988) = -10879,$

$\therefore b = 10879$

37. $a^2 - a - 3 = 0$ yields $a^2 - a = 3$. On the other hand,

$a^3 + 1 = (a + 1)(a^2 - a + 1),$
$a^5 - a^4 - a^3 + a^2 = a^2(a^3 - a^2 - a + 1)$
$= a^2[(a^2(a - 1) - (a - 1)]$
$= a^2(a - 1)(a^2 - 1) = a^2(a + 1)(a - 1)^2$
$= (a + 1)(a^2 - a)^2,$

It follows that

$$\frac{a^3+1}{a^5-a^4-a^3+a^2} = \frac{a^2-a+1}{(a^2-a)^2} = \frac{4}{3^2} = \frac{4}{9}.$$

38. The given equation has real roots so $D \geq 0$

$4(1 + a)^2 - 4(3a^2 + 4ab + 4b^2 + 2) \geq 0,$

$(1 + a)^2 - (3a^2 + 4ab + 4b^2 + 2) \geq 0,$

$2a^2 - 2a + 1 + 4ab + 4b^2 \leq 0,$

$(a - 1)^2 + (a + 2b)^2 \leq 0,$

This is possible only when both the perfect squares are zero.

$\therefore a = 1$, and $a + 2b = 0$,

$$a = 1, b = -\frac{1}{2}.$$

39. Let x_0 be the common root, then

$x_0^2 - kx_0 - 7 = 0$ and $x_0^2 - 6x_0 - (k + 1) = 0$

Then their difference gives

$(6 - k)x_0 - (6 - k) = 0$, or $(6 - k)(x_0 - 1) = 0$

Since $k \neq 6$ otherwise, the two equations are identical so that they have two common roots. Therefore $x_0 = 1$, and it implies that $k = -6$. By substituting back the value of k into the given equations, it follows that

$x^2 + 6x - 7 = 0$ and $x^2 - 6x + 5 = 0$

$\therefore (x - 1)(x + 7) = 0$ and $(x - 1)(x - 5) = 0.$

Thus the different roots are -7 and 5 respectively.

Exercis-B

Concept Deviator (CD)

(Single Correct MCQs)

1. (a) Let $x^2 - 2p_1x + 1 = 0$ has roots α, β

$\Rightarrow \alpha + \beta = 2p_1$ and $\alpha\beta = 1$

$x^2 - 4p_2x + 2 = 0$ has roots α, γ,

$\Rightarrow \alpha + \gamma = 4p_2$ and $\alpha\gamma = 2$

And $x^2 - 6p_3x + 3 = 0$ has roots β, γ

$\Rightarrow \beta + \gamma = 6p_3$ and $\beta\gamma = 3$

$(\alpha\beta)(\beta\gamma)(\gamma\alpha) = (1)(2)(3) = 6$

$\alpha\beta\gamma = \pm\sqrt{6}$

So we have two set of values of (α, β, γ)

$$\equiv \left(\frac{\sqrt{6}}{2}, \frac{\sqrt{6}}{3}, \sqrt{6}\right) \text{ or } \left(\frac{-\sqrt{6}}{2}, \frac{-\sqrt{6}}{3} - \sqrt{6}\right)$$

and so two sets of $(P_1\ P_2\ P_3)$ exist

2. (a) Let $f(x) = a\frac{(x-b)(x-c)}{(a-b)(a-c)} + b\frac{(x-c)(x-a)}{(b-c)(b-a)} + c\frac{(x-a)(x-b)}{(c-a)(c-b)} - x$

We have $f(a) = 0, f(b) = 0$ and $f(c) = 0.$

But, $f(x)$ is quadratic so it could not have more than two zeros.

$\Rightarrow f(x)$ is identity

Hence, $a\frac{(x-b)(x-c)}{(a-b)(a-c)}+b\frac{(x-a)(x-c)}{(b-a)(b-c)}+$

$$c\frac{(x-a)(x-b)}{(c-a)(c-b)}\equiv 0\forall x\in R$$

3. (a) Let $t=x+\sqrt{x^2+b^2}$

$\Rightarrow \frac{1}{t}=\frac{1}{x+\sqrt{x^2+b^2}}=\frac{\sqrt{x^2+b^2}-x}{b^2}$

$\therefore t-\frac{b^2}{t}=2x$ and $t+\frac{b^2}{t}=2\sqrt{x^2+b^2}$

Thus, $2(a-x)\left(x+\sqrt{x^2+b^2}\right)=\left(2a-t+\frac{b^2}{t}\right)(t)$

$=2at-t^2+b^2$

$=a^2+b^2-\left(a^2-2at+t^2\right)$

$=a^2+b^2-(a-t)^2$

So, $y=2(a-x)\left(x+\sqrt{x^2+b^2}\right)=$

$$a^2+b^2-(a-t)^2\le a^2+b^2$$

Hence, the maximum value of y is a^2+b^2 when $a=t=x+\sqrt{x^2+b^2}$

4. (a) Consider, LHS of the given equation, it is always positive

Now consider the RHS which is a quadratic function. $f(x)=-3+x-x^2$ since coefficient of $x^2<0$ and $D<0$ so RHS is always negative

The given equation has no solution.

5. (a) From the given condition ax^3+bx^2+cx+d and $(a+1)x^3+(b+1)x^2+(c+1)x+(d+1)=0$ must be identical.

$\Rightarrow \frac{a+1}{a}=\frac{b+1}{b}=\frac{c+1}{c}=\frac{d+1}{d}$

$\Rightarrow a=b=c=d\ne 0$

6. (a) Let $y=\frac{x^2-x}{1-ax}\Rightarrow x^2-x=y-axy$

$\Rightarrow x^2+x(ay-1)-y=0$

Since, 'x' is real so D of the above equation must be more than or equal to zero

$\Rightarrow (ay-1)^2+4y\ge 0$

$\Rightarrow a^2y^2+2y(2-a)+1\ge 0\forall y\in R$

$\Rightarrow a^2>0, 4(2-a)^2-4a^2\le 0$

$\Rightarrow 4+a^2-4a-a^2\le 0$

$\Rightarrow 4a\ge 4$

$\Rightarrow a\ge 1$

$\Rightarrow a\in[1,\infty)$

7. (a) Given that one root is less than –2 and the other root is more than 2

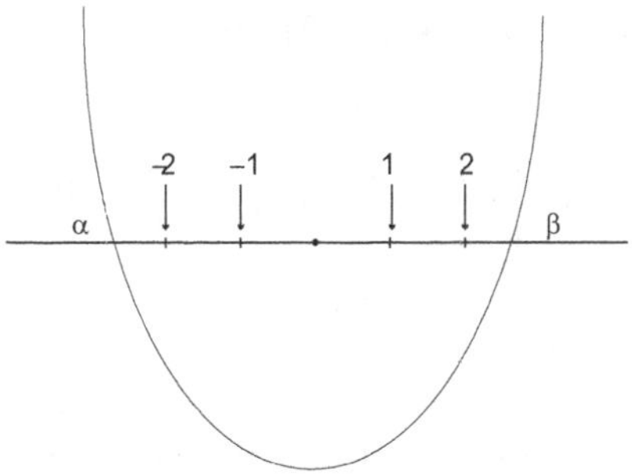

Since the equation has two distinct roots α and β. the Discriminant $b^2-4ac>0$, we must have $f(x)=ax^2+bx+c<0$ for $\alpha<x<\beta$

Since, $\alpha<0<\beta$ we, must have , $f(0)=c<0$

Also, as $\alpha<-1, 1<\beta$ we get, $f(-1)=a-b+c<0$ and $f(1)=a+b+c<0$.

$\Rightarrow a+|b|+c<0$.

Now , Since $<-2, 2<$.

$$F(-2)=4a-2b+c<0$$

And $f(2)=4a+2b+c<0$,

i. e., $4a+2|b|+c<0$.

All four statements are correct and none of them are incorrect

8. (c) Given equation is $x^2-(a-3)x+a=0$,

$D=(a-3)^2-4a=a^2-10a+9\ge 0$

$\Rightarrow (a-9)(a-1)\ge 0$

$\Rightarrow a\in(-\infty,1]\cup[9,\infty)$

Case 1: When both roots are positive.

$$D\ge 0$$

$$f(0)>0\Rightarrow a>0$$

$$-\frac{b}{2a}>0,(a-3)>0\Rightarrow a>3 \quad ....(i)$$

Combining all the above results we get $a \in [9, \infty)$

Case 2: When exactly one root is positive.

$\Rightarrow a \le 0$ ...(ii)

Combining (i) and (ii) $a \in (-\infty, 0] \cup [9, \infty)$

9. (d) $\log_2(ax^2 + x + a) \ge 1 \forall x \in R$

$\Rightarrow ax^2 + x + a \ge 2^1 = 2 \forall x \in R$

$\Rightarrow ax^2 + x + (a-2) \ge 0 \forall x \in R$

$\Rightarrow a > 0, 1 - 4a(a-2) \le 0$

$\Rightarrow 4a^2 - 8a - 1 \ge 0$

$\Rightarrow a \in \left(-\infty, 1 - \frac{\sqrt{5}}{2}\right] \cup \left[1 + \frac{\sqrt{5}}{2}, \infty\right)$ but $a > 0$,

$\Rightarrow a \in \left[1 + \frac{\sqrt{5}}{2}, \infty\right)$

10. (c) $x^n + ax + b = (x - x_1)(x - x_2)\ldots(x - x_n$

$\Rightarrow (x - x_2)(x - x_3)\ldots(x - x_n) = \frac{x^n + ax + b}{(x - x_1)}$

$\Rightarrow (x_1 - x_2)(x_1 - x_3)\ldots\ldots(x_1 - x_n) =$

$$\lim_{x \to x_1} \frac{x^n + ax + b}{x - x_1} = nx_1^{n-1} + a$$

11. (a) Here, $\log_{|\sin x|}|\cos x| + \frac{1}{\log_{|\sin x|}|\cos x|} = 2$

Let $y = \log_{|\sin x|}|\cos x|$ then $\Rightarrow y + \frac{1}{y} = 2$

$\Rightarrow y^2 - 2y + 1 = 0 \quad \Rightarrow y = 1$

$\therefore \log_{|\sin x|}|\cos x| = 1 \quad \Rightarrow |\cos x| = |\sin x|$

$\Rightarrow |\tan x| = 1 \qquad \Rightarrow x = \frac{\pi}{4}$

12. (d) Let $f(x) = \frac{a^3(x-b)(x-c)(x-d)}{(a-b)(a-c)(a-d)}$

$$+ \frac{b^3(x-c)(x-d)(x-a)}{(b-c)(b-d)(b-a)} + c^3\frac{(x-d)(x-a)(x-b)}{(c-d)(c-a)(c-b)} + d^3\frac{(x-a)(x-b)(x-c)}{(d-a)(d-b)(d-c)} - x^3$$

where $f(a) = a^3 - a^3 = 0$, similarly

$f(b) = 0, f(c) = 0, f(d) = 0$

$\therefore x = a, b, c, d$ are roots of $f(x) = 0$

But $f(x) = 0$ is a polynomial of degree 3 in x. So, it cannot have more than three roots. If it has more than three roots, it is identity in x. So, $f(x) = 0$ has infinity many solutions.

13. (a) Let $f(x) = ax^2 + bx + c$

Since, coefficients are integers and one root is irrational, then both the roots are irrational.

$\therefore$ For any $\lambda \in Q, f(\lambda) \ne 0$

$\Rightarrow f(\lambda) > 0$ or $f(\lambda) < 0$, but $f(\lambda) \ne 0$

$\therefore |f(\lambda)| > 0$

or $\left|a\frac{p^2}{q^2} + b\frac{p}{q} + c\right| > 0$

or $\frac{1}{q^2}|ap^2 + bpq + cq^2| > 0$ ... (i)

As, $a, b, c, p, q \in$ integers

$\therefore |ap^2 + bpq + cq^2| \ge 1$ ...(ii)

$\Rightarrow |f(\lambda)| = \frac{1}{q^2}|ap^2 + bpq + cq^2| \ge \frac{1}{q^2}$

14. (c) Let the polynomial be $P(x)$, when divided by $(x - 1)$ leaves remainder 2

$\Rightarrow P(1) = 2$ and when $P(x)$ divided by $(x - 2)$ leaves remainder 1

$\Rightarrow P(2) = 1$. $\therefore$ When $P(x)$ divided by $(x - 1)$ $(x - 2)$, could be given by

$P(x) = \{q(x)\}(x-1)(x-2) + r(x)$; where $q(x)$ is quotient and $r(x)$ is remainder

$(i.e., r(x) = ax + b)$

$\Rightarrow P(x) = (x-1)(x-2)q(x) + (ax+b)$ ...(i)

$\therefore P(1) \Rightarrow 0 + a + b = 2$ and $P(2) \Rightarrow 0 + 2a + b = 1$

On solving we get, $a = -1, b = 3$

Remainder $= -x + 3$

15. (a) Here, $\frac{(16)^{1/x}}{(2^{x+3})} > 1$

$\Rightarrow \dfrac{2^{4/x}}{2^{x+3}} > 1$ or $2^{\frac{4}{x}-x-3} > 1$

ie, $2^{\frac{4}{x}-x-3} > 2^0$

$\Rightarrow \dfrac{4}{x} - x - 3 > 0$

$\Rightarrow -\dfrac{(x^2+3x-4)}{x} > 0$ or $\dfrac{-(x+4)(x-1)}{x} > 0$

$\dfrac{(x+4)(x-1)}{x} < 0$

From method of interval

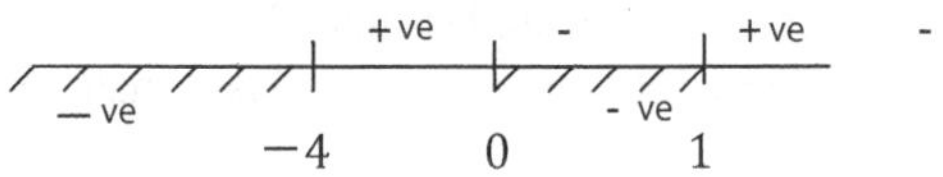

$\Rightarrow x \in (-\infty, -4) \cup (0, 1)$

16. (c) Given polynomial is,

$f(x) = (x-n)^m + (x-n^2)^m +$
$(x-n^3)^m + \ldots + (x-n^m)^m$

$f'(x) = m(x-n)^{m-1} + m(x-n^2)^{m-1} + \ldots + m$
$+ m(x-n^m)^{m-1}$ {as, m is odd $\Rightarrow (m-1)$ is even}

$\therefore f'(x) > 0$, as m is odd positive integer

So, $f'(x) = 0$ has no real root

$\Rightarrow f(x) = 0$ has one real and $(m-1)$ imaginary roots.

17. (a) Here, $x^2 - (a+b+c)x + (ab+bc+ca) = 0$ has imaginary root.

$\Rightarrow D < 0$ or $(a+b+c)^2 - 4(ab+bc+ca) < 0$

$\Rightarrow (a^2+b^2+c^2-2ab-2bc+2ac) < 4ac$

$\Rightarrow (a-b+c)^2 < 4ac$

$\Rightarrow -2\sqrt{ac} < a-b+c$

$\Rightarrow (a+c+2\sqrt{ac}) > b$

$\Rightarrow (\sqrt{a}+\sqrt{c})^2 > b$ or $\sqrt{a}+\sqrt{c} > \sqrt{b}$

Similarly, $\sqrt{b}+\sqrt{c} > \sqrt{a}$ and $\sqrt{a}+\sqrt{b} > \sqrt{c}$

$\Rightarrow \sqrt{a}, \sqrt{b}, \sqrt{c}$ can be sides of Δ.

18. (b) As; $x^2 = 2 + 2[x]$, where LHS is always positive. So, $2 + 2[x] \geq 0 \Rightarrow [x] \geq -1$

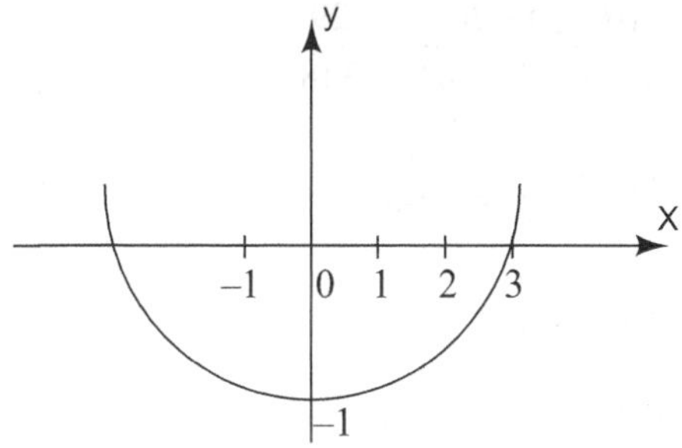

$\therefore$ If $[x] = -1$

$\Rightarrow x^2 - 2 = -2$

$\Rightarrow x = 0$, not possible

If $[x] = 0$ i.e. $0 \leq x < 1$

$\Rightarrow x^2 = 2 \Rightarrow x = \pm\sqrt{2}$, not possible

If $[x] = 1$ then $1 \leq x < 2$

$\Rightarrow x^2 = 4 \Rightarrow x = \pm 2$, not possible

If $[x] = 2$

$\Rightarrow x^2 = 6 \Rightarrow x = \pm\sqrt{6}$, only one possible value i.e., $x = \sqrt{6}$

For higher values of x, L.H.S will always be bigger than RHS

19. (c) The given expression can be written as

$3^{\sec^2 x - 1}\sqrt{y^2 - \dfrac{2y}{3} + \dfrac{2}{9}} \leq 1$

$\Rightarrow 3^{\sec^2 x}\sqrt{\left(y - \dfrac{1}{3}\right)^2 + \dfrac{1}{9}} \leq 1$ where $\sec^2 x \geq 1$

$\Rightarrow 3^{\sec^2 x} \geq 3$ and $\sqrt{\left(y - \dfrac{1}{3}\right)^2 + \dfrac{1}{9}} \geq \dfrac{1}{3}$

So, we should have and $\sec^2 x = 1$ and $y - \dfrac{1}{3} = 0$

$\Rightarrow x = 0, \pi, 2\pi, 3\pi$ and $y = \dfrac{1}{3}$

$\therefore$ Total number of solutions is four.

20. (c) $\sum_{r=1}^{n} s_r = (\alpha + \beta) + (\alpha^2 + \beta^2) + \ldots + (a^n + b^n)$

$= (\alpha + \alpha^2 + \ldots + \alpha^n) + (\beta + \beta^2 + \beta^n)$

$\therefore \lim\limits_{n\to\infty} \sum_{r=1}^{n} s_r = (\alpha + \alpha^2 + \alpha^3 + \ldots \text{ to } \infty) +$

$$\left(\beta+\beta^2+\beta^3+\ldots \text{to } \infty\right)=\frac{\alpha}{1-\alpha}+\frac{\beta}{1-\beta}$$

$$\frac{(\alpha+\beta)-2\alpha\beta}{1-(\alpha+\beta)+\alpha\beta}=\frac{29}{348}$$

$$\alpha+\beta=\frac{25}{375}=\frac{1}{15}$$

$$\alpha\beta=\frac{2}{375}$$

21. (b) $x = 2 + 2^{1/3} + 2^{2/3}$

$\Rightarrow x-2=2^{1/3}+2^{2/3}$

$\Rightarrow (x-2)^3=\left(2^{\frac{1}{3}}+2^{\frac{2}{3}}\right)^3$

$\Rightarrow x^3-6x^2+12x-8=6+6(x-2)$

$\Rightarrow x^3-6x^2+6x=2$

22. (d) Since $x(x-3)\geq 0$ *i.e.*, $x\leq 0$, or $x\geq 3$ ...(i)

The given equation can be written as $9|x|^2 - 19|x| + 2 = 0$

$\Rightarrow (9|x|-1)(|x|-2)=0 \;\Rightarrow |x|=2$ or $|x|=\frac{1}{9}$

$\therefore$ Solution of the given equation is $\pm 2, \pm\frac{1}{9}$

From condition (i), the required solutions are $-2,-\frac{1}{9}$

23. (d) Given that f(a) = f(b) = f(c) = – 1.

Or f(a) +1 = f(b) +1 = f(c) +1 = 0

So we can say that a, b, c are roots of f(x) + 1 = 0

Let f(x) + 1 = (x – a) (x – b) (x – c) g(x)

where g(x) is some polynomial with integral coefficients.

Let m be an integral root of f(x) then

f(m) + 1 = (m – a) (m – b) (m – c) g(m)

Or 1 = (m – a) (m – b) (m – c) g(m)

Since a, b, c are distinct at least one of m – a, m – b, m – c is different from 1 and – 1 therefore RHS can never be equal to 1. This is a contradiction, hence no such polynomial exist.

24. (c) Let α be the common root.

Then $\alpha^2+i\alpha+a=0, \alpha^2-2\alpha+ia=0$

$\Rightarrow \frac{\alpha^2}{a}=\frac{\alpha}{a(1-i)}=\frac{1}{-2-i}$

$\Rightarrow a^2(1-i)^2=a(-2-i)$

$\Rightarrow a=\frac{1}{2}-i$

25. (a) Discriminant D = $4a^2-4\left(a^2+a-3\right)>0$

$\Rightarrow a<3$ ...(i)

Since, the roots are less than 3.

$\therefore$ Sum of the roots is less than 6 ie, $\frac{2a}{1}<6\Rightarrow a<3$ ...(ii)

Now, $f(3)>0 \;\Rightarrow (a-2)(a-3)>0$

$\Rightarrow a<2$ or $a>3$ ...(iii)

From Eqs. (i), (ii) and (iii), we can conclude that $a<2$.

26. (a) Let the given function f(x) = $e^a x^2-e^{2a}x+e^a-1$

Then f(1).f(2) < 0

$\left(e^a-e^{2a}+e^a-1\right)\left(4e^a-2a^{2a}+e^a-1\right)<0$

$\Rightarrow \left(e^{2a}-2e^a+1\right)\left(2e^{2a}-5e^a+1\right)<0$

Let $x=e^a$

$\Rightarrow (x-1)^2\left(2x^2-5x+1\right)<0$

$\Rightarrow (x-1)^2\left(x-\frac{5-\sqrt{17}}{4}\right)\left(x-\frac{5+\sqrt{17}}{4}\right)<0$

$\Rightarrow \frac{5-\sqrt{17}}{4}<x<\frac{5+\sqrt{17}}{4}$

$\Rightarrow \log\left(\frac{5-\sqrt{17}}{4}\right)<a<\log\left(\frac{5+\sqrt{17}}{4}\right)$

27. (a) Let $f(x)=\frac{ax^3}{3}+\frac{bx^2}{2}+cx+d$, where $f(0)=f(1)=d$

$\therefore f'(x)$ has atleast one root in [0, 1] or ax^2+bx+c has atleast one root in [0, 1].

28. Given term is $\sum_{k=2}^{n}\left(x^2-\left(\frac{1}{k-1}+\frac{1}{k}\right)x+\frac{1}{k(k-1)}\right)$;

where constant term is given by $\frac{1}{k\cdot(k-1)}$

$$\therefore \sum_{k=2}^{\infty}\frac{1}{k(k-1)}=\left(\frac{1}{1\cdot 2}\right)+\left(\frac{1}{2\cdot 3}\right)+(\frac{1}{3\cdot 4})+\ldots$$

$$=\left(1-\frac{1}{2}\right)+\left(\frac{1}{2}-\frac{1}{3}\right)+\left(\frac{1}{3}-\frac{1}{4}\right)+\ldots\infty=1$$

29. (d) $y^{x^2+7x+12}=1$

$\Rightarrow x^2+7x+12=0 \Rightarrow x=-3,-4$

$\Rightarrow y=9,10(\text{when } y\neq 1)$

Again when y = 1, x = 5

$\therefore$ Solution set is (–3, 9)(– 4, 10), (5, 1)

30. Let $x=\cos\theta$, we get $4\cos^3\theta-3\cos\theta=p$

$\Rightarrow p=\cos 3\theta$

$\Rightarrow \theta=\frac{1}{3}\cos^{-1}(p)$ and $x=\cos\theta$

$\Rightarrow x=\cos\left(\frac{1}{3}\cos^{-1}(p)\right)$

Since, $-1\le p\le 1 \Rightarrow 0\le \cos^{-1}p\le \pi$

$0\le \frac{1}{3}\cos^{-1}p\le\frac{\pi}{3}$ or as we know cos x is decreasing

$\therefore \cos 0\ge \cos\frac{1}{3}\cos^{-1}(p)\ge\cos\frac{\pi}{3}$ ...(ii)

From Eqs. (i) and (ii), we get $\Rightarrow \frac{1}{2}\le x\le 1$;

$\Rightarrow x\in\left[\frac{1}{2},1\right]$

31. (b) Given condition is satisfied

if f(1) < 0 and f(2) < 0

f(1) = 1 – m + 1 < 0

$\Rightarrow m>2$

f(2) = 4 – 2m + 1 < 0

$\Rightarrow m>\frac{5}{2}$

least integral value of m is 3.

32. (b) Given that f(x) = $ax^2 + bx + c$,

Now consider f(0) = c< 0

$f(-\frac{1}{2})=\frac{a-2b+4c}{4}>0$

$\Rightarrow$ One root lies in the interval $(-\frac{1}{2}, 0)$

33. (b) |x + 2| vanishes at x = –2 and $|2^{x+1}-1|$ vanishes at x = –1, hence we divide the problem into three intervals:

(i) If x < –2, then |x +2| = – (x + 2)

Also, $x+1<-1 \Rightarrow 2^{x+1}<2^{-1}=$

$2^{x+1}<1 \Rightarrow |2^{x+1}-1|=-(2^{x+1}-1)$

$\therefore$ Equation is $2^{-x-2}+2^{x+1}-1=2^{x+1}+1$

$\Rightarrow 2^{-x-2}=2 \Rightarrow x=-3$

(ii) If $-2\le x<-1$, then |x +2| = x + 2

Also, $x+1<0 \Rightarrow 2^{x+1}<1$

$\Rightarrow |2^{x+1}-1|=-(2^{x+1}-1)$

$\therefore$ Equation is $2^{x+2}+2^{x+1}-1=2^{x+1}+1$

$\Rightarrow 2^{x+2}=2 \Rightarrow x=-1\notin[-1,\infty)$

(iii) If $x\ge -1$, then |x +2| = x + 2 and $|2^{x+1}-1|=2^{x+1}-1$

$\therefore$ Equation is $2^{x+2}-2^{x+1}+1=2^{x+1}+1$

$\Rightarrow 2^{x+2}=2^{x+2}$, which is identity.

$\therefore$ All x such that $x\ge -1$ satisfy the equation.

Hence, the solution set is $x\in\{-3\}\ [-1,\infty)$

So number of negative integers are –3 and –1 i.e 2 integers

34. (d) p (x) = $(2x+3)(x^{97}+x^{96}+\ldots+1)$.

$=(2x+3)(x+1)(x^{96}+x^{94}+\ldots+x^2+1)$

Also, $x^{96}+x^{94}+\ldots+x^2+1>1\ \forall x\in R$.

So required real roots are 1 and 3/2

Sum of roots is –5/2

35. (d) The equation is reciprocal so, f(x) =0 and

$f\left(\frac{1}{x}\right)=0$ represent same equation

$x^3+x^2+ax+b=0$ and $\frac{1}{x^3}+\frac{1}{x^2}+\frac{a}{x}+b=0$

bx^3+ax^2+x+1 0 are indentical

$\therefore \frac{1}{b}=\frac{1}{a}=\frac{a}{1}=\frac{b}{1}=<a^2=1$ and a = b

$=>a=b=\pm 1$

So number of ordered pairs is 2

36. (b) If x < 1 then

$$-x^3+1\ge 1-x$$
$$x^3-x\le 0 \text{ or } x(x+1)(x-1)\le 0$$

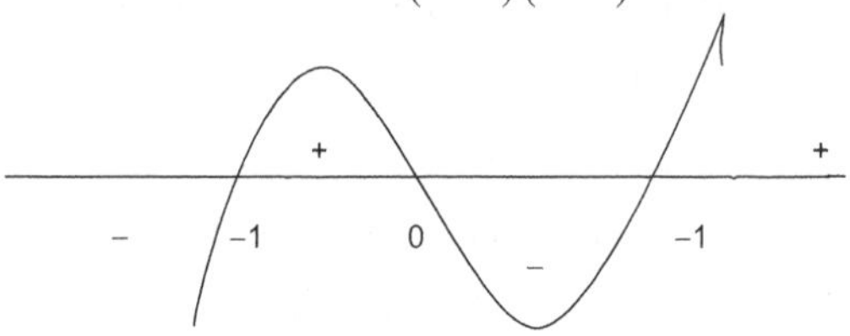

Solving by method of interval, we get

x ∈(– ∞, –1] ∪ [0,1]

Let x ≥ 1, then we get

$x^3-1\ge 1-x$

$x^3+x-2\ge 0$

⇒ (x – 1)(x^2 + x +2) ≥ 0

⇒ x ≥ 1 [since, x^2 + x +2 > 0 x ∈ R]

⇒ x ∈ [1, ∞)

Combining the two solutions, we get

x ∈ (∞, –1] ∪ [0,1] ∪ [1,∞)

or x ∈ (– ∞, –1] ∪ [0,∞)

37. (b) Put x^2 + x + 1 = y, the equation reduces to

$(y+1)^2 - (a-3)(y+1)y + (a-4)y^2 = 0$

$\Rightarrow (a-5)y - 1 = 0 \Rightarrow y = \dfrac{1}{a-5}$

$\therefore x^2+x+1 = \dfrac{1}{a-5}.$

Now $x^2+x+1 = \left(x+\dfrac{1}{2}\right)^2+\dfrac{3}{4}\ge\dfrac{3}{4}$

$\therefore \dfrac{1}{a-5}\ge\dfrac{3}{4} => \dfrac{4-3a+15}{4(a-5)}\ge 0$

$\Rightarrow \dfrac{3a-19}{a-5}\le 0 \Rightarrow (a-5)(3a-19)\le 0, a\ne 5$

$\Rightarrow 5 < a \le \dfrac{19}{3}$ since, $6<\dfrac{19}{3}<7$

∴ Only integral value of a such that a ∈ $(5, \dfrac{19}{3})$ is a = 6.

38. (b) Let f(x) = $ax^2+8x-(20+4a)$

f(2) = 4a + 16 – (20 + 4a) = – 4 < 0

f(0) < 0 ⇒ (20 +4a) > 0 => a > –5 ...(i)

If f(x) < 0 ∀ x ∈ [0, 2] ⇒ a > 0 and D > 0

D > 0 ⇒ 64 + 4a(20 + 4a) > 0

$\Rightarrow a^2+5a+4>0$

Or, a ∈ (– ∞, –4) ∪ (–1, – ∞)

From above results we get a ∈ (0, ∞)

Consider if a= 0 , f (x) = 8x – 20 < 0

$\Rightarrow x < \dfrac{5}{2}$

a ∈ [0,∞)

39. (d) From observation we find that 1 and (k+2) are roots of $x^2-(k+3)x+k+2=0$

Let f(x) = $x^2-(k-1)x+k(k+4)$

f(1). f(k +2) < 0

⇒ f(1) = 1 – (k –1) + k(k +4) = k^2 + 3k + 2

= (k + 1) (k + 2)

⇒ f(k+2) = $(k+2)^2$– (k – 1)(k +2) +k (k +4)

= k^2 + 7k + 6 = (k + 1) (k + 6)

$\Rightarrow f(1).f(k+2)=(k+1)^2(k+2)(k+6)<0$

⇒ k ∈ (–6, –2)

Number of integral values of ‘k’ is 3

Required sum is (– 5) + (– 4) + (–3) + (–2) = –14

40. (d) Since f(x) > 0 $\forall x\in R$ so a > 0

We have a > 0 and b^2 – 4ac < 0. So , f(–2) > 0

⇒ 4a – 2b + c > 0

f(–1) + a > 0

⇒ 2a – b + c > 0

f(3) + a > 0

⇒ 10a + 3b + c > 0

f(1) + a > 0

⇒ 2a + b + c > 0

All the statements are correct.

41. (c) Let us assume that the roots are rational and it is possible if discriminant is a perfect square. or b^2 – 4ac = d^2,d is an odd integer as a, b , c are odd. Let a = 2k +1, b = 2m +1, c = 2n + 1, d = 2p + 1. Then

$(2m+1)^2 - 4(2k+1)(2n+1) = (2p+1)^2$

$\Rightarrow (2m+1)^2 - (2p+1)^2 = 4(2k+1)(2n+1)$

$\Rightarrow$ (2m –2p)(2m +2p +2) = 4(2k + 1)(2n+ 1)

$\Rightarrow$ (m – p)(m +p +1) = (2k +1)(2n +1)

Now for any integral value of m and p, the LHS of the above is always even but RHS is always odd, which is a contradiction. Hence, the discriminant cannot be perfect square. So, the roots cannot be rational

42. (d) Domain: $x \in R$. Given inequality is equivalent to $\frac{3x^2-7x+8}{x^2+1}-2 \le 0$

$$\Rightarrow \frac{3x^2-7x+8-2x^2-2}{x^2+1} \le 0$$

$$\Rightarrow \frac{x^2-7x+6}{x^2+1} \le 0$$

$$\Rightarrow \frac{(x-1)(x-6)}{x^2+1} \le 0$$

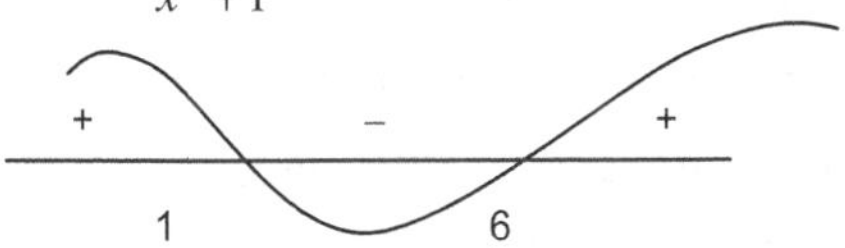

Since denominator is always positive so sign scheme depends on numerator which is shown above

$\Rightarrow x \in [1,6]$.

43. (a) Domain $x \in R-\{-2,0,2\}$ given inequality is equivalent to $\frac{1}{x-2}-\frac{1}{x}-\frac{2}{x+2} \le 0$

$$\frac{x^2+2x-(x^2-4)-2(x^2-2x)}{(x-2)(x+2)x} \le 0$$

$$\Rightarrow \frac{x^2-3x-2}{(x-2)(x+2)x} \ge 0$$

$$\Rightarrow \frac{\left(x-\frac{(3+\sqrt{17})}{2}\right)\left[x-\frac{(3-\sqrt{17})}{2}\right]}{(x-2)(x+2)x} \ge 0$$

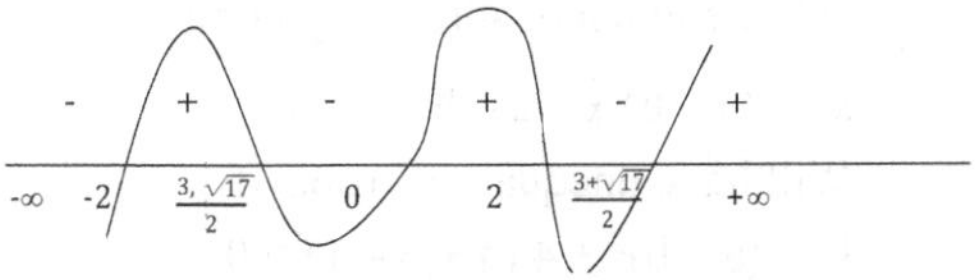

$$x \in \left(-2, \frac{(3-\sqrt{17})}{2}\right] \cup (0,2) \cup \left[\frac{(3-\sqrt{17})}{2}, +\infty\right)$$

44. (a) In order to solve this question one need to have complete command over complex numbers and Demoiver's theorem.

Roots of the equation $x=\frac{2\cos\theta \pm \sqrt{4\cos^2\theta-4}}{2}$

$=\cos\theta \pm i\sin\theta$

Let $\alpha=\cos\theta+i\sin\theta$, then

$\beta=\cos\theta-i\sin\theta$

$\Rightarrow \alpha^n=(\cos\theta+i\sin\theta)^n$

$=\cos n\theta+i\sin n\theta$

Also, $\beta^n=(\cos\theta-i\sin\theta)^n$

$=\cos n\theta-i\sin n\theta$

Now, $\alpha^n+\beta^n=2\cos n\theta$ and $\alpha^n\beta^n=1$

$\therefore$ Required equation is $x^2-(\alpha^n+\beta^n)x+\alpha^n\beta^n=0$

$\Rightarrow x^2-(2\cos n\theta)x+1=0$

45. (b) Solution: Let $f(x)=x^4+ax^3+bx^2+cx-1$

$(x-1)^3$ is a factor

$\Rightarrow f(1)=0, f'(1)=0, f''(1)=0$

$\Rightarrow a+b+c=0, 3a+2b+c=-4$ and $3a+b=-6$

$\Rightarrow a=-2, b=0, c=2$

$\Rightarrow$ Other factor is (x + 1)

Alternately :

Let roots of the equation are 1, 1, 1, and α

Then product of the roots are 1.1.1.α = –1 so α = –1

So (x + 1) is a factor.

46. (a) Consider the discriminant of the given equation

$D=4\left[({}^nC_r)^2-{}^nC_{r-1}\,{}^nC_{r+1}\right]$

$=4(p-q)$, where

$$p=({}^nC_r)^2, q={}^nC_{r-1}\cdot{}^nC_{r+1}$$

Now consider, $\frac{p}{q}=\frac{{}^nC_r\cdot{}^nC_r}{{}^nC_{r-1}\cdot{}^nC_{r+1}}=$

$$\frac{r+1}{r}\cdot\frac{n-r+1}{n-r}=\left(1+\frac{1}{r}\right)\left(1+\frac{1}{n-r}\right)>1$$

$\therefore p > q \Rightarrow D > 0$

$\Rightarrow$ Roots of given equation are real and distinct.

47. (d) Let $f(x) = 3x^4 + 4x^3 - 12x^2$

$f'(x) = 12x^3 + 12x^2 - 24x$

$= 12x(x^2 + x - 2) = 0 \Rightarrow x = 0, 1, -2$

$f(0) = 0, f(1) = 3 + 4 - 12 = -5$

$f(-2) = -32$

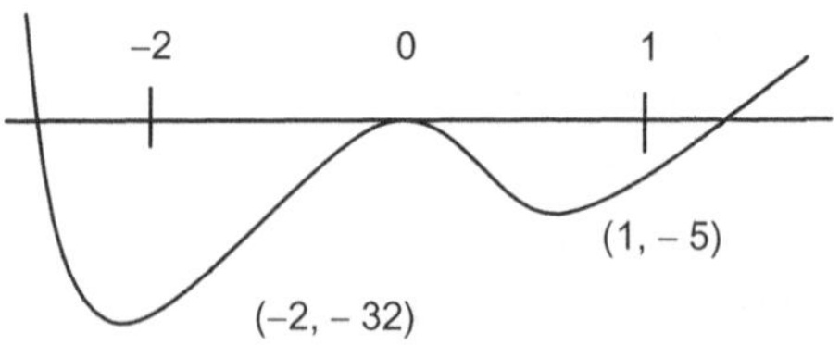

Since, $f(x) + 5a = 0$ has no real root.

$\Rightarrow 5a > 32$

$\Rightarrow a > \frac{32}{5}$

So, least +ve integral value of a = 7.

48. (a) Since $f(x) > 0\ \forall x \in R$ so $a > 0$ and $b^2 - 4ac < 0$.

Then $f'(x) = 2ax + b$ and $f''(x) = 2a$.

So, $g(x) = ax^2 + bx + c + 2ax + b + 2a$

$= ax^2 + (2a + b)x + (2a + b + c)$

We have $a > 0$ and $D = (2a+b)^2 - 4a(2a + b + c)$

$= 4a^2 + 4ab + b^2 - (8a^2 + 4ab + 4ac) = b^2 - 4ac - 4a^2 < 0$, [Since, $b^2 - 4ac < 0$]

So for g(x) =0, a > 0 and D< 0

Thus, $g(x) > 0\ \forall x \in R$.

49. (b) Since $y = 3^{x-1}$ and $y = 5^{x-1}$ are both increasing functions of x (exponential functions with base greater than 1). So their sum $y = 3^{x-1} + 5^{x-1}$ is also an increasing function so it will attain a value 34 only once, and the equation has only one root.

From observation we find that x = 3 is root of the equation.

50. (a) Given that $\tan\left(\frac{P}{3}\right)$ and $\tan\left(\frac{Q}{3}\right)$ are the roots of the equation $ax^2 + bx + c = 0$, So

$\tan\left(\frac{P}{3}\right) + \tan\left(\frac{Q}{3}\right) = -\frac{b}{a}$

$\tan\left(\frac{P}{3}\right)\tan\left(\frac{Q}{3}\right) = \frac{c}{a}$

Now, $\angle R = \pi/4 \Rightarrow P + Q = 3\pi/4$

Or $\frac{P}{3} + \frac{Q}{3} = \frac{\pi}{4}$

$\therefore 1 = \tan\left(\frac{\pi}{4}\right) = \frac{\tan(P/3) + \tan(Q/3)}{1 - \tan(P/3)\tan(Q/3)}$

$= \frac{-b/a}{1 - c/a} = \frac{-b}{a - c}$

$\Rightarrow a + b = c$

51. (a) Given equation is

$a_1x^4 + a_2x^3 + a_3x^2 + a_4x + a_5 = 0$

Here sum of its coefficients is

$a_1 + a_2 + a_3 + a_4 + a_5 = 0$

So x = 1 is always a root of the equation.

Therefore, the equation has at least two real roots

52. (c) $x + y = 4 - z$, $x^2 + y^2 = 6 - z^2$,

$\therefore 2xy = (x + y)^2 - (x^2 + y^2)$

$= (4 - z)^2 - (6 - z^2) = 2z^2 - 8z + 10$

The quadratic equation whose roots are x and y is

$t^2 - (x + y)t + xy = 0$

Or $t^2 - (4 - z)t + z^2 - 4z + 5 = 0$

Since x and y are real

$(4 - z)^2 - 4(z^2 - 4z + 5) \geq 0$

$\Leftrightarrow 3z^2 - 8z + 4 \leq 0 \Leftrightarrow (3z - 2)(z - 2) \leq 0$

$\Leftrightarrow 2/3 \leq z \leq 2$.

That z can take value 2 when x = y = 1.

53. (b) The given quadratic equation

$x^2 + (a - b)x - a - b + 1 = 0$

Will have unequal real roots if

$D = (a - b)^2 + 4(a + b - 1) > 0$

$\Rightarrow b^2 + (4 - 2a)b + a^2 + 4a - 4 > 0$

Now consider $b^2 + (4 - 2a)b + a^2 + 4a - 4$, which is a quadratic function of 'b' and $b^2 + (4 - 2a)b + a^2 + 4a - 4 > 0$ if its discriminant is negative

So

$(4 - 2a)^2 - 4(a^2 + 4a - 4) < 0$

$\Rightarrow 16 - 16a + 4a^2 - 4a^2 - 16a + 16 < 0$

$\Rightarrow 32 - 32a < 0 \Rightarrow a > 1$

54. (b) From observation we find that a, b, c are roots of $\frac{x}{t} + \frac{y}{t-p} + \frac{z}{t-q} = 1$

$\Rightarrow x(t-p)(t-q) + yt(t-q) + zt(t-p) = t(t-p)(t-q)$

$\Rightarrow t^3 - (x+y+z+p+q)t^2 + [(p+q)x + qy + pz]t - pqx = 0$

Sum of the roots $a+b+c = x+y+z+p+q$ or $x+y+z = a+b+c-p-q$ Statement I is correct

Product of the roots $abc = pq$ or $x = \frac{abc}{pq}$

From above result we get

$$y = \frac{(a-p)(b-p)(c-p)}{p(p-q)} \text{ and}$$

$$z = \frac{(a-q)(b-p)(c-q)}{q(q-p)}$$

So statement (iii) is correct and statement (iv) is incorrect.

55. (b) Let $\alpha = 4, \beta = 4\omega, \gamma = 4\omega^2$

$$\therefore \left(\frac{\alpha}{\beta}\right)^2 = \frac{1}{\omega^2} = \omega \text{ and } \left(\frac{\alpha}{\gamma}\right)^2 = \frac{1}{\omega^4} = \omega^2$$

$\therefore$ Required equation is $x^2 - (\omega + \omega^2)x + \omega^3 = 0$

$\Rightarrow x^2 + x + 1 = 0$

IIT CORNER

56. **(c)** Given that *l, m, n* are real, $l \neq m$

Given equation is

$(l - m)x^2 - 5(l + m)x - 2(l - m) = 0$

$D = 25(l + m)^2 + 8(l - m)^2 > 0, l, m \in R$

$\therefore$ Roots are real and unequal.

57. (a) $u = x^2 + 4y^2 + 9z^2 - 6yz - 3zx - 2xy$

$= \frac{1}{2}[2x^2 + 8y^2 + 18z^2 - 12yz - 6zx - 4xy]$

$= \frac{1}{2}[(x^2 - 4xy + 4y^2) + (4y^2 + 9z^2 - 12yz) + (x^2 + 9z^2 - 6zx)]$

$= \frac{1}{2}[(x - 2y)^2 + (2y - 3z)^2 + (3z - x)^2] \geq 0$

$\therefore$ u is always non – negative.

58. (c) Given $a, b, c > 0$,

Let α, β be the roots of $ax^2 + bx + c = 0$, then

$\alpha + \beta = -\frac{b}{a} = -\text{ve}, \qquad \alpha\beta = \frac{c}{a} = +\text{ve}$

$\Rightarrow$ Either both α, β are – ve (if roots are real) or both α, β have – ve real parts (if roots are complex conjugate).

59. (b) The given equation is

$(x - b)(x - c) + (x - a)(x - c) + (x - a)(x - b) = 0$

$\Rightarrow 3x^2 - 2(a + b + c)x + (ab + bc + ca) = 0$

Discriminant $= 4(a + b + c)^2 - 12(ab + bc + ca)$

$= 4[a^2 + b^2 + c^2 - ab - bc - ca]$

$= 2[(a - b)^2 + (b - c)^2 + (c - a)^2] \geq 0 \ \forall a, b, c$

$\therefore$ Roots of given equation are always real.

Alternately:

Consider $f(x) = (x - b)(x - c) + (x - a)(x - c) + (x - a)(x - b)$

Then $f(a) = (a - b)(a - c)$

$f(b) = (b - a)(b - c)$

$f(c) = (c - a)(c - b)$

now whatever the order of a, b and c at least one pair of f(a), f(b) and f(c) must be of opposite sign so f(x) =0 must have two real roots.

60. (c) Question is already discussed in Concept Deviator Question Number 7

61. (d) Let $f(x) = x^{12} - x^9 + x^4 - x + 1$

Now consider three cases

Case (i): $x < 0$

put $x = -y$ where $y > 0$

Then we get function as $= y^{12} + y^9 + y^4 + y + 1 > 0$ for $y > 0$

Case (ii): $0 < x < 1$

$x^9 < x^4 \Rightarrow -x^9 + x^4 > 0$

Also $1 - x > 0$ and $x^{12} > 0$

$\Rightarrow x^{12} - x^9 + x^4 + 1 - x > 0$

Case (iii): $x > 1$

$f(x) = x(x^3 - 1)(x^8 + 1) + 1 > 0$

So $f(x) > 0$ for $-\infty < x < \infty$.

62. (a) Given equation is $x - \frac{2}{x-1} = -\frac{2}{x-1}$

Clearly $x \neq 1$ for the given equation to be defined.

If $x - 1 \neq 0$, we can cancel the common term $\frac{-2}{x-1}$ on both sides to get $x = 1$, but at $x=1$ the equation is not defined. So given eq. has no roots.

63. (c) Given that $a^2 + b^2 + c^2 = 1$

We know $(a + b + c)^2 \geq 0$

$\Rightarrow a^2 + b^2 + c^2 + 2ab + 2bc + 2ca \geq 0$

$\Rightarrow 2(ab + bc + ca) \geq -1$

$\Rightarrow ab + bc + ca \geq -1/2$

Since $[(a - b)^2 + (b - c)^2 + (c - a)^2] \geq 0$

$\Rightarrow a^2 + b^2 + c^2 - ab - bc - ca \geq 0$

$\Rightarrow ab + bc + ca \leq 1$ [using (1)]

$\Rightarrow ab + bc + ca \leq 1$

Combining above result, we get

$-1/2 \leq ab + bc + ca \leq 1$

$\therefore ab + bc + ca \in [-1/2, 1]$

64. (d) Given α, β are the roots of

$x^2 + px + q = 0$

$\Rightarrow \alpha + \beta = -p$ and $\alpha\beta = q$

α^4, β^4 are the roots of $x^2 - rx + s = 0$

$\Rightarrow \alpha^4 + \beta^4 = r$ & $\alpha^4 . \beta^4 = s$

Then

$q = \alpha\beta$, $r = \alpha^4 + \beta^4$

product of the roots of the equation

$x^2 - 4qx + (2q^2 - r) = 0$

$= 2q^2 - r = 2\alpha^2\beta^2 - \alpha^4 - \beta^4 = -(\alpha^2 - \beta^2)^2$

⇒ Since product of the roots is negative so one root is positive and other is negative.

65. (d) This question is already discussed in Concept deviator (question number 24)

66. (a) Consider the given equation

$\sin(e^x) = 5^x + 5^{-x}$

We know 5^x and 5^{-x} both are + ve real numbers using A M ≥ G M

$\therefore 5^x + \frac{1}{5^x} \geq 2 \Rightarrow 5^x + 5^{-x} \geq 2$

∴ R. H. S. of given eq. ≥ 2

While $\sin e^x \in [-1, 1]$ i.e., L. H. S. $\in [-1, 1]$

So no real value of x exist

67. (b) For real roots $q^2 - 4pr \geq 0$

$\left(\frac{p+r}{2}\right)^2 - 4pr \geq 0$ (∴ p, q, r are in A. P.)

$\Rightarrow p^2 + r^2 - 14pr \geq 0 \Rightarrow \frac{p^2}{r^2} - 14\frac{p}{r} + 1 \geq 0$

$\Rightarrow \left(\frac{p}{r} - 7\right)^2 - 48 \geq 0 \left|\frac{p}{r} - 7\right| \geq 4\sqrt{3}.$

68. (c) This question is already discussed in theory part

69. (a) The given equation is

$$\sqrt{x+1} - \sqrt{x-1} = \sqrt{4x-1}$$

Squaring both sides, we get

$x + 1 + x - 1 - 2\sqrt{x^2 - 1} = 4x - 1$

$\Rightarrow -2\sqrt{x^2 - 1} = 2x - 1$

Again squaring both sides, we get

$\Rightarrow 4(x^2 - 1) = 4x^2 - 4x + 1 \Rightarrow -4x = -5$

$\Rightarrow x = 5/4$

Substituting this value of x in given equation, we get

$\sqrt{\frac{5}{4} + 1} - \sqrt{\frac{5}{4} - 1} = \sqrt{4 \times \frac{5}{4} - 1}$

$\Rightarrow \frac{3}{2} - \frac{1}{2} = 2$ (not satisfied)

∴ $x = 5/4$ is not a solution of given equation

⇒ the given equation has no solution.

∴ (a) is the correct option.

70. (a) We know that if both roots of a quadratic equation $ax^2 + bx + c = 0$ are less than k then $af(k) > 0$, $D \geq 0$, $\alpha + \beta < 2k$.

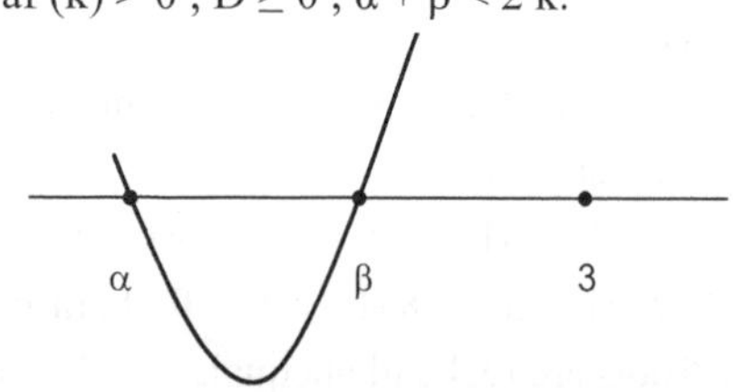

$f(x) = x^2 - 2ax + a^2 + a - 3 = 0,$

$f(3) > 0, \alpha + \beta < 6, a < 3, -4a = 12 \geq 0$

$\Rightarrow a^2 - 5a + 6 > 0, a < 3, -4a + 12 \geq 0$

$\Rightarrow a < 2$ or $a > 3, a < 3, a < 3 \Rightarrow a < 2.$

71. (b) Given that $c < 0 < b$ and

$\alpha + \beta = -b$ ...(i)

$\alpha\beta = c$ ...(ii)

from (ii), $c < 0 \Rightarrow \alpha\beta < 0$

$\Rightarrow$ either α is – ve and β is positive or β is – ve and α is +ve

From (i), $b > 0 \Rightarrow -b < 0 \Rightarrow \alpha + \beta < 0 \Rightarrow$ the sum is negative.

$\Rightarrow$ Therefore, it is clear that α is negative and β is positive and modulus of α is greater than modulus of

$\beta \Rightarrow \alpha < 0 < \beta < |\alpha|$

72. (d) Consider $(x - a)(x - b) = 0$

It has two roots a and b.

Now we need to find roots of $(x - a)(x - b) = 1$

This will have one root less than a and other more than b, as shown in the figure.

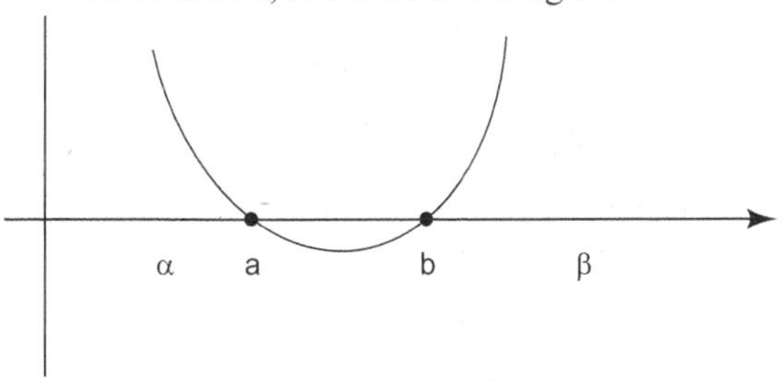

From graph it is clear that one root of the equation lies in $(-\infty, a)$ and other in (b, ∞).

73. (b) For $x < -2$, $|x + 2| = -(x + 2)$ and the inequality becomes

$x^2 + x + 2 + x > 0 \Rightarrow (x + 1)^2 + 1 > 0$

Which is valid $\forall x \in R$ but $x < -2$

$\therefore x \in (-\infty, -2)$

For $x \geq 2$, $|x + 2| = x + 2$ and the inequality becomes

$x^2 - x - 2 + x > 0 \Rightarrow x^2 > 2 \Rightarrow x > \sqrt{2}$ or $x < -\sqrt{2}$

i.e., $x \in (-\infty, -\sqrt{2}) \cup (\sqrt{2}, \infty)$

but $x \geq -2 \Rightarrow x \in [-2, -\sqrt{2}) \cup (\sqrt{2}, \infty)$

Combining above result we get

$x \in (-\infty, -2) \cup [-2, -\sqrt{2}) \cup (\sqrt{2}, \infty)$

$\Rightarrow x \in (-\infty, -\sqrt{2}) \cup (\sqrt{2}, \infty)$

74. (b) We know that $f(x) = ax^2 + bx + c$ has same sign as that of a if $D < 0$.

$x^2 + 2ax + 10 - 3a > 0 \ \forall x$

$\Rightarrow D < 0 \Rightarrow 4a^2 - 4(10 - 3a) < 0$

$\Rightarrow a^2 + 3a - 10 < 0$

$\Rightarrow (a + 5)(a - 2) < 0 \Rightarrow a \in (-5, 2)$

75. (a) Given equation is $x^2 + px + q = 0$

Let roots be α and α^2

$\Rightarrow \alpha + \alpha^2 = -p, \alpha\alpha^2 = q \Rightarrow \alpha = q^{1/3}$

$\Rightarrow (q)^{1/3} + (q^{1/3})^2 = -p$

Taking cube of both sides, we get

$q + q^2 + 3q(q^{1/3} + q^{2/3}) = -p$

$\Rightarrow q + q^2 - 3pq = -p^3$

$\Rightarrow p^3 + q^2 - q(3p - 1) = 0$

76. (a) $\because$ a, b, c are sides of a triangle and $a \neq b \neq c$

$\therefore |a - b| < |c| \Rightarrow a^2 + b^2 - 2ab < c^2$

Similarly

$b^2 + c^2 - 2bc < a^2$; $c^2 + a^2 - 2ac < b^2$

On adding

$a^2 + b^2 + c^2 < 2(ab + bc + ca)$

$\Rightarrow \dfrac{a^2 + b^2 + c^2}{ab + bc + ca} < 2$ ...(i)

$\because$ Roots of the given equation are real so

$\therefore (a + b + c)^2 - 3\lambda(ab + bc + ca) \geq 0$

$\Rightarrow \dfrac{a^2 + b^2 + c^2}{ab + bc + ca} \geq 3\lambda 2$ ...(ii)

From (i) and (ii), we get $3\lambda - 2 < 2 \Rightarrow \lambda < \dfrac{4}{3}$.

77. (d) As α, β are the roots of $x^2 - px + r = 0$

$\therefore \alpha + \beta = p$ ...(i)

And $\alpha\beta = r$ ...(ii)

Also, $\dfrac{\alpha}{2}, 2\beta$ are the roots of $x^2 - qx + r = 0$

$\therefore \dfrac{\alpha}{2} + 2\beta = q$ or $\alpha + 4\beta = 2q$ ...(iii)

Solving (i) and (iii) for α and β, we get

$\beta = \dfrac{1}{3}(2q - p)$ and $\alpha = \dfrac{2}{3}(2p - q)$

subtracting values of α and β, in equation (ii),

we get $\dfrac{2}{9}(2p - q)(2q - p) = r$.

78. (b) Given that $\alpha + \beta = -p$ and $\alpha^3 + \beta^3 = q$

$\Rightarrow (\alpha + \beta)^3 - 3\alpha\beta(\alpha + \beta) = q$

$\Rightarrow -p^3 - 3\alpha\beta(-p) = q \Rightarrow \alpha\beta = \dfrac{p^3+q}{3p}$

Now for required quadratic equation,

Sum of roots $= \dfrac{\alpha}{\beta} + \dfrac{\beta}{\alpha} = \dfrac{\alpha^2+\beta^2}{\alpha\beta}$

$$= \frac{p^2 - 2\left(\frac{p^3+q}{3p}\right)}{\frac{p^3+q}{3p}}$$

$$= \frac{3p^3 - 2p^3 - 2q}{p^3+q} = \frac{p^3-2q}{p^3+q}$$

and product of roots $= \dfrac{\alpha}{\beta}, \dfrac{\beta}{\alpha} = 1$

$\therefore$ Required equation is $x^2 - \left(\dfrac{p^3-2q}{p^3+q}\right)x + 1 = 0$

Or $(p^3 + q)x^2 - (p^3 - 2q)x + (p^3 + q) = 0$

79. (c) $\because \alpha, \beta$ are the roots of $x^2 - 6x - 2 = 0$

$\therefore \alpha^2 - 6\alpha - 2 = 0$

$\Rightarrow \alpha^{10} - 6\alpha^9 - 2\alpha^8 = 0$

$\Rightarrow \alpha^{10} - 2\alpha^8 = 6\alpha^9$ ... (i)

Similarly $\beta^{10} - 2\beta^8 = 6\beta^9$ ... (ii)

From equation (i) and (ii)

$\alpha^{10} - \beta^{10} - 2(\alpha^8 - \beta^8) = 6(\alpha^9 - \beta^9)$

$\Rightarrow a_{10} - 2a_8 = 6a_9$

$\Rightarrow \dfrac{a_{10} - 2a_g}{2a_9} = 3.$

80. (b) Let α be the common root of given equations, then

$\alpha^2 + b\alpha - 1 = 0$ ...(i)

and $\alpha^2 + \alpha + b = 0$ ... (ii)

subtracting (ii) from (i), we get

$(b - 1)\alpha - (b + 1) = 0$

Or $\alpha = \dfrac{b+1}{b-1}$

Subtracting this value of α in equation (1), we get

$\left(\dfrac{b+1}{b-1}\right)^2 + b\left(\dfrac{b+1}{b-1}\right) - 1 = 0$ or $b^3 + 3b = 0$

$\Rightarrow b = 0, i\sqrt{3}, -i\sqrt{3}.$

81. (d) Quadratic equation with real coefficients and purely imaginary roots can be considered as

$p(x) = x^2 + a = 0$ where $a > 0$ and $a \in R$

The $p[p(x)] = 0 \Rightarrow (x^2 + a)^2 + a = 0$

$\Rightarrow x^4 + 2ax^2 + (a^2 + a) = 0$

$\Rightarrow x^2 = \dfrac{-2a \pm \sqrt{4a^2 - 4a^2 - 4a}}{2}$

$\Rightarrow x^2 = -a \pm \sqrt{a}\, i$

$\Rightarrow x = \sqrt{-a \pm \sqrt{a}\, i} = \alpha \pm i\beta$ where $\alpha, \beta \neq 0$

$\therefore$ p[p(x)] = 0 has complex roots which are neither purely real nor purely imaginary.

82. (b) $x^2 - 2x\sec\theta + 1 = 0 \Rightarrow x = \sec\theta \pm \tan\theta$

and $x^2 + 2x\tan\theta - 1 = 0 \Rightarrow x = -\tan\theta \pm \sec\theta$

$\because -\dfrac{\pi}{6} < \theta < -\dfrac{\pi}{12}$

$\Rightarrow \sec\dfrac{\pi}{6} > \sec\theta > \sec\dfrac{\pi}{12}$

and $-\tan\dfrac{\pi}{6} < \tan\theta < -\dfrac{\tan\pi}{12}$

also $\tan\dfrac{\pi}{12} < -\tan\theta < \tan\dfrac{\pi}{6}$

α_1, β_1 are roots of $x^2 - 2x\sec\theta + 1 = 0$

and $\alpha_1 > \beta_1$

$\therefore \alpha_1 = \sec\theta - \tan\theta$ and $\beta_1 = \sec\theta + \tan\theta$

α_2, β_2 are roots of $x^2 + 2x\tan\theta - 1 = 0$ and $\alpha_2 > \beta_2$

$\therefore \alpha_2 = -\tan\theta + \sec\theta$, $\beta_2 = -\tan\theta - \sec\theta$

$\therefore \alpha_1 + \beta_2 = \sec\theta - \tan\theta - \tan\theta - \sec\theta$

$= -2\tan\theta$

(Passage Type)

For (Q. 83-84):

Let roots of $x^4 - 12x^3 + bx^2 + cx + 81 = 0$ be $\alpha, \beta, \gamma, \delta$

$\Rightarrow$ Sum of the roots is $\alpha + \beta + \gamma + \delta = 12$ and

Product of the roots is $\alpha\beta\gamma\delta = 81$

We know that AM $\geq$ GM and equality holds when all the elements are equal

$$\frac{\alpha+\beta+\gamma+\delta}{4} \geq (\alpha\beta\gamma\delta)^{1/4}$$

Since, $\frac{\alpha+\beta+\gamma+\delta}{4} = 3$ and $(\alpha\beta\gamma\delta)^{1/4} = 3$

$$\alpha = \beta = \gamma = \delta \Rightarrow \alpha = \beta = \gamma = \delta = 3$$

Now, $b = \Sigma \alpha \beta = 6 \times 9 = 54$

$c = -\Sigma \alpha \beta \gamma = -4 \times 27 = -108$

83. (a) So $b - c = 54 + 108 = 162$

84. (b) **Consider** $bx^2 + bx + c = 0$ sum of the co efficient is $b + b + c = 2b + c = 0$

$\therefore$ 1 is a root of equation $bx^2 + bx + c = 0$

For (Q. 85-86):

Let roots of $x^2 + b_1x + c_1 = 0$ be α & β then that of $x^2 + b_2x + c_2 = 0$ be $\alpha + k$ & $\beta + k$

$$(\beta - \alpha) = (\beta + k) - (\alpha + k)$$

$$(\beta + \alpha)^2 - 4\alpha\beta = [(\beta + k) + (\alpha + k)]^2 - 4(\beta + k)(\alpha + k)$$

$$(-b_1)^2 - 4c_1 = (-b_2)^2 - 4c_2$$

$$D_1 = D_2 \quad \text{..... (i)}$$

85. (c) Least value of $x^2 + b_1x + c_1$ is $-\frac{D_1}{4} = -\frac{1}{4}$

$\Rightarrow D_1 = 1$ and $D_2 = 1$

$\therefore$ Least value of is $x^2 + b_2x + c_2$ is $-\frac{D_2}{4} = -\frac{1}{4}$

86. (b) Least value of $x^2 + b_2x + c_2$ occurs at

$$x = -\frac{b_2}{2} = \frac{7}{2} \Rightarrow b_2 = -7$$

For (Q. 87-88):

Since λ is a root of $f(f(x)) = x$

$\Rightarrow f(f(\lambda)) = \lambda$

Let $f(\lambda) = t \Rightarrow f(t) = \lambda$

Means if point (λ, t) lie on the function then (t, λ) also lie on the function

$\Rightarrow$ There must be two points C and D on the parabola $y = ax^2 + bx + c$ which are image of each other in the line $y = x$

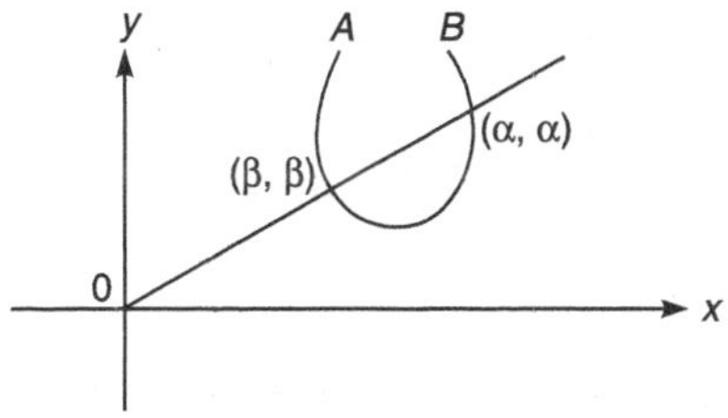

Now we have following cases-

Case (i): If α, β are real and distinct

In this case graph of f (x) will cut the graph of y = x twice

$\Rightarrow$ If α, β are real so are λ and δ as α, β will lie on line y = x

So number of roots of $f(f(x)) = x$ is 4

Case (ii): If α, β are real and equal

In this case graph of f (x) will touch the line y = x so λ and δ are not real.

Case (iii): If α, β are imaginary

In this case graph of f (x) will not meet the line y = x so λ and δ are not real

87. (b)

88. (a)

For (Q. 89-90):

89. (b) Since x_1, x_2, x_3, x_4 be the roots of the equation $x^4 + ax^3 + bx^2 + cx + d = 0$

So $x^4 + ax^3 + bx^2 + cx + d = (x - x_1)(x - x_2)(x - x_3)(x - x_4)$

Let $(x - x_1)(x - x_2) = x^2 + px + q$ and

$(x - x_3)(x - x_4) = x^2 + px + r$

$\therefore q = x_1x_2$ and $r = x_3x_4$

$\therefore x^4 + ax^3 + bx^2 + cx + d = x^4 + 2px^3 + (p^2 + q + r)x^2 + p(q + r)x + qr$

Comparing we get $a = 2p$ ... (i)

$b = p^2 + q + r$ ... (ii)

$c = p(q + r)$ ... (iii)

$d = qr$ ...(iv)

On solving we get, $a^3 - 4ab + 8c = 0$ ... (v)

If $a = 2 \Rightarrow b - c = 1$

90. (a) Substituting $c = 1 - b$ in Eq. (v) we get,

$(a + 2)[(a - 1)^2 + 3 - 4b] = 0 \quad \Rightarrow 4b - 3 > 0$

$\Rightarrow b > \frac{3}{4} \quad \Rightarrow c < \frac{1}{4}$

For (Q. 91-92):`

91. (a) Let $f(x) = 8x^3 - 20x^2 + 6x + 9$, then

$f'(x) = 24x^2 - 40x + 6.$

$f'(x) = 0$

$\Rightarrow 24x^2 - 40x + 6 = 0$ has roots $x = \frac{3}{2}, \frac{1}{6}$

$\Rightarrow$ But $f\left(\frac{3}{2}\right) = 0.$

We know that the equation $g(x) = 0$ has a repeated root $x = \alpha$, if $g(\alpha) = g'(\alpha) = 0$.

Therefore, $x = \frac{3}{2}$ is a repeated root of $f(x) = 0$

92. (c) This type of questions solved with some observations and use the property if f(a) . f(b) < 0 then f(x) =0 must have at least one real root between (a, b).

In other words if f(a) and f(b) are of opposite sign then f(x) = 0 must have at least one real root between (a, b).

We have $f(-2) = -1 < 0$ and $f(-1) = 3 > 0$, so there is one root (at least one) in $(-2, -1)$

Also $f(0) = 1 > 0$, and $f(1) = -1 < 0$, so there is one root (at least one) in (0, 1)

Also $f(1) = -1 < 0$, and $f(2) = 1 > 0$, so there is one root (at least one) in (1, 2)

Thus, the question f(x) = 0 has three distinct roots α, β, γ such that $-2 < \alpha < -1$, $0 < \beta < 1$, $1 < \gamma < 2$.

$[\alpha] = -2$, $[\beta] = 0$ and $[\gamma] = 1$;

so $[\alpha] + [\beta] + [\gamma] = -1$

For (Q. 93-94):`

93. (d) $y = \frac{x - 5}{x(x + 4)}$. Points of discontinuity are $x = 0, x = -4$

Thus, x = 0 and x = – 4 are vertical asymptote.

y = 0 is a horizontal asymptote.

94. (c) Since, $x^2 + 4x + 5 > 0$ for all x, so the curve has no vertical asymptote.

Now we can rewrite the function as,

$$y = \frac{x + 2}{x^2 + 4x + 5} = \frac{1}{(x + 2) + \frac{1}{x + 2}}$$

Clearly $y \to 0 + 0$ if $x \to +\infty$ and $y \to 0 - 0$ if $x \to -\infty$

Thus, y = 0 is a horizontal asymptote.

For (Q. 95-96):`

The given equations are

$x^2 + 2p_1x + 1 = 0$ ...(1)

Let α and β be roots of this equation

$x^2 + 4p_2x + 2 = 0$ ...(2)

β and γ be roots of this equation

$x^2 + 6p_3x + 3 = 0$ ...(3)

γ and α be roots of this equation

then

$\alpha + \beta = 2p_1$ and $\alpha\beta = 1$

$\beta + \gamma = 4p_2$ and $\beta\gamma = 2$

$\gamma + \alpha = 6p_3$ and $\gamma\alpha = 3$

From above $\alpha\beta^2\gamma = 2 \Rightarrow \beta^2 = \frac{2}{3}$;

$\alpha^2\beta\gamma = 3 \Rightarrow \alpha^2 = \frac{3}{2}$

$\alpha\beta\gamma^2 = 6 \Rightarrow \gamma^2 = 6$

Clearly, α, β and γ all three must have same signs. So, $\alpha = \sqrt{\frac{3}{2}}, \beta = \sqrt{\frac{2}{3}}, \gamma = \sqrt{6}$

Or, $\alpha = -\sqrt{\frac{3}{2}}, \beta = -\sqrt{\frac{2}{3}}, \gamma = -\sqrt{6}$

95. (b) Sum of all the three common roots is $\sqrt{\frac{3}{2}}$

$+\sqrt{\frac{2}{3}}+\sqrt{6}=\frac{11}{\sqrt{6}}$ or $=-\frac{11}{\sqrt{6}}$

96. (c) Required equation is $\left(x-\sqrt{\frac{3}{2}}\right)\left(x+\sqrt{\frac{3}{2}}\right)$

$\left(x-\sqrt{\frac{2}{3}}\right)\left(x+\sqrt{\frac{2}{3}}\right)\left(x-\sqrt{6}\right)\left(x+\sqrt{6}\right)=0$

So required degree of equation is 6

97. (a) Given that $\alpha+\beta+\gamma=2$,

$\frac{1}{\alpha}+\frac{1}{\beta}+\frac{1}{\gamma}=\frac{1}{2}$,

$\frac{1}{\alpha^2}+\frac{1}{\beta^2}+\frac{1}{\gamma^2}=\frac{9}{4}$

Consider $2\left(\frac{1}{\beta\gamma}+\frac{1}{\gamma\alpha}+\frac{1}{\alpha\beta}\right)=\left(\frac{1}{\alpha}+\frac{1}{\beta}+\frac{1}{\gamma}\right)^2$

$-\left(\frac{1}{\alpha^2}+\frac{1}{\beta^2}+\frac{1}{\gamma^2}\right)$

$\Rightarrow \frac{2(\alpha+\beta+\gamma)}{\alpha\beta\gamma}=-2 \Rightarrow \alpha\beta\gamma=-2$

Also, $\frac{\beta\gamma+\gamma\alpha+\alpha\beta}{\alpha\beta\gamma}=\frac{1}{2}$

$\Rightarrow \beta\gamma+\gamma\alpha+\alpha\beta=-1$

Equation whose roots are α, β & γ is given by

$x^3-(\alpha+\beta+\gamma)x^2+(\beta\gamma+\gamma\alpha+\alpha\beta)x-\alpha\beta\gamma=0$

$x^3-2x^2-x+2=0$

98. (b) Now consider $=\alpha^2+\beta^2+\gamma^2=(\alpha+\beta+\gamma)^2-2(\beta\gamma+\gamma\alpha+\alpha\beta)=6$

And $\alpha^3+\beta^3+\gamma^3-3\alpha\beta\gamma$

$=(\alpha+\beta+\gamma)(\alpha^2+\beta^2+\gamma^2-\beta\gamma-\gamma\alpha-\alpha\beta)$

So $\alpha^3+\beta^3+\gamma^3=-6+(2)(6+1)=8$

(Integer Type)

99. **(4)** Let $f(x)=x^3-6x^2+9x+\lambda$

$\therefore f'(x)=3x^2-12x+9=3(x-1)(x-3)$

$\therefore f'(x)=0$ at $(1, 3)$

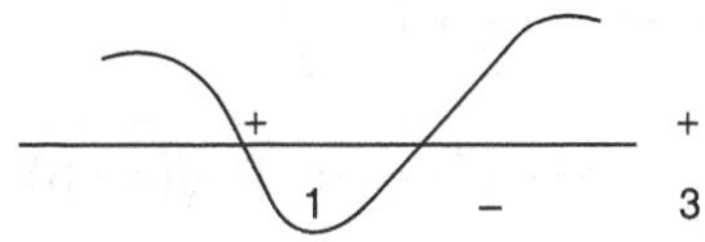

But $f(1)=4+\lambda$ and $f(3)=\lambda$ for $f(x)=0$ to have exactly one root in $(1, 3)$

$\Rightarrow f(1)$ and $f(3)$ should have opposite signs.

$\therefore f(1)f(3)<0$

$\Rightarrow \lambda(\lambda+4)<0$

$\Rightarrow [\lambda+1]=-3,-2,-1,0$

Number of integral values is 4.

100. (7) Since sum of the coefficents of the 2nd equation is zero so $x=1$ is one of the roots of the 2nd equation.

$\therefore$ Putting $x \Rightarrow 1$ in 1st equation $x^2+\lambda x-8=0$ then $1+\lambda-8=0$ or $\lambda=7$

101. **(4)** Given inequality is $-3\le\frac{x^2-\lambda x-2}{x^2+x+1}\le 2$

$\Rightarrow -3x^2-3x-3\le x^2-\lambda x-2\le 2x^2+2x+2$

(since $x^2+x+1>0$ for all $x\in R$)

From 1st inequality $4x^2-(\lambda-3)x+1\ge 0$

$\Rightarrow (\lambda-3)^2-4\cdot 1\cdot 4\cdot 1\le 0$

$\Rightarrow -4\le\lambda-3\le 4$

$\Rightarrow -1\le\lambda\le 7$

From 2nd inequality $x^2+(\lambda+2)x+4\ge 0$

$\therefore (\lambda+2)^2-4\cdot 1\cdot 4\cdot 1\le 0$

$\Rightarrow -4\le\lambda-3\le 4$

$\Rightarrow -6\le\lambda\le 2$

Combining the above two results we get $-1\le\lambda\le 2$

Number of integral value is 4

102. **(5)** Sum of the roots $p+q=\alpha-2$

Product of the roots $pq=-\alpha-1$

$\therefore p^2+q^2=(\alpha-2)^2-2(-\alpha-1)=\alpha^2-4\alpha+4+2\alpha+2=\alpha^2-2\alpha+6$

Minimum value of this is when $\alpha=1$

$\therefore$ Least value of $p^2+q^2=5$

103. **(6)** $\alpha+\beta=-\frac{7}{2}, \alpha\beta=\frac{k}{2}$

$$\therefore \frac{7}{4}=|\alpha^2-\beta^2|=|\alpha+\beta|\sqrt{(\alpha+\beta)^2-4\alpha\beta}$$

$$=\frac{7}{2}\sqrt{\frac{49}{4}-2k}=\frac{7}{4}\sqrt{49-8k}$$

$\Rightarrow 49-8k=1 \qquad \Rightarrow k=6$

104. **(0)** Given equation can be written as

$$x^4\left(x-\frac{1}{2}\right)^2+\frac{3}{4}x^2\left(x-\frac{2}{3}\right)^2+\frac{3}{8}=0$$

All the terms are positive

$\Rightarrow$ No real solution

Answer is [0]

105. **(3)** Roots of the given equation will be opposite sign, if $D \geq 0$ and product of roots < 0.

$\Rightarrow (a^3+8a-1)^2-8(a^2-4a) \geq 0$ and $\frac{a^2-4a}{2}<0$

$\Rightarrow a^2-4a<0$

[Since $a^2-4a<0 \Rightarrow (a^3+8a-1)^2-8(a^2-4a) \geq 0$]

$\Rightarrow 0<a<4$ integral values of a are $\{1, 2, 3\}$

106. **(6)** The given equation has negative roots of discriminant ≥ 0, $\alpha<0$, $\beta<0$ ie, $\alpha+\beta<0$ and $f(0)>0$

$\therefore 4(a+1)^2-36a+20>0$

$\Rightarrow a^2-7a+6 \geq 0$

$\Rightarrow a \leq 1$ or $a \geq 6$ ... (i)

$\alpha+\beta<0$

$\Rightarrow a+1>0 \Rightarrow a>-1$ ... (ii) and $f(0)>0$

$9a-5>0 \Rightarrow a>\frac{5}{9}$ ... (iii)

By combining equations (i), (ii) and (iii), we get $a \geq 6$

So minimum value of 'a' is 6

107. **(4)** For a complete polynomial equation with rational coefficients irrational roots occur in pairs.

Let $x=\sqrt{3}-\sqrt{2}$ or $x^2=5-2\sqrt{6}$

$$(x^2-5)^2=24$$

$x^4-10x^2+25=24 \Rightarrow x^4-10x^2+1=0$

108. **(1)** Since roots are of opposite sign so product of the roots are –ve

$a^2-3a+2<0 \Rightarrow 1<a<2$ or $[a]=1$

109. (2) Let α, β, γ, δ be the roots, then

$\alpha+\beta+\gamma+\delta=8$ and $\alpha\beta\gamma\delta=16$

$\Rightarrow \frac{\alpha+\beta+\gamma+\delta}{4}=(\alpha\beta\gamma\delta)^{1/4}=2$

A.M. = G. M.

$\therefore \alpha=\beta=\gamma=\delta=2$

Or, $x^4-8x^3+bx^2-cx+16=(x-2)^4$

110. **(6) From** the given condition $b^2-4ac<0$ and $a<0$

Hence $\cot\alpha<0$ i.e., $\alpha \in$ 2nd and 4th quadrant and $4\sin\alpha-2\tan\alpha\cot\alpha<0$

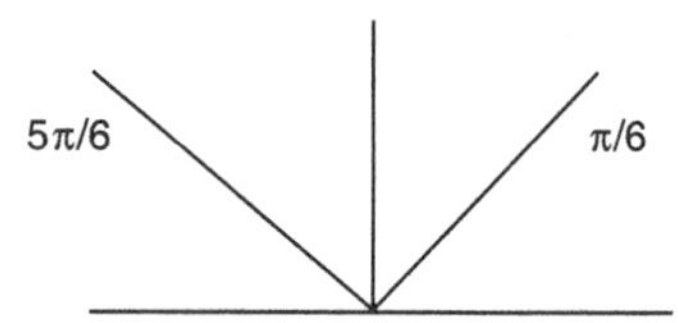

$2\sin\alpha<1 \Rightarrow \sin\alpha<\frac{1}{2}$

$\therefore \alpha \in 2^{nd}$ and 4th quadrant

$\sin\alpha \geq 0$, as $\sqrt{\sin\alpha}$ to exists

$\therefore \alpha \in$ 2nd quadrant

Hence, $\alpha \in \left(\frac{5\pi}{6}, \pi\right)$

111. **(3)** Let the original equation has two roots α & β

Then $\alpha\beta=\alpha^2\beta^2$ .... (i) and $\alpha^2+\beta^2=\alpha+\beta$ ... (ii)

Hence, $\alpha\beta(1-\alpha\beta)=0 \Rightarrow \alpha=0$ or $\beta=0$ or $\alpha\beta=1$

If $\alpha=0$, then from Eq. (ii) $\beta=0$ or $\beta=1$ or $\alpha\beta=1$ or (0, 1), if $\beta=0$ then $\alpha=0$ or $\beta=1$, If $\beta=\frac{1}{\alpha}$, then $\alpha^2+\frac{1}{\alpha^2}=\alpha+\frac{1}{\alpha}$

$$\Rightarrow \left[\alpha+\frac{1}{\alpha}\right]^2-2=\alpha+\frac{1}{\alpha}$$

Hence, $t^2-t-2=0 \Rightarrow (t-2)(t+1)=0$

$\Rightarrow t = 2$ or $t = -1$, if $t = 2 \Rightarrow \alpha = 1$ and $\beta = 1$, if $t = -1$ roots are imaginary (ω or ω^2)

$\therefore$ The number of quadratic equation is 3.

112. **(1)** $\log_4 (2x^2 + x + 1) - \log_2 (2x - 1) = 1$

$$\Rightarrow \frac{\log_e(2x^2+x+1)}{\log_e 4} - \frac{\log_e(2x-1)}{\log_e 2} = 1$$

$$\Rightarrow \log_e \frac{2x^2+x+1}{(2x-1)^2} = \log_e 4$$

$\Rightarrow (2x^2 + x + 1) = 4(4x^2 - 4x + 1)$

$\Rightarrow 14x^2 - 17x + 3 = 0 \Rightarrow (14x - 3)(x - 1) = 0$

$\Rightarrow x = \frac{3}{14}$, $x = 1$

But $x = \frac{3}{14}$ does not lie in the domain of function hence eliminated

Hence, x = 1 is the only solution.

113. **(1)** Here, $\left.\begin{matrix}\alpha+\beta=-p \\ \gamma+\delta=-p\end{matrix}\right\} \Rightarrow \alpha+\beta=\gamma+\delta$

Now, $(\alpha - \gamma)(\alpha - \delta) = \alpha^2 - \alpha(\gamma + \delta) + \gamma\delta$

$= \alpha^2 - \alpha(\alpha + \beta) + r = -\alpha\beta + r = -(-q) + r = q + r$

By symmetry of the results

$(\beta - \gamma)(\beta - \delta) = q + r$

Hence, the required ratio is 1

114. **(8)** The given inequality $\log_{\log_2 \frac{x}{2}} (x^2 - 10x + 22) > 0$

The given expression is defined when

(i) $x^2 - 10x + 22 > 0 \Rightarrow x < 5 - \sqrt{3}$ or $x > 5 + \sqrt{3}$

(ii) $\frac{x}{2} > 0 \Rightarrow x > 0$

We have two cases-

Case 1: $0 < \log_2\left(\frac{x}{2}\right) < 1 \Rightarrow 1 < \frac{x}{2} < 2$

$\Rightarrow 2 < x < 4$

Then $\log_{\log_2 \frac{x}{2}}(x^2 - 10x + 22) > 0$

$\Rightarrow x^2 - 10x + 22 < 1 \Rightarrow x^2 - 10x + 21 < 0$

$\Rightarrow 3 < x < 7$.

So $3 < x < 4$

Case 2: $\log_2\left(\frac{x}{2}\right) > 1 \Rightarrow \frac{x}{2} > 2 \Rightarrow x > 4$.

Then $\log_{\log_2 \frac{x}{2}}(x^2 - 10x + 22) > 0$

$\Rightarrow x^2 - 10x + 22 > 1 \Rightarrow x^2 - 10x + 21 > 0$

$\Rightarrow x < 3$ or $x > 7$

So $x > 7$.

$\therefore$ The values of x from two cases are

$x \in (3, 4) \cup (7, \infty)$

Combining above all result we get $x \in (3, 5 - \sqrt{3}) \cup (7, \infty)$

115. **(2)** The given equation can be written as $x^2 - x + a - 2 = 0$

Its roots will be real if its D is positive

So $1 - 4(a - 2) > 0 \Rightarrow a < 9/4$

Also, a – 2 product of roots of the equation ≥ 0

So $a - 2 \geq 0$

Thus, $2 \leq a < 9/4$

only integral value of 'a' is 2

[Answer is 2]

116. **(1)** Let $f(x) = x^3 - x^2 + \beta x + \gamma$

f(x) = 0 has three positive real roots in G.P.

$\Rightarrow f'(x) = 0$ will have two distinct real roots

$\Rightarrow 3x^2 - 2x + \beta = 0$ has two distinct roots

$\therefore D > 0$

$4 - 12\beta > 0 \Rightarrow \beta < \frac{1}{3}$

Since x_1, x_2, x_3 are roots of equation

$x^3 - x^2 + \beta x + \gamma = 0$

$x_1x_2 + x_2x_3 + x_3x_1 = \beta$

$x_1 + x_2 + x_3 = 1$

$x_1 x_2 x_3 = -\gamma$

$\Rightarrow x_2^3 = -\gamma$ (since, $x_2^2 = x_1x_3$)

$\Rightarrow \gamma < 0$

Also, $x_2(x_1 + x_3) + x_2^2 = \beta$

$\Rightarrow x_2(1 - x_2) + x_2^2 = \beta \Rightarrow x_2 = \beta > 0$

From above result we get $0 < \beta < \frac{1}{3}$ and $\gamma < 0$.

$[\beta]_{max} = 0$, $[\gamma]_{max} = -1$

Hence, $[\beta] + [\gamma] + 2 = 1$.

117. **(4)** Consider the upper limit $= \dfrac{x^2 + x + d}{x^2 + 2x + d} = \dfrac{3}{2}$

So $2(x^2 + x + d) = 3(x^2 + 2x + d)$

$x^2 + 4x + d = 0$

Above equation has equal roots $\Rightarrow d = 4$

Similarly we will find the same value of d, if $\dfrac{x^2 + x + d}{x^2 + 2x + d} = \dfrac{5}{6}$ has equal roots

118. **(9)** The given inequality $x^2 + |x - a| - 3 < 0$

Here we have two cases-

Case (i): Let $x \geq a$, then

$x^2 + x - (a + 3) < 0$.

It will have a solution if $D > 0$

$\Rightarrow 1 + 4(a + 3) > 0 \Rightarrow a > -\frac{13}{4}$ and the solution is $\dfrac{-1 - \sqrt{4a + 13}}{2} < x < \dfrac{-1 + \sqrt{4a + 13}}{2}$.

Clearly, it contains negative values of x.

Case (ii): Let $x < a$, then

$x^2 - x + a - 3 < 0$. It will have a solution, if $D > 0$

$\Rightarrow 1 - 4(a - 3) > 0 \Rightarrow a < \frac{13}{4}$ and the solution is $\dfrac{1 - \sqrt{13 - 4a}}{2} < x < \dfrac{1 + \sqrt{13 - 4a}}{2}$

It will have a negative value of x, if

$\dfrac{1 - \sqrt{13 - 4a}}{2} < 0$

$\Rightarrow 1 - \sqrt{13 - 4a} < 0 \Rightarrow \sqrt{13 - 4a} > 1$

$\Rightarrow 13 - 4a > 1 \Rightarrow a < 3$

Combining these three values of a, we get $-\frac{13}{4} < a < 3$.

So, the integral values of a are –3, –2, –1, 0,1,2. And the sum of absolute values of the integral values of 'a' is 3+2+1+0+1+2 = 9

119. **(3)** Given, $x^2 - 2x + \cos^{10}\alpha = 0$

$\Rightarrow \cos^{10}\alpha = 2x - x^2$

We know that $-1 \leq \cos\alpha \leq 1$

$\therefore 0 \leq \cos^{10}\alpha \leq 1$ or $0 \leq 2x - x^2 \leq 1$

If $2x - x^2 \geq 0$ then $0 \leq x \leq 2$

If $2x - x^2 \leq 1$ then $(x - 1)^2 \geq 0$, which is true

$\therefore 0 \leq x \leq 2$

So possible integral values are 0,1 and 2

120. **(6)** We Know, $(\alpha + \beta + \gamma)^2 = \alpha^2 + \beta^2 + \gamma^2 + 2(\beta\gamma + \gamma\alpha + \alpha\beta) \Rightarrow 4 = 6 + 2(\beta\gamma + \gamma\alpha + \alpha\beta)$

$\Rightarrow \beta\gamma + \gamma\alpha + \alpha\beta = -1$ .... (i)

Also, $\alpha^3 + \beta^3 + \gamma^2 - 3\alpha\beta\gamma = (\alpha + \beta + \gamma)(\alpha^2 + \beta^2 + \gamma^2 - \beta\gamma - \gamma\alpha - \alpha\beta)$

$\Rightarrow 8 - 3\alpha\beta\gamma = 2(6 + 1)$

$\Rightarrow 3\alpha\beta\gamma = 8 - 14 = -6$ or $\alpha\beta\gamma = -2$ ... (ii)

Now, $(\alpha^2 + \beta^2 + \gamma^2)^2 = \Sigma\alpha^4 + 2\Sigma\beta^2\gamma^2 = \Sigma\alpha^4 + 2[(\Sigma\beta\gamma)^2 - 2\alpha\beta\gamma(\Sigma\gamma)]$

$\Rightarrow \Sigma\alpha^4 = 36 - 2[(-1)^2 - 2(-2)(2)] = 18$

So required value is 18/3 = 6

121. **(2)** $(\lambda^2 + \lambda - 2)x^2 + (\lambda + 2)x - 1 < 0$ for all $x \in R$, if $D < 0$ and $a < 0$, ie, : $(\lambda + 2)^2 + 4(\lambda^2 + \lambda - 2) < 0$ and $\lambda^2 + \lambda - 2 < 0$

$5\lambda^2 + 8\lambda - 4 < 0$ and $\lambda^2 + \lambda - 2 < 0$

$\Rightarrow (5\lambda - 2)(\lambda + 2) < 0$ and $(\lambda + 2)(\lambda - 1) < 0$

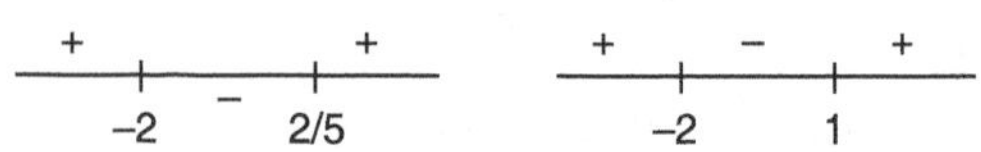

ie, $-2 < \lambda < \frac{2}{5}$ and $-2 < \lambda < 1$

ie $-2 < \lambda < \frac{2}{5}$ or $\lambda \in \left(-2, \frac{2}{5}\right)$. Integral values are –1 and 0

122. **(5)** Solution: $k-2>0$ and $64-4(k-2)(k+4)<0$

$k>2$ and $16-(k^2+2k-8)<0$

$k^2+2k-24>0$

$(k<-6$ or $k>4)$

$\Rightarrow k>4$

$\therefore$ Least integral value of $k=5$

123. **(6)** Consider the left side inequality as $1<\frac{3x^2-7x+8}{x^2+1}$

So, $\frac{3x^2-7x+8}{(x^2+1)}-1>0$

$\Rightarrow \frac{3x^2-7x+8-x^2-1}{(x^2+1)}>0$

$\Rightarrow \frac{2x^2-7x+7}{(x^2+1)}>0$

As $x^2+1>0$ is always positive, so consider numerator $2x^2-7x+7>0$

Discriminant of $y=2x^2-7x+7$

$D=(7)^2-4.2.7=49-56<0$

Thus, $y>0\ \forall\ x$

Thus, solution of this inequality will be R or $x\in(-\infty,\infty)$

Then, solve right side inequality

$\frac{3x^2-7x+8}{x^2+1}\le 2$

$\Rightarrow \frac{(3x^2-7x+8)}{(x^2+1)}-2\le 0$

$\Rightarrow \frac{(3x^2-7x+8)-2(x^2+1)}{(x^2+1)}\le 0$

$\Rightarrow \frac{x^2-7x+6}{(x^2+1)}\le 0$

$\Rightarrow \frac{(x-1)(x-6)}{(x^2+1)}\le 0 \Rightarrow 1\le x\le 6$

Thus, common solution is $x\in[1,6]$

124. **(8)** Substitute $y=x+\frac{3+5}{2}=x+4$

Then

$(y-1)^4+(y+1)^4=16$

$\Rightarrow 2\{y^4+6y^2+1\}=16 \Rightarrow y^4+6y^2-7=0$

$\Rightarrow (y^2+7)(y^2-1)=0$

$\Rightarrow y^2=-7$ or $y^2=1$

$\Rightarrow y=\pm\sqrt{7}i$ or $y=\pm 1$

$\Rightarrow x=-4\pm\sqrt{7}i$ or $x=-4\pm 1=-5$ or -3.

So sum of all the real roots is $-5-3=-8$

125. **(2)** Here we have two cases

Case (i) When $x\ge 1$,

$x^2+x-1=1$

$\Leftrightarrow (x+2)(x-1)=0$

$\Rightarrow x=-2$ or 1

Case (ii) $x<1$,

$x^2+1-x=1 \Rightarrow x=0, 1$

So the roots are 0 and 1

126. **(2)** The given equation is

$x^2-8kx+16(k^2-k+1)=0$

$\because$ Both the roots are real and distinct

$\therefore D>0 \Rightarrow (8k)^2-4\times16(k^2-k+1)>0$

$\Rightarrow k>1$ ...(i)

$\because$ Both the roots are greater than or equal to 4
$\alpha+\beta>8$ and $f(4)\ge 0$

$\Rightarrow k>1$ ...(ii)

and $16-32k+16(k^2-k+1)\ge 0$

$\Rightarrow k^2-3k+2\ge 0 \Rightarrow (k-1)(k-2)\ge 0$

$\Rightarrow k\in(-\infty,1]\cup[2,\infty)$ ...(iii)

Combining (i), (ii) and (iii), we get $k\ge 2$ or the smallest value of $k=2$.

127. **(2)** We have $x^4-4x^3+12x^2+x-1=0$

$\Rightarrow x^4-4x^3+6x^2-4x+1+6x^2+5x-2=0$

$\Rightarrow (x-1)^4+6x^2+5x-2=0$

$\Rightarrow (x-1)^4=-6x^2-5x+2$

To solve the above polynomial, it is equivalent to find the intersection points of the curves $y = (x - 1)^4$ and

$y = -6x^2 - 5x + 2$ or $y = (x - 1)^4$ and

$$= \left(x + \frac{5}{12}\right)^2 = -\frac{1}{6}\left(y - \frac{73}{24}\right)$$

The graph of above two curves as follows.

Clearly they have two points of intersection.

Hence the given polynomial has two real roots.

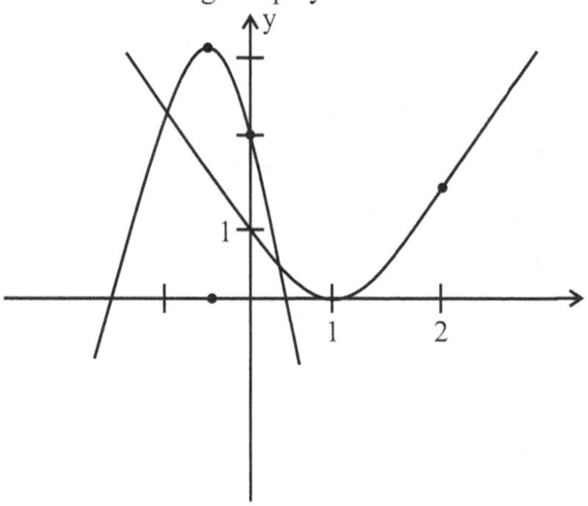

(Assertion and Reason)

128. (a) If quadratic equation $ax^2 + bx + c = 0$ have one root $x = 1$, then $a + b + c = 0$ Statement II is a fact and correct.

In Statement I sum of the coefficients is zero so it has one root 1 so it must have real roots

129. (d) Statement II is a fact and hence correct

let α is repeated root of $f(x) = 0$, then $f(\alpha) = 0$ and $f'(\alpha) = 0$.

$$\therefore f(\alpha) = 1 + \frac{\alpha}{1} + \frac{\alpha^2}{2!} + \frac{\alpha^3}{3!} + \frac{\alpha^4}{4!} + \ldots + \frac{\alpha^{100}}{100!} = 0 \quad \ldots (i)$$

$$\text{And } f'(\alpha) = 1 + \frac{\alpha}{1} + \frac{\alpha^2}{2!} + \frac{\alpha^3}{3!} + \ldots + \frac{\alpha^{99}}{99!} = 0 \quad \ldots (ii)$$

From equation (i) and (ii) we get

$$\Rightarrow \frac{\alpha^{100}}{100!} = 0 \Rightarrow \alpha = 0.$$

But $\alpha = 0$ does not satisfy $f(x) = 0$.

Hence statement (i) is incorrect

Statement I is False, Statement II is True.

130. (a) Statement II is a fact and correct.

Statement I: $x = \sqrt{5} - \sqrt{2}$

Squaring both sides twice we get

$x^4 - 14x^2 + 49 = 40$

131. (b) Statement II is a fact and hence true.

Statement II: Lets assume that each of the three equation has real roots.

Then $b^2 - 4ac \geq 0$, $c^2 - 4ab \geq 0$ and $a^2 - 4ab \geq 0$

$\Rightarrow a^2 b^2 c^2 \geq 64\, a^2 b^2 c^2$. A contradiction .

So statement II is correct.

Statement I is True, Statement II is True; Statement II is NOT a correct explanation for Statement I.

132.(b) Statement I: Given equation $x^2 + bx + c = 0$

Let α, β be two roots such that $|\alpha - \beta| = 1$

$\Rightarrow (\alpha + \beta)^2 - 4\alpha\beta = 1$

$\Rightarrow b^2 - 4c = 1$ so statement I is correct

Statement II Given equation $4abc\,x^2 + (b^2 - 4ac)\,x - b = 0$

$D = (b^2 + 4ac)^2 + 16ab^2c$

$D = (b^2 + 4ac)^2 > 0$

Hence, roots are real and unequal. Statement II is correct but not a correct explanation of statement I

133.(c) Consider statement I : Given that $a c \neq b c$, which implies $c \neq 0$ and $a \neq b$.

we have $x_1^2 + a x_1 + b c = 0$

$x_1^2 + b x_1 + c a = 0$

Subtracting, we get

$(a - b) x_1 + c (b - a) = 0$

$\Rightarrow x_1 = c$ as $a \neq b$.

Thus, $c^2 + a c + b c = 0 \Rightarrow a + b + c = 0$

Also, $x_1 = b$ and $x_3 = a$.

Equation whose roots are x_1 and x_3 is

$x^2 - (a + b) x + a b = 0$ or $x^2 + c x + a b = 0$.

So statement I is true

Statement – II is not true when a = b

134. (b) Statement I Let $f(x)=(x-p)(x-r)+k(x-q)(x-s)$

If $k>0$, then $f(p)>0, f(q)<0, f(r)<0$ and $f(s)>0$

$\Rightarrow f(x)=0$ has one real root between p and q and other real root between r and s. so statement I is correct Statement II is a fact and correct but not a correct explanation of statement I

135. (a) Statement I: Given expression is an identity then-

$a^2-3a+2=0 \Rightarrow a=1, 2, 2a^2-5a+6=0$

$\Rightarrow a=2,3, a^2-4=0 \quad \Rightarrow a=\pm 2$

$a=2$ is the only solution.

So, statement I is True.

Statement II is a fact and correct and a correct explanation of statement I

136. (a) Remainder = f(1)= 1+1+1+1+1+1+1 = 7

Statement II is a fact and correct and a correct explanation of statement I

137. (a) As $f'(\alpha)=0$, so y = f(x) has a tangent parallel to x – axis at $x=\alpha$ but $f''(\alpha)=0$ and $f'''(\alpha)\neq 0$, so $x=\alpha$ represents a point of inflexion

Statement II is a fact and correct and a correct explanation of statement I

138. (c) Consider statement II 1st

Let P(x) = $ax^2+bx+c=0$ and

Q(x) = $dx^2+ex+f=0$

Then P(x) + λQ(x) = $(ax^2+bx+c)+\lambda(dx^2+ex+f)=0$

Or $(a+\lambda d)x^2+(b+\lambda e)x+(c+\lambda f)=0$

Nature of roots of this equation depends on λ so statement II is incorr ect.

Consider statement I

Let f(x)= $(x^2-4x+3)+\lambda(x^2-6x+8)$

= (x – 1)(x – 3) + λ (x – 2)(x- 4)

If λ = 0 then f(X) = 0 has roots 1 and 3

Now f(1) = 3 λ , f(2) = –1 , f(3) = – λ, f(4) = 3

If λ> 0 then f(X) = 0 has a root in (1, 2) and other in (3, 4).

If λ< 0 then f(x) = 0 has root in (2, 3) and other in (-∞, 1) or (4, ∞)

139. (a) The equation can be rewritten as

(x + 1)[x(a +2)+ a] = 0

For both roots integer $\frac{a}{a+2}$ must be integer

So, a = 0, –1, –3, – 4

So statement II is correct and correct explanation of statement I

140. (c) If a ≠ 0, we rewrite f (x) as follows:

$$f(x)=a\left\{x^2+\frac{b}{a}x+\frac{c}{a}\right\}=a\left\{\left(x+\frac{b}{2a}\right)^2+\frac{4ac-b^2}{4a^2}\right\}$$

now consider $f\left(-\frac{b}{2a}+i\right)=a\left\{(i)^2+\frac{4ac-b^2}{4a^2}\right\}$

$$=a\left(-1+\frac{4ac-b^2}{4a^2}\right)=\text{real number}$$

So the given condition is true when, a = 0.

Statement – II is false, it is true when a, b, c ∈ R.

IIT CORNER

141. (b) As a, b, c, p, q, ∈ R and the two given equations have exactly one common root

⇒ Either both equations have real roots

Or both equations have imaginary roots

⇒ Either $\Delta_1\ge 0$ and $\Delta_2\ge 0$ or $\Delta_1\le 0$ and $\Delta_2\le 0$

⇒ $p^2-q\ge 0$, $b^2-ac\ge 0$

Or $p^2-q\le 0$ and $b^2-ac\le 0$

⇒ $(p^2-q)(b^2-ac)\ge 0$

⇒ Statement 1 is True .

Also we have $\alpha\beta=q$ and $\frac{\alpha}{\beta}=\frac{c}{a}$

$\therefore \frac{\alpha\beta}{\alpha/\beta}=\frac{q}{c}\times a \Rightarrow \beta=\frac{qa}{c}$

As $\beta\neq 1$ or $-1 \Rightarrow \beta^2\neq 1 \Rightarrow \frac{qa}{c}\neq 1$ or $c\neq qa$

Again, as exactly one root α is common, and $\beta\neq 1$

$\therefore \alpha+\beta \neq \alpha+\frac{1}{\beta} \Rightarrow \frac{-2b}{a} \neq -2p$

$\Rightarrow b \neq ap$

$\therefore$ Statement 2 is a fact and hence correct but is not a correct explanation of Statement 1 .

(Matrix Match Type)

142. (A) $y=-x^2+3x-2=\frac{9}{4}-\left(x^2-3x+\frac{9}{4}\right)-2$

$y=\frac{1}{4}-\left(x-\frac{3}{2}\right)^2 \leq \frac{1}{4}$

A→r

(B) Consider the given expression $x \;\; -4x-2$ it will attain minimum value at x =2 and the minimum value is -6 so a>6

B→p

(C) Given, $x^2-2(4a-1)x+15a^2-2a-7>0$ discriminant < 0 [since coefficient of $x^2>0$]

$4(4a-1)^2-4(15a^2-2a-7)<0$

$\Rightarrow a^2-6a+8<0 \Rightarrow 2<a<4$

C→s

(D) Let $y=\frac{x^2+2x+a}{x^2+4x+3a}$

$\Rightarrow x^2(y-1)+2(2y-1)x+a(3y-1)=0$

As, $x \in R, D \geq 0$

$\therefore 4(2y-1)^2-4(y-1)a(3y-1) \geq 0$

$(4-3a)y^2-4(1-a)y+1-a \geq 0$

$a<\frac{4}{3}$ and $a(a-1) \leq 0$

$\therefore 0 \leq a \leq 1$

D→q

143. (A) Since, $f(x)=x^2+2(k+1)x+(9k-5)$ have both negative roots, then condition for that is

$\Rightarrow$ (i) Discriminant ≥ 0 (ii) $\alpha<0, \beta<0$ (iii) $f(0)>0$ $ie, \alpha+\beta<0$

$\therefore$ Discriminant ≥ 0

$\Rightarrow 4(1+k)^2-36k+20>0$

$\Rightarrow k^2-7k+6 \geq 0$

$\Rightarrow (k-1)(k-6) \geq 0$

$\Rightarrow k \leq 1 or\ k \geq 6$ ...(i)

$\alpha+\beta<0 \Rightarrow -2(k+1)<0$

$\Rightarrow k+1>0 \Rightarrow k>-1$ ...(ii)

and $f(0)>0 \Rightarrow 9k-5>0$

$\Rightarrow k>\frac{5}{9}$

Combining the results of (i), (ii) and (iii), we get k $\geq$ 6

A→q

(B) Consider the equation $x^2-4x+k=0$ it will attain minimum value at x = 2 and at x=2 the value of quadratic expression is k-4 < 0 or k <4 or D <0

B→p

(C) $f(x)=x^2-2(k-1)x+(2k+1)$ has both roots positive, if

(i) Discriminant > 0

(ii) Sum of the roots > 0 and

(iii) $f(0)>0$

Now, Discriminant ≥ 0

$\Rightarrow 4(k-1)^2-4(2k+1) \geq 0$

$\Rightarrow k^2-4k \geq 0$

$\Rightarrow k \leq 0$ or $k \geq 4$

Sum of roots > 0

$\Rightarrow 2(k-1)>0 \Rightarrow k>1$ and $f(0)>0$

$\Rightarrow (2k+1)>0 \Rightarrow k>-\frac{1}{2}$

Combining the results of (i), (ii) and (iii), we get $k \geq 4$

C→s

(D) Given that $x^2-kx+16>0, \forall x \in R$ that means the corresponding equation has no real roots or $k^2-64<0$ or $-8<k<8$

D→r

144. (A) From the given information $|\alpha-\beta|>\sqrt{3k}$

$\Rightarrow(\alpha+\beta)^2-4\alpha\beta>3k$

$\Rightarrow k^2-3k-4>0$ and $k>0$

$\Rightarrow(k-4)(k+1)>0$ and $k>0$

$\Rightarrow k>4$

A→q

(B) $f(x)>0$ *for all* $x\Rightarrow$ Discriminant <0

$\Rightarrow 4(4k-1)^2-4(15k^2-2k-7)<0$

$\Rightarrow k^2-6k+8<0\Rightarrow 2<k<4$

B→p

(C) Given that $x^2+kx+k<0$

so $k^2-4k<0$

$\Rightarrow k\in(0,4)$

C→s

(D) Since, 'k' lies between the roots of $f(x)=2x^2-2(2k+1)x+k(k+1)$

$\Rightarrow$ Discriminant ≥ 0 and $f(k)<0$

$\Rightarrow 4(2k+1)^2-8k(k+1)>0$ and

$2k^2-2k(2k+1)+k(k+1)<0$

$\Rightarrow 8\left(k^2+k+-\right)\geq 0$ and $-k^2-k<0$

$\Rightarrow$ Required range of k is $k>0$ and $k<-1$

D→r

145. (A) Let the common difference of AP be k then

$\Rightarrow 2(-k)+x(-k)^2+(2k)^3=2(-3k)+(-2k)^2+(-k)^3$

$\Rightarrow 9k^3+(x-4)k^2+4k=0$ and $k\neq 0$

$\Rightarrow 9k^2+(x-4)k+4=0$

$\Rightarrow k\in R\Rightarrow (x-4)^2-144\geq 0$

$\Rightarrow x\leq -8$ or $x\geq 16$

A→p, q, s

(B) $\alpha+\beta=\dfrac{-b}{a}$, $\alpha\beta=\dfrac{c}{a}$

If the roots of other equation be γ and δ then

$\gamma+\delta=\left(\dfrac{-b}{a}\right)\left(\dfrac{c}{a}\right)^2=(\alpha+\beta)(\alpha^2\beta^2)$

$=\alpha^3\beta^2+\alpha^2\beta^3$

$\gamma\delta=\left(\dfrac{c}{a}\right)^5=(\alpha\beta)^5=(\alpha^3\beta^2)(\alpha^2\beta^3)$

So, the roots of the second equation are $\alpha^3\beta^2$ and $\alpha^2\beta^3$

$\Rightarrow(\alpha^3\beta^2)(\alpha^2\beta^3)=4*8\Rightarrow|\alpha\beta|=2$

B→p, r

(C) $\log_3(\log_5 x)+\log_{1/3}(\log_{1/5} y)=1$

$\Rightarrow \log_3\dfrac{\log_5 x}{\left(\log_{\frac{1}{5}} y\right)}=1$

$\Rightarrow\dfrac{\log_5 x}{\left(\log_{\frac{1}{5}} y\right)}=3\Rightarrow xy^3=1$

Also, $x^2y=1\Rightarrow x=y=1$

$\Rightarrow 5x+y=6$

C→p, s

(D) Since and x_1, x_2 and x_1x_2 are the real roots of the equation $x^3-ax^2+bx-c=0$

$x_1+x_2+x_1x_2=a$

$x_1x_2+x_1^2x_2+x_1x_2^2=b$

$x_1^2x_2^2=c$

So

$x_1x_2(1+x_1+x_2)=b\Rightarrow x_1x_2(1+a-x_1x_2)=b$

$\Rightarrow x_1x_2(1+a)-c=b\Rightarrow x_1x_2-\dfrac{b+c}{1+a}$, which is a rational number.

D→p

146. (A) We know that $(\sum\alpha)^2=\sum\alpha^2-2\sum\alpha\beta\Rightarrow\sum\alpha\beta=-1$

Also, $\sum\alpha^3-3\alpha\beta\gamma=\sum(\alpha)(\sum\alpha^2-2\sum\alpha\beta)\Rightarrow\alpha\beta\gamma=-2$

$(\sum\alpha^2)^2=\sum\alpha^4+2\sum\alpha^2\beta^2=\sum\alpha^4+2[(\sum\alpha\beta)^2-2\alpha\beta\gamma(\sum\alpha)]$

$\therefore\sum\alpha^4=18$

A→q, r

(B) α, β, γ are roots of the equation $x^3 - 2x^2 - x + 2 = 0$

So, $(x - \alpha)(x - \beta)(x - \gamma) = x^3 - 2x^2 - x + 2$

Now put $x = 1$, then $(1 - \alpha)(1 - \beta)(1 - \gamma) = 0$

B→t

(C) Since the roots of $x^3 - 2x^2 - x + 2 = 0$ are -1, 1 and 2. So, for $-1 < x < 1$,

$x^3 - 2x^2 - x + 2 > 0$

(Local max of $f(x) = x^3 - 2x^2 - x + 2$ is less than 18)

C→r

(D) As α, β, γ are equals to -1, 1 and 2. So,

$(1 + \alpha^2)(1 + \beta^2)(1 + \gamma^2) = 20$

D→p, r

147. (A) Since $\alpha, \beta, \gamma, \delta$ are in A.P.

$\Rightarrow \beta - \alpha = \delta - \gamma$

$\Rightarrow (\beta + \alpha)^2 - 4\beta\alpha = (\delta + \gamma)^2 - 4\delta\gamma$

$\Rightarrow \dfrac{b^2 - ac}{a^2} = \dfrac{q^2 - pr}{p^2}$

So **A→ s**

(B) Since, $\alpha, \beta, \gamma, \delta$ are in G.P.

$\Rightarrow \dfrac{\alpha}{\beta} = \dfrac{\gamma}{\delta}$

$\Rightarrow \dfrac{\alpha - \beta}{\alpha + \beta} = \dfrac{\gamma - \delta}{y + \delta}$

$\Rightarrow 1 - \dfrac{4\alpha\beta}{(\alpha + \beta)^2} = 1 - \dfrac{4\gamma\delta}{(\gamma + \delta)^2}$

$\Rightarrow \dfrac{b^2}{ac} = \dfrac{q^2}{pr}$

B→p

(C) Since, $\alpha, \beta, \gamma, \delta$ are in H.P so $\dfrac{1}{\alpha}, \dfrac{1}{\beta}, \dfrac{1}{\gamma}, \dfrac{1}{\delta}$ are in A.P

And $\dfrac{1}{\alpha}, \dfrac{1}{\beta}$ are roots of $cx^2 + 2bx + a = 0$

and $\dfrac{1}{\gamma}, \dfrac{1}{\delta}$, are roots of $rx^2 + 2qx + p = 0$

Since roots $\dfrac{1}{\alpha}, \dfrac{1}{\beta}, \dfrac{1}{\gamma}, \dfrac{1}{\delta}$ are in A.P so using result of A, we get,

$\therefore \dfrac{b^2 - ac}{c^2} = \dfrac{q^2 - pr}{r^2}$.

C→q

(D) Given that $\alpha - \gamma = \delta - \beta$ or $\alpha + \beta = \gamma + \delta$

Or $-\dfrac{2b}{a} = -\dfrac{2q}{p}$ or $\dfrac{b}{a} = \dfrac{q}{p}$

D→r

148. (A) Given that

$a(p+q)^2 + 2bpq + c = 0$ ...(i)

$a(p+r)^2 + 2bpr + c = 0$ ...(ii)

From above two results we can say that q and r are roots of

$a(p+x)^2 + 2bpx + c = 0$

or $ax^2 + 2(a+b)px + ap^2 + c = 0$

$\therefore$ Sum of the roots $q + r = \dfrac{-2(a+b)p}{a}$

A→r,

(B) Product of the roots $qr = p^2 + \dfrac{c}{a}$

B→q

(C) $(q - r)^2 = (q + r)^2 - 4qr$

$= \dfrac{4}{a^2}[(a + b)^2 p^2 - a^2 p^2 - ac]$

$= \dfrac{4}{a^2}[(2ab + b^2) p^2 - ac]$

$\dfrac{2}{|a|}\sqrt{(2a+b)bp^2 - ac} = |q - r|$

C→s

(D) We know that $q + r = -\dfrac{2(a+b)p}{a}$ and

$qr = p^2 + \dfrac{c}{a}$

So $\dfrac{1}{q} + \dfrac{1}{r} = \dfrac{q+r}{qr} = \dfrac{\left\{-\dfrac{2(a+b)p}{a}\right\}}{\left\{\dfrac{ap^2+c}{a}\right\}} = -\dfrac{2(a+b)p}{ap^2+c}$

D→p

149. Putting x = 0, 1, 1/2, we obtain

$|c| \le 1$, $|a + b + c| \le 1$, $|\frac{1}{4}a + \frac{1}{2}b + c| \le 1$

$\Rightarrow -1 \le c \le 1, -1 \le a + b + c \le 1.$

$-4 \le a + 2b + 4c \le 4$

$|a + 2b + 4c| \le 4$

D→r

$\Rightarrow -4 \le 4a + 4b + 4c \le 4$.

And $-4 \le -a - 2b - 4c \le 4$

Adding we get

$-8 \le 3a + 2b \le 8$.

Also, $-8 \le a + 2b \le 8$

$\therefore -16 \le 2a \le 16 \Rightarrow |a| \le 8$

A→q

Since $-1 \le -c \le 1, -8 \le -a \le 8$,

$|c|$ 1

B→p

We get $-16 \le 2b \le 16 \Rightarrow |b| \le 8$

Thus, $|a| + |b| + |c| \le 17$.

C→s

Combining above result we get

A→q; B→p; C→s; D→r

150. (A) Consider $f(x) = f_0(x) + f_1(x) + f_2(x)$

Or $f(x)=\frac{(x-x_1)(x-x_2)}{(x_0-x_1)(x_0-x_2)}+\frac{(x-x_0)(x-x_2)}{(x_1-x_0)(x_1-x_2)}$

$+\frac{(x-x_0)(x-x_1)}{(x_2-x_0)(x_2-x_1)}$

From observation we can say that $f(x) -1 = 0$ has three roots x_0, x_1, x_2 but f(x) is a quadratic function with three roots, that means it is an identity and $f(x) -1 =0$ for all x

So $f(x) =1$

A→ r

(B) let $g(x) = (x_1+ x_2)f_0(x) + (x_2+x_0)f_1(x) + (x_0 +x_1)f_2(x)$ then

$g(x) - (x_0 + x_1 + x_2 - x) = 0$ for $x = x_0, x_1, x_2$

So $g(x) = x_0 + x_1 + x_2 - x \ \forall x$

B→p

(C) Let $h(x) = \sum_{k=0}^{2} f_k(x)x^2 =$

$\frac{(x-x_1)(x-x_2)}{(x_0-x_1)(x_0-x_2)}x^2+\frac{(x-x_0)(x-x_2)}{(x_1-x_0)(x_1-x_2)}x^2+$

$\frac{(x-x_0)(x-x_1)}{(x_2-x_0)(x_2-x_1)}$

$h(x) -x^2 =0$ for $x_0 + x_1 + x_2$

So $h(x) = x^2 \ \forall x$

C→s

(D) Consider $f'(x) =(x - x_1)(x-x_2)+(x-x_0)(x-x_2)+(x-x_0)(x-x_1)$

This gives $\frac{f(x)}{(x-x_0)f'(x_0)} = f_0(x)$ and so on

$$\frac{f(x)}{(x-x_0)f'(x_0)}+\frac{f(x)}{(x-x_1)f'(x_1)}$$

$$+\frac{f(x)}{(x-x_2)f'(x_2)} = f_0(x)+ f_1(x)+ f_2(x) =1$$

D→q

151. From given condition $(x-a)(x-12)+2=0$ must have integer zeros.

If $x \in I$. Then $(x-a)(x-12) = -2$

Both the factors, (x–a) and (x–12) must be integers.

$\Rightarrow x - a = 1, x - 12 = -2$.

Or $x - a = -1, x - 12 = 2$.

Or $x - a = 2, x - 12 = -1$.

Or $x - a = -2, x - 12 = 1$

$\Rightarrow a = 9, 15$

when $a = 9$, b, c = – 10, – 11

When $a = 15$, b, c = – 13 , – 14

A→p, q , B→r, s, C→t, u

(More Than One Correct MCQs)

152.(a, c) If α is a common root of the two equations then

$\alpha^2 + b\alpha + c = 0$

$b\alpha^2 + c\alpha + 1 = 0$

Multiply the first by b and subtracting we get

$(b^2 - c)\alpha + bc - 1 = 0$

$\Rightarrow \alpha = \frac{1-bc}{b^2-c}$

Substituting this value we get,

$(1 - bc)^2 + b(1 - bc)(b^2 - c) + c(b^2 - c)^2 = 0$

$\Rightarrow b^3 + c^3 + 1 - 3bc = 0$

$\Rightarrow (b + c + 1)(b^2 + c^2 + 1 - bc - b - c) = 0$

$\Rightarrow \frac{1}{2}(b + c + 1)[(b - c)^2 + (b - 1)^2 + (c - 1)^2]$ $= 0$

So either $b + c + 1 = 0$

Or $(b - c)^2 + (b - 1)^2 + (c - 1)^2 = 0$ or $b = c = 1$

153. (a, b, c) According to the given conditions

$a > 0, b^2 < 4ac; b > 0, c^2 < 4ab; c > 0, a^2 < 4bc$

Adding above results we get

$\therefore a^2 + b^2 + c^2 < 4(bc + ca + ab)$

$\Rightarrow \frac{bc+ca+ab}{a^2+b^2+c^2} > \frac{1}{4}$

We know that, $a^2 + b^2 + c^2 - (bc + ca + ab)$

$= \frac{1}{2}[(b - c)^2 + (c - a)^2 + (a - b)^2] > 0$

$\Rightarrow \frac{bc+ca+ab}{a^2+b^2+c^2} < 1$

So range of $\frac{bc+ca+ab}{a^2+b^2+c^2}$ is $\left(\frac{1}{4}, 1\right)$

154. (b, c, d) The given equation can be written as,

$\pi^e(x-\pi)(x-\pi-e) + e^\pi(x-e)(x-\pi-e)$
$+(\pi^\pi + e^e)(x-e)(x-\pi) = 0$

Let, f (x) = $\pi^e(x-\pi)(x-\pi-e) + e^\pi(x-e)$ $(x-\pi-e)+(\pi^\pi+e^e)(x-e)(x-\pi)$

Then consider f(e) = $\pi^e(e-\pi)(-\pi) > 0$ [since, $e < \pi$]\

And $f(\pi) = e^\pi(\pi - e)(-e) < 0$

We know that if f(a) and f(b) are of opposite sign then f(x) =0 has at least one root in (a, b)

$\therefore$ Equation f(x) = 0 has a real root in (e, π)

Now $f(\pi + e) = (\pi^\pi + e^e)(\pi) > 0$/

$\therefore$ Equation f(x) = 0 has a real root in $(\pi, \pi + e)$.

$\therefore$ f(0) = 0 has a real roots in (e, π) and other in $(\pi, e + \pi)$.

Also, $\pi - e < e$

$\therefore$ Equation f(x) = 0 has real roots in $(\pi - e, \pi + e)$.

155. (a, d) Given equations are

$x^2 + y^2 - 4x - 4y = 60$ ...(i)

$xy + 2x + 2y = 20$ ...(ii)

Multiply (ii) by 2 an adding it to (i) we get,

$x^2 + y^2 + 2xy = 100$

Or, $(x + y)^2 = 100$

$\Rightarrow x + y = \pm 10 \Rightarrow |x + y| = 10$

156. (c, d) Let $f(x) = x^4 + ax^3 - 13x^2 + bx - 4 = 0$

If $2 + \sqrt{5}$ is one root then other has to be $2 - \sqrt{5}$.

Let α be repeated root, then

Product of the roots $(2 + \sqrt{5})(2 - \sqrt{5})\alpha^2 = -4$

$\Rightarrow \alpha^2 = 4$ or $\alpha = \pm 2$ ...(i)

But α can take only one value, so now consider

$\sum \alpha\beta = -13$

$\therefore -1 + (4 - 2\sqrt{5})\alpha + (4 + 2\sqrt{5})\alpha + \alpha^2 = -13$

$\alpha^2 + 8\alpha + 12 = 0$

$(\alpha + 6)(\alpha + 2) = 0$

$\alpha = -2$ or -6 ...(ii)

From result (i) and (ii), $\alpha = -2$

Sum of the roots, $2 + \sqrt{5} + 2 - \sqrt{5} + (2) + (-2)$ $= a = 0$

Similarly $b = -20$

157. (a, b, c, d) Consider $2x^3 + x^2 + 2x - 5 = 0$

Since sum of the co–efficient is zero, so x=1 is a root of the equation

$\Rightarrow 2x^3 + x^2 + 2x - 5 = (x - 1)(2x^2 + 3x + 5) = 0$

So, one root is 1 and other roots are imaginary.

Next , $ax^3 + (a + b)x^2 + (b + c)x + c = 0$

$\Rightarrow (x + 1)(ax^2 + bx + c) = 0$

So the two equation have either 1 as common root or both roots must be common. So, a +b +c $= 0$ or $\frac{a}{2} = \frac{b}{3} = \frac{c}{5} \Rightarrow a + b + c = 0$

Or, a +b +c = 2c or 5a.

158. (a, b, c, d) Let, the roots be x_1, x_2, x_3, x_4 then

$x_1 + x_2 + x_3 + x_4 = 4$ and $x_1x_2x_3x_4 = 1$

$\Rightarrow$ A.M. of x_1, x_2, x_3, x_4 = G.M. of x_1, x_2, x_3, x_4

$\Rightarrow \Leftrightarrow x_1 = x_2 = x_3 = x_4$

$\therefore x_1 = x_2 = x_3 = x_4 = 1 \Rightarrow x_1, x_2, x_3, x_4$ in A.P. as well in H.P.

159. (a,b, d) The given equation is $x^6 + 2x^3 + 5 + ax^3 + a = 0$ $(x^3 + 1)^2 + a(x^3 + 1) + 4 = 0$

Substitute $t = x^3 + 1$ then equation is

$(x^3 + 1)^2 + a(x^3 + 1) + 4 = 0 \Rightarrow t^2 + at + 4 = 0$

Now we have following cases

Case (i) $D < 0 \Rightarrow a^2 - 16 < 0$ then above equation has no real roots and all six roots are imaginary.

Case (ii) D> 0, then the equation has two roots. So, if $a^2 - 16 > 0$

Then the given equation has two real roots and other imaginary, except when t = 1. That is when a = –5, and equation is $x^3(x^3 - 3) = 0$, which has three repeated, one real and two imaginary roots.

160. (a, c) Equation has solution only if

$a^2 - 2a - 3 \geq 0$

$\Rightarrow a \leq -1$ or $a \geq 3$

Then equation becomes

$x^2 - 1 = \pm (a^2 - 2a - 3)$

$\Rightarrow x^2 = a^2 - 2a - 2$ or $x^2 = -a^2 + 2a + 4$

For real solution

$a^2 - 2a - 2 \geq 0$ or $-a^2 + 2a + 4 \geq 0$

$\Rightarrow a \leq 1 - \sqrt{3}$ or $a \geq 1 + \sqrt{3}$ or $1 - \sqrt{5} \leq a \leq 1 + \sqrt{5}$

$\therefore a \in [1 - \sqrt{5}, -1] \cup [3, 1 + \sqrt{5}]$

161. (c, d) Let, $x^2 + 4x + 4 = y$, then the equation becomes $ay^2 + by + c = 0$ ...(1)

Since, $b^2 - 4ac > 0 \Rightarrow$ equation (1) has two distinct real roots, say α and β.

Now consider $x^2 + 4x + 4 = \alpha$ or β

$\Rightarrow (x + 2)^2 = \alpha$ or β ...(2)

Equation (2) will give four distinct real values of x if α and β > 0.

That means both the roots of equation (1) is positive. For this a and c should have the same sign and the sign of b should be opposite.

162. (a, b) Consider $f(x) = x^2 + a^2x + b^2$ since both the roots exceed a number 'c' then–

(i) Discriminant $> 0 \Rightarrow a^4 > 4b^2$

(ii) $f(c) > 0 \Rightarrow c^2 + a^2c + b^2 > 0$

(iii) $c < -\frac{a^2}{2}$.

The condition is as shown in the figure below

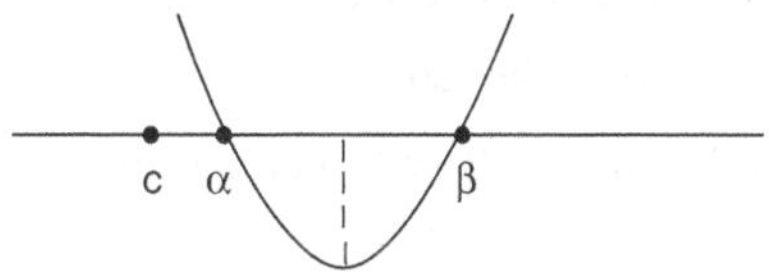

163. (a, c, d) Let $f(x) = (2x - a)(3x - b) + (3x - b)(4x - c) + (4x - c)(2x - a)$

$f\left(\frac{a}{2}\right) = \left(\frac{3a}{2} - b\right)(2a - c) > 0$ since $\frac{a}{2} < \frac{b}{3} < \frac{c}{4}$

$f\left(\frac{b}{3}\right) = \left(\frac{4b}{3} - c\right)\left(\frac{2b}{3} - a\right) < 0$

Since $f\left(\frac{c}{4}\right)$ and $f\left(\frac{b}{3}\right)$ are of opposite sign so there exists one real root between $\left(\frac{a}{2}, \frac{b}{3}\right)$

$\Rightarrow$ Both roots are real but not equal

Also, $f\left(\frac{c}{4}\right)$ and $f\left(\frac{b}{3}\right)$ are of opposite sign so there exist one real root between $\left(\frac{b}{3}, \frac{c}{4}\right)$.

164. (a, b, c) Let $f(x) = ax^2 + bx + c$, –3 and 3 are lying between the roots of f(x) = 0

$\Rightarrow f(-3) > 0$ and $f(3) > 0$

$\Rightarrow 9a + 3|b| + c > 0$

Also, 0, –2, 2 are lying between the roots.

Therefore $c > 0$ and $4a + 2|b| + c > 0$

165. (a, c, d) Let $f(x) = cx^2 + bx - 2a$

Since $b + c - 2a > 0 \Rightarrow f(1) > 0$

Given that f (x)=0 has no real root. That means f(x) > 0 for all x, hence c> 0

$\Rightarrow f(0) > 0 \Rightarrow a < 0$

$\Rightarrow ac < 0$. Also, $f(-1) > 0 \Rightarrow c - b > 2a$

Similarly, $f(1/2) > 0 \Rightarrow \frac{c + 2b}{8} > a$.

166. (a,b, c, d) Let f(x) = ax^2+bx+c

Discriminant D = $b^2 - 4ac < 0$

And f(–1) = a–b +c > 0

So, f(x) > 0 for all x ∈ R and it is possible when a > 0

f(0) > 0 ⇒ c> 0

f(1) > 0 ⇒ a +b +c > 0

f(–2) > 0 ⇒ 4a –2b + c > 0

f(1/2) = a/4 +b/2 +c > 0 or a + 2b > 4c

167. (b, c) Let $\frac{x^2+ax+3}{x^2+x+a}=y$

$\Rightarrow x^2(1-y)-x(y-a)+3-ay=0$

Since x ∈ R so descriminant of above equation must be +ve

$\Rightarrow (y-a)^2 - 4(1-y)(3-ay) \ge 0$

$(1-4a)y^2 + (2a+12)y + a^2 - 12 \ge 0$ ...(1)

Equation (1) is true for all y ∈ R, if 1 – 4a > 0 and D ≤ 0

$\Rightarrow a < \frac{1}{4}$ and $4(a+6)^2 - 4(a^2-12)(1-4a) \le 0$

$\Rightarrow a < \frac{1}{4}$ and $4a^3 - 36a + 48 \le 0 \Rightarrow a < \frac{1}{4}$ and

$4a^3 \le 36a - 48 \Rightarrow 4a^3 < 36\left(\frac{1}{4}\right)-48$

$\Rightarrow 4a^3 + 39 < 0$ [since $a<\frac{1}{4}$]

168. (a, c) From the given information

$a\alpha^2+b\alpha+c=0$ ...(1)

And $-a\beta^2+b\beta+c=0$ ...(2)

Now, Let f(x) = $ax^2+2bx+2c=0$

Then

$f(\alpha) = a\alpha^2+2b\alpha+2c = a\alpha^2-2a\alpha^2 = -a\alpha^2$ ...(4)

$f(\beta) = a\beta^2+2b\beta+2c = a\beta^2+2a\beta^2 = 3a\beta^2$ ...(5)

∴ f(α) and f(β) are of opposite sign so f(x) = 0 has a real root between α and β.

169. (b, c) Here we have two cases–

Case (i) If x ≥ a, then |x – a|= x – a, and the equation becomes $x^2-2ax-a^2=0$

$\Rightarrow x = a(1\pm\sqrt{2})$

Since a < 0; ∴ $a(1+\sqrt{2}) < a$,

Hence $x = a(1+\sqrt{2})$ cannot be the solution.

∴ $a(1-\sqrt{2}) > a$ is the solution, which is positive.

Case (ii) if x < a, then |x – a| = –(x– a) and the equation becomes $x^2+2ax-5a^2=0$

$\Rightarrow x = (-1\pm\sqrt{6})a$

Again as a< 0; $a(-1-\sqrt{6}) > 0 > a$,

hence $x = a(-1-\sqrt{6})$ cannot be the solution.

∴ $x = a(-1+\sqrt{6}) < a$ is the solution, which is negative

170. (b, c) Let f (x) = (x – a)(x – c) +λ(x – b)(x – d). Now, f (a) = λ(a – b)(a – d) and f(c) = λ(c – b) (c– d)

∴ $f(a)f(c) = \lambda^2(a-b)(a-d)(c-b)(c-d) < 0$ [since a < b< c< d]

Hence, f (x) = 0 has one real root between a and c.

Similarly,

f(b)f(d) = (b– a)(b – c) (d – a)(d– c)< 0. [since a < b< c< d]

Hence, f(x) = 0 has one real root between b and d.

171. (a,b, c, d) Since the sum of the coefficients of the equation = 0, so one root is 1 and other root is $\frac{c+a-2b}{a+b-2c}$, which is a rational number since a, b, c are rational. Hence, both the roots of the equation are rational .

Given that a > b > c ...(i)

And given equation is $(a+b-2c)x^2+(b+c-2a)x+(c+a-2b)=0$ ...(ii)

Since Equation (ii) has a root in the interval (–1, 0)

∴ f(–1)f(0) < 0

⇒ (2a – b – c)(c +a – 2b)< 0 ...(iii)

From (i), a> b ⇒ a – b > 0 and a > c ⇒ a – c > 0

∴ 2a – b – c > 0...(iv)

From (iii) and (4), c +a – 2b < 0 or c + a < 2b.

Option (a) is correct.

Now consider, the discriminant of equation $ax^2 + 2bx + c$ is $d = 4b^2 - 4ac$.

As, deduced ealier, $c + a < 2b$

$\Rightarrow 4b^2 > (c + a)^2$

$\Rightarrow 4b^2 > c^2 + a^2 + 2ac$

$\Rightarrow 4b^2 - 4ac > c^2 + a^2 - 2ac$

$\Rightarrow (c - a)^2 \Rightarrow 4b^2 - 4ac > 0 \Rightarrow D > 0$. Also, each of a, b, c are positive .

$\therefore$ The equation $ax^2 + 2bx + c$ has real and negative roots. So, (c) is also correct.

Similarly, (d) is also correct.

172. (a,b, c) Since, $\sin^{-1}\theta$ is defined only when $-1 \le \theta \le 1$

$\Rightarrow -1 \le x^2 - 4x + 5 \le 1$

$\Rightarrow x^2 - 4x + 4 \le 0$ and $x^2 - 4x + 6 \ge 0$

$\Rightarrow (x-2)^2 \le 0$ and $(x-2)^2 + 2 \ge 0$

$\Rightarrow x = 2$ is only solution

Putting x = 2, we get $4 + 2k + \frac{\pi}{2} = 0$

$\Rightarrow k = -2 - \frac{\pi}{4}$

173. (a, b) For $x > 0$, $f(x) > 0$ so there is no positive real root of $f(x) = 0$

Now, $f(0) = 1 \ne 0$ so $x = 0$ is not a root

$f(x) = 1 + 2x + 3x^2 + \ldots + (n+1)x^n$,

$xf(x) = x + 2x^2 + \ldots + nx^n + (n+1)x^{n+1}$

$\Rightarrow (1-x)f(x) = (1 + x + x^2 + \ldots + x^n) - (n+1)x^{n+1}$

$= \frac{1 - x^{n+1}}{1-x} - (n+1)x^{n+1}$

$$\Rightarrow f(x) = \frac{1 - x^{n+1}}{(1-x)^2} - \frac{(n+1)x^{n+1}}{(1-x)} = \frac{1 - (n+2)x^{n+1} + (n+1)x^{n+2}}{(1-x)^2} > 0$$

$\therefore$ The function $f(x) > 0$ for all x, hence, no solution.

174. (a, c) Given, equation is $x^2 + 9y^2 - 4x + 3 = 0$ ...(i) or $x^2 - 4x + 9y^2 + 3 = 0$

Since, x is realso $D \ge 0 \therefore (-4)^2 - 4(9y^2 + 3) \ge 0$

$\Rightarrow 16 - 4(9y^2 + 3) \ge 0 \Rightarrow 4 - 9y^2 - 3 \ge 0$

$\Rightarrow 9y^2 - 1 \le 0$

$\Rightarrow (3y-1)(3y+1) \le 0$

$\Rightarrow -\frac{1}{3} \le y \le \frac{1}{3}$

Now consider equation w.r.t y we get $9y^2 + 0y + x^2 - 4x + 3 = 0$

Now y is real if $\therefore 0^2 - 4 \cdot 9(x^2 - 4x + 3) \ge 0$

$\Rightarrow (x-1)(x-3) \le 0$

$\Rightarrow 1 \le x \le 3$

175. (a, b, d) we have

$ax + by = 1 \Rightarrow y = \frac{1 - ax}{b}$

Putting this value in the second equation, we get

$cx^2 + \frac{d}{b^2}(1 - ax)^2 = 1$

$\Rightarrow (b^2c + a^2d)x^2 - 2adx + d - b^2 = 0$

The quadratic equation will have equal roots then, D = 0 so

$D = 4a^2d^2 - 4(b^2c + a^2d)(d - b^2) = 0$

$\Rightarrow a^2d^2 + (b^2c + a^2d)b^2 - b^2cd - a^2d^2 - 0$

$\Rightarrow b^2[b^2c + a^2d - cd] = 0$

$\Rightarrow b^2c + a^2d = cd \Rightarrow \frac{b^2}{d} + \frac{a^2}{c} = 1$

So $x = \frac{2ad}{2(b^2c + a^2d)} = \frac{ad}{cd} = \frac{a}{c}$

And $y = \frac{1 - ax}{b} = \frac{1}{b}\left(1 - \frac{a^2}{c}\right) = \frac{1}{b}.\frac{b^2}{d} = \frac{b}{d}$

IIT CORNER

176. (c, d) Let us assume that $y = \frac{(x-a)(x-b)}{x-c}$

$\Rightarrow (x - c)y = x^2 - (a + b)x + ab$

$\Rightarrow x^2 - (a + b + y)x + ab + cy = 0$

Since x is real and y assumes all real values.

$\therefore \Delta \ge 0$ for all real values of y

Since x is real and y assumes all real values.

$\therefore \Delta \ge 0$ for all real values of y

Here , $\Delta = (a + b + y)^2 - 4(ab + cy)$

$= y^2 + 2y(a + b - 2c) + (a - b)^2$

$\Rightarrow y^2 + 2y(a + b - 2c) + (a - b)^2 \geq 0$

Now we know that the sign of a quadratic function is same as of coefficient of y^2 provided its Discriminant is negative.

So $4(a + b - 2c)^2 - 4(a - b)^2 < 0$

Or $4(a + b - 2c + a - b)(a + b - 2c - a + b) < 0$

$\Rightarrow 16(a - c)(b - c) < 0$

$\Rightarrow 16(c - a)(c - b) < 0$

Then if $a < b$ then we get $c \in (a, b)$

$\Rightarrow a < c < b$

Or if $a > b$ then we get, $c \in (a, b)$

$\Rightarrow b < c < a$ or $a > c > b$

177. (a, d) We have,

$$f(x) = \frac{2x-1}{2x^2+3x^2+x} = \frac{2x-1}{x(2x+1)(x+1)}$$

Critical points are $x = 1/2, 0, -1/2, -1$

On number line by wavy method, we have

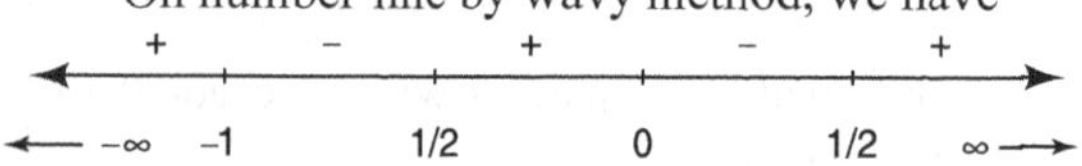

For $f(x) > 0$, when

$x \in (-\infty, 1) \cup (-1/2, 0) \cup (1/2, \infty)$

Clearly S contains $(-\infty, -3/2)$ and $(1/2, 3)$

178. (b) As the given expression is symmetric in a, b, c without loss of generality, we can assume that $a < b < c$.

Then $c - a = +$ ve and $c - b = +$ ve

$\therefore$ $b + c - a$ and $c + a - b = +$ ve

But $a + b - c$ may be + ve or – ve.

Now we have following cases

Case–I: if $a + b - c = +$ ve then we can say that a, b, c are such that sum of any two is greater than the 3rd. Consider $x = a + b - c$, $y = b + c - a$, $z = c + a - b$

Then x, y, z all are + ve.

And then $a = \frac{x+z}{2}, b = \frac{y+x}{2}, c = \frac{z+y}{2}$

Now we know that A. M. > G. M. for distinct real numbers

$\therefore \frac{x+y}{2} > \sqrt{xy}, \frac{y+z}{2} > \sqrt{yz}, \frac{z+x}{2} > \sqrt{zx}$

$\Rightarrow \left(\frac{x+y}{2}\right)\left(\frac{y+z}{2}\right)\left(\frac{z+x}{2}\right) > xyz$

$\Rightarrow abc > (a + b - c)(b + c - a)(c + a - b)$

$\Rightarrow (b + c - a)(c + a - b)(a + b - c) - abc < 0$

Case II: If $a + b - c = -$ ve then

$(b + c - a)(c + a - b)(a + b - c) - abc$

$= (+\text{ve})(+\text{ve})(-\text{ve}) - (+\text{ve})$

$= (-\text{ve}) - (+\text{ve}) = (-\text{ve})$

$\Rightarrow (b + c - a)(c + a - b)(a + b - c) < abc$

Hence in either case given expression is – ve.

179. (b) Given that a, b, c, d, p are real and distinct numbers such that

$(a^2 + bg^2 + c^2)p^2 - 2(ab + bc + cd)p + (b^2 + c^2 + d^2) \leq 0$

$\Rightarrow (a^2p^2 + b^2p^2) - (2abp + 2bcp + 2cdp) + (b^2 + c^2 + d^2) \leq 0$

$\Rightarrow (a^2p^2 - 2abp + b^2) + (b^2p^2 - 2bcp + c^2) + (c^2p^2 - 2cdp + d^2) \leq 0$

$\Rightarrow (ap - b)^2 + (bp - c)^2 + (cp - d)^2 \leq 0$

This is possible only when

$(ap - b)^2 + (bp - c)^2 + (cp - d)^2 = 0$

which is possible only when each term is zero individually i. e.,

$ap - b = 0$; $bp - c = 0$; $cp - d = 0$

$\Rightarrow \frac{b}{a} = p; \frac{c}{b} = p; \frac{d}{c} = p \Rightarrow \frac{b}{a} = \frac{c}{b} = \frac{d}{c} = p$

$\Rightarrow$ a, b, c, d aer in G. P.

180. (a) We know $240 = 2^4 . 3 . 5$.

Divisors of 240 are

$$\begin{matrix} 1,2, & 4,8, & 16 \\ 3,6, & 12,24, & 48 \\ 5,10, & 20,40, & 80 \\ 15,30, & 60,120, & 240 \end{matrix}$$

Out of these divisors just 4 divisors viz ., 2, 6, 10, 30 are of the form $4n + 2$.

181. (a, d) $\alpha x^2 - x + \alpha = 0$ has distinct real roots.

$\therefore D > 0 \Rightarrow 1 - 4\alpha^2 > 0$

$\Rightarrow \alpha \in \left(-\frac{1}{2}, \frac{1}{2}\right)$ ...(i)

Also $|x_1 - x_2| < 1$

$\Rightarrow (x_1 - x_2)^2 < 1 \Rightarrow (x_1 + x_2)^2 - 4x_1x_2 < 1$

$\Rightarrow \frac{1}{\alpha^2} - 4 < 1 \Rightarrow \frac{1}{\alpha^2} < 5$ or $a^2 > \frac{1}{5}$

$\Rightarrow \alpha \in \left(-\infty, -\frac{1}{\sqrt{5}}\right) \cup \left(\frac{1}{\sqrt{5}}, \infty\right)$ ...(ii)

Combining (i) and (ii)

$S = \left(-\frac{1}{2}, -\frac{1}{\sqrt{5}}\right) \cup \left(\frac{1}{\sqrt{5}}, \frac{1}{2}\right)$

$\therefore$ Subsets of S can be $\left(-\frac{1}{2}, -\frac{1}{\sqrt{5}}\right)$ and $\left(\frac{1}{\sqrt{5}}, \frac{1}{2}\right)$

(IIT JEE Subjective Questions)

182. $4^x - 3^{x-\frac{1}{2}} = 3^{x+\frac{1}{2}} - \frac{(2^2)^x}{2}$

$\Rightarrow 4^x - \frac{3^x}{\sqrt{3}} = 3^x\sqrt{3} - \frac{4^x}{2}$

$\Rightarrow \frac{3}{2} \cdot 4^x = 3^x\left(\sqrt{3} + \frac{1}{\sqrt{3}}\right)$

$\Rightarrow \frac{3}{2} \cdot 4^x = 3^x \frac{4}{\sqrt{3}}$

$\rightarrow \frac{4^{x-1}}{4^{1/2}} = \frac{3^{x-1}}{\sqrt{3}} \Rightarrow 4^{x-3/2} = 3^{x-3/2}$

$\Rightarrow \left(\frac{4}{3}\right)^{x-3/2} = 1 \Rightarrow x - \frac{3}{2} = 0 \Rightarrow x = 3/2$

183. Consider R H S = (m – 1 , n + 1) + x^{m-n-1}(m – 1 , n)

$= \frac{(1-x^{m-1})(1-x^{m-2})\ldots(1-x^{m-n-1})}{(1-x)(1-x^2)\ldots(1-x^{n+1})} + x^{m-n-1}$

$\left[\frac{(1-x^{m-1})(1-x^{m-2})\ldots(1-x^{m-n})}{(1-x)(1-x^2)\ldots(1-x^n)}\right]$

$= \frac{(1-x^{m-1})(1-x^{m-2})\ldots(1-x^{m-n})}{(1-x)(1-x^2)\ldots(1-x^n)}$

$\left[\frac{1-x^{m-n-1}}{1-x^{n+1}} + x^{m-n-1}\right]$

$= \frac{(1-x^m)(1-x^{m-1})\ldots(1-x^{m-n})}{(1-x)(1-x^2)\ldots(1-x^n)(1-x^{n+1})}$

= (m, n + 1) = L. H. S. Hence proved.

184. Given equation is $\sqrt{x+1} - \sqrt{x-1} = 1$

Squaring both sides, we get

$x + 1 = 1 + x - 1 + 2\sqrt{x-1} \Rightarrow 1 = 2\sqrt{x-1}$.

Again squaring both sides

$\Rightarrow 1 = 4(x - 1)$

$\Rightarrow x = 5/4$

185. Given $a > 0$, so we have to consider two cases: $a \neq 1$ and $a = 1$. Also it is clear that $x > 0$ and $x \neq 1$, $ax \neq 1$, $a^2x \neq 1$.

CASE I: If $a > 0, \neq 1$ then given equation can be simplified as

$\frac{2}{\log_a x} + \frac{1}{1+\log_a x} + \frac{3}{2+\log_a x} = 0$

Putting $\log_a x = y$, we get

$2(1 + y)(2 + y) + y(2 + y) + 3y(1 + y) = 0$

$\Rightarrow 6y^2 + 11y + 4 = 0 \Rightarrow y = -4/3$ and $-1/2$

$\Rightarrow \log_a x = -4/3$ and $\log_a x = -1/2$

$\Rightarrow x = a^{-4/3}$ and $x = a^{-1/2}$

CASE II: If a = 1 then equation becomes $2\log_x 1 + \log_x 1 + 3\log_x 1 = 6\log_x 1 = 0$ which is true $\forall x > 0, \neq 1$

Hence solution is if a = 1, $x > 0, \neq 1$

If $a > 0, \neq 1$; $x = a^{-1/2}, a^{-4/3}$

186. Let $x = \frac{\sqrt{26-15\sqrt{3}}}{5\sqrt{2} - \sqrt{38+5\sqrt{3}}}$

$\Rightarrow x^2 = \frac{26-15\sqrt{3}}{50+38+5\sqrt{3}-10\sqrt{76+10\sqrt{3}}}$

$\Rightarrow x^2 = \frac{26-15\sqrt{3}}{88+5\sqrt{3}-10\sqrt{75+1+10\sqrt{3}}}$

$\Rightarrow x^2 = \frac{26-15\sqrt{3}}{88+5\sqrt{3}-10\sqrt{(5\sqrt{3})^2+(1)^{20}+2\times5\sqrt{3}\times1}}$

$\Rightarrow x^2 = \frac{26-15\sqrt{3}}{88+5\sqrt{3}-10\sqrt{(5\sqrt{3}+1)^2}}$

$\Rightarrow x^2 = \frac{26-15\sqrt{3}}{88+5\sqrt{3}-10(5\sqrt{3}+1)}$

$$\Rightarrow x^2 = \frac{26-15\sqrt{3}}{88+5\sqrt{3}-50\sqrt{3}-10} = \frac{26-15\sqrt{3}}{78-45\sqrt{3}}$$

$$= \frac{26-15\sqrt{3}}{3(26-15\sqrt{3})} = \frac{1}{3}$$, which is a rational number.

187. There are two parts of this question

$(5x - 1) < (x + 1)^2$ and $(x + 1)^2 < (7x - 3)$

Taking first part

$(5x - 1) < (x + 1)^2 \Rightarrow 5x - 1 < x^2 + 2x + 1$

$\Rightarrow x^2 - 3x + 2 > 0 \Rightarrow (x - 1)(x - 2) > 0$

(using wavy method)

$\Rightarrow x < 1$ or $x > 2$ ...(i)

Taking second part

$(x + 1)^2 < (7x - 3) \Rightarrow x^2 - 5x + 4 < 0$

$\Rightarrow (x - 1)(x - 4) < 0$

(using wavy method)

$\Rightarrow 1 < x < 4$ ...(ii)

Combining (i) and (ii) [taking common solution], we get $2 < x < 4$ but x is an integer

Therefore $x = 3$

188. α, β are the roots of $x^2 + px + q = 0$

$\therefore \alpha + \beta = -p, \alpha\beta = q$

$\because \gamma, \delta$ are the roots of $x^2 + rx + s = 0$

$\therefore \gamma + \delta = -r, \gamma\delta = s$

Now, $(\alpha - \gamma)(\alpha - \delta)(\beta - \delta)(\beta - \gamma)$

$= [\alpha^2 - (\gamma + \delta)\alpha + \gamma\delta][\beta^2 - (\gamma + \delta)\beta\gamma\delta]$

$= [\alpha^2 + r\alpha + s][\beta^2 + r\beta + s]$

[$\because \alpha, \beta$ are roots of $x^2 + px + q = 0$

$\therefore \alpha^2 + p\alpha + q = 0$ and $\beta^2 + p\beta + q = 0$]

$= [(r - p)\alpha + (s - q)][(r - p)\beta + (s - q)]$

$= (r - p)^2\alpha\beta + (r - p)(s - q)(\alpha + \beta) + (s - q)^2$

$= q(r - p)^2 - p(r - p)(s - q) + (s - q)^2$

Now if the equations $x^2 + px + q = 0$ and $x^2 + rx + s = 0$ have a common root say α, then $\alpha^2 + p\alpha + q = 0$ and $\alpha^2 + r\alpha + s = 0$

$$\Rightarrow \frac{\alpha^2}{ps-qr} = \frac{\alpha}{q-s} = \frac{1}{r-p}$$

$$\Rightarrow \alpha^2 = \frac{ps-qr}{r-p} \text{ and } \alpha = \frac{q-s}{r-p}$$

$\Rightarrow (q - s)^2 = (r - p)(ps - qr)$ which is the required condition.

189. We know that for sides a, b, c of a Δ

$(a - b)^2 \geq 0$

$\Rightarrow a^2 + b^2 \geq 2ab$ ...(i)

Similarly $b^2 + c^2 \geq 2bc$ ...(ii)

$c^2 + a^2 \geq 2ca$ ...(iii)

Adding the three inequations, we get

$2(a^2 + b^2 + c^2) \geq 2(ab + bc + ca)$

$\Rightarrow a^2 + b^2 + c^2 \geq ab + bc + ca$

Adding $2(ab + bc + ca)$ to both sides, we get

$(a + b + c)^2 \geq 3(ab + bc + ca)$

Or $3(ab + bc + ca) \leq (a + b + c)^2$

Hence proved the 1st part.

Also $c < a + b$ (triangle inequality)

$\Rightarrow c^2 < ac + bc$ ...(iv)

Similarly $b^2 < ab + bc$ ...(v)

$a^2 < ab + ca$ ...(vi)

Adding above three inequalities, we get

$a^2 + b^2 + c^2 < 2(ab + bc + ca)$

Adding $2(ab + bc + ca)$ to both sides, we get

$\Rightarrow (a + b + c)^2 < 4(ab + bc + ca)$

Hence 2nd result is proved.

Combining above proved results we get,

$3(ab+bc+ca) \leq (a+b+c)^2 < 4(ab+bc+ca)$

First two expressions will be equal for $a = b = c$.

190. Proof : Since $n^4 < 10^n \Rightarrow 10n^4 < 10^{n+1}$

So it is sufficient to prove that $(n + 1)^4 < 10n^4$

Now $\left(\frac{n+1}{n}\right)^4 = \left(1+\frac{1}{n}\right)^4 \leq \left(1+\frac{1}{2}\right)^4$ [$\because n \geq 2$]

$= \frac{81}{16} < 10$

$\Rightarrow (n + 1)^4 < 10n^4$

Hence, $(n+1)^4 < 10^{n+1}$ Proved.

191. $y = \sqrt{\dfrac{(x+1)(x-3)}{(x-2)}}$

y will take all real values if $\dfrac{(x+1)(x-3)}{(x-2)} \geq 0$

By wavy method

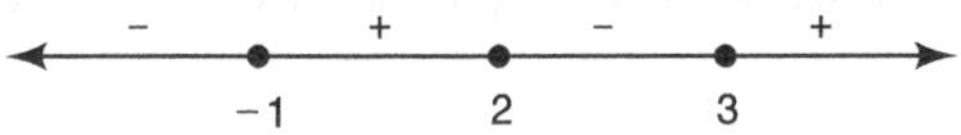

$x \in [-1, 2) \cup [3, \infty)$

[2 is not included as it makes denominator zero, and hence y as undefined number] .

192. The given equations are $3x + my - m = 0$ and $2x - 5y - 20 = 0$. Solving these equations by cross product method. We get

$$\frac{x}{-20m-5m} = \frac{y}{-2m+60} = \frac{1}{-15-2m}$$

$$\Rightarrow x = \frac{25m}{2m+15}, y = \frac{2m-60}{2m+15}$$

For $x > 0 \Rightarrow \dfrac{25m}{2m+15} > 0$

$\Rightarrow m < -\dfrac{15}{2}$ or $m > 0$

For $y > 0 \Rightarrow \dfrac{2(m-30)}{2m+15} > 0$

$\Rightarrow m < -\dfrac{15}{2}$ or $m > 30$

Combining above results we get the common values of m as follows:

$m < -\dfrac{15}{2}$ or $m > 30$

$\therefore m \in (-\infty,) \dfrac{15}{} (30, \infty)$

193. Solution : $e^{\sin x} - e^{-\sin x} - 4 = 0$

Let $e^{\sin x} = y$ then $e^{-\sin x} = 1/y$

$\therefore$ Equation becomes , $y - \dfrac{1}{y} - 4 = 0$

$\Rightarrow y^2 - 4y - 1 = 0$

$\Rightarrow y = 2 + \sqrt{5}, 2 - \sqrt{5}$

But y is real + ve number ,

$\therefore y \neq 2 - \sqrt{5} \Rightarrow y = 2 + \sqrt{5}$

$\Rightarrow e^{\sin x} = 2 + \sqrt{5} \Rightarrow \sin x = \log_e (2 + \sqrt{5})$

But $2 + \sqrt{5} > e \Rightarrow \log_e (2 + \sqrt{5}) > \log_e e$

$\Rightarrow \log_e (2 + \sqrt{5}) > 1$ Hence , $\sin x > 1$.

which is not possible .

$\therefore$ Given equation has no real solution.

194. For any square there can be at most 4 , neighboring squares,

		q				
	p	d	R			
		s				

Let for a square having largest number d, p, q, r, s be written then

According to the question,

$$p + q + r + s = 4d$$

$\Rightarrow (d - p) + (d - q) + (d - r) + (d - s) = 0$

Sum of four + ve numbers can be zero only if these are zero individually

$\therefore d - p = 0 = d - q = d - r = d - s$

$\Rightarrow p = q = r = s = d$

$\Rightarrow$ all the numbers written are same .

Hence proved.

195. Let α, β be the roots of eq. $ax^2 + bx + c = 0$

According to the question ,

$\beta = \alpha^n$

Also $\alpha + \beta = -b/a$; $\alpha\beta = c/a$

$\alpha\beta = \dfrac{c}{a} \Rightarrow \alpha . \alpha^n = \dfrac{c}{a} \Rightarrow \alpha = \left(\dfrac{c}{a}\right)^{\frac{1}{n+1}}$

then $\alpha + \beta = -b/a \Rightarrow \alpha + \alpha^n = \dfrac{-b}{a}$

or $\left(\dfrac{c}{a}\right)^{\frac{1}{n+1}} + \left(\dfrac{c}{a}\right)^{\frac{n}{n+1}} = \dfrac{-b}{a}$

$\Rightarrow a . \left(\dfrac{c}{a}\right)^{\frac{1}{n+1}} + a . \left(\dfrac{c}{a}\right)^{\frac{n}{n+1}} + b = 0$

$\Rightarrow (a^n c)^{\frac{1}{n+1}} + (a c^n)^{\frac{1}{n+1}} + b = 0$

Hence proved.

196. $x^2 - 3x + 2 > 0, \quad x^2 - 3x - 4 \leq 0$

$\Rightarrow (x - 1)(x - 2) > 0$ and $(x - 4)(x + 1) < 0$

$\Rightarrow x \in (-\infty, 1) \cup (2, \infty)$ and $x \in [-1, 4]$

$\therefore$ Common solution is $[-1, 1) \cup (2, 4]$

197. The given equation is

$$\left(5+2\sqrt{6}\right)^{x^2-3} + \left(5-2\sqrt{6}\right)^{x^2-3} = 10 \quad \text{...(i)}$$

Let $\left(5+2\sqrt{6}\right)^{x^2-3} = y$ ...(ii)

Then $\left(5-2\sqrt{6}\right)^{x^2-3} = \dfrac{\left(5-2\sqrt{6}\right)^{(x^2-3)}\left(5+2\sqrt{6}\right)^{x^2-3}}{\left(5+2\sqrt{6}\right)^{x^2-3}}$

$$\left(\frac{25-24}{5+2\sqrt{6}}\right)^{x^2-3} = \left(\frac{1}{5+2\sqrt{6}}\right)^{x^2-3} = \frac{1}{y}$$

$\therefore$ The given equation (i) becomes $y + \dfrac{1}{y} = 10$

$\Rightarrow y^2 - 10y + 1 = 0$

$\Rightarrow y = \dfrac{10 \pm \sqrt{100-4}}{2} = \dfrac{10 \pm 4\sqrt{6}}{2}$

$\Rightarrow y = 5 + 2\sqrt{6}$ or $5 - 2\sqrt{6}$

Consider, $y = 5 + 2\sqrt{6}$

$\Rightarrow \left(5+2\sqrt{6}\right)^{x^2-3} = \left(5+2\sqrt{6}\right)$

$\Rightarrow x^2 - 3 = 1 \Rightarrow x^2 - 4 \Rightarrow x = \pm 2$

Again consider

$y = 5 - 2\sqrt{6} = \dfrac{1}{5+2\sqrt{6}} = \left(5+2\sqrt{6}\right)^{-1}$

$\Rightarrow \left(5+2\sqrt{6}\right)^{x^2-3} = \left(5+2\sqrt{6}\right)^{-1} \Rightarrow x^2 - 2 = -1$

$\Rightarrow x^2 = 2$

$\Rightarrow x = \pm\sqrt{2}$

Hence the solutions are $2, -2, \sqrt{2}, -\sqrt{2}$

198. The given equation is,

$x^2 - 2a\,|x - a| - 3a^2 = 0$

Here two cases are possible .

Case I: $x - a > 0$ then $|x - a| = x - a$

$\therefore$ Equation becomes $x^2 - 2a(x - a) - 3a^2 = 0$

Or $x^2 - 2ax - a^2 = 0 \Rightarrow x = \dfrac{2a \pm \sqrt{4a^2 + 4a^2}}{2}$

$\Rightarrow x = a \pm a\sqrt{2}$

Case II: $x - a < 0$ then $|x - a| = -(x - a)$

$\therefore$ Equation becomes

$x^2 = 2a(x - a) - 3a^2 = 0$

Or $x^2 + 2ax - 5a^2 = 0$

$\Rightarrow x = \dfrac{2a \pm \sqrt{4a^2 + 20a^2}}{2}$

$\Rightarrow x = \dfrac{-2a \pm 2a\sqrt{6}}{2}$

$X = -a \pm a\sqrt{6}$

Thus the solution set is $\{a \pm a\sqrt{2}, -a \pm a\sqrt{6}\}$

199. Given that

$$\frac{2x}{\left(2x^2+5x+2\right)} > \frac{1}{(x+1)}$$

$$\Rightarrow \frac{2x}{\left(2x^2+5x+2\right)} - \frac{1}{x+1} > 0$$

$$\Rightarrow \frac{2x^2+2x-2x^2-5x-2}{\left(2x^2+5x+2\right)(x+1)} > 0$$

$$\Rightarrow \frac{-3x-2}{(2x+1)(x+1)(x+2)} > 0$$

$$\Rightarrow \frac{(3x+2)}{(x+1)(x+2)(2x+1)} < 0$$

$$\Rightarrow \frac{(3x+2)(x+1)(x+2)(2x+1)}{(x+1)^2(x+2)^2(2x+1)^2} < 0$$

$\Rightarrow (3x+2)(x+1)(x+2)(2x+1) < 0$ ...(i)

Critical points are $x = -2/3, -1, -2, -1/2$

On number line

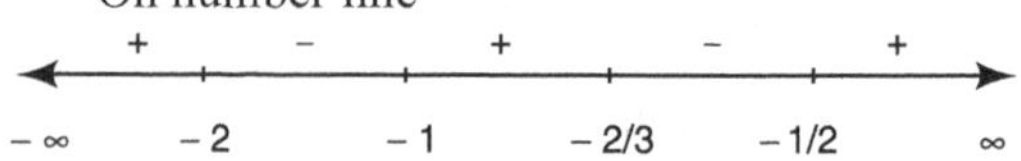

Inequality (i) holds for,

$x \in (-2, -1) \cup (-2/3, -1/2)$

[as $x \neq -2, -1, -2/3, -1/2$]

200. We are given that

α_1, α_2 be the roots of

$ax^2 + bx + c = 0$

$\therefore \alpha_1 + \alpha_2 = -\dfrac{b}{a};\ \alpha_1\alpha_2 = \dfrac{c}{a}$ ...(i)

And β_1, β_2 are the roots of $px^2 = qx + r = 0$

$\therefore \beta_1 + \beta_2 = -\dfrac{q}{p};\ \beta_1\beta_2 = \dfrac{r}{p}$ ...(ii)

The system of equations, $\alpha_1 y + \alpha_2 z = 0$

And $\beta_1 y + \beta_2 z = 0$ has a non trivial solution.

$\therefore$ We must have $\frac{\alpha_1}{\alpha_2} = \frac{\beta_1}{\beta_2} = 0$

So

$\Rightarrow \alpha_1 \beta_2 - \alpha_2 \beta_1 = 0 \Rightarrow \frac{\alpha_1}{\alpha_2} = \frac{\beta_1}{\beta_2}$

By componendo and dividendo, we get

$\frac{\alpha_1 + \alpha_2}{\alpha_1 - \alpha_2} = \frac{\beta_1 + \beta_2}{\beta_1 - \beta_2}$

$\Rightarrow (\alpha_1 + \alpha_2)(\beta_1 - \beta_2) = (\alpha_1 - \alpha_2)(\beta_1 + \beta_2)$

$\Rightarrow (\alpha_1 + \alpha_2)^2 [(\beta_1 + \beta_2)^2 - 4\beta_1 \beta_2]$

$= [(\alpha_1 + \alpha_2)^2 - 4\alpha_1 \alpha_2](\beta_1 + \beta_2)^2$

Using equations (i) and (ii) we get

$\frac{b^2}{a^2} = \left[\frac{q^2}{p^2} - \frac{4r}{p}\right] = \frac{q^2}{p^2}\left[\frac{b^2}{a^2} - \frac{4c}{a}\right]$

$\Rightarrow \frac{b^2q^2}{a^2p^2} - \frac{4b^2r}{a^2p} = \frac{q^2b^2}{p^2a^2} - \frac{4cq^2}{ap^2} \Rightarrow \frac{-4b^2r}{a^2p} = \frac{-4cq^2}{ap^2}$

$\Rightarrow \frac{b^2r}{a} = \frac{cq^2}{p} \Rightarrow \frac{b^2}{q^2} = \frac{ac}{pr}$ Hence proved.

201. The given equation is,

$|x^2 + 4x + 3| + 2x + 5 = 0$

Now there can be two cases.

CASE I: $x^2 + 4x + 3 \geq 0 \Rightarrow (x + 1)(x + 3) \geq 0$

$\Rightarrow x \in (-\infty, -3] \cup [-1, \infty)$ ...(i)

Then given equation becomes ,

$\Rightarrow x^2 + 6x + 8 = 0$

$\Rightarrow (x + 4)(x + 2) = 0 \Rightarrow x = -4, -2$

But $x = -2$ does not satisfy (i), hence rejected

$\therefore x = -4$ is one of the solution

CASE II: $x^2 + 4x + 3 < 0$

$\Rightarrow (x = 1)(x + 3) < 0$

$\Rightarrow x \in (-3, -1)$

...(ii)

Then given equitation becomes,

$-(x^2 + 4x + 3) + 2x + 5 = 0$

$\Rightarrow -x^2 - 2x + 2 = 0 \Rightarrow x^2 + 2x - 2 = 0$

$\Rightarrow x = \frac{-2 \pm \sqrt{4+8}}{2}$

$\Rightarrow x = -1 + \sqrt{3}, -1 - \sqrt{3}$

Out of which $x = -1 - \sqrt{3}$ is the other solution

Combining the two cases we get the solution of given equation as $x = -4, -1 - \sqrt{3}$.

202. Given that for $a, b, c \in R$, $ax^2 + bx + c = 0$ has two real roots α and β, where $\alpha < -1$ and $\beta > 1$. There may be two cases depending upon value of a, as shown below.

In each of the cases $af(1) < 0$

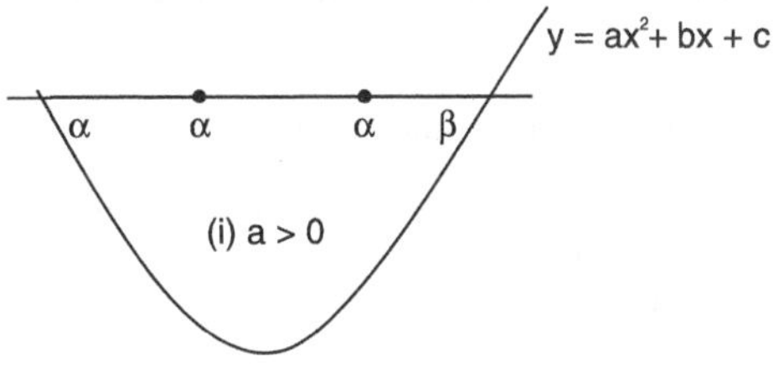

$\Rightarrow a(a - b + c) < 0$ and $a(a + b + c) < 0$

Dividing by a^2 (> 0), we get

$1 - \frac{b}{a} + \frac{c}{a} < 0$ ...(i)

And $1 + \frac{b}{a} + \frac{c}{a} < 0$ ...(ii)

Combining (i) and (ii) we get

$1 + \left|\frac{b}{a}\right| + \frac{c}{a} < 0$ or $1 + \frac{c}{a} + \left|\frac{b}{a}\right| < 0$ Hence Proved.

203. The given equation is,

$2^{|y|} - |2^{y-1} - 1| = 2^{y-1} + 1$

On the basis of absolute values involved here ($|y|$ and $2^{y-1} - 1|$) . there are two critical points 0 and 1. We have three cases here

CASE I: $y \in (-\infty, 0)$ then

$|y| = -y$ and $|2^{y-1} - 1| = 1 - 2^{y-1}$

$\therefore$ The given equation becomes

$2^{-y} - 1 + 2^{y-1} = 2^{y-1} + 1$

$\Rightarrow 2^{-y} = 2 \Rightarrow y = -1 \in (-\infty, 0)$

CASE II: $y \in [0, 1]$

204. $a^2 = p^2 + s^2$, $b^2 = (1 - p)^2 + q^2$

$c^2 = (1 - q)^2 + (1 - r)^2$, $d^2 = r^2 + (1 - s)^2$

$\therefore a^2 + b^2 + c^2 + d^2 = \{p^2 + (1 - p)^2\} + \{q^2$

$-(1-q)^2\} + \{r^2\} + (1-r)^2\} + \{s^2 + (1-s)^2\}$

where p, q, r, s all vary in the interval [0, 1].

Now consider the function

$y^2 = x^2 + (1-x)^2, 0 \le x \le 1,$

$2y\frac{dy}{dx} = 2x - 2(1-x) = 0$

$\Rightarrow x = \frac{1}{2}$ which $\frac{d^2y}{dx^2} = 4$ i.e. + ve

Hence y is minimum at $x = \frac{1}{2}$ and its minimum value is $= \frac{1}{4} + \frac{1}{4} = \frac{1}{2}$.

Clearly value is maximum at the end pts which is 1.

$\therefore$ Minimum value of $a^2 + b^2 + c^2 + d^2 = \frac{1}{2} + \frac{1}{2} + \frac{1}{2} + \frac{1}{2} = 2$ and maximum value is 1 + 1 + 1 + 1 = 4.

Hence proved.

If y = 0 we get $1 - \left|\frac{1}{2} - 1\right| = \frac{1}{2} + 1$

$1 - \frac{1}{2} = \frac{1}{2} + 1$ (not satisfied)

$\therefore$ y = 0 is not a solution

If y = 1 we get $2 - |2^0 - 1| = 2^0 + 1$

$\Rightarrow 2 = 2$ (satisfied)

$\therefore$ y = 1 is a solution

If $y \in (0, 1)$

then $|y| = y$ and $|2^{y-1} - 1| = 1 - 2^{y-1}$

$\therefore$ The equation becomes

$2^y - 1 + 2^{y-1} = 2^{y-1} + 1 \Rightarrow 2^y = 2$

$\Rightarrow y = 1 \notin (0, 1)$

$\therefore$ y = 1 is the only solution in the case.

CASE III: $y \in (1, \infty)$

Then $|y| = 1$, $|2^{y-1} - 1| = 2^{y-1} - 1$

The given equation becomes, $2^y - 2^{y-1} + 1 = 2^{y-1} + 1$

$\Rightarrow 2^y - 2^y = 0$

Which is satisfied for all real values of y but $y \in (1, \infty)$

$\therefore (1, \infty)$ is the solution in this case.

Combining all the cases, we get the final answer as as $y \in \{-1\} \cup [1, \infty)$.

205. We know that,

$(\alpha - \beta)^2 = [(\alpha + \delta) - (\beta + \delta)]^2$

$\Rightarrow (\alpha + \beta)^2 - 4\alpha\beta = (\alpha + \delta + \beta + \delta)^2 - 4(\alpha + \delta)(\beta + \delta)$

$\Rightarrow \frac{b^2}{a^2} - \frac{4c}{a} = \frac{B^2}{A^2} - \frac{4C}{A} \Rightarrow \frac{4ac - b^2}{a^2} = \frac{4AC - B^2}{A^2}$

[Here $\alpha + \beta = -\frac{b}{a}$, $\alpha\beta = \frac{c}{a}$, $(\alpha + \delta)(\beta + \delta) = -\frac{B}{A}$ and $(\alpha + \delta)(\beta + \delta) = \frac{C}{A}$]

Hence proved.

206. $ax^2 + bx + c = 0$ has roots α and β (given)

$\Rightarrow \alpha + \beta = -b/a$ and $\alpha\beta = c/a$

Now, $a^3x^2 + abcx + c^3 = 0$

Dividing the equation by c^2, we get

$\frac{a^3}{c^2}x^2 + \frac{abcx}{c^2} + \frac{c^3}{c^2} = 0, a\left(\frac{ax}{c}\right)^2 + b\left(\frac{ax}{c}\right) + c = 0$

$\Rightarrow \frac{ax}{c} = \alpha, \beta$ are the roots

$\Rightarrow x = \frac{c}{a}\alpha, \frac{c}{a}\beta$ are the roots

$\Rightarrow x = \alpha\beta\alpha, \alpha, \beta\beta$ are the roots

$\Rightarrow x = \alpha^2\beta, \alpha\beta^2$ are the roots

ALTERNATE SOLUTION

$\alpha + \beta = \frac{-b}{a}, \alpha\beta = \frac{c}{a}$

Roots of the equation

$a^3x^2 + abcx + c^3 = 0$ are

$$x = \frac{-abc \pm \sqrt{(abc)^2 - 4a^3c^3}}{2a^3}$$

$$= \frac{\left(-\frac{b}{a}\right)\left(\frac{c}{a}\right) \pm \sqrt{\left(\frac{b}{a}\right)^2\left(\frac{c}{a}\right)^2 - 4\left(\frac{c}{a}\right)^2}}{2}$$

$$= \frac{(\alpha+\beta)(\alpha\,\beta) \pm \sqrt{(\alpha+\beta)^2(\alpha\,\beta)^2 - 4(\alpha\,\beta)^3}}{2}$$

$$= (\alpha+\beta)\frac{((\alpha+\beta) \pm \sqrt{(\alpha-\beta)^2}}{}$$

$$= (\alpha+\beta)\frac{((\alpha+\beta) \pm (\alpha-\beta))}{2} = \alpha^2\,\beta, \alpha\beta^2$$

Let γ and δ be the required roots. Then γ = and δ = 2.

ALTERNATE SOLUTION:

Divide the equation by a^3, we get

$$x^2 + \frac{b}{c}\cdot\frac{c}{a}\cdot x + \left(\frac{c}{a}\right)^3 = 0$$

$$\Rightarrow x^2 - (\alpha+\beta)\cdot(\alpha\beta)\,X + (\alpha\beta)^3 = 0$$

$$\Rightarrow x^2 - \alpha^2\,\beta x - \alpha\,\beta^2\,x + (\alpha\beta)^3 = 0$$

$$\Rightarrow x\,(x - \alpha^2\beta)\,\alpha\,\beta^2\,(x - \alpha^2\beta) = 0$$

$$\Rightarrow (x - \alpha^2\beta)\,(x - \alpha\,\beta^2) = 0$$

$\Rightarrow x = \alpha^2\,\beta, \alpha\,\beta^2$ which is the required answer .

207. The given equation is,

$x^2 + (a - b)\,x + (1 - a - b) = 0$

$a, b \in R$

For this equation to have unequal real roots ∀ b

$D > 0$

$\Rightarrow (a - b)^2 - 4\,(1 - a - b) > 0$

$\Rightarrow a^2 + b^2 - 2\,a\,b - 4 + 4\,a + 4\,b > 0$

$\Rightarrow b^2 + b\,(4 - 2\,a) + a^2 + 4\,a - 4 > 0$

Which is a quadratic expression in b, and it will be true ∀ b ∈ R if D of above equation less than zero.

i.e., $(4 - 2\,a)^2 - 4\,(a^2 + 4\,a - 4) < 0$

$\Rightarrow \quad (2 - a)^2 - (a^2 + 4\,a - 4) < 0$

$\Rightarrow \quad 4 - 4\,a + a^2 - a^2 - 4\,a + 4 < 0$

$\Rightarrow \quad -8\,a + 8 < 0$

$\Rightarrow \quad a > 1.$

208. Roots of $x^2 - 10\,c\,x - 11\,d = 0$ are a and b

$\Rightarrow a + b = 10\,c$ and $a\,b = -11\,d$

Similarly c and d are the roots of

$x^2 - 10\,c\,x - 11\,d = 0$

$\Rightarrow c + d = 10\,a$ and $c\,d = -11\,b$

$\Rightarrow a + b + c + d = 10\,(a + c)$ and $a\,b\,c\,d = 121\,b\,d$

$\Rightarrow b + d = 9\,(a + c)$ and $a\,c = 121$

Also we have $a^2 - 10\,a\,c - 11\,d = 0$ and

$c^2 - 10\,a\,c - 11\,b = 0$

$\Rightarrow a^2 + c^2 - 20\,a\,c - 11\,(b + d) = 0$

$\Rightarrow (a + c)^2 - 22 \times 121 - 99\,(a + c) = 0$

$\Rightarrow a + c = 121$ or -22

For $a + c = -22$, we get $a = c$

⇒ rejecting this value we have $a + c = 121$

$\Rightarrow a + b + c + d = 10\,(a + c) = 1210.$

Concept Eliminator (CE)

1. If x_1, x_2, x_3 are roots of the given cubic equation then by the Viete formula , we have

$$\begin{cases} x_1 + x_2 + x_3 = 0 \\ x_1x_2 + x_2x_3 + x_3x_1 = -1, \\ x_1x_2x_3 = -1. \end{cases}$$

Since x_i where i ∈ (1, 2, 3) are roots of $x^3 - x + 1 = 0$

So, $x_i^3 - x_i + 1 = 0$ from it follows that

$$x_i^3 = x_i - 1,$$

$$x_i^5 = x_i^3 - .x_i^2 = (x_i - 1)x_i^2 = x_i^3 - x_i^2 = -x_i^2 + x_i - 1$$

$$x_i^8 = x_i^5.x_i^3 = (-x_i^2 + x_i - 1)(x_i - 1) = -x_i^3 + 2x_i^2 - 2x_i + 1 = 2x_i^2 - 3x_i + 2$$

Then $x_1^8 + x_2^8 + x_3^8 = 2(x_1^2 + x_2^2 + x_3^2) - 3(x_1 + x_2 + x_3) + 6.$

But $x_1^2 + x_2^2 + x_3^2 = (x_1 + x_2 + x_3)^2 - 2(x_1x_2 + x_2x_3 + x_3x_1) = 2.$

And so $x_1^8 + x_2^8 + x_3^8 = 4 - 0 + 6 = 10.$

2. (a) According to the condition given in question,

$\Delta_1 = a^2 - 4(3 - b) > 0$, or $a^2 + 4b - 12 > 0$, ...(i)

$\Delta_2 = (6 - a)^2 - 4(6 - b) = 0$ *or* $a^2 - 12a + 12 + 4b = 0$ or $a^2 + 4b = 12a - 12$ ...(ii)

$\Delta_3 = (4 - a)^2 - 4(5 - b) < 0.$ *or* $a^2 - 8a - 4 + 4b < 0.$ ...(iii)

From eqs (i) & (ii) we get

$$12a - 12 - 12 > 0, i.e.\, a > 2$$

And from eq (ii) and (iii) we get

$$12a-12-8a-4<0, i.e. a<4$$

So $2<a<4$ ...(iv)

Consider again equation (ii) $4b=24-(6-a)^2$

Since $2<a<4$ so

$24-(6-2)^2<4b<24-(6-4)^2$

$8<4b<10$, i.e. $2<b<5$

3. The given equation can be further sub divided into two groups

If $x \geq 0$, then equation is

$x^2+(a-5)x+1=0.$ ...(i)

If $x<0$, then equation is

$x^2+(a+1)x+1=0.$ ...(ii)

For these two equations to have exactly three distinct real solutions if one equation has two distinct real roots and other has repeated roots, and required condition for this is

Case (I) either $(a-5)^2>4$ and $(a+1)^2=4$;

Case (II) or $(a-5)^2=4$ and $(a+1)^2>4$.

CASE (I) From $(a+1)^2=4$ we get $a=1$ or -3.

But only $a=1$ satisfies $(a-5)^2>4$.

Thus $a=1$.

Check when $a=1$, equation (i) has solutions $x=2+\sqrt{3}$; and (ii) has solutions $x=-1,-1$.

As $2\pm\sqrt{3}>0$ and $-1<0$, we see that $a=1$ satisfy the condition.

CASE (II) From $(a-5)^2=4$, we have $a=3$ or 7.

Both these values of a satisfy the inequality $(a+1)^2>4$.

When $a=3$, equation (i) has solutions $x=1,1$ and (ii) has the solutions $x=-2\pm\sqrt{3}$.

As $1>0$ and $-2\pm\sqrt{3}<0$, we see that $a=3$ satisfy the condition.

When $a=7$, equation (i) has solutions $x=-1$, -1, which are negative contradicting $x\geq 0$.

Thus $a=1$, $a=3$ are the two desired values.

4. Let $P(x)=x^3+ax^2+bx+c$

Given that its roots are t, u, v, and

$Q(x)=x^3+a^3x^2+b^3x+c^3$,

Its roots are t^3, u^3, v^3.

By the Viete formula, we have

$t+u+v=-a$

$tu+uv+vt=b$

$tuv=c$, and

$t^3+u^3+u^3=-a^3$

and $(tu)^3+(uv)^3+(vt)^3=b^3$

$(tuv)^3=-c^3$

We know that

$$(t+u+v)^3=t^3+u^3+v^3+3(t+u+v)(tu+uv+vt)-3tuv,$$

From this we get

$a^3=-a^3-3ab+3c$, or $c=ab$.

In this case Q(x) has the form

$$Q(x)=x^3+a^3x^2+b^3x+(ab)^3=(x+a^3)(x^2+b^3)$$

So $Q(x)=0$ has a root $x=-a$, and for the other two roots we should have $b\leq 0$. Thus the conditions are $\begin{cases} ab=c, \\ b\leq 0. \end{cases}$

5. Let $x=\sqrt{2}+\sqrt[3]{3}$. Then squaring both the sides we get

$x^2=2+2\sqrt{2}\sqrt[3]{3}+\sqrt[3]{9}$ ...(i)

And cubing both the side we get

$x^3=2\sqrt{2}+6\sqrt[3]{3}+3\sqrt{2}\sqrt[3]{9}+3$ ...(ii)

From (i) $\sqrt[3]{3}=x-\sqrt{2}$ now squaring both the sides we get

$\sqrt[3]{9}=x^2-2-\sqrt{2}\sqrt[3]{3}$

$=x^2-2-2\sqrt{2}(x-\sqrt{2})$

$=x^2+2-2x\sqrt{2}$

Substituting two expressions for $\sqrt[3]{3}$ and $\sqrt[3]{9}$ into (ii), we get

$x^3=2\sqrt{2}+6(x-\sqrt{2})+3\sqrt{2}(x^2+2-2x\sqrt{2})+3,$

Or equivalently,

$x^3+6x-3=\sqrt{2}(3x^2+2).$

Squaring both sides yields

$x^6 - 6x^4 - 6x^3 + 12x^2 - 36x + 1 = 0,$

Which is the required polynomial of least degree and required degree is 6.

6. The given equation is defined if $|x| \le 1$ if so we can assume $x = \cos y$ where $y \in [0, \pi]$.

Now consider the left hand side of the equation

$$\sqrt{1+\sin y}(\sqrt{(1+\cos y)^3} - \sqrt{(1-\cos y)^3})$$

$$= \sqrt{\left(\sin\frac{y}{2}+\cos\frac{y}{2}\right)^2}$$

$$\left(\sqrt{\left(2\cos^2\frac{y}{2}\right)^3} - \sqrt{\left(2\sin^2\frac{y}{2}\right)^3}\right)$$

$$= 2\sqrt{2}\left(\sin\frac{y}{2}+\cos\frac{y}{2}\right)\left(\cos^3\frac{y}{2}-\sin^3\frac{y}{2}\right)$$

$$= 2\sqrt{2}\left(\sin\frac{y}{2}+\cos\frac{y}{2}\right)\left(\cos\frac{y}{2}-\sin\frac{y}{2}\right)$$

$$\left(\cos^2\frac{y}{2}+\sin^2\frac{y}{2}+\sin\frac{y}{2}\cos\frac{y}{2}\right)$$

$$= 2\sqrt{2}\left(\sin\frac{y}{2}+\cos\frac{y}{2}\right)$$

$$\left(\cos\frac{y}{2}-\sin\frac{y}{2}\right)\left(1+\frac{1}{2}\sin y\right)$$

$$= 2\sqrt{2}\left(\cos^2\frac{y}{2}-\sin^2\frac{y}{2}\right)\left(1+\frac{1}{2}\sin y\right)$$

$$= 2\sqrt{2}\cos^2\frac{y}{2}\sin^2\frac{y}{2}\left(1+\frac{1}{2}\sin y\right)$$

$$= 2\sqrt{2}\cos y\left(1+\frac{1}{2}\sin y\right)$$

Then consider the right hand side of the equation

$$2+\sqrt{1-\cos^2 y} = 2+\sin y$$

since $2 + \sin y \ne 0$

Therefore, the given equation is equivalent to

so $\cos y = \frac{1}{\sqrt{2}}$; that is

$\cos x = \frac{1}{\sqrt{2}}$, $x = \frac{\pi}{4}$

7. Consider the equation $x^2 + ax + b = 0$. It has two roots (not necessarily real), say α and β.

Either $\alpha = \beta$ or $\alpha \ne \beta$.

Case–(i) If $\alpha = \beta$, then is a repeated root. Given that $\alpha^2 - 2$ is a also a root, so $\alpha = \alpha^2 - 2$. or $(\alpha + 1)(\alpha + 2) = 0$. Hence $\alpha = -1$ or $\alpha = 2$.

When $\alpha = -1$, then (a, b) = (2, 1)

When $\alpha = 2$, then (a, b) = (–4, 4)

Case (ii): Suppose $\alpha \ne \beta$. Therefore are three possibilities ;

(I) $\alpha = \alpha^2 - 2$ and $\beta = \beta^2 - 2$;

Here $(\alpha,\beta) = (2,-1)$ or $(-1,2)$. Hence $(a,b) =$ $= (-(\alpha+\beta),\alpha\beta) = (-1,-2)$.

(II) $\alpha = \beta^2 - 2$ and $\beta = \alpha^2 - 2$;

Suppose $\alpha = \beta^2 - 2$ and $\beta = \alpha^2 - 2$. Then

$$\alpha - \beta = \beta^2 - \alpha^2 = (\beta-\alpha)(\beta+\alpha)$$

Since, $\alpha \ne \beta$, we get $\beta + \alpha = \beta^2 + \alpha^2 - 4$ $= (\alpha+\beta)^2 - 2\alpha\beta - 4$.

Thus $-1 = 1 - 2\alpha\beta - 4$, which implies that $\alpha\beta = -1$. Therefore $(a,b) = (-((\alpha+\beta),\alpha\beta)$ $= (1,-1)$.

(III) $\alpha = \alpha^2 - 2 = \beta^2 - 2$ and $\alpha \ne \beta$

If $\alpha = \alpha^2 - 2 = \beta^2 - 2$ and $\alpha \ne \beta$, then $\alpha = -\beta$ Thus $\alpha = 2, \beta = -2$ *or* (a, b) = (0, –4)

Thusweget6pairs; $(a,b) = -(4,4),(2,1),(-1,-2),$ $(1,-1),(0,-4),(0,1)$.

8. Since 0 is not a root of the equation so we can assume $t = x + \frac{1}{x}$ then $|t| \ge 2$

Then we can write the equation as

$$16x^2 - mx + (2m+17) - \frac{m}{x} + \frac{16}{x^2} = 0,$$

or $16t^2 - mt + (2m-15) = 0,$

From this it follows that the given equation has four distinct real roots if and only if the quadratic $16t^2 - mt + (2m-15) = 0$ has two distinct real roots t_1, t_2 which are not in, (–2, 2), as the equation $|t| = 2$ give two equal roots.

The given equation has two distinct real roots if,

$D = m^2 - 64(2m-15) > 0$

$\Leftrightarrow (m-8)(m-120) > 0 \Leftrightarrow$

$\Leftrightarrow m < 8, m > 120$

Sum of the roots is $t_1 + t_2 = \frac{m}{16}$

Either both the roots are less than –2 or both are more than 2

Consider the 1st case when both the roots are less than –2

$\frac{m}{16} = t_1 + t_2 < -4$

$\Leftrightarrow m < -64 \Rightarrow t_1 t_2 = \frac{2m-15}{6} < 0,$

Which is impossible.

So 1st case ruled out.

Consider the 2nd case when both the roots are greater than 2.

Thus we get $2 < t_1 < t_2$ and each of the two equations $x + \frac{1}{x} = t_1, x + \frac{1}{x} = t_2$, has two real positive distinct roots, which we denote by x_1, x'_1 and x_2, x'_2, respectively.

Then $x_1, x'_1 = x_2, x'_2 = 1$.

We can assume that $1 < x_1 < x_2$fform an increasing geometric progression. Therefore, $x_2 = (x_1)^3, x'_2 = (x'_1)^3$,

which implies that

$t_2 = x_2 + \frac{1}{x_2} = x_2 + x'_2$

$t_2 = x_2 + \frac{1}{x_2} = x_2 + x'_2$

$= (x_1)^3 + (x'_1)^3 = (x_1, x'_1)\left[(x_1)^2 - x_1 x'_1 + (x'_1)^2\right]$

$= (x_1 + x'_1)[(x_1 + x'_1)^2 - 3\, x_1 x'_1]$

$= t_1[(t_1)^2 - 3]$.

Then, $\frac{m}{16} = t_1 + t_2 = t_1[(t_1)^2 - 2]$,

Andhence $m = 16t_1[(t_1)^2 - 2\,] = t_1[16(t_1)^2 - 32\,]$

$= t_1[mt_1 - 2m + 15) - 32] = m(t_1)^2 - (2m+17)t_1$,

Which gives $= \frac{17t_1}{(t_1)^2 - 2t_1 - 1}$

Substituting this value of m into the equation $f(t_1) = 16(t_1)^2 - mt_1 + 2m - 15 = 0$ we obtain $16(t_1)^4 - 31(t_1)^3 - 48(t_1)^2 + 64t_1 + 15 = 0$.

Substitute $y = 2t_1$, we have

$y^4 - 4y^3 - 12y^2 + 32y + 15 = 0 \Leftrightarrow (y-5)$

$(y+3)(y^2 - 2y - 1) = 0$.

From this it follows that the unique possible value of y for which $t_1 > 2 is y = 5$. Hence $t_1 = \frac{5}{2}$, and so $m = 170$.

Conversely, for $m = 170$ the equation $16\,x^4 - 170x^3 + 357x^2 - 170x + 6 = 0$ has four distinct roots $\frac{1}{8}, \frac{1}{2}, 2, 8$ which obviously form a geometric progression with the ratio $r = 4$. Thus the only solution to the problem is $m = 170$.

9. Note that if $u\sqrt[3]{3} + v\sqrt[3]{9} = 0$ with $u, v \in \mathbb{Q}$ then $u = v = 0$.

(1) Since we need to find out polynomial with lowest degree so we will start with linear function.

Consider a polynomial $P(x) = ax + b$ with $a, b \in \mathbb{Q}$. with If $\underline{P}$ (x) satisfies the problem, that is, $a\left(\sqrt[3]{3} + \sqrt[3]{9}\right) + b = 3 + \sqrt[3]{3}$, then $(a-1)\sqrt[3]{3} + a\sqrt[3]{9} = 3 - b \in \mathbb{Q}$ and hence $a = 0 = a - 1$. This is impossible, and so there does not exist such a linear polynomial. Now consider a quadratic $P(x) = ax^2 + bx + c$ with $a, b, c \in \mathbb{Q}$.

From $f\,(\sqrt[3]{3} + \sqrt[3]{9}) = 3 + \sqrt[3]{3}$ which is equivalent to

$(a+b)\sqrt[3]{9} + (3a+b)\sqrt[3]{3} + 6a + c = 3 + \sqrt[3]{3}$,

If follows that

$a + b = 0,$

$3a + b = 1,$

$6a + c = 3,$

On solving above equations we get, $a = \frac{1}{2}$,

$b = -\frac{1}{2}, c = 0.$

Thus $P(x) = \frac{1}{2}(x^2 - x)$ is the required polynomial of lowest degree.

(2) Put $s = \sqrt[3]{3} + \sqrt[3]{9}$ we have $s^3 = 9s + 12$. This shows that the polynomial $G(x) = x^3 - 9x - 12$ has s as its root.

Assume that there exists a polynomial $P(x)$ of degree $n \geq 3$ with integer coefficients satisfying $P(s) = 3 + \sqrt[3]{3}$. Then by long division, we have

$P(x) = G(x).\, Q(x) + R(x),$

Where $Q(x)$ and $R(x)$ are polynomials with integer coefficients, and deg $(R) < 3$.

In this case

$$P(s) = G(s).\, Q(s) + R(s) = 0.\, Q(s) + R(s) = R(s),$$

That is p $(s) = 3 + \sqrt[3]{3}$. This is impossible, because by (i) there exists a unique polynomial of degree ≤ 2 with rational coefficients satisfying the given condition, which is $\frac{1}{2}(x^2 - x)$

So there does not exist other a polynomial $P(x)$.

10. We have three relations:

$a\alpha^2 - b\alpha - c = \lambda,$

$b\alpha^2 - c\alpha - a = \lambda,$

$c\alpha^2 - a\alpha - b = \lambda,$

where λ is the common value. Eliminating α^2 from these, taking these equations pair– wise, we get three relations:

$(ca - b^2)\alpha - (bc - a^2) = \lambda(b - a),$

$(ab - c^2)\alpha - (ca - b^2) = \lambda(c - b),$

$(bc - a^2) - (ab - c^2) = \lambda(a - c).$

Adding these three, we get

$(ab + bc + ca - a^2 - b^2 - c^2)(\alpha - 1) = 0$.

(Alternatively, multiplying above relations respectively by $b - c, c - a$ and $a - b$, and adding also leads to this). Thus either $ab + bc + ca - a^2 - b^2 - c^2 = 0$ or $\alpha = 1$. In the first case

$$0 = ab + bc + ca - a^2 - b^2 - c^2 = \frac{1}{2}\left((a - b)^2 + (b - c)^2 + (c - a)^2\right)$$

Shows that $a = b = c$. If $\alpha = 1$, then we obtain $a - b - c = b - c - a = c - a - b.$

And once again we obtain a + b + c

11. $x = \sqrt{19 - 8\sqrt{3}} = \sqrt{(4 - \sqrt{3})^2} = 4 - \sqrt{3}$ *or* $4 - x = \sqrt{3}.$

Squaring both the sides we get $x^2 - 8x + 13 = 0$.

Hence, by long division,

$$x^4 - 6x^3 - 2x^2 + 18x + 23 = (x^2 - 8x + 13)(x^2 - 2x + 1) + 10 = 10,$$

And $x^2 - 8x + 15 = 2$

So $\frac{x^4 - 6x^3 - 2x^2 + 18x + 23}{x^2 - 8x + 15} = \frac{10}{2} = 5$

12. When $m = 2$, the equation becomes $-5x - 11 = 0$, the solution is $x = -1$.

When $m \neq 2, \Delta = [-(m + 3)]^2 + 4(2m + 1)(m - 2) = 9m^2 - 6m + 1$

$= (3m - 1)^2 \geq 0$, and roots are

$$x_1 = \frac{(m + 3) - (3m - 1)}{2(m - 2)} = -1,$$

$$x_2 = \frac{(m + 3) + (3m - 1)}{2(m - 2)} = \frac{2m + 1}{m - 2}.$$

So there exist only one roots independent of m which is x = – 1

13. (i) When $m = 1$, then $-6x + 1 = 0$, so there is a real root $x = 1/6$.

(ii) When $m = -1$, then $-2x + 1 = 0$, so there is a real root $x = 1/2$.

(iii) When $m^2 - 1 \neq 1$, then $\Delta = 4(m + 2)^2 - 4(m^2 - 1) = 16m + 20 \geq 0$ implies that $m \geq -\frac{5}{4}$ and $m \neq 1$.

Thus the range of $m \geq -\frac{5}{4}$.

If condition could have been that the equation has two distinct roots then range of m is

$(-5/4, 1) \cup (1, \infty)$

14. (d) The Discriminant of the equation is given by

$$\Delta = (a+b+c)^2 - 4(a^2+b^2+c^2)$$
$$= -3a^2 - 3b^2 - 3c^2 + 2ab + 2bc + 2ca$$
$$= -\left[(a-b)^2 + (b-c)^2 + (c-a)^2 + (a^2+b^2+c^2)\right] < 0$$

So the equation has no real roots

15. Mr. Fat has the winning strategy. A set of three distinct rational non – zero numbers *a*, *b*, and *c*, such that $a + b + c = 0$,

Since sum of coefficients is zero then 1 is a root of the equation and since coefficients are rational

16. Notice that $x_0 \neq 1$. Otherwise, if $x_0 = 1$, then above two equalities becomes $a = 1 = b$ and the given equation wont be a quadratic equation.

From the assumptions $a > 1$ and $b > 1$ and $a \neq b$.

Let x_0 be the common root, then

$$(a-1)x_0^2 - (a^2+2)x_0 + (a^2+2a) = 0,$$
$$(b-1)x_0^2 - (b^2+2)x_0 + (b^2+2b) = 0.$$

After eliminating the term of x_0^2 and doing simplification we get

$$(a-b)(ab-a-b-2)(x_0-1) = 0.$$

Since $a - b \neq 0$ and $x_0 \neq 1$, so $ab - a - b - 2 = 0$, i.e. $ab = a + b + 2$.

(i) When $a > b > 1$, then $b = 1 + \frac{b}{a} + \frac{2}{a} < 3$, so $b = 2$, $a = \frac{4}{b-1} = 4$.

(ii) When $b > a > 1$, then by symmetry, $a = 2, b = 4$.

Thus in each of above two cases,

$$\frac{a^b + b^a}{a^{-b} + b^{-a}} = (a^b + b^a).\frac{a^b b^a}{a^b + b^a} = a^b b^a = 256.$$

17. Since the left hand side cannot be zero, so $m \neq 0$

We have two cases

Case (i) When $m > 0$, then $x > 0$.

For $0 < x \le 1$, then $1 - x^2 + 4 - x^2 = mx$, i.e. $2x^2 + mx - 5 = 0$, so

$$x = \frac{-m + \sqrt{m^2+40}}{4} \le 1.$$

It implies that

$$m^2 + 40 \le m^2 + 8m + 16, i.e. 3 \le m.$$

For $1 < x \le 2$, then $x^2 - 1 + 4 - x^2 = mx$, so $1 < x = \frac{3}{m} \le 2$, i.e. $\frac{3}{2} \le m < 3$.

for $2 < x$, then $2x^2 - mx - 5 = 0$, so $x = \frac{m + \sqrt{m^2+40}}{4} > 2$, it implies that

$$m^2 + 40 > m^2 - 16m + 64, i.e.\ m > \frac{3}{2}$$

Case (ii) When $m < 0$, then $x < 0$, so the case be converted to the case (*i*) if use $-m$, $-x$ to replace *m*, *x* respectively. Thus the solutions are

$$x = \begin{cases} \frac{3}{m} or \frac{m+\sqrt{m^2+40}}{4} & if\ \frac{3}{2} \le m < 3 \\ \frac{-m+\sqrt{m^2+40}}{4} & if\ m \ge 3. \\ \frac{3}{m} or \frac{m-\sqrt{m^2+40}}{4} & if\ -3 < m \le -\frac{3}{2} \\ \frac{-m+\sqrt{m^2+40}}{4} & if\ m \le -3, \end{cases}$$

And has no solution for other values of.

18. Suppose that the first equation has no real roots, then

$$\Delta_1 = 1 - 4q_1 \le 0, i.e.\ q_1 \ge \frac{1}{4}.$$

Now consider the Discriminant of the second equation

$$\Delta_2 = p^2 - 4q_2 = (q_1 + q_2 + 1)^2 - 4q_2$$
$$= q_2^2 + 2q_2(q_1+1) + (q_1+1)^2 - 4q_2$$
$$= q_2^2 + 2q_2(q_1-1) + (1+q_1)^2.$$

To show Δ_2 is always positive, consider the last expression as a quadratic function of q_2, then its Discriminant is

$$\Delta_3 = 4(q_1-1)^2 - 4(1+q_1)^2 = -16q_1 \le -4,$$

so, $\Delta_2 > 0$ for any value of q_2, i.e. the second equation has two distinct real roots,

Hence proved.

19. By Viete Theorem, $x_1 + x_2 = -1, x_1x_2 = -3$.

Let $A = x_1^3 - 4x_2^2 + 19$,

$B = x_2^3 - 4x_1^2 + 19$. Then

$$A + B = \left(x_1^3 + x_2^3\right) - 4\left(x_1^2 + x_2^2\right) + 38$$
$$= (x_1 + x_2)\left[(x_1 + x_2)^2 - 3x_1x_2\right] - 4\left[(x_1 + x_2)^2 - 2x_1x_2\right] + 38.$$
$$= -[1+9] - 4[1+6] + 38 = 0.$$
$$A - B = \left(x_1^3 - x_2^3\right) + 4\left(x_1^2 - x_2^2\right)$$
$$= (x_1 - x_2)\left[(x_1 + x_2)^2 - x_1x_2 + 4(x_1 + x_2)\right]$$
$$= (x_1 - x_2)[1 + 3 - 4] = 0.$$

Thus, $2A = (A + B) + (A - B) = 0, i.e. A = 0$.

20. (i) The Viete Theorem yields $\alpha + \beta = p, \alpha\beta = q$, so

$$(\alpha)^3.(\beta)^3 = (\alpha\beta)^3 = q^3,$$
$$\alpha^3 + \beta^3 = (\alpha + \beta)^3 - 3\alpha\beta(\alpha + \beta)$$
$$= p^3 - 3pq = p\left(p^2 - 3q\right),$$

So the required equation is

$$x^2 - p\left(p^2 - 3q\right)x + q^3 = 0.$$

(ii) If the new equation is the same as the original equation, then by comparing the coefficients we get

$$p\left(p^2 - 3q\right) = p, \quad ...(i)$$
$$q^3 = q. \quad ...(ii)$$

From equation (ii) we get, $q = 0, 1$ or -1.

When $q = 0$, then from (i), $p^3 = p$, so $p = 0$, 1 or -1.

when $q = 1$, then $p^3 = 4p$, so $p = 0, 2$ or -2.

when $q = -1$ then $p^3 = -2p$, so $p = 0$.

corresponding to the obtained seven pairs of (p, q) there are seven equations

$x^2 = 0, x^2 - x = 0, x^2 + x = 0, x^2 + 1 = 0,$

$x^2 - 2x + 1 = 0, x^2 + 2x + 1 = 0, x^2 - 1 = 0.$

Among them only $x^2 + 1 = 0$ has no real roots, so the rest 6 pairs satisfy the requirements.

Thus, the solutions for (p, q) are

$(0,0), (1,0), (-1,0), (2,1), (-2,1), (0,-1)$.

21. The equation has two distinct real roots implies that $\Delta > 0$,

so $\Delta = 4(m-2)^2 - 4\left(m^2 - 3m + 3\right) = -4m + 4 > 0.$

Thus, $-1 \le m < 1$.

(i) By Viete Theorem,

$x_1 + x_2 = -2(m-2), x_1x_2 = m^2 - 3m + 3$, so

$$6 = x_1^2 + x_2^2 = (x_1 + x_2)^2 - 2x_1x_2$$
$$= 4(m-2)^2 - 2\left(m^2 - 3m + 3\right) = 2m^2 - 10m + 10,$$

i.e. $m^2 - 5m + 2 = 0$, so $m = \frac{5 \pm \sqrt{17}}{2}$. Since $-1 \le m < 1$, $m = \frac{5 - \sqrt{17}}{2}$.

(ii) $$\frac{mx_1^2}{1 - x_1} + \frac{mx_2^2}{1 - x_2} = \frac{m\left[x_1^2(1 - x_2) + x_2^2(1 - x_1)\right]}{(1 - x_1)(1 - x_2)}$$

$$= \frac{m\left[x_1^2 + x_2^2 - x_1x_2(x_1 + x_2)\right]}{x_1x_2 - (x_1 + x_2) + 1}$$

$$= \frac{m\left[\left(2m^2 - 10m + 10\right) + 2\left(m^2 - 3m + 3\right)(m - 2)\right]}{m^2 - 3m + 3 + 2(m-2) + 1}$$

$$= \frac{m\left(2m^3 \quad 8m^2 + 8m - 2\right)}{m^2 - m} = 2\left(m^2 - 3m + 1\right)$$

Since $-1 \le m < 1$.

$$= 2\left(m - \frac{3}{2}\right)^2 - \frac{5}{2} \le 2\left(-1 - \frac{3}{2}\right)^2 - \frac{5}{2} = 10$$

Thus, the maximum value of $\frac{mx_1^2}{1 - x_1} + \frac{mx_2^2}{1 - x_2}$ is 10.

22. Let us consider the given equations

$$x^2 + px + q = 0 \quad ...(i)$$
$$x^2 + 2qx + \frac{1}{2}p = 0 \quad ...(ii)$$

And let α, β be the roots of (i), then the roots of (ii) are $\alpha - 1, \beta - 1$.

By Viete Theorem,

$$\alpha + \beta = -p, \quad ...(iii)$$
$$\alpha\beta = q, \quad ...(iv)$$
$$\alpha + \beta - 2 = -2q, \quad ...(v)$$
$$(\alpha - 1)(\beta - 1) = \frac{1}{2}p. \quad ...(vi)$$

By (iii) – (v)

$$-p + 2q = 2. \quad ...(vii)$$

By, (iv) – (v) – (vi)

$-p+6q=2.$ ...(viii)

Then from (viii) – (vii) we get α = 0, β = – 2 hence the equation $x^2-2x=0$, its roots are $\alpha=0,\beta=2$.

Similarly, the equation $x^2-1=0$, its roots are $x_1=-1=\alpha-1, x_2=1=\beta-1$

23. Let (x_1,x_2) be an integer solution of the given solution. Then by Viete Theorem

$x_1+x_2=-p, x_1x_2=q,$

so $198=p+q=-(x_1+x_2)+x_1x_2=(x_1-1)(x_2-1)-1,$

$\therefore (x_1-1)(x_2-1)=199=(1)(199)=(-1)\times(-199).$

Letting $x_1 \le x_2$, then $x_1-1=1, x_2-1=199$ or $x_1-1=-199, x_2-1=-1$, so the solutions are $(x_1,x_2)=(2,200)$ or $(-198,0)$.

24. Given that α & β are the real roots of $x^2+19x-97=0$

So $\alpha+\beta=-19, \alpha\beta=-97.$

Then $-\frac{m}{n}=\frac{1+\alpha}{1-\alpha}+\frac{1+\beta}{1-\beta}=\frac{(1+\alpha)(1-\beta)+(1-\alpha)(1+\beta)}{(1-\alpha)(1-\beta)}$

$$=\frac{2(1-\alpha\beta)}{1-(\alpha+\beta)+\alpha\beta}=\frac{2.98}{77}=-\frac{28}{11},$$

So $m=28, n=11$ and $m+n=39$.

25. α & β are the roots of the equation $x^2+px+1=0$.

So $\alpha+\beta=-p,$ & $\alpha\beta=1,$

and if γ and δ are the roots of the equation $x^2+qx+1=0,$

so $\gamma+\delta=-q$ & $\gamma\delta=1$.

So $(\alpha-\gamma)(\beta-\gamma)(\alpha+\delta)(\beta+\delta)$

$$=[(\alpha-\gamma)(\beta+\delta)][(\beta-\gamma)(\alpha+\delta)]$$

$$=(\alpha\beta+\alpha\delta-\beta\gamma-\gamma\delta)(\alpha\beta+\beta\delta-\alpha\gamma-\gamma\delta)$$

$$=(\alpha\delta-\beta\gamma)(\beta\delta-\alpha\gamma)=\alpha\beta\delta^2-\alpha^2\gamma\delta-\beta^2\gamma\delta+\alpha\beta\gamma^2$$

$$=\delta^2-\alpha^2-\beta^2+\gamma^2=\left[(\delta+\gamma)^2-2\delta\gamma\right]-[(\alpha+\beta)^2-2\alpha\beta$$

$$=(q^2-2)-(p^2-2)=q^2-p^2.$$

26. Let $A=\frac{2}{\alpha}+3\beta^2, B=\frac{2}{\beta}+3\alpha^2.$

Since $\alpha+\beta=7, \alpha\beta=8,$ and $\alpha-\beta=\sqrt{\Delta}$

$=\sqrt{49-32}=\sqrt{17}.$

It follows that

$$A+B=\frac{2(\alpha+\beta)}{\alpha\beta}+3(\alpha^2+\beta^2)=\frac{14}{8}+3(7^2-16)=\frac{403}{4},$$

$$A-B=-\frac{2(\alpha-\beta)}{\alpha\beta}-3(\alpha^2-\beta^2)=-(\alpha-\beta)\left[\frac{2}{\alpha\beta}+3(\alpha+\beta)\right]$$

$$=-\sqrt{\Delta}\left[\frac{1}{4}+21\right]=-\frac{85\sqrt{17}}{4}.$$

By adding above two results, we get

$$\frac{2}{\alpha}+3\beta^2=A=\frac{1}{2}\left[\frac{403}{4}-\frac{85\sqrt{17}}{4}\right]=\frac{403-85\sqrt{17}}{8}$$

27. The given equation can be written as

$$\frac{x(13-x)}{x+1}\left(x+\frac{13-x}{x+1}\right)=42$$

Let $y=\frac{13-x}{x+1}$, then the given equation becomes

$$xy(x+y)=42.$$

The technique of solving the problem is to solve xy and $x+y$ first by applying the inverse of Viete Theorem. For this the value of $(xy)+(x+y)$ is needed.

Since $xy+(x+y)=\frac{13x-x^2}{x+1}+\frac{x^2+x+13-x}{x+1}=\frac{13+13x}{x+1}=13$

By using inverse Viete Theorem, xy and $x+y$ are the roots of the equation

$z^2-13z+42=0.$

Since $z=6$ or 7, by solving the systems $xy=6, x+y=7$ and $xy=7, x+y=6$ respectively, the solutions for x are

$x_1=1, x_2=6, x_3=3+\sqrt{2}, x_4=3-\sqrt{2}.$

By checking, they are all the solution of the original equation.

28. This equation can be solved by factorization method.

Let $n^2+19n+48 = m^2$ for some $m \in \mathbb{N}$, then $4n^2+76n+192=4m^2$,

so $(2n+19)^2-(2m)^2=169$.

$(2n-2m+19)(2n+2m+19)$

$= 13^2 = 1.169 = 13.13.$

Now we have following Cases

Case (i) 169 = 1 x 169

Then $2n-2m+19=1$ and

$2n+2m+19=169$

On solving these two equations we get n = 33, m= 42

Case (ii) 169 = 169 x 1

Then $2n-2m+19=169$ and

$2n+2m+19=1$

On solving these two equations we get n = 33, m= – 42

Case (iii) 169 = 13x13

Then $2n-2m+19=13$ and

$2n+2m+19=13$

on solving these two equations we get n = –3, m= 0, this answer ruled out as n is negative.

So only possible value of n is n = 33

29. Since the equation has integer solutions,

$\Delta = 4\left[(m+5)^2-(100m+9)\right]=4(m^2-90m+16)\geq 0$

and $m^2-90m+16=n^2$ for some non–negative integer n. Then

$(m-45)^2+16-2025=n^2$, so

$(m-n-45)(m+n-45)=2009$

$=1.2009=41.49=(-49)(-41)=(-2009)(-1).$

When $m-n-45=1, m+n-45=2009$,

then $m=1050, n=1004$.

When $m-n-45=41, m+n-45=49$, then $m=90, n=4$.

When $m-n-45=-2009, m+n-45=-1$, then $m=-960, n=1004$

When $m-n-45=-49, m+n-45=-41$, then $m=0, n=4$.

Since $x=-(m+5)\pm n$, all above values of (m,n) give integral roots. Thus, $m=1050, 90, -960, 0,$ and the smallest positive value of m is 90.

30. From Viete Theorem we get

$p+q=\dfrac{p^2+11}{9}$,

$pq=\dfrac{15}{4}(p+q)+16.$

Since here both $p+q>0$ and $pq>0$ so p, q are both positive integers.

Above result can be written as

$16pq-60(p+q)=16^2$

$\therefore (4p-15)(4q-15)=256+225=481.$

Since $481=1\times 481=13\times 37=(-1)\times(-481)=(-13)\times(-37)$,

and $4p-15$ *or* $4q-15$ cannot be $-37 \text{ or } -481$, so the pair $(4p-15, 4q-15)$ has the following four possible cases :

$(1,481),(481,1),(13,37),(37,37)$.

Corresponding to them the pairs of (p, q) are

$(4,124),(124,4),(7,13),(13,7)$.

By checking only the pair (13, 7) satisfies the original equation : the equation becomes

$$x^2-20x+91=0,$$

And its roots are $\{13,7\}$.

31. Let (a,b,c) be an integer solution. We show that a, b and c must all even by taking modulo 4 to both side of the equation. There are three possible cases to be considered :

Case 1: When a, b, c are all odd, then $a^2+b^2+c^2\equiv 3(\text{mod } 4)$, whereas $a^2b^2\equiv 2$ (mod 4), so its impossible.

Case 2: When two of a, b, c are odd and the other is even, then $a^2+b^2+c^2\equiv 2$ (mod 4), whereas $a^2b^2\equiv 0$ or 1 (mod 4), so it is impossible also.

Case 3: When two of a, b, c are even and the other is odd, then $a^2+b^2+c^2\equiv 1$ (mod 4), whereas $a^2b^2\equiv 0$ (mod 4), so it is impossible also.

Thus a, b, c are all even. Let $a=2a_1, b=2b_1$, $c=2c_1$, this leads to the relation

$a_1^2+b_1^2+c_1^2=4a_1^2b_1^2$.

Since $4\,a_1^2b_1^2\equiv 0$ (mod 4) and each of a_1, b_1, c_1 has remainder 0 or 1 modulo 4 a_1, b_1, c_1, must all be even also, then $a_1=2a_2$, $b_1=2b_2$ $c_1=2c_2$.

This leads to the relation

$a_2^2+b_2^2+c_2^2=16a_2^2b_2^2$.

Again we can conclude that a_2, b_2, c_2 are all even , so it leads to the relation

$a_3^2+b_3^2+c_3^2=64a_3^2b_3^2$.

The process can be continued to any times, Since we have the relation

$a_n^2+b_n^2+c_n^2=4^n a_n^2 b_n^2$.

for any natural number n. Hence $a_n=\frac{a}{2^n}$, $b_n=\frac{b}{2^n}, c_n=\frac{c}{2^n}$ are integers for any nautral number n, i.e., $a=b=c=0$.

Thus the equation has only zero solution.

NOTE: This example is one of Fermat's method of infinite decent.

It is sometimes useful to use substitution to simplify the equation first, this is such an example.

32. Since $a^2+a-6=(a-2)(a+3)$

Given that $\frac{1260}{a^2+a-6}$ is a positive integer it is possible when $(a-2)$ and $(a+3)$ are both factors of 1260 , and their difference is 5.

Since $1260=2^2\times3^2\times5\times7$, where the pairs of two factors with the difference 5 are (1, 6), (2, 7), (4, 9), (7, 12) and (9; 14). Thus, $a-2$ = 1, 2, 4, 7, 9, i.e. a = 3, 4, 6, 9, 11.

33. Suppose that the two roots of the given equation are m and n. Then, by Viete Theorem,

$m+n=p,$ ...(i)

$mn=-580p,$ ...(ii)

Therefore one of m and n is divisible by p. Without loss of generality, we assume that $p|m$. Then for $m=kp$ some integer k. then from (i) we get $n=(1-k)p$. and from (*ii*) we get $(k-1)kp^2=580p, i.e. (k-1)kp=580$ = 4 × 5 × 29. Thus, $p=29$.

34. The Discriminant of the quadratic equation must be a perfect square number implies that $(5p)^2-20(66p-1)=n^2$ for some integer $n\geq0$, so

$25p^2-1320p+20=n^2$ or $(5p-132)^2-n^2$ $=17404=2^2.19.229$.

Note that $5|n$, and the equation gives

$p-1\equiv0 \pmod 5$,

so $p=5k+1$ for some positive integer k, then

$(25k-127-n)(25k-127+n)=38.458$

$=(-38)(-458)$.

The system $25k-127-n=38, 25k-127+n$ $=458$ yields the solution

$=n=\frac{1}{2}(458-38)=210, k=\frac{1}{25}(38+127+210)$

$=15, \therefore p=76$.

However the system

$25k-127-n=-458$

$25-127+n=-38$ has no integer solution since the left hand sides are 3 modulo 5, but right hand sides are 2 modulo 5.

By checking, $p=76$ yields equation $5x^2-380x+5015=0$ or

$5(x-17)(x-59)=0$, its two roots are 17 and 59. Thus, $p=76$ is the answer.

35. Let x_1, x_2 be the integer roots of the given equation.

From Viete Theorem $x_1+x_2=pq$ and $x_1x_2=p+q$,

i.e.

$(x_1-1)(x_2-1)+(p-1)(q-1)=2$.

If $(x_1-1)(x_2-1)=1$ then $(p-1)(q-1)=1$

$p-1=1, q-1=1, i.e. (p,q)=(2,2)$,

So the equation is $x^2-4x+4=0$, which has two equal integer solutions $x_1=x_2=2$.

If $(x_1-1)(x_2-1)=2$ then $(p-1)(q-1)=0$

$p-1=0$ or $q-1=0$ and $(x_1x_2)=(2,3) or (3,2)$,

so the equation is

$x^2-5x+6=0$ which implies that $(p,q)=(1,6) or (5,1)$.

Thus possible values of, $(p,q)=(2,3),(3,2),(2,2),(1,5),(5,1)$.

36. (c) Let $y=x-[x]$. Then, equation $(a-2)(x-[x])^2+2(x-[x])+a^2=0$ .... (i)

can be written as

$f(y)=(a-2)y^2+2y+a^2=0$ ...(ii)

As x cannot be an integer, $y=x-[x]\neq0$

Thus, $a\neq0$. When $2<x<3, [x]=2$

$\Rightarrow 0<x-[x]<1$ ie, $0<y<1$

Since, Eq. (i) has exactly one solution in the interval (2, 3), Eq. (ii) has exactly one solution in the interval (0, 1). This is possible, if $f(0)f(1)<0$ otherwise the Eq. (ii) has either no or two solutions in (0, 1).

$\Rightarrow a^2\{a-2+2+a^2\}<0$

$\Rightarrow a^3(a+1)<0 \quad [\because a^2>0]$

$\Rightarrow -1<a<0 \, or \; a\in(-1,0)$

37. (b) The equation

$2\log_x a+\log_{ax} a+3\log_{a^2x} a=0$

can be written as

$$\frac{2\log a}{\log x}+\frac{\log a}{\log(ax)}+\frac{3\log a}{\log(a^2x)}=0 \qquad \text{...(i)}$$

As $a>0\ a\neq 1, \log a\neq 0$, Eq. (i) can be written as $\frac{2}{y}+\frac{1}{b+y}+\frac{3}{2b+y}=0$ where $b=\log a$ and $y=\log x$

$\Rightarrow 2(b+y)(2b+y)+y(2b+y)+3y(b+y)=0$

$\Rightarrow 4b^2+11by+6y^2=0$

$\Rightarrow y=\frac{-11b\pm\sqrt{121b^2-96b^2}}{12}=-\frac{4b}{3},-\frac{b}{2}$

$\Rightarrow \log x=-\frac{4}{3}\log a$ or $-\frac{1}{2}\log a$

$\Rightarrow x=a^{-4/3},a^{-1/2}$

38. (a) $ax^2+bx+6=0$, roots are not real

$\Rightarrow D<0$

$\Rightarrow b^2-24a<0$

$\Rightarrow a>\frac{b^2}{24} ie, a$ is +ve ...(i)

Also, $f(-1)>0\Rightarrow a-b+6>0$

$\Rightarrow a>b-6$ ...(ii)

$\Rightarrow a>\max\left\{\frac{b^2}{24},b-6\right\}$; using Eqs. (i) and (ii)

39. (c) Here, $x^2\cdot 2^{|x-3|+4}+2^{x-1}=x^2\cdot 2^{x+1}+2^{|x-3|+2}$

If $x-3\geq 0$

$\Rightarrow x^2\cdot 2^{x-3+4}+2^{x-1}=x^2\cdot 2^{x+1}+2^{x-3+2}$

$\Rightarrow x^2\cdot 2^{x+1}+2^{x-1}=x^2\cdot 2^{x+1}+2^{x-1}$ which is always true for $x\geq 3$.

Again, if $x-3<0$

$x^2\cdot 2^{-(x-3)+4}+2^{x-1}=x^2\cdot 2^{x+1}+2^{-(x-3)+2}$

$\Rightarrow x^2\cdot 2^{7-x}+2^{x-1}=x^2\cdot 2^{x+1}+2^{5-x}$

$\Rightarrow x^2\cdot 2^{7-x}-2^{5-x}=x^2\cdot 2^{x+1}-2^{x-1}$

$\Rightarrow 2^{5-x}(x^2\cdot 2^2-1)=2^{x-1}(x^2\cdot 2^2-1)$

$\Rightarrow (2^{5-x}-2^{x-1})(4x^2-1)=0$

$\Rightarrow x^2=\frac{1}{4}$ or $5-x=x-1$

$\Rightarrow x=\pm\frac{1}{2}$ or $x=3$

Solution set is $x=\pm\frac{1}{2}$ and $x\geq 3$.

40. (c) Here, $\frac{x-a}{b}+\frac{x-b}{a}=\frac{b}{x-a}+\frac{a}{x-b}$

$\Rightarrow \frac{(x-a)^2-b^2}{b(x-a)}+\frac{(x-b)^2-a^2}{a(x-b)}=0$

$\Rightarrow (x-a-b)\left\{\frac{x-a+b}{b(x-a)}+\frac{x-(b-a)}{a(x-b)}\right\}=0$

$\Rightarrow (x-a-b)\{a(x-a)(x-b)+ab(x-b)+b(x-a)(x-b)+ab(x-a)\}=$

$\Rightarrow x(x-a-b)\{a(x-a)+b(x-b)\}=0$

$\Rightarrow x=0,a+b,\frac{a^2+b^2}{a+b}$

So, $x_1=a+b, x_2=\frac{a^2+b^2}{a+b}, x_3=0$

Now, $x_1-x_2-x_3=c \Rightarrow c=\frac{2ab}{a+b} \Rightarrow a,c,b$ are in HP.

41. (a) Here let $t=x^2+3x$

So, $(t+1)(t-3)\geq 5$

$t^2-2t-3-5\geq 0$

$t^2-2t-8\geq 0$

$(t-4)(t+2)\geq 0$

So, $t\leq -2 or\ t\geq 4$

Let us consider $t\leq -2$

So, $x^2+3x\leq -2\Rightarrow x^2+3x+2\leq 0$

$\Rightarrow (x+1)(x+2)\leq 0\Rightarrow -2\leq x\leq -1$

Now consider $t\geq 4$

$x^2+3x\geq 4\Rightarrow x^2+3x-4\geq 0$

$(x+4)(x-1)\geq 0\Rightarrow x\leq -4 or\ x\geq 1$

Thus, final solution is

$(-\infty,-4]\cup[-2,-1]\cup[1,\infty)$

42. Since a, b, c, d are four consecutive terms of an increasing AP so, $a<b<c<d$

$\therefore f(x)=(x-a)(x-c)+2(x-b)(x-d)$

$\Rightarrow f(a)=(+ve), f(b)=(-ve), f(c)=(-ve)$

And $f(d)=(+ve)$

$\Rightarrow$ One root lies in (a, b) and other in (c, d).

43. Since roots are in AP so let the roots of the equation be a – d, a, a + d then

Sum of the roots $(a-d)+a+(a+d)=1 \Rightarrow a=\frac{1}{3}$

$(a-d)a+(a-d)(a+d)+a(a+d)=\beta$

$\Rightarrow 3a^2-d^2=\beta \Rightarrow d^2=\frac{1}{3}-\beta$

$\because d^2 \geq 0$ So, $\beta \leq \frac{1}{3}$

Product of the roots $(a-d)\,a\,(a+d)=-\gamma$

$\Rightarrow a(a^2-d^2)=-\gamma$

Or $\frac{1}{3}(\frac{1}{9}-d^2)=-\gamma \Rightarrow \frac{1}{27}+\gamma=\frac{1}{3}d^2$

So, $\gamma+\frac{1}{27}\geq 0 \Rightarrow \gamma \geq -\frac{1}{27}$

44. Roots of $x^2-10cx-11d=0$ are a and b

$\Rightarrow a+b=10c$ and $ab=-11d$

Similarly, c and d are the roots of $x^2-10ax-11b=0 \Rightarrow c+d=10a$ and $cd=-11b$

$\Rightarrow a+b+c+d=10(a+c)$ and $abcd=121bd$

$\Rightarrow b+d=9(a+c)$ and $ac=121$

Also, we have $a^2-10ac-11d=0$ & c

$c^2-10ac-11b=0$

$\Rightarrow a^2+c^2-20ac-11(b+d)=0$

$\Rightarrow (a+c)^2-22\times 121-99(a+c)=0$

$\Rightarrow a+c=121 or -22$

For $a+c=-22$ we get $a=c$

$\therefore$ rejecting this value we have $a+c=121$

$\therefore a+b+c+d=10(a+c)=10\times 121=1210$

45. (d) Here, $(x-2)^6+(x-4)^6=64$ ...(i)

Put $y=AM$ of $(x-2)$ and $(x-4)=x-3$ in Eq. (i)

The Eq. (i) becomes $(y+1)^6+(y-1)^6=64$

$\Rightarrow 2\left[y^6+15y^4+15y^2+1\right]=64$

$\Rightarrow y^6+15y^4+15y^2-31=0$ ...(ii)

Now substitute $y^2=u,$ in Eq. (ii)

$u^3+15^2+15u-31=0$ ...(iii)

It has a root $u=1,$

$u^3+15^2+15u-31=(u-1)(u^2+16u+31)=0$

$\Rightarrow u=1, u=\frac{-16\pm\sqrt{256-124}}{2}=-8\pm\sqrt{33}$

$\Rightarrow y^2=1,-8+\sqrt{33},-8-\sqrt{33}$

$\Rightarrow y=\pm 1,\pm\sqrt{8-\sqrt{33}}i,\pm\sqrt{8+\sqrt{33}}i$

$\Rightarrow x=2,4,3\pm\sqrt{8-\sqrt{33}}i,\pm\sqrt{8+\sqrt{33}}i.$

46. (d) Here, $x=\frac{y^4}{8-6y^2}$ but is non–negative.

So $\frac{y^4}{8-6y^2}\geq 0$ or

$\Rightarrow 8-6y^2>0$ or $y^2<\frac{4}{3}$

ie, $y=0,1$ (as is non–negative integer)

For $y=0$ we get $x=0$

For $y=1$ we get $x=\frac{1}{2}\notin$ integer

None of the points lie in 1st quadrant

47. Let $y=-5+4i \Rightarrow y^2+10y+41=0$

$\therefore f(x)=x^4+9x^3+35x^2-x+4$

$=x^2(x^2+10x+41)-x(x^2+10x+41)+$

$4(x^2+10x+41)-160=-160$

$\therefore f(-5+4i)=-160$

CHAPTER

5

SEQUENCE & SERIES

PART-A

5.1 ARITHMETIC PROGRESSION

5.1.1 Introduction

A sequence or series is defined as a function whose domain is the set of natural numbers and the range of a sequence is a subset of real numbers (if range is a subset of complex numbers then it is called complex sequence). Terms are denoted by f(1), f(2), f(3), … f(n) or If t_n = f(n), then the sequence is written as t_1, t_2, t_3, …

Here t_1, t_2, t_3, are called 1st, 2nd, 3rd term respectively. A series is finite or infinite according as the number of terms in the corresponding sequence is finite or infinite.

Progressions: It is not necessary that the terms of a sequence always follow a certain pattern or they are described by some explicit formula for the nth term. Those sequences whose terms follow certain patterns are called progressions.

Eaxmple: 1, 5, 4, 11, 17, 8, 1 is not a progression.

Example: 2, 3, 5, 7, 11, 13 is a progression with nth term is nth prime number but its nth term can not be represented by an algebraic formula.

Illustration 1. If nth term of a series is defined as $t_n = 3n + 4$ then find the 10th term.

Solution: Since given that $t_n = 3n + 4$ so

$$t_1 = 3(1) + 4 = 7$$

$$t_2 = 3(2) + 4 = 10 \text{ and so on..}$$

$$t_{10} = 3(10) + 4 = 34$$

Illustration 2. What is the 15th term of the sequence defined by $t_n = \dfrac{1}{3n-1}$

Solution: Substituting $n = 15$ in $t_n = \dfrac{1}{3n-1}$, we get

$$t_{15} = \frac{1}{3(15)-1} = \frac{1}{44}$$

So the 15th term of the given sequence is $\dfrac{1}{44}$

Illustration 3. Find the 5th term of the sequence for which $t_1 = 1$, $t_2 = 2$ and $t_{n+2} = t_n + t_{n+1}$

Solution: Given that, $t_1 = 1$, $t_2 = 2$, $t_{n+2} = t_n + t_{n+1}$

Now substituting:

$n = 1$, we get $t_3 = t_1 + t_2 = 1 + 2 = 3$

$n = 2$, we get $t_4 = t_2 + t_3 = 2 + 3 = 5$

$n = 3$, we get $t_5 = t_3 + t_4 = 3 + 5 = 8$

So 5th term is 8

Types of Progression/Series

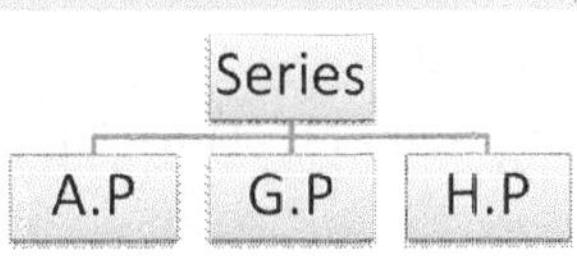

5.1.2 Arithmetic Progression (AP)

A sequence of numbers is said to be in arithmetic progression (A.P.) when the difference $t_n - t_{n-1}$ is a constant for all $n \in N$, $n > 1$. This constant is called the common difference of the A. P. and it is usually denoted by the letter '*d*'.

Let a is the first term (i.e $t_1 = a$) and d the common difference of the A.P. then:

1st term $\quad \mathbf{a = a + (1 - 1)d}$

2nd term $\quad \mathbf{a + d = a + (2 - 1)d}$

3rd term $\quad \mathbf{a + 2d = a + (3 - 1)d}$ and so on

nth term $\quad \mathbf{t_n = a + (n - 1)d.}$

Illustration 4. If for an AP 1st term is 5 and common difference is –3, then find the 10th term

Solution: Given that 1st term = a = 5 and common difference is $d = -3$

We know that $\quad t_n = a + (n - 1)d$

so $\quad t_{10} = a + 9d = 5 + 9(-3) = -22$

Here common difference is negative so it is called decreasing AP.

Illustration 5. If for an AP 5th term is 24 and 10th term is 49, then find the 13th term of the AP

Solution: As per the given information:

5th term is 24 or $a + 4d = 24$ ...(i)

10th term is 49 or $a + 9d = 49$ ...(ii)

From above two equations we get $a = 4$ and $d = 5$

So 13th term is $a + 12d = 4 + 60 = 64$

Illustration 6. The 11th term of an A.P. is 80 and the 16th term is 110. Find the 31st term.

Solution : The nth term of an A.P. whose first term is 'a' and common difference is 'd', is given by

$$t_n = a + (n - 1)\, d$$

Given, $\quad t_{11} = 80$

$\therefore \quad 80 = a + 10d$ ...(i)

And $\quad t_{16} = 110$

$\therefore \quad 110 = a + 15d$ ...(ii)

Solving equations (i) and (ii), we get $30 = 5d$

$\therefore \quad d = 6$ and $a = 20$

The 31st term, $t_{31} = 20 + (31-1).\,6 = 20 + 30 \times 6 = 200$

Illustration 7. Is 65 a term of the sequence 1, 3, 5, 7,...? If yes, find which term it is.

Solution : If possible let n^{th} term of the sequence be 65.

Here a = 1, d = 2

And $t_n = a + (n-1)d$.

Here $t_n = 65$,

$\therefore \quad 65 = 1 + (n - 1)2$

or $\quad 2n = 66, n = 33$

Hence 65 is 33rd term of the given sequence.

Illustration 8. Prove that 5th, 10th and 15th term of an AP is in AP

Solution: Consider an AP with 1st term **a** and common difference **d** then

5th term $\quad t_5 = a + 4d$

10th term $\quad t_{10} = a + 9d$

15th term $\quad t_{15} = a + 14d$

Since $2t_{10} = t_5 + t_{15}$, so 5th, 10th, and 15th term of an AP is in AP.

5.1.3 Sum of Arithmetic Progression

Consider an AP with 1st term as 'a' and common difference as 'd'

The AP is a, (a + d), (a + 2d),... {a + (*n* – 1)d}

Sum to *n* terms is –

$$S_n = (a) + (a + d) + (a + 2d) + \ldots \{a + (n - 2)d\} + \{a + (n - 1)d\} \quad \ldots(i)$$

Now consider the same AP from its last term

If last term is a_n then we can write same AP as

$$S_n = (a_n) + (a_n - d) + (a_n - 2d) + \ldots \{a_n - (n - 2)d\} + (a_n - (n - 1)d\} \quad \ldots(ii)$$

Adding equations (*i*) and (*ii*), we get

$$2S_n = n\{a + a_n\}$$
$$= n\{a + a + (n - 1)d\}$$
$$= n\{2a + (n - 1)d\}$$

or the sum S_n of the first n terms of such an A.P. is given by:

$$S_n = \frac{n}{2}\left[2a + (n-1)d\right] = \frac{n}{2}(t_1 + t_n)$$
$$= \frac{n}{2}(t_2 + t_{n-1}) = \frac{n}{2}(t_3 + t_{n-2})$$

5.1.4 Properties of Arithmetic Progression

(i) If a fixed number is added to (or subtracted from) each term of a given A.P (with common difference **d**)., then the resulting sequence is also an A.P., and it has the same common difference as that of the given A.P.

Consider an A.P $a_1, a_2, \ldots, a_n$ if a fixed number (say k) is added to (or subtracted from) each term of a given term then resultant $(a_1 \pm k)$, $(a_2 \pm k)$, $(a_3 \pm k)$, $\ldots(a_n \pm k)$ is also an AP with common difference d

(ii) If each term of an A.P. is multiplied by a fixed constant (or divided by a non–zero fixed constant), then the resulting sequence is also an A.P.

Consider an A.P $a_1, a_2, \ldots, a_n$ if a fixed number (say k) is multiplied to (or divided by a non–zero fixed constant say **k**) each term of a given series then resultant (a_1/k), (a_2/k), (a_3/k), … (a_n/k) is also an AP with common difference (d/k)

(iii) If $a_1, a_2, a_3, \ldots$ and $b_1, b_2, b_3, \ldots$ are two arithmetic progressions with common difference d_1, and d_2 respectively, then $(a_1 \pm b_1), (a_2 \pm b_2), \ldots(a_n \pm b_n)$ is also an A.P with common difference $(d_1 \pm d_2)$

(iv) If $a_1, a_2, a_3, \ldots$ are in A.P. then $a_p - a_q = (p-q)d$ where d is the common difference of the A.P.

(v) If we have to take three terms in A.P. whose sum is known, it is convenient to take them as $a - d$, a, $a + d$. Similarly if sum of 5 terms is known then we can take these 5 terms as $a-2d,\ a-d,\ a,\ a+d,\ a+2d$.

(vi) If we have to take four terms in A.P., whose sum is known, we take them as a – 3d, a – d, a + d, a + 3d. in general, if we have to take 2r terms in A.P., we take them as $a-(2r-1)d, a-(2r-3)d, \ldots, a-d, a+d, \ldots, a+(2r+1)d, a+(2r+3)d$.

Important properties of AP

(i) If p^{th} term of an AP is q and q^{th} term of the same AP is p then $(p + q)^{th}$ term is 0

or if $a_p = q$ and $a_q = p$, then

$$a_{p+q} = 0, T_r = p+q-r$$

(ii) If p times the pth term of an AP is equal to q times the qth term of the same AP then $(p + q)^{th}$ term is 0

if $pT_p = qT_p, a_{p+q} = 0$

(iii) If $a_p = \frac{1}{q}$ and $a_q = \frac{1}{p}$, then $a_{pq} = 1$

(iv) If $S_p = q$ and $S_q = p$, then $S_{p+q} = -(p+q)$

(v) If $S_p = S_q$, then $S_{p+q} = 0$

When sum of AP is given:

(i) If sum of 3 terms of an AP is given then assume these terms as (a – d), a and (a + d).

(ii) If sum of 5 terms of an AP is given then assume these terms as (a – 2d), (a – d), a, (a + d) and (a + 2d).

(iii) If sum of (2n + 1) terms of an AP is given then assume these terms as (a – nd) … (a – 2d), (a – d), a, (a + d) and (a + 2d)…(a + nd).

(iv) If sum of 4 terms of an AP is given then assume these terms as (a – 3d),(a – d), (a + d) and (a + 3d).

(v) If sum of (2n) terms of an AP is given then assume these terms as (a – (2n – 1)d) … (a – 3d), (a – d), a, (a + d) and (a + 2d) … (a + (2n – 1)d).

Illustration 9. Find the sum up to 100 terms of the series 4, 7, 10, 13.

Solution: In the given AP, 1st term is a = 4 and common difference $d = 3$

Then
$$S_n = \frac{n}{2}[2a+(n-1)d]$$
$$= \frac{100}{2}[8+(99)(3)]$$
$$= \frac{100 \times 305}{2} = 15250$$

Illustration 10. Consider an AP 400, 393, 386, 379… Find the summation of all the +ve terms of the series.

Solution: For this AP a = 400 and d = –7

Here 1st we have to find the number of terms let nth term is positive then $T_n = 400 + (n-1)(-7) > 0$, then $400 + 7 - 7n > 0$ or $n < 407/7$ or $n < 58.14$ so largest value of n is 58 and last positive term is

$$400 + (57)(-7) = 1$$

Required summation

$$S_n = \frac{n}{2}\left[2a + (n-1)d\right]$$

$$= \frac{n}{2}(t_1 + t_n) = \frac{58(400+1)}{2}$$

$$= 29 \times 401 = 11629$$

Illustration 11. Consider an AP 400, 393, 386, 379... If the summation of 1st 'n' terms is positive then find the maximum value of 'n'.

Solution: For this AP a = 400 and d = –7

Since $S_n = \frac{n}{2}\left[2a + (n-1)d\right] = \frac{n}{2}(t_1 + t_n)$

so $(t_1 + t_n) > 0$

$800 + (n-1)(-7) > 0$

807 –7n > 0 or $n < 807/7$ or $n < 115.28$ or largest value of n is 115.

Illustration 12. If sum of 5 consecutive term of an AP is 65 while ratio of 5th and 1st term is 21:5 then find the 5th term.

Solution: Let the 5 terms be (a – 2d), (a – d), a, (a + d) and (a + 2d)

Sum of these 5 terms is 5a = 65 so a = 13

Now ratio of 5th and 1st term is 21:5 so

$$\frac{a+2d}{a-2d} = \frac{13+2d}{13-2d} = \frac{21}{5} \text{ or } d = 4$$

so 5th term is 21

5.1.5 Arithmetic Mean

Arithmetic Mean: It is defined as sum of the terms divided by number of terms.

$$\textbf{A.M} = \frac{S_n}{n} = \frac{t_1 + t_n}{2} = \frac{t_2 + t_{n-1}}{2}$$

Arithmetic mean is also known as '**Mean**' or '**Average**'

Some more properties of AM

(i) The arithmetic mean A of any two numbers 'a' and 'b' is given by $A = \frac{a+b}{2}$.

Here sequence a, A, b is in A.P.

(ii) The arithmetic mean A of any three numbers a, b and c is given by $A = \frac{a+b+c}{3}$.

(iii) If $a_1, a_2, \ldots, a_n$ are n numbers, then the arithmetic mean A, of these numbers is given by

$$A = \frac{S_n}{n} = \frac{1}{n}(a_1 + a_2 + \ldots + a_n).$$

5.1.6 Insertion of Arithmetic Means

Insertion of *n* Arithmetic Means between *a* and *b*

Let $A_1, A_2, \ldots, A_n$ be n arithmetic means between two quantities a and b. Then, $a, A_1, A_2, \ldots, A_n, b$ is an AP. Let d be the common difference of this AP. Clearly, it contains $(n + 2)$ terms.

$\therefore \quad b = (n+2)\text{th term}$

$\Rightarrow \quad b = a + (n+1)d$

$\Rightarrow \quad d = \frac{b-a}{n+1}$

Now, $\quad A_1 = a + d$

$\Rightarrow \quad A_1 = \left(a + \frac{b-a}{n+1}\right)$

$A_2 = a + 2d$

$\Rightarrow \quad A_2 = \left(a + \frac{2(b-a)}{n+1}\right)\ldots$, and so on

$$A_n = a + nd \Rightarrow A_n = \left(a + \frac{n(b-a)}{n+1}\right)$$

These are the required arithmetic means between a and b.

Illustration 13. Consider an AP series 5, 11, 17, 23... Up to 50 terms. Find the AM of this series

Solution: The last term is 5 + 49 × 6 = 299

$$\text{Required AM} = \frac{t_1 + t_n}{2} = \frac{5 + 299}{2} = 152$$

Illustration 14. Insert 10 AMs between 20 and 75, find the 8th AM

Solution: Let the 10 AMs be $t_1, t_2, \ldots t_{10}$

From the given information we can say that 20, $t_1, t_2, \ldots t_{10}$, 75 are 12 consecutive terms of an AP, let common difference of AP is d then 75 = 20 + 11d or d = 5

So 8th AM is 20 + 8(5) = 60

Illustration 15. If n A.M.'s are inserted between 1 and 31 such that the ratio of the 7th and (n – 1)th means is 5 : 9. Find n.

Solution: Let $A_1, A_2, \ldots, A_n$ be the n arithmetic means between 1 and 31.

$\therefore$ 1, $A_1, A_2, \ldots, A_n$, 31 are in A.P. with a = 1, b = 31 and common difference

$$d = \frac{31-1}{n+1} = \frac{30}{n+1} \qquad \left[\because d = \frac{b-a}{n+1}\right]$$

Now, consider A_7 = 8th term of A.P.

$$= a + 7d$$

$$= 1 + \frac{7.30}{n+1} = \frac{n+211}{n+1}$$

and A_{n-1} = *nth* term of *A.P.*

$$= a + (n-1)d$$

$$= 1 + (n-1)\cdot\frac{30}{n+1} = \frac{31n-29}{n+1}$$

From the given information, $\frac{A_7}{A_{n-1}} = \frac{5}{9}$

$$\Rightarrow \quad \frac{\frac{n+211}{n+1}}{\frac{31n-29}{n+1}} = \frac{5}{9}$$

$$\Rightarrow \quad \frac{n+211}{31n-29} = \frac{5}{9}$$

$$\therefore \quad n = 14$$

Concept Applicator (CA)-1

1. In an AP, the pth term is $-q$ and the $(p+q)$th term is 0. Then the qth term is
   (a) $-p$ (b) p
   (c) $p+q$ (d) $p-q$
2. Consider two A.Ps (3, 7, 11, ...) and (1, 6, 11, ...) what is the 20th common term between these two series
   (a) 371 (b) 391
   (c) 411 (d) None of these
3. Find the number of common term in the sequences 1, 11, 21, 31, ... to 100 terms and 31, 36, 41, 46, ... to 100 terms is
   (a) 49 (b) 51
   (c) 50 (d) None of these
4. In an AP, if S_r denotes the sum of the first r terms, $S_p = q, S_q = p$. Then find S_{p+q}
   (a) 0 (b) $-(p+q)$
   (c) $p+q$ (d) pq
5. The product of four numbers in A.P, such that their sum is 60 and greatest of them is 4 times the least, is
   (a) 31104 (b) 30720
   (c) 31416 (d) None of these
6. If the sum of four integers in A.P is 24 and their product is 945, then find the difference between largest and smallest terms.
   (a) 11 (b) 6
   (c) 8 (d) None of these
7. The maximum sum of the AP 40, 38, 36, 34, … is
   (a) 390 (b) 420
   (c) 460 (d) None of these
8. In a certain AP, 11 times the 11th term is equal to 13 times the 13th term, then the 24th term is
   (a) – 13 (b) – 12
   (c) – 1 (d) 0
9. If the sum of n terms of an AP is $3n^2 - n$ and its common difference is 6, then find its 10th term
   (a) 62 (b) 63
   (c) 64 (d) None of these
10. Find the difference between AM of 1st 50 terms and 1st 100 terms of AP 3, 8, 13, 18…
    (a) 250 (b) 125
    (c) 120 (d) None of these

5.2 GEOMETRIC PROGRESSION (GP)

5.2.1 Geometric Progression (GP)

Geometric progression is a series when its first term (say $a \neq 0$) is non–zero and each of its succeeding terms is r times the preceding term, where r is some fixed non–zero number and is known as the common ratio of the G. P.

Let a is the first term (i.e $t_1 = a$) and

r the common ratio of the G.P. then:

1st term **a**

2nd term **ar**

3rd term ar^2 and so on

n^{th} term $t_n = ar^{n-1}$

Illustration 1. Consider a GP, 2, 6, 18, 54, 162... find its 10 term.

Solution: From the given series we can say that 1st term is 2 and common ratio is 3 so its nth term is $2(3^{n-1})$ so 10th term is 2×3^9

Illustration 2. If r^{th} term of a GP is $5(7^r)$ then find the value of common ratio of the GP

Solution: Since r^{th} term of a GP is $5(7^r)$ so its common ratio is 7.

Illustration 3. If 7, a, 21 and b are in GP then find a and b.

Solution: Since these four are in GP so a = 7r and 21 = $7(r^2)$ or $r = \sqrt{3}$

So a = $7\sqrt{3}$ and b = $21\sqrt{3}$

Illustration 4. Let f(x) = 2x + 1. Then, the numbers of real values of x for which the three unequal numbers f(x), f(2x), f(4x) are in GP is

(a) 1 (b) 2

(c) 0 (d) None of these

Solution: (c) Given that $f(x), f(2x), f(4x)$ are in GP

$\therefore \quad (4x+1)^2 = (2x+1)(8x+1)$

$\Rightarrow \quad 16x^2 + 8x + 1 = 16x^2 + 10x + 1$

$\Rightarrow \quad x = 0 \quad \ldots(i)$

Since, $f(x), f(2x), f(4x)$ are three unequal numbers but if, $x = 0$

$\Rightarrow f(x) = f(2x) = f(4x) = f(0) = 1$, which is not possible

$\therefore$ No solution.

5.2.2 Sum of GP Series

Sum to *n* Terms of a GP

(i) The sum of n terms of a GP with first term 'a' and common ratio 'r' is given by

$$S_n = \begin{cases} a\dfrac{(1-r^n)}{1-r}, & |r| < 1 \\ a\dfrac{(r^n-1)}{r-1}, & |r| > 1 \end{cases}$$

(ii) If l is the nth term of the GP, then $l = ar^{n-1}$

$$\Rightarrow S_n = \frac{a-lr}{1-r} \text{ or } S_n = \frac{lr-a}{r-1}, r \neq 1$$

Sum to infinite Terms of a GP

If $|r| < 1$, then $\lim_{n\to\infty} r^n = 0$. Therefore, the sum S of an infinite GP with common ratio r satisfying $|r| < 1$ is given by $S = \lim_{n\to\infty} S_n = \lim_{n\to\infty} a\left(\frac{1-r^n}{1-r}\right)$

$$\Rightarrow \quad S = \frac{a}{1-r}$$

Thus, the sum S of an infinite GP with first term a and common ratio $r(-1 < r < 1)$ is given by $S = \frac{a}{1-r}$

Illustration 5. Find the sum of the series

1 + 2 + 4 + 8 + ... + 1024

Solution: Here 1st term is a = 2^0 = 1 and common ratio is $r = 2$ and number of terms is 12

So sum of the series is given by $1\dfrac{(2^{12}-1)}{2-1}$

= 2048 – 1 = 2047

Illustration 6. Find the sum of infinite terms of the series 1 + 1/2 + 1/4 + 1/8 + ...up to ∞

Solution: Since the sum S of an infinite GP with first term a and common ratio $r(-1<r<1)$ is given by

$S=\frac{a}{1-r}$ here in this example $a = 1$ and $r = 1/2$, so

required sum is $S=\frac{1}{1-\frac{1}{2}}=2$

5.2.3 Properties of G.P.

(i) If each term of a G.P. with common ratio r is multiplied by a fixed non–zero constant (say k), then the resulting sequence is also a G.P. with common ratio r

Consider a G.P $a, ar, ar^2, ar^3...$ if each term is multiplied by a fixed constant 'k' then resultant ka, kar, kar^2, kar^3 is also in G.P with common ratio r

(ii) If each term of a G.P. with common ratio r is divided by a fixed non–zero constant (say k), then the resulting sequence is also a G.P. with common ratio r

Consider a G.P a, ar, ar^2, ar^3...if each term is divided by a fixed constant 'k' then resultant $\frac{a}{k},\frac{ar}{k},\frac{ar^2}{k},\frac{ar^3}{k}$ is also in G.P with common ratio r

(iii) If each term of a G.P. with common ratio r is squared (raised by power 2), then the resulting sequence is also a G.P. with common ratio r^2

Consider a G.P a, ar, ar^2, ar^3... if each term is squared then resultant

$a^2, a^2r^2, a^2r^4, a^2r^6$ ... is also in G.P with common ratio r^2

(iv) If each term of a G.P. with common ratio r is raised by power k, then the resulting sequence is also a G.P. with common ratio r^k

Consider a G.P a, ar, ar^2, ar^3...if each term is squared then resultant

$a^k, a^kr^k, a^kr^{2k}, a^kr^{3k}$ ... is also in G.P with common ratio r^k

(v) If a_1, a_2, a_3, ... and b_1, b_2, b_3, ... are two geometric progressions with common ratio r_1 and r_2 respectively, then the sequence a_1b_1, a_2b_2, a_3b_3, ... is also a G.P. with common ratio r_1r_2

(vi) If a_1, a_2, a_3, ... are in G.P., then $\frac{a_p}{a_q}=r^{p-q}$ where r is the common ratio of the G.P.

(vii) If we have to take four terms in G.P., we take them as a/r, a, ar and four terms as $\frac{a}{r^3},\frac{a}{r},ar,ar^3$

(viii) If a_1, a_2, a_3, ... is a G.P. $(a_i>0\forall i)$, then $\log a_1$, $\log a_2$, $\log a_3$, ... is an A.P. in this case the converse also holds true.

Selection of Terms in GP if product is given

If product of GP series is given and it is required to select a finite number of terms in GP. It is always convenient, if we select the terms in the following manner.

No. of terms	Terms	Common ratio
3 (odd)	$\frac{a}{r},a,ar$	r
4 (Even)	$\frac{a}{r^3},\frac{a}{r},ar,ar^3$	r^2
5 (odd)	$\frac{a}{r^2},\frac{a}{r},a,ar,ar^2$	r
6 (Even)	$\frac{a}{r^5},\frac{a}{r^3},\frac{a}{r},ar,ar^3,ar^5$	r^2

If the product of the numbers is not given, then the numbers are taken as a, ar, ar^2, ar^3, ...

Illustration 7. The number of terms common between the series 1, 2, 4, 8, ... to 100 terms and 1, 4, 7, 10, ... to 100 terms is

Solution: For GP, $t_n=2^{n-1}$, for AP, $T_m=1+(m-1)3$.

They are common, if $2^{n-1}=3m-2$

or $$2^{n-2}+1=\frac{3m}{2}\le 150$$

$\Rightarrow$ $$n\le 9, m\le 100$$

So we can conclude that $2^{n-1} = 3m - 2$

$\Rightarrow$ So the possible pairs of (n, m) are (1, 1), (3, 2), (5, 6), (7, 22), (9, 86), and for $n = 2, 4, 6, 8, m \in$ fraction, which is not possible. So total 5 common terms exists.

Illustration 8. If $\log\left(\frac{5c}{a}\right), \log\left(\frac{3b}{5c}\right)$ and $\log\left(\frac{a}{3b}\right)$ are in AP, where a, b, c are in GP, then a, b, c are the lengths of sides of which type of triangle?

Solution: $2\log\frac{3b}{5c} = \log\frac{5c}{a} + \log\frac{a}{3b}$

$$\Rightarrow \quad \left(\frac{3b}{5c}\right)^2 = \frac{5c}{a}\cdot\frac{a}{3b}$$

$$\Rightarrow \quad 3b = 5c$$

Also, $b^2 = ac$. So, $9ac = 25c^2$ or $9a = 25c$

$$\therefore \quad \frac{9a}{5} = 5c = 3b$$

$$\Rightarrow \frac{a}{5} = \frac{b}{3} = \frac{c}{9/5} \Rightarrow b + c < a$$

Since, sum of two sides is smaller than the third.

$\therefore$ Triangle is not formed.

5.2.4 Geometric Means

Geometric mean of two numbers 'a' and 'b' is defined as $\sqrt{ab}$. To understand AM and GM let us compare these two

Numbers	AM	GM
a, b	$\frac{a+b}{2}$	$(ab)^{\frac{1}{2}}$
a, b, c	$\frac{a+b+c}{3}$	$(abc)^{\frac{1}{3}}$
a, b, c, d	$\frac{a+b+c+d}{4}$	$(abcd)^{\frac{1}{4}}$
$a_1, a_2, \ldots a_n$	$\frac{a_1 + a_2 + \ldots + a_n}{n}$	$(a_1 a_2 \ldots a_n)^{\frac{1}{n}}$

Insertion of Geometric Means between Two Given Numbers

Geometric means: Let a and b be two given numbers. If n numbers $G_1, G_2, \ldots, G_n$ are inserted between a and b such that the sequences $a, G_1, G_2, \ldots, G_n, b$ is a GP. Then, $G_1, G_2, \ldots, G_n$ are known as n geometric means (GM's) between a and b.

Insertion of n Geometric Means Between Two Given Numbers a and b

Let $G_1, G_2, \ldots, G_n$ be n geometric means between two given numbers a and b. Then,

$a, G_1, G_2, \ldots, G_n, b$ is a GP consisting of $(n + 2)$ terms of a GP

Let r be the common ratio of this GP. Then,

$b = (n + 2)$th term of the GP $= ar^{n+1}$

$$\Rightarrow \quad r^{n+1} = \frac{b}{a}$$

$$\Rightarrow \quad r = \left(\frac{b}{a}\right)^{\frac{1}{n+1}}$$

$$\therefore \quad G_1 = ar = a\left(\frac{b}{a}\right)^{\frac{1}{n+1}}$$

$$G_2 = ar^2 = a\left(\frac{b}{a}\right)^{\frac{2}{n+1}}$$

$$G_n = ar^n = a\left(\frac{b}{a}\right)^{\frac{n}{n+1}}$$

Illustration 9. Find the GM of three numbers 10, 20 and 40 and compare it with AM.

Solution: AM of 10, 20 and 40 is $\frac{10 + 20 + 40}{3} = \frac{70}{3}$

GM of 10, 20 and 40 is $(10 \times 20 \times 40)^{\frac{1}{3}} = 20$

From comparison we get A.M $\geq$ G.M

Illustration 10. Consider the AP $a_1, a_2, \ldots, a_n \ldots$ the GP $b_1, b_2, \ldots, b_n \ldots$ such that $a_1 = b_1 = 1; a_9 = b_9$ and $\sum_{r=1}^{9} a_r = 369$, then find b_7

Solution: $a_1 = b_1 = 1; a_9 = 1 + 8d = b_9 = 1 \cdot r^8$

Now, $\sum_{r=1}^{9} a_r = \frac{9}{2}(1 + a_9) = \frac{9}{2}(1 + r^8) = 369 \Rightarrow r = \sqrt{3}$

$$\therefore \quad b_7 = b \cdot r^6 = 1(\sqrt{3})^6 = 27$$

Concept Applicator (CA)-2

1. Find the sum of the infinite series $1+\frac{2}{3}+\frac{4}{9}+\frac{8}{27}+......$

   (a) 2 (b) 3
   (c) 5/3 (d) None of these

2. Find the value of $2\times2^{\frac{1}{2}}\times2^{\frac{1}{4}}\times2^{\frac{1}{8}}\times...$ up to infinite

   (a) 4 (b) $\sqrt{2}$
   (c) $\sqrt{8}$ (d) None of these

3. If x, $2y$, $3z$ are in AP, and x, y, z are distinct terms of a GP, then the common ratio of the GP is

   (a) 3 (b) $\frac{1}{3}$
   (c) 2 (d) $\frac{1}{2}$

4. The lengths of three unequal edges of a rectangular solid block are in GP. The volume of the block is 216 cm^3 and the total surface area is 252 cm^2. The sum of the length of the three edges

   (a) 12 cm (b) 21 cm
   (c) 18 cm (d) 30 cm

5. The sum of an infinite geometric series is 2 and the sum of the geometric series made from the cubes of this infinite series is 24. Then, find the sum of the geometric series made from the squares of this infinite series.

   (a) 6 (b) 9
   (c) 15 (d) None of these

6. The AM of two given positive numbers is 2. If the larger number is increased by 1, the GM of the numbers becomes equal to the AM of the given numbers. Then, find the difference between the two numbers.

   (a) $\frac{3}{2}$ (b) $\frac{2}{3}$
   (c) $\frac{1}{2}$ (d) None of these

7. Three numbers a, b and c are in GP, whose sum is 70, if 4a, 5b, 4c are in AP then the numbers are

   (a) 10, 20, 40 (b) 10, 30, 40
   (c) 20, 30, 40 (d) None of these

8. If the roots of the cubic $x^3-px^2+qx-r=0$ are in GP, then

   (a) $q^3=p^3r$ (b) $p^3=q^3r$
   (c) $pq=r$ (d) $pr=q$

9. If one AM, A and two GM's p and q are inserted between two given numbers, then $\frac{p^2}{q}+\frac{q^2}{p}$ is equal to

   (a) A (b) 2A
   (c) 3A (d) 4A

10. If sum up to 'n' terms of the series is $S_n=3(2^n)-3$, then find the 5th term of the series

    (a) 24 (b) 16
    (c) 48 (d) None of these

5.3 HARMONIC PROGRESSION & INEQUALITY

5.3.1 Harmonic Progression

The sequence $a_1, a_2, \ldots, a_n, \ldots$ where $a_i \neq 0$ for each i is said to be in harmonic progression (H.P.) if the sequence $\frac{1}{a_1}, \frac{1}{a_2}, \ldots, \frac{1}{a_n}, \ldots$ is in A.P.

So in this A.P 1st term is $a = \frac{1}{a_1}$ 2nd term is $\frac{1}{a_2}$ and nth term is $\frac{1}{a_n}$

So common difference of the A.P is $d = \frac{1}{a_2} - \frac{1}{a_1}$

So we can conclude that a_n, the nth term of the H.P. is given by $a_n = \frac{1}{a+(n-1)d}$ where $a = \frac{1}{a_1}$ and $d = \frac{1}{a_2} - \frac{1}{a_1}$

The sequence $1, \frac{1}{3}, \frac{1}{5}, \frac{1}{7}, \ldots$ is an HP, because the sequence 1, 3, 5, 7 is an AP.

If three numbers a, b and c are in H.P then

$$\frac{2}{b} = \frac{1}{a} + \frac{1}{c} \text{ or } b = \frac{2ac}{a+c}$$

Illustration 1. If consecutive terms of a series is 2, 3 and 6 then is the series in HP?

Solution: Consider 1/2, 1/3, 1/6

Since $\frac{1}{2} - \frac{1}{3} = \frac{1}{3} - \frac{1}{6} = \frac{1}{6}$ so 1/2, 1/3, 1/6 in AP so 2, 3, and 6 are in HP.

Illustration 2. If a, b, c are in HP, then prove that $\frac{a}{b+c}, \frac{b}{c+a}, \frac{c}{a+b}$ are in HP

Solution: Since, a, b, c are in HP.

$\therefore \quad \frac{1}{a}, \frac{1}{b}$ and $\frac{1}{c}$ are in AP

$\Rightarrow \quad \frac{a+b+c}{a}, \frac{a+b+c}{b}, \frac{a+b+c}{c}$ are in AP

$\Rightarrow \quad 1+\frac{b+c}{a}, 1+\frac{c+a}{b}, 1+\frac{a+b}{c}$

$\Rightarrow \quad \frac{b+c}{a}, \frac{c+a}{b}, \frac{a+b}{c}$ are in AP

$\Rightarrow \quad \frac{a}{b+c}, \frac{b}{c+a}, \frac{c}{a+b}$ are in HP

Illustration 3. If $x > 1, y > 1, z > 1$ are in GP, then prove that $\frac{1}{1+\ln x}, \frac{1}{1+\ln y}, \frac{1}{1+\ln z}$ are in HP

Solution: Since, x, y, z are in GP

$\therefore \quad y^2 = xz$

Taking log on both sides, we get

$\Rightarrow \quad 2\ln y = \ln x + \ln z$

$\Rightarrow \quad 2 + 2\ln y = (1+\ln x) + (1+\ln z)$

$\Rightarrow \quad 2(1+\ln y) = (1+\ln x) + (1+\ln z)$

$\therefore \quad 1+\ln x, 1+\ln y, 1+\ln z$ are in AP

$\Rightarrow \quad \frac{1}{1+\ln x}, \frac{1}{1+\ln y}, \frac{1}{1+\ln z}$ are in HP

Illustration 4. If the 7th term of a HP is 8 and 8th term is 7, then find its 15th term

Solution: T_7 of HP = 8

$\Rightarrow \quad T_7$ of AP $= \frac{1}{8}$

And $\quad T_8$ of HP = 7

$\Rightarrow \quad D = \frac{1}{7} - \frac{1}{8} = \frac{1}{56}$ and $a = \frac{1}{56}$

$\therefore \quad T_{15}$ of AP = a + 14d $= \frac{15}{56}$

$\Rightarrow \quad T_{15}$ of HP $= \frac{56}{15}$

5.3.2 Harmonic Mean

Let H is the harmonic mean between a and b then a, H and b are in HP

Harmonic mean of two numbers 'a' and 'b' is defined as $\frac{2}{\frac{1}{a}+\frac{1}{b}}$. To understand AM, GM and HM let us compare these

Numbers	AM	GM	HM
a, b	$\frac{a+b}{2}$	$(ab)^{\frac{1}{2}}$	$\frac{2}{\frac{1}{a}+\frac{1}{b}}$
a, b, c	$\frac{a+b+c}{3}$	$(abc)^{\frac{1}{3}}$	$\frac{3}{\frac{1}{a}+\frac{1}{b}+\frac{1}{c}}$
a, b, c, d	$\frac{a+b+c+d}{4}$	$(abcd)^{\frac{1}{4}}$	$\frac{4}{\frac{1}{a}+\frac{1}{b}+\frac{1}{c}+\frac{1}{d}}$
a_1, a_2, ... a_n	$\frac{a_1+a_2+\ldots+a_n}{n}$	$(a_1a_2\ldots a_n)^{\frac{1}{n}}$	$\frac{n}{\frac{1}{a_1}+\frac{1}{a_2}+..+\frac{1}{a_n}}$

(i) If a and b are two non–zero numbers, then the harmonic mean of a and b is a number H such that the sequence a, H, b is an H.P. we have

$$\frac{1}{H}=\frac{1}{2}\left(\frac{1}{a}+\frac{1}{b}\right) \text{ or } H=\frac{2ab}{a+b}$$

(ii) If $a_1, a_2, \ldots, a_n$ are n non–zero numbers, then the harmonic mean H of these numbers is given by $\frac{1}{H}=\frac{1}{n}\left(\frac{1}{a_1}+\frac{1}{a_2}+\ldots+\frac{1}{a_n}\right)$

(iii) The n numbers $H_1, H_2, \ldots, H_n$ are said to be harmonic means between a and b if $a, H_1, H_2, \ldots, H_n, b$ are in H.P., that is, if $\frac{1}{a},\frac{1}{H_1},\frac{1}{H_2},\ldots,\frac{1}{H_n},\frac{1}{b}$ are in A.P. let d be the common difference of this A.P. then

$$\frac{1}{b}=\frac{1}{a}+(n+2-1)d$$

$$\Rightarrow \quad d=\frac{1}{n+1}\left(\frac{1}{b}-\frac{1}{a}\right)=\frac{a-b}{(n+1)ab},$$

Thus $\frac{1}{H_1}=\frac{1}{a}+\frac{a-b}{(n+1)ab}$,

$$\frac{1}{H_2}=\frac{1}{a}+\frac{2(a-b)}{(n+1)ab},\ldots,\frac{1}{H_n}=\frac{1}{a}+\frac{n(a-b)}{(n+1)ab}$$

From here, we can get the values of $H_1, H_2, \ldots, H_n$

Illustration 5: If the harmonic mean between P and Q be H, then find the value of $H\left(\frac{1}{P}+\frac{1}{Q}\right)$ is equal to

Solution: Since $H=\frac{2PQ}{P+Q}$

or $H\left(\frac{P+Q}{PQ}\right)=2$

i.e., $H\left(\frac{1}{P}+\frac{1}{Q}\right)=2$

Illustration 6. Let the positive numbers a, b, c, d be in A.P, then prove that abc, abd, acd, bcd are in HP

Solution: a, b, c, d be in A.P.

$\Rightarrow$ d, c, b, a are in A.P.

Also, $\frac{1}{abc},\frac{1}{abd},\frac{1}{acd},\frac{1}{bcd}$ are in A.P.

[Divide by abcd]

$\Rightarrow$ abc, abd, acd, bcd are in H.P.

5.3.3 AM, GM, HM Inequality

(i) If a and b are two positive real numbers and A, G, H are arithmetic mean, geometric mean and harmonic mean between a and b respectively, then $A \geq G \geq H$ and the equality holds if and only if a = b.

(ii) If $a_1, a_2, \ldots, a_n$ are n positive real numbers, and

$$A=\frac{1}{n}(a_1+a_2+\ldots+a_n)$$

$$G_n=(a_1\, a_2 \ldots a_n)^{1/n}$$

and $\frac{1}{H}=\frac{1}{n}\left(\frac{1}{a_1}+\frac{1}{a_2}+\ldots+\frac{1}{a_n}\right)$, then $A \geq G \geq H$

Illustration 7. If $a > 1, b > 1$, then find the minimum value of $\log_b a+\log_a b$

Solution: Here,

$$\frac{\log_b a+\log_a b}{2} \geq (\log_b a \cdot \log_a b)^{1/2}=1$$

$\Rightarrow \log_b a+\log_a b \geq 2$

So required minimum value is 2.

Illustration 8. Find the minimum value of $4x+4^{1-x}$, $x \in R$ is

Solution: $AM \geq GM$ for positive numbers

So, $\frac{4^x+\frac{4}{4^x}}{2} \geq \sqrt{4^x \cdot \frac{4}{4^x}}=2 \Rightarrow 4^x+4^{1-x} \geq 4$

So required minimum value is 4

Illustration 9. If x, y, z are three positive real numbers, then find the minimum value of

$$\frac{(x+y)(y+z)(z+x)}{xyz}$$

Solution: Since, AM $\geq$ GM

$$\Rightarrow \quad \frac{x+y}{2} \geq \sqrt{xy}$$

$$\Rightarrow \quad x+y \geq 2\sqrt{xy}$$

Similarly, $\quad y+z \geq 2\sqrt{yz}$ and $z+x \geq 2\sqrt{zx}$

On multiplication, we get

$$\frac{(x+y)(y+z)(z+x)}{xyz} \geq 8$$

So required minimum value is 8.

Illustration 10. $\log(x+z) + \log(x+z-2y) = 2\log(x-z)$, then prove that x, y, z are in HP

Solution: $\log(x+z) + \log(x+z-2y) = 2\log(x-z)$

$$\Rightarrow \log[(x+z-2y)(x+z)] = \log(x-z)^2$$

$$\Rightarrow (x+z-2y)(x+z) = (x-z)^2$$

$$\Rightarrow x^2 + 2xz + z^2 - 2yz - 2xy = x^2 + z^2 - 2xz$$

$$\Rightarrow 4xz = 2yz + 2xy$$

$$\Rightarrow 2xz = y(x+z)$$

$$\therefore \quad y = \frac{2xz}{x+z}$$

$\therefore \quad x, y, z$ are in HP

Concept Applicator (CA)-3

1. If $a > 0$, $b > 0$, $c > 0$ and the minimum value of $a(b^2+c^2) + b(c^2+a^2) + c(a^2+b^2)$ is λabc, then find the value of λ
   (a) 2 (b) 1
   (c) 6 (d) 3

2. If $-\frac{\pi}{2} < \theta < \frac{\pi}{2}$, then the minimum value of $\cos^3\theta + \sec^3\theta$ is
   (a) 1 (b) 2
   (c) 0 (d) None of these

3. The three terms, $\log_3 2, \log_6 2, \log_{12} 2$ are in
   (a) AP (b) GP
   (c) HP (d) None of these

4. If $H_n = 1 + \frac{1}{2} + \ldots + \frac{1}{n}$, then value of $1 + \frac{3}{2} + \frac{5}{3} + \ldots + \frac{2n-1}{n}$ is
   (a) $H_n + n$ (b) $2n - H_n$
   (c) $(n-1) + H_n$ (d) $H_n + 2n$

5. If a, b, c and d are in HP, then $ab + bc + cd$ is equal to
   (a) ad (b) $2ad$
   (c) $3ad$ (d) None of these

6. If $(p+q)$, $(q+r)$ and $(r+p)$ are in H.P, then
   (a) p, q, r are in A.P
   (b) q^2, p^2, r^2 are in A.P
   (c) p^2, q^2, r^2 are in A.P
   (d) p, q, r are in G.P

7. If x, y, z are in AP then e^{-x}, e^{-y} and e^{-z} are in
   (a) AP (b) GP
   (c) HP
   (d) no definite sequence

8. If $\frac{a^n + b^n}{a^{n-1} + b^{n-1}}$ is the HM between a and b, if n is
   (a) 0 (b) $\frac{1}{2}$
   (c) $-\frac{1}{2}$ (d) 1

9. If the first and the $(2n-1)$th terms of an AP, a GP and a HP are equal and their *nth* terms are a, b and c respectively, then
   (a) $a = b = c$ (b) $a < b < c$
   (c) $a + c = b$ (d) $ac - b^2 = 0$

10. If the first two terms of harmonic progression be $\frac{1}{2}$ and $\frac{1}{4}$, then the harmonic mean of the first four numbers is
    (a) 5 (b) $\frac{1}{5}$
    (c) 10 (d) 12

5.4 MISCELLANEOUS SERIES

5.4.1 Sum of squares of the 1st 'n' natural numbers

To find the sum of the squares of the first N natural numbers or to find Σn^2

Proof: Let $s = \Sigma n^2 = 1^2 + 2^2 + 3^2 + ... + n^2$ ...(i)

Now, $r^3 - (r-1)^3 = r^3 - (r^3 - 3r^2 + 3r - 1)$

$= 3r^2 - 3r + 1$

$\therefore \quad r^3 - (r-1)^3 = 3r^2 - 3r + 1$ ...(ii)

Now in (ii), When

$r = 1, \quad 1^3 - 0^3 = 3 \cdot 1^2 - 3 \cdot 1 + 1$

$r = 2, \quad 2^3 - 1^3 = 3 \cdot 2^2 - 3 \cdot 2 + 1$

$r = 3, \quad 3^3 - 2^3 = 3 \cdot 3^2 - 3 \cdot 3 + 1$

... ...

... ...

$r = n, \quad n^3 - (n-1)^3 = 3n^2 - 3n + 1$

Adding all these we get

$$n^3 = 3(1^2 + 2^2 + ... + n^2) - 3(1 + 2 + ... + n) + (1 + 1 ... \text{to } n \text{ terms})$$

or $\quad n^3 = 3S - 3 \cdot \dfrac{n(n+1)}{2} + n$

$$3S = n^3 - n + \frac{3n(n+1)}{2} = n(n-1)(n+1) + \frac{3n(n+1)}{2}$$

$$= n(n+1)\left(n - 1 + \frac{3}{2}\right)$$

$$= \frac{n(n+1)(2n+1)}{2}$$

$$\therefore \quad S = \frac{n(n+1)(2n+1)}{6}$$

2nd Proof:

Let $S = 1^2 + 2^2 + 3^2 + ... + n^2$ consider the identify

$(2r+1)^3 - (2r-1)^3 = 24r^2 + 2$ ...(1)

putting $r = 1, 2, 3, ..., (n-1), n$, We get

$3^3 - 1^3 = 24 \cdot 1^2 + 2$

$5^3 - 3^3 = 24 \cdot 2^2 + 2$

$7^3 - 5^3 = 24 \cdot 3^2 + 2$

... ...

... ...

$(2n-1)^3 - (2n-3)^3 = 24 \cdot (n-1)^2 + 2$

$(2n+1)^3 - (2n-1)^3 = 24 \cdot n^2 + 2$

Adding, We get

$$(2n+1)^3 - 1^3 = 24(1^2 + 2^2 + ... + n^2) + 2n$$

$$\therefore \quad 24S = (2n+1)^3 - 1 - 2n = (2n+1)^3 - (2n+1)$$

$$= (2n+1)\{(2n+1)^2 - 1\}$$

$$= (2n+1)(4n^2 + 4n) = 4n(n+1)(2n+1)$$

$$\therefore \quad S = \frac{4n(n+1)(2n+1)}{24} = \frac{n(n+1)(2n+1)}{6}$$

$$\therefore \Sigma n^2 = 1^2 + 2^2 + ... + n^2 = \frac{n(n+1)(2n+1)}{6}$$

5.4.2 Sum of cubes of the 1st 'n' natural numbers

Let $S = 1^3 + 2^3 + 3^3 + ... + n^3$

We consider the identify

$$r^4 - (r-1)^4 = 4r^3 - 6r^2 + 4r - 1$$

Putting $r = 1, 2, 3, ... n$ successively, we get

$1^4 - 0^4 = 4 \cdot 1^3 - 6 \cdot 1^2 + 4 \cdot 1 - 1$

$2^4 - 1^4 = 4 \cdot 2^3 - 6 \cdot 2^2 + 4 \cdot 2 - 1$

$3^4 - 2^4 = 4 \cdot 3^3 - 6 \cdot 3^2 + 4 \cdot 3 - 1$

... ...

... ...

$(n-1)^4 - (n-2)^4 = 4(n-1)^3 - 6(n-1)^2 + 4(n-1) - 1$

$n^4 - (n-1)^4 = 4 \cdot n^3 - 6 \cdot n^2 + 4 \cdot n - 1$

Adding we get

$$n^4 = 4(1^3 + 2^3 + 3^3 + ... + n^3) - 6(1^2 + 2^2 + 3^2 + ... + n^2) + 4(1 + 2 + 3 + ... + n) - n$$

$$= 4S - \frac{6n(n+1)(2n+1)}{6} + \frac{4n(n+1)}{2} - n$$

$\therefore \quad 4S = n^4 + n(n+1)(2n+1) - 2n(n+1) + n$

$= n\left\{n^3 + (n+1)(2n+1) - 2(n+1) + 1\right\}$

$= n\left(n^3 + 2n^2 + 3n + 1 - 2n - 2 + 1\right)$

$= n\left(n^3 + 2n^2 + n\right)$

$= n^2\left(n^2 + 2n + 1\right) = n^2(n+1)^2$

$\therefore \quad S = \dfrac{n^2(n+1)^2}{4} = \left[\dfrac{n(n+1)}{2}\right]^2$

Second proof:

$\because (r+1)^2 - (r-1)^2 = 4r$

$\therefore r^2(r+1)^2 - r^2(r-1)^2 = 4r^3$

Putting $r = 1, 2, \ldots, n$ successively, we get

$1^2 \cdot 2^2 - 1^2 \cdot 0^2 = 4 \cdot 1^3$

$2^2 \cdot 3^2 - 2^2 \cdot 1^2 = 4 \cdot 2^3$

$3^2 \cdot 4^2 - 3^2 \cdot 2^2 = 4 \cdot 3^3$

... ...

... ...

$n^2(n+1)^2 - n^2(n-1)^2 = 4n^3$

Adding, we get $n^2(n+1)^2 = 4\left(1^3 + 2^3 + \ldots + n^3\right)$

$\therefore \quad 1^3 + 2^3 + \ldots. + n^3 = \dfrac{n^2(n+1)^2}{4} = \left[\dfrac{n(n+1)}{2}\right]^2$

Note: $\sum_{n=1}^{n} n^3 = \left(\sum_{n=1}^{n} n\right)^2$

5.4.3 Power Series

Power Series		
Summation	**Expansion**	**Equivalent Value**
$\sum_{k=1}^{n} k$	$= 1 + 2 + 3 + 4 + .. + n$	$= (n^2 + n)/2$ $= \left(\frac{1}{2}\right)n^2 + \left(\frac{1}{2}\right)n$
$\sum_{k=1}^{n} k^2$	$1^2 + 2^2 + 3^2 + 4^2 + .. + n^2$	$= (1/6)n(n+1)(2n+1)$ $= \left(\frac{1}{3}\right)n^3 + \left(\frac{1}{2}\right)n^2 + \left(\frac{1}{6}\right)n$
$\sum_{k=1}^{n} k^3$	$1^3 + 2^3 + 3^3 + 4^3 + \ldots + n^3$	$= \left(\frac{1}{4}\right)n^4 + \left(\frac{1}{2}\right)n^3 + \left(\frac{1}{4}\right)n^2$
$\sum_{k=1}^{n} k^4$	$1^4 + 2^4 + 3^4 + 4^4 + .. + n^4$	$= \left(\frac{1}{5}\right)n^5 + \left(\frac{1}{2}\right)n^4 + \left(\frac{1}{3}\right)n^3 - \left(\frac{1}{30}\right)n$
$\sum_{k=1}^{n} k^5$	$1^5 + 2^5 + 3^5 + 4^5 + .. + n^5$	$= \left(\frac{1}{6}\right)n^6 + \left(\frac{1}{2}\right)n^5 + \left(\frac{5}{12}\right)n^4 - \left(\frac{1}{12}\right)n^2$
$\sum_{k=1}^{n} k^6$	$1^6 + 2^6 + 3^6 + 4^6 + .. + n^6$	$\left(\frac{1}{7}\right)n^7 + \left(\frac{1}{2}\right)n^6 + \left(\frac{1}{2}\right)n^5 - \left(\frac{1}{6}\right)n^3 + \left(\frac{1}{42}\right)n$
$\sum_{k=1}^{n} k^7$	$1^7 + 2^7 + 3^7 + 4^7 + .. + n^7$	$\left(\frac{1}{8}\right)n^8 + \left(\frac{1}{2}\right)n^7 + \left(\frac{7}{12}\right)n^6 - \left(\frac{7}{24}\right)n^4 + \left(\frac{1}{12}\right)n^2$
$\sum_{k=1}^{n} k^8$	$1^8 + 2^8 + 3^8 + 4^8 + .. + n^8$	$\left(\frac{1}{9}\right)n^9 + \left(\frac{1}{2}\right)n^8 + \left(\frac{2}{3}\right)n^7 - \left(\frac{7}{15}\right)n^5 + \left(\frac{2}{9}\right)n^3 - \left(\frac{n}{30}\right)$

Power Series		
Summation	**Expansion**	**Equivalent Value**
$\sum_{k=1}^{n} k^9$	$1^9 + 2^9 + 3^9 + 4^9 + .. + n^9$	$\left(\frac{n^{10}}{10}\right)+\left(\frac{n^9}{2}\right)+\left(\frac{3}{4}\right)n^8-\left(\frac{7}{10}\right)n^6+\left(\frac{1}{2}\right)n^4-\left(\frac{3}{20}\right)n^2$
$\sum_{k=1}^{n} k^{10}$	$1^{10} + 2^{10} + 3^{10} + 4^{10} + .. + n^{10}$	$\left(\frac{1}{11}\right)n^{11}+\left(\frac{1}{2}\right)n^{10}+\left(\frac{5}{6}\right)n^9-n^7+n^5-\left(\frac{1}{2}\right)n^3+\left(\frac{5}{66}\right)n$

Approach to problems based on Sum of a series when n^{th} term of the series is a polynomial in n.

Step 1. Write down the nth term t_n of the given series as a polynomial in n.

Step 2. Then put $n = 1, 2, 3, \ldots n$ and add or use Σ notation.

Step 3. Use the results of power series whichever are required:

> **Some more properties of Natural number**
>
> Sum of squares of 1st n Odd natural number i.e.,
>
> $1^2 + 3^2 + 5^2 + \ldots + (2n-1)^2 = \frac{n(2n-1)(2n+1)}{3}$
>
> Sum of squares of 1st n Even natural number i.e.,
>
> $2^2 + 4^2 + 6^2 + \ldots + (2n)^2 = \frac{2n(n+1)(2n+1)}{3}$
>
> Sum of cubes of 1st n Odd natural number i.e.,
>
> $1^3 + 3^3 + 5^3 + \ldots.. + (2n-1)^3 = n^2\left(2n^2-1\right)$
>
> Sum of cubes of 1st n Even natural number i.e.,
>
> $2^3 + 4^3 + 6^3 + \ldots.. + 2n^3 = 2\left[n(n+1)\right]^2$

Illustration 1. Find the sum of 1st n odd natural numbers

Solution: Since rth odd natural number is defined as $t_r = (2r-1)$

Sum to n terms is defined as

$$s_n = \sum_{r=1}^{n} t_r = \sum_{r=1}^{n}(2r-1) = \frac{2n(n+1)}{2} - n = n^2$$

Illustration 2. Find the sum to 1st 50 terms of $2^2 + 4^2 + 6^2 + \ldots..$

Solution: Given series is

$2^2 + 4^2 + 6^2 + \ldots.. = 2^2(1^2 + 2^2 + \ldots 50^2)$

$$= 4\frac{(50)(51)(101)}{6} = 171700$$

Illustration 3.

If $\sum_{k=1}^{n}\left(\sum_{m=1}^{k} m^2\right) = an^4 + bn^3 + cn^2 + dn + e$, then

(a) $a = \frac{1}{12}$ (b) $b = \frac{1}{2}$

(c) $d = \frac{1}{6}$ (d) $e = 0$

Solution: (a) $\sum_{k=1}^{n}\left(\sum_{m=1}^{k} m^2\right) = \sum_{k=1}^{n}\frac{k(k+1)(2k+1)}{6}$

$$= \frac{1}{6}\sum_{k=1}^{n}\left(2k^3 + 3k^2 + k\right)$$

$$= \frac{1}{3}\cdot\left\{\frac{n(n+1)}{2}\right\}^2 + \frac{1}{2}\frac{n(n+1)(2n+1)}{6} + \frac{1}{6}\frac{n(n+1)}{2}$$

$a =$ coefficient of $n^4 = \frac{1}{3}\cdot\frac{1}{4}$

5.4.4 Type 1, when $T_n = an + b$

If rth term of the series is $T_r = ar + b$ then series is in AP with common difference 'a'.

Sum of the n terms of the series S_n is defined as

$$S_n = \sum_{r=1}^{n} t_r = \sum_{r=1}^{n}(ar+b) = a\sum_{r=1}^{n} r + \sum_{r=1}^{n} b$$

$$= \frac{an(n+1)}{2} + bn = \left(\frac{a}{2}\right)n^2 + \left(\frac{a}{2}+b\right)n = An^2 + Bn$$

Observations:

1. If $T_r = ar + b$ then series is in AP with common difference 'a' and sum to n terms is in the form of $S_n = An^2 + Bn$
2. If sum to n terms is in the form of $S_n = An^2 + Bn$ then series must be in AP

5.4.5 Type 2, when $T_r = ar^2 + br + c$ (Arithmetico Arithmetic Progression)

If r[th] term of the series is $T_r = ar^2 + br + c$ then series is in AAP

Sum of the n terms of the series S_n is defined as

$$S_n = \sum_{r=1}^{n} t_r = \sum_{r=1}^{n}\left(ar^2 + br + c\right) = a\sum_{r=1}^{n} r^2 + b\sum_{r=1}^{n} r + \sum_{r=1}^{n} c$$

$$= \frac{an(n+1)(2n+1)}{6} + \frac{bn(n+1)}{2} + cn$$

Sum to n terms is in the form of $S_n = An^3 + Bn^2 + Cn$

In this type of series 2nd order difference is constant.

Consider a series 2, 6, 13, 23, 36, 52 …

1st order difference is 4, 7, 10, 13, 16…(Which is in AP)

2nd order difference is 3 (a constant)

Since 2nd order difference is constant so r[th] term is a 2nd order expression ($T_r = ar^2 + br + c$) and sum to n terms is a 3rd order expression ($S_n = An^3 + Bn^2 + Cn$)

Imp Concept

If r[th] term is a polynomial of degree k then
(i) Its kth order difference must be constant
(ii) Its summation to n terms must be a polynomial of degree (k + 1) with constant term zero

Illustration 4. Find the 10th term of the sequence 1, 3, 6, 10, 15 ...

(a) 16 (b) 13
(c) 15 (d) None of these

Solution: (d) 1st order difference 2, 3, 4, 5,...

2nd order difference 1, 1, 1,...

Second order of differences is constant.

$$\therefore \quad t_n = a(n-1)(n-2) + b(n-1) + c$$

$n = 1, 2, 3$ gives, c = 1, b = 2 and a = 1/2

$$\therefore \quad t_n = \frac{1}{2}n^2 + \frac{n}{2} = \frac{n(n+1)}{2}$$

$$\Rightarrow \quad t_{10} = 55$$

Illustration 5. The next term of the sequence 1, 5, 14, 30, 55, ... is

(a) 91 (b) 85
(c) 90 (d) 95

Solution: (a) Given series is 1, 5, 14, 30, 55…

1st order difference 4, 9, 16, 25,... which is in the form of n^2

So next difference is 36, so next term is 55 + 36 = 91

Illustration 6. If the sum of any number of terms in a sequence is always three times the squared number of these terms, then the sequence is

(a) an A.P (b) a G.P
(c) an H.P (d) None of these

Solution: Given $S_n = 3n^2$

$$\therefore \quad t_n = S_n - S_{n-1} = 3n^2 - 3(n-1)^2$$
$$= 6n - 3$$

Putting $n = 1, 2, 3$ …, The sequence is 3, 9, 15, 21,... which is clearly in an AP.

5.4.6 Arithmetic–Geometric Sequence

Consider two sequences $a_1, a_2, \ldots, a_n, \ldots$ which is an A.P. and $b_1, b_2, \ldots, b_n, \ldots$ which is in a G.P. then the sequence $a_1b_1, a_2b_2, \ldots, a_nb_n, \ldots$ is said to be an arithmetic-geometric sequence (or AGP),

An arithmetic-geometric sequence is of the form $ab, (a+d)br, (a+2d)br^2, (a+3d)br^3, \ldots$

Sum to n term of an Arithmetic-Geometric Series.

$$\text{Let } S_n = ab + (a+d)br + (a+2d)br^2 + \ldots$$
$$+\left[a+(n-2)d\right]br^{n-2} + \left[a+(n-1)d\right]br^{n-1} \quad \ldots(1)$$

Multiply each term by r and write the first term below the second, the second below the third and so on..

$$rS_n = abr + (a+d)br^2 + \ldots + \left[a+(n-3)d\right]br^{n-2}$$
$$+\left[a+(n-2)d\right]br^{n-1} + \left[a+(n-1)d\right]br^n$$

Subtracting we get,

$$(1-r)S_n = ab + dbr + dbr^2 + \ldots + dbr^{n-2} + dbr^{n-1}$$
$$-\left[a+(n-1)d\right]br^n$$

$$= ab + \frac{dbr\left(1-r^{n-1}\right)}{1-r} - \left[a+(n-1)d\right]br^n \quad (r \neq 1)$$

$$\Rightarrow S_n = \frac{ab}{1-r} + \frac{dbr\left(1-r^{n-1}\right)}{(1-r)^2} - \frac{\left[a+(n-1)d\right]br^n}{(1-r)}$$

$(r \neq 1)$

if $-1 < r < 1$, then $\lim_{n\to\infty} r^n = 0$ and $\lim_{n\to\infty} nr^n = 0$

Hence, the sum of an infinite number of terms of the sequence is $S = \frac{ab}{1-r} + \frac{dbr}{(1-r)^2} (-1 < r < 1)$

Illustration 7. The sum of the series $1 + 2x + 3x^2 + 4x^3 + \ldots$ upto infinity when x lies between 0 and 1 ie, $0 < x < 1$ is

(a) $\frac{1}{1+x}$ (b) $\frac{1}{1-x}$

(c) $\frac{1}{1-2x}$ (d) $\frac{1}{(1-x)^2}$

Solution: (d) Let $S = 1 + 2x + 3x^2 + 4x^3 + \ldots$

$xS = x + 2x^2 + 3x^3 + \ldots$

On subtraction,

$$(1-x)S = 1 + x + x^2 + x^3 + \ldots = \frac{1}{1-x}$$

$$\Rightarrow \quad S = \frac{1}{(1-x)^2}$$

Illustration 8. Sum of the series $(1+x) + (1+x+x^2) + (1+x+x^2+x^3) + \ldots$ Upto n terms is

(a) $\frac{1}{1-x}\left[n - \frac{x^2\left(1-x^n\right)}{1-x}\right]$

(b) $\frac{1}{1-x}\left[n - \frac{x^3\left(1-x^n\right)}{1-x}\right]$

(c) $\frac{1}{1-x}\left[n - \frac{x\left(1-x^n\right)}{1-x}\right]$

(d) None of these

Solution: (a) $(1+x) + (1+x+x^2) + (1+x+x^2+x^3) + \ldots$ Upto n terms

$$= \frac{1-x^2}{1-x} + \frac{1-x^3}{1-x} + \frac{1-x^4}{1-x} + \ldots \text{to } n \text{ terms}$$

$$= \frac{1}{1-x}[(1+1+1+\ldots n \text{ terms}) - (x^2 + x^3 + x^4 + \ldots \text{ to } n \text{ terms})]$$

$$= \frac{1}{1-x}\left[n - \frac{x^2\left(1-x^n\right)}{1-x}\right]$$

Illustration 9. Sum to infinity of the series $1 + \frac{4}{5} + \frac{7}{5^2} + \frac{10}{5^3} + \ldots$ is

(a) $\frac{16}{25}$ (b) $\frac{11}{8}$

(c) $\frac{35}{16}$ (d) $\frac{8}{11}$

Solution: (c) Let $S = 1 + \frac{4}{5} + \frac{7}{5^2} + \frac{10}{5^3} + \ldots$

$$\Rightarrow \quad \frac{1}{5}S = \frac{1}{5} + \frac{4}{5^2} + \frac{7}{5^3} + \ldots$$

On subtraction,

$$\Rightarrow \quad \frac{4}{5}S = 1 + \frac{3}{5} + \frac{3}{5^2} + \frac{3}{5^3} + \ldots$$

$$\Rightarrow \quad S = \frac{7}{4} \times \frac{5}{4} = \frac{35}{16}$$

Illustration 10. (a) The sum of first n terms of the series $1^2 + 2\cdot 2^2 + 3^2 + 2\cdot 4^2 + 5^2 + 2\cdot 6^2 + \ldots$ is $\frac{n(n+1)^2}{2}$ where n is even. When n is odd the sum is

(a) $\frac{n^2(n+1)}{2}$ (b) $\frac{n(n+1)^2}{2}$

(c) $\left[\frac{n(n+1)}{2}\right]^2$ (d) $\frac{n(n+1)}{2}$

Solution: when n is odd, last term will be n^2

$\therefore$ Then, the sum is

$$1^2 + 2\cdot 2^2 + 3^2 + 2\cdot 4^2 + 5^2 + 2\cdot 6^2 + \ldots + 2(n-1)^2 + n^2$$

$$= \frac{(n-1)n^2}{2} + n^2 \text{ [Replacing } n \text{ by } n\text{–}1 \text{ in } \frac{n(n+1)^2}{2}\text{]}$$

$$= \frac{n^3 - n^2 + 2n^2}{2} = \frac{n^3 + n^2}{2} = \frac{n^2(n+1)}{2}$$

5.4.7 Some Special Series

In this case we will discuss the type of questions when in r^{th} term numerator is constant and denominator is a polynomial function of r.

If $t_n = \dfrac{1}{\{(an+b)(cn+d)(en+f)\}}$ then use partial fraction and write

$$t_n = \frac{1}{\{(an+b)(cn+d)(en+f)\}}$$

$$= \frac{A}{an+b} + \frac{B}{cn+d} + \frac{C}{en+f}$$

$$= \frac{\left\{\begin{matrix} A(cn+d)(en+f) + B(an+b)(en+f) \\ +C(an+b)(cn+d) \end{matrix}\right\}}{\{(an+b)(cn+d)(en+f)\}}$$

In order to find the value of A, put $an+b=0$ or $n = -\frac{b}{a}$ put this value of 'n' in $\dfrac{1}{\{(en+f)(cn+d)\}}$

So $A = \dfrac{1}{\{(en+f)(cn+d)\}}$ & put $n = -\dfrac{b}{a}$

$B = \dfrac{1}{\{(an+b)(en+f)\}}$ & put $n = -\dfrac{d}{c}$

$C = \dfrac{1}{\{(an+b)(cn+d)\}}$ & put $n = -\dfrac{f}{e}$

Illustration 11. Find to $\frac{1}{1.2} + \frac{1}{2.3} + \frac{1}{3.4} + \ldots\ldots n$ terms

Solution: Here $t_r = \dfrac{1}{\{r.(r+1)\}} = \dfrac{1}{r} - \dfrac{1}{r+1}$

$$t_r = \frac{1}{r} - \frac{1}{r+1}$$

$$t_1 = \frac{1}{1} - \frac{1}{2}$$

$$t_2 = \frac{1}{2} - \frac{1}{3}$$

$$t_3 = \frac{1}{3} - \frac{1}{4}$$

$\ldots\ \ldots\ \ldots$

$\ldots\ \ldots\ \ldots$

$$t_{n-1} = \frac{1}{n-1} - \frac{1}{n}$$

$$t_n = \frac{1}{n} - \frac{1}{n+1}$$

On adding these terms we will get

$$S_n = \frac{1}{1} - \frac{1}{n+1} = \frac{n}{n+1}$$

Illustration 12. Find to $\frac{1}{1.3} + \frac{1}{3.5} + \frac{1}{5.7} + \ldots\ldots n$ terms

Solution: Here

$$t_r = \frac{1}{\{(2r-1).(2r+1)\}} = \frac{1}{2}\left[\frac{1}{2r-1} - \frac{1}{2r+1}\right]$$

$$t_r = \frac{1}{2}\left[\frac{1}{2r-1} - \frac{1}{2r+1}\right]$$

$$t_1 = \frac{1}{2}\left[\frac{1}{1} - \frac{1}{3}\right]$$

$$t_2 = \frac{1}{2}\left[\frac{1}{3} - \frac{1}{5}\right]$$

$$t_3 = \frac{1}{2}\left[\frac{1}{5} - \frac{1}{7}\right]$$

$\ldots\ \ldots\ \ldots$

$\ldots\ \ldots\ \ldots$

$$t_{n-1} = \frac{1}{2}\left[\frac{1}{2n-3} - \frac{1}{2n-1}\right]$$

$$t_n = \frac{1}{2}\left[\frac{1}{2n-1} - \frac{1}{2n+1}\right]$$

On adding these terms we will get

$$S_n = \frac{1}{2}\left[1 - \frac{1}{2n+1}\right] = \frac{n}{2n+1}$$

Illustration 13. Find the sum to n terms of the series whose nth term is $t_n = n^2 + 2^n$

Solution: Given $t_n = n^2 + 2^n$

$$\therefore\ S_n = \sum_{n=1}^{n} t_n = \sum_{n=1}^{n}\left(n^2 + 2^n\right)$$

$$= \sum_{n=1}^{n} n^2 + \sum_{n=1}^{n} 2^n$$

$$= \frac{n(n+1)(2n+1)}{6} + \left(2^1 + 2^2 + 2^3 + \ldots + 2^n\right)$$

$$= \frac{n(n+1)(2n+1)}{6} + \frac{2\left(2^n - 1\right)}{2-1}$$

$$= \frac{n(n+1)(2n+1)}{6} + 2\left(2^n - 1\right)$$

Illustration 14. Find $(3^3 - 2^3) + (5^3 - 4^3) + (7^3 - 6^3) + \ldots$ to n terms

Solution: Let $S_n = (3^3 - 2^3) + (5^3 - 4^3) + (7^3 - 6^3) + \ldots$to n terms ...(i)

Now nth terms of sequence 3,5,7,...

$= 3 + (n-1)2 = 2n + 1$

$\therefore$ nth term of the series $3^3 + 5^3 + 7^3 + ...$ is $(2n+1)^3$

nth terms of sequence 2,4,6,... is $2 + (n - 1)\,2 = 2n$

$\because$ n^{th} term of the series $2^3 + 4^3 + 6^3 + ...$ is $(2n)^3 = 8n^3$

From (i), n^{th} term of the given series,

$$t_n = (2n + 1)^3 - 8n^3$$
$$= 8n^3 + 3\cdot 4n^2 \cdot 1 + 3\cdot 2n\cdot 1^2 + 1^3 - 8n^3$$
$$= 12n^2 + 6n + 1$$
$$\therefore\ S_n = \sum 12n^2 + \sum 6n + \sum 1 = 12\sum n^2 + 6\sum n + n$$
$$= \frac{12n(n+1)(2n+1)}{6} + \frac{6n(n+1)}{2} + n$$
$$= 2n(n + 1)(2n + 1) + 3n(n + 1) + n$$
$$= n[2(2n^2 + 3n + 1) + 3n + 3 + 1]$$
$$= n(4n^2 + 9n + 6) = 4n^2 + 9n^2 + 6n$$

Illustration 15. Find the sum $(n^2 + 1) + 2(n^2 - 2^2) + 3(n^2 - 3^2) + ...$ to n terms

Solution:

$$t_r = r(n^2 - r^2)$$
$$\Rightarrow\ n^2\left(\sum_{r=1}^{n-1} r\right) - \sum_{r=1}^{n-1} r^3 = \frac{n^2}{2}(n-1)n - \frac{1}{4}n^2(n-1)^2$$
$$= \frac{1}{4}n^2(n-1)(2n-n+1)$$
$$= \frac{1}{4}n^2(n^2-1).$$

Concept Applicator (CA)-4

1. Find the sum of the series $1^3 + 3^3 + 5^3 + ...19^3$
   (a) 19900 (b) 19700
   (c) 19500 (d) None of these
2. The sum of the first n natural number is one – fifth of the sum of their squares, then n is
   (a) 5 (b) 6 (c) 7 (d) 8
3. The 10th term of the series 5, 17, 35,59,89,... is
   (a) 271 (b) 268 (c) 329 (d) 269
4. A series whose nth term is $(\frac{n}{x}) + y$, then sum of r terms will be
   (a) $\left\{\frac{r(r+1)}{2x}\right\} + ry$ (b) $\left\{\frac{r(r-1)}{2x}\right\}$
   (c) $\left\{\frac{r(r-1)}{2x}\right\} - xy$ (d) $\left\{\frac{r(r+1)}{2x}\right\} - ry$
5. The sum of all the numbers of the form n^3 which lie between 100 and 10000 is
   (a) 43261 (b) 53261
   (c) 63261 (d) None of these
6. The sum of n terms of $1\cdot 2\cdot 3 + 2\cdot 3\cdot 4 + ...$ will be
   (a) $\frac{n(n+1)(n+2)(n+3)}{4}$
   (b) $\frac{2n(n+1)(n+2)(n+3)}{3}$
   (c) $\frac{(n+1)(n+2)(n+3)}{4}$
   (d) $\frac{n(n-1)(n-2)(n-3)}{4}$
7. Find the 10th term of the series, 2, 8, 32, 86, 182, 332
   (a) 1709 (b) 1712
   (c) 1717 (d) None of these
8. Sum of the series $1 + 2.\,2 + 3.2^2 + 4.\,2^3 + ... + 100.2^{99}$ is
   (a) $100.\,2^{100} + 1$ (b) $99.\,2^{100} + 1$
   (c) $99.\,2^{99} - 1$ (d) $100.\,2^{100} - 1$
9. The positive integer n for which $2 \times 2^2 + 3 \times 2^3 + 4\times 2^4 + ... + n\times 2^n = 2^{n+10}$ is
   (a) 510 (b) 511 (c) 512 (d) 513
10. Let S_n denote the sum of the cubes of the first n natural numbers and s_n denote the sum of the first n natural numbers. Then, $\sum_{r=1}^{n}\frac{S_r}{s_r}$ is equal to
    (a) $\frac{n(n+1)(n+2)}{6}$ (b) $\frac{n(n+1)}{2}$
    (c) $\frac{n^2+3n+2}{6}$ (d) None of these

Exercise A

Concept Builder (CB)

1. In an arithmetic progression of 16 distinct terms with $a_1 = 16$, the sum is equal to square of the last term. The common difference of the A. P. is
   (a) 8/15 (b) – 4/5
   (c) – 8/5 (d) – 8/15
2. Find the number of numbers lying between 100 and 500 that are divisible by 7 but not by 21
   (a) 57 (b) 38
   (c) 39 (d) None of these
3. If the first, second and the last terms of an A.P are a, b, c respectively, then find the sum of the series
   (a) $\frac{(a+b)(a+c-2b)}{2(b-a)}$
   (b) $\frac{(b+c)(a+b-2c)}{2(b-a)}$
   (c) $\frac{(a+c)(b+c-2a)}{2(b-a)}$
   (d) None of these
4. The sum of all natural numbers less than 200, that are divisible neither by 3 nor by 5, is
   (a) 10730 (b) 10735
   (c) 10732 (d) None of these
5. Find the sum of all two digit numbers which are in the form of $4k + 1$
   (a) 1100 (b) 1200
   (c) 1210 (d) None of these
6. Hasan took a loan of ₹ 3600 and agrees to pay off by 40 annual instalments which are in AP. It is observed that after 30 of the instalments are paid, the due amount is ₹ 1200. Find the value of the 1st instalment
   (a) ₹ 51 (b) ₹ 50
   (c) ₹ 65 (d) None of these
7. The sum of the series 2, 5, 8, 11,...upto 'n' terms is 60100, then find the sum of the digits of n.
   (a) 9 (b) 8
   (c) 6 (d) None of these
8. Find the sum of all 2 digit odd numbers.
   (a) 2475 (b) 2530
   (c) 4905 (d) 5049
9. A polygon has 25 sides, the lengths of which starting from the smallest side are in AP. If the perimeter of the polygon is 2100 cm and the length of the largest side 20 times that of the smallest, then the find the difference between largest and smallest side.
   (a) 171 (b) 152
   (c) 133 (d) None of these.
10. A club consists of members whose ages are in AP, the common difference being 3 months. If the youngest member of the club is just 7 yr old and the sum of the ages of all the members is 975 yr, then the number of members in the club are
    (a) 45 (b) 120
    (c) 65 (d) 60
11. If the sum of first n terms of a series be $5n^2 + 2n$, then its second term is
    (a) 107 (b) 97
    (c) 87 (d) None of these
12. Three numbers are in GP. If we double the middle term, we get an AP. Then, find the common ration of the GP
    (a) $2\pm\sqrt{3}$ (b) $3\pm\sqrt{2}$
    (c) $3\pm\sqrt{5}$ (d) none of these
13. If in a GP, the $(m + n)^{th}$ term is p and the $(m - n)^{th}$ term is q, then find its mth term
    (a) 0 (b) pq
    (c) $\sqrt{pq}$ (d) $\frac{1}{2}(p+q)$
14. The fourth, seventh and tenth terms of a GP are p, q and r respectively, then p, q and r are in
    (a) AP
    (b) GP
    (c) HP
    (d) No specific sequence

15. The 6th term of a GP is 32 and its 8th term is 128, then find the value of the common ratio if all the terms are positive.
(a) 3 (b) 2
(c) 4 (d) 3/2

16. If $a^x = b^y = c^z$ and a, b, c are in GP, then x, y, z are in
(a) AP (b) GP
(c) HP (d) None of these

17. If p, q, r are in AP and x, y, z are in GP, then find the value of $x^{q-r} . y^{r-p} .z^{p-q}$
(a) 1 (b) 2
(c) –1 (d) None of these

18. If the ratio between HM and GM of two positive numbers is 12: 13, then the number are in the ratio
(a) 4:9 (b) 1:3
(c) 2:9 (d) 9:2

19. If $a^x = b^y = c^z = d^u$ where a, b, c, d are in GP, then x, y, z, u are in
(a) AP (b) GP
(c) HP (d) None of these

20. If $x = 1 + y + y^2 + \ldots\infty$, where $|y| < 1$, then y is
(a) $\frac{x}{x-1}$ (b) $\frac{x}{1-x}$
(c) $\frac{x-1}{x}$ (d) $\frac{1-x}{x}$

21. Find the sum of the series $S = 1^2 - 2^2 + 3^2 - 4^2 + \ldots - 2002^2 + 2003^2$ is
(a) 2007006 (b) 1005004
(c) 2000506 (d) None of these

22. If a, b, c are positive unequal numbers in HP, then find the minimum value of $a^5 + c^5$
(a) b^5 (b) $2b^5$
(c) $3b^5$ (d) None of these

23. Find the value of $(0.2)^{\log_{\sqrt{5}}\left(\frac{1}{4}+\frac{1}{8}+\frac{1}{16}+\ldots\infty\right)}$ is
(a) 1 (b) 2
(c) $\frac{1}{2}$ (d) None of these

24. The sum of the first three terms of an A. P. is 9 and sum of their squares is 35. The sum to n terms of the series is S_n then
(a) $S_{10} = 40$ (b) $S_5 = 15$
(c) $S_5 = 9$ (d) $S_{10} = -40$

25. The sum of the products of the 2n numbers $\pm1, \pm2, \pm3, \ldots, \pm n$ taking two at a time is
(a) $-\frac{n(n+1)}{2}$
(b) $\frac{n(n+1)(2n+1)}{6}$
(c) $-\frac{n(n+1)(2n+1)}{6}$
(d) None of these

26. The largest value of the positive integer k for which $n^k + 1$ divides $1 + n + n^2 + \ldots + n^{127}$ is
(a) 128 (b) 16
(c) 32 (d) 64

27. If $A = 1 + r^a + r^{2a} + r^{3a} + \ldots\infty$ and and $B = 1 + r^b + r^{2b} + r^{3b} + \ldots\infty$, then $\frac{a}{b}$ is equal to, here $|r| < 1$
(a) $\log_B A$ (b) $\log_{1-B}(1-A)$
(c) $\log_{\frac{B-1}{B}}\left(\frac{A-1}{A}\right)$ (d) None of these

28. Find the number of terms common to two APs 3, 7, 11, ..., 407 and 2, 9, 16, ..., 709
(a) 21 (b) 28
(c) 14 (d) None of these

29. If a, b, c are in AP. a, kb, c are in GP, then a, k^2b, c are in
(a) HP (b) GP
(c) AP (d) None of these

30. If a, b, c are in AP, then $7^a, 7^b, 7^c$ are in
(a) GP (b) HP
(c) AP (d) None of these

31. a, b, c are three distinct real numbers in a GP such that $a + b + c = xb$, then
(a) $x < -1$ or $x > 1$ (b) $x < -3$ or $x > 2$
(c) $x < -4$ or $x > 3$ (d) None of these

32. If three positive real numbers a, b, c are in AP such that abc = 4, then the minimum possible value of b is
(a) $2^{3/2}$ (b) $2^{2/3}$
(c) $2^{1/3}$ (d) $2^{5/2}$

33. If a, b, c are positive numbers in AP, then roots of the equation $ax^2 - 4bx + c = 0$ are
(a) Always real (b) always rational
(c) unreal (d) data insufficient

34. Along a road lies an odd number of stones placed at intervals of 10m. These stones have to be assembled around the middle stone. A person can carry only one stone at a time. A man carried out the job starting with the stone in the middle, carrying stones in succession, thereby covering a distance of 7.6 km. Then, the number of stones is
(a) 39 (b) 19
(c) 31 (d) 35

35. if for an AP $a_1, a_2, a_3, ..., a_n,...$
$a_1 + a_3 + a_5 = -12$ and $a_1a_2a_3 = 8$,
then find the value of $\frac{a_2 + a_4 + a_6}{a_2a_4a_6}$
(a) 3/13 (b) –3/13
(c) –1/3 (d) None of these

36. If a, b, c, d are non–zero real numbers such that $\left(a^2+b^2+c^2\right)\left(b^2+c^2+d^2\right) \le (ab+bc+cd)^2$,
then a, b, c, d are in
(a) AP (b) GP
(c) HP (d) None of these

37. Let $t_r = 2^{r/2} + 2^{-r/2}$. Then $\sum_{r=1}^{n} t_r^2$ is equal to
(a) $\frac{2^{2n+1}-1}{2^n} + 2n$ (b) $\frac{2^{2n+1}-1}{2^n} + 2n - 1$
(c) $\frac{2^{2n+1}-1}{2^n} - 1$ (d) None of these

38. If $a_n > 1$ for all $n \in N$, then find the minimum value of $\log_{a_2} a_1 + \log_{a_3} a_2 + \ldots + \log_{a_n} a_{n-1} + \log_{a_1} a_n$
(a) n (b) 2
(c) 0 (d) None of these

39. If x, y, z are three negative real numbers, then the value of $\frac{x}{y} + \frac{y}{z} + \frac{z}{x}$ lies in the interval
(a) $[2, +\infty)$ (b) $[3, +\infty)$
(c) $(3, +\infty)$ (d) $(-\infty, 3)$

40. Find the least value of $2 \log_{100} a - \log_a 0.0001$, if $a > 1$ is
(a) 2 (b) 3
(c) 4 (d) None of these

41. If $0 < x < \frac{\pi}{2}$, then the minimum value of $(\sin x + \cos x + \operatorname{cosec} 2x)^3$ is
(a) 27 (b) 13.5
(c) 6.75 (d) None of these

42. Let $t_n = n \cdot (n!)$. Then find the value of $\sum_{n=1}^{15} t_n$.
Here $n! = 1.2.3..n$
(a) 15! – 1 (b) 15! + 1
(c) 16! – 1 (d) None of these

43. Let x, y, z be three positive prime numbers. The progression in which $\sqrt{x}, \sqrt{y}, \sqrt{z}$ can be three terms (not necessarily consecutive) is
(a) AP (b) GP
(c) HP (d) None of these

44. If $4a^2 + 9b^2 + 16c^2 = 2(3ab + 6bc + 4ca)$, where a, b, c are non–zero numbers, then a, b, c are in
(a) AP (b) GP
(c) HP (d) None of these

45. If the sum of n terms of an AP is $3n^2 + 5n$, then find the ratio of 10th term to 7th term
(a) 4:3 (b) 11:8
(c) 13:10 (d) none of these

46. The first and last term of an AP are 1 and 15. If the sum of its terms is 64, then the number of terms will be
(a) 5 (b) 6
(c) 7 (d) 8

47. Find the sum of the series 0.4 + 0.004 + 0.00004 + …∞
(a) $\frac{11}{25}$ (b) $\frac{41}{100}$
(c) $\frac{40}{99}$ (d) $\frac{2}{5}$

48. If log(a + c), log(c – a), log(a –2b + c) are in AP, then
(a) a, b, c are in AP (b) a^2, b^2, c^2 are in AP
(c) a, b, c are in GP (d) a, b, c are in HP

49. The sum of an infinite geometric series is 3. A series which is formed by squares of its terms also have sum equal to 3. Then the series will be
(a) 1st term = $\frac{3}{2}$, common ratio = $\frac{1}{2}$
(b) 1st term = $\frac{3}{4}$, common ratio = $\frac{1}{2}$
(c) 1st term = $\frac{1}{4}$, common ratio = $-\frac{1}{2}$
(d) 1st term = $\frac{3}{2}$, common ratio = $\frac{1}{4}$

50. If $s_n = \sum_{n=1}^{n} \frac{1+2+2^2+\ldots \text{ to } n \text{ terms}}{2^n}$, then S_n is equal to
(a) $2^n-(n+1)$ (b) $1-\frac{1}{2^n}$
(c) $n-1+\frac{1}{2^n}$ (d) 2^n-1

51. Sum of all terms of an infinite GP is $\frac{1}{5}$ times the sum of odd terms. The common ratio is
(a) 2 (b) 3
(c) $-\frac{4}{5}$ (d) 5

52. The three terms, $\log_3 2, \log_6 2, \log_{12} 2$ are in
(a) AP (b) GP
(c) HP (d) None of these

53. If a, b, c are in HP, then which one of the following is correct.
(a) $a^2+c^2>b^2$ (b) $a^2+c^2>2b^2$
(c) $a^2+c^2<2b^2$ (d) $a^2+c^2=2b^2$

54. the sum of first four terms of GP is $12(1-\sqrt{5})$. If the common ratio is $-\sqrt{5}$, then the first term of the GP is,
(a) 1 (b) 2
(c) 3 (d) 4

55. The sum to infinity of the series
$S=1+\frac{2}{3}+\frac{6}{3^2}+\frac{10}{3^3}+\frac{14}{3^4}+\ldots$ is
(a) 4 (b) 6
(c) 2 (d) 3

56. The sum to infinity of the series
$S=\frac{4}{19}+\frac{44}{19^2}+\frac{444}{19^3}+\ldots$ is
(a) 38/81 (b) 4/19
(c) 36/171 (d) none of these

57. The sum to infinity of the series
$1+\frac{4}{5}+\frac{7}{5^2}+\frac{10}{5^3}+\ldots$ is
(a) $\frac{16}{35}$ (b) $\frac{11}{8}$
(c) $\frac{35}{16}$ (d) $\frac{17}{6}$

58. The interior angles of a polygon are in A. P. if the smallest angle is 100° and the common difference is 4°. Then the number of sides is– [we know that sum of the interior angles of a polygon of n sides is $(n-2)(180°)$]
(a) 5 (b) 7
(c) 36 (d) 4

59. If a and x are positive integers such that $x < a$ and $\sqrt{a-x}, \sqrt{x}, \sqrt{a+x}$ are in A. P., then least possible value of a is
(a) 5 (b) 7
(c) 11 (d) none of these

60. If S_r denotes the sum of the first r terms of an AP, and the value of $\frac{S_{3r}-S_{r-1}}{S_{2r}-S_{2r-1}} = \text{pr}+q$ then find the value of $p+q$
(a) –1 (b) 1
(c) 3 (d) None of these

Concept Cracker (CC)

1. Find the sum of 1st 20 terms of the series $\sqrt{2}+\sqrt{8}+\sqrt{18}+\sqrt{32}+\ldots$
(a) $210\sqrt{2}$ (b) $\frac{210}{\sqrt{2}}$
(c) $420\sqrt{2}$ (d) None of these

2. The sum of 100 terms of the series, $\frac{1}{\sqrt{1}+\sqrt{3}}+\frac{1}{\sqrt{3}+\sqrt{5}}+\frac{1}{\sqrt{5}+\sqrt{7}}+\ldots$
(a) $\frac{\sqrt{201}}{2}$ (b) $\frac{-1+\sqrt{201}}{2}$
(c) $\frac{1}{\sqrt{201}}$ (d) $\sqrt{199}$

3. If $a_1, a_2, \ldots, a_{24}$ are in AP and $a_1+a_5+a_{10}+a_{15}+a_{20}+a_{24}=999$, then find the sum of 1st 24 terms of AP
(a) 2372 (b) 3996
(c) 7212 (d) None of these

4. Let f (n) = $\sum_{r=1}^{n} r^4$ then find the value of $\sum_{r=1}^{n}(2r-1)^4$
(a) f(2n)– 16f(n) (b) f(2n)– 7f(n)
(c) f(2n –1)– 8 f(n) (d) None of these

5. If $\frac{a+be^y}{a-be^y} = \frac{b+ce^y}{b-ce^y} = \frac{c+de^y}{c-de^y}$, then a, b, c, d are in
(a) AP (b) GP
(c) HP (d) none of these

6. If $\frac{1}{2}\text{cosec}^2\theta$, $2\cot\theta$, $\sec\theta$, $0 < \theta < \frac{\pi}{2}$ are in GP, if θ is equal to
(a) $\frac{\pi}{6}$ (b) $\frac{\pi}{4}$
(c) $\frac{\pi}{3}$ (d) $\frac{\pi}{2}$

7. If the mth, nth and pth terms of an AP and GP be equal and be respectively x, y, z then which one of the following is correct?
(a) $x^{y-z}y^{z-x}z^{x-y} = 1$
(b) $(x-y)^x(y-z)^y = (z-x)^z$
(c) $(x-y)^z(y-z)^x = (z-x)^y$
(d) None of the above

8. What will be quadratic equation in x when the roots have arithmetic mean A and the geometric mean G?
(a) $x^2 + 2Ax + G^2 = 0$ (b) $x^2 + G^2x + A = 0$
(c) $x^2 - 2Ax + G^2 = 0$ (d) $x^2 + Gx + A = 0$

9. If x_1, x_2, x_3 as well as y_1, y_2, y_3 are in GP with same common ratio, then the points $(x_1, y_1)(x_2, y_2)$ and (x_3, y_3)
(a) lie on a straight line
(b) are vertices of a triangle
(c) lie on a circle
(d) None of these

10. If the roots of the equation $a(b-c)x^2 + b(c-a)x + c(a-b) = 0$ are equal, then a, b, c are in
(a) AP (b) GP
(c) HP (d) None of these

11. Three non-zero numbers a, b and c are in AP. Increasing a by 1 or increasing c by 2, the numbers are in GP, then find the sum of the three numbers.
(a) 32 (b) 34
(c) 12 (d) None of these

12. If a, b, c are in AP and a^2, b^2, c^2 are in HP, then
(a) $a = b = c$ (b) $2b = 3a + c$
(c) $b^2 = \sqrt{ac/8}$ (d) None of these

13. If a, b, c, d are positive real numbers, such that $a + b + c + d = 18$, then find the maximum value of $(a + b)(c + d)$
(a) 27 (b) 81
(c) 51 (d) None of these

14. If each term of a GP is positive and each term is the sum of its two succeeding terms, then the common ratio of the GP is,
(a) $\frac{\sqrt{5}-1}{2}$ (b) $\frac{\sqrt{5}+1}{2}$
(c) $-\frac{(\sqrt{5}+1)}{2}$ (d) $\frac{1-\sqrt{5}}{2}$

15. If one geometric mean G and two arithmetic means p and q be inserted between two number, then which one of the following is correct?
(a) $G^2 = (3p-q)(3q-p)$
(b) $G^2 = (2p-q)(2q-p)$
(c) $G^2 = (4p-q)(4q-p)$
(d) None of theses

16. The AM between two numbers b and c is 'a' and the two GMs between them are g_1 and g_2 then find the value of $\frac{g_1^3 + g_2^3}{abc}$
(a) 1 (b) 4
(c) 3 (d) None of these

17. If $3 + \frac{1}{4}(3 + d) + \frac{1}{4^2}(3 + 2d) + \ldots$ upto $\infty = 8$, then find the value of d
(a) 9 (b) 5
(c) 1 (d) None of these

18. The harmonic mean of 2 numbers is 4 . Their arithmetic mean is A and Geometric mean is G. If $2A + G^2 = 27$, then sum of the two numbers is
(a) 9 (b) 12 (c) 6 (d) 8

19. Find the value of n for which $704 + \frac{1}{2}(704) + \frac{1}{4}(704) + \ldots$ Upto n terms $= 1984 - \frac{1}{2}(1984) + \frac{1}{4}(1984)$ ... Upto n
(a) 5 (b) 3 (c) 4 (d) 10

37. (c) $\because$ x, y, z are the pth, q^{th} and r^{th} terms of an AP.

$\therefore x = A + (p-1)D; y = A + (q-1)D;$
$z = A + (r-1)D$

$\Rightarrow x - y = (p-q)D;\ y - z = (q-r)D$
$z - x = (r-p)D$ ....(1)

where A is the first term and D is the common difference.

Also x, y, z are the pth, q^{th}, and r^{th} terms of a GP.

$\therefore x = AR^{p-1}, y = AR^{q-1}, z = AR^{r-1}$

$\therefore x^{y-z}\,y^{z-x}\,z^{x-y} = (AR^{p-1})^{y-z}(AR^{q-1})^{z-x}(AR^{r-1})^{x-y}$

$= A^{y-z+z-x+x-y}R^{(p-1)(y-z)+(q-1)(z-x)+(r-1)(x-y)}$

$= A^0 r^{(p-1)(q-r)D+(q-1)(r-p)D+(r-1)(p-q)D}$

$= A^0 R^0 = 1$

38. (b) $ar^2 = 4$ ....(1)

$a.\,ar.\,ar^2.ar^3.ar^4 = a^5r^{10} = (ar^2)^5 = 4^5.$

39. (c) $2.\overline{357} = 2 + .357 + 0.000357 + ...\infty$

$= 2 + \frac{357}{10^3} + \frac{357}{10^6} + ...\infty$

$= 2 + \frac{\frac{357}{10^3}}{1 - \frac{1}{10^3}} = 2 + \frac{357}{999} = \frac{2355}{999}$

40. (a) a, b, c are in G.P.

$b^2 = ac$ ....(1)

$ax^2 + 2bx + c = 0$

and $dx^2 + 2ex + f = 0$ have a common root

Let it be α, then $a\alpha^2 + 2b\alpha + c = 0$

$d\alpha^2 + 2e\alpha + f = 0$

$\Rightarrow \frac{\alpha^2}{2(bf-ec)} = \frac{\alpha}{cd-af} = \frac{1}{2(ae-bd)}$

$\Rightarrow \alpha^2 = \frac{bf-ce}{ae-bd};\ \alpha = \frac{cd-af}{2(ae-bd)}$

Substituting the value of α, we get

$\frac{(cd-af)^2}{4(ae-bd)^2} = \frac{bf-ce}{ae-bd}$

$\Rightarrow (cd-af)^2 = 4(ae-bd)(bf-ce)$

Dividing both sides by a^2c^2 we get

$\left(\frac{d}{a} - \frac{f}{c}\right)^2 = 4\left(\frac{e}{c} - \frac{bd}{ac}\right)\left(\frac{bf}{ac} - \frac{e}{a}\right)$

$\left(\frac{d}{a} - \frac{f}{c}\right)^2 = 4\left(\frac{e}{c} - \frac{d}{b}\right)\left(\frac{f}{b} - \frac{e}{a}\right)$

[Using eq. (1)]

$\Rightarrow \frac{d^2}{a^2} + \frac{f^2}{c^2} - \frac{2df}{ac} = \frac{4ef}{cb} - \frac{4e^2}{ac} - \frac{4df}{b^2} + \frac{4de}{ab}$

$\Rightarrow \frac{d^2}{a^2} + \frac{f^2}{c^2} + \frac{4e^2}{b^2} + 2\frac{d}{a}.\frac{f}{c} - 4\frac{e}{b}.\frac{f}{c} - 4.\frac{d}{a}.\frac{e}{b} = 0$

[Using eq.(1)]

$\Rightarrow \left(\frac{d}{a} + \frac{f}{c} - 2\frac{e}{b}\right)^2 = 0 \Rightarrow \frac{d}{a} + \frac{f}{c} = \frac{2e}{b}$

$\Rightarrow \frac{d}{a}, \frac{e}{b}, \frac{f}{c}$ are in A.P.

41. (c) Let $S = \frac{1}{2} + \frac{3}{4} + \frac{7}{8} + \frac{15}{16} + ...n\ terms$

$= \left(1 - \frac{1}{2}\right) + \left(1 - \frac{1}{4}\right) + \left(1 - \frac{1}{8}\right) + \left(1 - \frac{1}{16}\right)$....n terms

$= (1 + 1 + 1 + ...n$ terms$)$
$-\left(\frac{1}{2} + \frac{1}{2^2} + \frac{1}{2^3} + \frac{1}{2^4} + ... + \frac{1}{2^n}\right)$

$= n - \left[\frac{\frac{1}{2}\left(1 - \frac{1}{2^n}\right)}{1 - \frac{1}{2}}\right] = n - 1 + 2^{-n}$

42. (c) We know that $\log_2 4 = 2$ and $\log_2 8 = 3$

$\therefore \log_2 7$ lies between 2 and 3

$\therefore \log_2 7$ is either rational or irrational but not integer or prime number.

If possible let $\log_2 7 = \frac{p}{q}$ (a rational number)

$\Rightarrow 2^{p/q} = 7 \Rightarrow 2^p = 7^q$

$\Rightarrow$ even number = odd number

$\therefore$ We get a contradiction, so assumption is wrong.

Hence $\log_2 7$ must be an irrational number.

43. (d) $\ln(a+c), \ln(a-c), \ln(a-2b+c)$ are in A.P.

$\Rightarrow a + c, a - c, a - 2b + c$ are in G.P.

$\Rightarrow (c-a)^2 = (a+c)(a-2b+c)$

$\Rightarrow \quad (c-a)^2 = (a+c)^2 - 2b(a+c)$

$\Rightarrow \quad 2b(a+c) = (a+c)^2 - (c-a)^2$

$\Rightarrow \quad 2b(a+c) = 4ac \Rightarrow b = \dfrac{2ac}{a+c}$

$\Rightarrow \quad a, b, c$ are in H.P.

44. (d) $a_1 = h_1 = 2, a_{10} = h_{10} = 3$

$3 = a_{10} = 2 + 9d \quad \Rightarrow \quad d = 1/9$

$\therefore \quad a_4 = 2 + 3d = 7/3$

$3 = h_{10} \Rightarrow \dfrac{1}{3} = \dfrac{1}{h_{10}} = \dfrac{1}{2} + 9D$

$\therefore \ D = -\dfrac{1}{54}$

$\dfrac{1}{h_7} = \dfrac{1}{2} + 6D = \dfrac{1}{2} - \dfrac{1}{9} = \dfrac{7}{18}$

$\therefore \ a_4 h_7 = \dfrac{7}{3} \times \dfrac{18}{7} = 6.$

45. (b) $\dfrac{1}{H} = \dfrac{1}{2}\left(\dfrac{1}{\alpha} + \dfrac{1}{\beta}\right) = \dfrac{\alpha+\beta}{2\alpha\beta}$

$= \dfrac{(4+\sqrt{5})}{(5+\sqrt{2})} \cdot \dfrac{(5+\sqrt{2})}{(8+2\sqrt{5})} \cdot \dfrac{1}{2} = \dfrac{1}{4}$

$\therefore \quad H = 4.$

46. (d) Sum = 4 and second term = 3/4, it is given that first term is a and common ratio r

$\Rightarrow \quad \dfrac{a}{1-r} = 4$ and $ar = 3/4 \Rightarrow r = \dfrac{3}{4a}$

Therefore, $\dfrac{a}{1-\dfrac{3}{4a}} = 4 \Rightarrow \dfrac{4a^2}{4a-3} = 4$

or $\quad a^2 - 4a + 3 = 0 \Rightarrow (a-1)(a-3) = 0$

$\Rightarrow \quad a = 1$ or 3

When $a = 1, r = 3/4$ and when $a = 3, r = 1/4$

47. (a) α, β are the roots of $x^2 - x + p = 0$

$\therefore \quad \alpha + \beta = 1 \quad$ ....(1)

$\alpha\beta = p \quad$ ...(2)

γ, δ are the roots of $x^2 - 4x + q = 0$

$\therefore \quad \gamma + \delta = 4 \quad$ ....(3)

$\gamma\delta = q \quad$ ....(4)

$\alpha, \beta, \gamma, \delta$ are in G.P.

$\therefore$ Let $\alpha = a;\ \beta = ar,\ \gamma = ar^2,\ \delta = ar^3.$

Substituting these values in equations (1), (2), (3) and (4), we get

$a + ar = 1 \quad$ ....(5)

$a^2 r = p \quad$ ....(6)

$ar^2 + ar^3 = 4 \quad$ ....(7)

$a^2 r^5 = q \quad$ ....(8)

Dividing (7) by (5) we get

$\dfrac{ar^2(1+r)}{a(1+r)} = \dfrac{4}{1} \Rightarrow r^2 = 4 \Rightarrow r = 2, -2$

(5)

$\Rightarrow a = \dfrac{1}{1+r} = \dfrac{1}{1+2}$ or $\dfrac{1}{1-2} = \dfrac{1}{3}$ or -1

As p is an integer (given), r is also an integer (2 or – 2).

$\therefore$ (6)

$\Rightarrow a \neq \dfrac{1}{3}$. Hence $a = -1$ and $r = -2$.

$\therefore \quad p = (-1)^2 \times (-2) = -2$

$q = (-1)^2 \times (-2)^5 = -32$

48. (d) a, b, c, d are in A.P.

$\therefore \quad d, c, b, a$ are also in A.P.

$\Rightarrow \quad \dfrac{d}{abcd}, \dfrac{c}{abcd}, \dfrac{b}{abcd}, \dfrac{a}{abcd}$ are also in A.P.

$\Rightarrow \quad \dfrac{1}{abc}, \dfrac{1}{abd}, \dfrac{1}{acd}, \dfrac{1}{bcd}$ are in A.P.

$\Rightarrow \quad abc, abd, acd, bcd$ are in H.P.

49. (c) ATQ $2 + 5 + 8 + ... 2n$ terms $= 57 + 59 + 61 + ...\ n$ terms

$\Rightarrow \quad \dfrac{2n}{2}[4 + (2n-1)3] = \dfrac{n}{2}[114 + (n-1)2]$

$\Rightarrow \quad 6n + 1 = n + 56 \Rightarrow 5n = 55 \Rightarrow n = 11$

50. (d) Given that a, b, c are in A.P.

$\Rightarrow \quad 2b = a + c$

$\Rightarrow \quad$ but given $a + b + c = 3/2 \Rightarrow 3b = 3/2$

$\Rightarrow \quad b = 1/2$ and then $a + c = 1$

Again a^2, b^2, c^2, are in G.P. $\Rightarrow b^4 = a^2 c^2$

$\Rightarrow \quad b^2 = \pm ac \Rightarrow ac = \dfrac{1}{4}$ or $-\dfrac{1}{4}$

and $\quad a + c = 1 \quad$ ....(1)

Considering $a + c = 1$ and $ac = 1/4$

$\Rightarrow \quad (a-c)^2 = 1 - 1 = 0 \Rightarrow a = c$ but $a \neq c$ as given that $a < b < c$

$\therefore$ We consider $a + c = 1$ and $ac = -1/4$

$\Rightarrow \quad (a-c)^2 = 1+1=2 \Rightarrow a-c = \pm\sqrt{2}$

but $a < c \Rightarrow a-c = -\sqrt{2}$ ....(2)

Solving (1) and (2) we get $a = \frac{1}{2} - \frac{1}{\sqrt{2}}$

51. (c) $\frac{x}{1-r} = 5 \Rightarrow r = 1 - \frac{x}{5}$

Since G.P. contains infinite terms

$\therefore \quad -1 < r < 1$

$\Rightarrow \quad -1 < 1 - \frac{x}{5} < 1 \Rightarrow -2 < -\frac{x}{5} < 0$

$\Rightarrow \quad -10 < x < 0. \Rightarrow 0 < \frac{x}{5} < 2$

$\Rightarrow \quad 0 < x < 10.$

52. (c) In the quadratic equation $ax^2 + bx + c = 0$

$\Delta = b^2 - 4ac$ and $\alpha + \beta = -\frac{b}{a}, \alpha\beta = \frac{c}{a}$

$\alpha^2 + \beta^2 = (\alpha+\beta)^2 - 2\alpha\beta$

$= \frac{b^2}{a^2} - \frac{2c}{a} = \frac{b^2 - 2ac}{a^2}$

and $\alpha^3 + \beta^3 = -\frac{b^3}{a^3} - \frac{3c}{a}\left(-\frac{b}{a}\right)$

$= -\left(\frac{b^3 - 3abc}{a^3}\right)$

Given $\quad \alpha+\beta, \alpha^2+\beta^2, \alpha^3+\beta^3$ are in G.P.

$\Rightarrow$

$-\frac{b}{a}, -\frac{b^2-2ac}{a^2}, -\frac{(b^3-3\,abc)}{a^3}$ are in G.P.

$\Rightarrow \left(\frac{b^2-2\,ac}{a^2}\right)^2 = \frac{b}{a}\left(\frac{b^3-3\,abc}{a^3}\right)$

$\Rightarrow \quad b^4 + 4a^2c^2 - 4ab^2c = b^4 - 3ab^2c$

$\Rightarrow \quad 4a^2c^2 - ab^2c = 0 \Rightarrow ac\,\Delta = 0$

$\Rightarrow \quad c\,\Delta = 0 \quad (\because$ In quadratic $a \neq 0)$

53. (c) Given that for an A.P, $S_n = cn^2$

Then $T_n = S_n - S_{n-1} = cn^2 - c(n-1)^2$

$= (2n-1)c$

$\therefore$ Sum of squares of n terms of this A.P

$= \sum T_n^2 == \sum (2n-1)^2.c^2$

$= c^2\left[4\sum n^2 - 4\sum n + n\right]$

$= c^2\left[\frac{4n(n+1)(2n+1)}{6} - \frac{4n(n+1)}{2} + n\right]$

$= c^2 n\left[\frac{2(2n^2+3n+1) - 6(n+1) + 3}{3}\right]$

$= c^2 n\left[\frac{4n^2-1}{3}\right] = \frac{n(4n^2-1)c^2}{3}$

54. (d) $\because a_1, a_2, a_3, ...$ are in H.P.

$\therefore \frac{1}{a_1}, \frac{1}{a_2}, \frac{1}{a_3}$ .... are in A.P.

$\therefore \quad \frac{1}{a_1} = \frac{1}{5}$ and $\frac{1}{a_{20}} = \frac{1}{25}$

$\frac{1}{a_1} + 19d = \frac{1}{a_{20}}$

$\Rightarrow \frac{1}{5} + 19d = \frac{1}{25} \Rightarrow d = \frac{-4}{475}$

Now $\frac{1}{a_n} = \frac{1}{5} + (n-1)\left(\frac{-4}{475}\right)$

Clearly $a_n < 0$ if $\frac{1}{a_n} < 0$

$\Rightarrow \frac{1}{5} - \frac{4n}{475} + \frac{4}{475} < 0$

$\Rightarrow -4n < -99$ or $n > \frac{99}{4} = 24\frac{3}{4}$

$\therefore n \geq 25$

Hence least value of n is 25.

55. (b) $\log_c b_1, \log_c b_2, ----, \log_c b_{101}$ are in A.P.

$\Rightarrow \quad b_1, b_2, ---- b_{101}$ are in G.P.

Also $a_1, a_2, ---- a_{101}$ are in A.P.

where $a_1 = b_1$ are $a_{51} = b_{51}$.

$\therefore \quad b_2, b_3, ---, b_{50}$ and GM's and $a_2, a_3, ----, a_{50}$ are AM's between b_1 and b_{51}.

$\because$ GM < AM $\Rightarrow b_2 < a_2, b_3 < a_3, ----, b_{50} < a_{50}$

$\because \quad b_1 + b_2 + ---- + b_{51} < a_1 + a_2 + ---- + a_{51}$

$\Rightarrow \quad t < s$

Also a_1, a_{51}, a_{101} are in AP

b_1, b_{51}, b_{101} are in GP

$\because \quad a_1 = b_1$ and $a_{51} = b_{51}$

$\therefore \quad b_{101} > a_{101}$

Passage Type

56. (b, c) Consider for AM,

$$A_1 + A_2 + A_3 + \ldots + A_{m-1} + A_m = 1025 \times 171$$

$$\therefore m\left(\frac{-2+1027}{2}\right) = 1025 \times 171$$

$$\therefore m = 342$$

Now for GM,

$$G_1G_2 \ldots G_n = \left(\sqrt{1 \times 1024}\right)^n = 2^{5n}$$

$$\therefore 2^{5n} = 2^{45}$$

$$\therefore n = 9$$

57. (b, d) Common difference of sequence

$A_1, A_2, \ldots, A_m$ is $\dfrac{1027+2}{342+1} = 3$

Similarly common ratio is 2

58. (a, c) Since, a, b, c, d are in GP (say with common ratio r), ln a, ln b, ln c, ln d, are in AP with common difference ln r. Now, $a^x = b^y = c^z = d^v$

$$\Rightarrow x\ln a = z\ln c = v\ln d = y\ln b$$

$$\Rightarrow x\ln a = y(\ln a + \ln r) = z(\ln a + 2\ln r)$$

$$= v(\ln a + 3\ln r)$$

$$\Rightarrow \frac{x-z}{z} = 2\frac{x-y}{y} \text{ and } \frac{y-v}{v} = 2\frac{y-z}{z}$$

$$\Rightarrow \frac{1}{x}+\frac{1}{z} = \frac{2}{y} \text{ and } \frac{1}{y}+\frac{1}{v} = \frac{2}{z}$$

$\Rightarrow x, y, x, v$ are in HP

For options c and d

Since, p, q, r are in AP

b^p, b^q, b^r are in GP,

$\Rightarrow b^{p-1}, b^{q-1}, b^{r-1}$ are in GP

$\Rightarrow ab^{p-1}, ab^{q-1}, ab^{r-1}$ are in GP

$\Rightarrow$ The pth, the qth and the rth terms of any GP are in GP.

59. (a, c) Since, x, y, z are terms of a GP (say with common ratio t). ln x, ln y, ln z are in AP with common difference ln t. Also, x, y, z are in AP (say with common difference d).

So, $x - y = (p - q)$d etc. And ln x – ln $y = (p - q)$ ln t, and so on,

Let us assume that the required product as $P = x^{y-z}\, y^{z-x}\, z^{x-y}$ then taking log both the side

$$\ln P = (y-z)\ln x + (z-x)\ln y + (x-y)\ln z$$

$$= (q-r)d\ln x + (r-p)d\ln y + (p-q)d\ln z$$

$$= d[p(\ln z - \ln y) + q(\ln x - \ln z) + r(\ln y - \ln x)$$

$$= d\ln t\left[p(r-q) + q(p-r) + r(q-p)\right] = 0$$

$$\Rightarrow P = 1$$

60. (a, b, c) $\dfrac{a+b+c}{3} > b\ (HM)$

$$\Rightarrow a + c > 2b$$

Also, $\dfrac{a^2+c^2}{2} > \sqrt{a^2c^2}$

$$\Rightarrow a^2 + c^2 > 2ac > 2b^2 (GM > HM)$$

$$\Rightarrow a^2 + c^2 > 2b^2$$

61. (a) b is the HM of a and c and their GM $= \sqrt{ac}$

GM > HM $\Rightarrow \sqrt{ac} > b$ and AM of a^n and c^n > GM of a^n and c^n

$$\frac{a^n + c^n}{2} > \sqrt{a^nc^n}$$

$$c^n + c^n > 2\left(\sqrt{ac}\right)^n > 2b^n$$

62. (a, b, c, d) Given that a + b + c = n

$a^ab^bc^c$ = (aa…a times) (bb…b times) (cc…c times)

Applying AM–GM inequality

$$(a^ab^bc^c)^{1/n} \le \frac{(a+a+\ldots+a\text{ times}) + (b+b+\ldots+b\text{ times}) + (c+c+\ldots+c\text{ times})}{n}$$

$$= \frac{a^2+b^2+c^2}{n}$$

Similarly,

$$(a^bb^cc^a)^{1/n} \le \frac{ab+bc+ca}{n} \le \frac{a^2+b^2+c^2}{n}$$

$$\text{Also, } (a^cb^ac^b)^{1/n} \le \frac{ac+ba+cb}{n} \le \frac{a^2+b^2+c^2}{n}$$

So, $(a^ab^bc^c)^{1/n} + (a^bb^cc^a)^{1/n} + (a^cb^ac^b)^{1/n}$

$$\le \frac{(a+b+c)^2}{n} = n$$

Further, $a < n \Rightarrow a^a < a^n \ldots$

$\therefore \quad (a^a b^b c^c) < (abc)^n \Rightarrow (a^a b^b c^c)^{1/n} < (abc)$

63. (a, c, d)

64. (b) Here a, b, c, are in H.P.

$\Rightarrow a^{-1}, b^{-1}, c^{-1}$ are in A.P.

$\Rightarrow e^{(-a)^{-1}}, e^{(-b)^{-1}}, e^{(-c)^{-1}}$ are in G.P.

65. (a) Since x, y, z are respectively the pth, q^{th} and the rth terms of a G.P so $\ell n\, x$, $\ell n\, y$, $\ell n\, z$ are in A.P. with common difference ℓnt. Here t is the common ratio

Also, x, y, z are in A.P.(say with common difference d.)

Hence, $x - y = (p-q)d$ etc.

and $\ell n\, x - \ell n\, y = (p-q)\ell n\, t$.

Let $S = (x^{y-z}), (y^{z-x}), (z^{x-y})$

so that $\ell n\, S = (y-z)\ell n\, x + (z-x)\ell n\, y + (x-y)\ell n\, z$

$= (q-r)d\ell n\, x + (r-p)d\ell n\, y + (p-q)d\ell n\, z$

$= d[p(\ell n\, z - \ell n\, y) + q(\ell n\, x - \ell n\, z) + r(\ell n\, y - \ell n\, x)]$

$= d\ell nt[p(r-q) + q(p-r) + r(q-p)] = 0$

$\Rightarrow S = 1$

66. (c) Since $t_n = S_n - S_{n-1} = a.2^{n-1}, n > 1$

$$\therefore \sum_{r=2}^{\infty} \frac{1}{t_r} = \frac{1}{a} \sum_{r=2}^{\infty} \frac{1}{2^{r-1}} = \frac{1}{a} \frac{\frac{1}{2}}{1-\frac{1}{2}} = \frac{1}{a}$$

67. (a) Given that $S_n = \frac{1}{4} n(n+1)(n+2)(n+3)$

$\therefore t_n = S_n - S_{n-1}$

$$= \frac{1}{4} n(n+1)(n+2)(n+3) - \frac{1}{4}(n-1)\, n(n+1)(n+2)$$

$$= \frac{1}{4} n(n+1)(n+2)\{n+3-n+1\}$$

$$= n(n+1)(n+2)$$

68. (b) $V_1 + V_2 + \ldots + V_n = \sum_{r=1}^{n} V_r = \sum_{r=1}^{n}\left(r^3 - \frac{r^2}{2} + \frac{r}{2}\right)$

$$= \Sigma n^3 - \frac{\Sigma n^2}{2} + \frac{\Sigma n}{2}$$

$$= \frac{n^2(n+1)^2}{4} - \frac{n(n+1)(2n+1)}{12} + \frac{n(n+1)}{4}$$

$$= \frac{n(n+1)}{4}\left[n(n+1) - \frac{2n+1}{3} + 1\right]$$

$$= \frac{n(n+1)(3n^2+n+2)}{12}$$

69. (d) $T_r = V_{r+1} - V_r - 2$

$$= \left[(r+1)^3 - \frac{(r+1)^2}{2} + \frac{r+1}{2}\right] - \left[r^3 - \frac{r^2}{2} + \frac{r}{2}\right] - 2$$

$$= 3r^2 + 2r + 1$$

$T_r = (r+1)(3r-1)$

For each r, T_r has two different factors other than 1 and itself.

$\therefore \quad T_r$ is always a composite number.

70. (b) $\therefore Q_{r+1} - Q_r = T_{r+2} - T_{r+1} - (T_{r+1} - T_r)$

$$= T_{r+2} - 2T_{r+1} + T_r$$

$$= (r+3)(3r+5) - 2(r+2)(3r+2) + (r+1)(3r-1)$$

$\because Q_{r+1} - Q_r = 6(r+1) + 5 - 6r - 5 = 6$ (constant)

$\therefore Q_1, Q_2, Q_3, \ldots$ are in AP with common difference 6.

71. (c) Given $A_1 = \frac{a+b}{2}, G_1 = \sqrt{ab}, H_1 = \frac{2ab}{a+b}$

also $A_n = \frac{A_{n-1} + H_{n-1}}{2}, G_n = \sqrt{A_{n-1}H_{n-1}}$

$$H_n = \frac{2A_{n-1}H_{n-1}}{A_{n-1} + H_{n-1}}$$

$\Rightarrow G_n^2 = A_n H_n \Rightarrow A_n H_n = A_{n-1} H_{n-1}$

Similarly we can prove

$A_n H_n = A_{n-1} H_{n-1} = A_{n-2} H_{n-2} = \ldots = A_1 H_1$

$\Rightarrow A_n H_n = ab$

$\therefore G_1^2 = G_2^2 = G_3^2 \ldots = ab$

$\rightarrow G_1 = G_2 = G_3 \ldots = \sqrt{ab}$

72. (a) We have $A_n = \frac{A_{n-1} + H_{n-1}}{2}$

$\therefore A_n - A_{n-1} = \frac{A_{n-1} + H_{n-1}}{2} - A_{n-1}$

$= \frac{H_{n-1} - A_{n-1}}{2} < 0 \quad (\because A_{n-1} > H_{n-1})$

$\Rightarrow A_n < A_{n-1}$ or $A_{n-1} > A_n$

$\therefore$ We can conclude that $A_1 > A_2 > A_3 > \ldots$

73. (b) We have $A_n H_n = ab \Rightarrow H_n = \frac{ab}{A_n}$

$\because \quad \frac{1}{A_{n-1}} < \frac{1}{A_n} \Rightarrow H_{n-1} < H_n$

$\therefore \quad H_1 < H_2 < H_3 < ....$

Integer Type

74. (8) Given, length of a side of S_n = Length of a diagonal of S_{n+1}

$\Rightarrow$ For a square $\frac{\text{Length of a side}}{\text{length of diagonal}} = \frac{1}{\sqrt{2}}$

$\Rightarrow$ Length of a side of $S_n = \sqrt{2}$ (Length of a side of S_{n+1})

$$\Rightarrow \frac{\text{Length of a side of } S_{n+1}}{\text{length of a side of } S_n} = \frac{1}{\sqrt{2}} \text{ for all } n \geq 1$$

$\Rightarrow$ Side of $S_1, S_2, \ldots, S_n$ form a GP with common ratio $\frac{1}{\sqrt{2}}$ and first term 10.

$$\therefore \text{Side of } S_n = 10\left(\frac{1}{\sqrt{2}}\right)^{n-1} = \frac{10}{2^{(n-1)/2}}$$

$$\Rightarrow \text{Area of } S_n = (S_n)^2 = \frac{100}{2^{n-1}}$$

As per the given condition Area of S_n<1 (given)

$$\frac{100}{2^{n-1}} < 1$$

$$\Rightarrow \quad 2^{n-1} > 100$$

$$\Rightarrow \quad n - 1 \geq 7 => n \geq 8$$

75. (1) $k = (32)(32)^{1/6}(32)^{1/36} \ldots \infty$

$$= 32^{1+\frac{1}{6}+\frac{1}{36}+\ldots\infty} = 32^{\frac{1}{1-\frac{1}{6}}}$$

$$= 32^{\frac{6}{5}} = (2^5)^{\frac{6}{5}} = 2^6 = 64$$

When 64 divided by 9, remainder is 1

76. (6) Given that $a_{10} = 3 \quad \Rightarrow a_1 + 9d = 3$

$$\Rightarrow \quad 2 + 9d = 3 \; [a_1 = 2]$$

$$\Rightarrow \quad d = \frac{1}{9}$$

$$\therefore \quad a_4 = a_1 + 3d = 2 + \frac{3}{9} = \frac{7}{3}$$

$$h_{10} = 3$$

$$\Rightarrow \quad \frac{1}{h_{10}} = \frac{1}{3}$$

$\Rightarrow$ Common difference of corresponding AP is $D = -\frac{1}{54}$

$$\therefore \frac{1}{h_7} = \frac{1}{h_1} + 6D = \frac{1}{2} - \frac{1}{9} = \frac{7}{18}$$

$$\Rightarrow \quad h_7 = \frac{18}{7}$$

so, $\quad a_4 h_7 = \frac{7}{3} \times \frac{18}{7} = 6$

77. (9) $2^{n+10} = 2.2^2 + 3.2^3 + 4.2^4 + .. + n.2^n$

$\Rightarrow \quad 2(2^{n+10})$

$= 2.2^3 + 3.2^4 + .. + (n-1).2^n + n.2^{n+1}$

From above two results, we get

$$-2^{n+10} = 2.2^2 + 2^3 + 2^4 + .. + 2^n - n.2^{n+1}$$

$$= 8 + \frac{8(2^{n-2}-1)}{2-1} - n.2^{n+1}$$

$$= 8 + 2^{n+1} - 8 - n.2^{n+1}$$

$$= 2^{n+1} - (n)\, 2^{n+1}$$

$$\Rightarrow \quad 2^{10} = 2n - 2$$

$\Rightarrow \quad n = 513$ sum of the digits is $5 + 1 + 3 = 9$

78. (4) The given equation can be written as

$$\log_2 x^{1+\frac{1}{2}+\frac{1}{4}+\frac{1}{8}+\frac{1}{16}+\ldots}$$

$$\Rightarrow \quad \log_2(x^2) = 4$$

$$\Rightarrow \quad x^2 = 2^4$$

$$\Rightarrow \quad x = 4$$

79. (3) Using, $AM \geq GM$

$$\therefore \left|\frac{a^{\log b - \log c} + b^{\log c - \log a} + c^{\log a - \log b}}{3}\right|$$

$$\geq \sqrt[3]{a^{\log b - \log c} \cdot b^{\log c - \log a} \cdot c^{\log a - \log b}} \geq \sqrt[3]{x} \ldots \text{(i)}$$

Where $x = a^{\log b - \log c} \cdot b^{\log c - \log a} \cdot c^{\log a - \log b}$

$\Rightarrow \log x = (\log b - \log c)\log a$

$+(\log c - \log a)\log b + (\log a - \log b)\log c = 0$

$\Rightarrow \quad x = 1$

$\therefore$ From Eq. (i),

$\left|a^{\log b - \log c} + b^{\log c - \log a} + c^{\log a - \log b}\right| \geq 3$

80. (9) Let d be the common difference of the AP, then

$a_{2r} = a_{2r-1} + d$

$\Rightarrow \sum_{r=1}^{10^{99}} a_{2r} = \sum_{r=1}^{10^{99}} (a_{2r-1} + d)$

$\Rightarrow \sum_{r=1}^{10^{99}} a_{2r} = \sum_{r=1}^{10^{99}} a_{2r-1} + 10^{99} d$

$\Rightarrow 10^{100} = 10^{99} + 10^{99} d$

$\Rightarrow d = \frac{10^{100} - 10^{99}}{10^{99}} = \frac{10^{99}(10-1)}{10^{99}} = 9$

81. (4) Since, $\sum_{n=1}^{n} n, \frac{\sqrt{10}}{3} \cdot \sum_{n=1}^{n} n^2$ & $\sum_{n=1}^{n} n^3$ are in GP

$\Rightarrow \frac{n(n+1)}{2}, \frac{\sqrt{10}}{3} \cdot \frac{n(n+1)(2n+1)}{6}, \frac{n^2(n+1)^2}{4}$

are in GP

$\Rightarrow \frac{10 \cdot n^2(n+1)^2(2n+1)^2}{9 \cdot 36} = \frac{n(n+1)}{2} \cdot \frac{n^2(n+1)^2}{4}$

$\Rightarrow 20(2n+1)^2 = 81n(n+1)$

$\Rightarrow n^2 + n - 20 = 0$ or $n = 4$

82. (4) Given $b^2 = ac, x = \frac{a+b}{2}, y = \frac{b+c}{2}$

Now, $\frac{a}{x} + \frac{c}{y} = \frac{2a}{a+b} + \frac{2c}{b+c}$

$= \frac{2(ab + ac + ac + bc)}{ab + ac + b^2 + bc} = 2 \quad [\because b^2 = ac]$

Again $\frac{b}{x} + \frac{b}{y} = 2b\left[\frac{1}{a+b} + \frac{1}{b+c}\right]$

$= \frac{2b(b + c + a + b)}{ab + ac + b^2 + bc} = 2$

$\therefore \left(\frac{a}{x} + \frac{c}{y}\right)\left(\frac{b}{x} + \frac{b}{y}\right) = 4.$

83. (5) $(1+x)(1+x^2)(1+x^4)\ldots(1+x^{128})$

$= 1 + x + x^2 + \ldots + x^n$

$\Rightarrow (1-x)\{(1+x)(1+x^2)(1+x^4)\ldots(1+x^{128})\}$

$= 1 - x^{n+1}$

$\Rightarrow (1-x^2)\{(1+x^2)\ldots(1+x^{128})\} = 1 - x^{n+1}$

$\Rightarrow (1-x^4)(1+x^4)\ldots(1+x^{128}) = 1 - x^{n+1}$

$\Rightarrow 1 - x^{256} = 1 - x^{n+1}$

$\therefore n + 1 = 256$ or $n = 255$

84. (5) $S_k = 1 + \frac{1}{k+1} + \frac{1}{(k+1)^2} + \ldots\infty = \frac{1}{1 - \frac{1}{k+1}}$

$= \frac{k+1}{k}$

$\therefore k\,S_k = k + 1$

$\therefore \sum_{k=1}^{n} kS_k = \sum_{k=1}^{n} (k+1) = 2 + 3 + \ldots + (n+1)$

$= \frac{n}{2}(2 + (n+1)) = \frac{n(n+3)}{2} = \frac{n^2}{2} + \frac{3n}{2}$

So a = ½, b = 3/2 and c = 0

Then 4a + 2b + 3c = 2 + 3 = 5

85. (5) Let the GP be $a, ar, ar^2, \ldots (0 < r < 1)$

From the question, $\frac{a}{1-r} = 3^3 + 3 \cdot 3 - 9$ ... (i)

{since $f'(x) = 3x^2 + 3 > 0$.

So, $f(x)$ is monotonically increasing.

$\therefore f(3)$ is the greatest value in $[-2, 3]$

Also, $f'(0) = 3$. So, $a - ar = 3$ ... (ii)

Solving Eqs. (i) and (ii), i.e., $a = 27(1-r)$ and

$a(1-r) = 3$, we get $r = \frac{2}{3}, \frac{4}{3}$. but $r < 1$

So $p + q = 2 + 3 = 5$

86. (5)

$S = (1+3^{-1})(1+3^{-2})(1+3^{-4})(1+3^{-8})\ldots(1+3^{-2n})$

$\Rightarrow (1-3^{-1})S = (1-3^{-1})(1+3^{-1})(1+3^{-2})$

$(1+3^{-4})(1+3^{-8})\ldots(1+3^{-2n})$

$$\Rightarrow \frac{2}{3}S = \left(1-3^{-2}\right)\left(1+3^{-2}\right)\left(1+3^{-4}\right)\left(1+3^{-8}\right)\ldots\left(1+3^{-2n}\right)$$

$$= \left(1-3^{-4}\right)\left(1+3^{-4}\right)\left(1+3^{-8}\right)\ldots\left(1+3^{-2n}\right)$$

$$= \left(1-3^{-8}\right)\left(1+3^{-8}\right)\ldots\left(1+3^{-2n}\right)$$

$$= \left(1-3^{-2n}\right)\left(1+3^{-2n}\right) = 1-\left(3^{-2n}\right)^2 = 1-3^{-(2n+1)}$$

$$\Rightarrow S = \frac{3}{2}\left(1-3^{-(2n+1)}\right)$$

$$\therefore \lim_{n\to\infty} S = \lim_{n\to\infty}\frac{3}{2}\left(1-3^{-(2n+1)}\right) = \frac{3}{2}(1-0) = \frac{3}{2}$$

$$= \frac{p}{q} \text{ so } p+q=5$$

87. (7) $\frac{1}{a_1}, \frac{1}{a_2}, \frac{1}{a_3}, \ldots$ are in A. P. Let d be the common difference of this A. P., then

$$\frac{1}{a_{20}} - \frac{1}{a_1} = 19d$$

$$\Rightarrow \frac{1}{25} - \frac{1}{5} = 19d \Rightarrow d = -\frac{4}{475}$$

Now, $a_n < 0 \Rightarrow \frac{1}{5} - \frac{(n-1)4}{475} < 0$

$$\Rightarrow n-1 > \frac{95}{4} \Rightarrow n \geq 25.$$

Thus, least value of n is 25.

88. (4) We have $t_r = \frac{(r-1)!}{(r+4)!}$

And $t_{r+1} = \frac{r!}{(r+5)!}$

Now, $rt_r - (r+5)t_{r+1} = \frac{r!}{(r+4)!} - \frac{r!}{(r+4)!} = 0$

$$\Rightarrow \quad rt_r - (r+1)t_{r+1} = 4t_{r+1}$$

$$\Rightarrow \quad 4\sum_{r=1}^{n-1} t_{r+1} = \sum_{r=1}^{n-1}\left[rt_r - (r+1)t_{r+1}\right]$$

$$\Rightarrow 4\left(t_2 + t_3 + \ldots + t_n\right) = 1t_1 - nt_n$$

$$\Rightarrow 4\left(t_1 + t_2 + \ldots + t_n\right) = 5t_1 - nt_n$$

$$= 5\left(\frac{0!}{5!}\right) - \frac{n(n-1)!}{(n+4)!} = \frac{1}{4!} - \frac{n!}{(n+4)!}$$

$$\Rightarrow t_1 + t_2 + \ldots + t_n = \frac{1}{4}\left[\frac{1}{4!} - \frac{n!}{(n+4)!}\right]$$

So a = 4, b = 4, c = 0 and d = 4 and a + b–c–d = 4

89. (3) We have $t_r = \frac{2r-1}{r(r+1)(r+2)}$ and

$$t_{r+1} = \frac{2r+1}{(r+1)(r+2)(r+3)}$$

$$\Rightarrow \quad rt_r = \frac{2r-1}{(r+1)(r+2)},$$

$$(r+3)t_{r+1} = \frac{2r+1}{(r+1)(r+2)}$$

$$\Rightarrow \quad rt_r - (r+3)t_{r+1} = \frac{-2}{(r+1)(r+2)}$$

$$\Rightarrow \quad rt_r - (r+1)t_{r+1} + 2\left(\frac{1}{r+1} - \frac{1}{r+2}\right) = 2t_{r+1}$$

$$\Rightarrow \quad \sum_{r=1}^{n-1}\left[rt_r - (r+1)t_{r+1}\right] + 2\sum_{r=1}^{n-1}\left(\frac{1}{r+1} - \frac{1}{r+2}\right) = 2\sum_{r=1}^{n-1} t_{r+1}$$

$$\Rightarrow \quad t_1 - nt_n + 1 - \frac{2}{n+1} = 2\left(t_2 + t_3 + \ldots + t_n\right)$$

$$\Rightarrow \quad 3t_1 - nt_n + 1 - \frac{2}{n+1} = 2\sum_{r=1}^{n} t_r$$

$$\frac{1}{2} - \frac{n(2n-1)}{n(n+1)(n+2)} + \frac{n-1}{n+1} = 2\sum_{r=1}^{n} t_r$$

$$\therefore 2\sum_{r=1}^{n} t_r = \frac{(n+1)(n+2) - 2(2n-1) + 2(n-1)(n+2)}{2(n+1)(n+2)}$$

$$\Rightarrow \quad \sum_{r=1}^{n} t_r = \frac{n(3n+1)}{4(n+1)(n+2)}$$

So a = 3, b = 1, c = 1, d = 1, e = 1, f = 2 so a + b + c + d–e–f = 3

90. (2) $\frac{s}{10\Delta^2}$ $[(s-a)^2 + (s-a) + 1]$ $[(s-b)^2 + (s-b) + 1]$ $[(s-c)^2 + (s-c) + 1]$

$$= \frac{1}{10}\frac{\left[(s-a)^2+(s-a)+1\right]}{(s-a)}\frac{\left[(s-b)^2+(s-b)+1\right]}{(s-b)}$$

$$\frac{\left[(s-c)^2+(s-c)+1\right]}{(s-c)}$$

$$= ((s-a)+\frac{1}{s-a}+1)\ ((s-b)+\frac{1}{s-b}+1)$$

$$((s-c)+\frac{1}{s-c}+1),$$

Now consider $((s-a)+\frac{1}{s-a}+1) \geq 3$

$((s-b)+\frac{1}{s-b}+1) \geq 3$

$((s-c)+\frac{1}{s-c}+1) \geq 3$

$((s-a)+\frac{1}{s-a}+1)\ ((s-b)+\frac{1}{s-b}+1)$

$((s-c)+\frac{1}{s-c}+1) \geq 3 \times 3 \times 3 = 27$

The required minimum value is

$27/10 = 2.7 = k$

Then $[k] = 2$

Equality holds if $a = b = c = 2$

91. (6) $c > 4b - 3a \Rightarrow ar^2 > 4ar - 3a$

$\Rightarrow r^2 - 4r + 3 > 0 \Rightarrow r < 1$ or $r > 3$. But the terms are positive so $r \in (0,1) \cup (3, \infty)$

So the values that r can assume $\{4,5,6,7,8,9\}$ or 6 values

92. (1) $\tan^{-1}(\frac{1}{2r†}) = \tan^{-1}(\frac{2}{4r†})$

$$= \tan^{-1}\frac{(2r+1)-(2r-1)}{1+(2r+1)(2r-1)}$$

$= \tan^{-1}(2r+1) - \tan^{-1}(2r-1)$

$\therefore \sum_{r=1}^{n} tan^{-1}(\frac{1}{2r†}) = \tan^{-1}(2n+1) - \tan^{-1}(1)$

$= \tan^{-1}(2n+1) - \frac{\pi}{4}$

$\therefore \tan\{\sum_{r=1}^{n} tan^{-1}(\frac{1}{2r†})\}$

$$= \frac{(2n+1)-1}{1+(2n+1).1} = \frac{n}{n+1}$$

$$\therefore \lim_{n\to\infty} tan\ [\sum_{r=1}^{n} tan^{-1}(\frac{1}{2r†})] = \lim_{n\to\infty}\frac{n}{n+1} = 1.$$

93. (1) $\sum_{k=1}^{n}\left(\sum_{m=1}^{k} m^2\right) = \sum_{k=1}^{n}\left(1^2+2^2+3^2+\ldots\ldots+k^2\right)$

$$\sum_{k=1}^{n}\frac{k(k+1)(2k+1)}{6} = \frac{1}{6}\sum_{k=1}^{n}\left(2k^3+3k^2+k\right)$$

$$= \frac{1}{3}\left\{\frac{n(n+1)}{2}\right\}^2 + \frac{1}{2}\left\{\frac{n(n+1)(2n+1)}{6}\right\} + \frac{1}{6}\left\{\frac{n(n+1)}{2}\right\}$$

$$= \frac{1}{12}\left\{n^4+4n^3+5n^2+2n\right\}$$

$\therefore a = \frac{1}{12}, b = \frac{1}{3}, c = \frac{5}{12}, d = \frac{1}{6}$ and $e = 0$.

So, $a+b+c+d+e = 1$.

94. (3) Since, a_1, a_2, a_3, a_4 are in H.P.,

$\frac{1}{a_1}, \frac{1}{a_2}, \frac{1}{a_3}, \frac{1}{a_4}$ are in A.P.

Let common difference of this AP is d then

$\frac{1}{a_2} - \frac{1}{a_1} = d \Rightarrow a_1 - a_2 = da_1a_2$

Similarly, $a_2 - a_3 = da_2a_3$

Add $a_3 - a_4 = da_3a_4$

$a_1 - a_4 = d\left[a_1a_2 + a_2a_3 + a_3a_4\right]$

$\Rightarrow a_1a_2 + a_2a_3 + a_3a_4 = \frac{a_1 - a_4}{d}$

also $\frac{1}{a_4} = \frac{1}{a_1} + 3d \Rightarrow \frac{a_1 - a_4}{d} = 3a_1a_4$

$\Rightarrow a_1a_2 + a_2a_3 + a_3a_4 = 3a_1a_4$ of

$$\frac{1}{a_1a_4}\sum_{r=1}^{3} a_ra_{r+1} = 3$$

95. (3) Using $S_\infty = \frac{a}{1-r}$, we get

$$S_k = \begin{cases} \dfrac{\frac{k-1}{k!}}{1-\frac{1}{k!}}, & k \neq 1 \\ 0, & k=1 \\ \dfrac{1}{(k-1)!}, & k \geq 2 \end{cases}$$

Now

$$\sum_{k=1}^{100} |(k^2-3k+1)S_k| = \sum_{k=2}^{100} |(k^2-3k+1)| \frac{1}{(k-1)!}$$

$$= |-1| + \sum_{k=3}^{100} \frac{(k^2-1)+1-3(k-1)-2}{(k-1)!} \text{ as } k^2-3k+1>0 \forall k \geq 3$$

$$= 1+\sum_{k=3}^{100}\left(\frac{1}{(k-3)!}-\frac{1}{(k-1)!}\right)$$

$$= 1+\left(1-\frac{1}{2!}\right)+\left(\frac{1}{1!}-\frac{1}{3!}\right)+\left(\frac{1}{2!}-\frac{1}{4!}\right)+....$$

$$....+\left(\frac{1}{96!}-\frac{1}{98!}\right)+\left(\frac{1}{97!}-\frac{1}{99!}\right)$$

$$= 3-\frac{1}{98!}-\frac{1}{99!} = 3-\frac{9900}{100!}-\frac{100}{100!}$$

$$= 3-\frac{10000}{100!} = 3-\frac{(100)^2}{100!}$$

$$\therefore \quad \frac{100^2}{100!}+3\sum_{k=1}^{100}\left|\left(k^2-3k+1\right)S_k\right| = 3.$$

96. (0) Given that $a_k = 2a_{k-1} - a_{k-2}$

$$\Rightarrow \frac{a_{k-2}+a_k}{2} = a_{k-1}, 3 \leq k \leq 11$$

$\Rightarrow a_1, a_2, a_3, ..., a_{11}$ are in AP.

If a is the first term and D the commen difference then

$$a_1^2+a_2^2+...+a_{11}^2 = 990$$

$$\Rightarrow 11a^2+d^2\left(1^2+2^2+...+10^2\right)+2ad\left(1+2+...+10\right)$$

$$= 990$$

$$\Rightarrow 11a^2+\frac{10\times 11\times 21}{6}d^2+2ad\times\frac{10\times 11}{2} = 990$$

$\Rightarrow a^2 + 35d^2 + 150d = 90$

Using $a = 15$, we get

$35d^2 + 150d + 135 = 0$ or $7d^2 + 30d + 27 = 0$

$\Rightarrow (d+3)(7d+9) = 0 \Rightarrow d = -3$ or $-9/7$

then $a_2 = 15 - 3 = 12$ or $15-\frac{9}{7}=\frac{96}{7}>\frac{27}{2}$

$\therefore d \neq -9/7$

Hence

$$\frac{a_1+a_2+...+a_{11}}{11} = \frac{\frac{11}{2}\left[2\times 15+10(-3)\right]}{11} = 0$$

97. (9) We have $\dfrac{S_m}{S_n} = \dfrac{S_{5n}}{S_n} = \dfrac{\frac{5n}{2}[2\times 3+(5n-1)d]}{\frac{n}{2}[6+(n-1)d]}$

$$= \frac{5[(6-d)+5nd]}{(6-d)+nd]}$$

which will be independent of n if $d = 6$ or $d = 0$

For a proper A.P. we take $d = 6$

then $a_2 = 3 + 6 = 9$

98. (5) Let k, $k + 1$ be removed from pack.

$\therefore (1 + 2 + 3 + ... + n) - (k + k + 1) = 1224$

$$\frac{n(n+1)}{2} - 2k = 1225$$

$$k = \frac{n(n+1)-2450}{4}$$

for $n = 50$, $k = 25$

$\therefore \quad k - 20 = 5$

99. (4) $\because$ a, b, c are in G.P

$\therefore b = ar$ and $c = ar^2$

Also $\dfrac{b}{a}$ is an integer

$\Rightarrow r$ is an integer

$\because$ A.M. of a, b, c is $b + 2$

$$\Rightarrow \frac{a+b+c}{3} = b+2$$

$\Rightarrow \quad a + ar + ar^2 = 3ar + 6$

$\Rightarrow \quad a(r^2 - 2r + 1) = 6$

$\Rightarrow \quad a(r - 1)^2 = 6$

$\because$ a and r are integers

$\therefore$ The only possible values of a and r can be 6 and 2 respectively.

Then $\dfrac{a^2+a-14}{a+1} = \dfrac{36+6-14}{6+1} = \dfrac{28}{7} = 4$

100. (9) $\dfrac{\frac{7}{2}[2a+6d]}{\frac{11}{2}[2a+10d]} = \dfrac{6}{11} \Rightarrow a = 9d$

$a_7 = a + 6d = 15d$

$\because$ $130 < 15d < 140 \Rightarrow d = 9$

($\because$ All terms are natural numbers $\therefore$ $d \in N$)

101. (8) In expansion of $(1 + x)(1 + x^2)(1 + x^3)$ .... $(1 + x^{100})$

x^9 can be found in the following ways

$x^9, x^{1+8}, x^{2+7}, x^{3+6}, x^{4+5}, x^{1+2+6}, x^{1+3+5}, x^{2+3+4}$

The coefficient of x^9 in each of the above 8 cases is 1.

$\therefore$ Required coefficient = 8.

Assertion & Reason

102. (a) Let $T_{k+1} = ar^k$ and $T_{k+1} = br^k$

Since, $T_{k+1} = ar^k + br^k = (a+b)r^k$

$\therefore T_{k+1}$ is a general term of a GP

103. (c) Consider the given expression

$$3(a^2 + b^2 + c^2 + 1) - 2(a + b + c + ab + bc + ca) = 0$$

$$\Rightarrow (a-1)^2 + (b-1)^2 + (c-1)^2 + (a-b)^2 + (b-c)^2 + (c-a)^2 = 0$$

$\Rightarrow a = b = c = 1$

Statement I is correct and statement II is incorrect

104. (a) We can use $AM \ge GM$ to have

$$\frac{\frac{3a}{3} + 2b + \frac{c}{1/3}}{6} \ge \left(\frac{a^3}{27} \cdot b^2 \cdot 3c\right)^{1/6} \left(\frac{1}{6}\right)^6 \ge \frac{a^3b^2c}{9}$$

or $a^3b^2c < \dfrac{9}{6^6}$

or $a^3b^2c \le \dfrac{1}{5184}$,

Statement I is true and Statement II is true and correct explanation of statement I

105. (a) Statement I a, A, b are in AP

$\Rightarrow 2A = a + b$

a, p, q, b are in GP $\Rightarrow$ $pq = ab$

and let common ratio of GP be r

$\therefore b = ar^3$

$$\Rightarrow r = \left(\frac{b}{a}\right)^{\frac{1}{3}}$$

$$p = ar \Rightarrow p = a\left(\frac{b}{a}\right)^{\frac{1}{3}} \Rightarrow p^3 = a^2b$$

$$q = ar^2 \Rightarrow q = a\left(\frac{b}{a}\right)^{\frac{2}{3}} \Rightarrow q^3 = ab^2$$

From above relations we get Statement I is true, and statement II is correct explanation

106. (d) Statement I is obviously false

Statement II, if we solve the equations given in statement II

$$\sqrt{3} - \sqrt{5} = (p-q)d, \quad \sqrt{5} - \sqrt{7} = (q-r)d$$

On dividing, we get

$$\frac{\sqrt{3}-\sqrt{5}}{\sqrt{5}-\sqrt{7}} = \frac{p-q}{q-r}$$

$\Rightarrow$ Rational = Irrational which is not possible, so there exist no solution

$\Rightarrow$ Statement I is false and Statement II is true

107. (d) Statement II is a fact.

Now consider statement I

$a > 0$, $ar > 0$, $ar^2 > 0$

We have two cases here

Case (i) If $r > 1$ then ar^2 is the largest side

So $ar^2 > ar + a$

$r^2 - r - 1 > 0$,

or $r \in \left(-\infty, \dfrac{1-\sqrt{5}}{2}\right) U \left(\dfrac{1+\sqrt{5}}{2}, \infty\right)$,

but $r > 1$ so $r \in \left(\dfrac{1+\sqrt{5}}{2}, \infty\right)$

Case (ii) If $r < 1$ then a is the largest side

So $a > ar + ar^2$

$r^2 + r - 1 < 0$,

or $r \in \left(\dfrac{-1-\sqrt{5}}{2}, \dfrac{-1+\sqrt{5}}{2}\right)$, but $0 < r < 1$

so $r \in \left(0, \dfrac{-1+\sqrt{5}}{2}\right)$

108. (d) From first two equations we get $x = -2$

$\Rightarrow \beta = -m$ and $\gamma = m$

3rd equation, $\frac{1}{x+p} + \frac{1}{x+q} = \frac{1}{r}$

$\Rightarrow x^2 + (p+q-2r)x + pq - r(p+q) = 0$

$\therefore \quad \beta + \gamma = p + q - 2r = 0$

$\Rightarrow \quad p, r, q,$ are in A.P.

109. (b) Statement II is correct as $a^2 + b^2 + c^2 - ab - bc - ca \geq 0$

Now consider statement I

$f(x) = (a_1x - a_2)^2 + (a_2x - a_3)^2 + \ldots + (a_{n-1}x - a_n)^2$

Thus, $f(x) \leq 0$

$\Rightarrow a_1x - a_2 = a_2x - a_3 = \ldots = a_{n-1}{}^{x-a}{}_n = 0$

$\Rightarrow x = \frac{a_2}{a_1} = \frac{a_3}{a_2} = \ldots = \frac{an}{an_{-1}}$

$\Rightarrow a_1, a_2, \ldots, a_n$ are in G.P.

110. (c) Let us consider the GP 1st, let a be the first term and r be the common ratio of the G.P. As four numbers are distinct and positive $a > 0, r > 0$ and $r \neq 1$

$b_1 = a, b_2 = a(1+r), b_3 = a(1+r+r^2)$

and $b_4 = a(1+r+r^2+r^3)$

As $b_2 - b_1 = ar \neq ar^2 = b_3 - b_2$

$\Rightarrow b_1, b_2, b_3, b_4$ are not in A.P.

$\frac{b_2}{b_1} = 1 + r \neq \frac{1+r+r^2}{1+r} = \frac{b_3}{b_2}$

$\Rightarrow b_1, b_2, b_3, b_4$ are not in G.P.

As $\frac{1}{b_2} - \frac{1}{b_1} = \frac{-r}{a(1+r)} \neq \frac{-r^2}{a(1+r)(1+r+r^2)}$

$= \frac{1}{b_3} - \frac{1}{b_2}$

b_1, b_2, b_3, b_4 cannot be in H.P.

111. (c) 1st consider statement II

$\frac{1}{(1)(3)} + \frac{1}{(3)(5)} + \ldots + \frac{1}{(2n-1)(2n+1)}$

Here

$t_r = \frac{1}{(2r-1)(2r+1)} = \frac{1}{2}\left[\frac{1}{2r-1} - \frac{1}{2r+1}\right]$

So $\frac{1}{(1)(3)} + \frac{1}{(3)(5)} + \ldots + \frac{1}{(2n-1)(2n+1)}$

$= \frac{1}{2}\left[\left\{\frac{1}{1} - \frac{1}{3}\right\} + \left\{\frac{1}{3} - \frac{1}{5}\right\} + \ldots + \left\{\frac{1}{2n-1} - \frac{1}{2n+1}\right\}\right]$

$= \frac{n}{2n+1}$

So statement II is false

Now consider Statement I

$t_r = \frac{r^2}{(2r-1)(2r+1)}$

$\Rightarrow 4t_r = \frac{4r^2 - 1 + 1}{(2r-1)(2r+1)}$

$= 1 + \frac{1}{(2r-1)(2r+1)}$

$= 1 + \frac{1}{2}\left(\frac{1}{2r-1} - \frac{1}{2r+1}\right)$

$\Rightarrow 4\sum_{r=1}^{n} t_r = n + \frac{1}{2}\sum_{r=1}^{n}\left(\frac{1}{2r-1} - \frac{1}{2r+1}\right)$

$= n + \frac{1}{2}\left(1 - \frac{1}{2n+1}\right)$

$= n + \frac{n}{2n+1} = \frac{2n(n+1)}{2n+1}$

$\Rightarrow \sum_{r=1}^{n} t_r = \frac{n(n+1)}{2(2n+1)}$

$\therefore$ statement I is true.

112. (d) We know that If $0 < y < 1$, then sum of the AGP series $1 + 3y + 5y^2 + 7y^3 + \ldots$ is $\frac{1+y}{(1-y)^2}$

So statement II is true

Putting $y = 1 - \frac{1}{x}$, we get

$S = \frac{1 + 1 - 1/x}{(1/x)^2} = 2x^2 - x.$

So statement I false and Statement II is true

113. (c) Given a_1, a_2, a_3, a_4 are in GP.
Then b_1, b_2, b_3, b_4 are the numbers
$a_1, a_1 + a_2, a_1 + a_2 + a_3,\ a_1 + a_2 + a_3 + a_4$

or a, a + ar, $a + ar + ar^2$, $a + ar + ar^2 + ar^3$
Clearly above numbers are neither in AP nor in G.P. and hence statement 1 is true.

Also $\frac{1}{a}, \frac{1}{a+ar}, \frac{1}{a+ar+ar^2}, \frac{1}{a+ar+ar^2+ar^3}$

are not in A.P. b_1, b_2, b_3, b_4 are not in H.P.

∴ Statement 2 is false.

Matrix Match

114. Let number in set A be $a - D,\ a,\ a + D$ and in set B be $b - d,\ b, b + d$

$3a = 3b = 15 \Rightarrow a = b = 5$

Set $A = \{5 - D, 5, 5 + D\}$

Set $B = \{5 - d, 5, 5 + d\}$ where D = d + 1

$$\frac{p}{q} = \frac{5(25 - D^2)}{5(25 - d^2)} = \frac{7}{8}$$

$\Rightarrow d = -17, 1$ but $d > 0 \Rightarrow d = 1$ and so D = 2

So, numbers in set A are 3, 5, 7

Now, $p = 3 \times 5 \times 7 = 105$

$q = 4 \times 5 \times 6 = 120$

Value of $D + d = 3$

(A) → (*s*), (B) → (*p*), (C) → (*q*), (D) → (*r*)

115. (A) Let the 3 consecutive terms are a – d, a, a + d, d > 0

Hence, $a^2 - 2ad + d^2 = 36 + N$ ...(i)

$a^2 = 300 + N$ ...(ii)

$a^2 + 2ad + d^2 = 596 + N$ ...(iii)

Now subtracting Eq(i) from Eq (ii), gives

$d(2a - d) = 264$ ...(iv)

On subtracting Eq(ii) from Eq(iii), gives

$d(2a + d) = 296$ ...(v)

On subtracting Eq(iv) from Eq(v), gives

$2d^2 = 32$

$\Rightarrow\quad d^2 = 16$

$\Rightarrow\quad d = 4$ (d = –4 rejected)

Hence, from Eq (iv),

$4(2a - 4) = 264$

$\Rightarrow\quad 2a - 4 = 66$

$\Rightarrow\quad 2a = 70$

$\Rightarrow\quad a = 35$

$\therefore\quad N = 35^2 - 300$

$= 1225 - 300 = 925$

So A→ q

(B) Since, $a_1, a_2 \ldots, a_n \ldots$, are in AP. And $b_1, b_2 \ldots, b_n \ldots$, are in GP

$\Rightarrow 1, (1 + d), \ldots, (1 + (n - 1)d), \ldots$ in AP

$\Rightarrow 1, r, r^2, \ldots, r^{n-1}, \ldots$ In GP

Also given, $a_9 = b_9$

$\Rightarrow\quad 1 + 8d = r^8$

$\therefore S_9 = \frac{9}{2}[2 + (9 - 1)d] = 369$

$\Rightarrow\quad 2 + 8d = 41 \times 2$

$\Rightarrow\quad d = 10$

$\Rightarrow\quad r^8 = 81$

$\Rightarrow\quad b_9 = 81$

$\Rightarrow\quad 9b_9 = 729$

B→r

(C) Sum of all two digit numbers $= \frac{90}{2}(10 + 99)$

Sum of all two digit numbers divisible by 2 is

$$\frac{45}{2}(10 + 98) = (45)(54)$$

Sum of all two digit numbers divisible by 3 is

$$\frac{30}{2}(12 + 99) = 15(111)$$

Sum of all two digit numbers divisible by 6 is

$$\frac{15}{2}(12 + 96) = 15(54)$$

The required sum is

$$45(109) + 15(54) - (45)(54) - 15(111) = 1620$$

C→p

(D) $S = 1 + 5 + 13 + 29 + \ldots + t_{10}$

$S = 1 + 5 + 13 + \ldots + t_9 + t_{10}$

Subtracting $t_{10} = 1 + 4 + 8 + 16 + \ldots$ upto 10 terms

$= 1 + (4 + 8 + 16 + \ldots \text{ upto 9 terms})$

$= 2045$

D→s

A→q, B→r, C→p, D→s

116. (A) $a_1 + 2d + a_1 + 4d + a_1 + 10d + a_1 + 16d + a_1 + 18d$

$= 5a_1 + 50d$

$= 5(a_1 + 10d) = 10$ i.e., $a_1 + 10d = 2$

Now,

$$\sum_{i=1}^{21} a_i = \frac{21}{2}[2a_1 + 20d] = 21(a_1 + 10d) = 42$$

(A) $\to (s)$,

(B) $S = \frac{5}{1^2 \cdot 4^2} + \frac{11}{4^2 \cdot 7^2} + \frac{17}{7^2 \cdot 10^2} + \ldots$

$$\Rightarrow 3S = \frac{3 \cdot 5}{1^2 \cdot 4^2} + \frac{3 \cdot 11}{4^2 \cdot 7^2} + \frac{3 \cdot 17}{7^2 \cdot 10^2} + \ldots$$

$$\Rightarrow 3S = \frac{(4-1)\cdot(4+1)}{1^2 \cdot 4^2} + \frac{(7-4)(7+4)}{4^2 \cdot 7^2} + \frac{(10-7)(10+7)}{7^2 \cdot 10^2} + \ldots$$

$$\Rightarrow 3S = \frac{4^2 - 1^2}{1^2 \cdot 4^2} + \frac{7^2 - 4^2}{4^2 \cdot 7^2} + \frac{10^2 - 7^2}{7^2 \cdot 10^2} + \ldots$$

$$\Rightarrow 3S = 1 - \frac{1}{4^2} + \frac{1}{4^2} - \frac{1}{7^2} + \frac{1}{7^2} - \frac{1}{10^2} + \ldots$$

$\Rightarrow 3S = 1 \Rightarrow S = \frac{1}{3}$ so 99S = 33

(B) $\to (q)$,

(C) $T_n = \frac{n^2}{500 + 3n^3}$

Let $U_n = \frac{1}{T_n} = \frac{500}{n^2} + 3n$

For the condition of maximum value of T_n

$$\frac{dT_n}{dn} = \frac{\left(500 + 3n^2\right) \cdot 2n - n^2 9n^2}{\left(500 + 3n^2\right)^2}$$

$$= \frac{n\left(1000 - 3n^3\right)}{\left(500 + 3n^2\right)^2}$$

$$= 0 \Rightarrow n = \left(\frac{1000}{3}\right)^{1/3}$$

Now, $6 < \left(\frac{1000}{3}\right)^{1/3} < 7$

Hence, T_7 is the largest term, so largest term in the given sequence is $\frac{49}{1529}$

(C) $\to (p)$,

(D) $F(n+1) = \frac{2f(n)+1}{2} = F(n) + \frac{1}{2}$

$\therefore F(1), F(2), F(3), \ldots$ is an AP with common difference $\frac{1}{2}$

$F(101) = 2 + 50 = 52$

D→r

(A) → (s), (B) → (q), (C) → (p), (D) → (r)

117. (a) $a + b = 12$

$ab + \frac{6ab}{a+b} = 48$

$ab + \frac{ab}{2} = 48$

$\therefore ab = 32$

$(A) \to (q)$

(B) As $n = 39$ is odd, the value of the given expression $= 1^3 - 2^3 + 3^3 - \ldots + n^3$

$$= \left(1^3 + 2^3 + 3^3 + \ldots + n^3\right) - 2\left\{2^3 + 4^3 + \ldots + (n-1)^3\right\}$$

$$= \left\{\frac{n(n+1)}{2}\right\}^2 - 16\left\{1^3 + 2^3 + \ldots + \left(\frac{n-1}{2}\right)^3\right\}$$

$$= \frac{n^2(n+1)^2}{4} - 16 \cdot \left\{\frac{\frac{n-1}{2} \cdot \frac{n+1}{2}}{2}\right\}^2$$

$$= \frac{(n+1)^2 \cdot (2n-1)}{4}$$

On putting the value we get $\frac{S_{39}}{100} = \frac{30800}{100} = 308$

$(B) \rightarrow (p)$

(C) HM of $\frac{1}{2}, \frac{1}{3}, \frac{1}{4}, \frac{1}{5}$ is

$$\frac{4}{\frac{1}{2}+\frac{1}{3}+\frac{1}{4}+\frac{1}{5}} = \frac{240}{77}$$

$(C) \rightarrow (r)$

(D) The numbers between 100 and 500 that are divisible by 7 are 105, 112, 119, 126, 133, 140, 147, ..., 483, 490, 497.

Let such numbers be n.

Then, $497 = 105 + (n-1) \times 7$ or $n = 57$

So there are 57 number of numbers lying between 100 and 500 that are divisible by 7

The number between 100 and 500 that are divisible by 21 are 105, 126, 147, ..., 483.

Let such number be m.

Then, $483 = 105 + (m-1) \times 21$ or $n = 19$

So there are 19 number of numbers lying between 100 and 500 that are divisible by 21

Hence, the required number = $n - m = 57 - 19 = 38$

$(D) \rightarrow (s)$

118. **(A) → (s), (B) → (p), (C) → (q), (D) → (r)**

(A) Given, $\sum n = \frac{1}{5}\left(\sum n^2\right)$

or $\frac{n(n+1)}{2} = \frac{1}{5}\frac{n(n+1)(2n+1)}{6}$

$\Rightarrow \quad 2n + 1 = 15$

$\Rightarrow \quad 2n = 14$

$\Rightarrow \quad n = 7$

(A) → (s)

(B) Let α and β be the roots of the given equation,

$$\alpha + \beta = \frac{4+\sqrt{3}}{5+\sqrt{2}} \text{ and } \alpha\beta = \frac{8+2\sqrt{3}}{5+\sqrt{2}}$$

Hence, required harmonic mean

$$= \frac{2\alpha\beta}{\alpha+\beta}$$

$$= \frac{2\left(\frac{8+2\sqrt{3}}{5+\sqrt{2}}\right)}{\frac{4+\sqrt{3}}{5+\sqrt{2}}} = 4$$

(B) → (p),

(C) x, y, z are in HP.

$$y = \frac{2xz}{x+z}$$

$$\Rightarrow \quad x - 2y + z = x + z - \frac{4xz}{x+z}$$

$$= \frac{(x+z)^2 - 4xz}{x+z} = \frac{(z-x)^2}{x+z}$$

$\Rightarrow \quad (x+z)(x-2y+z) = (z-x)^2$

$\Rightarrow \quad \log(x+z) + \log(x-2y+z) = 2\log(z-x)$

(C) → (q)

(D) $\frac{128r - a}{r-1} = 255$

$\Rightarrow \quad \frac{256-a}{2-1} = 255 \qquad$ [since $r = 2$]

$\Rightarrow \quad 256 - a = 225$

$\Rightarrow \quad a = 1$

(D) → (r)

119. **(A) → (q), (B) → (s), (C) → (p), (D) → (r)**

(A) Given that a, b, c are in A.P so 2b = a + c and $(2+b)^2 = (1+a)(1+c)$

$\Rightarrow \quad 4 + 4b + b^2 = 1 + a + c + ac$

$\Rightarrow \quad 4 + 4b + b^2 = 1 + 2b + ac$

$\Rightarrow \quad ac = b^2 + 2b + 3$

$= (b+1)^2 + 2$

$\Rightarrow \quad ac > 2 = k$

(A) → (q)

(B) $(1+x)(1+x^2)(1+x^4)(1+x^8)\ldots(1+x^{128})$

Multiply by $(1-x)$ in numerator and denominator we get

$$\frac{1-x^{256}}{1-x} = \sum_{r=0}^{255} x^r$$

$\Rightarrow \quad n = 255$

So $255-250 = 5$

(B) → (s)

(C) $a_2 = a_1 + 2^2 = 1^2 + 2^2$

$a_3 = a_2 + 3^2 = 1^2 + 2^2 + 3^2$

$\ldots \quad \ldots \quad \ldots$

$a_i = 1^2 + 2^2 + \ldots i^2$

$= \frac{i(i+1)(2i+1)}{6}$

$S = \sum_{i=1}^{10} a_I$

$= \sum_{i=1}^{10} \frac{i(i+1)(2i+1)}{6} = 1210$

$$\sqrt{\left\{\frac{\left(\sum_{i=1}^{10} a_i\right)}{10}\right\}} = 11$$

(C) → (p)

(D) Consider n^{th} term $T_n = \frac{n†}{500 + 3n†}$

Let $U_n = \frac{1}{T_n} = \frac{500}{n†} + 3n$

$$\frac{dT^n}{dn} = \frac{(500+3n^2).2n - n^2 9n^2}{(500+3n^2)^2}$$

$$= \frac{n(1000-3n^3)}{(500+3n^2)†} = 0$$

$\Rightarrow \quad n = (-\frac{1000}{3})^{1/3}$

Now, $6 < (\frac{1000}{3})^{1/3} < 7$.

Hence, T_7 is the largest term, so largest term in the given sequence is $\frac{49}{1529}$.

(C) → (r)

120. **(A) → (r), (B) → (s), (C) → (p), (D) → (q)**

(A) Let $x = I + f$ then from the given information

$\because \quad [x] = \frac{2x\{x\}}{x+\{x\}}$

$\Rightarrow \quad I = \frac{2f(1+f)}{1+2f}$

$I^2 + 2If = 2If + 2f^2$

$\Rightarrow \quad I^2 = 2f^2$

$\Rightarrow \quad I = \pm\sqrt{2}f$

$f = 0, \frac{1}{\sqrt{2}}$

If $f = 0$, $I = 0 \Rightarrow x = 0$ which is not possible

$\therefore \quad f = \frac{1}{\sqrt{2}}$

$\therefore \quad I = \pm\sqrt{2}\frac{1}{\sqrt{2}} = \pm 1$

$\therefore \quad x = 1 + \frac{1}{\sqrt{2}}, -1 + \frac{1}{\sqrt{2}}$.

So total 2 values of x exist

(A) → (r)

(B) Let $x = a^2, y = b^2, z = c^2$. Then $2b^2 = a^2 + c^2$

Since b is at least 2, consider the values of b from 2 onwards.

The first such value occurs when b = 5 (and then a = 7, c = 1)

So, the least possible value of $x = a^2 = 49$.

$\sqrt{k} - 4 = 7 - 4 = 3$

(B) → (s),

(C) Apply A.M, G.M inequality

A.M. ≥ H.M.

$\Rightarrow \frac{a+b+c}{3} \geq \frac{3}{\frac{1}{a}+\frac{1}{b}+\frac{1}{c}}$

$\Rightarrow (a+b+c)(\frac{1}{a}+\frac{1}{b}+\frac{1}{c}) \geq 9$

(C) → (p),

(D) $h = \frac{2ab}{a+b}, g = \sqrt{ab}$

$\Rightarrow \frac{h}{g} = \frac{2ab}{(a+b)\sqrt{ab}} \Rightarrow \frac{4}{5} = \frac{2\sqrt{ab}}{a+b}$

$\Rightarrow \frac{9}{1} = \frac{(\sqrt{a}+\sqrt{b})†}{(\sqrt{a}-\sqrt{b})†} \Rightarrow \left|\frac{\sqrt{a}+\sqrt{b}}{\sqrt{a}-\sqrt{b}}\right| = 3$

or, $\frac{\sqrt{a}+\sqrt{b}}{\sqrt{a}-\sqrt{b}} = \pm 3$

or, a : b = 1 : 4 or 4 : 1 .

(D) → (q)

More Than One Correct MCQs

121. (a, b, d) Let four integers be $a-d, a, a+d$ and $a+2d$ where a and d are integers and $d>0$

Since $a+2d=(a-d)^2+a^2+(a+d)^2$

$\Rightarrow 2d^2-2d+3a^2-a=0$

$\therefore d=\frac{1}{2}[1\pm\sqrt{1+2a-6a^2}$

Since, d is positive integer

$\therefore 1+2a-6a^2>0$ or $6a^2-2a-1<0$

$\Rightarrow \frac{1-\sqrt{7}}{6}<a<\frac{1+\sqrt{7}}{6}$

Since a is an integer.

$\therefore a=0,\ \therefore d=1$ or 0 but since $d>0 \therefore d=1$

$\therefore$ The four numbers are – 1, 0, 1, 2

122. (a, b, c) Consider a_N and assume that N is not a prime say $N=pq$ then

Consider

$$a_N=a_{pq}=\frac{1}{9}(999\ldots pq\,times)=\frac{1}{9}\left(10^{pq}-1\right)$$

$$=\frac{1}{9}\left[(10^p)^q-1\right]=\left[\frac{(10^p)^q-1}{10^p\quad 1}\right]\left[\frac{10^p-1}{10-1}\right]$$

$$=\left[(10^p)^{q-1}+(10^p)^{q-2}+\ldots+10^p+1\right]$$

$$\times\left[10^{p-1}+10^{p-2}+\ldots+10+1\right]$$

$\Rightarrow a_N$ is not prime if N is not a prime

123. (a, b, d) Since three numbers in AP so $2b=a+c$ and $b^2=\frac{2a^2c^2}{a^2+c^2}$

Eliminating b, we get $\left(\frac{a+c}{2}\right)^2=\frac{2a^2c^2}{a^2+c^2}$

$\Rightarrow \left(a^2+c^2\right)^2+2ac\left(a^2+c^2\right)-8a^2c^2=0$

$\Rightarrow \left(a^2+c^2+4ac\right)\left(a^2+c^2-2ac\right)=0$

$\Rightarrow \left[(a+c)^2+2ac\right](a-c)^2=0$

$\Rightarrow 4\left(b^2+\frac{1}{2}ac\right)(a-c)^2=0$

$\Rightarrow a-c=0\ or\, b^2=-\frac{1}{2}ac$

If $a=c$, we get $a=b=c$

If $b^2=-\frac{1}{2}ac$, then either $a, b, -\frac{1}{2}c$ are in GP

or $-\frac{1}{2}a, b, c$ are in GP

124. (a, b, c, d) Given that a, b, c are in HP

$\Rightarrow \frac{1}{a},\frac{1}{b},\frac{1}{c}$ are in AP.

$\Rightarrow \frac{a+b+c}{a},\frac{a+b+c}{b},\frac{a+b+c}{c}$ are in AP

$\Rightarrow \frac{b+c}{a},\frac{c+a}{b},\frac{a+b}{c}$ are in AP

Subtracting 1 from each term

$\Rightarrow \frac{b+c}{a}-1,\frac{c+a}{b}-1,\frac{a+b}{c}-1$ are in AP

$\Rightarrow \frac{b+c-a}{a},\frac{c+a-b}{a},\frac{a+b-c}{c}$ are in AP

Thus, $\frac{a}{b+c},\frac{b}{c+a},\frac{c}{a+b}$ are in HP

And $\frac{a}{b+c-a},\frac{b}{a+a-b},\frac{c}{a+b-c}$ are also in HP

Also, $b=\frac{2ac}{a+c}$, so

$$\frac{1}{b-a}+\frac{1}{b-c}=\frac{2b-(a+c)}{(b-a)(b-c)}$$

$$=\frac{2b-(a+c)}{b^2-b(a+c)+ac}$$

$$=\frac{2b-\frac{2ac}{b}}{b^2-b\frac{2ac}{b}+ac}=\frac{2}{b}\cdot\frac{b^2-ac}{b^2-ac}=\frac{2}{b}$$

Lastly, $\left(a-\frac{b}{2}\right)\left(c-\frac{b}{2}\right)=ac-\frac{b}{2}(a+c)+\frac{b^2}{4}$

$$=ac-\frac{b}{2}\cdot\frac{2ac}{b}+\frac{b^2}{4}=\frac{b^2}{4}$$

$\therefore a-\frac{b}{2},\frac{b}{2},c-\frac{b}{2}$ are in GP

125. (c, d) Let $b=a+p, c=a+2p, d=a+3p$

$$\frac{\frac{1}{a}+\frac{1}{d}}{\frac{1}{b}+\frac{1}{c}}=\frac{\frac{1}{a}+\frac{1}{a+3p}}{\frac{1}{a+p}+\frac{1}{a+2p}}=\frac{(a+p)(a+2p)}{a(a+3p)}$$

$$= \frac{a^2 + 3ap + 2p^2}{a^2 + 3ap} > 1$$

$$\therefore \frac{1}{a} + \frac{1}{d} > \frac{1}{b} + \frac{1}{c}$$

$$\left(\frac{1}{b} + \frac{1}{c}\right)(a + d) = \left(\frac{1}{a+p} + \frac{1}{a+2p}\right)(a + a + 3p)$$

$$= \frac{(2a+3p)^2}{a^2 + 3ap + 2p^2} = 4 + \frac{p^2}{a^2 + 3ap + 2p^2} > 4$$

126. (a, b, c, d) We have,

$$a(n) = 1 + \left(\frac{1}{2} + \frac{1}{3}\right) + \left(\frac{1}{4} + \ldots + \frac{1}{7}\right)$$

$$+ \left(\frac{1}{8} + \ldots + \frac{1}{15}\right) + \ldots + \left(\frac{1}{2^{n-1}} + \frac{1}{2^{n-1}+1} + \ldots\right.$$

$$\left. + \frac{1}{(2^n) - 1}\right) < 1 + \frac{2}{2} + \frac{4}{4} + \frac{8}{8} + \ldots + \frac{2^{n-1}}{2^{n-1}} = n$$

Thus, $a(n) < n$

$\Rightarrow a(2n) < 2n$

Next,

$$a(n) = 1 + \frac{1}{2} + \left(\frac{1}{3} + \frac{1}{4}\right) + \left(\frac{1}{5} + \ldots + \frac{1}{8}\right) + \ldots$$

$$+ \left(\frac{1}{2^{n-1}+1} + \ldots + \frac{1}{2^n}\right) - \frac{1}{2^n}$$

$$> \left(1 + \frac{1}{2} + \frac{2}{4} + \frac{4}{8} + \ldots + \frac{2^{n-1}}{2^n} - \frac{1}{2^n}\right)$$

$$\geq \frac{1}{2} + \frac{1}{2} + \ldots + \frac{1}{2} = \frac{n}{2}$$

Thus, $a(n) > \frac{n}{2} \Rightarrow a(2n) > n$

From above two results we can conclude that

$\frac{n}{2} < a(n) < n$

127. (a, c) Let the AP series be a, a + d, a + 2d, a + 3d, …

Given that,

$$3a + 3d = 9$$

$$a + d = 3$$

And $a^2 + (a + d)^2 + (a + 2d)^2 = 35$

$\Rightarrow a^2 + (3)^2 + (3 + d)^2 = 35$

$\Rightarrow (3 - d)^2 + 3^2 + (3 + d)^2 = 35$

$\Rightarrow d = \pm 2$

$\Rightarrow a = 1, 5$

$\therefore S_n = \frac{n}{2}[2a + (n-1)d]$

when $d = 2$, $a = 1$ $S_n = n^2$

when $d = -2$, $a = 5$, then $S_n = n(6 - n)$

128. (a, c) Since sin β is the GM of sin α and cosα so,

$$\sin\beta = \sqrt{\sin\alpha\cos\alpha}$$

$$\Rightarrow \sin^2\beta = \sin\alpha\cos\alpha$$

Now, $\cos 2\beta = 1 - 2\sin^2\beta$

$$= 1 - 2\sin\alpha\cos\alpha$$

$$= (\sin\alpha - \cos\alpha)^2$$

$$= 2\sin^2\left(\frac{\pi}{4} - \alpha\right)$$

or $2\cos^2\left(\frac{\pi}{4} + \alpha\right)$

129. (a, c) Given $\cot^{-1}\left[\frac{1 + a_1a_2}{d}\right] + \cot^{-1}\left[\frac{1 + a_2a_3}{d}\right]$

$$+ \cot^{-1}\left[\frac{1 + a_3a_4}{d}\right] + \ldots + \cot^{-1}\left[\frac{1 + a_na_{n-1}}{d}\right]$$

$$= \cot^{-1}a_1 - \cot^{-1}a_2 + \cot^{-1}a_2 - \cot^{-1}a_3$$

$$+ \ldots + \cot^{-1}a_{n-1} - \cot^{-1}a_n$$

$$= \cot^{-1}a_1 - \cot^{-1}a_n = \tan^{-1}a_n - \tan^{-1}a_1$$

130. (a, c) Let the roots be a_1, a_2, a_3, a_4, a_5, then

$$\frac{1}{a_1} + \frac{1}{a_2} + \frac{1}{a_3} + \frac{1}{a_4} + \frac{1}{a_5} = 10$$

$$\frac{\sum a_1a_2a_3a_4}{a_1a_2a_3a_4a_5} = 10$$

$$\Rightarrow 10S = \sum a_1a_2a_3a_4$$

Since, a_1, a_2, a_3, a_4, a_5, are in GP, then a, ar, ar^2, ar^3, ar^4

$10S = 10\,a^5r^{10} = a^3r^6\,[a + ar + ar^2 + ar^3 + ar^4]$

$= a^3r^6\,[40]$

[since sum of root = 40]

$\Rightarrow a^2r^4 = 4 \Rightarrow ar^2 = \pm 2$

$\Rightarrow S = a^3r^6\,[4] = (\pm 2)^3\,[4] = \pm 32$

131. (a, b) Given that, $\frac{1}{16}$, a, b are in GP

$$\Rightarrow \qquad a^2 = \frac{b}{16} \qquad \ldots(i)$$

Also, a, b, $\frac{1}{6}$ are in HP.

$$\Rightarrow \qquad b = \frac{2 \times a \times \frac{1}{6}}{a + \frac{1}{6}} \qquad \ldots(ii)$$

Solving eq (i) and (ii), we get

$$a = \frac{1}{12}, -\frac{1}{4} \text{ and } b = \frac{1}{9}, 1$$

132. (a, b, c, d)Let the three numbers in GP be a, ar, ar^2

$\therefore$ $a + ar + ar^2 = 56$ (given) ...(i)

Subtracting 1, 7, 21 from the numbers, we get $a - 1$, $ar - 7$, $ar^2 - 21$ which are given to be in AP

$\therefore$ $(ar - 7) - (a - 1) = (ar^2 - 21) - (ar - 7)$

$\Rightarrow$ $ar - a - 6 = ar^2 - ar - 14$

$\Rightarrow$ $a - 2ar + ar^2 = 8$ ...(ii)

Subtracting Eq. (ii) from Eq.(i), we get

$$3ar = 48$$

$$\Rightarrow \qquad a = \frac{16}{r} \qquad \ldots(iii)$$

Substituting $a = \frac{16}{r}$ in Eq . (i), we have,

$$\frac{16}{r} + 16 + 16r = 56$$

$\Rightarrow 16r^2 - 40r + 16 = 0$

$\Rightarrow$ $2r^2 - 5r + 2 = 0$

$\Rightarrow (r - 2)(2r - 1) = 0$

$\therefore$ $r = 2, \frac{1}{2}$

Where $r = 2$, then from eq (iii) a = 8 and the number are 8, 16, 32 .

Where $r = \frac{1}{2}$, then from eq (iii) a = 32 and the number are 32, 16, 8 .

And corresponding APs are 7, 9, 11 and 31, 9, −13

133. (a, d) α, β, γ and δ are in HP

$\Rightarrow \frac{1}{\alpha}, \frac{1}{\beta}, \frac{1}{\gamma}, \frac{1}{\delta}$ are in AP

Let d be the common difference of this AP.

Now, α, γ are roots of $Ax^2 - 4x + 1 = 0$

$$\therefore \frac{\alpha + \gamma}{\alpha\gamma} = \frac{\frac{4}{A}}{\frac{1}{A}} = 4$$

$$\text{or } \frac{1}{\alpha} + \frac{1}{\gamma} = 4 ie, \frac{1}{\alpha} + \frac{1}{\alpha} + 2d = 4$$

$$or \frac{1}{\alpha} + d = 2 \qquad \ldots(i)$$

β, δ are roots of $Bx^2 - 6x + 1 = 0$

$$\therefore \frac{\beta + \delta}{\beta\delta} = \frac{1}{\beta} + \frac{1}{\delta} = \frac{6/B}{1/B} = 6$$

$$\text{or } \frac{1}{\alpha} + d + \frac{1}{\alpha} + 3d = 6$$

$$\text{or } \frac{1}{\alpha} + 2d = 3 \qquad \ldots(ii)$$

From Eqs. (i) and (ii), on solving, we get

$$\frac{1}{\alpha} = 1, d = 1 \therefore \frac{1}{\alpha} = 1, \frac{1}{\beta} = 2, \frac{1}{\gamma} = 3, \frac{1}{\delta} = 4$$

$$Since, \frac{1}{\alpha\gamma} = A \Rightarrow A = 3$$

$$\text{Also, } \frac{1}{\beta\delta} = B, \Rightarrow B = 8$$

Hence, A = 3 and B = 8

134. (a, b)

$$T_r = \cot^{-1}\left(\frac{4r\dagger + 3}{4}\right)$$

$$= \tan^{-1} \frac{1}{1 + r\dagger - \frac{1}{4}}$$

$$= \tan^{-1}\left(\frac{4r\dagger + 3}{4}\right)$$

$$= \tan^{-1} \frac{1}{1 + r\dagger - \frac{1}{4}}$$

$$= \tan^{-1} \frac{\left(r + \frac{1}{2}\right) - \left(r - \frac{1}{2}\right)}{1 + \left(r + \frac{1}{2}\right)\left(r - \frac{1}{2}\right)}$$

$$= \tan^{-1}\left(r + \frac{1}{2}\right) - \tan^{-1}\left(r - \frac{1}{2}\right)$$

$$S_n = \sum T_r$$

$$= \tan^{-1}\left(n + \frac{1}{2}\right) - \tan^{-1}\frac{1}{2}$$

$$= \tan^{-1}\left(\frac{4n}{2n+5}\right)$$

$$S_\infty = \lim_{n\to\infty} S_n = \tan^{-1} 2 = \cot^{-1}\frac{1}{2}.$$

135. (a, b, c) $y = 2\sqrt{x}$, being in the first quadrant.

The sequence of x-coordinates 1, 2, 4, 8, ...

$\therefore$ The sequence of y-coordinates $2, 2\sqrt{2}, 2\sqrt{4}, 2\sqrt{8}$, ..., where the common ratio is $\sqrt{2}$.

$$\therefore y_n = 2\cdot\left(\sqrt{2}\right)^{n-1} = 2^{\frac{n+1}{2}}$$

136. (a, b, c, d) Let us add one more number, a_{n+1} to the given sequence. The number a_{n+1} is such that $|a_{n+1}| = |a_n + 1|$.

Squaring all the numbers, we have

$$a_1^2 = 0$$
$$a_2^2 = a_1^2 + 2a_1 + 1$$
$$a_3^2 = a_2^2 + 2a_2 + 1$$
$$a_4^2 = a_3^2 + 2a_3 + 1$$
$$\ldots$$
$$\ldots$$
$$\ldots$$
$$a_n^2 = a_{n-1}^2 + 2a_{n-1} + 1$$
$$a_{n+1}^2 = a_n^2 + 2a_n + 1$$

Adding the above equalities, we get

$$a_1^2 + a_2^2 + \ldots + a_n^2 + a_{n+1}^2$$
$$= a_1^2 + a_2^2 + \ldots + a_n^2 + 2(a_1 + a_2 + \ldots + a_n) + n$$
$$\Rightarrow \quad 2(a_1 + a_2 + \ldots + a_n) = -n + a_{n+1}^2 \geq -n$$
$$\Rightarrow \quad \frac{a_1 + a_2 + \ldots + a_n}{n} \geq -\frac{1}{2} = -\frac{\lambda}{\mu}$$

so $\lambda = 1$ and $\mu = 2$

$$\therefore \quad \lambda + \mu = 3$$
$$\lambda\mu = 2$$
$$\lambda^\mu + \mu^\lambda = 3$$

λ & μ can be roots of a quadratic equation with rational co-efficient

137. (a, b, c) We have

$$1^4 + 2^4 + 3^4 + \ldots + n^4 = an^5 + bn^4 + cn^3 + dn^2 + en + f \quad \ldots(1)$$

and $1^4 + 2^4 + \ldots + n^4 + (n+1)^4$

$$= a(n+1)^5 + b(n+1)^4 + c(n+1)^3 + d(n+1)^2 + e(n+1) + f \quad \ldots(2)$$

Subtracting (1) from (2), we get

$$(n+1)^4 = a\left[(n+1)^5 - n^5\right] + b\left[(n+1)^4 - n^4\right] + c\left[(n+1)^3 - n^3\right] + d\left[(n+1)^2 - n^2\right] + e\left[(n+1) - n\right]$$

Comparing coefficient for n^4, we get

$1 = 5a \Rightarrow a = 1/5$, $b = ½$ and $c = 1/3$

138. (a, c, d) We have $t_r = \sum_{k=1}^{r} t_k - \sum_{k=1}^{r-1} t_k$

$$= \frac{1}{12}r(r+1)(r+2) - \frac{1}{12}(r-1)(r)(r+1)$$

$$= \frac{1}{4}r(r+1)$$

Now, $\frac{1}{t_r} = \frac{4}{r(r+1)} = 4\left(\frac{1}{r} - \frac{1}{r+1}\right)$

$$\Rightarrow \sum_{r=1}^{n}\frac{1}{t_r} = 4\sum_{r=1}^{n}\left(\frac{1}{r} - \frac{1}{r+1}\right) = 4\left(1 - \frac{1}{n+1}\right) = \frac{4n}{n+1}$$

139. (a, c) We have

$$\sin^2 x + \sin^4 x + \sin^6 x + \ldots = \frac{\sin^2 x}{1 - \sin^2 x} = \tan^2 x$$

therefore,

$$\alpha = e^\wedge\left\{\tan^2 x \ln 2\right\} = e^\wedge\left\{\ln 2^{\tan^2 x}\right\}$$

$$= 2^{\tan^2 x}$$

As α satisfies the equation $x^2 - 17x + 16 = 0$ we get $\alpha = 1$ or $\alpha = 16$

Since $0 < x < \pi/2$,

$\tan^2 x > 0 \Rightarrow \alpha = 2^{\tan^2 x} > 1$.

Therefore, $2^{\tan^2 x} = 16 = 2^4$

$$\Rightarrow \quad \tan^2 x = 4 \Rightarrow \tan x = 2 \left[\because \tan x > 0\right]$$

Thus, $\frac{2\cos x}{\sin x + 2\cos x} = \frac{2}{\tan x + 2} = \frac{2}{2+2} = \frac{1}{2}$

140. (a, c) If t_r denotes the rth term of the series, then

$$xt_r = \frac{x}{(1+rx)(1+(r+1)x)}$$

$$= \frac{1}{1+rx} - \frac{1}{1+(r+1)x}$$

$$\Rightarrow x\sum_{r=1}^{n} t_r = \sum_{r=1}^{n}\left[\frac{1}{1+rx} - \frac{1}{1+(r+1)x}\right]$$

$$= \frac{1}{1+x} - \frac{1}{1+(n+1)x} = \frac{nx}{(1+x)(1+(n+1)x)}$$

$$\Rightarrow S_n = \sum_{r=1}^{n} t_r = \frac{n}{(1+x)[1+(n+1)x]}$$

141. (b) Let t_r denote the rth term of the given series, then

$$t_r = \frac{1}{r(r+1)(r+2)(r+3)}$$

and $t_{r+1} = \frac{1}{(r+1)(r+2)(r+3)(r+4)}$

$$\Rightarrow rt_r = \frac{1}{(r+1)(r+2)(r+3)}$$

and $(r+4)t_{r+1} = \frac{1}{(r+1)(r+2)(r+3)}$

Thus, $rt_r - (r+4)t_{r+1} = 0$

$rt_r - (r+1)t_{r+1} = 3t_{r+1}$

Putting $r = 1, 2, \ldots, n-2, n-1,$ we get

$1t_1 - 2t_2 = 3t_2$

$2t_2 - 3t_3 = 3t_3$

$3t_3 - 4t_4 = 3t_4$

...

$(n-2)t_{n-2} - (n-1)t_{n-1} = 3t_{n-1}$

$(n-1)t_{n-1} - nt_n = 3t_{n-1}$

Adding above expressions, we get

$t_1 - nt_n = 3(t_2 + t_3 + \ldots + t_n)$

$\Rightarrow 4t_1 - nt_n = 3(t_2 + t_3 + \ldots + t_n)$

$$\Rightarrow \frac{1}{1\cdot 2\cdot 3} - \frac{1}{(n+1)(n+2)(n+3)} = 3S_n$$

$$\Rightarrow S_n = \frac{1}{3}\left[\frac{1}{6} - \frac{1}{(n+1)(n+2)(n+3)}\right]$$

142. (a,b,d) Let x be the first term and y the $(2n-1)$ th terms of AP, GP and HP whose nth terms are a, b, c respectively.

For AP, $y = x + (2n-2)\,d$

$$\Rightarrow \quad d = \frac{y-x}{2(n-1)}$$

$$\therefore \quad a = x + (n-1)d = x + \frac{1}{2}(y-x) = \frac{1}{2}(x+y) \quad \text{....(1)}$$

$$\Rightarrow r = \left(\frac{y}{x}\right)^{\frac{1}{2n-2}}$$

For G.P. $y = xr^{2n-2}$

$$\therefore \quad b = xr^{n-1} = x.\left(\frac{y}{x}\right)^{1/2} = \sqrt{xy} \quad \text{....(2)}$$

For H.P. $\frac{1}{y} = \frac{1}{x} + (2n-2)d_1$

$$\Rightarrow \quad d_1 = \frac{x-y}{2xy\,(n-1)}$$

$$\therefore \quad \frac{1}{c} = \frac{1}{x} + (n-1)\,d_1 = \frac{1}{x} + \frac{x-y}{2xy}$$

$$\frac{1}{c} = \frac{x+y}{2xy} \Rightarrow c = \frac{2xy}{x+y} \quad \text{....(3)}$$

Thus from (1), (2) and (3), a, b, c are A.M., G.M. and H.M. respectively of x and y.

143. (b, c) We have for $0 < \phi < \frac{\pi}{2}$

$$x = \sum_{n=0}^{\infty} \cos^{2n}\phi = 1 + \cos^2\phi + \cos^4\phi + \ldots.\infty$$

$$\frac{1}{1-\cos^2\phi} = \frac{1}{\sin^2\phi} \quad \text{....(1)}$$

[Using sum of infinite G.P. $\cos^2\alpha$ being < 1]

$$y = \sum_{n=0}^{\infty} \sin^{2n}\phi = 1 + \sin^2\phi + \sin^4\phi + \ldots.\infty$$

$$= \frac{1}{1-\sin^2\phi} = \frac{1}{\cos^2\phi} \quad \text{....(2)}$$

$$z = \sum_{n=0}^{\infty} \cos^{2n}\phi\sin^{2n}\phi$$

$$= 1 + \cos^2\phi\sin^2\phi + \cos^4\phi\sin^4\phi + \ldots\infty$$

$$= \frac{1}{1-\cos^2\phi\sin^2\phi} \quad \text{....(3)}$$

Substituting the values of $\cos^2\phi$ and $\sin^2\phi$ in (3), from (1) and (2), we get

$$z = \frac{1}{1-\frac{1}{x}.\frac{1}{y}} \Rightarrow z = \frac{xy}{xy-1}$$

$\Rightarrow xyz - z = xy \Rightarrow xyz = xy + z.$

Also, $x + y + z = \frac{1}{\cos^2\phi} + \frac{1}{\sin^2\phi} + \frac{1}{1-\cos^2\phi\sin^2\phi}$

$$\frac{[\sin^2\phi(1-\cos^2\phi\sin^2\phi)+\cos^2\phi(1-\cos^2\phi\sin^2\phi)+\cos^2\phi\sin^2\phi]}{\cos^2\phi\sin^2\phi\,(1-\cos^2\phi\sin^2\phi)}$$

$$=\frac{(\sin^2\phi+\cos^2\phi)\,(1-\cos^2\phi\sin^2\phi)+\cos^2\phi\sin^2\phi}{\cos^2\phi\sin^2\phi\,(1-\cos^2\phi\sin^2\phi)}$$

$$=\frac{1}{\cos^2\phi\sin^2\phi\,(1-\cos^2\phi\sin^2\phi)}=xyz$$

Thus (*b*) and (*c*) both are correct.

144. (b) Putting $\theta = 0$, we get $b_0 = 0$

$$\therefore \quad \sin n\theta = \sum_{r=1}^{n} b_r \sin^r\theta$$

$$\Rightarrow \quad \frac{\sin n\theta}{\sin\theta} = \sum_{r=1}^{n} b_r (\sin\theta)^{r-1}$$

$$= b_1 + b_2\sin\theta + b_3\sin^2\theta + \dots\dots + b_n\sin^{n-1}\theta$$

Taking limit as $\theta \to 0$, we obtain

$$\lim_{\theta\to 0}\frac{\sin n\theta}{\sin\theta} = b_1 \Rightarrow b_1 = n.$$

145. (c) $T_m = a+(m-1)\,d = 1/n$ and $t_n = a+(n-1)\,d = 1/m$

$\Rightarrow \quad (m-n)\,d = 1/n - 1/m = (m-n)/mn$

$\Rightarrow \quad d = 1/mn$

$$\Rightarrow \quad a = \frac{1}{mn}$$

$$\therefore \quad t_{mn} = a + (mn-1)\,d = \frac{1}{mn} + (mn-1)\frac{1}{mn}$$

$$= \frac{1}{mn} + 1 - \frac{1}{mn} = 1$$

146. (b) If x, y, z are in G.P. ($x, y, z > 1$); $\log x$, $\log y$, $\log z$ will be in A.P.

$\Rightarrow$ $1 + \log x$, $1 + \log y$, $1 + \log z$ will also be in A.P.

$$\Rightarrow \quad \frac{1}{1+\log x}, \frac{1}{1+\log y}, \frac{1}{1+\log z} \text{ will be in H.P.}$$

147. (a, d) We have

$$a(n) = 1+\frac{1}{2}+\frac{1}{3}+\frac{1}{4}+\frac{1}{5}+\dots\dots+\frac{1}{2^n-1}$$

$$=1+\left(\frac{1}{2}+\frac{1}{3}\right)+\left(\frac{1}{4}+\frac{1}{5}+\frac{1}{6}+\frac{1}{7}\right)+\left(\frac{1}{8}+\dots\dots+\frac{1}{15}\right)+\dots+\left(\frac{1}{2^{n-1}}+\dots\dots+\frac{1}{2^n-1}\right)$$

$$<1+\left(\frac{1}{2}+\frac{1}{2}\right)+\left(\frac{1}{4}+\frac{1}{4}+\dots\dots+\frac{1}{4}\right)+\left(\frac{1}{8}+\frac{1}{8}+\dots+\frac{1}{8}\right)+\dots+\left(\frac{1}{2^{n-1}}+\dots+\frac{1}{2^{n-1}}\right)$$

$$<1+\frac{2}{2}+\frac{4}{4}+\frac{8}{8}+\dots\dots+\frac{2^{n-1}}{2^{n-1}}=1+1+\dots\dots+1=n$$

Thus, $a(100) < 100$

Also

$$a(n)=1+\frac{1}{2}+\left(\frac{1}{3}+\frac{1}{4}\right)+\left(\frac{1}{5}+\frac{1}{6}+\frac{1}{7}+\frac{1}{8}\right)+\dots\dots+\left(\frac{1}{2^{n-1}+1}+\dots+\frac{1}{2^n}\right)-\frac{1}{2^n}$$

$$>1+\frac{1}{2}+\left(\frac{1}{4}+\frac{1}{4}\right)+\left(\frac{1}{8}+\frac{1}{8}+\dots+\frac{1}{8}\right)+\dots+\left(\frac{1}{2^n}+\dots+\frac{1}{2^n}\right)-\frac{1}{2^n}$$

$$=1+\frac{1}{2}+\frac{2}{4}+\frac{4}{8}+\dots+\frac{2^{n-1}}{2^n}-\frac{1}{2^n}$$

$$=1+\underbrace{\left(\frac{1}{2}+\frac{1}{2}+\frac{1}{2}+\dots\dots+\frac{1}{2}\right)}_{n\text{ times}}-\frac{1}{2^n}$$

$$=1+\frac{n}{2}-\frac{1}{2^n}=\left(1-\frac{1}{2^n}\right)+\frac{n}{2}$$

Thus, $a(200) > \left(1-\frac{1}{2^{200}}\right)+\frac{200}{2} > 100$,

i.e. $a(200) > 100$.

148. (b, d)

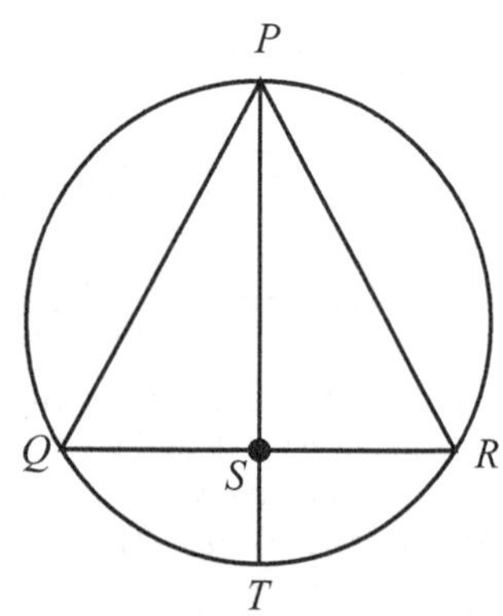

We know by geometry $PS \times ST = QS \times SR$ ...(1)

$\because$ S is not the centre of circulm circle, $PS \neq ST$

And we know that for two unequal real numbers.
H.M. < G.M.

$$\Rightarrow \frac{2}{\frac{1}{PS}+\frac{1}{ST}} < \sqrt{PS \times ST}$$

$$\Rightarrow \frac{1}{PS}+\frac{1}{ST} > \frac{2}{\sqrt{PS \times ST}}$$

$$\Rightarrow \frac{1}{PS}+\frac{1}{ST} > \frac{2}{\sqrt{QS \times SR}} \text{ [using eqn (1)] } ...(2)$$

$\therefore$ (b) is the correct option.

Also $\sqrt{QS \times SR} < \frac{QS+SR}{2}$ ($\because$ GM < AM)

$$\Rightarrow \frac{1}{\quad} > \frac{2}{\quad} \quad ...(3)$$

From equations (2) and (3) we get $\frac{1}{PS}+\frac{1}{ST} > \frac{4}{QR}$

$\therefore$ (d) is also the correct option.

149. (a, d) We have $S_n = \sum_{k=1}^{n} \frac{n}{n^2+kn+k^2}$

and $T_n = \sum_{k=0}^{n-1} \frac{n}{n^2+kn+k^2}$ $n = 1, 2, 3.........$

For $n = 1$ we get

$S_1 = \frac{1}{1+1+1} = \frac{1}{3} = 0.3$ and $T_1 = \frac{1}{1+0} = 1$

Also $\frac{\pi}{3\sqrt{3}} = \frac{\pi\sqrt{3}}{9} = \frac{3.14 \times 1.73}{9}$

$= 0.34 \times 1.73 = 0.58$

$\therefore S_1 < \frac{\pi}{3\sqrt{3}} < T_1$, $\therefore S_n < \frac{\pi}{3\sqrt{3}}$ and $T_n > \frac{\pi}{3\sqrt{3}}$

150. (a,d) $S_n = -1^2 - 2^2 + 3^2 + 4^2 - 5^2 - 6^2 + ...$

$= (3^2 + 7^2 + 11^2 + ...) + (4^2 + 8^2 + 12^2 + ...) - (1^2 + 5^2 + 9^2 + ...) - (2^2 + 6^2 + 10^2 + ...)$

$$= \sum_{r=1}^{n}(4r-1)^2 + \sum_{r=1}^{n}(4r)^2 - \sum_{r=1}^{n}(4r-3)^2 - \sum_{r=1}^{n}(4r-2)^2$$

$$= \left[\sum_{r=1}^{n}(4r-1)^2-(4r-3)^2\right] + 4\left[\sum_{r=1}^{n}(2r)^2-(2r-1)^2\right]$$

$$= 8\sum_{r=1}^{n}(2r-1) + 4\sum_{r=1}^{n}(4r-1)$$

$$= 8\left[2\frac{n(n+1)}{2}-n\right] + 4\left[4\frac{n(n+1)}{2}-n\right]$$

$= 8n^2 + 8n^2 + 4n = 16n^2 + 4n$

For $n = 8$, $16n^2 + 4n = 1056$

and for $n = 9$, $16n^2 + 4n = 1332$

Concept Eliminator (CE)
(Olympiad Based Questions)

1. (a) Let the set $S = \frac{1}{2^a}, \frac{1}{2^{a+b}}, \frac{1}{2^{a+2b}}....$ its 1st term is $\frac{1}{2^a}$ and common ratio is $\frac{1}{2^b}$

So required sum is $\frac{\frac{1}{2^a}}{1-\frac{1}{2^b}} = \frac{2^{b-a}}{2^b-1}$

If $\frac{2^{b-a}}{2^b-1} = \frac{1}{7}$ then b = 3 and a = 3

Similarly it cant attain 1/5, 1/9 and 1/11

2. Consider the given expression is

1st term

$$\left\{\frac{2^3-1}{2^3+1}\right\} = \left\{\frac{2-1}{2+1}\right\}\left\{\frac{2^2+2+1}{2^2-2+1}\right\} = \left(\frac{1}{3}\right)\left(\frac{7}{3}\right)$$

2nd term

$$\left\{\frac{3^3-1}{3^3+1}\right\} = \left\{\frac{3-1}{3+1}\right\}\left\{\frac{3^2+3+1}{3^2-3+1}\right\} = \left(\frac{2}{4}\right)\left(\frac{13}{7}\right)$$

3rd term

$$\left\{\frac{4^3-1}{4^3+1}\right\} = \left\{\frac{4-1}{4+1}\right\}\left\{\frac{4^2+4+1}{4^2-4+1}\right\} = \left(\frac{3}{5}\right)\left(\frac{21}{13}\right)$$

...

Product of the 1st part is

$$\left(\frac{1}{3}\right)\left(\frac{2}{4}\right)\left(\frac{3}{5}\right)...\left(\frac{n-3}{n-1}\right)\left(\frac{n-2}{n}\right)\left(\frac{n-1}{n+1}\right) = \frac{2}{n(n+1)}$$

Product of 2nd part is

$$\left(\frac{7}{3}\right)\left(\frac{13}{7}\right)\left(\frac{21}{13}\right)..\left(\frac{n^2-n+1}{n^2-3n+3}\right)\left(\frac{n^2+n+1}{n^2-n+1}\right)$$

$$= \frac{n^2+n+1}{3}$$

So product is $\frac{2(n^2+n+1)}{3n(n+1)}$

Required product is $\lim_{n\to\infty} \frac{2(n^2+n+1)}{3n(n+1)} = \frac{2}{3}$

3. Without loss of generality we can assume that a < b< c then we have two cases

Case (i) (a + b – c) < a < (c + a – b) < b < c < (b + c–a) < (a + b + c)

Let the common difference of AP is 'd' then a = (a + b + c) – 5d, b = (a + b + c) – 3d and c = (a + b + c) – 2d

adding these we get (a + b + c) = 5d or a = 0 but given that a, b, and c are positive real number so this case ruled out.

or

Case (ii) (a + b – c) < a < b < (c + a – b) < c < (b + c – a) < (a + b + c)

Let the common difference of AP is 'd' then a = (a + b + c) – 5d, b = (a + b + c) – 4d and c = (a + b + c) – 2d

From this we get (a + b + c) = $\frac{11}{2}d$

or $a = \frac{d}{2}, b = \frac{3d}{2}, c = \frac{7d}{2}$ ...(i)

Since (a + b – c) is the 1st term and (a + b + c) is the 7th term of AP so

$a + b - c = a + b + c - 6d = \frac{-d}{2}$

From result (i) we get

$a + b - c = d\left(\frac{1}{2} + \frac{3}{2} - \frac{7}{2}\right) = -\frac{3d}{2}$

which gives d = 0

so this case also ruled out .

So no such triplets (a, b, c) exist

4. Lets divide the given series in groups $(1),(2,2),(3,3,3),...,\underbrace{(k,k,...,k)}_{k\ times},...$

Here in k^{th} group we have 'k' elements.

Let nth term t_n is in the kth group, then $t_n = k$ since each element in k^{th} group is equal to k

Note that there are $1+2+...+(k-1) = \frac{k(k-1)}{2}$ terms preceding the kth group.

And total number of terms in the series till k^{th} group is $\frac{k(k-1)}{2}$

Therefore, if $\frac{(k-1)k}{2} + 1 \le n < \frac{k(k+1)}{2} + 1$

and $t_n = k$.

Solve the 1st inequality we get $\frac{(k-1)k}{2} + 1 \le n$

$\Rightarrow \frac{1-\sqrt{8n-7}}{2} \le k \le \frac{1+\sqrt{8n-7}}{2}$,

Now from the 2nd inequality

$n > \frac{k(k-1)}{2} + 1 = \frac{1-\sqrt{8n-7}}{2} > k,$

and $\frac{1-\sqrt{8n-7}}{2} < k.$

Thus $\frac{1-\sqrt{8n-7}}{2} < k \le \frac{1+\sqrt{8n-7}}{2}$,

or $k \le \frac{1+\sqrt{8n-7}}{2} < k+1,$

From this we can conclude that k = the integral part of $\frac{1+\sqrt{8n-7}}{2} = t_n$,

So $t_n = \left[\frac{1+\sqrt{8n-7}}{2}\right]$

Here [.] is greatest integer function

5. If we observe then we find that the series is formed as follows. The first term is 1. Then take the next two even numbers 2, 4. Then take the next three odd numbers 5, 7, 9. Then take the next four even numbers 10, 12, 14, 16, and so on. Lets divide the given sequence in the group (1), (2, 4), (5, 7, 9), (10, 12, 14, 16), (17, 19, 21, 23, 25)...

If we observe the we find that the nth term of the series $a_n = 2n - t_n$

Where t_n is the nth term of the series 1, 2, 2, 3, 3, 3, 4, 4, 4, 4......

So $a_n = 2n - t_n = 2n - \left[\frac{1+\sqrt{8n-7}}{2}\right]$

6. Lets define

$$f(n) = \sum_{i=1}^{n}(a_i - a_{n+1}) = a_0 + a_1 + a_2 ... + a_n - na_{n+1}$$

Here $f(0) = a_0 > 0$

Now consider

$$f(n+1) = \sum_{i=1}^{n+1}(a_i - a_{n+2}) = a_0 + a_1 + a_2 ... + a_{n+1} - (n+1)a_{n+2}$$

Since $a_0 < a_1 < a_2 < ...$ so $f(n) > f(n+1)$

So f is a monotonic decreasing function and $f(0) = a_0 > 0$

So there must exist a unique N such that

$$f(N-1) > 0 \geq f(N)$$

or $a_0 + a_1 + a_2 \ldots + a_{n-1} - (n-1)a_n > 0$ ...(i)

$a_0 + a_1 + a_2 \ldots + a_n - na_{n+1} \leq 0$ ...(ii)

From above two results we get for a unique N

$$a_n < \frac{a_0 + a_1 + a_2 \ldots + a_n}{n} < a_{n+1}$$

So only one integer will satisfy the condition.

7. Since 20, b, c are in harmonic progression, so $\frac{1}{20} + \frac{1}{c} = \frac{2}{b}$, or $bc + 20b - 40c = 0$

or $bc + 20b - 40c - 800 = -800$

or $(40 - b)(c + 20) = 800.$

Now we have to factorize 800 in such a way that one factor is more than 20.

$800 = 1 \times 800 = 2 \times 400 = 4 \times 200 = 5 \times 160$

$= 8 \times 100 = 10 \times 80 = 16 \times 50$

$(40 - b)(c + 20) = 800 = 1 \times 800 = 2 \times 400$

$= 4 \times 200 = 5 \times 160 = 8 \times 100 = 10 \times 80 = 16 \times 50.$

We thus get the pairs

$(b, c) = (39, 780), (38, 380), (36, 180),$

$(35, 140), (32, 80), (30, 60), (24, 30).$

Among these 7 pairs, only 5 pairs $(39, 780)$, $(38, 380), (36, 180), (35, 140)., (30, 60)$

Satisfy condition of divisibility: b divides c. thus there are 5 triples exist.

8. (b) For each positive integer n,

$$\frac{1}{(n+1)\sqrt{n} + n\sqrt{n+1}} = \frac{1}{\sqrt{n(n+1)}\left(\sqrt{n+1} + \sqrt{n}\right)}$$

$$= \frac{\sqrt{n+1} - \sqrt{n}}{\sqrt{n}(n+1)} = \frac{1}{\sqrt{n}} - \frac{1}{\sqrt{n+1}},$$

Hence,

$$p = \left(1 - \frac{1}{\sqrt{2}}\right) + \left(\frac{1}{\sqrt{2}} - \frac{1}{\sqrt{3}}\right) + \ldots + \left(\frac{1}{\sqrt{99}} - \frac{1}{\sqrt{100}}\right)$$

$$= 1 - \frac{1}{\sqrt{100}} = 1 - \frac{1}{10} = \frac{9}{10}.$$

9. Let $x_k = \max\{x_1, \ldots, x_n\}$. Then

$$\sum_{i=1}^{n-1} x_i x_{i+1} = \sum_{i=1}^{k-1} x_i x_{i+1} + \sum_{i=k}^{n-1} x_i x_{i+1}$$

$$\leq x_k \sum_{i=1}^{k-1} x_i + x_k \sum_{i=k}^{n-1} x_{i+1}$$

$$= x_k (p - x_k) \leq \frac{p^2}{4}.$$

The equality occurs if say $x_1 = x_2 = \frac{P}{2}$, $x_3 = \ldots = x_n = 0$. Thus the answer is $\frac{p^2}{4}$.

10. Since $\frac{1}{2n-2k+1} - \frac{1}{2n-2k+1}$

$$= \frac{k}{(2n-2k+1)(2n-k+1)},$$

We have $S_n = \sum_{k=1}^{n} \frac{k}{(2n-2k+1)(2n-k+1)}$

$$= \sum_{k=1}^{n} \frac{1}{2n-2k+1} - \sum_{k=1}^{n} \frac{1}{2n-2k+1}$$

$$= \left(\frac{1}{1} + \frac{1}{3} + \frac{1}{5} + \ldots + \frac{1}{2n-1}\right)$$

$$- \left(\frac{1}{n+1} + \frac{1}{n+2} + .. + \frac{1}{2n-1} + \frac{1}{2n}\right).$$

Then $T_n - S_n = \left(\frac{1}{1} + \frac{1}{3} + \ldots + \frac{1}{n} + \frac{1}{n+1} + \frac{1}{n+2} + \ldots + \frac{1}{2n-1} + \frac{1}{2n}\right)$

$$- \left(\frac{1}{1} + \frac{1}{3} + \frac{1}{5} + \ldots + \frac{1}{2n-1}\right)$$

$$= \frac{1}{2}\left(\frac{1}{1} + \frac{1}{2} + \frac{1}{3} + .. + \frac{1}{n}\right) = \frac{1}{2} T_n,$$

And hence $T_n = 2S_n$.

11. Let the six faces are assigned with a, b, c, d, e and f. Then sum of the numbers at the vertices is abc + acd + ade + aeb + fbc + fcd + feb + fde = (a + b)(c + d)(e + f)

From AM, GM inequality we know that this product is maximum when all the terms are equal i.e., (a + b) = (c + d) = (e + f) which is possible if (2 + 7) = (3 + 6) = (4 + 5)

So required maximum value is 9 × 9 × 9 = 729.

Here point to note that 2 and 7 are on the opposite faces, similarly 3 and 6 are on the opposite faces and same way 4 and 5.

12. Here x is the smallest one so maximum value of x is 33. Consider the cases $x = 0, x = 1, \ldots,$ $x = 33$.

Case (i) $x = 0$ then

If $y = 0$ then $z \in \{0, 1, 2, \ldots, 100\}$; 101 cases

if $y = 1$, then $z \in \{1, 2, \ldots, 99\}$; 99 cases

if $y = 2$, then $z \in \{2, 3, \ldots, 98\}$; 97 cases

and so on. Finally if $y = 50$, then $z \in \{50\}$.

1 case, So total number of cases

$101 + 99 + 97 + \ldots + 1 = 51^2$.

Case (ii) $x = 1$ then

If $y = 1$, then $z \in \{1, 2, \ldots, 98\}$; 98 cases

if $y = 2$, then $z \in \{2, 3, \ldots, 97\}$; 96 cases

if $y = 3$, then $z \in \{3, 4, \ldots, 96\}$; 94 cases

and so on. Finally if $y = 49$ then $z \in \{49, 50\}$.

2 cases

So total number of cases

$98 + 96 + 94 + \ldots + 2 = 49.50$.

Now lets generalize it, here also we will take two cases

Case (i) when x is even (let $x = 2k$)

Let, $x = 2k, 0 \le k \le 16$.

If $y = 2k$, then $z \in \{2k, 2k+1, \ldots, 100 - 4k\}$

total (101–6k) cases, if $y = 2k + 1$,

then $z \in \{2k+1, 2k+2, \ldots, 99 - 4k\}$;

total (99–6k) cases, if $y = 2k + 2$,

then $z \in \{2k+2, 2k+3, \ldots, 99 - 4k\}$;

total (97–6k) cases and so on.

Finally, if $y = 50 - k$, then $z \in \{50 - k\}$.

Adding all the above cases we get

$$(101 - 6k) + (99 - 6k) + (97 - 6k) + \ldots + 1 = (51 - 3k)^2$$

possibilities.

Let x be odd, say, $x = 2k + 1, 0 \le k \le 16$.

If $y = 2k + 1$,

then $z \in \{2k+2, 2k+3, \ldots, 97 - 4k\}$;

(98–6k) cases

If $y = 2k + 3$,

then $z \in \{2k+3, 2k+4, \ldots, 96 - 4k\}$;

(96–6k) cases and so on.

Finally, if

$y = 49 - k$, then $z \in \{49 - k, 50 - k\}$.

So total number of cases

$(98 - 6k) + (96 - 6k) + (94 - 6k) + \ldots + 2$

$= (49 - 3k)(50 - 3k)$ possibilities.

Thus the total number of triples, say T, is given by,

$$T = \sum_{k=0}^{16} (51 - 3k)^2 + \sum_{k=0}^{16} (49 - 3k)(50 - 3k).$$

$$\text{or } T = \sum_{k=1}^{17} (3k)^2 + \sum_{k=0}^{17} (3k - 2)(3k - 1)$$

$$= 18 \sum_{k=1}^{17} k^2 - 9 \sum_{k=0}^{17} k + 34$$

$$= 18\left(\frac{17.18.35}{6}\right) - 9\left(\frac{17.18}{2}\right) + 34 = 30{,}787.$$

Alternately

Using the property that the number of ways in which a given positive integer $n \ge 3$ can be expressed as a sum of three positive integers x, y, z (that is, $x + y + z = \pi$), subject to the condition $x \le y \le z$ is $\left\{\frac{n^2}{12}\right\}$, where {a} represents the integer closest to a. If zero values are allowed for x, y, z then the corresponding count is $\left\{\frac{(n+3)^2}{12}\right\}$, where now $n \ge 0$.

In the given question $n = x + y + z \in \{0, 1, 2, \ldots, 100\}$, so the number of ways is

$$\sum_{n=0}^{100} \left\{\frac{(n+3)^2}{12}\right\}.$$

By using above formula and taking necessary correction we get total number of ways is 30787.

CHAPTER 6

COMPLEX NUMBERS

6.1 INTRODUCTION

6.1.1 Introduction of Imaginary Number

A linear equation is always solvable in real numbers, but not all quadratic equations has this property. The simplest of such equation is $x^2 + 1 = 0$. Until the 18th century, mathematicians avoided such quadratic equation. It was Leonhard Euler who first introduced the number 'i' as $\sqrt{-1}$ and called it as 'imaginary unit' or 'iota'. After introducing 'i' now mathematicians are able to solve the equation $x^2 + 1 = 0$, as $x^2 = -1$ or $x = \pm\sqrt{\ } = \pm i$.

Consider one more equation $x^2 + 4 = 0$, or $x^2 = -4 = 4(i^2)$ or $x = \pm 2i$

Now with the introduction of 'i' we are able to solve the quadratic equation whose discriminant is less than zero.

6.1.2 Properties of iota

In this section we will discuss properties of imaginary number 'i'. From definition we know that $i^2 = -1$, now see the different exponents of 'i'

$$i^1 = i$$
$$i^2 = -1$$
$$i^3 = -i$$
$$i^4 = 1$$
$$i^5 = i$$
$$i^6 = -1$$
$$i^7 = -i$$
$$i^8 = 1 \text{ and so on}$$

Thus for any integer n

$$i^{4n} = 1$$
$$i^{4n+1} = i$$
$$i^{4n+2} = -1$$
$$i^{4n+3} = -i$$

Illustration 1. Find the value of i^{103},

Solution: First of all divide 103 by 4,

Since $\quad 103 = 4 \cdot (25) + 3$

So, $\quad i^{103} = i^{4\cdot(25)+3} = i^{4\cdot 25} i^3$

$$= (i^4)^{25} i^3 = (1)^{25} i^3$$
$$= i^3 = i^2 . i = -i$$

Illustration 2. Find the value of i^{97},

Solution: Divide 97 by 4. We get $97 = 4(24) + 1$

$$i^{97} = i^{4\cdot(24)+1} = i^{4\cdot 24} i^1 = (i^4)^{24} i^1 = i$$

Illustration 3. Find the value of $\frac{1}{i}$

Solution: Consider $\frac{1}{i}$ now multiplying by 'i' in numerator and denominator we get

$$\frac{1}{i} = \frac{1(i)}{i(i)} = \frac{i}{i^2} = \frac{i}{-1} = -i$$

Illustration 4. Find the value of

$\sqrt{-4}+2\sqrt{-9}+3\sqrt{-16}+4\sqrt{-25}$

Solution: Consider $\sqrt{-4}=\sqrt{4i^2}=2i$

$$\sqrt{-9}=3i$$
$$\sqrt{-16}=4i$$
$$\sqrt{-25}=5i$$
$$\sqrt{-4}+2\sqrt{-9}+3\sqrt{-16}+4\sqrt{-25}$$
$$=(2i)+2(3i)+3(4i)+4(5i)$$
$$=(2+6+12+20)\,i=40\,i$$

Illustration 5. Find the value of S

$= \mathbf{1 + i + i^2 + i^3 + ...+ i^{101}}$

Solution: $S = 1 + i + i^2 + i^3 + ...+ i^{101}$

(It is a G.P.)

$$=\frac{i^{102}-1}{i-1}\left[\because 1+r+r^2+....+r^n=\frac{r^{n+1}-1}{r-1}\right]$$
$$=\frac{-1-1}{i-1}\left(\because i^{102}=i^{4\times 25+2}=i^{4\times 25}\,i^2=i^2=-1\right)$$
$$=\frac{2}{1-i}$$
$$=\frac{2}{(1-i)}\frac{(1+i)}{(1+i)}=\frac{2(1+i)}{1+1}=1+i$$

Illustration 6. Find the value of

$i^n + i^{n-1} + i^{n-2} + i^{n-3}$ for all $n \in N$.

Solution: Let $S = i^n + i^{n-1} + i^{n-2} + i^{n-3}$

$$= i^n(1 + i^{-1} + i^{-2} + i^{-3})$$
$$= i^n\left[1+\frac{1}{i}+\frac{1}{i^2}+\frac{1}{i^3}\right]$$
$$= i^n\left[1+\frac{1}{i}-1-\frac{1}{i}\right]$$
$$(\because i^2=-1 \text{ and } i^3=-i)$$
$$=0$$

6.1.3 Complex Number

Now, we have already discussed about the real number and the imaginary number, so we can discuss complex number

Any number which is expressed in the form as $z = x + iy$ where $x, y \in R$ and $i^2 = -1$ or $i = \sqrt{-1}$

or $z = \{x + iy, x, y \in R, i = \sqrt{-1}\}$.

For example, $(2 + 6i)$, $(2 + i)$, $(1+2i)$, $(1 - 2i)$, all are complex numbers.

For any complex number $z = x + iy$, x is called as real part of z or Re(z) and y is called as imaginary part of z or Im (z).

Thus, for $z = 5 - 3i$.

Re (z) = 5 and Im (z) = –3.

For any complex number z, if Re (z) is zero then z is called as purely imaginary. Similarly if Im(z) is zero then z is called as purely real number.

Example of purely imaginary numbers:- $z = 2i$, $z = -4i$ etc

6.1.4 Algebra of Complex Number

Let us consider two complex numbers z_1 and z_2 as

$$z_1 = x_1 + iy_1 \text{ where } x_1, y_1 \in R$$
$$z_2 = x_2 + iy_2 \text{ where } x_2, y_2 \in R.$$

Now, we will discuss the different algebraic operations.

(a) Addition of two complex numbers

Real term is added to real term and imaginary term is added to imaginary term

$$z_1 + z_2 = (x_1 + iy_1) + (x_2 + iy_2)$$
$$= (x_1 + x_2) + i(y_1 + y_2)$$

Example if

$$z_1 = 5 + 3i \text{ and } z_2 = 7 + 2i$$
$$z_1 + z_2 = (5 + 3i) + (7 + 2i)$$
$$= (5 + 7) + i(3 + 2) = 12 + 5i$$

similarly, if $z_1 = 5 + 3i$ and $z_2 = 6 - 2i$

$$z_1 + z_2 = (5 + 3i) + (6 - 2i)$$
$$= (5 + 6) + i(3 - 2) = 11 + i$$

(b) Subtraction of two complex numbers

Let $z_1 = x_1 + iy_1$ and $z_2 = x_2 + iy_2$

$$z_1 - z_2 = (x_1 + iy_1) - (x_2 + iy_2)$$
$$= (x_1 - x_2) + i(y_1 - y_2)$$

For example,

let $z_1 = 3 + 3i$ and $z_2 = 5 - i$

$$z_1 - z_2 = (3 + 3i) - (5 - i)$$
$$= (3 - 5) + i[3 - (-1)]$$
$$= -2 + 4i$$

(c) Multiplication of two complex numbers z_1 and z_2

Here $z_1 = x_1 + iy_1$ and $z_2 = x_2 + iy_2$

$$z_1 z_2 = (x_1 + iy_1)(x_2 + iy_2)$$
$$z_1 z_2 = x_1 x_2 + iy_1 x_2 + ix_1 y_2 + i^2 y_1 y_2$$
$$= (x_1 x_2 - y_1 y_2) + i(x_1 y_2 + x_2 y_1)$$
$$(\because i^2 = -1)$$

Let $z_1 = 3 + 4i$ and $z_2 = 3 + 4i$

$$z_1 z_2 = (3 + 4i)(3 + 4i)$$
$$= 9 + 12i + 12i + 16i^2$$
$$= (9 - 16) + i(12 + 12) \ (\because i^2 = -1)$$
$$= -7 + 24i$$

(d) Division of two complex numbers

$$\frac{z_1}{z_2} = \frac{x_1 + i y_1}{x_2 + i y_2}$$

Now, similar to the concept of rationalization

$$= \frac{(x_1 + i y_1)(x_2 - i y_2)}{(x_2 + i y_2)(x_2 - i y_2)}$$

$$= \frac{x_1 x_2 + i x_2 y_1 - i x_1 y_2 - i^2 y_1 y_2}{(x_2^2 + y_2^2)}$$

$$= \frac{(x_1 x_2 + y_1 y_2) + i(x_2 y_1 - x_1 y_2)}{(x_2^2 + y_2^2)}$$

$$= \frac{(x_1 x_2 + y_1 y_2)}{(x_2^2 + y_2^2)} + i\frac{(x_2 y_1 - x_1 y_2)}{(x_2^2 + y_2^2)}$$

Illustration 7. Simplify $\frac{1+3i}{2-4i}$

Solution: Multiply numenator as well as denominator by (2 + 4i).

$$z = \frac{(1+3i)(2+4i)}{(2-4i)(2+4i)}$$

$$= \frac{2 + 6i + 4i + 12i^2}{2^2 + 4^2}$$

$$= \frac{(2-12) + 10i}{4 + 16}$$

$$= \frac{-10 + 10i}{20} = \frac{-1 + i}{2} = -\frac{1}{2} + \frac{i}{2}$$

6.1.5 Equality of Two Complex Numbers

Two complex numbers are said to be equal if and only if their real part as well as imaginary parts are equal.

Let us consider two complex numbers z_1 and z_2 as follows

let $z_1 = x_1 + i y_1$ and $z_2 = x_2 + i y_2$.

If $z_1 = z_2$, so $x_1 + i y_1 = x_2 + i y_2$

Thus, $x_1 = x_2$ and $y_1 = y_2$

Here real part is equated to real part and imaginary part is equated to imaginary part.

Illustration 8. If $(2x_1 + 3x_2) + i(-3x_1 + 5x_2) = 13 + 9i$ then find x_1 and x_2.

Solution: Comparing real and imaginary parts we get

$$2x_1 + 3x_2 = 13$$
$$-3x_1 + 5x_2 = 9$$

Solving above two equations we get $x_1 = 2$ and $x_2 = 3$

6.1.6 Square root of a Complex Number

Let us discuss the method to find the square root of a complex number step by step.

Suppose, we have to evaluate square root of (a + *i* b), then

Step – I: $\sqrt{(a + i b)} = x + i y$

Step – II: Square both sides and equate the real and imaginary parts we get,

$$a + i b = (x^2 - y^2) + i\,2xy$$
$$x^2 - y^2 = a \text{ and } 2xy = b$$

Step – III: $(x^2 + y^2)^2 = (x^2 - y^2)^2 + (2xy)^2 = a^2 + b^2$

Step – IV: $(x^2 + y^2) = \pm\sqrt{(a^2 + b^2)}$

(Ignoring negative sign) $= \sqrt{(a^2 + b^2)}$

Step – V: Solve two equations simultaneously.

$$x^2 - y^2 = a \text{ and } x^2 + y^2 = \sqrt{(a^2 + b^2)}$$

$$x^2 = \frac{a + \sqrt{(a^2 + b^2)}}{2};$$

$$y^2 = \frac{\sqrt{(a^2 + b^2)} - a}{2}$$

Step – VI:

$$x = \pm\sqrt{\frac{a + \sqrt{(a^2 + b^2)}}{2}};$$

$$y = +\sqrt{\frac{\sqrt{(a^2 + b^2)} - a}{2}}$$

Step – VII: Now consider 2xy = b, we have two cases

Case (i) b > 0

So, xy > 0

⇒ x > 0, y > 0

or x < 0, y < 0

Case (ii) b < 0

So, xy < 0

⇒ x > 0, y < 0

or x < 0, y > 0

Formula

$$\sqrt{z} = \pm\left(\sqrt{\frac{|z|+a}{2}} + i\sqrt{\frac{|z|-a}{2}}\right)$$

Illustration 9. Find the square root of $(3+4i)$

Solution: Let us assume that $\sqrt{(3+4i)}$ = x + *i* y where x, y ∈ R.

Now square both sides,

$$3 + 4i = x^2 - y^2 + i\,2xy.$$

Equate real and imaginary parts.

$$x^2 - y^2 = 3 \text{ and } 2xy = 4$$

$$(x^2 + y^2)^2 = (x^2 - y^2)^2 + (2xy)^2$$

$$= 3^2 + 4^2 = 5^2$$

$$x^2 + y^2 = \pm 5$$

$$x^2 + y^2 = 5 \quad \text{(Neglecting –ve sign)}$$

Now use $x^2 - y^2 = 3$

So, $2x^2 = 8$ thus $x^2 = 4$ and $y^2 = 1$

So, $x = \pm 2, y = \pm 1$

Now, consider 2xy = 4 so, xy = 2

As xy = 2 (A positive quantity)

Here, x > 0, y > 0 or x < 0, y < 0

So, either x = 2, y = 1 or x = –2, y = –1

Thus, $\sqrt{(3+4i)} = 2 + i$ or $(-2 - i)$

Illustration 10. Find the square root of 3 – 4i

Solution: Let $\sqrt{3-4i} = x + iy$

Now square both sides,

$$3 - 4i = x^2 - y^2 + i\,2xy$$

Equating real and imaginary parts, we get

$$x^2 - y^2 = 3 \quad \text{...(i)}$$

$$2xy = -4 \quad \text{...(ii)}$$

$$(x^2 + y^2)^2 = (x^2 - y^2)^2 + (2xy)^2$$

$$= 3^2 + (-4)^2 = 25$$

$$x^2 - y^2 = 5 \text{ (Ignoring – ve number)} \quad \text{...(iii)}$$

From equations (i) and (iii).

$$x^2 = 4 \text{ and } y^2 = 1$$

So, $x = \pm 2$ and $y = \pm 1$

From equation (ii), xy = – 2 (Negative)

So, if x > 0, then y < 0 or if x < 0, then y > 0

So, x = 2, y = – 1 or x = – 2, y = 1

Thus $\sqrt{3-4i} = 2 - i$ or $= -2 + i$

Illustration 11. $i^{57} + \frac{1}{i^{125}}$, when simplified has the value

(a) 0 (b) 2i
(c) – 2i (d) 2

Solution: (a)

$$i^{57} + \frac{1}{i^{125}} = \left(i^4\right)^{14} i + \frac{1}{\left(i^4\right)^{31} i} = i + \frac{1}{i} = i - i = 0$$

Illustration 12. If n is any positive integer, then the value of $\frac{i^{4n+1} - i^{4n-1}}{2}$ equals

(a) 1 (b) –1
(c) i (d) – i

Solution: (c) We have,

$$\frac{i^{4n+1} - i^{4n-1}}{2} = \frac{i^{4n}\cdot i - i^{4n}\cdot i^{-1}}{2} = \frac{i - \frac{1}{i}}{2}$$

[since $i^{4n} = (i^4)^n = (1)^n] = 1$

$$= \frac{i^2 - 1}{2i} = \frac{-1-1}{2i} = \frac{-2}{2i} = \frac{1}{-i} = i$$

Illustration 13. For a positive integer –n, the expression $(1-i)^n\left(1-\frac{1}{i}\right)^n$ equals

(a) 0 (b) $2i^n$
(c) 2^n (d) 4^n

Solution: (c) $(1-i)^n\left(1-\frac{1}{i}\right)^n = (1-i)^n(1+i)^n$ 1 1

$$= (1 + 1)^n = 2^n$$

Illustration 14. $\left(\frac{1+i}{1-i}\right)^2 + \left(\frac{1-i}{1+i}\right)^2$ is equal to

(a) $2i$ (b) $-2i$
(c) -2 (d) 2

Solution: (c) We have,

$$\left(\frac{1+i}{1-i}\right)^2 + \left(\frac{1-i}{1+i}\right)^2 = \frac{2i}{-2i} + \frac{-2i}{2i}$$

[Since $(1 + i) = 1 + i^2 + 2i =$ and $(1 - i)^2$

$= 1 + i^2 - 2i = -2i]$

$= -1 - 1 = -2$

Illustration 15. If $b + ic = (1+a)z$ and 1,
and $a^2 + b^2 + c^2 = 1$,

then $\frac{1+iz}{1-iz}$ is equal to

(a) $\frac{a-ib}{1-c}$ (b) $\frac{a-ib}{1+c}$

(c) $\frac{a+ib}{1-c}$ (d) $\frac{a+ib}{1+c}$

Solution: (d) We have, $b+ic=(1+a)z$

$$\Rightarrow z=\frac{b+ic}{1+a} \Rightarrow iz=\frac{ib-c}{1+a}$$

$$\Rightarrow \frac{1+iz}{1-iz}=\frac{1+a-c+ib}{1+a+c-ib}$$

$$=\frac{(1+a-c)+ib}{(1+a+c)-ib}\times\frac{(1+a+c)+ib}{(1+a+c)+ib}$$

$$=\frac{a^2+a+iab+ib}{1+a+ac+c}=\frac{(a+1)(a+ib)}{(a+1)(1+c)}=\frac{a+ib}{1+c}$$

Illustration 16. $\sqrt{2i}$ equals

(a) $1+i$ (b) $1-i$

(c) $-\sqrt{2}i$ (d) None of these

Solution: (a) $\sqrt{2i}=\sqrt{1+i^2+2i}=\sqrt{(1+i)^2}=1+i$

Illustration 17. The value of $\left[i^{19}+\left(\frac{1}{i}\right)^{25}\right]^2$ is

(a) 4 (b) – 4

(c) 2 (d) – 2

Solution: **(b)** We have,

$$\left[i^{19}+\left(\frac{1}{i}\right)^{25}\right]^2=\left[i^{16}\cdot i^3+\frac{1}{i^{24}\cdot i}\right]^2=\left[i^2\cdot i+\frac{1}{i}\right]^2$$

$$=(-i-i)^2=(-2i)=4i^2=-4$$

Illustration 18: The smallest positive integral value of n for which $\left(\frac{1-i}{1-i}\right)^n$ is purely imaginary with positive imaginary part is

(a) 1 (b) 3

(c) 5 (d) None of these

Solution: (b) $\because \frac{1-i}{1+i}=\frac{(1-i)^2}{1-i^2}=\frac{-2i}{2}=-i$

$\therefore \left(\frac{1-i}{1+i}\right)^n=(-i)^n=$ imaginary $\Rightarrow n=1,3,5.....$

But for +ve imaginary n = 3, 7,.......

∴ n = 3 is smallest positive integer

Illustration 19. If the number $\frac{(1-i)^n}{(1+i)^{n-2}}$ is real and positive, then n is

(a) any integer (b) 2λ

(c) 4λ + 1 (d) None of these

Solution: (c)

$$\frac{(1-i)^n}{(1+i)^{n-2}}=\frac{(1-i)^n}{(1+i)^{n-2}}\times\frac{(1-i)^{n-2}}{(1-i)^{n-2}}=2(-i)^{n-1}=2(-1)^{\frac{n-1}{2}}$$

This is positive and real if $\frac{n-1}{2}$ is even. Let. $\frac{n-1}{2}=2\lambda.\ \therefore n=4\lambda+1$

Illustration 20. The smallest positive number n for which $(1+i)^{2n}=(1-i)^{2n}$

(a) 4 (b) 8

(c) 2 (d) 12

Solution: (c) $(1+i)^{2n}=(1-i)^{2n}\Rightarrow(2i)^n=(-2i)^n$

∴ The smallest value of is 2.

Concept Applicator (CA)-1

1. Find the value of i^{315}
2. Find the value of $S=1+i+i^2+i^3+..........+i^{30}$
3. Find the value of $i^n+i^{n+1}+i^{n+2}+i^{n+3}$ for all n ∈ N.
4. Simplify $\frac{1}{-2+i}$
5. Find the value of the sum $\sum_{n=0}^{13}(i^n+i^{n+1})$, where $i=\sqrt{-1}$,
6. Simplify the complex number $$\frac{2^n}{(1+i)^{2n}}+\frac{(1+i)^{2n}}{2^n}, n\in I,$$
7. If n is an odd integer, $i=\sqrt{-1}$ then $(1+i)^{6n}+(1-i)^{6n}$ is equal to
8. The number $\frac{(1-i)^3}{1-i^3}$ is equal to
9. The smallest positive integer n for which $\left(\frac{1+i}{1-i}\right)^n=-1$
10. Find the square root of $-3-4i$

6.2 MODULUS & CONJUGATE

6.2.1 Geometric Representation of a Complex Number

Consider a complex number z = x + iy, here real part is x and imaginary part is y. We are aware about the Cartesian co-ordinate plane, it is also called Argand plane. The x-axis of Cartesian co-ordinate plane is called Real axis in argand plane and y-axis is called imaginary axis in argand plane.

The plane, in which complex number is presented, is called as Gaussian or Argand plane.

We will try to understand this with an example

Example : Suppose we have been given a complex number $Z = 3 + 4i$

Here Re (z) = 3 and Im (z) = 4

Then find a point P on the xy-plane (Argand plane) such that its x-coordinate = Real part of z = 3 and y – coordinate = Imaginary part of z = 4. Thus, point P represents complex number, where $p = (3, 4)$.

Now we have following cases-

Case (i) : If R(z) > 0, Im (z) > 0, then the complex number lies in 1st quadrant.

Case (ii) : If R(z) < 0, Im (z) > 0, then the complex number lies in 2nd quadrant.

Case (iii) : If R(z) < 0, Im (z) < 0, then the complex number lies in 3rd quadrant.

Case (iv) : If R(z) > 0, Im (z) < 0, then the complex number lies in 4th quadrant.

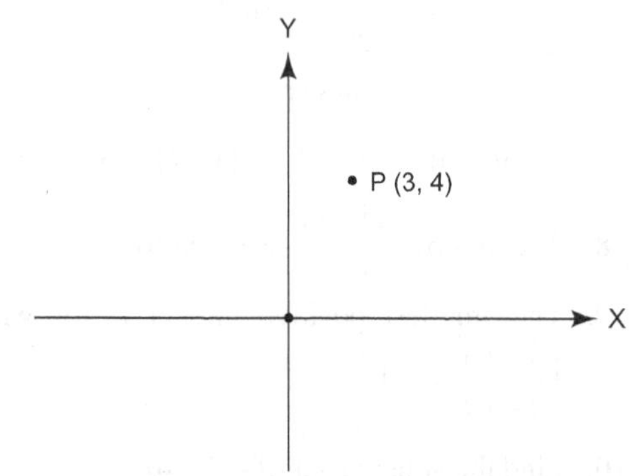

6.2.2 Modulus Complex Number

Modulus of a complex number is defined as distance of that point from origin on the Argand Plane.

When we join origin 'O' to the given point 'P', then OP is called the modulus of z and is denoted by | z |.

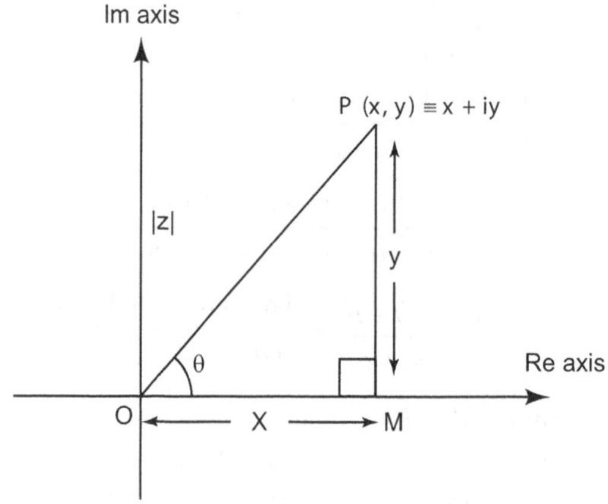

Thus, $|z| = \text{distance } OP = \sqrt{(x^2 + y^2)}$

$$|z| = \sqrt{(x^2 + y^2)}$$

$$= \sqrt{\left[(Re\, z)^2 + (Im\, z)^2\right]}$$

Example: Let $z_1 = 3 + 4i$ then $|z_1|$

$$= \sqrt{(3^2 + 4^2)} = 5$$

If $z_2 = 4 - 3i$, $|z_2|$

$$= \sqrt{(4)^2 + (-3)^2} = 5$$

Illustration 1. If $z_1 = 3 - 4i$ and $z_2 = -3 + 4i$ then find the value of $|z_1 - z_2|$

Solution: Given that $z_1 = 3 - 4i$ and $z_2 = -3 + 4i$ so $z_1 - z_2 = 6 - 8i$ then find the value of

$$|z_1 - z_2| = \sqrt{6^2 + 8^2} = 10$$

6.2.3 Properties of Conjugate of Complex Number

Let z = a + ib be a complex number. We define conjugate of z, denoted by $\overline{z}$ to be the complex number a – ib. i.e., if z = a + ib, then $\overline{z} = a - ib$.

Consider a complex number, $z = x + i\,y$; $x, y \in R$. Now replace i by $-i$ and we will get a new complex number called as conjugate of given complex number z. It is denoted by $\bar{z}$. Thus $\bar{z} = x - i\,y$.

So, if $z = 3 + 2\,i$, then $\bar{z} = 3 - 2\,i$

$z = -3 + 4\,i$, then $\bar{z} = -3 - 4\,i$

$z = 3 - 4\,i$, then $\bar{z} = 3 + 4\,i$

$z = 3 + 5\,i$, then $\bar{z} = 3 - 5\,i$

Geometrical understanding

$$z = x + i\,y;\ \bar{z} = x - i\,y.$$

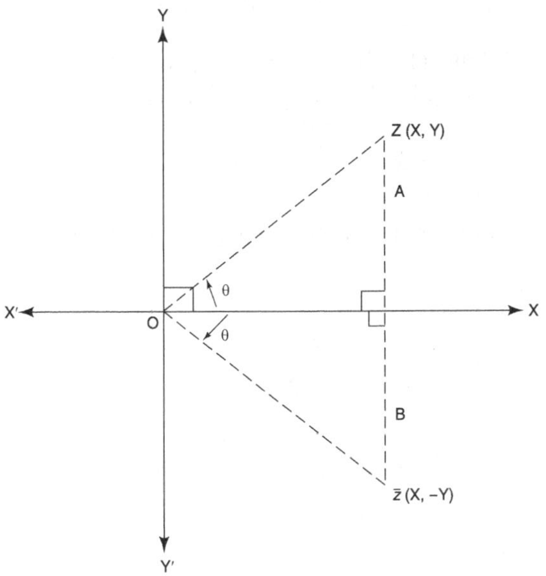

Here A and B are mirror image of each other with respect to x–axis.

Properties of Conjugate

(i) $\overline{(\bar{z})} = z$

Explanation: Let $z = x + i\,y$

It conjugate, $\bar{z} = x - i\,y$

Again conjugate, $\bar{\bar{z}} = x + i\,y = z$

Thus, conjugate of a conjugate gives original complex number, z.

(ii) $z + \bar{z} = 2\,\text{Re}\,(z)$

Explanation: Let $z = x + i\,y$, so $\bar{z} = x - i\,y$.

Thus, $z + \bar{z} = 2x = 2\,\text{Re}\,(z)$

(iii) $z - \bar{z} = 2\,i\,\text{Im}\,(z)$

Explanation: Here $z = x + i\,y$, $\bar{z} = x - i\,y$

$z - \bar{z} = 2\,i\,y = 2\,i\,\text{Im}\,(z)$

(iv) If $z = \bar{z}$, then z is purely real.

Explanation: Purely real means imaginary part of the complex number is zero. Let $z = x + i\,y$, so $\bar{z} = x - i\,y$

Thus, $z = \bar{z}$

$x + i\,y = x - i\,y$

$2\,i\,y = 0$

$\Rightarrow y = 0$

As imaginary part is zero, hence z is purely real.

(v) If $z + \bar{z} = 0$ then z is purely imaginary.

Explanation: Purely imaginary means real part of the complex number is zero. Let $z = x + i\,y$, and $\bar{z} = x - i\,y$.

$z + \bar{z} = 2x = 0$

Thus $x = 0$

As real part is zero, z is purely imaginary.

(vi) $z\,\bar{z} = |z|^2$

Explanation:

Here $z = x + i\,y$, and $\bar{z} = x - i\,y$

$z\,\bar{z} = (x + i\,y)\,(x - i\,y)$

$= x^2 - i^2 y^2 = x^2 + y^2 \quad (\because i^2 = -1)$

$= |z|^2 \quad \left(\because |z| = \sqrt{(x^2 + y^2)}\right)$

(vii) $\overline{z_1 + z_2} = \bar{z}_1 + \bar{z}_2$

Explanation: Let $z_1 = x_1 + i\,y_1$ and $z_2 = x_2 + i\,y_2$

$z_1 + z_2 = (x_1 + x_2) + i\,(y_1 + y_2)$

$\overline{z_1 \quad z_2} = (x_1 + x_2) - i\,(y_1 + y_2)$

$= (x_1 - i\,y_1) + (x_2 - i\,y_2)$

$= \bar{z} + \bar{z}_2$

(viii) $\overline{z_1 z_2} = \bar{z}_1 \cdot \bar{z}_2$

Explanation: Let $z_1 = x_1 + i\,y_1$ and $z_2 = x_2 + i\,y_2$

$z_1 - z_2 = (x_1 - x_2) + i\,(y_1 - y_2)$

$\overline{z_1 - z_1} = (x_1 - x_2) - i\,(y_1 - y_2)$

$= (x_1 - i\,y_1) - (x_2 - i\,y_2)$

$= \bar{z} - \bar{z}_2$

(ix) $\overline{z_1\,z_2} = \bar{z}_1\,\bar{z}_2$

Explanation: Let $z_1 = x_1 + i\,y_1$ and $z_2 = x_2 + i\,y_2$

$\bar{z}_1 = x_1 - i\,y_1$; $\bar{z}_2 = x_2 - i\,y_2$

$z_1\,z_2 = (x_1 + i\,y_1)\,(x_2 + i\,y_2)$

$= (x_1\,x_2 - y_1 y_2) + i\,(x_1 y_2 + x_2 y_1)$

$\overline{z_1\,z_2} = (x_1\,x_2 - y_1 y_2) - i\,(x_1 y_2 + x_2 y_1)$

$\bar{z}_1\,\bar{z}_2 = (x_1 - iy_1)(x_2 - iy_2)$
$= (x_1x_2 - y_1y_2) - i(x_1y_2 + x_2y_1)$
Thus, $\overline{z_1z_2} = \bar{z}_1 \cdot \bar{z}_2$

Properties of Conjugate of a Complex Number

(i) $z_1 = z_2 \Rightarrow \bar{z}_1 = \bar{z}_2$
(ii) $z + \bar{z} = 2Re(z)$
(iii) $z - \bar{z} = 2i\,Im(z)$
(iv) $z = \bar{z} \Rightarrow z$ is purely real
(v) $z + \bar{z} = 0 \Rightarrow z$ is purely imaginary
(vi) $z\bar{z} = [\text{Re}(z)]^2 + [Im(z)]^2$
(vii) $\overline{z_1 + z_2} = \bar{z}_1 + \bar{z}_2$
(viii) $\overline{z_1 - z_2} = \bar{z}_1 - \bar{z}_2$
(ix) $\overline{z_1z_2} = \bar{z}_1\bar{z}_2$
(x) $\overline{\left(\frac{z_1}{z_2}\right)} = \frac{\bar{z}_1}{\bar{z}_2}$ if $z_2 \neq 0$
(xi) Polynomial function of complex number:- If P(z) $= a_0 + a_1z + a_2z^2 + \ldots + a_nz^n$ where $a_0, a_1, \ldots a_n$ and z are complex number, then

$$\overline{P(z)} = \bar{a}_0 + \bar{a}_1(\bar{z}) + \bar{a}_2(\bar{z})^2 + \cdots + \bar{a}_n(\bar{z})^n = \bar{P}(\bar{z})$$

where $\bar{P}(z) = \bar{a}_0 + \bar{a}_1z + \bar{a}_2z^2 + \cdots + \bar{a}_nz^n$

(xii) Rational expression of complex number:- If R (z) = $\frac{P(z)}{Q(z)}$ where P (z) and Q (z) are polynomials in z, and Q (z), $\overline{R(z)} = \frac{\overline{P(z)}}{\overline{Q(z)}}$

(xiii) Determinant of complex number:- If

$$z = \begin{vmatrix} a_1 & a_2 & a_3 \\ b_1 & b_2 & b_3 \\ c_1 & c_2 & c_3 \end{vmatrix}, \text{ then } \bar{z} = \begin{vmatrix} \bar{a}_1 & \bar{a}_2 & \bar{a}_3 \\ \bar{b}_1 & \bar{b}_2 & \bar{b}_3 \\ \bar{c}_1 & \bar{c}_2 & \bar{c}_3 \end{vmatrix}$$

where $a_i, b_i, c_i\ (i = 1, 2, 3)$ are complex numbers.]

Illustration 2. Which one of the following is equal to Im (z)

(a) $\frac{1}{2}(z + \bar{z})i$ (b) $\frac{1}{2}(z - \bar{z})$
(c) $\frac{1}{2}(\bar{z} - z)i$ (d) None of these

Solution: If $z = x + iy$ and $\bar{z} = x - iy$

$\therefore\ 2iy = z - \bar{z}$ or $Im(z)$

$$= \frac{z - \bar{z}}{2i} = \frac{-i(z - \bar{z})}{2} = \frac{i(\bar{z} - z)}{2}$$

Option (c)

Illustration 3. If $(1 + i)z = (1 - i)\bar{z}$, then z is

(a) $x(1 - i), x \in R$ (b) $x(1 + i), x \in R$
(c) $\frac{x}{1 + i}, x \in R^+$ (d) None of these

Solution: $(1 + i)(x + iy) = (1 - i)(x - iy)$
$\Rightarrow (x - y) + i(x + y) = (x - y) - i(x + y)$
$\therefore\ x + y = 0$
$\therefore\ z = x - ix = x(1 - i)$

Option (a)

Illustration 4. Find the conjugate complex number of $\frac{2 - i}{(1 - 2i)^2}$

Solution: Let $z = \frac{2 - i}{(1 - 2i)^2}$

$$\therefore z = \frac{2 - i}{-3 - 4i} = \frac{-2 + i}{3 + 4i} \times \frac{3 - 4i}{3 - 4i} = -\frac{2}{25} + \frac{11}{25}i$$

$$\therefore \bar{z} = -\frac{2}{25} - \frac{11}{25}i$$

6.2.4 Properties of Modulus of Complex Number

Let z = a + ib be a complex number then modulus or the absolute value of z of a complex number is defined as to be the real number $\sqrt{a^2 + b^2}$ and denote it by $|z|$. And $|z| \geq 0\ \forall z \in c$

Properties of Modulus

(i) $|z| = 0 \Rightarrow z = 0$
(ii) $|z| = |\bar{z}| = |-z|$
(iii) $-|z| \leq \text{Re}(z) \leq |z|$
(iv) $-|z| \leq \text{Im}(z) \leq |z|$
(v) $z\bar{z} = |z|^2$
(vi) $|z_1z_2| = |z_1||z_2|$
(vii) $\left|\frac{z_1}{z_2}\right| = \frac{|z_1|}{|z_2|}$, if $z_2 \neq 0$

(viii) $|z_1+z_2|^2=|z_1|^2+|z_2|^2+\bar{z}_1z_2+z_1\bar{z}_2$
$=|z_1|^2+|z_2|^2+2\,\text{Re}(z_1\bar{z}_2)$

(ix) $|z_1-z_2|^2=|z_1|^2+|z_2|^2-\bar{z}_1z_2-z_1\bar{z}_2$
$=|z_1|^2+|z_2|^2+2\,\text{Re}(z_1\bar{z}_2)$

(x) $|z_1+z_2|^2+|z_1-z_2|^2=2\left(|z_1|^2+|z_2|^2\right)$

(xi) if a and b are real numbers and z_1, z_2 are complex numbers, then

$$|az_1+bz_2|^2-|bz_1-az_2|^2=(a^2+b^2)\left(|z_1|^2+|z_2|^2\right)$$

(xii) If $z_1, z_2 \neq 0$, then $|z_1+z_2|^2=|z_1|^2+|z_2|^2$

$\Rightarrow \frac{z_1}{z_2}$ is purely imaginary.

(xiii) If z_1 and z_2 are two complex numbers, then $|z_1+z_2|\le |z_1|+|z_2|$. The equality holds if and only if $z_1\bar{z}_2\ge 0$.

In general, $|z_1+z_2+...+z_n|\le|z_1|+|z_2|+...+|z_n|$ and the equality sign holds if and only if the ratio of any two non-zero terms is positive.

(xiv) $|z_1-z_2|\le|z_1|+|z_2|$

(xv) $\big||z_1|-|z_2|\big|\le|z_1|+|z_2|$

(xvi) $|z_1-z_2|\ge\big||z_1|-|z_2|\big|$

(xvii) If a_1, a_2, a_3, a_4 are four complex numbers, then $|z-a_1|+|z-a_2|+|z-a_3|+|z-a_4|\ge$ max $\{|a_l-a_l|+|a_m-a_n| : l, m,$ are distinct integers lying in $\{1, 2, 3, 4\}$ and m < n $\}$.

Illustration 5. If is a complex number satisfying the relation $|z+1|=z+2(1+i)$ then find z

Solution Let $z=x+iy$

$\therefore\quad |x+iy+1|=x+iy+2(1+i)$

$\sqrt{(x+1)^2+y^2}=(x+2)+i(y+2)$

$\Rightarrow\quad \sqrt{(x+1)^2+y^2}=x+2$ and $y+2=0$

$\Rightarrow\quad (x+1)^2+4=(x+2)^2$ and $y=-2$

$\Rightarrow\quad 2x+5=4x+4$ and $y=-2$

$\Rightarrow z=\frac{1}{2}(1-4i)$

Illustration 6. If $z = 1 + i\tan\alpha$, where $\pi<\alpha<3\pi/2$, then $|z|$ is equal to

Solution: As $z = 1 + i\tan\alpha$

$\therefore\quad |z|=\sqrt{1+\tan^2\alpha}=|\sec\alpha|$

Since $\pi<\alpha<\frac{3\pi}{2}$ so $|z|=-\sec\alpha$,

Illustration 7. Let z be a complex number of constant modulus such that z^2 is purely imaginary then the number of value of z is

(a) 2 (b) 1
(c) 4 (d) infinite

Solution: (c) Since z^2 is purely imaginary

$\text{Re}\{(x+iy)^2\}=0$ i.e., $x^2-y^2=0$

Solving these, $x=\pm k, y=\pm k$ so possible number of values of z is 4

Illustration 8. For a complex number z find the minimum value of $|z|+|z-2|$

Solution: $|z|+|z-2|$ using $|z_1|+|z_2|\ge|z_1-z_2|$

We have $|z|+|z-2|\ge|z-(z-2)|$

$\therefore\ |z|+|z-2|\ge 2$

So required minimum value is 2

Illustration 9. If $|z_1-1|<1, |z_2-2|<2, |z_3-3|<3$, then find the maximum value of $|z_1+z_2+z_3|$

Solution: $|z_1+z_2+z_3|$
$=|(z_1-1)+(z_2-2)+(z_3-3)+6|\le|z_1-1|+|z_2-2|+|z_3-3|+6<1+2+3+6=12$

So required maximum value is 12.

6.2.5 Multiplicative Inverse

Consider a complex number $z=x+iy$ its multiplicative inverse is defined as $z^{-1}=\frac{1}{z}$

$$\frac{1}{z}=\frac{1}{x+iy}=\left\{\frac{1}{x+iy}\right\}\left\{\frac{x-iy}{x-iy}\right\}=\frac{x-iy}{x^2+y^2}=\frac{\bar{z}}{|z|^2}$$

Illustration 10. Find the multiplicative inverse of $(6+5i)^2$ is

Solution: Let $z = (6 + 5i)^2 = 36 + 2(6)(5i) + 25i^2$
$= 36 + 60i - 25 = 11 + 60i$

Then, $\bar{z} = 11 - 60i$ and

$|z| = \sqrt{(11)^2 + (60)^2} = \sqrt{121 + 3600} = \sqrt{3721} = 61$

Multiplicative inverse of

$z = \frac{\bar{z}}{|z|^2} = \frac{11 - 60i}{(61)^2} = \frac{11}{(61)^2} - \frac{60}{(61)^2}i$

Concept Applicator (CA)-2

1. If $z = 1 + i$, then the multiplicative inverse of z^2 is
2. Find the multiplicative inverse of $\sqrt{2i}$
3. Find the value of $\left|\frac{(3+2i)^2}{4-3i}\right|$
4. Find the multiplicative inverse of $\frac{3+4i}{4-5i}$
5. If α, β are two different complex numbers such that $|\alpha| = 1, |\beta| = 1,$ then find the value of expression $\left|\frac{\beta - \alpha}{1 - \bar{\alpha}\beta}\right|$
6. If $|z| = 1$, then $\frac{1+z}{1+\bar{z}}$ is equal to (Here $z \neq -1$)
   (a) z (b) $\bar{z}$
   (c) $z + \bar{z}$ (d) None of these
7. If $z = x + iy$ and $w = \frac{1 - iz}{z - i}$, then $|w| = 1$ implies that in the complex plane
   (a) z lies on the imaginary axis
   (b) z lies on the real axis
   (c) z lies on the unit circle
   (d) None of the above
8. If z = x + iy, x, y real, then then k is equal to
   (a) $\sqrt{2}$ (b) $\sqrt{3}$
   (c) $\sqrt{5}$
   (d) None of these
9. Find the real values of x and y for which the complex numbers $-2 + i\,xy$ and $x + y - i$ are conjugate of each other.
10. Find the modulus of multiplicative inverse of $\frac{1}{a - ib} - \frac{1}{a - ic}$

6.3 AMPLITUDE & ARGUMENT

6.3.1 Argument of a Complex Number

Consider a complex number $z = x + iy$, 1st locate it on the Argand plane and let that point be P. When we join point P to the origin, then the angle θ which OP makes with positive direction of x–axis, is called argument (amplitude) of z. It is denoted by arg(*z*) or *amp*(*z*)

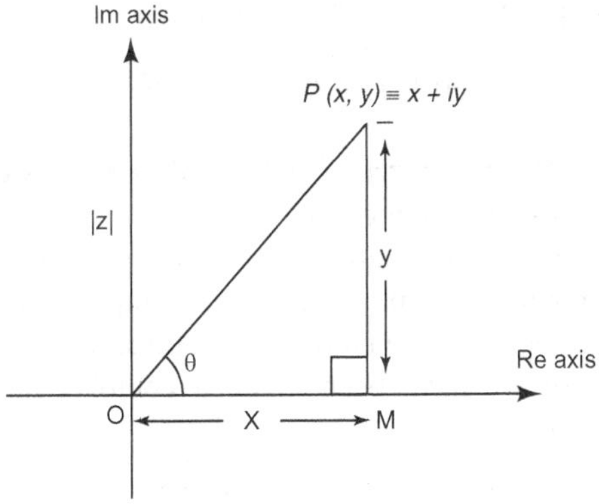

Here in above figure $\tan\theta = \frac{y}{x}$, so $\theta = \tan^{-1}(\frac{y}{x})$

or $$\theta = \tan^{-1}\left(\frac{Im(z)}{\operatorname{Re}(z)}\right)$$

> θ can have multiple values but all θ such that $-\pi < \theta \le \pi$ is called as principle value of argument or principle argumrnt.

Now we have four different cases which depends on location or quadrant in which complex number lies.

Case (i) : If complex number $z = x + iy$ is in first quadrant, then x > 0, y > 0

Here $|z| = \sqrt{(x^2 + y^2)}$

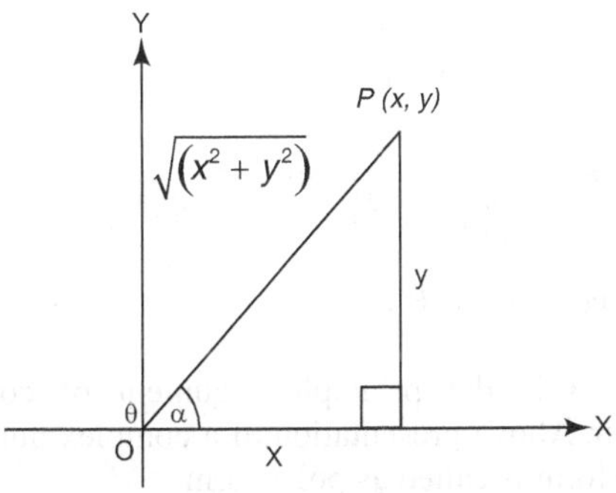

Here we define $\tan\alpha = |\frac{y}{x}|$

So, $\alpha = \tan^{-1}(|\frac{y}{x}|)$

Thus, argument, $\theta = \alpha = \tan^{-1}(|\frac{y}{x}|)$

Illustration 1. Find the argument of z = 3 + 4 i

Solution: Modulus of z, $|z| = \sqrt{(3^2 + 4^2)} = 5$

Here $\alpha = \tan^{-1}(|\frac{4}{3}|) = \tan^{-1}(\frac{4}{3})$

Thus, argument, $\theta = \alpha = \tan^{-1}(\frac{4}{3})$

Case (ii) : If complex number $z = x + iy$ is in 2nd quadrant, then x < 0, y > 0

Modulus of z is $|z| = \sqrt{(x^2 + y^2)}$

Then $\alpha = \tan^{-1}(|\frac{y}{x}|)$

Argument (*z*), $= \pi - \tan^{-1}(|\frac{y}{x}|)$

Illustration 2. Find the argument of z = –3 + 4i

Solution: Modulus of z, $|z| = \sqrt{(-3)^2 + (4^2)} = 5$

Here $\alpha = \tan^{-1}(|\frac{y}{x}|)$, $\alpha = \tan^{-1}(\frac{|3|}{|-4|}) = \tan^{-1}\frac{4}{3}$

Thus, argument, $(z) = \pi - \tan^{-1}(\frac{4}{3})$

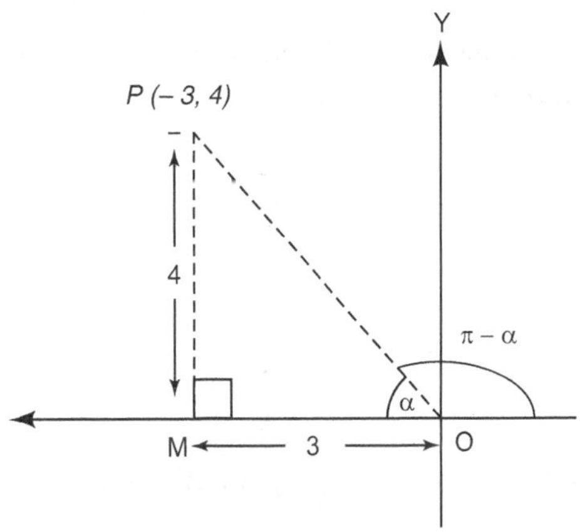

Case (iii) : If complex number $z = x + iy$ is in 3rd quadrant, then x < 0, y < 0

Here $\alpha = \tan^{-1}[|\frac{y}{x}|]$

And argument, $z = -(\pi - \alpha) = -[\pi - \tan^{-1}\left|\frac{y}{x}\right|]$

($\because$ Here angle is measured along clockwise direction, so we have taken –ve sign)

Illustration 3. Find the argument of z = –3 + 4i

Solution:

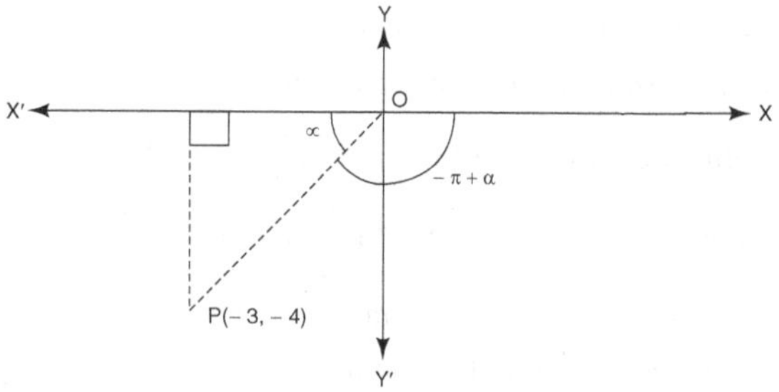

Modulus of z, $|z| = \sqrt{(-3)^2 + (-4^2)} = 5$

Here $\qquad \alpha = \tan^{-1}[|\frac{y}{x}|]$

$$= \tan^{-1}[\left|\frac{-4}{-3}\right|] = \tan^{-1}\left(\frac{4}{3}\right)$$

And argument, $\quad z = -(\pi - \alpha) = -[\pi - \tan^{-1}\left(\frac{4}{3}\right)]$

Case (iv) : If complex number $z = x + iy$ is in 4th quadrant, then $x > 0, y < 0$

$$1^{st} \text{ find } \propto = \tan^{-1}[|\frac{y}{x}|]$$

Argument, $\qquad z = -\alpha = -\tan^{-1}[|\frac{y}{x}|]$

Illustration 4. Find the argument of z = 3 – 4i

Solution. Given, z = 3 – 4*i*

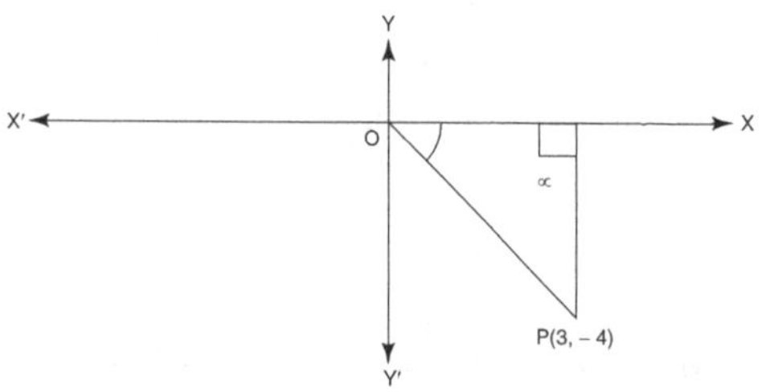

Modulus of z, $|z| = \sqrt{(3)^2 + (-4^2)} = 5$

Here $\alpha = \tan^{-1}|\frac{y}{x}| = \tan^{-1}\left|\frac{-4}{3}\right| = \tan^{-1}\left(\frac{4}{3}\right)$

$\arg(z) = -\alpha = -\tan^{-1}(\frac{4}{3})$

Summary: In any case 1st find $\alpha = \tan^{-1}\left|\frac{y}{x}\right|$ then –

Case (i): If z lies in 1st quadrant then

$$\arg(z) = \alpha = \tan^{-1}\left|\frac{y}{x}\right|$$

Case (ii): If z lies in 2nd quadrant then

$$\arg(z) = \pi - \alpha = \pi - \tan^{-1}\left|\frac{y}{x}\right|$$

Case (iii): If z lies in 3rd quadrant then

$$\arg(z) = -\pi + \alpha = -\pi + \tan^{-1}\left|\frac{y}{x}\right|$$

Case (iv): If z lies in 4th quadrant then

$$\arg(z) = -\alpha = -\tan^{-1}\left|\frac{y}{x}\right|$$

6.3.2 Polar Form of a Complex Number

Generally we represent a complex number as z = x + iy and in argand plane it represents co–ordinate of a point. Same point (or same complex number) can be represented in terms of |z| and arg(z).

Suppose we have been given a complex number $z = x + iy$

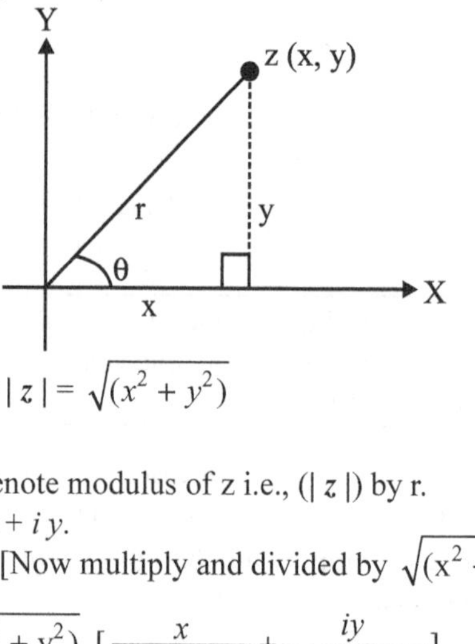

Then $r = |z| = \sqrt{(x^2 + y^2)}$

Let us denote modulus of z i.e., ($|z|$) by r.

So, $z = x + iy$.

[Now multiply and divided by $\sqrt{(x^2 + y^2)}$]

$$z = \sqrt{(x^2 + y^2)}\left[\frac{x}{\sqrt{(x^2 + y^2)}} + \frac{iy}{\sqrt{(x^2 + y^2)}}\right]$$

Here, $\sqrt{(x^2 + y^2)} = r$

and $\cos\theta = \frac{x}{\sqrt{(x^2 + y^2)}}$; $\sin\theta = \frac{y}{\sqrt{(x^2 + y^2)}}$

$z = r(\cos\theta + i\sin\theta)$

where θ is the principle argument of complex number. Above presentation of a complex number z in r, θ form is called as polar form.

Illustration 5. Write the complex number z = 1 + i in polar form

Solution: Given $z = 1 + i$

Here, z is in first quadrant.

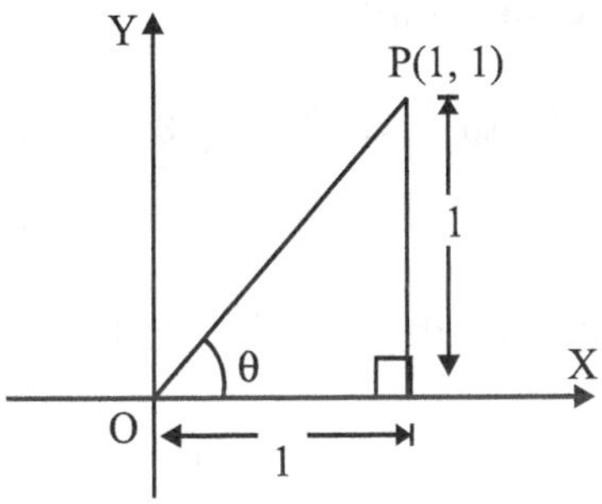

$$|z| = \sqrt{(1)^2 + (1)^2} = \sqrt{2}$$

$$\tan\theta = \frac{1}{1} = \tan\frac{\pi}{4} \text{ or } \arg(z) = \theta = \frac{\pi}{4}$$

$$\text{Polar form of } z = \sqrt{2}\left(\cos\left(\frac{\pi}{4}\right) + i\sin\left(\frac{\pi}{4}\right)\right)$$

Illustration 6. Write the complex number z = – 3 + 4i in polar form

Solution: Given complex number is z = – 3 + 4i and it lies in the 2nd quadrant.

$$|z| = \sqrt{(-\sqrt{3})^2 + (4)^2} = 5$$

$$\tan\alpha = \frac{4}{3}$$

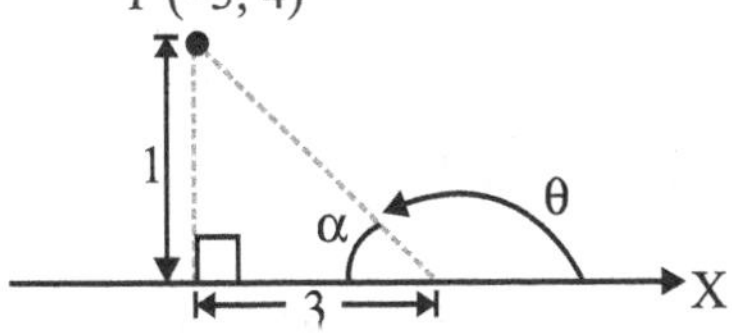

$$\arg z = \theta = (\pi - \alpha) \text{ where } \tan\alpha = \frac{4}{3},$$

Polar form of z,

$z = 5\,(\cos\theta + i\sin\theta)$

$$= 5\,[\cos(\pi-\alpha) + i\sin(\pi-\alpha)] \text{ where } \tan\alpha = \frac{4}{3}$$

6.3.3 Euler's Form of a Complex Number

We can represent $\cos\theta + i\sin\theta$ as $e^{-i\theta}$ and for $\cos\theta - i\sin\theta$, as $e^{-i\theta}$.

Thus
$$e^{i\theta} = \cos\theta + i\sin\theta$$
$$e^{-i\theta} = \cos\theta - i\sin\theta$$

Adding above two results we get $\dfrac{e^{i\theta} + e^{-i\theta}}{2} = \cos\theta$,

and on subtraction we get $\dfrac{e^{i\theta} - e^{-i\theta}}{2i} = \sin\theta$

Thus, for
$$z = x + iy \text{ if } |z| = r$$
$$z = r(\cos\theta + i\sin\theta) \text{ (Polar form)}$$
$$= re^{i\theta} \text{(Euler form) or } z(r, \theta)$$

Thus, $z = re^{i\theta}$, where $r = \sqrt{(x^2 + y^2)}$ and θ is argument, is called Euler form.

Sometime it is also represented as $z = |z|e^{i\theta}$.

It is very helpful in finding product and division of two complex numbers.

Product of Two Complex Numbers

Let us consider two complex numbers z_1, z_2 with modulus r_1, r_2 and argument θ_1, θ_2 respectively.

So $z_1 = r_1\, e^{i\theta_1}$ and $z_2 = r_2\, e^{i\theta_2}$

Product of z_1 and z_2,

$$z_1 z_2 = r_1 r_2\, e^{i(\theta_1+\theta_2)}$$
$$= r_1 r_2 [\cos(\theta_1 + \theta_2) + i\sin\theta(\theta_1 + \theta_2)]$$

It can also be verified by polar forms.

$$z_1 = r_1(\cos\theta_1 + i\sin\theta_1)\ ;\ z_2 = r_2(\cos\theta_2 + i\sin\theta_2)$$
$$z_1 z_2 = r_1 r_2 (\cos\theta_1 + i\sin\theta_1)(\cos\theta_2 + i\sin\theta_2)$$
$$= r_1 r_2 [(\cos\theta_1 \cos\theta_2 - \sin\theta_1 \sin\theta_2)$$
$$+ i(\cos\theta_1 \sin\theta_2 + \cos\theta_2 \sin\theta_1)]$$
$$= r_1 r_2 [\cos(\theta_1 + \theta_2) + i\sin\theta(\theta_1 + \theta_2)]$$
$$= r_1 r_2\, e^{i(\theta_1+\theta_2)}$$

Division by a Complex Numbers

Let
$$\mathbf{z_1 = r_1(\cos\theta_1 + i\sin\theta_1)} = r_1 e^{i\theta_1}$$
$$\mathbf{z_2 = r_2(\cos\theta_2 + i\sin\theta_2)} = r_2 e^{i\theta_2}$$

$$\frac{z_1}{z_2} = \frac{r_1 e^{i\theta_1}}{r_2 e^{i\theta_2}} = \frac{r_1}{r_2} e^{i(\theta_1-\theta_2)}$$

Illustration 7. Write $z = 1 - \sqrt{3}$ i in Euler's form

Solution: Given that $z = 1 - \sqrt{3}$ i

So,
$$|z| = \sqrt{(1)^2 + (-\sqrt{3})^2} = \sqrt{4} = 2$$

$$\tan\alpha = \sqrt{3} = \tan\frac{\pi}{3}$$

So, $\alpha = \frac{\pi}{3}$

$$\theta = -\alpha = -\frac{\pi}{3}$$

$\arg z = -\frac{\pi}{3}$ and $|z| = 2$

so $z = 2\left(\cos\left(-\frac{\pi}{3}\right) + i\sin\left(-\frac{\pi}{3}\right)\right) = 2e^{-\frac{i\pi}{3}}$

6.3.4 De-Moivre's Theorem

Consider a complex number $z = \cos\theta + i\sin\theta = e^{i\theta}$

Now consider

$$z^2 = (\cos\theta + i\sin\theta)^2$$

$$= (\cos^2\theta - \sin^2\theta) + i(2\cos\theta\sin\theta)$$

$$= \cos 2\theta + i\sin 2\theta = e^{i2\theta}$$

Similarly

$$z^3 = (\cos\theta + i\sin\theta)^3 = \cos 3\theta + i\sin 3\theta = e^{i3\theta}$$

And so on

$$z^n = (\cos\theta + i\sin\theta)^n = \cos n\theta + i\sin n\theta = e^{in\theta}$$

Consider n to be any integer then

$$(\cos\theta + i\sin\theta)^n = \cos(n\theta) + i\sin(n\theta)$$

Thus,

$$(\cos\theta + i\sin\theta)^n = (e^{i\theta})^n = e^{in\theta}$$

$$= \cos(n\theta) + i\sin(n\theta)$$

Here we must note that $n \in Z$ (set of integers).

Special Cases:

Case (i): Consider n to be any rational number of the form, $n = \frac{p}{q}$ where $q \neq 0$ and $p, q \in I$.

$$z^n = (\cos\theta + i\sin\theta)^n = \cos n\theta + i\sin n\theta = e^{in\theta}$$

or $z^{\frac{p}{q}} = (\cos\theta + i\sin\theta)^{\frac{p}{q}}$

$$= \cos\left(\frac{p}{q}\right)\theta + i\sin\left(\frac{p}{q}\right)\theta = e^{i\left(\frac{p}{q}\right)\theta}$$

Case (ii): $(\cos\theta - i\sin\theta)^n = \cos(n\theta) - i\sin(n\theta)$

We know $e^{-i\theta} = \cos\theta - i\sin\theta$ (Euler's form) thus

$(\cos\theta - i\sin\theta)^n = (e^{-i\theta})^n = e^{-in\theta} = \cos(n\theta) - i\sin(n\theta)$

Case (iii): If $z = \cos\theta + i\sin\theta$,

then $\frac{1}{z} = \cos\theta - i\sin\theta$.

Here, $z = e^{i\theta}$, so $\frac{1}{z} = e^{-i\theta} = \cos\theta - i\sin\theta$.

Case (iv):

$$(\sin\theta + i\cos\theta)^n = \left(\cos\left(\frac{\pi}{2} - \theta\right) + i\sin\left(\frac{\pi}{2} - \theta\right)\right)^n$$

$$= \left(e^{i\left(\frac{\pi}{2} - \theta\right)}\right)^n = e^{i\left(\frac{n\pi}{2} - n\theta\right)}$$

Illustration 8. Let $z = \frac{\cos\theta + i\sin\theta}{\cos\theta - i\sin\theta}, \frac{\pi}{4} < \theta < \frac{\pi}{2}$.

Then find arg z is

Solution: $z = (\cos\theta + i\sin\theta)^2 = \cos 2\theta + i\sin 2\theta$,

where $\frac{\pi}{2} < 2\theta < \pi$

$\therefore$ z is a complex number in the second quadrant.

So, $\frac{\pi}{2} < \arg z < \pi$

$\therefore$ $\arg z = \tan^{-1}(\tan 2\theta) = 2\theta - \pi$

Illustration 9. If $z = \cos\alpha + i\sin\alpha, 0 < \alpha < \pi/6$, then the argument of $1 + z^3$

Solution: $1 + z^3 = 1 + \cos 3\alpha + i\sin 3\alpha$

$= 2\cos^2\frac{3\alpha}{2} + 2i\sin\frac{3\alpha}{2}\cos\frac{3\alpha}{2}$

$= 2\cos\left(\frac{3\alpha}{2}\right)\left[\cos\left(\frac{3\alpha}{2}\right) + i\sin\left(\frac{3\alpha}{2}\right)\right]$

$\Rightarrow \arg(1 + z^3) = 3\alpha/2$

Illustration 10. If $z = \cos\alpha + i\sin\alpha, 0 < \alpha < \pi/6$, then the argument of $\frac{z^4 - 1}{z^3 + 1}$

Solution: From previous solution

$1 + z^3 = 2\cos\left(\frac{3\alpha}{2}\right)\left[\cos\left(\frac{3\alpha}{2}\right) + i\sin\left(\frac{3\alpha}{2}\right)\right]$

Similarly, $z^4 - 1 = -2\sin 2\alpha(\sin 2\alpha - i\cos 2\alpha)$

Consider

$$\frac{z^4-1}{z^3+1} = \frac{-\sin 2\alpha}{\cos(3\alpha/2)}\left[\cos\left(\frac{\alpha}{2}-\frac{\pi}{2}\right)+i\sin\left(\frac{\alpha}{2}-\frac{\pi}{2}\right)\right]$$

$$= \frac{-\sin 2\alpha}{\cos(3\alpha/2)}\left[\cos\left(\frac{\pi}{2}+\frac{\alpha}{2}\right)+i\sin\left(\frac{\pi}{2}+\frac{\alpha}{2}\right)\right]$$

Thus, $\arg\left(\frac{z^4-1}{z^3+1}\right) = \frac{\pi}{2}+\frac{\alpha}{2}$

Concept Applicator (CA)-3

1. Find the modulus and argument of $z = 1+i\sqrt{3}$
2. Find the modulus and argument of $z = \sqrt{3}+i$
3. Find the modulus and argument of $z = -1+i\sqrt{3}$
4. Write the complex number $z = \sqrt{3}-i$ in polar form
5. Write the complex number $z = -1-i$ in polar form
6. If $e^{i\theta} = \cos\theta + i\sin\theta$, then for the ΔABC, find $e^{iA}.e^{iB}.e^{iC}$
7. If $z = \frac{\sqrt{3}+i}{\sqrt{3}-i}$, then find the fundamental amplitude of z
8. If $z = x+iy$ satisfies $amp\,(z-1) = amp\,(z+3i)$, then find the value of $\frac{(x-1)}{y}$
9. If $z(2-i2\sqrt{3})^2 = i(\sqrt{3}+i)^4$, then find the amplitude of z
10. If $\left|\frac{z_1}{z_2}\right| = 1$ and $arg(z_1z_2) = 0$, then prove that $|z_2|^2 = z_1z_2$

6.4 APPLICATION OF MODULUS, CONJUGATE AND ARGUMENT

6.4.1 Properties of Modulus

Property 1. $|z| = 0$ **then** $\text{Re}(z) = \text{Im}(z) = 0$

Explanation: $z = x + iy$

So, $|z| = \sqrt{(x^2 + y^2)}$

If $|z| = 0$, so, $\sqrt{(x^2 + y^2)} = 0$

Thus, $x = 0$ as well as $y = 0$

Property 2. $|z| = |\bar{z}| = |-z| = |-\bar{z}|$

Explanation: $z = x + iy;\ \bar{z} = x - iy$

$-z = -x - iy;\ -\bar{z} = -x + iy$

$|z| = \sqrt{(x^2 + y^2)} = |\bar{z}| = |-z| = |-\bar{z}|$

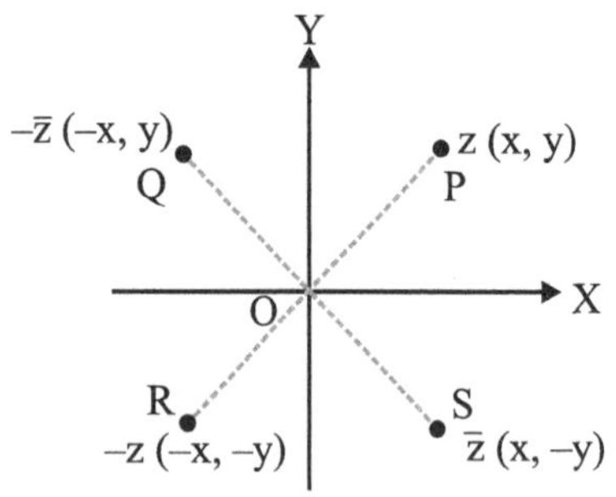

From the geometrical presentation given above also, it is clear that $OP = OQ = OR = OS$.

Property 3. $z\bar{z} = |z|^2$

Explanation: Let $z = x + iy$ so,

$$\bar{z} = x - iy$$

$$z\bar{z} = (x + iy)(x - iy)$$

$$= x^2 + y^2 = |z|^2$$

$$(\because |z| = \sqrt{(x^2 + y^2)})$$

Alternately:

$$z = re^{i\theta} \text{ (Euler's form)}$$

$$z = re^{i\theta};\ \bar{z} = re^{-i\theta}$$

$$z\bar{z} = r^2 = |z|^2$$

Property 4. $|z_1 z_2| = |z_1||z_2|$

Explanation: Let $z_1 = r_1 e^{i\theta_1}$ and $z_2 = r_2 e^{i\theta_2}$

$$z_1 z_2 = r_1 r_2\, e^{i(\theta_1 + \theta_2)}$$

$$|z_1 z_2| = r_1 r_2$$

$$= |z_1||z_2|$$

Property 5. $|z_1 z_2 z_3 \ldots\ldots z_n| = |z_1||z_2||z_3|\ldots\ldots|z_n|$

Special case If $z_1 = z_2 = z_3 = \ldots = z_n = z$

$$|z.z.z\ldots\ldots z| = |z||z||z|\ldots\ldots\ldots|z|$$

$\downarrow$ n times $\qquad$ $\downarrow$ n times

$$|z^n| = |z|^n$$

Property 6. $\left|\dfrac{z_1}{z_2}\right| = \dfrac{|z_1|}{|z_2|}$

Explanation: Let $z_1 = r_1 e^{i\theta_1}$ and $z_2 = r_2 e^{i\theta_2}$

$$\frac{z_1}{z_2} = \frac{r_1}{r_2} e^{i(\theta_1 - \theta_2)}$$

$$\left|\frac{z_1}{z_2}\right| = \frac{r_1}{r_2} = \frac{|z_1|}{|z_2|}$$

Property 7.

$$|z_1 + z_2|^2 = |z_1|^2 + |z_2|^2 + z_1\bar{z}_2 + \bar{z}_1 z_2$$

$$= |z_1|^2 + |z_2|^2 + 2\,\text{Re}(z_1\bar{z}_2)$$

Explanation: As $|z|^2 = z\bar{z}$

$|z_1 + z_2|^2 = (z_1 + z_2)\left(\overline{z_1 + z_2}\right)$

$= (z_1 + z_2)(\bar{z}_1 + \bar{z}_2)$

$= z_1\bar{z}_1 + z_2\bar{z}_2 + z_1\bar{z}_2 + z_2\bar{z}_1$

$$(\because z_1\bar{z}_1 = |z_1|^2,\ z_2\bar{z}_2 = |z_2|^2)$$

$= |z_1|^2 + |z_2|^2 + 2\,\text{Re}(z_1\bar{z}_2)\ [\because z + \bar{z} = 2Re(z)]$

Property 8.

$$|z_1 - z_2|^2 = |z_1|^2 + |z_2|^2 - z_1\bar{z}_2 - \bar{z}_1 z_2$$

$$= |z_1|^2 + |z_2|^2 - 2\,\text{Re}(z_1\bar{z}_2)$$

Explanation: Use $|z|^2 = z\bar{z}$

$|z_1 - z_2|^2 = (z_1 - z_2)\left(\overline{z_1 - z_2}\right)$

$= (z_1 - z_2)(\bar{z}_1 - \bar{z}_2)$

$= z_1\bar{z}_1 + z_2\bar{z}_2 - (z_1\bar{z}_2 + \bar{z}_1 z_2)$

$= |z_1|^2 + |z_2|^2 - 2\,\text{Re}(z_1\bar{z}_2)\quad [\because z + \bar{z} = 2Re(z)]$

Property 9.

$$|z_1+z_2|^2+|z_1-z_2|^2=2(|z_1|^2+|z_2|^2)$$

Explanation: As we have proved

$$|z_1+z_2|^2=|z_1|^2+|z_2|^2+2\,\text{Re}(z_1\overline{z_2}) \quad \ldots(i)$$

$$|z_1-z_2|^2=|z_1|^2+|z_2|^2-2\,\text{Re}(z_1\overline{z_2}) \quad \ldots(ii)$$

Adding equations (i) and (ii)

$$|z_1+z_2|^2+|z_1-z_2|^2=2(|z_1|^2+|z_2|^2)$$

Property 10. $|az_1-bz_2|^2+|bz_1+az_2|^2=(a^2+b^2)(|z_1|^2+|z_2|^2)$ **where a, b ∈ R.**

Explanation: $|az_1-bz_2|^2=(az_1-bz_2)(\overline{az_1-bz_2})$

$=(az_1-bz_2)(a\overline{z_1}-b\overline{z_2})$

$=a^2z_1\overline{z_1}-ab\,\overline{z_1}\,z_2-ab\,z_1\,\overline{z_2}+b^2z_2\,\overline{z_2}$

$=a^2|z_1|^2+b^2|z_2|^2-ab(z_1\overline{z_2}+\overline{z_1}z_2) \quad \ldots(i)$

$|bz_1+az_2|^2=(bz_1+az_2)(\overline{bz_1+az_2})$

$=(bz_1+az_2)(b\overline{z_1}+a\overline{z_2})$

$=b^2z_1\overline{z_1}+ab\,z_2\,\overline{z_1}+ab\,z_1\,\overline{z_2}+a^2z_2\overline{z_2}$

$=b^2|z_1|^2+a^2|z_2|^2+ab(z_1\overline{z_2}+\overline{z_1}z_2) \quad \ldots(ii)$

Adding eqns. (i) and (ii)

$|bz_1-az_2|^2+|bz_1+az_2|^2=(a^2+b^2)(|z_1|^2)+(a^2+b^2)(|z_2|^2)$

$=(a^2+b^2)(|z_1|^2+|z_2|^2)$

Property 11. Distance between two points $P(z_1)$ and $Q(z_2)$ on the plane, PQ is given by $|z_1-z_2|$.

Explanation: Let $z_1=x_1+iy_1$ and $z_2=x_2+iy_2$

$$z_1-z_2=(x_1+iy_1)-(x_2+iy_2)$$
$$=(x_1-x_2)+i(y_1-y_2)$$
$$|z_1-z_2|=\sqrt{[(x_1-x_2)^2+(y_1-y_2)^2]}$$

Which is the distance formula. So,

$$PQ=\sqrt{[(x_1-x_2)^2+(y_1-y_2)^2]}$$

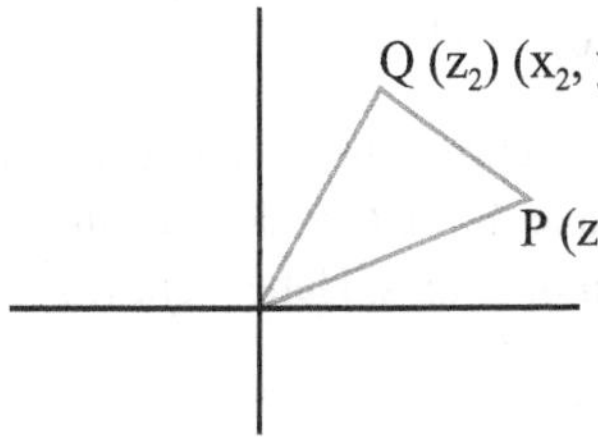

Property 12. $|z_1+z_2|\le|z_1|+|z_2|$

Explanation: Consider a triangle having sides $|z_1|$, $|z_2|$ then 3rd side is $|z_1+z_2|$ we know that some of the two sides is more than the third side hence $|z_1+z_2|\le|z_1|+|z_2|$

Property 13.

$$|z_1z_2z_3+\ldots+z_n|\le|z_1|+|z_2|+|z_3|+\ldots+|z_n|$$

Property 14. $|z_1-z_2|\le|z_1|+|z_2|$

Property 15. $|z_1+z_2|\ge|z_1|-|z_2|$

Property 16. $|z_1-z_2|\le|z_1|-|z_2|$

Property 17. $\big||z_1|-|z_2|\big|\le|z_1|+|z_2|$

Property 18. $|z_1-z_2|\ge\big||z_1|-|z_2|\big|$

Property 19. If $a_1, a_2, a_3,$ are four complex numbers, then $|z-a_1|+|z-a_2|+|z-a_3|+|z-a_4|\ge$ max $\{|a_1-a_l|+|a_m-a_n|: l, m, n$ are distinct integers lying in $\{1, 2, 3, 4\}$ and $m<n\}$.

Illustration 1. If z is any complex number such that $|z+3|\le 10$, then find the greatest value of $|z|$

Solution:

$$|z|=|(z+3)+(-3)|\le|z+3|+|-3|\le 10+3\le 13$$

Thus, greatest value is 13.

Illustration 2. For any complex number z. Find the minimum value of $|z|+|z-10i|$.

Solution: Use $|z_1+z_2|\le|z_1|+|z_2|$

$$|10i|=|z+10i-z|$$
$$10\le|z|+|10i-z|$$
$$10\le|z|+|z-10i|$$

Thus, $|z|+|z-10i|\ge 10$

Minimum value is 10.

Illustration 3. Find the greatest and the least value of $|z_1+z_2|$ if $z_1=3+4i$ and $|z_2|=2$.

Solution: $|z_1+z_2|\le|z_1|+|z_2|$

Here, $|z_1|=\sqrt{(3^2+4^2)}=5$

$|z_2|=2$

$|z_1 + z_2| \le 5 + 2$

$|z_1 + z_2| \le 7$

Now, use $|z_1 + z_2| = |z_1 - (-z_2)| \ge ||z_1| - |z_2||$

$|z_1 + z_2| \ge |5 - 2| \ge 3$

$3 \le |z_1 + z_2| \le 7$

So required minimum value is 3 and maximum value is 7.

Illustration 4. If z_1 and z_2 are two complex numbers and c > 0. Then prove that $|z_1 + z_2|^2 \le (1 + c)$ $|z_1|^2 + \left(1 + \frac{1}{c}\right) |z_2|^2$.

Solution: $|z_1 + z_2|^2 \le (1 + c)\, |z_1|^2 + \left(1 + \frac{1}{c}\right) |z_2|^2$

$|z_1|^2 + |z_2|^2 + z_1\overline{z_2} + \overline{z_1}\, z_2 \le |z_1|^2 + |z_2|^2 + c\, |z_1|^2$

$+ \frac{1}{c}\, |z_2|^2$

$z_1\overline{z_2} + \overline{z_1}\, z_2 \le c\, |z_1|^2 + \frac{1}{c}\, |z_2|^2$

So, $c\, |z_1|^2 + \frac{1}{c}\, |z_2|^2 - z_1\overline{z_2} - \overline{z_1}\, z_2 \ge 0$

$\left|\sqrt{c}z_1 - \frac{1}{\sqrt{c}} z_2\right|^2 \ge 0$

6.4.2 Properties of Argument

Property 1. Argument (z_1z_2) = argument (z_1) + argument (z_2)

Explanation: Let $z_1 = r_1\, e^{i\theta_1}$ and $z_2 = r_2\, e^{i\theta_2}$

So, $z_1z_2 = r_1\, r_2\, e^{i(\theta_1+\theta_2)}$

Argument $(z_1z_2) = \theta_1 + \theta_2$

$=$ argument (z_1) + argument (z_2)

or, $arg(z_1z_2) = \arg(z_1) + \arg(z_2) + 2k\pi$ where

$$k = \begin{cases} 0 \text{ if } -\pi < \arg(z_1) + \arg(z_2) \le \pi \\ 1 \text{ if } -2\pi < \arg(z_1) + \arg(z_2) \le -\pi \\ -1 \text{ if } \pi < \arg(z_1) + \arg(z_2) \le 2\pi \end{cases}$$

Property 2. Argument $(z_1\, z_2\, z_3 \ldots z_n)$ = argument z_1 + argument z_2 +... + argument z_n + $2k\pi$

Property 3. Arg (z^n) = n arg z

Property 4. arg $\left(\frac{z_1}{z_2}\right)$ = arg z_1 − arg z_2 + $2k\pi$

Explanation: Let $z_1 = r_1\, e^{i\theta_1}$ and $z_2 = r_2\, e^{i\theta_2}$

$$\frac{z_1}{z_2} = \frac{r_1 e^{i\theta_1}}{r_2 e^{i\theta_2}} = \frac{r_1}{r_2} e^{i(\theta_1 - \theta_2)}$$

$$\arg\left(\frac{z_1}{z_2}\right) = \theta_1 - \theta_2 = \arg z_1 - \arg z_2$$

Property 5. $\arg \bar{z} = -\arg(z)$

Explanation: Let $z = r\, e^{i\theta}$ and $\bar{z} = r e^{-i\theta}$

$\arg \bar{z} = \arg(r e^{-i\theta}) = -\theta = -\arg z$

Property 6. If z is purely real, then arg z = 0 or π

Explanation: Let $z = x + iy$

(Here $y = 0$ because purely real)

$z = x$

So, complex number z must lie on real axis. Hence argument will be either 0 or π. Let us consider following example.

$z = 3 + 0.\, i$ (purely real and Re $(z) > 0$) plot it on the graph

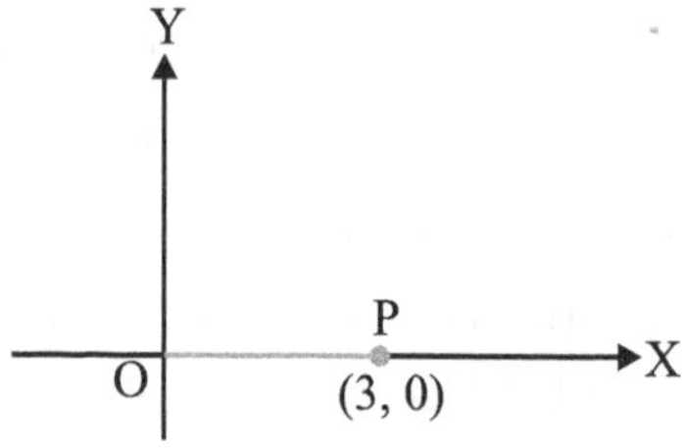

OP makes with positive direction of x-axis an angle θ, which is definitely zero.

So, arg z = 0, now take $z = -3 + i \cdot 0$ (Purely real but < 0)

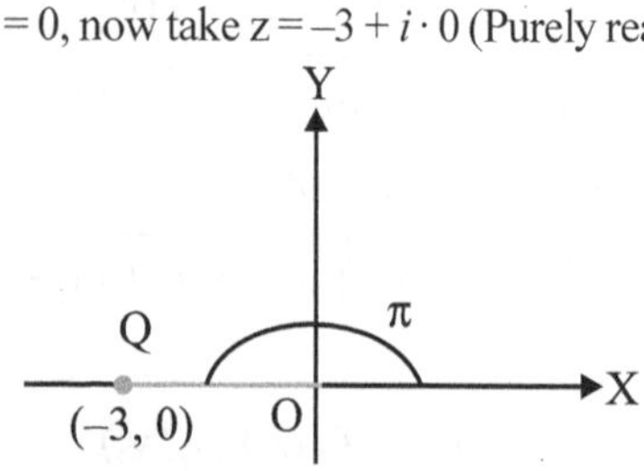

Let say $Q(-3, 0)$, represents this complex number z on the argand plane. So, OQ makes an angle π with positive direction of x–axis. Hence, arg (z) = π.

Property 7. If z is purely imaginary then

$$\arg z = \pm\frac{\pi}{2}$$

Explanation: Let us consider an example $\mathbf{z} = 0 + 3i$ (purely imaginary but $y > 0$). Plot it on argand plane.

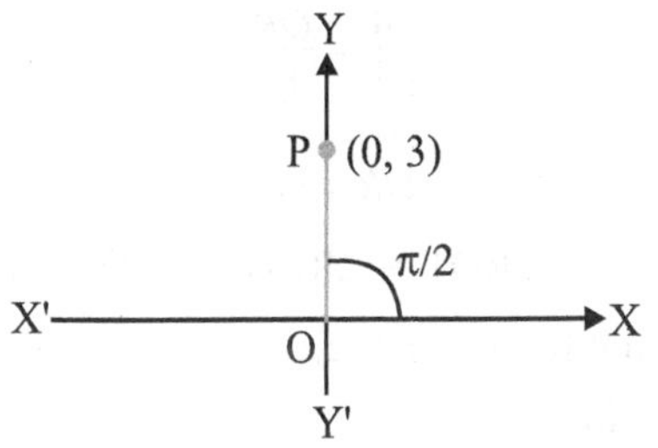

Here P (0, 3) represents z on argand plane. Note that OP makes an angle $\frac{\pi}{2}$ with the positive direction of x–axis. Here $\arg z = \frac{\pi}{2}$.

Now let us take $z = 0 - 3i$ (purely imaginary but $y < 0$).

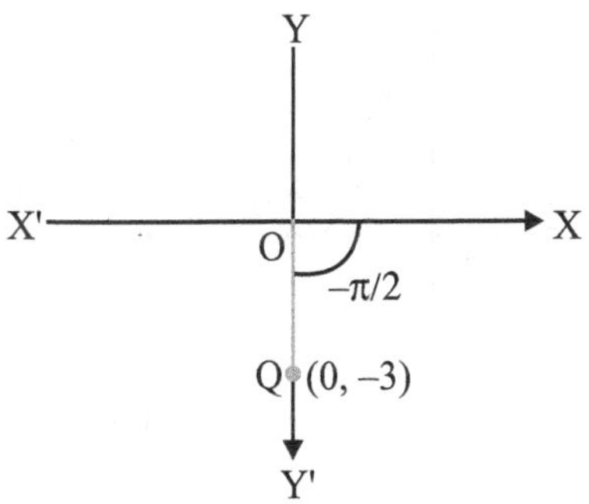

Here Q (0, –3) represents z on argand plane. Note that OQ makes an angle $-\frac{\pi}{2}$ with the positive direction of x–axis.

Here $\arg z = -\frac{\pi}{2}$.

Property 8. $\mathbf{arg}\left(\frac{1}{\bar{z}}\right) = \mathbf{arg\ z}$

Explanation: As $\arg\left(\frac{z_1}{z_2}\right) = \arg z_1 - \arg z_2$

$$\arg\left(\frac{1}{\bar{z}}\right) = \arg(1) - \arg(\bar{z})$$

$$= 0 - (-\arg z)$$

[$\because \arg(1) = 0$ discussed above]

$= \arg z$ [$\because \arg(\bar{z}) = -\arg z$]

Illustration 5. If $z_1 z_2$ are two non-zero complex number such that $|z_1 + z_2| = |z_1| + |z_2|$, then amp $\frac{z_1}{z_2}$ is equal to 0.

Solution $|z_1 + z_2| = |z_1| + |z_2|$ can hold when $0, z_1, z_2$ are collinear with 0 at one end

$$\therefore \quad \arg\left(\frac{z_1}{z_2}\right) = \arg z_1 - \arg z_2 = 0$$

6.4.3 Geometrical Representation of Algebraic Operations

(i) Addition of two complex numbers

Let us take two complex numbers

$$z_1 = x_1 + iy_1;\ z_2 = x_2 + iy_2$$

Let $\quad z = z_1 + z_2$

Lets denote z_1 on argand plane as P and z_2 on argand plane as Q then, first complete the triangle *POQ*. Then complete the parallelogram with adjacent sides as OP and OQ, then we get the other vertex of the parallelogram.

Now represents $\quad z = z_1 + z_2$.

$$z = (x_1 + x_2) + i(y_1 + y_2)$$

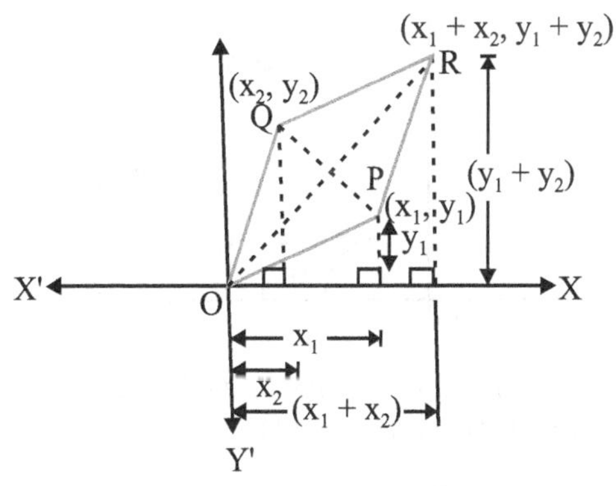

(ii) Subtraction

Let us take two complex numbers

$$z_1 = x_1 + iy_1;\ z_2 = x_2 + iy_2$$

$$z = z_1 - z_2 = ((x_1 - x_2)) + i(y_1 - y_2)$$

Lets denote z_1 on argand plane as P and z_2 on argand plane as Q then $-z_2$ will be is represented by Q′. So, complete the parallelogram with adjacent sides as OP and OQ′. The vertex R of the parallelogram denotes $(z_1 - z_2)$.

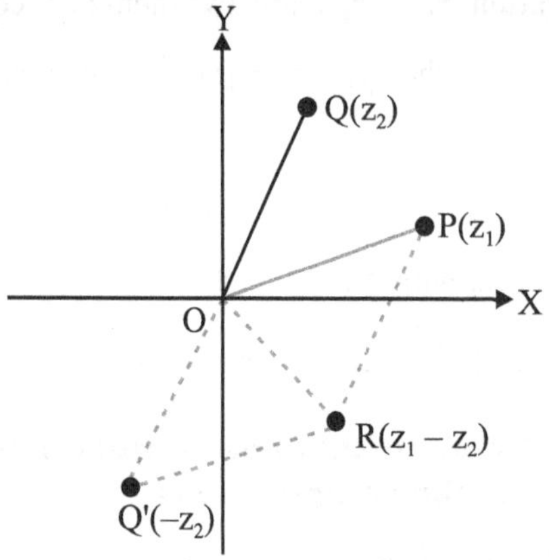

(iii) Multiplication

Let $z_1 = x_1 + iy_1 = r_1 e^{i\theta_1}$ (Euler's form)

$z_2 = x_2 + iy_2 = r_2 e^{i\theta_2}$ (Euler's form)

$$\mathbf{z = z_1 z_2 = r_1 r_2 e^{i(\theta_1+\theta_2)}}$$

Thus, $|z| = r_1 r_2$ and argument $z = (\theta_1 + \theta_2)$

Thus, z should be on that straight line passing through origin, which makes an angle $(\theta_1 + \theta_2)$ with positive direction of x–axis.

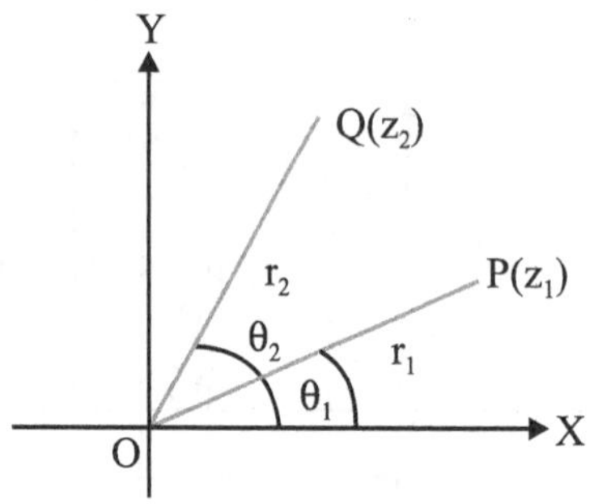

In the given figure two similar angle triangle ΔPOT and ΔROQ and assume that OT is of 1 unit.

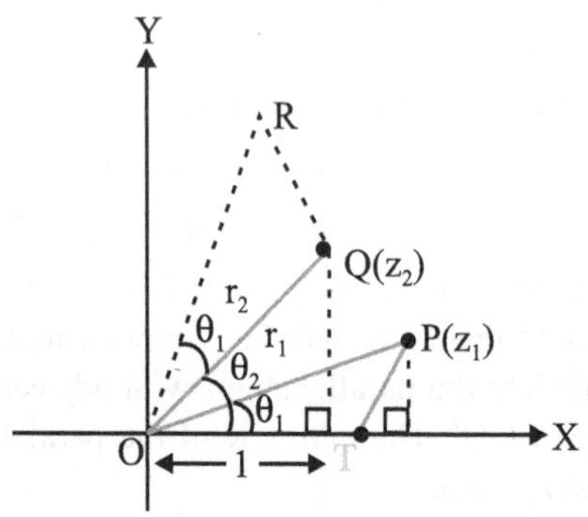

Thus, OT = 1 and OP = r_1, OQ = r_2

$$\angle POT = \theta_1 = \angle ROQ$$

So, $$\frac{OP}{OT} = \frac{OR}{OQ}$$

$$\frac{r_1}{1} = \frac{OR}{r_2}$$

so, $$OR = r_1 r_2$$

$$\angle ROT = \angle ROQ + \angle QOT = \theta_1 + \theta_2$$

Thus, point R represents $z_1 z_2$

(iv) Division

Let $z_1 = r_1 e^{i\theta_1}$ and $z_2 = r_2 e^{i\theta_2}$

$$\frac{z_1}{z_2} = \frac{r_1}{r_2} e^{i(\theta_1-\theta_2)}$$

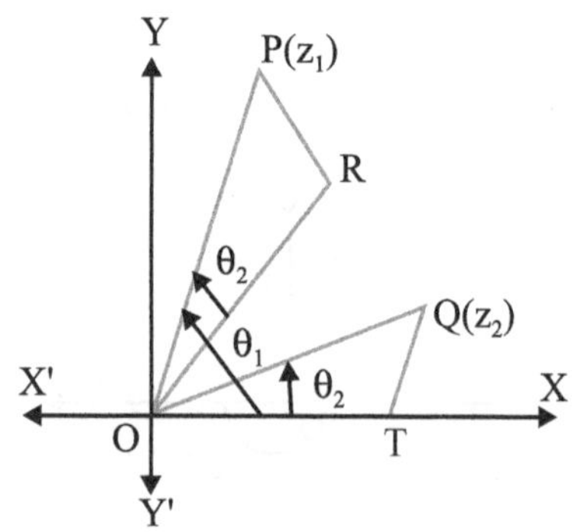

Thus, if $z = \frac{z_1}{z_2}$ then $|z| = \frac{r_1}{r_2}$ and argument $z = \theta_1 - \theta_2$.

Now consider similar triangle with $OT = 1$, $OQ = r_2$ and $OP = r_1$.

Δ QOT and ΔPOR are similar angle triangle.

$$\frac{OQ}{OT} = \frac{OP}{OR} \text{ so } \frac{r_2}{1} = \frac{r_1}{OR}$$

Thus, $$OR = \frac{r_1}{r_2'}$$

and $$\angle ROT = \angle POT - \angle POR = \theta_1 - \theta_2$$

Thus, R represents $z = \frac{z_1}{z_2}$

6.4.4 Rotation Theorem

Case (i) Rotation theorem for two complex numbers

Let us consider two complex numbers z_1 and z_2 such as $z_1 = r_1 e^{i\theta_1}$ and $z_2 = r_2 e^{i\theta_2}$.

$$\frac{z_2}{z_1} = \frac{r_2}{r_1} e^{i(\theta_2-\theta_1)}$$

$$\frac{z_2}{z_1} = \frac{|z_2|}{|z_1|} e^{i(\theta_2-\theta_1)}$$

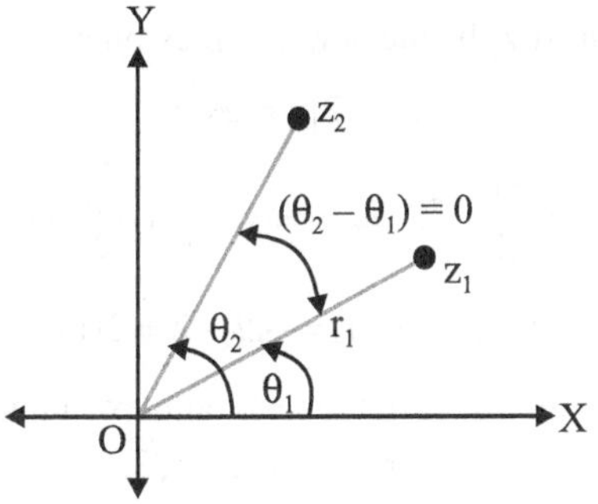

Let $\theta = (\theta_2 - \theta_1)$

$$\frac{z_2}{z_1} = \frac{|z_2|}{|z_1|} e^{i\theta}$$

Illustration 6. Find a complex number which makes an angle $\frac{\pi}{6}$. (anti clockwise direction) and magnitude $\sqrt{2}$ times the magnitude of complex number 1 + i

Solution: Let

$$z_1 = 1 + i = \sqrt{2}\left(\frac{1}{\sqrt{2}} + i\frac{1}{\sqrt{2}}\right) = \sqrt{2}\left(\cos\frac{\pi}{4} + i\sin\frac{\pi}{4}\right)$$

$$= \sqrt{2}\ e^{i\pi/4}$$

Let us plot $z_1 = 1 + i$ on argand plane.

Now, we have to find z_2 with $|z_2| = \sqrt{2}\ |z_1|$

and $\arg z_2 - \arg z_1 = \frac{\pi}{6}$.

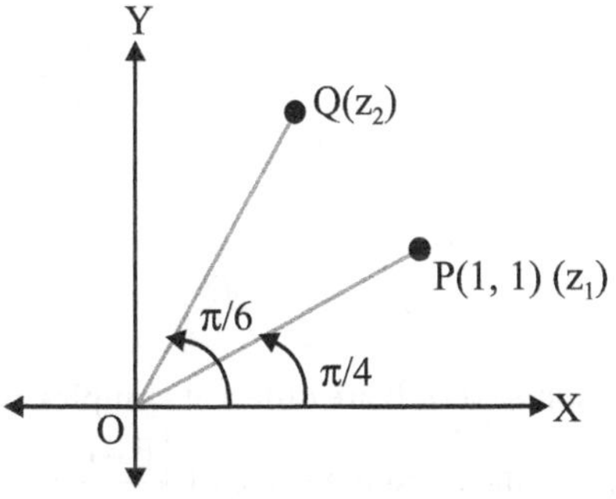

Here let $z_2 = r_2 e^{i\theta_2}$ with $r_2 = |z_2|$ and arg $z_2 = \theta_2$ using concept if rotation.

$$\frac{z_2}{z_1} = \frac{|z_2|}{|z_1|} e^{i\pi/6} = \sqrt{2}\ e^{i\pi/6}$$

$$\frac{z_2}{z_1} = \sqrt{2}\ e^{i\pi/6} \qquad (\because \frac{|z_2|}{|z_1|} = \sqrt{2})$$

$$z_2 = \sqrt{2}\ e^{i\pi/6}\ z_1$$

$$= \sqrt{2}\ e^{i\pi/6}\ \sqrt{2}\ e^{i\pi/4}$$

$$(\because z_1 = \sqrt{2}\ e^{i\pi/4})$$

$$= 2e^{\frac{i5\pi}{12}} = 2\left(\cos\frac{5\pi}{12} + i\sin\frac{5\pi}{12}\right)$$

Case (ii): Rotation theorem for three complex numbers

Consider three complex numbers z_1, z_2 and z_3 and represent them on argand plane as Q, P, & R respectively.

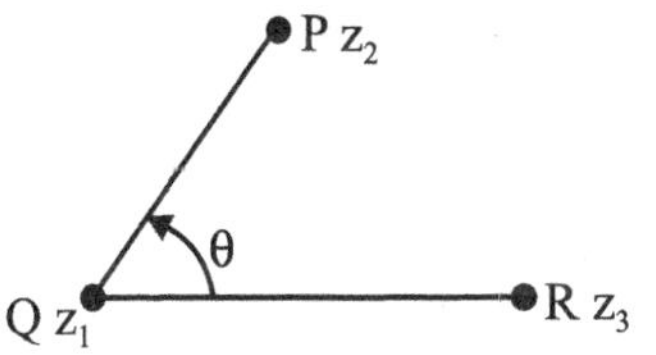

Here, $PQ = |z_2 - z_1|$

$$RQ = |z_3 - z_1|$$

$$\frac{z_2 - z_1}{z_3 - z_1} = \frac{|z_2 - z_1|}{|z_3 - z_1|} e^{i\theta}$$

where θ should be taken in anti– clockwise sense.

Special Case: When angle is π/2

Suppose z_2, z_1 and z_3 are situated on the vertices of a right angle tringle isosceles triangle.

Here, $PQ = RQ$

$$|z_2 - z_1| = |z_3 - z_1|$$

$$\frac{z_2 - z_1}{z_3 - z_1} = \frac{|z_2 - z_1|}{|z_3 - z_1|} e^{i\theta}$$

or $$\frac{z_2 - z_1}{z_3 - z_1} = i$$

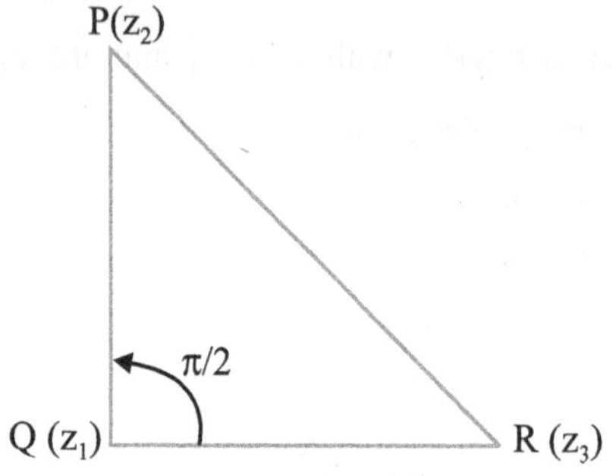

Thus, $z_2 - z_1 = i(z_3 - z_1)$

Suppose out of z_1, z_2 and z_3 any two complex numbers are known then third complex number can be evaluated using above relation.

Special Case: For equilateral triangle

Application of the concept of rotation on equilateral triangle.

Suppose, z_1, z_2 and z_3 are vertices of an equilateral triangle *ABC*.

$$AB = BC = CA$$

$$|z_1 - z_2| = |z_2 - z_3| = |z_3 - z_1|$$

$$\angle ABC = 60°$$

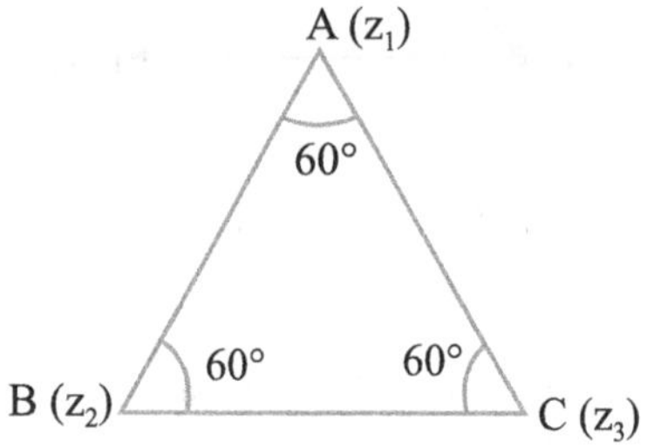

$$\frac{z_1 - z_2}{z_3 - z_2} = \frac{|z_1 - z_2| e^{i\pi/3}}{|z_3 - z_2|} = e^{i\pi/3}$$

$$\frac{z_2 - z_3}{z_1 - z_3} = \frac{|z_2 - z_3|}{|z_1 - z_3|} e^{i\pi/3} = e^{i\pi/3}$$

$$\frac{z_1 - z_2}{z_3 - z_2} = \frac{z_2 - z_3}{z_1 - z_3}$$

$$(z_1 - z_2)(z_1 - z_3) = -(z_2 - z_3)^2$$

$$z_1^2 - z_1 z_2 - z_1 z_3 + z_2 z_3 = -z_2^2 - z_3^2 + 2z_2 z_3$$

$$z_1^2 + z_2^2 + z_3^2 = z_1 z_2 + z_2 z_3 + z_3 z_1$$

Illustration 7. The complex number z = 1 + i is rotated through an angle $\frac{3\pi}{2}$ in anti–clockwise direction about the origin and whose modulus is $2|z|$ units, then find the new complex number

Solution. If z_1 be the new complex number then

$$|z_1| = 2|z| = 2\sqrt{2}$$

Also, $\frac{z_1}{z} = \frac{|z_1|}{|z|} e^{i3\pi/2} \Rightarrow z_1 = z \cdot 2(\cos\frac{3\pi}{2} + i\sin\frac{3\pi}{2})$

$$= 2(1+i)(0-i) = -2i + 2 = 2(1-i)$$

Illustration 8. Find the complex number 'z' satisfying $\text{Re}(z^2 + z) = 0$ and $|z| = \sqrt{3}$.

Solution: Let $z = x + iy$

$$z^2 = x^2 - y^2 + i\,2xy$$

$$z^2 + z = x^2 - y^2 + x + i(2xy + y)$$

$$\text{Re}(z^2 + z) = x^2 - y^2 + x = 0 \quad \ldots\text{(i)}$$

$$|z| = \sqrt{(x^2 + y^2)} = \sqrt{3}$$

So, $x^2 + y^2 = 3$ ... (ii)

Adding (i) and (ii).

$$2x^2 + x = 3$$

$$2x^2 + x - 3 = 0$$

$$(x - 1)(2x + 3) = 0$$

So, $x = 1,\ x = -\frac{3}{2}$

If $x = 1$, $y^2 = 3 - x^2 = 3 - 1 = 2$

$y = \pm\sqrt{2}$, thus $z = 1 \pm \sqrt{2}i$

If $x = -\frac{3}{2}$,

$$y^2 = 3 - \left(-\frac{3}{2}\right)^2 = 3 - \frac{9}{4} = \frac{3}{4}$$

$$y = \pm\frac{\sqrt{3}}{2}$$

Thus, $z = -\frac{3}{2} \pm \frac{\sqrt{3}}{2} i$

Illustration 9. If α, β are different complex numbers with |β| = 1, then find the value of $\left|\frac{\beta - \alpha}{1 - \bar{\alpha}\beta}\right|$

Solution: Here given to us, $|\beta| = 1$

Thus, $|\beta|^2 = 1$ so, $\beta\bar{\beta} = 1$

$$\left|\frac{\beta-\alpha}{1-\bar{\alpha}\beta}\right| = \frac{|\beta-\alpha|}{|(1-\bar{\alpha}\beta)|} = \frac{|\beta-\alpha|}{\left|\beta\left(1-\frac{\bar{\alpha}}{\bar{\beta}}\right)\right|}$$

$$= |\bar{\beta}|\frac{|(\beta-\alpha)|}{|(\bar{\beta}-\bar{\alpha})|} = \frac{|\bar{\beta}||(\beta-\alpha)|}{|\overline{(\beta-\alpha)}|}$$

$$\left(\because \bar{\beta}-\bar{\alpha} = \overline{\beta-\alpha} \text{ and } |\bar{\beta}| = |\beta|\right)$$

$$= |\beta|\frac{|\beta-\alpha|}{|\beta-\alpha|}|\beta|$$

$$\left(\because |\overline{\beta-\alpha}| = |\beta-\alpha|\right)$$

$$= 1 \qquad (\because |\beta| = 1 \text{ given})$$

6.4.5 Computation of Polynomial

In this type of question, a complex number x is given in complex form i.e $x = a + ib$, and then we have to form a polynomial in terms 'x'.

In order to solve this type of question 1st bring all the real parts together.

Step 1: $ib = x - a$

Step 2: Squaring both the side

$$-b^2 = (x-a)^2 = x^2 - 2ax + a^2$$

Step 3: Substitute this value of polynomial.

Illustration 10. Find the value of $x^4 - x^3 + x^2 + 1$. If $x = 1 + i$

Solution : Here $x = 1 + i$

Now squaring both sides,

$$(x-1)^2 = i^2$$

$$x^2 - 2x + 1 = -1$$

Thus, $x^2 - 2x + 2 = 0$

$$x^4 - x^3 + x^2 + 1 = x^2(x^2-2x+2) + x(x^2-2x+2) + 1(x^2-2x+2) + 1$$

$$= x^2(0) + x(0) + 1(0) + 1$$

$$= 1$$

Concept Applicator (CA)-4

1. If $|z_1 - a| \le a$, $|z_2 - b| \le b$, $|z_3 - c| \le c$ then, find the greatest value of $|z_1 + z_2 + z_3|$ (where a, b, c > 0).
2. If $|z_1| = |z_2|$ and $\arg(z_1) + \arg(z_2) = \pi/2$ then prove that $z_1 z_2$ is purely imaginary
3. If z_1, z_2 are complex numbers such that $\frac{z_1}{z_2}$ is purely imaginary number, then find the value of $\left|\frac{z_1 - z_2}{z_1 + z_2}\right|$.
4. If $x + iy = (a + ib)(b + ic)(c + id)(d + ie)$ then prove that $x^2 + y^2 = (a^2 + b^2)(b^2 + c^2)(c^2 + d^2)(d^2 + e^2)$
5. If $|\omega| = 1$ and $\omega = \frac{(1 - iz)}{(z - i)}$, then show that z is purely real.
6. If $z + \frac{1}{z} = 2\cos\theta$, then prove $|z| = 1$, and $\left|\frac{z^{2n}-1}{z^{2n}+1}\right| = |\tan n\theta|$.
7. Simplify $\left[\frac{\cos\theta + i\sin\theta}{\sin\theta + i\cos\theta}\right]^{100}$
8. Simplify $$\left[\frac{(\cos\theta + i\sin\theta)^{-1}(\cos 2\theta + i\sin 2\theta)^{-2}}{(\cos 3\theta + i\sin 3\theta)^{-3}(\cos 4\theta - i\sin 4\theta)^{-4}}\right]$$
9. Write complex number $z = \frac{(\sqrt{3}+i)^{99}}{(1-i)^{101}}$, in polar form
10. Find the value of $x^3 + 8x^2 + 2x + 1$, when $x = 1 + i$.

6.5 ROOTS OF UNITY

6.5.1 Cube Roots of Unity

Cube roots of unity means all the complex numbers z satisfying $z = 1^{1/3}$.

Cubing both sides $\mathbf{z^3 = 1}$

Thus, $\quad \mathbf{z^3 - 1 = 0}$

$\Rightarrow (z - 1)(z^2 + z + 1) = 0$

Hence, $\quad z - 1 = 0, z^2 + z + 1 = 0$

$z = 1; z^2 + z + 1 = 0$

$$z^2 + z + 1 = 0$$

$$= \frac{-1 \pm \sqrt{(1-4)}}{2} = \frac{-1 \pm \sqrt{3}i}{2}$$

Thus, $\quad z_1 = \frac{-1+\sqrt{3}i}{2}$

$z_2 = \frac{-1-\sqrt{3}i}{2}$

and $\quad z_3 = 1$

are called as cube root of unity.

Now consider $z_1{}^2 = \left(\frac{-1+\sqrt{3}i}{2}\right)^2 = \frac{-1-\sqrt{3}i}{2} = z_2$

So the cube roots of unity are denoted as –

$$z_1 = \frac{-1+\sqrt{3}i}{2} = \omega$$

$$z_2 = \frac{-1-\sqrt{3}i}{2} = \omega^2$$

6.5.2 Properties of Cube Roots of Unity

Property 1. $\boldsymbol{\omega^3 = 1}$

Since ω is a cube root of unity so $\omega^3 = 1$

Property 2. We denote

$$\omega = -\frac{1}{2} + \frac{\sqrt{3}}{2}i \text{ and } \omega^2 = -\frac{1}{2} - i\frac{\sqrt{3}}{2}$$

So three cube roots of unity are $1, \omega, \omega^2$.

Property 3. $1 + \omega + \omega^2 = 0$

Since $1, \omega \,\&\, \omega^2$ are the roots of $z^3 - 1 = 0$, so sum of three cube roots is zero

Also

$$1 + \omega + \omega^2 = (1) + \left(-\frac{1}{2} + \frac{\sqrt{3}}{2}i\right) + \left(-\frac{1}{2} - \frac{\sqrt{3}}{2}i\right) = 0$$

Property 4. $\bar{\omega} = -\frac{1}{2} - i\frac{\sqrt{3}}{2} = \omega^2$ and $\bar{\omega}^2 = \omega$

Property 5. $\omega = \frac{1}{\omega^2}$ and $\omega^2 = \frac{1}{\omega}$

Property 6. Suppose $n \in N$ and $1, \omega, \omega^2$ are cube roots of unity.

$1 + \omega^n + \omega^{2n} = 0$, when n is not multiple of 3.

$= 3$, when n is a multiple of 3

When n is a multiple of 3.

$$\omega^n = 1$$

$$\omega^{2n} = (\omega^n)^2 = 1$$

Thus, $1 + \omega^n + \omega^{2n} = 1 + 1 + 1 = 3$

When n is not a multiple of 3, then n will be either 3m + 1 or 3m + 2

Let n = 3m + 1

$$\text{So,} \quad 1 + \omega^n + \omega^{2n} = 1 + \omega^{3m+1} + \left(\omega^2\right)^{3m+1}$$

$$= 1 + \omega^{3m}\omega^1 + \omega^{6m}\omega^2$$

$$= 1 + (\omega^3)^m\omega^1 + (\omega^3)^{2m}\omega^2$$

$$= 1 + \omega + \omega^2 = 0$$

Let $\quad n = 3m + 2$

$$\text{So,} \quad 1 + \omega^n + \omega^{2n} = 1 + \omega^{3m+2} + \left(\omega^2\right)^{3m+2}$$

$$= 1 + \omega^{3m}\omega^2 + \omega^{6m}\omega^4$$

$$= 1 + \omega + \omega^2 = 0$$

Property 7. Cube roots of – 1 are $-1, -\omega, -\omega^2$.

Suppose, we have to solve $(x - 1)^3 + 8 = 0$

$$(x - 1)^3 = -8 = (-2)^3 (1)$$

$$(x - 1) = -2 (1)^{1/3}$$

So, $\quad x - 1 = -2 . 1$

$= -2\omega$

$= -2\omega^2$

So, $\quad x = 1 - 2, 1 - 2\omega, 1 - 2\omega^2$

Thus, $\quad x = -1, 1 - 2\omega, 1 - 2\omega^2$

6.5.3 Geometrical Representation of Cube Roots of Unity

We know that cube roots of unity are 1, $\frac{-1+\sqrt{3}i}{2}$, $\frac{-1-\sqrt{3}i}{2}$ lets represent them on the argand plane.

In the figure below 1 means (1, 0) is represented by P or z_1

$\omega = \frac{-1+\sqrt{3}i}{2}$ means $\left(-\frac{1}{2}, \frac{\sqrt{3}}{2}\right)$ is represented by Q

or z_2 and $\omega^2 = \frac{-1-\sqrt{3}i}{2}$ means $\left(-\frac{1}{2}, -\frac{\sqrt{3}}{2}\right)$ is represented by R or z_3

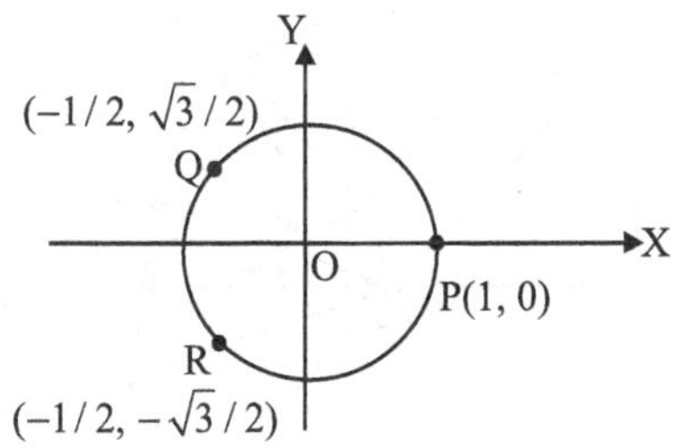

$$z_2 = -\frac{1}{2} + \frac{\sqrt{3}}{2}i; |z_2| = \sqrt{\left(\frac{1}{4}+\frac{3}{4}\right)} = 1$$

$$z_2 = \cos\frac{2\pi}{3} + i\sin\frac{2\pi}{3} = e^{i2\pi/3}$$

$$z_3 = -\frac{1}{2} - \frac{\sqrt{3}}{2}i$$

$$|z_3| = \sqrt{\left(\frac{1}{4}+\frac{3}{4}\right)} = 1$$

and $z_3 = \cos\frac{2\pi}{3} - i\sin\frac{2\pi}{3} = e^{-i2\pi/3}$

Important Points:

1. z_1, z_2, z_3 will form an equilateral triangle with side length of $\sqrt{3}$.
2. A circle with centre at origin and radius 1 unit, circumscribe the equilateral triangle.
3. Length of in-radius of equilateral triangle is $\frac{1}{2}$
4. Area of equilateral triangle is $\frac{3\sqrt{3}}{4}$.

Illustration 1. If ω is an imaginary cube root of unity then find the value of $(1+\omega-\omega^2)^7$

Solution: We know, $1+\omega+\omega^2 = 0$

$\therefore \quad 1+\omega-\omega^2 = -2\omega^2$

or $(1+\omega-\omega^2)^7 = -128\omega^{14} = -128\omega^2 \quad (\because \omega^3 = 1)$

Illustration 2. If $i = \sqrt{-1}$, then find the value of

$$4+5\left(-\frac{1}{2}+i\frac{\sqrt{3}}{2}\right)^{334} - 3\left(\frac{1}{2}+i\frac{\sqrt{3}}{2}\right)^{365}$$

Solution: Since,

$$4+5\left(-\frac{1}{2}+i\frac{\sqrt{3}}{2}\right)^{334} - 3\left(\frac{1}{2}+i\frac{\sqrt{3}}{2}\right)^{365}$$

$\Rightarrow \quad 4+5(\omega)^{334} - 3(-\omega^2)^{365}$

$\Rightarrow \quad 4+5\omega+3\omega$

$\Rightarrow \quad \frac{1}{2}\{8-5+5i\sqrt{3}-3+3i\sqrt{3}\}$

$\Rightarrow \quad \frac{1}{2}\{8i\sqrt{3}\} = 4\sqrt{3}i$

Illustration 3. If $z = \left(\frac{\sqrt{3}+i}{2}\right)^5 + \left(\frac{\sqrt{3}-i}{2}\right)^5$, then find Im(z)

Solution: $z = \left(\frac{\sqrt{3}+i}{2}\right)^5 + \left(\frac{\sqrt{3}-i}{2}\right)^5$

$$= \left(e^{i\pi/6}\right)^5 + \left(e^{-i\pi/6}\right)^5$$

$$= e^{i5\pi/6} + e^{-i5\pi/6}$$

$$= 2\cos\frac{5\pi}{6} = -2\cos\frac{\pi}{6} = -\sqrt{3}$$

Thus, $Im(z) = 0$

Illustration 4. The value of $\text{amp}(i\omega) + \text{amp}(i\omega^2)$ where $i = \sqrt{-1}$ and ω is non-real cube root of unity.

Solution: $amp(i\omega) + amp(i\omega^2) = amp(i^2.\omega^3)$

$= amp(-1) = \pi$

Illustration 5. If $z^2 - z + 1 = 0$, then find $z^n - z^{-n}$, where n is a multiple of 3.

Solution: As $z^2 - z + 1 = 0$

$\Rightarrow z = -\omega, -\omega^2$

$\therefore z^n - z^{-n} = \omega^n - \omega^{-n}$, if n is a multiple of 3.

$= 1 - 1 = 0.$

6.5.4 n^{th} Roots of Unity

All the complex numbers z, satisfying $z^n = 1$ represents the nth Roots of Unity.

$$z^n = 1$$
$$z^n = \cos(0) + i\sin(0)$$
$$z^n = \cos(2k\pi + 0) + i\sin(2k\pi + 0)$$
$$z^n = e^{i2k\pi}$$
$$z = e^{\frac{i2k\pi}{n}}$$

Here k can assume n value for 0, 1, 2, 3,…., n – 1.

For $k = 0$, $z_0 = e^{i.0} = 1$

For $k = 1$, $z_1 = e^{\frac{i2\pi}{n}} = \alpha$ (say)

For $k = 2$, $z_2 = e^{\frac{i4\pi}{n}} = \alpha^2$

For $k = 3$, $z_3 = e^{\frac{i6\pi}{n}} = \alpha^3$

⋮ ⋮ ⋮

For $k = n-1$, $z_{n-1} = e^{\frac{i2(n-1)\pi}{n}} = \alpha^{n-1}$

Thus, if $\alpha = e^{\frac{i2\pi}{n}}$ then all the nth roots of unity are as follows $1, \alpha, \alpha^2, \alpha^3, \ldots\ldots, \alpha^{n-1}$.

Properties of nth roots of unity:

1. Sum of all the roots of unity is zero:
$$1 + \alpha + \alpha^2 + \alpha^3 + \ldots\ldots + \alpha^{n-1}$$
$$= \frac{\alpha^n - 1}{\alpha - 1} = \frac{1-1}{\alpha - 1} = 0$$
$$[\because \alpha^n = \left(e^{\frac{i2\pi}{n}}\right)^n = e^{i2\pi} = 1]$$

2. Product of all the nth roots of unity:
$$1 \cdot \alpha \cdot \alpha^2 \cdot \alpha^3 \cdot \ldots\ldots \cdot \alpha^{n-1} = \alpha^{1+2+3+\ldots+n-1}$$
$$= \alpha^{\frac{(n-1)n}{2}} = \left(e^{\frac{i2\pi}{n}}\right)^{\frac{n(n-1)}{2}} = e^{i\pi(n-1)}$$
$$= \cos[(n-1)\pi] + i\sin[(n-1)\pi]$$
$$= \cos[(n-1)\pi] + 0 \quad [\because \sin((n-1)\pi) = 0]$$
Product = 1, if n is odd.
–1, if n is even.

We can also prove above two results by sum of roots/ product of roots of a polynomial.

6.5.5 Geometrical Representation of nth Roots of Unity

Construct a circle of radius unity and mark point A = 1 on the x–axis, then create a regular polygon of n sides with internal angle $A_1 0 A_0 = \frac{2\pi}{n}$.

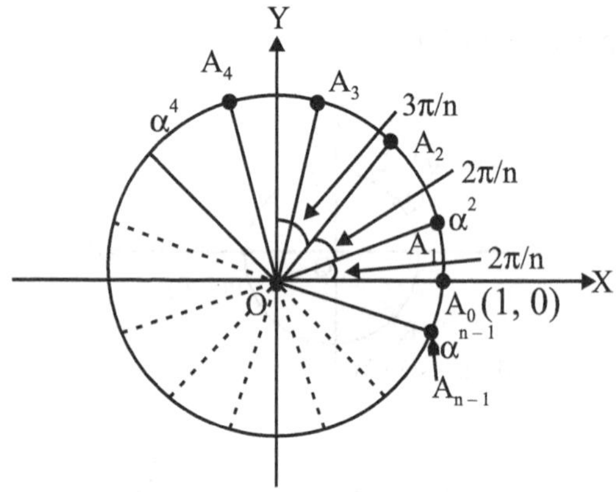

Here vertices A_1, A_2, A_3 …denotes α, α^2, α^3, ….. respectively.

Illustration 6. If $(x-1)^4 - 16 = 0$, then the sum of non-real complex value of x is

Solution: As $(x-1)^4 = 16$

$\Rightarrow \quad x - 1 = 2, -2, 2i, -2i$

Thus, the sum of non–real complex values is 2.

Illustration 7. If the cube roots of unity are $1, \omega, \omega^2$ then find the roots of the equation $(x-1)^3 + 8 = 0$

Solution: $(x-1)^3 + 8 = 0$

$\Rightarrow \quad (x-1)^3 = -8$

$\Rightarrow \quad x - 1 = (-8)^{1/3} = -2, -2\omega - 2\omega^2$

$\Rightarrow \quad x = -1, 1-2\omega, 1-2\omega^2$

Illustration 8. If the fourth roots of unity are z_1, z_2, z_3 and z_4, then find the value of $z_1^2 + z_2^2 + z_3^2 + z_4^2$

Solution: $1^{1/4} = (\cos 2\pi r + i\sin 2\pi r)^{1/4}$

$$= \cos\frac{\pi r}{2} + i\sin\frac{\pi r}{2}; r = 0, 1, 2, 3$$

$\therefore \; 1^{1/4} = 1, i, -1, -i$

$\therefore \; z_1^2 + z_2^2 + z_3^2 + z_4^2 = 1^2 + i^2 + (-1)^2 + (-i)^2 = 0$

Illustration 9. If 1, $z_1, z_2, z_3, \ldots, z_{n-1}$ are the nth roots of unity, then prove that $(1 - z_1)(1 - z_2) \ldots (1 - z_{n-1}) = n$.

Solution: $z^n = 1$, it has roots as $1, z_1, z_2, z_3, \ldots, z_{n-1}$

So, $z^n - 1 = (z-1)(z-z_1)(z-z_2)\ldots(z-z_{n-1})$

$$\frac{z^n-1}{z-1} = (z-z_1)(z-z_2)\ldots(z-z_{n-1})$$

But $\frac{z^n-1}{z-1} = 1 + z + z^2 + z^3 + \ldots\ldots + z^{n-1}$

$$1 + z^1 + z^2 + z^3 + \ldots\ldots + z^{n-1} = (z-z_1)(z-z_2)\ldots(z-z_{n-1})$$

Put z = 1,

$n = (1-z_1)(1-z_2)\ldots(1-z_{n-1})$. Hence proved.

Illustration 10. $\alpha_1, \alpha_2, \alpha_3, \ldots, \alpha_{100}$ are all the 100th roots of unity. Then find the numerical value of $\sum_{1 \le i < j \le 100} \sum (\alpha_i \alpha_j)^5$ is

Solution: $\sum_{1 \le i < j \le 100} \sum \alpha_i^5 \alpha_j^5$

$$= \left(\alpha_1^5 + \alpha_2^5 + \alpha_3^5 + \alpha_4^5 + \alpha_5^5 + \ldots\right)^2 - \left(\alpha_1^{10} + \alpha_2^{10} + \alpha_3^{10} + \alpha_4^{10} + \alpha_5^{10} + \ldots\right)$$

$= 0 - 0 = 0$

Because,

$$\left(\alpha_1^r + \alpha_2^r + \ldots + \alpha_{100}^r\right) = \begin{cases} 100, \text{ if } r = 100k \\ 0, \text{ if } r \ne 100k \end{cases}$$

Concept Applicator (CA)-5

1. If $3^{49}(x+iy) = \left(\frac{3}{2} + \frac{\sqrt{3}}{2}i\right)^{100}$, then find $x + iy$
2. If $i = \sqrt{-1}, \omega =$ non-real cube root of unity then find the value of $\frac{(1+i)^{2n} - (1-i)^{2n}}{\left(1+\omega^4-\omega^2\right)\left(1-\omega^4+\omega^2\right)}$ if n is even
3. The values of $(16)^{1/4}$ are
4. If $x^2 - x + 1 = 0$, then find the value of $\sum_{n=1}^{5}(x^n + \frac{1}{x^n})^2$ is
5. If $z = -2 + 2\sqrt{3}i$, then analyze the value of $z^{2n} + 2^{2n} z^n + 2^{4n}$
6. Find the value of $\left(-1+\sqrt{-3}\right)^{62} + \left(-1-\sqrt{-3}\right)^{62}$
7. If z is any complex number such that $z + \frac{1}{z} = 1$, then the value of $z^{99} + \frac{1}{z^{99}}$ is
8. If ω is a cube root of unity, then $\left(3+5\omega+3\omega^2\right)^2 + \left(3+3\omega+5\omega^2\right)^2$ is equal to
9. Find the value of xyz if $x = a + b, y = a\omega + b\omega^2, z = a\omega^2 + b\omega$
10. If 1, ω, ω^2 are the cube roots of unity, then find the value of $(1+\omega)^3 - \left(1+\omega^2\right)^3$

6.6 GEOMETRY OF COMPLEX NUMBER

6.6.1 Distance & Section Formula

1. Distance formula

Distance between $A(z_1)$ and $B(z_2)$ is given by

$$AB = |z_2 - z_1|$$

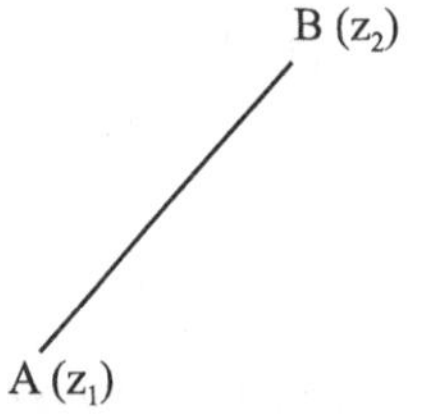

2. Section formula

Consider line segment AB, with A represented by z_1 and B by z_2. Let us consider a point $C(z)$ dividing the line segment AB in the ratio $m : n$.

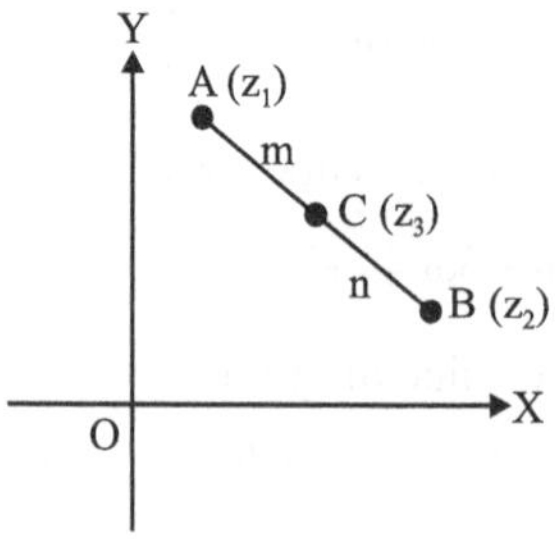

Then $z = \dfrac{nz_1 + mz_2}{m+n}$

3. Mid–point formula

Mid–point M(z) of the line segment AB is given by

$$z = \frac{1}{2}(z_1 + z_2)$$

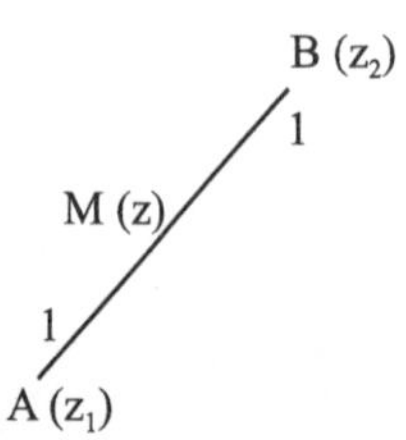

In section formula if we put m = n then we will get mid point formula.

6.6.2 Properties of Straight Lines in Complex Numbers

1. Equation of straight line passing through the two points z_1 and z_2

Let A (z_1) and B (z_2) are two points as shown in the figure below.

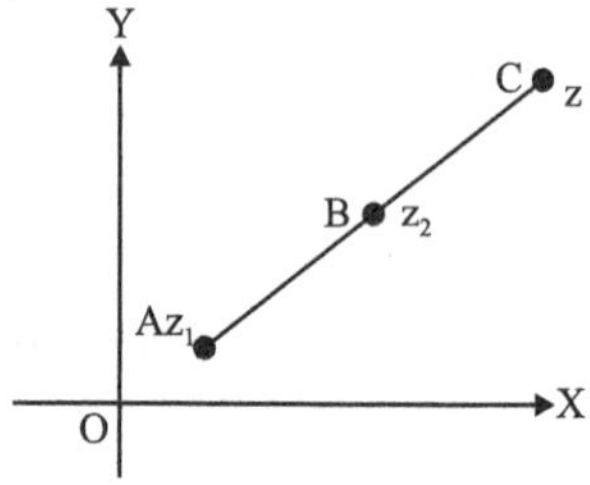

Let us consider a point C (z), which is on the straight line AB. Thus, the angle between AB and BC is zero.

or $\quad \arg\left(\dfrac{z - z_2}{z_2 - z_1}\right) = 0$

We know that if arg(z) = 0, then Im(z) = 0

Thus, $\dfrac{z - z_2}{z_2 - z_1}$ is purely real.

For any complex number z to be purely real, $z = \bar{z}$.

$$\frac{z - z_2}{z_2 - z_1} = \overline{\left(\frac{z - z_2}{z_2 - z_1}\right)}$$

$$\frac{z - z_2}{z_2 - z_1} = \frac{\bar{z} - \bar{z}_2}{\bar{z}_2 - \bar{z}_1} \quad \ldots(i)$$

Equation (i) can also be written as

$$\begin{vmatrix} z & \bar{z} & 1 \\ z_1 & \bar{z}_1 & 1 \\ z_2 & \bar{z}_2 & 1 \end{vmatrix} = 0 \quad \ldots(ii)$$

Thus, for three points z_1, z_2 and z_3 to be lie on a line (to be collinear points) the condition should be

$$\begin{vmatrix} z_1 & \bar{z}_1 & 1 \\ z_2 & \bar{z}_2 & 1 \\ z_3 & \bar{z}_3 & 1 \end{vmatrix} = 0$$

We can write equation (i) as

$$(z - z_2)(\bar{z}_2 - \bar{z}_1) = (z_2 - z_1)(\bar{z} - \bar{z}_2)$$

$$(\bar{z}_1 - \bar{z}_2)z - (z_1 - z_2)\bar{z} + z_1\bar{z}_2 - z_2\bar{z}_1 = 0$$

Multiply it with i

$\Rightarrow i(\bar{z}_1 - \bar{z}_2)z - i(z_1 - z_2)\ \bar{z} + i(z_1\bar{z}_2 - z_2\bar{z}_1) = 0$

Let a = $-i(z_1 - z_2)$ so $\bar{a} = i(\bar{z}_1 - \bar{z}_2)$

$\Rightarrow \bar{a}z + a\bar{z} + b = 0$ where b = $i(z_1\bar{z}_2 - z_2\bar{z}_1)$

$$b = i\left(z_1\bar{z}_2 - \overline{z_1\bar{z}_2}\right)$$

$$= i(2i\ \text{Im}(z_1\bar{z}_2)) = -2\text{Im}(z_1\bar{z}_2)$$

= a real number

Now lets discuss the slope of the given line, $\bar{a}z + a\bar{z} + b = 0$ put $z = x + iy, \bar{z} = x - iy$.

$\bar{a}(x+iy) + a(x-iy) + b = 0$

$x(a + \bar{a}) + iy(\bar{a} - a) + b = 0$

Thus, slope = $\dfrac{(a+\bar{a})}{i(a-\bar{a})} = \dfrac{2Re(a)}{i[2iIm(a)]} = -\dfrac{Re(a)}{Im(a)}$

Complex Angular Co-efficient (or Complex Slope)

Consider a linear equation

$$\alpha z + \bar{\alpha}\bar{z} + \beta = 0$$

The ratio $m_d = -\left(\dfrac{\bar{\alpha}}{\alpha}\right)$ is called the complex angular coefficient of the line.

Consider two equations–

$\alpha_1 z + \bar{\alpha}_1\bar{z} + \beta = 0$

$\alpha_2 z + \bar{\alpha}_2\bar{z} + \beta = 0$

Now we have following conditions

(i) Two lines are parallel if and only if

$$\left(\frac{\bar{\alpha}_1}{\alpha_1}\right) = \left(\frac{\bar{\alpha}_2}{\alpha_2}\right)$$

(ii) Two lines are perpendicular if and only if

$$\left(\frac{\bar{\alpha}_1}{\alpha_1}\right) + \left(\frac{\bar{\alpha}_2}{\alpha_2}\right) = 0$$

(iii) Two lines are concurrent if and only if

$$\left(\frac{\bar{\alpha}_1}{\alpha_1}\right) \neq \left(\frac{\bar{\alpha}_2}{\alpha_2}\right)$$

Complex angular coefficient satisfy the properties of slope so it is also called as Complex Slope of the line.

Summary:

(i) Non-parametric form:

An equation of a straight line joining the two points A (z_1) and B (z_2) is $\begin{vmatrix} z & \bar{z} & 1 \\ z_1 & \bar{z}_1 & 1 \\ z_2 & \bar{z}_2 & 1 \end{vmatrix} = 0$

or $\dfrac{z - z_1}{z_2 - z_1} = \dfrac{\bar{z} - \bar{z}_1}{\bar{z}_2 - \bar{z}_1}$

or $z(\bar{z}_1 - \bar{z}_2) - \bar{z}(z_1 - z_2) + z_1\bar{z}_2 - z_2\bar{z}_1 = 0$

(ii) Parametric form: An equation of the straight line joining the points A(z_1) and B(z_2) is $z = tz_1 + (1-t)z_2$ where t is a real parameter.

(iii) General equation of a straight line: The general equation of a straight line is $\bar{a}z + a\bar{z} + b = 0$ where a is a non–zero complex number and b is a real number.

(iv) Equation of line parallel to $a\bar{z} + \bar{a}z + b = 0$ is $a\bar{z} + \bar{a}z + \lambda = 0$

(v) Equation of line perpendicular to $a\bar{z} + \bar{a}z + b = 0$ is $\bar{z} + \bar{a}z + i\lambda = 0$, where $\lambda \in R$.

Equation of perpendicular bisector

Any point P(z) on the perpendicular bisector of A and B is at equal distance from A and B.

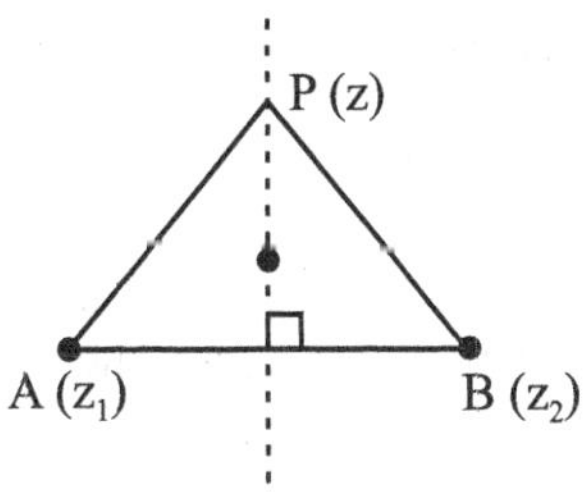

Let A = z_1 and B = z_2

$$|z - z_1| = |z - z_2|$$

Squaring both sides

$|z - z_1|^2 = |z - z_2|^2$

$\Rightarrow (z - z_1)(\bar{z} - \bar{z}_1) = (z - z_2)(\bar{z} - \bar{z}_2)$

$\Rightarrow z\bar{z} + z_1\bar{z}_1 - z_1\bar{z} - z\bar{z}_1 = z\bar{z} - z_2\bar{z} - z\bar{z}_2 + z_2\bar{z}_2$

$$\Rightarrow z\left(\bar{z}_2 - \bar{z}_1\right) + \bar{z}\left(z_2 - z_1\right) + z_1\bar{z}_1 - z_2\bar{z}_2 = 0$$

Let $a = z_2 - z_1$ and $b = z_1\bar{z}_1 - z_2\bar{z}_2$

Thus, equation of perpendicular bisector will be $\bar{a}z + a\bar{z} + b = 0$

Distance of a given point from a given line

Consider a point P (Z_0) and a line $\bar{a}z + a\bar{z} + b = 0$

Put $z = x + iy$

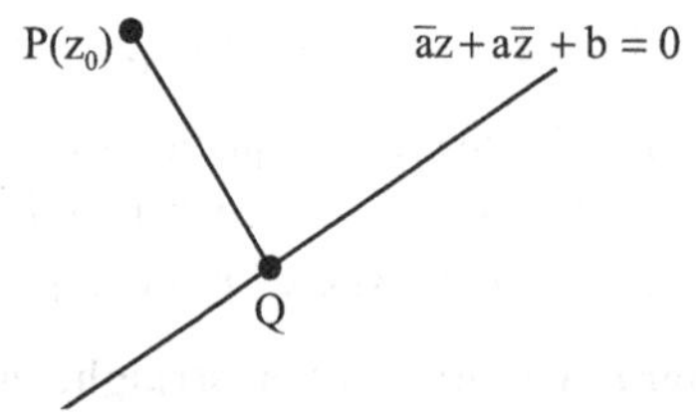

$$\bar{a}z + a\bar{z} + b = 0$$

$$a(x - iy) + \bar{a}\,(x + iy) + b = 0$$

$$x(a + \bar{a}) + iy(-a + \bar{a}) + b = 0$$

Let us take, $z_0 = x_0 + i\,y_0$

$$\text{Distance } PQ = \frac{\left|x_0(a+\bar{a}) + iy_0(\bar{a}-a) + b\right|}{\sqrt{\left[(a+\bar{a})^2 - (\bar{a}-a)^2\right]}}$$

$$= \frac{\left|z_0\bar{a} + a\bar{z}_0 + b\right|}{\sqrt{4[Re(a)]^2 + 4[Im(a)]^2}} = \frac{\left|z_0\bar{a} + a\bar{z}_0 + b\right|}{2|a|}$$

6.6.3 Properties of Triangle in Complex Numbers

- **Centroid of a triangle with vertices $A(z_1)$, $B(z_2)$ and $C(z_3)$.**

Centroid G(z) of the ΔABC is the point of concurrence of medians of ΔABC and is given by

$$z = \frac{1}{3}\left(z_1 + z_2 + z_3\right)$$

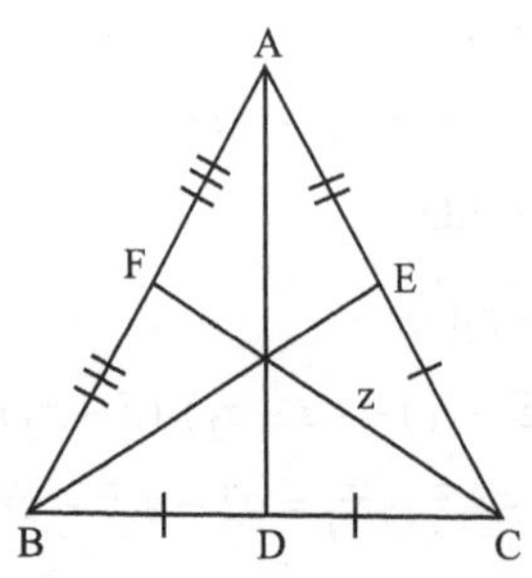

- **Incentre of a triangle with vertices $A(z_1)$, $B(z_2)$ and $C(z_3)$.**

Incentre I(z) of the ΔABC is the point of concurrence of internal bisectors of angles of ΔABC and is given by $z = \dfrac{az_1 + bz_2 + cz_3}{a+b+c}$.

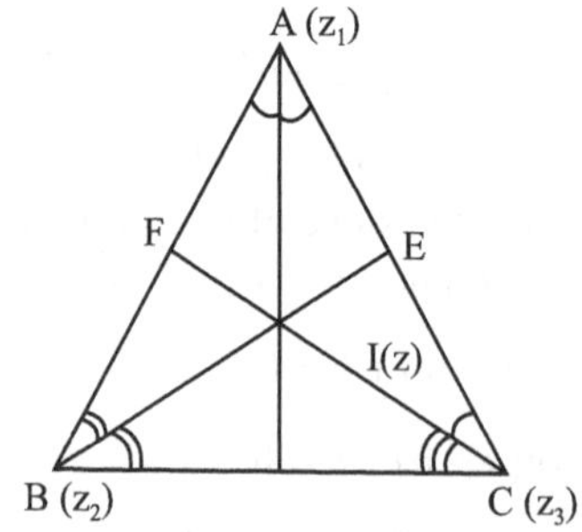

- **Circumcentre of a triangle with vertices $A(z_1)$, $B(z_2)$ and $C(z_3)$**

S(z) of the ΔABC is the point of concurrence of perpendicular bisectors of sides ΔABC and is given by

$$z = \frac{(z_2 - z_3)|z_1|^2 + (z_3 - z_1)|z_2|^2 + (z_1 - z_2)|z_3|^2}{\bar{z}_1(z_2 - z_3) + \bar{z}_2(z_3 - z_1) + \bar{z}_3(z_1 - z_2)}$$

$$z = \frac{\begin{vmatrix} |z_1|^2 & z_1 & 1 \\ |z_2|^2 & z_2 & 1 \\ |z_3|^2 & z_3 & 1 \end{vmatrix}}{\begin{vmatrix} \bar{z}_1 & z_1 & 1 \\ \bar{z}_2 & z_2 & 1 \\ \bar{z}_3 & z_3 & 1 \end{vmatrix}}$$

Circum centre in terms of angle

$$z = \frac{z_1(\sin 2A) + z_2(\sin 2B) + z_3(\sin 2C)}{\sin 2A + \sin 2B + \sin 2C}$$

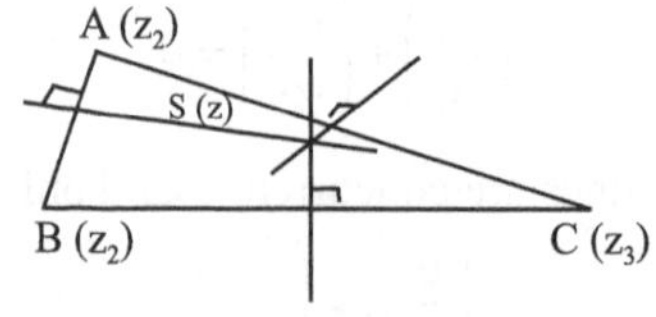

- **Orthocentre of a triangle with vertices $A(z_1)$, $B(z_2)$ and $C(z_3)$**

H(z) of the ΔABC is the point of concurrence of altitudes of ΔABC and is given by

$$z=\frac{\begin{vmatrix} z_1^2 & \overline{z_1} & 1 \\ z_2^2 & \overline{z_2} & 1 \\ z_3^2 & \overline{z_3} & 1 \end{vmatrix}+\begin{vmatrix} |z_1|^2 & z_1 & 1 \\ |z_2|^2 & z_2 & 1 \\ |z_3|^2 & z_3 & 1 \end{vmatrix}}{\begin{vmatrix} \overline{z_1} & z_1 & 1 \\ \overline{z_2} & z_2 & 1 \\ \overline{z_3} & z_3 & 1 \end{vmatrix}}$$

Orthocentre in terms of angle

$$\text{or } z=\frac{(\tan A)z_1+(\tan B)z_2+(\tan C)z_3}{\tan A+\tan B+\tan C}$$

$$\text{or } z=\frac{(a\sec A)z_1+(b\sec B)z_2+(c\sec C)z_3}{a\sec A+b\sec B+c\sec C}$$

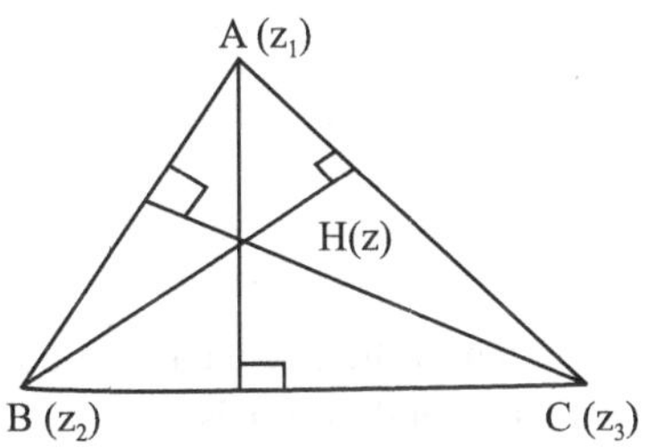

☛ **Euler's Line:** The centroid G of a triangle lies on the segment joining the Orthocentre H and the Circumcentre S of the triangle. G divides the line joining H and S in the ratio 2 : 1

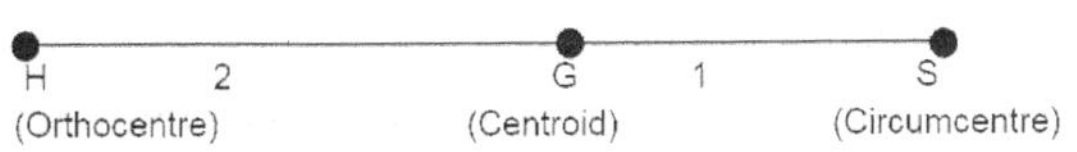

Thus, $z_G=\frac{1}{3}(z_H+2z_s)$

☛ **Area of a triangle – Area of ΔABC with vertices $A(z_1)$, $B(z_2)$ and $C(z_3)$ is given by**

$$\Delta=\left|\frac{1}{4i}\begin{vmatrix} z_1 & \overline{z_1} & 1 \\ z_2 & \overline{z_2} & 1 \\ z_3 & \overline{z_3} & 1 \end{vmatrix}\right|=\left|\frac{1}{2}Im\left(\overline{z_1}z_2+\overline{z_2}z_3+\overline{z_3}z_1\right)\right|$$

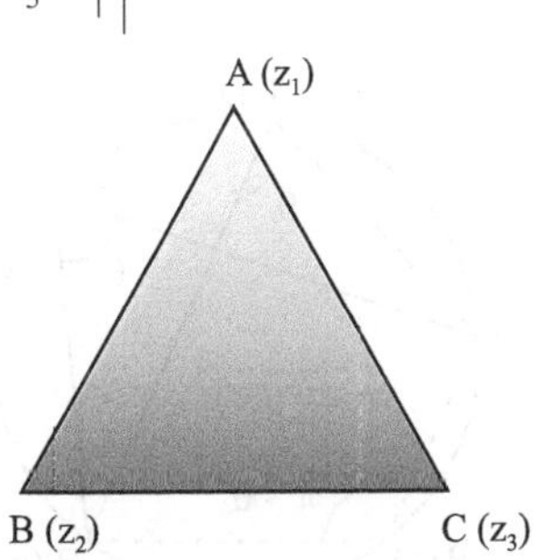

$$\Delta=\sqrt{s(s-a)(s-b)(s-c)}$$

☛ **Equilateral triangle:** Triangle ABC with vertices $A(z_1), B(z_2)$ and $C(z_3)$ is equilateral if and only if

$$\frac{1}{z_2-z_3}+\frac{1}{z_3-z_1}+\frac{1}{z_1-z_2}=0$$

$$\Rightarrow\quad z_1^2+z_2^2+z_3^2=z_2z_3+z_3z_1+z_1z_2$$

$$\Rightarrow\quad z_1\overline{z_2}=z_2\overline{z_3}=z_3\overline{z_1}$$

$$\Rightarrow\quad z_1^2=z_2z_3 \text{ and } z_2^2=z_1z_3$$

$$\Rightarrow\quad \begin{vmatrix} 1 & z_2 & z_3 \\ 1 & z_3 & z_1 \\ 1 & z_1 & z_2 \end{vmatrix}=0$$

$$\Rightarrow\quad \frac{1}{z-z_1}+\frac{1}{z-z_2}+\frac{1}{z-z_3}=0$$

where $z=\frac{1}{3}(z_1+z_2+z_3)$

Illustration 1. Let $z_1, z_2, z_3 \in C$ are such that $\text{Re}\left(\frac{z_3-z_1}{z_2-z_1}\right)=0$, then discuss the type of all the triangles with vertices z_1, z_2, z_3

Solution: As $\frac{z_3-z_1}{z_2-z_1}$ is purely imaginary,

So from rotation theorem angle between the segment joining $z_3\ z_1$ and $z_2\ z_1$ is a right angle .
$\Rightarrow z_1$ lies on a circle with the segment joining $z_2\ z_3$ as diameter .

Illustration 2. Find the value of $\frac{z_3-z_1}{z_2-z_1}$ for which triangle with vertices z_1, z_2, z_3 is equilateral,
Solution: Since triangle is equilateral,

$$\frac{z_3-z_1}{z_2-z_1}=\frac{|z_3-z_1|}{|z_2-z_1|}\left(\cos\frac{\pi}{3}+i\sin\frac{\pi}{3}\right)$$

$$=\frac{1}{2}\,(\,1+\sqrt{3}i\,)$$

Illustration 3. Let $z_1, z_2, z_3 \in C$ are such that $\text{Im}\left(\frac{z_3-z_1}{z_2-z_1}\right)=0$, then z_1, z_2, z_3 lie on

Solution: Given that $z_3-z_1=\lambda\,(\,z_2-z_3\,)$
$\Rightarrow z_3=(\,1-\lambda)\,z_1+z_2$
$\Rightarrow z_3$ lies on the line joining z_1 and z_2 .

6.6.4 Properties of Circle in Complex Numbers

We know that locus of circle is the moving points whose distance from a fixed point (called center) is always constant (Called radius).

Let us consider a circle with centre $Q\ (z_0)$ and radius r and any point on its periphery (z).

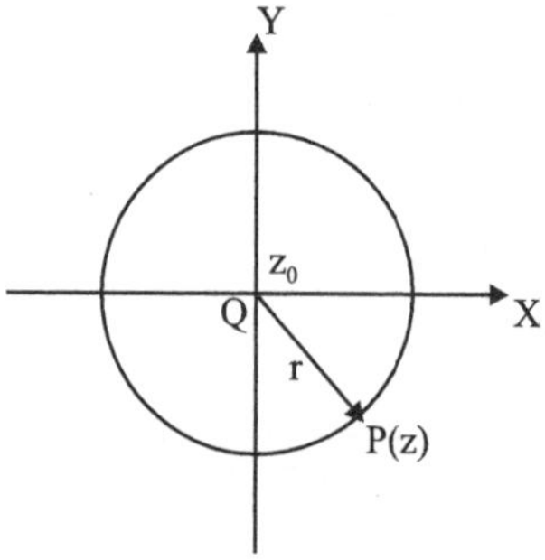

$|z - z_0| = r$

Square both side $|z - z_0|^2 = r^2$

$$(z - z_0)(\bar{z} - \bar{z}_0) = r^2$$

$z\bar{z} - z_0\bar{z} - z\bar{z}_0 + z_0\bar{z}_0 - r^2 = 0$

$z\bar{z} + \bar{a}z + a\bar{z} + b = 0$ ...(i)

where $\alpha = -z_0$ and $\bar{\alpha} = -\bar{z}_0$

$$b = z_0\bar{z}_0 - r^2$$

Thus, centre of circle represented by equation is $-\alpha$ (i) and radius

$r = \sqrt{(z_0\bar{z}_0 - b)} = \sqrt{(|z_0|^2 - b)} = \sqrt{(a^2 - b)}$

- **Equation of circle:**

(i) An equation of the circle with centre at z_0 and radius r is given by following methods-

(a) $|z - z_0| = r$

(b) or $z = z_0 + re^{i\theta}, 0 \le \theta < 2\pi$ (parametric form)

(c) or $z\bar{z} - z_0\bar{z} - \bar{z}_0 z + z_0\bar{z}_0 - r^2 = 0$

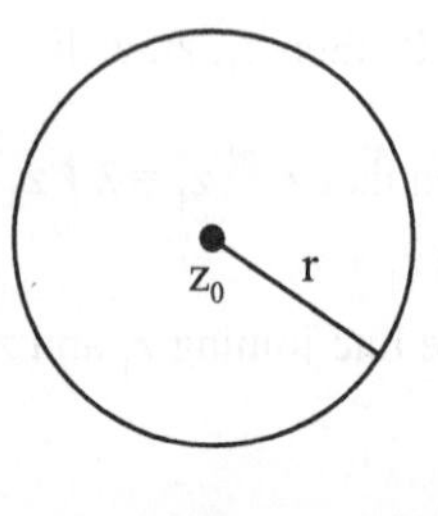

(d) General equation of a circle is

$$z\bar{z} + a\bar{z} + \bar{a}z + b = 0$$

Here a is a complex number and b is a real number such that $a\bar{a} - b \ge 0$.

Centre of the above circle is –a and its radius is $\sqrt{a\bar{a} - b}$.

(e) Diameter form of a circle: An equation of the circle whose diameter is AB where $A(= z_1)$ and $B(= z_2)$ is

$$(z - z_1)(\bar{z} - \bar{z}_2) + (\bar{z} - \bar{z}_1)(z - z_2) = 0$$

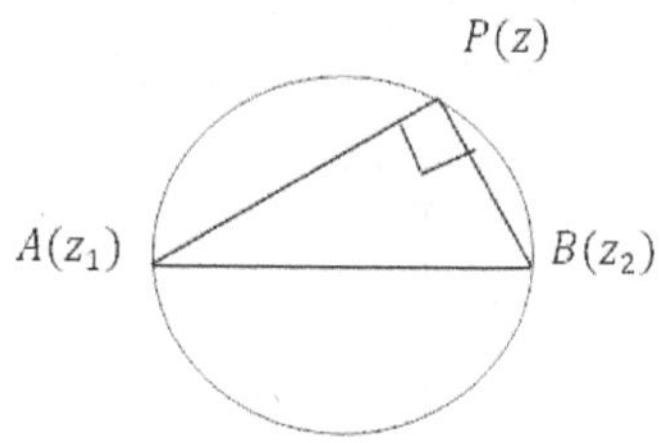

(f) An equation of the circle passing through two points $A(z_1)$ and $B(z_2)$ is

$$(z - z_1)(\bar{z} - \bar{z}_2) + (\bar{z} - \bar{z}_1)(z - z_2) + k\begin{vmatrix} z & \bar{z} & 1 \\ z_1 & \bar{z}_1 & 1 \\ z_2 & \bar{z}_2 & 1 \end{vmatrix} = 0$$

where k is a parameter.

(g) Equation of a circle passing through three non-collinear points.

Let three non-collinear points be $A(z_1), B(z_2)$ and $C(z_3)$. Let P(z) be any point on the circle.

Then either $\angle ACB = \angle APB$ [when these angles are in the same segment] or $\angle ACB + \angle APB = \pi$ [when these angles are in the opposite segment]

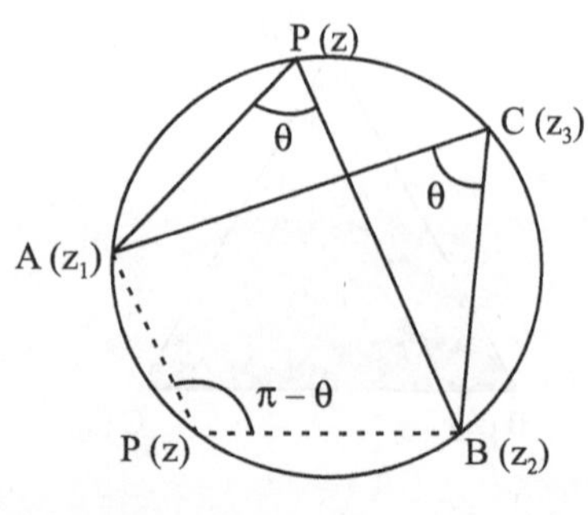

$$\Rightarrow \arg\left(\frac{z_3 - z_2}{z_3 - z_1}\right) - \arg\left(\frac{z - z_2}{z - z_1}\right) = 0$$

$$\text{or } \arg\left(\frac{z_3 - z_2}{z_3 - z_1}\right) + \arg\left(\frac{z - z_1}{z - z_2}\right) = \pi$$

$$\Rightarrow \arg\left[\left(\frac{z_3 - z_2}{z_3 - z_1}\right)\left(\frac{z - z_1}{z - z_2}\right)\right] = 0$$

$$\text{or } \arg\left[\left(\frac{z - z_1}{z - z_2}\right)\left(\frac{z_3 - z_2}{z_3 - z_1}\right)\right] = \pi$$

In both the cases, $\frac{(z - z_1)(z_3 - z_2)}{(z - z_2)(z_3 - z_1)}$ is purely real.

$$\Rightarrow \frac{(z - z_1)(z_3 - z_2)}{(z - z_2)(z_3 - z_1)} = \frac{(\bar{z} - \bar{z}_1)(\bar{z}_3 - \bar{z}_2)}{(\bar{z} - \bar{z}_2)(\bar{z}_3 - \bar{z}_1)}$$

Condition for four points to be concyclic.

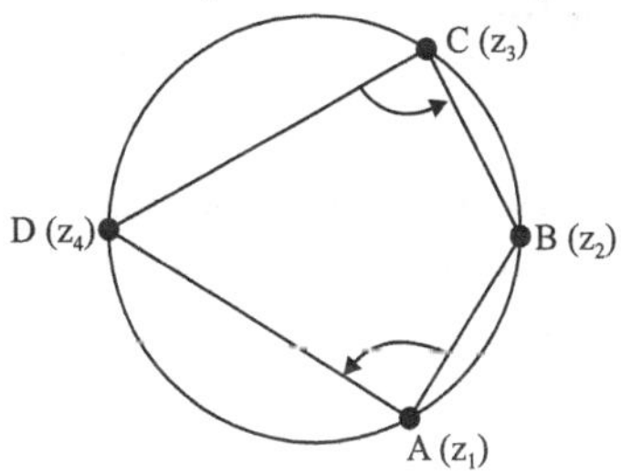

Four points z_1, z_2, z_3 and z_4 will lie on the same circle if and only if $\frac{(z_4 - z_1)(z_3 - z_2)}{(z_4 - z_2)(z_3 - z_1)}$ is purely real.

$\Rightarrow \frac{(z_1 - z_3)(z_2 - z_4)}{(z_1 - z_4)(z_2 - z_3)}$ is purely real.

Sum of angles $BAD + BCD = \pi$

$$\arg\left(\frac{z_4 - z_1}{z_2 - z_1}\right) + \arg\left(\frac{z_2 - z_3}{z_4 - z_3}\right) = \pi$$

Inverse points w.r.t a circle

Two points P & Q are said to be inverse w. r. t. A circle with centre 'O' and radius p, if :

(i) the point O, P, Q are collinear and on the same side of O .

(ii) OP . OQ = p^2.

N.B: Two points z_1 and z_2 will be the inverse points w. r. t. The circle $z\bar{z} + \bar{\alpha}z + r = 0$ if and only if $z_1\bar{z}_2 + \bar{\alpha}z_1 + \alpha\bar{z}_2 + r = 0$.

Illustration 1. Find the centre and radius of the circle $|z-2|^2 + |z-4i|^2 = 20$

Solution: The given equation represents a circle with the segment joining 2 and 4i as diameter so its centre is 1 + 2i and radius is $\frac{\sqrt{2^2+4^2}}{2} = \sqrt{5}$

Illustration 2. Find the centre and radius of the circle $\left|\frac{z-1}{z+1}\right| = \frac{1}{2}$

Solution: $4|z-1|^2 = |z+1|^2$

$\Rightarrow 4(z\bar{z} - z - \bar{z} + 1) = z\bar{z} + z + \bar{z} + 1$

$\Rightarrow 3z\bar{z} - 5z - 5\bar{z} + 3 = 0$

It represents a circle with centre at 5/3 + 0i

One Important Property

The equation $|z - z_1|^2 + |z - z_2|^2 = k$ (where k is a real number) will represent a circle then find the condition on k and center and radius of circle.

$|z - z_1|^2 + |z - z_2|^2 = k$ implies

$$|z|^2 + |z_1|^2 - 2Re(z\bar{z}_1) + |z|^2 + |z_2|^2 - 2Re(z\bar{z}_2) = k$$

$$\Rightarrow |z|^2 - Re\left[z\left(\overline{\frac{z_1 + z_2}{2}}\right)\right] + \frac{1}{4}|z_1 + z_2|^2$$

$$= \frac{k}{2} + \frac{1}{4}\left[|z_1 + z_2|^2 - 2|z_1|^2 - 2|z_2|^2\right]$$

$$\left|z - \frac{z_1 + z_2}{2}\right|^2 = \frac{1}{2}\left[k - \frac{1}{2}\left\{|z_1|^2 + |z_2|^2 - 2Re(z_1\bar{z}_2)\right\}\right]$$

$$\rightarrow \left|z - \frac{z_1 + z_2}{2}\right|^2 = \frac{1}{2}\left[k - \frac{1}{2}|z_1 - z_2|^2\right]$$

This will represent a circle with centre at $\frac{1}{2}(z_1 - z_2)$ and radius $\frac{1}{2}\sqrt{2k - |z_1 - z_2|^2}$ provided

$$k \geq \frac{1}{2}|z_1 - z_2|^2$$

6.6.5 Properties of Quadrilaterals in Complex Numbers

- **Parallelogram:**

***A*(z_1), *B*(z_2), *C*(z_3) and *D*(z_4) to represent vertices of a Parallelogram:** The diagonals AC and BD must bisect each other

$\Rightarrow \quad \frac{1}{2}(z_1 + z_3) = \frac{1}{2}(z_2 + z_4)$

$\Rightarrow \quad z_1 + z_3 = z_2 + z_4$

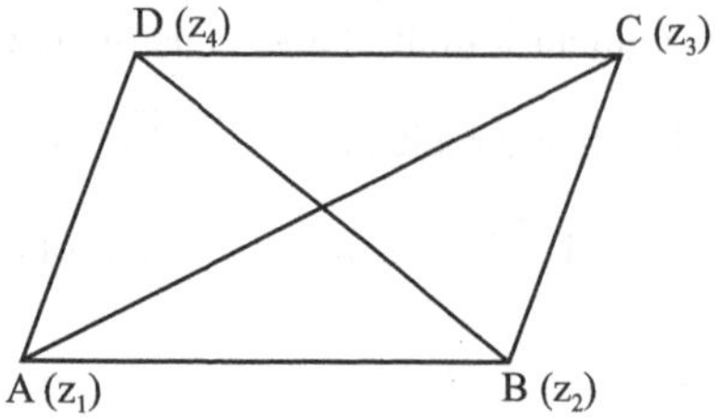

Rhombus:-

***A*(z_1), *B*(z_2), *C*(z_3) and *D*(z_4) to represent vertices of a Rhombus –**

(a) the diagonals AC and BD bisect each other $\Rightarrow z_1 + z_3 = z_2 + z_4$, and

(b) a pair of two adjacent sides are equal, for instance, AD = AB $\Rightarrow |z_4 - z_1| = |z_2 - z_1|$

or $|z_4 - z_3| = |z_2 - z_3|$

(c) Diagonals AC and BD bisect each other at right angle.

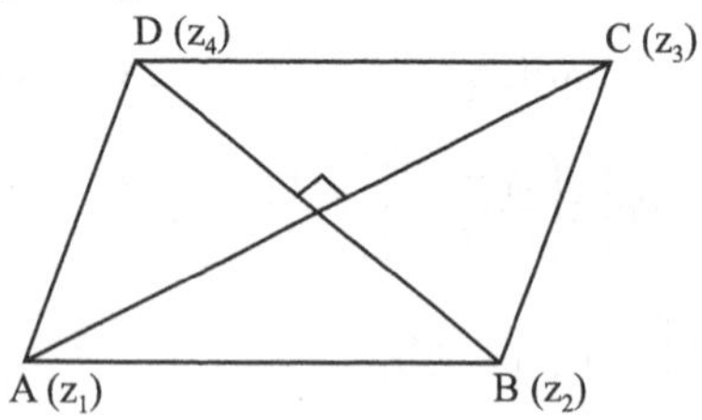

Square –

***A*(z_1), *B*(z_2), *C*(z_3) and *D*(z_4) to represent vertices of a Square –**

(a) the diagonals AC and BD bisect each other $\Rightarrow z_1 + z_3 = z_2 + z_4$

(b) a pair of adjacent sides are equal; for instance, AD = AB $\Rightarrow |z_4 - z_1| = |z_2 - z_1|$

(c) the two diagonals are equal, that is, AC ≈ BD $\Rightarrow |z_3 - z_1| = |z_4 - z_2|$

(d) Adjacent sides are perpendicular to each other.

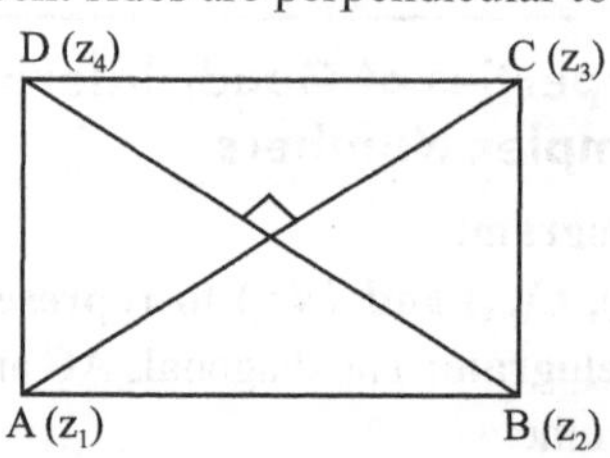

Rectangle –

***A*(z_1), *B*(z_2), *C*(z_3) and *D*(z_4) to represent vertices of a Rectangle –**

(a) the diagonals AC and BD bisect each other $\Rightarrow z_1 + z_3 = z_2 + z_4$

(b) the diagonals AC and BD are equal $\Rightarrow |z_3 - z_1| = |z_4 - z_2|$

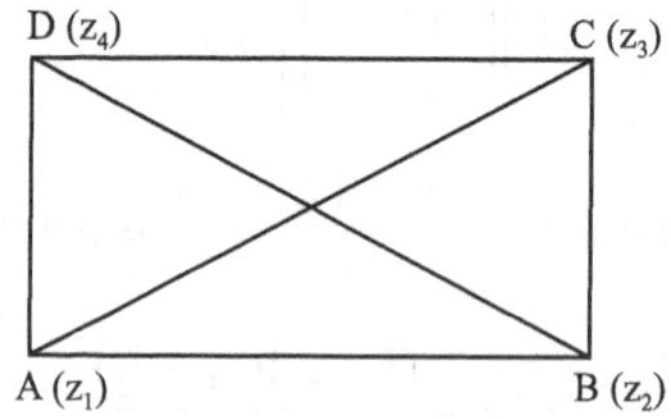

6.6.6 Standard Locus in Complex Numbers

1. $|z - z_1| = |z - z_2|$

It represents a straight line. In fact, a perpendicular bisector joining the line between two points z_1 and z_2

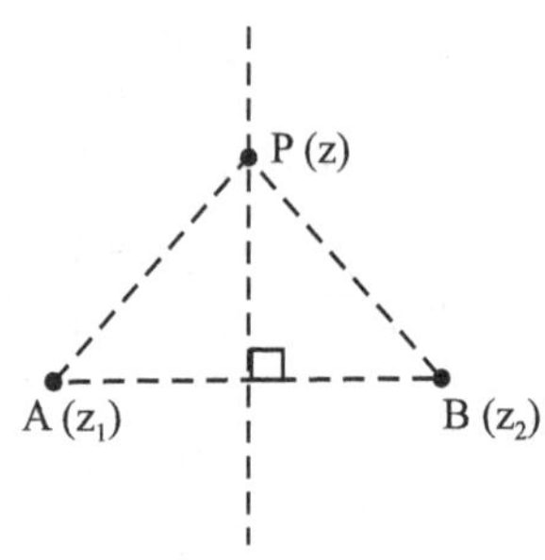

2. $|z - z_1| + |z - z_2| = |z_1 - z_2|$

Here points *A* and *B* represents complex numbers z_1 and z_2 respectively. Point P represents any complex number, z on the line segment AB

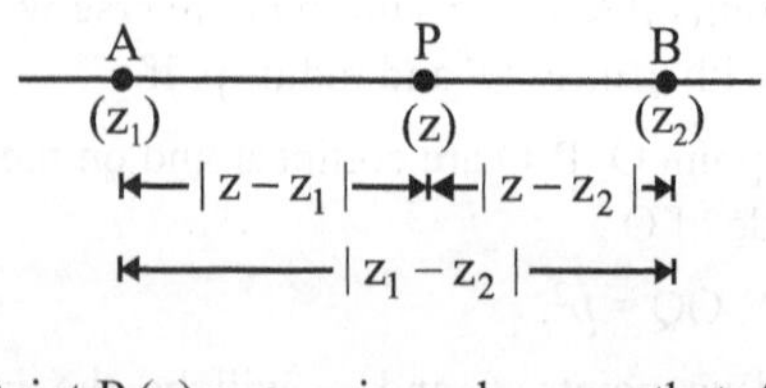

Point P (z) moves in such a way that *AP* + *PB* = *AB*. Thus locus of z is a line segment joining z_1 and z_2.

3. $||z - z_1| - |z - z_2|| = |z_1 - z_2|$

Thus, locus of z is a straight line joining z_1 and z_2 but excluding the line segment AB.

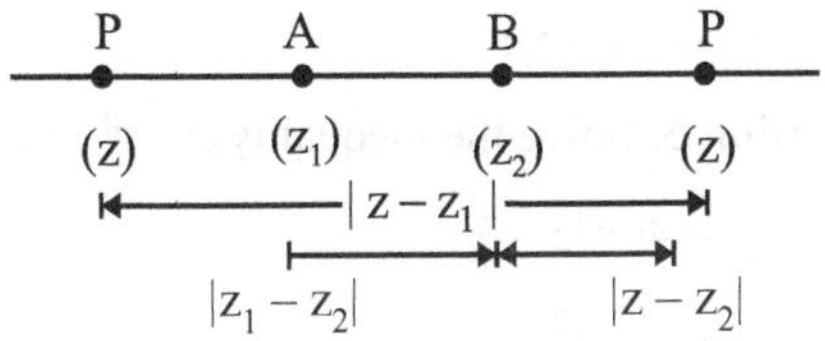

4. $|z - z_1| - |z - z_2| = |z_1 - z_2|$

$PA - PB = AB$

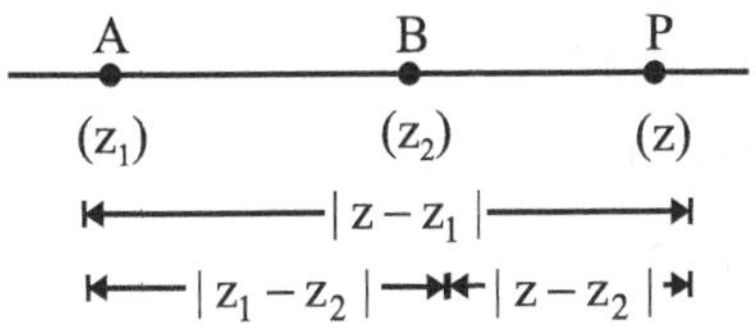

Thus, locus of z is a straight line but P will always lie on the right side of the point B.

5. $|z - z_1| + |z - z_2| = k$ **where** $k > |z_1 - z_2|$.

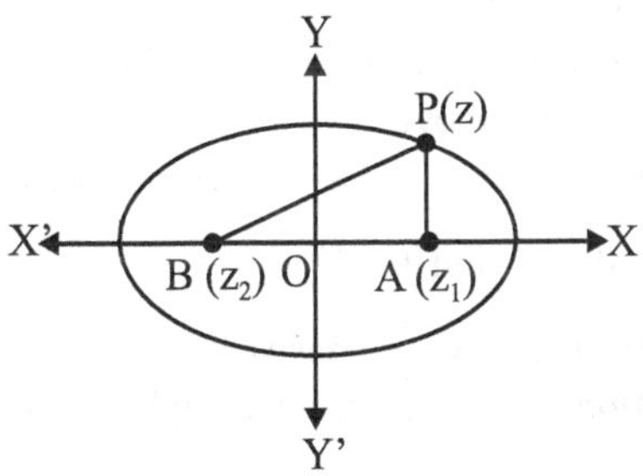

Let z be the point P.

$$PA = |z - z_1|$$

$$PB = |z - z_2|$$

$$AB = |z_1 - z_2|$$

If $PA + PB = k$ and $k > |z_1 - z_2|$

Then locus of z will be an ellipse.

6. $||z - z_1| - |z - z_2|| = k$

where $k > |z_1 - z_2|$

Then locus of z is a hyperbola.

So, $|AP - BP|$ = Constant.

7. $|z - z_1|^2 + |z - z_2|^2 = |z_1 - z_2|^2$

$PA^2 + PB^2 = AB^2$

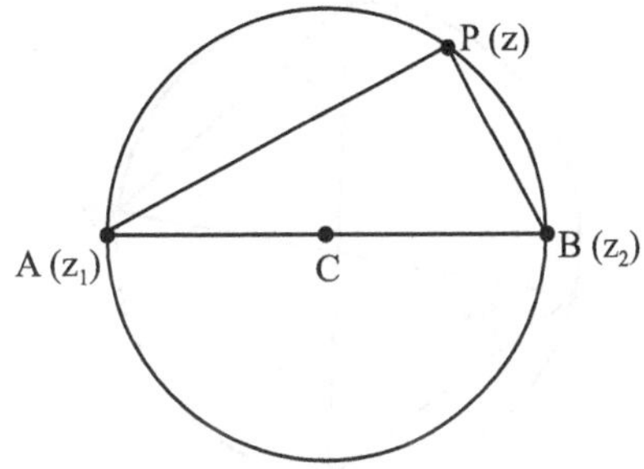

Thus locus of P is a circle with z_1 and z_2 as ends of any diameter of circle.

$$PA^2 + PB^2 = AB^2$$

8. $|z - z_1| = k\,|z - z_2|$ where $k \neq 1$.

If a point moves in such a way that its distance from a given point A is in a constant ratio to its distance from another point B. Then locus of P will be a circle.

9. $\arg\left(\dfrac{z - z_1}{z - z_2}\right) = 0$ **or** π

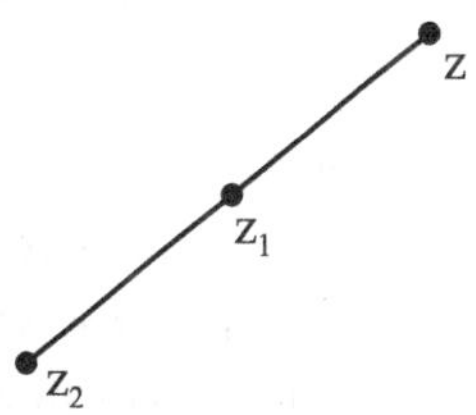

Then locus of z is a straight line joining z_1 and z_2.

10. $\arg\left(\dfrac{z - z_1}{z - z_2}\right) = \pm\dfrac{\pi}{2}$

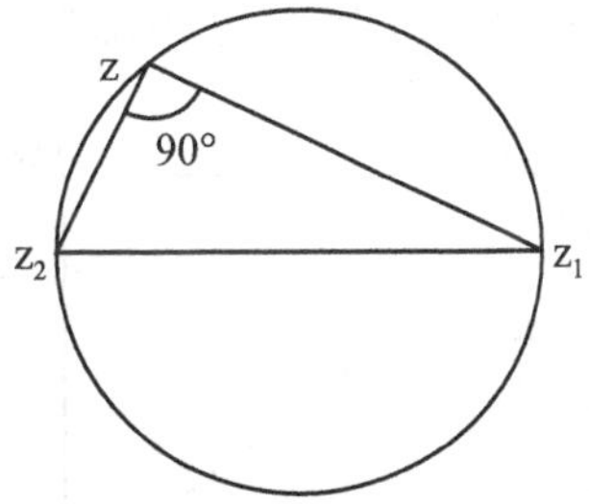

Here z represents a circle with diametrical end points as z_1 and z_2.

11. $\arg\left(\frac{z-z_1}{z-z_2}\right) = \alpha(\textit{fixed})$

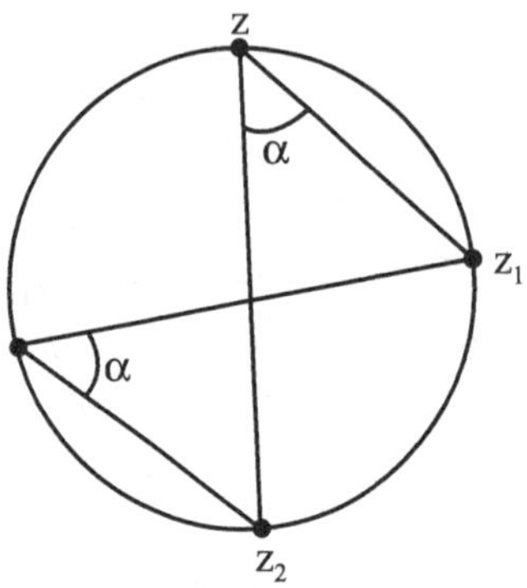

Here locus of z is a segment of a circle passing through the points z_1 and z_2.

Illustration 3. If A, B, C, are three points in the Argand plane representing the complex number z_1, z_2, z_3 such that $z_1 = \frac{\lambda z_2 + z_3}{\lambda + 1}, \lambda \in R$, then the distance of A from the line BC is

Solution: $z_1 = \frac{\lambda z_2 + z_3}{\lambda + 1}$

which show z_1, divides the line joining z_2 & z_3 in the ratio of $1 : \lambda$.

Thus, the point are collinear.

$\therefore$ Distance of A from line BC is zero.

Illustration 4. Consider a square ABCD such that z_1, z_2, z_3 and z_4 represent its vertices A, B, C and D respectively. Express 'z_3' and 'z_4' in term of z_1 and z_2.

Solution: Consider the rotation of AB about A through an angle $\frac{\pi}{4}$, we get

$$\frac{z_3 - z_1}{z_2 - z_1} = \left|\frac{z_3 - z_1}{z_2 - z_1}\right| e^{i\pi/4} = \sqrt{2}\left(\cos\frac{\pi}{4} + i\sin\frac{\pi}{4}\right)$$

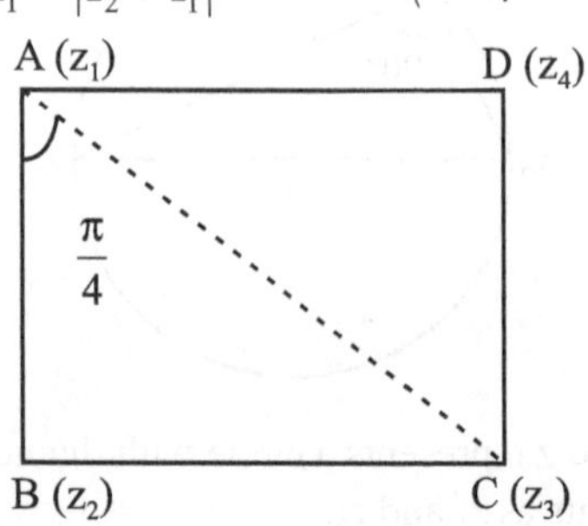

$$\Rightarrow z_3 = z_1 + (z_2 - z_1)(1 + i)$$

Similarly, $\frac{z_4 - z_1}{z_2 - z_1} = \left|\frac{z_4 - z_1}{z_2 - z_1}\right| e^{i\pi/2} = i$

$$\Rightarrow z_4 = z_1 + i(z_2 - z_1)$$

Illustration 5. Solve the inequality $|z-4| < |z-2|$.

Solution $|z-4| < |z-2|$

$\Rightarrow |z-4|^2 < |z-2|^2$

$\Rightarrow (z-4)(\bar{z}-4) < (z-2)(\bar{z}-2)$

$\Rightarrow z\bar{z} - 4(z+\bar{z}) + 16 < z\bar{z} - 2(z+\bar{z}) + 4$

$\Rightarrow 2(z+\bar{z}) > 12$

$\Rightarrow 4Re(z) > 12$

$\Rightarrow Re(z) > 3$

Illustration 6. The complex number $z = x + iy$ which satisfy the equation $\left|\frac{z-5i}{z+5i}\right| = 1$, then z lie on

(a) x-axis
(b) the straight line $y = 5$
(c) a circle passing through the origin
(d) None of the above

Solution: $|z-5i| = |z+5i|$, represents perpendicular bisector of $(0,5)$ and $(0,-5)$

Hence, (a) is the correct answer.

Illustration 7. If z_1, z_2, z_3 are complex number such that $\frac{2}{z_1} = \frac{1}{z_2} + \frac{1}{z_3}$, show that the points represented by z_1, z_2, z_3 lies on a circle passing through the origin.

Solution: Let $P(z_1), Q(z_2), R(z_3)$ and $S(z_4)$ are concylic points, $\angle PSQ = \angle PRQ$.

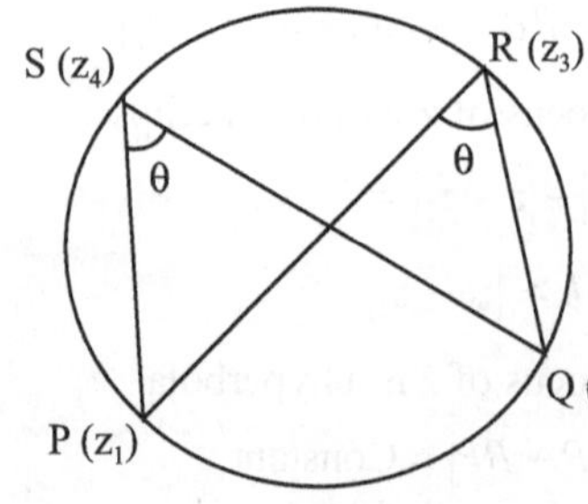

$$\Rightarrow \arg\left(\frac{z_2 - z_4}{z_1 - z_4}\right) = \arg\left(\frac{z_2 - z_3}{z_1 - z_3}\right)$$

$$\Rightarrow \arg\left[\left(\frac{z_2 - z_4}{z_1 - z_4}\right)\left(\frac{z_1 - z_3}{z_2 - z_3}\right)\right] = 0$$

$$\Rightarrow \left(\frac{z_2 - z_4}{z_1 - z_4}\right)\left(\frac{z_1 - z_3}{z_2 - z_3}\right) = \text{real}$$

If $z_4 = 0 + i0$, then $\left(\frac{z_2}{z_1}\right)\left(\frac{z_1 - z_3}{z_2 - z_3}\right) = \text{real} \quad \ldots(i)$

Now consider,

$\left(\frac{2}{z_1}\right) = \left(\frac{1}{z_2}\right) + \left(\frac{1}{z_3}\right)$ from which $z_3 = \frac{z_1 z_2}{2z_2 - z_1} \ldots(ii)$

From eqs. (i) and (ii) $\frac{z_2}{z_1} \times \frac{z_1 - \frac{z_1 z_2}{2z_2 - z_1}}{z_2 - \frac{z_1 z_2}{2z_2 - z_1}} = \text{real}$

$$\Rightarrow \frac{z_2 - z_1}{2(z_2 - z_1)} = \text{real} \Rightarrow \frac{1}{2} = \text{real, which is true.}$$

Therefore z_1, z_2, z_3 and the origin are concyclic.

Illustration 8. If $|z_1| = |z_2| = |z_3| = |z_4|$, then prove that the point representing z_1, z_2, z_3, z_4 are concyclic

Solution: $|z_1|$ = the distance of the point representing z_1 from the origin. So, the distances of the four points from the origin are equal shown as below.

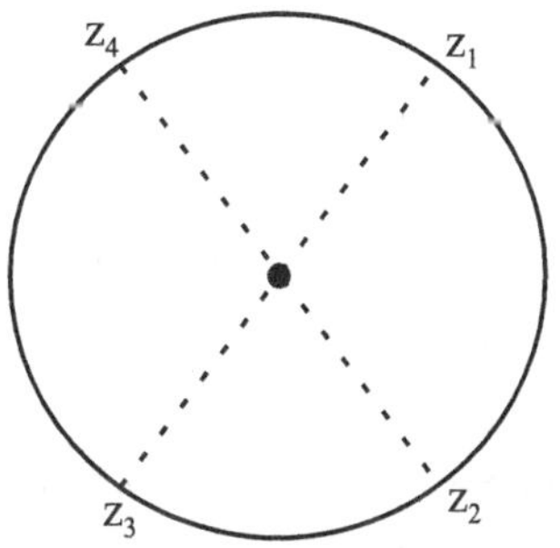

Illustration 9. Complex numbers z_1, z_2, z_3, are the vertices A, B, C respectively of an isosceles right angled triangle with right angle at C. Prove that $(z_1 - z_2)^2 = 2(z_1 - z_3)(z_3 - z_2)$.

Solution: In the isosceles triangle ABC, $AC = BC$ and BC perpendicular to AC. It means that AC is rotated through angle $\frac{\pi}{2}$ to occupy the position BC.

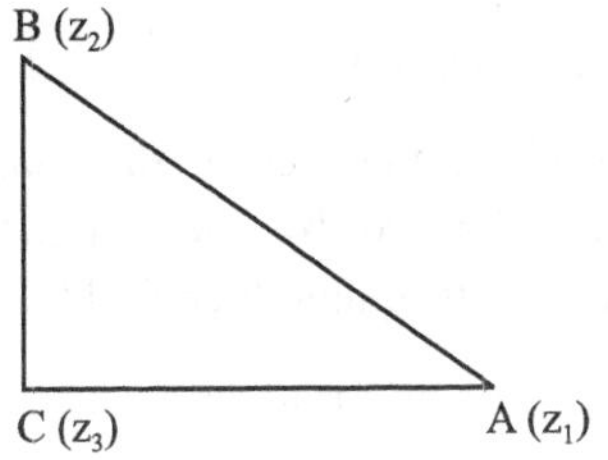

Hence, we have $\frac{z_2 - z_3}{z_1 - z_3} = e^{+i\pi/2} = +i$

$$\Rightarrow z_2 - z_3 = +i(z_1 - z_3)$$

$$\Rightarrow z_2^2 + z_3^2 - 2z_2 z_3 = -\left(z_1^2 + z_3^2 - 2z_1 z_3\right)$$

$$\Rightarrow (z_1 - z_2)^2 = 2(z_1 - z_3)(z_3 - z_2)$$

Illustration 10. Consider an 'n' sided regular polygon with the origins its centre. If z_1 be the complex number representing a vertex A_1 or the polygon, find the complex number associated with the vertex that is adjacent A_1.

Solution: Vertex adjacent to A_1 is either A_2 or A_n. Let z_2 and z_n represent the vertices A_1 and A_2 respectively.

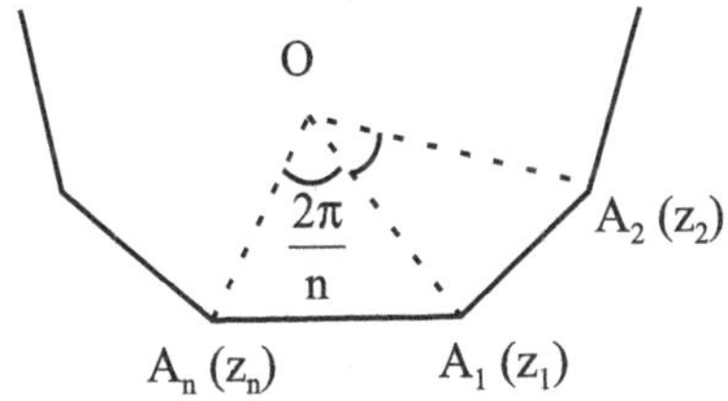

Considering the rotation about the origin, we get

$$\frac{z_2 - 0}{z_1 - 0} = \frac{|z_2|}{|z_1|} . e^{i2\pi/n} = e^{i2\pi/n}$$

$$\Rightarrow z_2 = z_1 e^{i2\pi/n}$$

Similarly $\frac{z_n}{z_1} = e^{-i2\pi/n} \Rightarrow z_n = z_1 e^{-i2\pi/n}$

Illustration 11. If the equation $|z - z_1|^2 + |z - z_2|^2 = k$ represent the equation of a circle, where $z_1 = 2 + 3i$, $z_2 = 4 + 3i$ are the extremities of a diameter, then find the value of k

Solution: As z_1 and z_1 are the extremities of diameter

$\Rightarrow |z-z_1|^2 + |z-z_2|^2 = |z_1 - z_2|^2$

$\Rightarrow k = |z_1 - z_2|^2 = |2+3i-4-3i|^2 = |-2|^2 = 4$

Illustration 12. If the area of the triangle on the Argand plane formed by the complex numbers $-z, iz, z-iz$ is 150 sq unit, then find the value of $|z|$

Solution: Area of the triangle on the argand plane formed by the complex numbers $-z, iz, z-iz$ is

$\frac{3}{2}|z|^2$

$\therefore \frac{3}{2}|z|^2 = 150 \qquad \Rightarrow |z| = 10$

Concept Applicator (CA)-6

1. The equation $z\bar{z} + (4-3i)z + (4+3i)\bar{z} + 5 = 0$ represents a circle find the radius of that circle.
2. Let z_1 and z_2 be two non-real complex cube roots of unity and $|z-z_1|^2 + |z-z_2|^2 = \lambda$ be the equation of a circle z_1 z_2 as ends of a diameter, then the value of λ is
3. For all complex numbers z_1, z_2 satisfying $|z_1| = 12$ and $|z_2 - 3 - 4i| = 5$, the minimum value of $|z_1 - z_2|$ is
   (a) 0 (b) 2
   (c) 7 (d) 17
4. If $Im\left(\frac{2z+1}{iz+1}\right) = -2$, then find the locus of the point representing z, in the complex plane.
5. If z_1, z_2, z_3, z_4 are the four complex numbers represented by the vertices of a quadrilateral taken in order such that $z_1 - z_4 = z_2 - z_3$ and $\text{amp}\frac{z_4 - z_1}{z_2 - z_1} = \frac{\pi}{2}$, then the quadrilateral is a
   (a) square
   (b) rhombus
   (c) a cyclic quadrilateral
   (d) None of these
6. The triangle formed by the points 1, $\frac{1+i}{\sqrt{2}}$ and i as vertices in the Argand diagram is
   (a) scalene
   (b) equilateral
   (c) isosceles
   (d) right angled
7. Area of the triangle formed by complex numbers $1+i, i-1, 2i$ in the Argand plane is
   (a) $\frac{1}{2}$ (b) 1
   (c) $\sqrt{2}$ (d) 2
8. Locus of the point z satisfying the equation $|iz-1| + |z-i| = 2$ is
   (a) a straight line
   (b) a circle
   (c) an ellipse
   (d) a pair of straight lines
9. If z_1, z_2, z_3 are vertices of an equilateral triangle inscribed in the circle $|z| = 2$ and if $z_1 = 1 + i\sqrt{3}$, then which one of the following says correctly about z_2, z_3
   (a) $z_2 = -2, z_3 = 1 - i\sqrt{3}$
   (b) $z_2 = 2, z_3 = 1 - i\sqrt{3}$
   (c) $z_2 = -2, z_3 = -1 - i\sqrt{3}$
   (d) $z_2 = 1 - i\sqrt{3}, z_3 = -1 - i\sqrt{3}$
10. If $Re\left(\frac{z+2i}{z+4}\right) = 0$, then locus of z is
   (a) A circle with centre (–2, 1) and radius $\sqrt{3}$
   (b) A circle with centre (–2, –1) and radius $\sqrt{3}$
   (c) A circle with centre (–2, –1) and radius $\sqrt{5}$
   (d) None of these

Exercise A

Concept Builder

1. The locus of z satisfying $Im\left(z^2\right)=4$ is
   (a) $x-y=2$ (b) $x+y=2$
   (c) $xy=2$ (d) None of these

2. If $amp\left(z_1z_2\right)=0$ and $|z_1|=|z_2|=1$, then
   (a) $z_1z_2=0$ (b) $z_1z_2=1$
   (c) $z_1z_2=-1$ (d) None of these

3. The value of $(1+i)^3+(1-i)^6$ is
   (a) $i+5$ (b) $2(-1+5i)$
   (c) $1-5i$ (d) None of these

4. If $z=x+iy$ such that $|z-4i|=|z+4i|$, then locus of z is equal to
   (a) x-axis (b) y-axis
   (c) $x=y$ (d) $x^2+y^2=1$

5. The curve represented by $Re\left(z^2\right)=4$ is
   (a) $x-y=2$ (b) $x+y=2$
   (c) $xy=2$ (d) None of these

6. If $\sqrt{3}+i=(a+ib)(c+id)$, then
   $$\tan^{-1}\frac{b}{a}+\tan^{-1}\frac{d}{c}$$
   (a) $\frac{\pi}{2}+2n\pi$ for some integer n
   (b) $-\frac{\pi}{3}+n\pi$ for some integer n
   (c) $\frac{\pi}{6}+n\pi$ for some integer n
   (d) $-\frac{\pi}{6}+2n\pi$ for some integer n

7. If z is any complex number satisfying $|z-1|=1$, then find the value of $\frac{\arg(z-1)}{\arg(z)}$
   (a) 2 (b) −1
   (c) 1 (d) None of these

8. The locus represented by $|z-1|=|z+i|$ is
   (a) a circle of radius 1
   (b) an ellipse with foci at (1, 0) and (0, −1)
   (c) a straight line through the origin
   (d) a circle on the line joining (1, 0), (0, 1) as diameter.

9. Which one of the following represents polar form of $i+\sqrt{3}$ in
   (a) $\frac{1}{\sqrt{2}}\left(\sin\frac{\pi}{6}+i\cos\frac{\pi}{6}\right)$
   (b) $2\left(\cos\frac{\pi}{6}+i\sin\frac{\pi}{6}\right)$
   (c) $\frac{1}{2}\left(\sin\frac{\pi}{6}+i\cos\frac{\pi}{6}\right)$
   (d) $4\left(\cos\frac{\pi}{6}+i\sin\frac{\pi}{6}\right)$

10. If $|z_1+z_2|=|z_1|+|z_2|$ then
    (a) $z_2=\overline{z_1}$
    (b) $z_2=\frac{1}{z_1}$
    (c) $\arg(z_1)=\arg(z_2)$
    (d) $|z_1|=|z_2|$

11. If $-1+\sqrt{-3}=re^{i\theta}$, then r and θ is
    (a) $\theta=-\frac{\pi}{3}, r=2$
    (b) $\theta=\frac{\pi}{3}, r=2$
    (c) $\theta=-\frac{\pi}{3}, r=1$
    (d) None of these

12. The value of $\left(\frac{1+i\sqrt{3}}{1-i\sqrt{3}}\right)^6+\left(\frac{1-i\sqrt{3}}{1+i\sqrt{3}}\right)^6$ is
    (a) 2 (b) −2
    (c) 1 (d) 3

13. The equation $|z+1-i| = |z+i-1|$ represents
(a) a straight line (b) a circle
(c) a parabola (d) a hyperbola

14. If $z_1 = \sqrt{3} + i\sqrt{3}$ and $z_2 = \sqrt{3} + i$, then the complex number $\left(\frac{z_1}{z_2}\right)^{50}$ lies in which quadrant
(a) I (b) II
(c) III (d) IV

15. The roots of $(z-1)^n = i(z+1)^n$ lies on
(a) on a parabola (b) y-axis
(c) x-axis (d) a circle

16. For any complex number z, find the minimum value of $|z| + |z-1|$
(a) 1 (b) 0
(c) $\frac{1}{2}$ (d) $\frac{3}{2}$

17. Find the value of $\frac{\left(\sin\frac{\pi}{8} + i\cos\frac{\pi}{8}\right)^8}{\left(\sin\frac{\pi}{8} - i\cos\frac{\pi}{8}\right)^8}$
(a) −1 (b) 0
(c) 1 (d) $2i$

18. If $a = \cos\frac{4\pi}{3} + i\sin\frac{4\pi}{3}$, then the value of $\left(\frac{1+a}{2}\right)^{3n}$ is
(a) $(-1)^n$ (b) $\frac{(-1)^n}{2^{3n}}$
(c) $\frac{1}{2^{3n}}$ (d) $(-1)^n + 1$

19. $\frac{(\cos\theta + i\sin\theta)^4}{(i\cos\theta + \sin\theta)^5}$ is equal to
(a) $\cos\theta - i\sin\theta$ (b) $\cos 9\theta - i\sin 9\theta$
(c) $\sin\theta - i\cos\theta$ (d) $\sin 9\theta - i\cos 9\theta$

20. If $z = \frac{1-i\sqrt{3}}{1+i\sqrt{3}}$, then find $\arg(z)$
(a) $-\frac{2\pi}{3}$ (b) $\frac{\pi}{3}$
(c) $-\frac{\pi}{3}$ (d) None of these

21. In a geometrical progression, first term and common ratio are both $\frac{1}{2}(\sqrt{3}+i)$. Then, find the absolute value of the nth term of the progression
(a) 2^n (b) 4^n
(c) 1 (d) None of these

22. If z_0 is an imaginary 7th root of unity, then find the value of $\log_2\left|1 + z_0 + z_0^2 + z_0^3 + z_0^4 + z_0^5 - \frac{1}{z_0}\right|$.
(a) 1 (b) 2
(c) 4 (d) None of these

23. If $Im\left(\frac{z+2i}{z+2}\right) = 0$, then locus of z is
(a) A circle which passes through origin
(b) Straight line
(c) Acircle which touches x-axis
(d) None of the above

24. For $x_1, x_2, y_1, y_2 \in R$, if $0 < x_1 < x_2, y_1 = y_2$ and $z_1 = x_1 + iy_1$, $z_2 = x_2 + iy_2$ and $z_3 = \frac{1}{2}(z_1 + z_2)$, then z_1, z_2, z_3 satisfy
(a) $|z_1| = |z_2| = |z_3|$ (b) $|z_1| < |z_2| < |z_3|$
(c) $|z_1| > |z_2| >| z_3$ (d) $|z_1| < |z_3| < |z_2|$

25. If z_1, z_2, z_3 represent the vertices of an equilateral triangle such that $|z_1| = |z_2| = |z_3|$, then
(a) $z_1 + z_2 = z_3$ (b) $z_1 + z_2 + z_3 = 0$
(c) $z_1 z_2 = \frac{1}{z_3}$ (d) $z_1 - z_2 = z_3 - z_2$

26. If 1, ω, ω^2 are the three cube roots of unity and α, β and γ are the cube roots of p, $p < 0$, then for any x, y and z the expression $\frac{x\alpha + y\beta + z\gamma}{x\beta + y\gamma + z\alpha}$ equals
(a) 1 (b) ω
(c) ω^2 (d) None of these

27. In the Argand plane, all the complex numbers satisfying $|z-4i|+|z+4i|=10$ lie on

(a) a straight line (b) a circle

(c) an ellipse (d) a parabola

28. For how many values of 'x' the complex numbers $\sin x + i\cos 2x$ and $\cos x - i\sin 2x$ are conjugate to each other

(a) 5 (b) 7

(c) more than 7 (d) 0

29. Let z and w be two non-zero complex numbers such that $|z|=|w|$ and $\arg(z)+\arg(w)=\pi$, then

(a) $\bar{w}=-zi$ (b) $z=-\bar{w}$

(c) $z=\bar{w}$ (d) $z=-\bar{w}i$

30. The area of the triangle whose vertices are i, ω, ω^2 where ω, ω^2 are the non-real cube roots of unity, is

(a) $\frac{3\sqrt{3}}{2}$ (b) $\frac{3\sqrt{3}}{4}$

(c) 0 (d) $\frac{\sqrt{3}}{4}$

31. If $A(z_1), B(z_2)$ and $C(z_3)$ be the vertices of a triangle ABC in which $\angle ABC = \frac{\pi}{4}$ and $\frac{AB}{BC}=\sqrt{2}$, then find the value of z_2

(a) $z_3+i(z_1+z_3)$ (b) $z_3-i(z_1-z_3)$

(c) $z_3+i(z_1-z_3)$ (d) None of these

32. Locus of z, if

$$\arg[z-(1+i)] = \begin{cases} \frac{3\pi}{4}, \text{when } |z| \le |z-2| \\ -\frac{\pi}{4}, \text{when } |z| > |z-2| \end{cases} \text{ is}$$

(a) straight line passing through (2, 0)

(b) straight lines passing through (2, 0), (1, 1)

(c) a line segment

(d) a set of two rays

33. Consider a function of 'z' such that dividing $f(z)$ by $z-i$, we obtain the remainder i and dividing it by $z+i$, we get the remainder $1+i$. The remainder upon the division of $f(z)$ by z^2+1, is

(a) $\frac{1}{2}(z+1)+i$ (b) $\frac{1}{2}(iz+1)+i$

(c) $\frac{1}{2}(iz-1)+i$ (d) $\frac{1}{2}(z+i)+1$

34. Find $x+y$ if the values of x and y which satisfy the equation $\frac{(1+i)x-2i}{3+i}+\frac{(2-3i)y+i}{3-i}=i$ are

(a) 2 (b) 0

(c) 4 (d) None of these

35. If z is a complex number such that $|z| \ne 0$ and Re $(z)=0$, then

(a) $\text{Re}(z^2)=0$ (b) $\text{Im}(z^2)=0$

(c) $\text{Re}(z^2)=\text{Im}(z^2)$ (d) None of these

36. If $(x+iy)^{1/3}=a+ib$, then $\frac{x}{a}+\frac{y}{b}$ is equal to

(a) $2(a^2-b^2)$ (b) $4(a^2-b^2)$

(c) $8(a^2-b^2)$ (d) None of these

37. $\frac{1-ix}{1+ix}=a-ib$ and $a^2+b^2=1$, where a and b are real, then x is equal to

(a) $\frac{b}{1+a}$ (b) $\frac{2b}{1+a}$

(c) $\frac{b}{1-a}$ (d) None of these

38. If $z=x+iy, x, y$ real, then $|x|+|y| \le \lambda|z|$, find the value of λ^2

(a) 1 (b) 4

(c) 9 (d) None of these

39. If α, β are the roots of the equation $x^2-2x+4=0$, then the value of $\alpha^6+\beta^6$ is

(a) 64 (b) 128

(c) 256 (d) None of these

40. If z is any complex number such that $|z+4| \le 3$, then $|z+1|$ can take how many integral values?

(a) 6 (b) 7
(c) 5 (d) None of these

41. If $z = x + iy = \frac{u+iv}{u-iv}$, then find the value of $|z|$

(a) 1 (b) 2
(c) 0 (d) None of these

42. If $\left(\frac{1+i}{1-i}\right)^3 - \left(\frac{1-i}{1+i}\right)^3 = x + iy$, then (x, y) is equal to

(a) (0, 2) (b) (–2, 0)
(c) (0, –2) (d) None of these

43. If $(1+i)(1+2i)(1+3i)\dots(1+ni) = \alpha + i\beta$, then $2 \cdot 5 \cdot 10 \dots (1+n^2)$ is equal to

(a) $(\alpha+\beta)^2$ (b) $\alpha^2 - \beta^2$
(c) $\alpha^2 + \beta^2$ (d) None of these

44. If $P^2 + Q^2 = 1$, then $\frac{1+P+iQ}{1+P-iQ}$ is equal to

(a) $P + iQ$ (b) $P - iQ$
(c) $Q + iP$ (d) $Q - iP$

45. The imaginary part of $(z-1)(\cos\alpha - i\sin\alpha) + (z-1)^{-1} \times (\cos\alpha + i\sin\alpha)$ is zero, if

(a) $|z-1| = 2$ (b) $\arg(z-1) = 2\alpha$
(c) $\arg(z-1) = \alpha$ (d) $|z| = 1$

Concept Cracker

1. Let z and w be two complex numbers such that $|z| \le 1, |w| \le 1, |z - iw| = |z - i\overline{w}| = 2$, then how many complex numbers z can satisfy

(a) 2 (b) 3
(c) 4 (d) None of these

2. Find the square roots of $-2 + 2\sqrt{3}i$

(a) $\pm(1+\sqrt{3}i)$ (b) $\pm(1-\sqrt{3}i)$
(c) $\pm(-1+\sqrt{3}i)$ (d) None of these

3. If $1, \alpha_1, \alpha_2, \dots, \alpha_{n-1}$ are the nth roots of unity, then the value of $(1+\alpha_1)(1+\alpha_2)\dots(1+\alpha_{n-1})$ is

(a) n (b) $-n$
(c) $\frac{1}{2}\{1-(-1)^n\}$ (d) None of these

4. If α, β be two complex numbers, then $|\alpha|^2 + |\beta|^2$ is equal to

(a) $\frac{1}{2}(|\alpha+\beta|^2 - |\alpha-\beta|^2)$

(b) $\frac{1}{2}(|\alpha+\beta|^2 + |\alpha-\beta|^2)$

(c) $|\alpha+\beta|^2 + |\alpha-\beta|^2$

(d) None of these

5. If α is non-real such that $\alpha = \sqrt[5]{1}$, then the value of $2^{|1+\alpha+\alpha^2+\alpha^{-2}-\alpha^{-1}|}$ is equal to

(a) 4 (b) 2
(c) 1 (d) None of these

6. Find the locus of z in the Argand plane if $z = 1 - t + i\sqrt{t^2+t+2}$, where t is a real parameter

(a) a hyperbola (b) an ellipse
(c) a straight line (d) None of these

7. The equation $|z+i| - |z-i| = k$ represents a hyperbola if

(a) $|k| < 1$ (b) $|k| < 2$
(c) $|k| > 2$ (d) None of these

8. The equation $|z-i| + |z+i| = k, k > 0$, can represent an ellipse if

(a) $|k| < 1$ (b) $|k| < 2$
(c) $|k| > 2$ (d) None of these

9. If $|z-i| \le 2$ and $z_0 = 5 + 3i$, then the maximum value of $|iz + z_0|$ is

(a) $2 + \sqrt{31}$ (b) 7
(c) $\sqrt{31} - 2$ (d) None of these

10. Find the value of i^i

(a) $e^{-\pi/2}$ (b) $e^{\pi/2}$

(c) $e^{\pi/4}$ (d) None of these

11. If $|z| = \max\{|z-1|, |z+1|\}$, then find the value of $|z+\bar{z}|$

(a) $1\frac{1}{2}$ (b) 1

(c) $\frac{1}{2}$ (d) None of these

12. The roots z_1, z_2, z_3 of the equation $x^3 + 3px^2 + 3qx + r = 0$ (p, q, r are complex numbers) correspond to points A, B and C, then triangle ABC is equilateral, if

(a) $p = q^2$ (b) $p^2 = 3q$

(c) $p^2 = q$ (d) $q^2 = 3p$

13. Find the value of $\sum_{k=2}^{n}(k-1)(k-\omega)(k-\omega^2)$ where ω is an imaginary cube root of unity

(a) $\left(\frac{n(n+1)}{2}\right)^2$ (b) $\left(\frac{n(n+1)}{2}\right)^2 - n$

(c) $\left(\frac{n(n+1)}{2}\right)^2 + n$ (d) None of these

14. If $\log_{1/2}\frac{|z|^2 + 2|z| + 4}{2|z|^2 + 1} < 0$, then the region traced by z is

(a) $|z| < 3$ (b) $1 < |z| < 3$

(c) $|z| > 1$ (d) $|z| < 2$

15. If $z_1 = 8 + 4i$, $z_2 = 6 + 4i$ and $\arg\left(\frac{z-z_1}{z-z_2}\right) = \frac{\pi}{4}$, then which one of the following is true about the locus of z

(a) Locus of z is a circle with centre (–7, 5) and radius 2

(b) Locus of z is a circle with centre (–7, –5) and radius $\sqrt{2}$

(c) Locus of z is a circle with centre (7, 5) and radius $\sqrt{2}$

(d) Locus of z is a circle with centre (7, –5) and radius $\sqrt{2}$

16. For the complex numbers z_1 and z_2 find the value of $\frac{\left\{|1-\bar{z}_1 z_2|^2 - |z_1 - z_2|^2\right\}}{\left\{\left(1-|z_1|^2\right)\left(1-|z_2|^2\right)\right\}}$

(a) 1 (b) – 1

(c) 2 (d) 4

17. If $z = i\log_e\left(2-\sqrt{3}\right)$, then the value of cos z

(a) 2 (b) –2

(c) $2i$ (d) $-2i$

18. If $|z| = 2$ and locus of $5z - 1$ is the circle having radius a and $z_1^2 + z_2^2 - 2z_1 z_2\cos\theta = 0$, then $|z_1| : |z_2|$ is equal to

(a) $a:1$ (b) $2a:1$

(c) $a:10$ (d) None of these

19. Let α and β be two fixed non-zero complex numbers and 'z' a variable complex number. If the lines $\alpha\bar{z} + \bar{\alpha}z + 1 = 0$ and $\beta\bar{z} + \bar{\beta}z - 1 = 0$ are mutually perpendicular, then

(a) $\alpha\beta + \bar{\alpha}\,\bar{\beta} = 0$ (b) $\alpha\beta - \bar{\alpha}\bar{\beta} = 0$

(c) $\bar{\alpha}\beta - \alpha\bar{\beta} = 0$ (d) $\alpha\bar{\beta} + \bar{\alpha}\beta = 0$

20. The expression $\tan\left[i\ln\left(\frac{a-ib}{a+ib}\right)\right]$ reduces to

(a) $\frac{ab}{a^2+b^2}$ (b) $\frac{ab}{a^2-b^2}$

(c) $\frac{2ab}{a^2+b^2}$ (d) $\frac{2ab}{a^2-b^2}$

21. If z is a complex number $z^2 + \bar{z}^2 = 2$ represents

(a) a circle (b) a straight line

(c) h hyperbola (d) an ellipse

22. If $z = x + iy$ and $|z-1| + |z+1| \le 4$ then

(a) $\frac{x^2}{4} + \frac{y^2}{3} \le 1$ (b) $\frac{x^2}{4} + \frac{y^2}{3} > 1$

(c) $\frac{x^2}{3} + \frac{y^2}{4} \le 1$ (d) None of these

23. If z_1, z_2 and z_3 are the vertices of an equilateral triangle then $z_1^2 + z_2^2 + z_3^2$ is equal to
(a) 1
(b) –1
(c) $z_1z_2 + z_1z_3 + z_2z_3$
(d) None of these

24. Which one the following is true about the expression

$$S_n = 2\left(1+\frac{1}{\omega}\right)\left(1+\frac{1}{\omega^2}\right)+3\left(2+\frac{1}{\omega}\right)\left(2+\frac{1}{\omega^2}\right)$$
$$+4\left(3+\frac{1}{\omega}\right)\left(3+\frac{1}{\omega^2}\right)+\ldots+(n+1)\left(n+\frac{1}{\omega}\right)\left(n+\frac{1}{\omega^2}\right)$$

where ω is an imaginary cube root of unity.
(a) $S_{20} = 44100$ (b) $S_{10} = 44100$
(c) $S_{10} = 3025$ (d) $S_{20} = 44120$

25. Let 'z' be a complex number and 'a' be a real parameter such that $z^2 + az + a^2 = 0$, then how many of the following statements are true?
(i) Locus of z is a pair of straight lines
(ii) $|z| = |a|$
(iii) $\arg(z) = \pm\frac{2\pi}{3}$
(a) 0 (b) 1
(c) 2 (d) 3

26. For how many complex number z satisfying the equations $|z| - 4 = |z - i| - |z + 5i| = 0$,
(a) 0 (b) 1
(c) 2 (d) None of these

27. If $2\cos\theta = x + \frac{1}{x}$ and $2\cos\phi = y + \frac{1}{y}$, then how many of the following statements are correct?
(i) $\frac{x}{y} + \frac{y}{x} = 2\cos(\theta + \phi)$
(ii) $x^m y^n + \frac{1}{x^m y^n} = 2\cos(m\theta + \phi)$
(iii) $\frac{x^n}{y^n} + \frac{y^n}{x^n} = 2\cos(\theta + \phi)$
(iv) $xy + \frac{1}{xy} = 2\cos(\theta - \phi)$
(a) 0 (b) 1
(c) 2 (d) 4

28. Let a be a complex number such that $|a| < 1$ and $z_1, z_2, \ldots, z_n$ be the vertices of a polygon such that $z_k = 1 + a + a^2 + \ldots a^k$, then the vertices of the polygon lie within the circle whose
(a) Radius is $\frac{1}{|1+a|}$ and centre is $\frac{1}{1-a}$
(b) Radius is $\frac{1}{|1+a|}$ and centre is $\frac{1}{1+a}$
(c) Radius is $\frac{1}{|1-a|}$ and centre is $\frac{1}{1-a}$
(d) None of these

29. If $z = x + iy$ is a complex number satisfies the equation $|z| - z = 1 + 2i$, then find $2|x - y|$
(a) 7 (b) 6
(c) 8 (d) None of these

30. The complex numbers z_1 and z_2 are such that $z_1 \neq z_2$ and $|z_1| = |z_2|$. If z_1 has positive real part and z_2 has negative imaginary part then $\left(\frac{z_1 + z_2}{z_1 - z_2}\right)$ may be
(a) zero (b) real and positive
(c) real and negative (d) puerly imaginary

31. If $\left|\frac{z_1 z - z_2}{z_1 z + z_2}\right| = k, (z_1, z_2 \neq 0)$, then
(a) for k ≠ 1, locus z is a straight line
(b) for $k \notin \{1, 0\}$, z lies on a circle
(c) for $k \neq 0$, z represents a point
(d) for $k \neq 1$, z lies on the perpendicular bisector of the line segment joining $\frac{z_2}{z_1}$ and $-\frac{z_2}{z_1}$.

32. If $z = x + iy$ and $\alpha + i\beta = \tan^{-1} z$, and α is constant, then the locus of z is

(a) $x^2 + y^2 + 2x\cot 2\alpha = 1$

(b) $\left(x^2 + y^2\right)\cot 2\alpha = 1 + x$

(c) $x^2 + y^2 + 2y\tan 2\alpha = 1$

(d) $x^2 + y^2 + 2x\sin 2\alpha = 1$

33. $\cos\left(i\log\dfrac{a-ib}{a+ib}\right)$ is equal to

(a) ab (b) $\dfrac{a^2-b^2}{a^2+b^2}$

(c) $\dfrac{a^2-b^2}{2ab}$ (d) $\dfrac{2ab}{a^2+b^2}$

34. If $z_1 z_2 \in C, z_1^2 + z_2^2 \in R, z_1\left(z_1^2 - 3z_2^2\right) = 2$ and $z_2\left(3z_1^2 - z_2^2\right) = 11$, then find the value of $z_1^2 + z_2^2$

(a) 5 (b) 6

(c) 10 (d) 12

35. If $\sqrt{1-c^2} = nc - 1$ and $z = e^{i\theta}$, then $\left(\dfrac{c}{2n}\right)(1+nz)\left(1+\dfrac{n}{z}\right)$ is equal to

(a) $1 - c\cos\theta$ (b) $1 + 2c\cos\theta$

(c) $1 + c\cos\theta$ (d) $1 - 2c\cos\theta$

36. If $|z - 2 - i| = |z|\left|\sin\left(\dfrac{\pi}{4} - \arg z\right)\right|$, then find the locus of z

(a) a pair of straight lines

(b) circle

(c) parabola

(d) ellipse

37. If $\omega(\neq 1)$ is cube root of unity satisfying

$$\frac{1}{a+\omega} + \frac{1}{b+\omega} + \frac{1}{c+\omega} = 2\omega^2$$

and $\dfrac{1}{a+\omega^2} + \dfrac{1}{b+\omega^2} + \dfrac{1}{c+\omega^2} = 2\omega$,

then the value of $\dfrac{1}{a+1} + \dfrac{1}{b+1} + \dfrac{1}{c+1}$ is

(a) 2 (b) –2

(c) $-1+\omega^2$ (d) None of these

38. If the points are represented by complex numbers $z_1 = a + ib, z_2 = a' + ib'$ and $z_1 - z_2$ are collinear, then

(a) $ab' + a'b = 0$ (b) $ab' - a'b = 0$

(c) $ab + a'b' = 0$ (d) $ab - a'b' = 0$

39. If z_1, z_2, z_3 are the affixes points A, B and C are lying on the circle centred at origin. If the altitude is drawn from vertex A to base BC such that it meets the circumcircle at P(z), then $zz_1 + z_2z_3$ is

(a) 0 (b) 1

(c) –1 (d) None of these

40. If the complex numbers z_1, z_2 and z_3 are the vertices of a triangle such that $\dfrac{z_1 - z_3}{z_2 - z_3} = \dfrac{1 - i\sqrt{3}}{2}$ then the triangle is

(a) non right angled isosceles

(b) right angled isosceles

(c) equilateral

(d) obtuse angled isosceles

IIIT JEE Mains/AIEEE Corner

41. If z and ω are two non-zero complex numbers such that $|z\omega| = 1$ and $Arg(z) - Arg(\omega) = \dfrac{\pi}{2}$, then $\bar{z}\omega$ is equal to (2003)

(a) $-i$ (b) 1

(c) – 1 (d) i

42. Let Z_1 and Z_2 be two roots of the equation $Z^2 + aZ + b = 0, Z$ being complex. Further, assume that the origin, Z_1 and Z_2 form an equilateral triangle. Then (2003)

(a) $a^2 = 4b$ (b) $a^2 = b$

(c) $a^2 = 2b$ (d) $a^2 = 3b$

43. If $\left(\frac{1+i}{1-i}\right)^x = 1$, then (2006)

(a) $x = 2n + 1$, where n is any positive integer
(b) $x = 4n$, where n is any positive integer
(c) $x = 2n$, where n is any positive integer
(d) $x = 4n + 1$, where n is any positive integer

44. Let z and w be complex numbers such that $\bar{z} + i\bar{w} = 0$ and arg zw = π. Then arg z equals (2004)

(a) $\frac{5\pi}{4}$ (b) $\frac{\pi}{2}$
(c) $\frac{3\pi}{4}$ (d) $\frac{\pi}{4}$

45. If $z = x + iy$ and $z^{\frac{1}{3}} = p + iq$, then $\frac{\left(\frac{x}{p} + \frac{y}{q}\right)}{\left(p^2 + q^2\right)}$ is equal to (2004)

(a) -2 (b) -1
(c) 2 (d) 1

46. If $|z^2 - 1| = |z|^2 + 1$, then z lies on (2004)

(a) an ellipse (b) the imaginary axis
(c) a circle (d) the real axis

47. If z_1 and z_2 are two non-zero complex numbers such that $|z_1 + z_2| = |z_1| + |z_2|$, then arg $z_1 - \arg z_2$ is equal to (2005)

(a) $\frac{\pi}{2}$ (b) π
(c) 0 (d) $\frac{-\pi}{2}$

48. If $\omega = \frac{z}{z - \frac{1}{3}i}$ and $|\omega| = 1$, then z lies on (2005)

(a) an ellipse (b) a circle
(c) a straight line (d) a parabola

49. The value of $\sum_{k=1}^{10}\left(\sin\frac{2k\pi}{11} + i\cos\frac{2k\pi}{11}\right)$ is (2006)

(a) i (b) 1
(c) -1 (d) $-i$

50. If $z^2 + z + 1 = 0$, where z is complex number, then the value of

$$\left(z + \frac{1}{z}\right)^2 + \left(z^2 + \frac{1}{z^2}\right)^2 + \left(z^3 + \frac{1}{z^3}\right)^2 + \ldots + \left(z^6 + \frac{1}{z^6}\right)^2$$ (2006)

(a) 18 (b) 54
(c) 6 (d) 12

51. If $|z + 4| \le 3$, then the maximum value of $|z + 1|$ is (2007)

(a) 6 (b) 0
(c) 4 (d) 10

52. The conjugate of a complex number is $\frac{1}{i-1}$ then that complex number is (2008)

(a) $\frac{-1}{i-1}$ (b) $\frac{1}{i+1}$
(c) $\frac{-1}{i+1}$ (d) $\frac{1}{i-1}$

53. The number of complex numbers z such that $|z - 1| = |z + 1| = |z - i|$ (2010)

(a) 1 (b) 2
(c) ∞ (d) 0

54. Let α, β be real and z be a complex number. If $z^2 + \alpha z + \beta = 0$. If has two distinct roots on the line Re $z = 1$, then it is necessary that (2011)

(a) $\beta \in (-1, 0)$ (b) $|\beta| = 1$
(c) $\beta \in (1, \infty)$ (d) $\beta \in (0, 1)$

55. If z ≠ 1 and $\frac{z^2}{z-1}$ is real, then the point represented by the complex number z lies: (2012)

(a) either on the real axis or on a circle passing through the origin.
(b) on a circle with centre at the origin
(c) either on the real axis or on a circle not passing through the origin
(d) on the imaginary axis.

56. If z is a complex number of unit modulus and argument θ, then arg $\left(\frac{1+z}{1+\bar{z}}\right)$ equals:

(JEE M 2013)

(a) $-\theta$ (b) $\frac{\pi}{2}-\theta$

(c) θ (d) $\pi-\theta$

57. If z is a complex number such that $|z| \geq 2$, then the minimum value of $\left|z+\frac{1}{2}\right|$:

(JEE M 2014)

(a) is strictly greater than $\frac{5}{2}$

(b) is strictly greater than $\frac{3}{2}$ but less than $\frac{5}{2}$

(c) is equal to $\frac{5}{2}$

(d) lie in the interval (1, 2)

58. A complex number z is said to be unimodular if $|z| = 1$. Suppose z_1 and z_2 are complex numbers such that $\frac{z_1-2z_2}{2-z_1\bar{z}_2}$ is unimodular and z_2 is not unimodular. Then the point z_1 lies on a:

(JEE M 2015)

(a) circle of radius 2.

(b) circle of radius $\sqrt{2}$.

(c) straight line parallel to x-axis

(d) straight line parallel to y-axis.

59. A value of θ for which $\frac{2+3i\sin\theta}{1-2i\sin\theta}$ is purely imaginary, is:

(JEE M 2016)

(a) $\sin^{-1}\left(\frac{\sqrt{3}}{4}\right)$ (b) $\sin^{-1}\left(\frac{1}{\sqrt{3}}\right)$

(c) $\frac{\pi}{3}$ (d) $\frac{\pi}{6}$

Board Level Questions

Level 1

1. The closest distance of the origin from a curve $a\bar{z} + \bar{a}z + a\bar{a} = 0$ (a is a complex number) is

(a) $\frac{|a|}{3}$ (b) $\frac{|a|}{2}$

(c) $\frac{Re(a)}{|a|}$ (d) $\frac{Im(a)}{|a|}$

2. A point 'z' is equidistant from three distinct points z_1, z_2, z_3 in Argand plane. If z, z_1 and z_2 are collinear, then arg$\left(\frac{z_3-z_1}{z_3-z_2}\right)$ will be (z_1, z_2, z_3 in anti-clockwise sense)

(a) $\frac{\pi}{2}$ (b) $\frac{\pi}{4}$

(c) $\frac{\pi}{6}$ (d) $\frac{2\pi}{3}$

3. For the equation $3x^2 + px + 3 = 0, p > 0$, if one of the roots is square of the other, then find the value of p

(a) $\frac{1}{3}$ (b) 1

(c) 3 (d) $\frac{2}{3}$

4. If z_1, z_2, z_3 are vertices of an equilateral triangle with z_0 its circum centre, then find the value of $\frac{z_1^2 + z_2^2 + z_3^2}{z_0^2}$

(a) 1 (b) 3

(c) 9 (d) None of these

5. Find the value of $\frac{\cos 75° + i\sin 75°}{\cos 30° + i\sin 30°}$

(a) $\frac{1}{\sqrt{2}}(1+i)$ (b) $\frac{1}{\sqrt{2}}(1-i)$

(c) $-\frac{1}{\sqrt{2}}(1+i)$ (d) None of these

6. Find the value of

$$\frac{(\cos 2\theta - i\sin 2\theta)^7 (\cos 3\theta + i\sin 3\theta)^{-5}}{(\cos 4\theta + i\sin 4\theta)^{12} (\cos 5\theta + i\sin 5\theta)^{-6}}$$

(a) $\cos 33\theta + i\sin 33\theta$

(b) $\cos 33\theta - i\sin 33\theta$

(c) $\cos 47\theta + i\sin 47\theta$

(d) $\cos 47\theta - i\sin 47\theta$

7. If P represents $z = x + iy$ in the Argand plane and $|z-1|^2 + |z+1|^2 = 4$, then the locus of P is
   (a) Circle with radius 1
   (b) Circle with radius 2
   (c) Circle with centre (1, 1)
   (d) None of these

8. For a complex number z if
   $$\log_{1/2}\left(\frac{|z-1|+4}{3|z-1|-2}\right) > 1 \left(\text{where } |z-1| \neq \frac{2}{3}\right)$$
   then
   (a) $|z-1| < 3$ (b) $|z-1| > 3$
   (c) $|z-1| > 4$ (d) None of these

9. Find the value of $\left(\frac{\sqrt{3}}{2}+\frac{i}{2}\right)^5 + \left(\frac{\sqrt{3}}{2}-\frac{i}{2}\right)^5$
   (a) i (b) ω
   (c) $i\omega$ (d) None of these

10. Find the value of $(a+b+c)(a\alpha+b\beta+c)$ $(a\beta+b\alpha+c)$ if α and β are complex cube roots of unity,
    (a) $2(a^3+b^3+c^3)$ (b) $2(a^3-b^3-c^3)$
    (c) $a^3+b^3+c^3$ (d) None of these

11. Which one of the following is equal to $(a+b)^3 + (a\omega+b\omega^2)^3 + (a\omega^2+b\omega)^3$
    (a) a^3+b^3 (b) $3(a^3+b^3)$
    (c) a^3-b^3 (d) $3(a^3-b^3)$

12. Which one of the following is correct about the complex number $z = \frac{(1+\cos\theta+i\sin\theta)^5}{(\cos\theta+i\sin\theta)^3}$
    (a) $\text{Arg(z)} = -\frac{\theta}{2}, |z| = 32\cos^5\frac{\theta}{2}$
    (b) $\text{Arg(z)} = \frac{\theta}{2}, |z| = 32\cos^5\frac{\theta}{2}$
    (c) $\text{Arg}(z) = -\frac{\theta}{2}, |z| = 16\cos^4\frac{\theta}{2}$
    (d) None of these

13. Find the number of common roots of the equation $z^3+2z^2+2z+1=0$ and $z^{1985}+z^{100}+1=0$
    (a) 0 (b) 1
    (c) 2 (d) 3

14. If $z = \cos\theta + i\sin\theta$, then
    (a) $z^n + \frac{1}{z^n} = 2\cos n\theta$
    (b) $z^n + \frac{1}{z^n} = 2^n \cos n\theta$
    (c) $z^n - \frac{1}{z^n} = 2^n i \sin n\theta$
    (d) $z^n - \frac{1}{z^n} = (2i)^n \sin n\theta$

15. If $z = a+ib$ where $a > 0, b > 0$, then
    (a) $|z| \geq \frac{1}{\sqrt{3}}(a+b)$ (b) $|z| \geq \frac{1}{\sqrt{2}}(a+b)$
    (c) $|z| < \frac{1}{\sqrt{5}}(a+b)$ (d) None of these

16. If $z = \cos\theta + i\sin\theta$, then $\frac{z^{20}-1}{z^{20}+1}$ is equal to (n is an integer)
    (a) $i\cot 10\theta$ (b) $i\tan 10\theta$
    (c) $\tan 10\theta$ (d) $\cot 10\theta$

17. If $z_r = \cos\frac{\pi}{2^r} + i\sin\frac{\pi}{2^r}$, then $z_1z_2z_3\ldots\infty$ is equal to
    (a) 1 (b) –1
    (c) 0 (d) None of these

18. If p, q, r are positive integers and ω be an imaginary cube root of unity then find the value of $\omega^{3p} + \omega^{3q+1} + \omega^{3r+2}$
    (a) ω (b) $-\omega^2$
    (c) 0 (d) None of these

19. Find the maximum value of $|z|$ if $\left|z+\frac{2}{z}\right| = 2$
    (a) $\sqrt{3}-1$ (b) $\sqrt{3}+1$
    (c) $\sqrt{3}$ (d) $\sqrt{2}+\sqrt{3}$

20. The complex numbers given by $1-3i, 4+3i$ and $3+i$ represent the vertices of
(a) a right angled triangle
(b) an isosceles triangle
(c) an equilateral triangle
(d) None of these

Level 2

21. If $z_1 \neq -z_2$ and $|z_1+z_2| = \left|\frac{1}{z_1}+\frac{1}{z_2}\right|$, then find the value of $|z_1 z_2|$

22. If n_1, n_2 are positive integers, then $(1+i)^{n_1} + (1+i^3)^{n_1} + (1+i^5)^{n_2} + (1+i^7)^{n_2}$ is a real number then prove that n_1 n_2 are any two positive integers

23. Prove that the roots of $(z-1)^n = 2\omega(z+1)^n$ (where $n \geq 3$ and ω is a complex cube root of unity) lie on a circle

24. If $z \neq 0$ is defined as $z = \frac{3}{2+\cos\theta + i\sin\theta}$ then find the value of $|z|^2 - 2(z+\bar{z})$.

25. Let $a, b, c, d \in R$ and $a^2+b^2=4$, $c^2+d^2=2$ and $a+ib = (c+id)\sqrt{x+iy}$, find the value of x^2+y^2.

26. If $1, x_1, x_2, x_3$ are the roots of $x^4-1=0$ and ω is a complex cube root of unity, find the value of
$$\frac{(\omega^2-x_1)(\omega^2-x_2)(\omega^2-x_3)}{(\omega-x_1)(\omega-x_2)(\omega-x_3)}.$$

27. If a, b, c are distinct integers and $\omega \neq 1$ is a cube root of unity, then find the minimum value of $|a+b\omega+c\omega^2|^2$.

28. If z_1 and z_2 are two non-zero complex numbers such that $|z_1+z_2|^2 = |z_1|^2+|z_2|^2$, then how many of the following statements are correct?
(i) $z_1\bar{z}_2$ is purely imaginary
(ii) $z_1/\bar{z}_2$ is purely real
(iii) $z_1\bar{z}_2 + \bar{z}_1 z_2 = 0$
(iv) O, z_1, z_2 are the vertices of an equilateral triangle.

29. Let $z_1 = 6+i$ and $z_2 = 4-3i$. Let z be a complex number such that $\arg\left(\frac{z-z_1}{z_2-z}\right) = \frac{\pi}{2}$; then find the locus of z

30. If $|z_1-1|<1$, $|z_2-2|<2$, $|z_3-3|<3$, then find the maximum value of $|z_1+z_2+z_3|$

31. If $\omega \neq 1$ and ω is a nth root of unity, then find the value of $1+4\omega+9\omega^2+16\omega^3+\ldots+n^2\omega^{n-1}$

32. If it is given that the equation $|z|^2 - 2iz + 2c(a+i) = 0$ possesses solution for all $c \in R$, then find the number of integral values of 'a' for which it is true.

33. If a and b are two real numbers lying between 0 and 1 such that $z_1 = a+i$, $z_2 = 1+ib$, $z_3 = 0$ from an equilateral triangle then find a and b

34. If z_1, z_2, z_3 are non-zero complex numbers representing the points A, B, C such that $\frac{2}{z_1} = \frac{1}{z_2}+\frac{1}{z_3}$. Then prove that circle through A, B, C passes through origin.

35. If $z = \frac{1}{2}(\sqrt{3}-i)$, find the smallest value of positive integer n for which $(z^{89}+i^{97})^{94} = z^n$

36. Find the number of complex numbers satisfying $|z+\bar{z}| + |z-\bar{z}| = 2$ and $|z+i| + |z-i| = 2$ is/are

37. Consider a complex number $z = \frac{b+ic}{1+a}$ here a, b, c are real numbers such that $a^2+b^2+c^2=1$, then prove that $\frac{1+iz}{1-iz} = \frac{a+ib}{1+c}$

38. If $\theta_1, \theta_2, \theta_3$ and θ_4 are four real numbers and α is a root of the equation $\sin\theta_1 z^3 + \sin\theta_2 z^2 + \sin\theta_3 z + \sin\theta_4 = 3$.
Such that $|\alpha| < 1$, then find the minimum value of $|\alpha|$

39. If $\log_{(0.8)}|z+1| > \log_{(0.8)}|z-1|$, then prove that $\text{Re}(z) < 0$.

6.7 ADVANCE MATERIAL FOR JEE ADVANCED, KVPY AND OLYMPIAD

6.7.1 Trigonometry of Complex Number

Property 1.

If $\sin\alpha + \sin\beta + \sin\gamma = \cos\alpha + \cos\beta + \cos\gamma = 0$, then

(a) $\cos 2\alpha + \cos 2\beta + \cos 2\gamma = 0$

(b) $\sin 2\alpha + \sin 2\beta + \sin 2\gamma = 0$

(c) $\cos 4\alpha + \cos 4\beta + \cos 4\gamma = 0$

(d) $\sin 4\alpha + \sin 4\beta + \sin 4\gamma = 0$

(e) $\cos(\beta+\gamma) + \cos(\gamma+\alpha) + \cos(\alpha+\beta) = 0$

(f) $\sin(\beta+\gamma) + \sin(\gamma+\alpha) + \sin(\alpha+\beta) = 0$

(g) $\cos 3\alpha + \cos 3\beta + \cos 3\gamma = 3\cos(\alpha+\beta+\gamma)$

(h) $\sin 3\alpha + \sin 3\beta + \sin 3\gamma = 3\sin(\alpha+\beta+\gamma)$

Explanation.

Let $a = \cos\alpha + i\sin\alpha$, $b = \cos\beta + i\sin\beta$ and $c = \cos\gamma + i\sin\gamma$.

$$\frac{1}{a} = \frac{1}{\cos\alpha + i\sin\alpha} = \cos\alpha - i\sin\alpha$$

Similarly,

$$\frac{1}{b} = \frac{1}{\cos\beta + i\sin\beta} = \cos\beta - i\sin\beta$$

and $\frac{1}{c} = \frac{1}{\cos\gamma + i\sin\gamma} = \cos\gamma - i\sin\gamma$

Now $a + b + c = 0$ ...(*i*)

And $\frac{1}{a} + \frac{1}{b} + \frac{1}{c} = 0$

so $(\frac{1}{a} + \frac{1}{b} + \frac{1}{c})abc = 0$

$\Rightarrow\ bc + ca + ab = 0$ ...(*ii*)

$(a+b+c)^2 = a^2 + b^2 + c^2 + 2(ab+bc+ca)$

$a^2 + b^2 + c^2 = 0$ ...(*iii*)

or $(\cos\alpha + i\sin\alpha)^2 + (\cos\beta + i\sin\beta)^2 + (\cos\gamma + i\sin\gamma)^2 = 0$

or $(\cos 2\alpha + \cos 2\beta + \cos 2\gamma) + i(\sin 2\alpha + \sin 2\beta + \sin 2\gamma) = 0$

Equating the real and imaginary part we get

$\cos 2\alpha + \cos 2\beta + \cos 2\gamma = 0$

$\sin 2\alpha + \sin 2\beta + \sin 2\gamma = 0$

$(a^2+b^2+c^2)^2 = a^4 + b^4 + c^4 + 2(a^2b^2 + b^2c^2 + c^2a^2)$

Consider $(ab+bc+ca)^2 = (a^2b^2 + b^2c^2 + c^2a^2) + 2abc(a+b+c)$

From equations (i) and (ii), we get

$$a^2b^2 + b^2c^2 + c^2a^2 = 0$$

So $(a^2+b^2+c^2)^2 = a^4 + b^4 + c^4 + 2(a^2b^2 + b^2c^2 + c^2a^2)$

or $a^4 + b^4 + c^4 = 0$ ...(*iv*)

or $(\cos\alpha + i\sin\alpha)^4 + (\cos\beta + i\sin\beta)^4 + (\cos\gamma + i\sin\gamma)^4 = 0$

or $(\cos 4\alpha + \cos 4\beta + \cos 4\gamma) + i(\sin 4\alpha + \sin 4\beta + \sin 4\gamma) = 0$

Equating the real and imaginary part we get

$\cos 4\alpha + \cos 4\beta + \cos 4\gamma = 0$

$\sin 4\alpha + \sin 4\beta + \sin 4\gamma = 0$

Now from equation (ii) $bc + ca + ab = 0$

Consider $ab = (\cos\alpha + i\sin\alpha)(\cos\beta + i\sin\beta)$

$= (\cos\alpha.\cos\beta - \sin\alpha.\sin\beta) + i(\sin\alpha\cos\beta + \cos\alpha\sin\beta)$

$= \cos(\alpha+\beta) + i\sin(\alpha+\beta)$

$ab + bc + ca = \cos(\beta+\gamma) + \cos(\gamma+\alpha) + \cos(\alpha+\beta)$

$$+i\left[\sin(\beta+\gamma)+\sin(\gamma+\alpha)+\sin(\alpha+\beta)\right]=0$$

Equating real and imaginary part we get

$$\cos(\beta+\gamma)+\cos(\gamma+\alpha)+\cos(\alpha+\beta)=0$$

$$\sin(\beta+\gamma)+\sin(\gamma+\alpha)+\sin(\alpha+\beta)=0$$

Now consider $a^3+b^3+c^3-3abc$

$$=(a+b+c)(a^2+b^2+c^2-bc-ca-ab)=0$$

So $a^3+b^3+c^3=3abc$

or

$$(\cos\alpha+i\sin\alpha)^3+(\cos\beta+i\sin\beta)^3+(\cos\gamma+i\sin\gamma)^3$$

$$=3(\cos\alpha+i\sin\alpha)(\cos\beta+i\sin\beta)(\cos\gamma+i\sin\gamma)$$

or

$$(\cos 3\alpha+\cos 3\beta+\cos 3\gamma)+i(\sin 3\alpha+\sin 3\beta+\sin 3\gamma)$$

$$=3\cos(\alpha+\beta+\gamma)+i3\sin(\alpha+\beta+\gamma)$$

Equating real imaginary part we get

$$\cos 3\alpha+\cos 3\beta+\cos 3\gamma=3\cos(\alpha+\beta+\gamma)$$

$$\sin 3\alpha+\sin 3\beta+\sin 3\gamma=3\sin(\alpha+\beta+\gamma)$$

Property 2. If $z=\cos x+i\sin x$ then

$$Re\left(\frac{1}{1-z}\right)=Re\left(\frac{1}{1-z^2}\right)=Re\left(\frac{1}{1-z^n}\right)=\frac{1}{2}$$

Explanation. Consider

$$\left(\frac{1}{1-z^n}\right)=\frac{1}{1-(\cos x+i\sin x)^n}$$

$$=\frac{1}{(1-\cos nx)-i\sin nx}$$

$$=\frac{1}{(1-\cos nx)-i\sin nx}\,x\,\frac{(1-\cos nx)+i\sin nx}{(1-\cos nx)+i\sin nx}$$

$$-\frac{(1-\cos nx)+i\sin nx}{(1-\cos nx)^2+\sin^2 nx}$$

So $$\boldsymbol{Re}\left(\frac{1}{1-z^n}\right)=\frac{(1-\cos nx)}{(1-\cos nx)^2+\sin^2 nx}$$

$$=\frac{2\sin^2\frac{nx}{2}}{\left(2\sin^2\frac{nx}{2}\right)^2+\left(2\sin\frac{nx}{2}\cos\frac{nx}{2}\right)^2}$$

$$=\frac{2\sin^2\frac{nx}{2}}{4\sin^4\frac{nx}{2}+4\sin^2\frac{nx}{2}\cos^2\frac{nx}{2}}$$

$$=\frac{2\sin^2\frac{nx}{2}}{\left(4\sin^2\frac{nx}{2}\right)\left(\sin^2\frac{nx}{2}+\cos^2\frac{nx}{2}\right)}$$

or $$\boldsymbol{Re}\left(\frac{1}{1-z^n}\right)=\frac{2\sin^2\frac{nx}{2}}{4\sin^2\frac{nx}{2}}=\frac{2}{4}=\frac{1}{2}$$

Hence $$Re\left(\frac{1}{1-z}\right)=Re\left(\frac{1}{1-z^2}\right)=Re\left(\frac{1}{1-z^n}\right)=\frac{1}{2}$$

Illustration 1. If $\sin^6 x-\cos^6 x=A+B\cos 4x$, then find A and B

Solution: We have $A+B\cos 4x$

$$=\frac{1}{64}\left[\left(e^{ix}+e^{-ix}\right)^6-\left(e^{ix}-e^{-ix}\right)^6\right]$$

$$=\frac{1}{32}\left[{}^6C_1(e^{ix})^5(e^{-ix})+{}^6C_3(e^{ix})^3(e^{-ix})^3+{}^6C_5(e^{ix})(e^{-ix})^5\right]$$

$$=\frac{1}{32}[6(e^{4ix}+e^{-4ix})+20]$$

$$=\frac{1}{32}[(6)(2\cos 4x)+20]=\frac{3}{8}\cos 4x+\frac{5}{8}$$

Thus, $A=5/8$ and $B=3/8$

Illustration 2. Value of

$$2^{n-1}\sin\left(\frac{\pi}{n}\right)\sin\left(\frac{2\pi}{n}\right)\ldots\sin\left(\frac{n-1}{n}\pi\right) \text{ is}$$

Solution: Let $\alpha_0,\alpha_1,\alpha_2,\ldots\alpha_{n-1}$ are nth roots of unity, then

$$\alpha_k=\cos\frac{2k\pi}{n}+i\sin\frac{2k\pi}{n}$$

where $0\le k\le n-1$. Also

$$x^n-1=(x-\alpha_0)(x-\alpha_1)\ldots(x-\alpha_{n-1})$$

Now consider $$|1-\alpha_k|^2=\left(1-\cos\frac{2k\pi}{n}\right)^2+\sin^2\frac{2k\pi}{n}$$

Since $$\left(1-\cos\frac{2k\pi}{n}\right)^2=4\sin^2\frac{k\pi}{n}$$ and

$$|(1-\alpha_1)(1-\alpha_2)\ldots(1-\alpha_{n-1})|=n$$

$$\Rightarrow 2^{n-1}\sin\left(\frac{\pi}{n}\right)\sin\left(\frac{2\pi}{n}\right)\ldots\sin\left(\frac{n-1}{n}\pi\right)=n$$

Illustration 3. Find the value of

$$\cos\frac{\pi}{11}+\cos\frac{3\pi}{11}+\cos\frac{5\pi}{11}+\cos\frac{7\pi}{11}+\cos\frac{9\pi}{11}$$

Solution: Let $z = \cos\frac{\pi}{11} + i\sin\frac{\pi}{11}$ then

Then $z^{11} = -1$

$$z + z^3 + z^5 + z^7 + z^9 = \frac{z\left(z^{10}-1\right)}{z^2-1}$$

$$= \frac{z^{11}-z}{z^2-1} = \frac{-1-z}{z^2-1} = \frac{1}{1-z}$$

But $1-z = 1-\left(\cos\frac{\pi}{11} + i\sin\frac{\pi}{11}\right) = 2\sin^2\frac{\pi}{22} - i\sin\frac{\pi}{11}$

So $\frac{1}{1-z} = \frac{1}{2\sin^2\frac{\pi}{22} - i\sin\frac{\pi}{11}}$

$$= \frac{1}{2\sin^2\frac{\pi}{22} - i\sin\frac{\pi}{11}} x \frac{2\sin^2\frac{2\pi}{11} + i\sin\frac{\pi}{11}}{2\sin^2\frac{\pi}{22} + i\sin\frac{\pi}{11}}$$

$$= \frac{2\sin^2\frac{2\pi}{11} + i\sin\frac{\pi}{11}}{4\sin^4\frac{\pi}{22} - \sin^2\frac{\pi}{11}}$$

So $Re\left(\frac{1}{1-z}\right) = \frac{1}{2}$

So $Re\left(z + z^3 + z^5 + z^7 + z^9\right) = \frac{1}{2}$

$$\cos\frac{\pi}{11} + \cos\frac{3\pi}{11} + \cos\frac{5\pi}{11} + \cos\frac{7\pi}{11} + \cos\frac{9\pi}{11} = \frac{1}{2}$$

Illustration 4. Find the value of

$$\cos^2 10° + \cos^2 50° + \cos^2 70°$$

Solution: Let $z = \cos 10° + i\sin 10°$, then

$$z^9 = (\cos 10° + i\sin 10°)^9 = \cos 90° + i\sin 90° = i$$

Since $z = \cos 10° + i\sin 10°$

So $\frac{1}{z} = \frac{1}{\cos 10° + i\sin 10°} = \cos 10° - i\sin 10°$

So $z + \frac{1}{z} = 2\cos 10°$

or $\cos 10° = \frac{z^2+1}{2z}$

Similarly $\cos 50° = \frac{z^{10}+1}{2z^5}$

And $\cos 70° = \frac{z^{14}+1}{2z^7}$

Then $\cos^2 10° + \cos^2 50° + \cos^2 70°$

$$= \left(\frac{z^2+1}{2z}\right)^2 + \left(\frac{z^{10}+1}{2z^5}\right)^2 + \left(\frac{z^{14}+1}{2z^7}\right)^2$$

$$= \frac{\left\{z^{12}\left(z^2+1\right)^2 + z^4\left(z^{10}+1\right)^2 + \left(z^{14}+1\right)^2\right\}}{4z^{14}}$$

$$= \frac{z^{12}\left(z^4+2z^2+1\right) + z^4\left(z^{20}+2z^{10}+1\right) + \left(z^{28}+2z^{14}+1\right)}{4z^{14}}$$

Now put $z^{18} = -1$ then we get

$$\frac{\left(\left(z^{16}+z^{12}-z^{10}-z^6+z^4+1\right)+6z^{14}\right)}{4z^{14}}$$

$$= \frac{\left(z^4+1\right)\left(z^{12}-z^6+1\right)+6z^{14}}{4z^{14}}$$

Now consider

$$\left(z^4+1\right)\left(z^{12}-z^6+1\right) = \frac{\left(z^4+1\right)\left(z^{18}+1\right)}{z^6+1} = 0$$

So given expression is

$$\frac{\left(z^4+1\right)\left(z^{12}-z^6+1\right)+6z^{14}}{4z^{14}} = \frac{6z^{14}}{4z^{14}} = \frac{3}{2}$$

Illustration 5. Find the value of

$$\cos\frac{\pi}{7} - \cos\frac{2\pi}{7} + \cos\frac{3\pi}{7}$$ [Based on IMO]

Solution: Let $z = \cos\frac{\pi}{7} + i\sin\frac{\pi}{7}$

Then $z^7 = \cos\pi + i\sin\pi = -1$

or $z^7 + 1 = 0$

$$(z+1)\left(z^6 - z^5 + z^4 - z^3 + z^2 - z + 1\right) = 0$$

But $z \neq -1$ so $z^6 - z^5 + z^4 - z^3 + z^2 - z + 1 = 0$

$$z^4\left(z^2-z+1\right) - z\left(z^2-z+1\right) = -1$$

or $\left(z^4-z\right)\left(z^2-z+1\right) = -1$

or $\left(z^3-z^2+z\right) = \frac{1}{1-z^3}$

Now consider $\left(z^3-z^2+z\right)$

$$= \left(\cos\frac{\pi}{7} + i\sin\frac{\pi}{7}\right)^3 - \left(\cos\frac{\pi}{7} + i\sin\frac{\pi}{7}\right)^2$$

$$+\left(\cos\frac{\pi}{7}+i\sin\frac{\pi}{7}\right)$$

$$=\left(\cos\frac{3\pi}{7}-\cos\frac{2\pi}{7}+\cos\frac{\pi}{7}\right)$$

$$+i\left(\sin\frac{3\pi}{7}-\sin\frac{2\pi}{7}+\sin\frac{\pi}{7}\right)$$

So $Re\left(z^3-z^2+z\right)=\left(\cos\frac{3\pi}{7}-\cos\frac{2\pi}{7}+\cos\frac{\pi}{7}\right)$

or $\left(\cos\frac{3\pi}{7}-\cos\frac{2\pi}{7}+\cos\frac{\pi}{7}\right)$

$$= Re\left(z^3-z^2+z\right)=Re\left(\frac{1}{1-z^3}\right)$$

We have already discussed in theory that

$$\text{Re}\left(\frac{1}{1-z^n}\right)=\frac{1}{2}$$

So $\cos\frac{\pi}{7}-\cos\frac{2\pi}{7}+\cos\frac{3\pi}{7}=\frac{1}{2}$

6.7.2 Application of De-Moiver's Theorem

Here we will discuss expansion of $\cos n\theta$ and $\sin n\theta$ when n is a positive integer and θ is real.

By De Moivre's theorem, we have

$$\cos n\theta+i\sin n\theta=\left(\cos\theta+i\sin\theta\right)^n.$$

Expanding the right-hand side by the binomial theorem, we get

$\cos n\theta + i\sin n\theta =$

$$\cos^n\theta+{}^nC_1\cos^{n-1}\theta(i\sin\theta)+{}^nC_2\cos^{n-2}\theta\left(i\sin\theta\right)^2$$

$$+{}^nC_3\cos^{n-3}\theta\left(i\sin\theta\right)^3+\ldots+{}^nC_r\cos^{n-r}\theta\left(i\sin\theta\right)^r$$

$$+\ldots+(i\sin\theta)^n$$

$$=(\cos^n\theta-{}^nC_2\cos^{n-2}\theta\sin^2\theta+{}^nC_4\cos^{n-4}\theta$$

$$\sin^4\theta-\ldots)+i({}^nC_1\cos^{n-1}\theta\sin\theta-{}^nC_3\cos^{n-3}\theta$$

$$\sin^3\theta+{}^nC_5\cos^{n-5}\theta\sin^5\theta-\ldots)$$

Equating the real and imaginary parts, we get

$$\mathbf{\cos n\theta = \cos^n\theta - {}^nC_2\cos^{n-2}\theta\sin^2\theta + {}^nC_4\cos^{n-4}\theta}$$

$\mathbf{\sin^4\theta-\ldots}$ and

$$\mathbf{\sin n\theta = {}^nC_1\cos^{n-1}\theta\sin\theta - {}^nC_3\cos^{n-3}\theta\sin^3\theta}$$

$$\mathbf{+{}^nC_5\cos^{n-5}\theta\sin^5\theta-\ldots}$$

Illustration 6. If $a_1, a_2, \ldots, a_n$ are real numbers and $\cos\alpha + i\sin\alpha$ is a root of $x^n + a_1x^{n-1} + a_2x^{n-2} + \ldots + a_{n-1}x + a_n = 0$. Then find the value of $|a_1\cos\alpha + a_2\cos 2\alpha + \ldots + a_n\cos(n\alpha)|$

Ans. 1.

Solution: As $a_1, a_2, \ldots a_n$ are real numbers and $z = \cos\alpha + i\sin\alpha$ is a root

$\bar{z}=\cos\alpha-i\sin\alpha$ is also a root of the equation

Thus, $\cos\alpha - i\sin\alpha = \dfrac{1}{\cos\alpha+i\sin\alpha} = \dfrac{1}{z}$ is a root of the equation

Therefore,

$$\frac{1}{z^n}+\frac{a_1}{z^{n-1}}+\frac{a_2}{z^{n-2}}+\ldots+a_n=0$$

$$\Rightarrow\quad 1+a_1z+a_2z^2+\ldots+a_nz^n=0$$

$$\Rightarrow\quad 1+a_1\left(\cos\alpha+i\sin\alpha\right)+a_2\left(\cos2\alpha+i\sin2\alpha\right))$$

$$+\ldots+a_n(\cos n\alpha+i\sin n\alpha)=0$$

[using De Moivre's Theorem]

Equating real parts, we get

$$1+a_1\cos\alpha+a_2\cos2\alpha+\ldots+a_n\cos n\alpha=0.$$

$$\Rightarrow\quad |a_1\cos\alpha+a_2\cos2\alpha+\ldots+a_n\cos n\alpha|=1.$$

Illustration 7. Write $\cos 6\theta$ in terms of powers of $\cos\theta$.

Solution: $\cos 6\theta = \text{Re}(\cos 6\theta + i\sin 6\theta)$

$$=\text{Re}(\cos\theta+i\sin\theta)^6$$

Expand by Binomial theorem, we get

$$\text{Re}(\cos^6\theta+6\cos^5\theta\cdot i\sin\theta$$

$$+{}^6C_2\cos^4\theta\cdot i^2\sin^2\theta+{}^6C_3\cos^3\theta\cdot i^3\sin^3\theta$$

$$+{}^6C_4\cos^2\theta\cdot i^4\sin^4\theta+{}^6C_5\cos\theta\cdot i^5\sin^5\theta$$

$$+i^6\sin^6\theta)$$

$$=\cos^6\theta-{}^6C_2\cos^4\theta\sin^2\theta+{}^6C_4\cos^2\theta\sin^4\theta$$

$$-\sin^6\theta$$

$$=\cos^6\theta-15\cos^4\theta\left(1-\cos^2\theta\right)$$

$$+15\cos^2\theta\left(1-\cos^2\theta\right)^2-\left(1-\cos^2\theta\right)^3$$

$$=32\cos^6\theta-48\cos^4\theta+16\cos^2\theta-1.$$

Expansion of $\cos^n\theta$

Let $z=\cos\theta+i\sin\theta$.

$\therefore\ z^n = \cos n\theta + i\sin n\theta$ and $z^{-n} = \cos n\theta - i\sin n\theta$.

$z^n + \frac{1}{z^n} = 2\cos n\theta$.

For $n = 1, z + \frac{1}{z} = 2\cos\theta$.

Now, $2^n cos^n\theta = \left(z + \frac{1}{z}\right)^n$

$= z^n + {}^nC_1 z^{n-2} + {}^nC_2 z^{n-4} + \ldots + {}^nC_{n-1}\frac{1}{z^{n-2}} + \frac{1}{z^n}$

$= \left(z^n + \frac{1}{z^n}\right) + {}^n C_1\left(z^{n-2} + \frac{1}{z^{n-2}}\right) + {}^n C_2\left(z^{n-4} + \frac{1}{z^{n-4}}\right) + \ldots$

$= 2\cos n\theta + {}^n C_1 2\cos(n-2)\theta + {}^n C_2 2\cos(n-4)\theta + \ldots$

or, $2^{n-1}\cos^n\theta$

$= \cos n\theta + {}^n C_1\cos(n-2)\theta + {}^n C_2\cos(n-4)\theta + \ldots$

Expansion of $\sin^n\theta$

Let z = $\cos\theta + i\sin\theta$, then $z - \frac{1}{z} = 2i\sin\theta$.

We have $(2i\sin\theta)^n = \left(z - \frac{1}{z}\right)^n$.

or, $2^n i^n sin^n\theta = z^n - {}^n C_1 z^{n-2} + {}^n C_2 z^{n-4} - \ldots$

$+(-1)^{n-1}\,{}^nC_{n-1}\frac{1}{z^{n-2}} + (-1)^n\frac{1}{z^n}$ ...(1)

Now we have following cases

CASE I: n is even .

From $(1), (-1)^{n/2}\, 2^n sin^n\theta$

$= y\left(z^n + \frac{1}{z^n}\right) - {}^nC_1\left(z^{n-2} + \frac{1}{z^{n-2}}\right)$

$+ {}^nC_2\left(z^{n-4} + \frac{1}{z^{n-4}}\right) - \ldots + (-1)^{n/2n} C_{n/2} = 2$

$\cos n\theta - {}^n C_1 2\cos(n-2)\theta + {}^n C_2\cos(n-4)\theta -$

$\ldots + (-1)^{n/2n} C_{n/2},$

or, $(-1)^{n/2}\, 2^{n-1} sin^n\theta = \cos n\theta - {}^n C_1\cos(n-2)\theta$

$+ {}^n C_2\cos(n-4)\theta - .. + (-1)^{n/2}.\frac{1}{2}.{}^nC_{n/2}$ ... (2)

CASE – II : n is odd

From (1), $2^n i(1)^{(n-1)/2}\ \sin^n\theta$

$= \left(z^n + \frac{1}{z^n}\right) - {}^n C_1\left(z^{n-2} + \frac{1}{z^{n-2}}\right) +$

${}^nC_2\left(z^{n-4} + \frac{1}{z^{n-4}}\right) - \ldots + (1)^{(n-1)/2}\,{}^nC_{(n-1)/2}\left(z - \frac{1}{z}\right)$

$= 2i\sin n\theta - {}^n C_1 2i\sin(n-2)\theta + {}^n C_2 2i\sin(n-4)\theta$

$- \ldots + (1)^{(n-1)/2n} C_{(n-1)/2} 2i\sin\theta,$

Or, $(1)^{(n-1)/2}\, 2^{n-1} sin^n\theta = \sin n\theta - {}^n C_1\sin(n-2)\theta$

$+ {}^n C_2\sin(n-4)\theta - \ldots (1)^{(n-1)/2n} C_{(n-1)/2}\sin\theta \ldots(3)$

Illustration 8. Show that $\sin^4\theta\cos^5\theta$

$= \frac{1}{256}(\cos 9\theta + \cos 7\theta - 4\cos 5\theta - 4\cos 3\theta + 6cos\theta)$

Solution. Let z = $\cos\theta = i\sin\theta$. Then

$z^n = \cos n\theta + i\sin n\theta$ and $z^{-n} = \cos n\theta - i\sin n\theta$

$\therefore\ z^n + \frac{1}{z^n} = 2\cos n\theta.$

For $n = 1, z + \frac{1}{z} = 2\cos\theta$.

Again

$z - \frac{1}{z} = \cos\theta + i\sin\theta - \cos\theta + i\sin\theta = 2i\sin\theta$.

Thus, $(2i\sin\theta)^4.(2\cos\theta)^5$

$= \left(z - \frac{1}{z}\right)^4\left(z + \frac{1}{z}\right)^5 = \left(z^2 - \frac{1}{z^2}\right)^4\left(z + \frac{1}{z}\right)$

$= \left(z^8 - 4z^4 + 6 - \frac{4}{z^4} + \frac{1}{z^8}\right)\left(z + \frac{1}{z}\right)$

$= \left(z^9 + \frac{1}{z^9}\right) + \left(z^7 + \frac{1}{z^7}\right) - 4\left(z^5 + \frac{1}{z^5}\right) - 4\left(z^3 + \frac{1}{z^3}\right)$

$+6\left(z + \frac{1}{z}\right)$

$= 2\cos 9\theta + 2\cos 7\theta - 8\cos 5\theta - 8\cos 3\theta + 12\cos\theta.$

$\therefore\ \sin^4\theta\cos^5\theta = \frac{1}{256}$

$(\cos 9\theta + \cos 7\theta - 4\cos 5\theta - 4\cos 3\theta + 6\cos\theta)$.

Sum of series

$\cos\theta + \cos 2\theta + \cos 3\theta + \ldots + \cos n\theta$ for $0 < \theta < 2\pi$

Consider the complex number $\cos\theta + i\sin\theta$

We have $(\cos\theta + i\sin\theta)^k = \cos k\theta + i\sin k\theta$, for any integer k. If we put $z = \cos\theta + i\sin\theta$ then

$\cos k\theta$ = Real part of z^k.

$S_n = \cos\theta + \cos 2\theta + \ldots + \cos n\theta$

$= Re\left(z + z^2 + \ldots z^n\right)$

$= Re\left(1 + z + ..z^n - 1\right).$

If $0 < \theta < 2\pi$, then $\neq 1$.

Hence, we get $S_n = Re\left(\dfrac{z^{n+1} - 1}{z - 1} - 1\right)$

Thus

$$S_n = Re\left[\frac{\cos(n+1)\theta + i\sin(n+1)\theta - 1}{\cos\theta + i\sin\theta - 1}\right] - 1$$

$$= Re\left[\frac{[(\cos\theta - 1) - i\sin\theta]\,[\cos(n+1)\theta - 1 + i\sin(n+1)\theta]}{[(\cos\theta - 1) + i\sin\theta]\,[(\cos\theta - 1) - i\sin\theta]}\right] - 1$$

$$= \frac{(\cos\theta - 1)(cos(n+1)\theta - 1) + \sin\theta\sin(n+1)\theta}{(\cos\theta - 1)^2 + sin^2\theta} - 1$$

$$= \frac{\cos\theta\cos(n+1)\theta + \sin\theta\sin(n+1)\theta - \cos(n+1)\theta - \cos\theta + 1}{(\cos\theta - 1)^2 + sin^2\theta} - 1$$

$$= \frac{\cos n\theta - \cos(n+1)\theta - \cos\theta + 1}{2(1 - \cos\theta)} - 1$$

$$= \frac{\cos n\theta - \cos(n+1)\theta}{2(1 - \cos\theta)} - \frac{1}{2}$$

$$\text{or } S_n = \frac{2\sin\left(n + \frac{1}{2}\right)\theta\sin(\theta/2)}{4sin^2(\theta/2)} - \frac{1}{2}$$

$$= \frac{\sin\left(n + \frac{1}{2}\right)\theta}{2\sin(\theta/2)} - \frac{1}{2}$$

$$= \frac{\sin\left(n + \frac{1}{2}\right)\theta - \sin\frac{\theta}{2}}{2\sin(\theta/2)} = \frac{\sin\frac{n\theta}{2}\cos\left(\frac{n+1}{2}\right)\theta}{\sin(\theta/2)}$$

Note:

(i) $\cos\theta + \cos 2\theta + \cos 3\theta + \ldots + \cos n\theta$

$$= \frac{\sin\frac{n\theta}{2}\cos\left(\frac{n+1}{2}\right)\theta}{\sin(\theta/2)}$$

(ii) $\sin\theta + \sin 2\theta + \sin 3\theta + .. + \sin n\theta$

$$= \frac{\sin\frac{n\theta}{2}\sin\left(\frac{n+1}{2}\right)\theta}{\sin(\theta/2)}.$$

if, $\theta = (2\pi/n)$ then the sum of the above series vanish.

Illustration 9. Find the sum of the infinite series,

$$\sin\alpha + \frac{1}{2}\sin 2\alpha + \frac{1}{2^2}\sin 3\alpha + \frac{1}{2^3}\sin 4\alpha + \ldots\infty.$$

Solution : Let $S = \sin\alpha + \frac{1}{2}\sin 2\alpha + \ldots$

and $C = \cos\alpha + \frac{1}{2}\cos 2\alpha + \ldots$

$$r = \left|\frac{1}{2}e^{i\alpha}\right| = \frac{1}{2} < 1$$

using sum of a infinite G.P., we get

$$C + iS = \frac{e^{i\alpha}}{1 - \frac{1}{2}e^{i\alpha}} = \frac{2e^{i\alpha}}{2 - e^{i\alpha}}$$

$$\text{or } C + iS = \frac{4\cos\alpha - 2}{5 - 4\cos\alpha} + i\frac{4\sin\alpha}{5 - 4\cos\alpha}$$

Equating the real and imaginary parts we have

$$C = \frac{4\cos\alpha - 2}{5 - 4\cos\alpha} \text{ and } s = \frac{4\sin\alpha}{5 - 4\cos\alpha}$$

6.7.3 Exponential, Logarithmic and Trigonometric Operations of Complex Number

1. Complex exponential

- The exponential of a complex number z = x + i y is defined as

$$e^z = e^{x+iy} = e^x e^{iy} = e^x(\cos y + i\sin y)$$

- As for real numbers, the exponential function is equal to its derivative, i. e.

$$\frac{d}{dz}\exp(z) = \exp(z) \qquad \ldots(1)$$

Illustration 10. Find the value of e^{2+3i}

Solution: $e^{2+3i} = e^2 e^{i3} = e^2(\cos 3 + i\sin 3)$

Properties of the exponential function

Property 1. The complex exponential function is periodic with period $2\pi i$: indeed, for any integer $k \in Z$,

$$e^{z+2k\pi i} = e^{x+i(2k\pi+y)}$$

$$= e^x(\cos(y+2k\pi)+i\sin(y+2k\pi))$$

$$= \exp(x)(\cos(y)+i\sin(y)) = \exp(z) = e^z$$

So e^z is periodic with period $2\pi i$

Property 2:

$$|e^z| = |e^x e^{iy}| = |e^x||e^{iy}| = |e^x||\cos y + i\sin y|$$

$$= |e^x| = |e^{Re(z)}|$$

So $|e^z| = |e^{Re(z)}|$

- As with real numbers, it satisfy for complex numbers

  $e^{(z_1+z_2)} = e^{(z_1)}.e^{(z_2)}$

  $e^z \neq 0$

2. Trigonometric functions

- The complex sine and cosine functions are defined in a way similar to their real counterparts,

  $$\cos(z) = \frac{e^{iz}+e^{-iz}}{2}, \sin(z) = \frac{e^{iz}-e^{-iz}}{2i}.$$

- The tangent, cotangent, secant and cosecant are defined as usual . For instance,

  $$\tan(z) = \frac{\sin(z)}{\cos(z)}, \sec(z) = \frac{1}{\cos(z)}, \text{etc}.$$

3. Hyperbolic functions

The complex hyperbolic sine and cosine are defined in a way similar to their real counterparts,

$$\cos h(z) = \frac{e^{iz}+e^{-iz}}{2}, \sin(z) = \frac{e^{iz}-e^{-iz}}{2i} \quad ...(3)$$

4. Complex logarithm

- The logarithm of a complex number is defined as

  $$e^w = z.$$

  Since the exponential is $2\pi i$-periodic, the complex logarithm is multi-valued.

- Solving the above equation for $w = w_r + iw_i$ and $z = re^{i\theta}$ gives

  $$e^w = e^{w_r}e^{iw_i} = re^{i\theta}$$

  $$\Rightarrow \begin{cases} e^{w_r} & = r \\ w_i & = \theta + 2p\pi \end{cases},$$

  So we can conclude $w_r = \ln(r)$ and $w_i = \theta + 2k\pi$, $k \in Z$.

- Therefore,

  $\ln(z) = \ln(|z|) + i\arg(z)$

Principal value of ln (z)

We know that

$\ln(z) = \ln(|z|) + i\,\mathrm{Arg}(z)$.

Properties of the logarithm

- For complex logarithms we have to be careful when you use identities like

  $\ln(z_1z_2) = \ln(z_1) + \ln(z_2)$,

  or $\ln = \left(\frac{z_1}{z_2}\right) \ln(z_1) - \ln(z_2)$.

  They are only true up to multiples of $2\pi i$.

- For example, if $z_1 = i = \exp(i\pi/2)$ and $z_2 = -1 = \exp(i\pi)$,

  $\ln(z_1) = i\frac{\pi}{2} + 2k_1\pi$, $\ln(z_2)$

  $= i\pi + 2k_2 i\pi$, $k_1, k_2 \in Z$.

  and $\ln(z_1z_2) = i\frac{3\pi}{2} + 2k_3 i\pi$ $k_3 \in Z$, but k_3 is not necessarily equal to $k_1 + k_2$.

- However, every branch of the logarithm (i.e. each expression of $\ln(z)$ with a given value of $k \in Z$) is analytic except at the branch point $z = 0$ and on the branch cut of $\ln(z)$. In the domain of analyticity of $\ln(z)$,

  $$\frac{d}{dz}(\ln(z)) = \frac{1}{z}.$$

Illustration 11. Find $\log(1 + i)$

Solution. Let $\log(1+i) = x + iy$

$\Rightarrow 1 + i = e^{x+iy}$

$$\Rightarrow \sqrt{2}\left(\cos\frac{\pi}{4}+i\sin\frac{\pi}{4}\right) = e^x(\cos y + i\sin y)$$

$$\Rightarrow e^x = \sqrt{2} \qquad \Rightarrow x = \ln\sqrt{2} = \frac{1}{2}\ln 2$$

$y = \frac{\pi}{4}$. So $\log(1+i) = \frac{1}{2}\ln 2 + \frac{\pi}{4}$.

6.7.4 Properties of Roots of Unity

Property 1. Let n be an odd positive integer and $\varepsilon_0, \varepsilon_1, \ldots \varepsilon_{n-1}$ the complex roots of unity of order n . Prove that $\prod_{k=0}^{n-1}(A+B\varepsilon_k)=A^n+B^n$ for all complex numbers A and B .

Proof: From the identity

$$x^n-1=(x-\varepsilon_0)(x-\varepsilon_1)\ldots(x-\varepsilon_0)=\prod_{k=0}^{n-1}(x-\varepsilon_k)$$

Now put x = $-\frac{A}{B}$ then we get

$$-\left(\frac{A^n}{B^n}+1\right)=-\prod_{k=0}^{n-1}\left(\frac{A}{B}+\varepsilon_k\right).$$

or $\prod_{k=0}^{n-1}(A+B\varepsilon_k)=A^n+B^n$

Property 2. Let n be an odd positive integer and ε_0, ε_1, … ε_{n-1} the complex roots of unity of order n . Prove that $\prod_{k=0}^{n-1}(A+B\varepsilon_k^2)=A^n+B^n$ for all complex numbers A and B . **[Romanian Maths Olympiad]**

Proof: If ε_k is one of the nth root of unity then we get $\varepsilon_j^2=\varepsilon_k$ another n[th] root of unit.

From the Property 1 we have

$$\prod_{k=0}^{n-1}(A+B\varepsilon_k^2)=\prod_{k=0}^{n-1}(A+B\varepsilon_j)=A^n+B^n .$$

Property 3. Let n be an even positive integer such that $\frac{n}{2}$ is odd and let ε_0, ε_1 …, ε_{n-1} be the complex roots of unity of order n.

Prove that $\prod_{k=0}^{n-1}(a+b\varepsilon_k^2)=\left(a^{\frac{n}{2}}+b\right)^2$ for any complex numbers a and b.

Proof: Let $n = 2(2s + 1)$. Consider a complex number α such that $\alpha^2=\frac{a}{b}$

The polynomial.

$$f(x) = X^n - 1 = (X-\varepsilon_0)(X-\varepsilon_1)\ldots(X-\varepsilon_{n-1})$$

Then $f(\frac{\alpha}{i})=\left(\frac{1}{i}\right)^{\alpha}(\alpha-i\varepsilon_0)\ldots(\alpha-i\varepsilon_{n-1})$ …(i)

and $f(-\frac{\alpha}{i})=\left(\frac{-1}{i}\right)^{\alpha}(\alpha+i\varepsilon_0)\ldots(\alpha+i\varepsilon_{n-1})$ …(ii)

From equations (i) and (ii)

$$f(\frac{\alpha}{i})f\left(-\frac{\alpha}{i}\right)=(\alpha^2+\varepsilon_0^2)\ldots(\alpha^2+\varepsilon_{n-1}^2)$$

Therefore $\prod_{k=0}^{n-1}(a+b\varepsilon_k^2)=b^n\prod_{k=0}^{n-1}\left(\frac{a}{b}+\varepsilon_k^2\right)$

$$=b^n\prod_{k=0}^{n-1}(\alpha^2+\varepsilon_k^2)$$

$$=b^n f(\frac{\alpha}{i})f(-\frac{\alpha}{i})=b^n[(\alpha^2)^{2s+1}+1]^2$$

$$=b^n[(\frac{a}{b})^{2s+1}+1]^2$$

$$=b^{2(2s+1)}\left(\frac{a^{2s+1}+b^{2s+1}}{b^{2s+1}}\right)^2=\left(a^{\frac{n}{2}}+b^{\frac{n}{2}}\right)^2$$

The Nine-Point Circle of Euler

Consider a triangle ABC with its circumcenter at Origin. Let A_1, B_1, C_1 be the mid points of sides BC, CA and AB, while A', B', and C', be the feet of altitudes and A", B", C" be the mid points of the segment AH, BH and CH respectively Then 9-points A_1, B_1, C_1, A', B', C', A", B", C" lie on same circle named as **Nine-Point Circle of Euler**

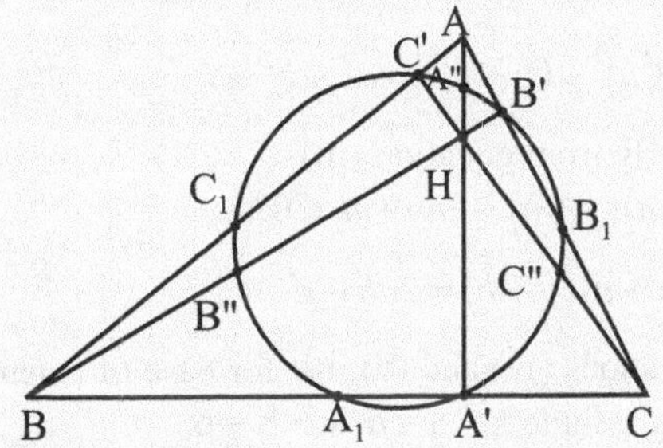

Properties of Nine-Point Circle of Euler

(1) Centre of nine-point circle is the mid point of OH (Mid point of circum center and ortho centre)

(2) Radius of nine-point circle is R/2, where R is the circum radius of triangle ABC

(3) If Circum circle and nine-point circle touch each other then triangle must be right-angled.

(4) Consider a point P on the circum circle of triangle ABC, then nine-point centers of triangles PBC, PCA and PAB are the vertices of a triangle which is similar to triangle ABC.

(5) In any triangle ABC, the points O, G and H are lie on a line called as Euler Line.

6.7.5 The General Quadratic Equation with Complex Coefficients

In this section we will discuss the quadratic equation with complex co-efficients. The formulae for determination of roots, sum of roots and product of roots hold true as before.

Consider the equation $pz^2+qz+r=0$ where $\boldsymbol{p}$, $\boldsymbol{q}$ and $\boldsymbol{r}$ are complex numbers.

or this equation can be re-written as

$$z^2+\left(\frac{q}{p}\right)z+\left(\frac{r}{p}\right)=0, \text{ or } z^2+sz+t=0$$

where ***s*** and ***t*** are complex numbers.
We can write the equation as

$z^2 + 2(a + bi)z + (c + di) = 0$ ...(i)

Here ***a***, ***b***, ***c*** and ***d*** are real numbers

Putting z = x + yi, and equating real and imaginary parts, we obtain a pair of simultaneous equations for x and y,

$$x^2-y^2+2(ax-by)+c=0 \quad ...(ii)$$

$$2xy+2(ay+bx)+d=0 \quad ...(iii)$$

Consider equation (ii), we can re-write it as

$$(x+a)^2-(y+b)^2-a^2+b^2+c=0$$

or $(x+a)^2-(y+b)^2=a^2-b^2-c$ ...(iv)

Similarly from equation (iii)

$$2(x+a)(y+b)-2ab+d=0$$

or $2(x+a)(y+b)=2ab-d$ ...(v)

In equations (iv) and (v) just for ease of calculation we can assume $x + a = \alpha$, $y + b = \beta$,

$$a^2-b^2-c=p, 2ab-d=q,$$

these equations becomes $\alpha^2-\beta^2=p$, $2\alpha\beta=q$.

Squaring and adding we obtain

$$\alpha^2+\beta^2=\sqrt{(p^2+q^2)},$$

$$\alpha=\pm\sqrt{\frac{1}{2}\left\{\sqrt{(p^2+q^2)}+p\right\}},$$

$$\beta=\pm\sqrt{\frac{1}{2}\left\{\sqrt{(p^2+q^2)}-p\right\}}.$$

or $x=\alpha-a=-a\pm\sqrt{\frac{1}{2}\left\{\sqrt{(p^2+q^2)}+p\right\}}$

And $y=\beta-b=-b\pm\sqrt{\frac{1}{2}\left\{\sqrt{(p^2+q^2)}-p\right\}}$

Since $2\alpha\beta = k$, so if k > o then both $\alpha>0,\beta>0$ or $\alpha<0,\beta<0$ then two roots are

$$\left(-a+\sqrt{\frac{1}{2}\left\{\sqrt{(p^2+q^2)}+p\right\}}\right)+i\left(-b+\sqrt{\frac{1}{2}\left\{\sqrt{(p^2+q^2)}-p\right\}}\right) \text{ and}$$

$$\left(-a-\sqrt{\frac{1}{2}\left\{\sqrt{(p^2+q^2)}+p\right\}}\right)+i\left(-b-\sqrt{\frac{1}{2}\left\{\sqrt{(p^2+q^2)}-p\right\}}\right)$$

if $k<0$ then both $\alpha>0,\beta<0$ or $\alpha<0,\beta>0$

$$\left(-a+\sqrt{\frac{1}{2}\left\{\sqrt{(p^2+q^2)}+p\right\}}\right)+i\left(-b-\sqrt{\frac{1}{2}\left\{\sqrt{(p^2+q^2)}-p\right\}}\right) \text{ and}$$

$$\left(-a-\sqrt{\frac{1}{2}\left\{\sqrt{(p^2+q^2)}+p\right\}}\right)+i\left(-b+\sqrt{\frac{1}{2}\left\{\sqrt{(p^2+q^2)}-p\right\}}\right).$$

So for the equation (i)

$$z^2+2(a+bi)z+(c+di)=0$$

Sum of the roots = $-2(a + ib)$

Product of the roots= $c + di$

or Consider the equation $z^2+Az+B=0$ here A and B are complex numbers then-

Roots of the equation is given by

$$z=\frac{-A\pm\sqrt{A^2-4B}}{2},$$

here A and B are complex numbers

Condition for equal roots

The two roots can only be equal if both the square roots above vanish and it is possible when if p = 0, q = 0 or if c = a^2-b^2, C = 2 ab. These conditions are equivalent to the single condition c + d i = $(a+bi)^2$

Consider the original equation i.e equation (i)

$$z^2+2(a+bi)z+(c+di)=0$$

Now this will transform to

$$z^2+2(a+bi)z+(a+bi)^2=0$$

or $(z-(a+bi))^2=0$

And both the roots are $(a + bi)$

So required condition is ***c* + *d i* = (*a* + *b i*)²**

Condition for a real root

For real roots y = 0

So equation (ii) and (iii) will become

$$x^2 + 2ax + c = 0, 2bx + d = 0$$

Eliminating x we find that the required condition is

$$d^2 - 4abd + 4cb^2 = 0.$$

Condition for a purely imaginary root

For purely imaginary roots x = 0

So equations (ii) and (iii) will become

$$y^2 + 2by - c = 0, 2ay + d = 0$$

Eliminating y we get.

And the required condition is $d^2 - 4abd + 4a^2c = 0$

Illustration 12. If $(-1 + 2i)$ is one of the roots of the equation $z^2 - (2+i)z - 1 - 7i = 0$ then find the other root.

Solution:

For the equation $z^2 + 2(a+bi)z + (c+di) = 0$

Sum of the roots = $-2(a + ib)$

Product of the roots = $c + di$

Let other root be z then

Sum of the roots $z + (-1 + 2i) = (2 + i)$

Then the other root $z = 3 - i$

Illustration 13. Find the roots of the equation

$$z^2 + (5+5i)z + 13i = 0$$

Solution: From the formula roots is given by

$$z = \frac{-(5+5i) \pm \sqrt{(5+5i)^2 - 4(13i)}}{2}$$

$$= \frac{-(5+5i) \pm \sqrt{-2i}}{2} = \frac{-(5+5i) \pm (1-i)}{2}$$

So required roots are $-2 - 3i$ and $-3 - 2i$

Exercise B

Concept Deviator (CD) (Single Correct MCQs)

1. If $f(x)$ and $g(x)$ are two polynomials such that the polynomial $h(x) = x f(x^3) + x^2 g(x^6)$ is divisible by $x^2 + x + 1$, then, how many of the following statements are correct?
   (i) $f(1) = g(1)$ (ii) $f(1) = -g(1)$
   (iii) $h(1) = 0$
   (a) 0 (b) 1
   (c) 2 (d) 3

2. If z_1, z_2, z_3 lie on a circle with centre origin and radius 1 then find the sum of maximum and minimum value of $|z_2 - z_3|^2 + |z_3 - z_1|^2 + |z_1 - z_2|^2$
   (a) 9 (b) 7
   (c) 12 (d) None of these

3. z_1, z_2, z_3, z_4 correspond to the points A, B, C, D on a circle $|z| = 1$. If $z_1 + z_2 + z_3 + z_4 = 0$, then how many of the following statements are correct?
   (i) ABCD is necessarily rectangle
   (ii) ABCD is necessarily a rhombus
   (iii) Angle ABC must be right angle.
   (a) 0 (b) 1
   (c) 2 (d) None of these

4. Find the condition for a quadratic equations $az^2 + bz + c = 0$, where a, b, c are complex numbers, to have one purely imaginary root
   (a) $(b\bar{c} + c\bar{b})(a\bar{b} + \bar{a}b) + (c\bar{a} - a\bar{c})^2 = 0$
   (b) $(b\bar{c} + c\bar{b})(c\bar{a} + a\bar{c}) + (a\bar{b} - \bar{a}b)^2 = 0$
   (c) $(a\bar{b} + \bar{a}b)(c\bar{a} + \bar{c}a) + (b\bar{c} - \bar{b}c)^2 = 0$
   (d) None of these

5. If $|z-1| + |z+3| \le 8$, then the number of integral values that $|z-4|$ can assume is
   (a) 9 (b) 8
   (c) 7 (d) None of these

6. Find the number of integral solution of the equation $(3 + 4i)^x = 5^x$
   (a) 0 (b) 1
   (c) more than 5 (d) None of these

7. If the complex number z satisfy the equation $(i - z)(1 + 2i) + (1 - iz)(3 - 4i) = 1 + 7i$, then
   (a) There are two values of z exist
   (b) At least one value of z lies in 1st quadrant
   (c) Exactly one value of z lies in 3rd quadrant
   (d) None of these

8. If α is a complex n^{th} root of unity and if z_1 and z_2 are two complex numbers, then $\sum_{r=0}^{n-1} |z_1 + \alpha^r z_2|^2 =$
   (a) $n^2 |z_1 + z_2|^2$ (b) $\left(\frac{z_1}{n} + \frac{z_2}{n}\right)^2$
   (c) $n\left(|z_1|^2 + |z_2|^2\right)$ (d) $n^2\left(|z_1|^2 + |z_2|^2\right)$

9. If $\frac{1}{a+\omega}+\frac{1}{b+\omega}+\frac{1}{c+\omega}+\frac{1}{d+\omega}=\frac{1}{\omega}$, where a, b, c, d $\in$ R and ω is a cube root of unity then $\Sigma \frac{3}{a^2-a+1}$ is equal to
(a) 1 (b) 2
(c) 3 (d) $\frac{1}{3}$

10. Let the altitudes from the vertices A, B, and C of the triangle ABC meet its circumcircle at D, E, and F respectively and z_1, z_2, and z_3 represent the points D, E and F respectively. If $\frac{z_3-z_1}{z_2-z_1}$ is purely real then the triangle ABC is
(a) acute angled (b) right angled
(c) obtuse angled (d) None of these

11. If the complex number z satisfies the equation $\left|z-\frac{4}{z}\right|=2$, then $|z|$ can assume how many integral values?
(a) 2 (b) 1
(c) 0 (d) None of these

12. At how many points curves arg $(z-3i)=\frac{3\pi}{4}$ and arg $(2z+1-2i)=\frac{\pi}{4}$ intersect each other
(a) 0 (b) 1
(c) 2 (d) None of these

13. The value of $\sum_{k=1}^{10}\left(\sin\frac{2\pi k}{11}-i\cos\frac{2\pi k}{11}\right)$ is
(a) -1 (b) 0
(c) $-i$ (d) i

14. z_1 and $\bar{z}_1$ represent adjacent vertices of a regular polygon of n sides whose centre is original and if $\frac{\text{Im}(z_1)}{\text{Re}(z_1)}=\frac{\sqrt{3}-1}{\sqrt{3}+1}$, then n is equal to
(a) 12 (b) 16
(c) 24 (d) 32

15. Let ω and ω^2 be two non-real complex cube roots of unity and $|z-\omega|^2+|z-\omega^2|^2=a$ be the equation of a circle with ω, ω^2 as ends of a diameter, then the value of a is
(a) 4 (b) 3
(c) 2 (d) $\sqrt{2}$

16. If z_1, z_2, z_3 are three distinct complex numbers such that $\frac{1}{|z_2-z_3|}=\frac{2}{|z_3-z_1|}=\frac{3}{|z_1-z_2|}$, then
$\frac{1}{(z_2-z_3)}+\frac{4}{(z_3-z_1)}+\frac{1}{(z_1-z_2)}=$
(a) 0 (b) 6
(c) 14 (d) None of these

17. The locus of the centre of a circle, which touches the circles $|z-z_1|=a$ and $|z-z_2|=b$ externally can be
(a) an ellipse (b) a hyperbola
(c) a circle (d) parabola

18. Find the smallest value of $|z|^2+|z-3|^2+|z-6i|^2$
(a) 10 (b) 30
(c) 16 (d) None of these

19. If ω be an imaginary 10th root of unity, then $\sum_{r=1}^{10}(ar+b)\,\omega^{r-1}$ is equal to
(a) $\frac{10a}{\omega-1}$ (b) $\frac{11a}{\omega-1}$
(c) $\frac{9a}{\omega-1}$ (d) None of these

20. If z_1, z_2, z_3 lie on a circle with center origin and radius 1 unit and $\frac{z_1^2}{z_2z_3}+\frac{z_2^2}{z_3z_1}+\frac{z_3^2}{z_1z_2}=-1$, then sum of all the possible values of $|z_1+z_2+z_3|$ is
(a) 1 (b) 2
(c) 3 (d) 4

21. If z is a complex number with modulus 1, then the equation $\left(\frac{1+ia}{1-ia}\right)^4=z$, in unknown a, has
(a) all roots real and distinct
(b) two roots real and two roots imaginary
(c) all roots are imaginary
(d) None of these

22. If z_1, z_2, z_3 are the points lying on the circle $|z|=2$, then the minimum value of $|z_1+z_2|^2+|z_2+z_3|^2+|z_3+z_1|^2$ is equal to

(a) 6 (b) 12
(c) 15 (d) 24

23. If z, z_1, z_2, z_3 are four complex number such that

$$2|z - z_1| = |z - z_2| + |z - z_3|$$
$$2|z - z_2| = |z - z_1| + |z - z_3|$$
$$2|z - z_3| = |z - z_1| + |z - z_2|$$

If complex numbers z_1, z_2 and z_3 are represented by the points A, B and C then the complex number z, is
(a) the circumcentre of ΔABC
(b) the incentre of ΔABC
(c) the orthocenter of ΔABC
(d) the centroid of ΔABC

24. The reflection of the complex number $\frac{4+3i}{1+2i}$ in the straight line $iz = \bar{z}$, is
(a) $1 - 2i$ (b) $4 - 3i$
(c) $3 + 4i$ (d) $2 + i$

25. The roots of the equation $x^4 - 2x^2 + 4 = 0$ are the vertices of a
(a) square inscribed in a circle of radius 2
(b) rectangle inscribed in a circle of radius 2
(c) square inscribed in a circle of radius $\sqrt{2}$
(d) rectangle inscribed in a circle of radius $\sqrt{2}$

26. Two complex numbers z_1 and z_2 lie on the curve $2\ |z + \bar{z}| = 32 + (z - \bar{z})^2$ find the maximum value of $|z_1 - z_2|$
(a) 8 (b) 16
(c) 24 (d) 4

27. Find the number of complex numbers satisfying $z^2 = \bar{z}^3$
(a) 0 (b) 5
(c) 3 (d) None of these

28. If $\left|\frac{z-25}{z-1}\right| = 5$, then the locus of $|z|$ is
(a) A straight line with –ve slope
(b) A circle with radius more than 5
(c) A parabola whose focus is on X-axis
(d) None of these

29. Find the range of $|z|$ if $|z - 1| \le 1$ and $|z - 2| = 1$
(a) $[0, \sqrt{5}]$ (b) $[0, \sqrt{3}]$
(c) $[1, \sqrt{5}]$ (d) None of these

IIT Corner

30. If the cube roots of unity are $1, \omega, \omega^2$, then the roots of the equation $(x - 1)^3 + 8 = 0$ are **(1979)**
(a) $1, 1 + 2\omega, 1 + 2\omega^2$ (b) $1, 1 - 2\omega, 1 - 2\omega^2$
(c) $-1, -1, -1$ (d) None of these

31. The smallest positive integer n for which $\left(\frac{1+i}{1-i}\right)^n = 1$ is **(1980)**
(a) n = 8 (b) n = 16
(c) n = 12 (d) None of these

32. The complex numbers $z = x + iy$ which satisfy the equation $\left|\frac{z-5i}{z+5i}\right| = 1$ lie on ***(1981)***
(a) The x-axis
(b) The straight line y = 5
(c) a circle passing though the origin
(d) none of these

33. If $z = \left(\frac{\sqrt{3}}{2} + \frac{i}{2}\right)^5 + \left(\frac{\sqrt{3}}{2} - \frac{i}{2}\right)^5$, then **(1982)**
(a) Re (z) = 0
(b) Im (z) = 0
(c) Re (z) > 0, Im (z) > 0
(d) Re (z) > 0, Im (z) < 0

34. The Complex number $|z - 4| < |z - 2|$ represents the region given by **(1982)**
(a) $Re(z) \ge 0$ (b) $Re(z) < 0$
(c) $Re(z) > 0$ (d) None of these

35. If $z = x + iy$ and $\omega = (1 - iz)(z - 1)$, then $|\omega| = 1$ implies that, in the complex plane, **(1983)**
(a) z lies on the imaginary axis
(b) z lies on the real axis
(c) z lies on the unit circle
(d) None of these

36. The point z_1, z_2, z_3, z_4 in the complex plane are the vertices of a parallelogram taken in order if and and only if, **(1983)**
(a) $z_1 + z_4 = z_2 + z_3$ (b) $z_1 + z_3 = z_2 + z_4$
(c) $z_1 + z_2 = z_3 + z_4$ (d) None of these

37. If a, b, c and u, v, w are complex numbers representing the vertices of two triangles such

that c = $(1-r)a+rb$ and $w=(1-r)u+rv$, where r is a complex number, then the two triangles **(1983)**

(a) have the same area
(b) are similar
(c) are congruent
(d) none of these

38. If $\omega(\neq 1)$ is a cube root of unity and $(1+\omega)^7 = A + B\omega$ then A and B are respectively ***(1995)***

(a) 0, 1 (b) 1, 1
(c) 1, 0 (d) – 1, 1

39. Let z and ω be two non zero complex numbers such that $|z| = |\omega|$ and Arg z + Arg ω = π, then z equals **(1995)**

(a) ω (b) $-\omega$
(c) $\bar{\omega}$ (d) $-\bar{\omega}$

40. Let z and ω be two non zero complex numbers suchthat $|z| < 1, |\omega| \le 1$ and $|z+i\omega| = |z - i\bar{\omega}| = 2$ then z equals **(1995)**

(a) 1 or i (b) i or $-i$
(c) 1 or – 1 (d) i or -1

41. For positive integers $n_1, n^2$the value of expression
$$(1+i)^{n_1} + (1+i^3)^{n_1} + (1+i^5)^{n_2} + (1+i^7)^{n_2},$$
where $i = \sqrt{-1}$ is a real number if and only *if* **(1996)**

(a) $n_1 = n_2 + 1$ (b) $n_1 = n_2 - 1$
(c) $n_1 = n_2$ (d) $n_1 > 0, n_2 > 0$

42. If $= \sqrt{-1}$, then
$$4 + 5\left(-\frac{1}{2} + \frac{i\sqrt{3}}{2}\right)^{334} + 3\left(-\frac{1}{2} + \frac{i\sqrt{3}}{2}\right)^{365}$$
is equal to **(1999)**

(a) $1 - i\sqrt{3}$ (b) $-1 + i\sqrt{3}$
(c) $i\sqrt{3}$ (d) $-i\sqrt{3}$

43. If arg (z) < 0, then arg (– z) – arg (z) = **(2000)**

(a) π (b) –π
(c) $-\frac{\pi}{2}$ (d) $\frac{\pi}{2}$

44. If z_1, z_2 and z_3 are complex numbers such that
$$|z_1| = |z_2| = |z_3| = \left|\frac{1}{z_1} + \frac{1}{z_2} + \frac{1}{z_3}\right| = 1,$$
then $|z_1 + z_2 + z_3|$ is **(2000)**

(a) equal to 1 (b) less than 1
(c) greater than 3 (d) equal to 3

45. Let z_1 and z_2 be n[th] roots of unity which subtend a right angle at the origin. Then n must be of the form **(2001)**

(a) $4k + 1$ (b) $4k + 2$
(c) $4k + 3$ (d) $4k$

46. The complex numbers z_1, z_2 and z_3 satisfying $\frac{z_1 - z_3}{z_2 - z_3} = \frac{1 - i\sqrt{3}}{2}$ are the vertices of a triangle which is **(2001)**

(a) of area zero
(b) right-angled isosceles
(c) equilateral
(d) obtuse-angled isosceles

47. For all complex numbers z_1, z_2 satisfying $|z_1| =$ 12 and $|z_2 - 3 - 4i| = 5$, the minimum value of $|z_1 - z_2|$ is **(2002)**

(a) 0 (b) 2
(c) 7 (d) 17

48. $|z_1| = 1$ and $\omega = \frac{z-1}{z+1}$ (where z ≠ –1), then Re(ω) is **(2003)**

(a) 0 (b) $-\frac{1}{|z+1|^2}$
(c) $\left|\frac{z}{z+1}\right| \cdot \frac{1}{|z+1|^2}$ (d) $\frac{\sqrt{2}}{|z+1|^2}$

49. If ω(≠1) is a cube root of unity and $(1+\omega^2)^n = (1+\omega^4)^n$, then the least positive value of n is **(2004)**

(a) 2 (b) 3
(c) 5 (d) 6

50. The locus of z which lies in shaded region (excluding the boundaries) is best represented by **(2005S)**

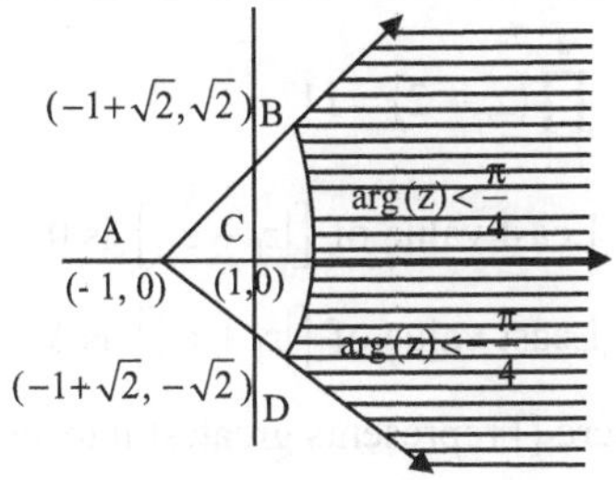

(a) z : $|z+1|>2$ and $|\arg(z+1)|<\pi/4$

(b) z : $|z-1|>2$ and $|\arg(z-1)|<\pi/4$

(c) z : $|z+1|>2$ and $|\arg(z+1)|<\pi/2$

(d) z : $|z-1|>2$ and $|\arg(z+1)|<\pi/2$

51. a, b, c are integers, not all simultaneously equal and ω is cube root of unity (ω ≠1), then minimum value of $\left|a+b\omega+c\omega^2\right|$ is **(2005)**

(a) 0 (b) 1

(c) $\frac{\sqrt{3}}{2}$ (d) $\frac{1}{2}$

52. Let $\omega=-\frac{1}{2}+\frac{i\sqrt{3}}{2}$, then the value of the

$$\begin{vmatrix} 1 & 1 & 1 \\ 1 & -1-\omega^2 & \omega^2 \\ 1 & \omega^2 & \omega^4 \end{vmatrix}$$ **(2002)**

(a) 3ω (b) 3ω (ω −1)

(c) 3ω² (d) 3ω (1 − ω)

53. If $\frac{w-\overline{w}z}{1-z}$ is purely real where w = $\alpha+i\beta,\beta\neq 0$ and z ≠ 1, then the set of the values of z is *(2006)*

(a) $\{z:|z|=1\}$ (b) $\{z:z=\overline{z}\}$

(c) $\{z:z\neq 1\}$ (d) $\{z:|z|=1,z\neq 1\}$

54. A man walks a distance of 3 units from the origin towards the north-east (N 45°E) direction. Form there, he walks a distance of 4 units towards the north-east (N 45°W) direction to reach a point P. Then the position of p in the Argand plane is *(2007)*

(a) $3e^{i\pi/4}+4i$ (b) $(3-4)\,e^{i\pi/4}$

(c) $(4+3i)e^{i\pi/4}$ (d) $(4-3i)e^{i\pi/4}$

55. If $|z|=1\, and\ z\neq\pm 1$, then all the values of $\frac{z}{1-z^2}$ lie on *(2007)*

(a) a line not passing through the origin

(b) $|z|=\sqrt{2}$

(c) the x-axis

(d) the y-axis

56. A particles P starts from the point $z_0=1+2i$, where $i=\sqrt{-1}$. It moves horizontally away from origin by 5 units and then vertically away from origin by 3 units to reach a point z_1. From z_1 the part ticle moves $\sqrt{2}$ unit in the direction of the vector $i+j$ and the it moves through an angle $\frac{\pi}{2}$ in anticlockwise direction on a circle with centre at origin, to reach a point z_2. The point z_2 is given by **(2008)**

(a) 6 + 7*i* (b) −7 + 6*i*

(c) 7 + 6*i* (d) −6 + 7*i*

57. Let z = cos θ + *i*sin θ. Then the value of $\sum_{m=1}^{15} Im\left(z^{2m-1}\right)$ at θ = 2° is **(2009)**

(a) $\frac{1}{\sin 2°}$ (b) $\frac{1}{3\sin 2°}$

(c) $\frac{1}{2\sin 2°}$ (d) $\frac{1}{4\sin 2°}$

58. Let $z=x+iy$ be a complex number where x and y are integers. Then the area of the rectangle whose vertices are the roots of the equation $z\overline{z}^3+\overline{z}z^3=350$ is **(2009)**

(a) 48 (b) 32

(c) 40 (d) 80

59. Let z be a complex number such that the imaginary part of *z* is non-zero and $a=z^2+z+1$ is real. Then a cannot take the value **(2012)**

(a) − 1 (b) $\frac{1}{3}$

(c) $\frac{1}{2}$ (d) $\frac{3}{4}$

60. Let complex number α and $\frac{1}{\bar{\alpha}}$ lie on circles

$(x-x_0)^2+(y-y_0)^2=r^2$ and

$(x-x_0)^2+(y-y_0)^2=4r^2$, respectively.

If $z_0=x_0+iy_0$ satisfies the equation

$2|z_0|^2=r^2+2$, then $|\alpha|=$ **(JEE Adv. 2013)**

(a) $\frac{1}{\sqrt{2}}$ (b) $\frac{1}{2}$

(c) $\frac{1}{\sqrt{7}}$ (d) $\frac{1}{3}$

(Passage Type)

Passage-I

If $a=\cos\alpha+i\sin\alpha$, $b=\cos\beta+i\sin\beta$

$c=\cos\gamma+i\sin\gamma$ and $\frac{a}{b}+\frac{b}{c}+\frac{c}{a}=1$, then

61. $4(\cos(\alpha-\beta)+\cos(\beta-\gamma)+\cos(\gamma-\alpha))$ is equal to

(a) 4 (b) −4

(c) 8 (d) 1

62. $4(\sin(\alpha-\beta)+\sin(\beta-\gamma)+\sin(\gamma-\alpha))$ is equal to

(a) 4 (b) −4

(c) 8 (d) None of these

Passage-II

If z_1, z_2, z_3, z_4 are the roots of the equation $z^4+z^3+z^2+z+1=0$, then

63. If $P=\left|\sum_{i=1}^{4} z_i^4\right|$ and $Q=\sum_{i=1}^{4} z_i^5$ then

(a) $P+Q=5$ (b) $|P-Q|=3$

(c) $|P-Q|=2$ (d) $|P+Q|=0$

64. Identify the correct statements

(a) $\prod_{i=1}^{4}(z_i+2)=9$

(b) $\prod_{i=1}^{4}(z_i+2)=11$

(c) Least value of $[|z_1+z_2|]$ is 0

(d) Least value of $[|z_1+z_2|]$ is 1

(where [] represents greatest integer function)

Passage-III

Let $A_1, A_2, \ldots, A_n$ be the vertices of a regular polygon of n sides in a circle $|z|=1$ then complex numbers A_1, $A_2, \ldots, A_n$ represents location of roots of equation $|z|^n=1$

65. Which all of the following is/are true about $S_n=|A_1A_2|^2+|A_1A_3|^2+\ldots|A_1A_n|^2$,

(a) Is independent of 'n'

(b) S_n can never be a prime

(c) $S_{12}>20$

(d) S_n can never be a perfect square

66. Which all of the following is/are false about $P_n=|A_1A_2||A_1A_3|\ldots\ldots|A_1A_n|$,

(a) Is independent of 'n'

(b) S_n can never be a prime

(c) $S_{12}>20$

(d) S_n can never be a perfect square

Passage-IV

The equation $f(x)=a_0x^n+a_1x^{n-1}+a_2x^{n-2}+\ldots+a_{n-1}x+a_n=0$ where $a_0, a_1, \ldots, a_{n-1}, a_n \in C$, the set of complex numbers, and $a_0\neq 0$, is said to be an equation of degree n. An equation of degree n has exactly n roots. Let $\alpha_1, \alpha_2, y\ldots, \alpha_n \in C$ be the n roots of. Then

$$f(x)=a_0(x-\alpha_1)(x-\alpha_2)\ldots(x-\alpha_n)$$

$$\alpha_1+\alpha_2+\ldots+\alpha_n=-\frac{a_1}{a_0}$$

$$\alpha_1\alpha_2\ldots\alpha_n=(-1)^n\frac{a_n}{a_0}$$

Now consider the equation $z^6 - 6z + 20 = 0$,

67. The number of roots in the 2nd quadrant can be
 (a) 3 (b) 4
 (c) 1 (d) 5

68. The number of roots in the 4th quadrant can not be
 (a) 2 (b) 3
 (c) 1 (d) None of these

Passage-V

Consider a triangle having vertices at the points $A\left(2e^{\frac{i\pi}{4}}\right)$, $B\left(2e^{11i\pi/12}\right)$ and $C\left(2e^{-5i\pi/12}\right)$. Let the incircle of ΔABC touches the sides BC, CA and AB at D, E and F, respectively which are represented by the complex number z_d, z_e, z_f respectively. If $P(z)$ be any point on the incircle, then

69. $AP^2 + BP^2 + CP^2$ is equal to
 (a) 14 (b) 15
 (c) 18 (d) None of these

70. If the altitude through vertex A cuts the circumcircle of ΔABC at Q, then the complex number representing Q is
 (a) $-\sqrt{2}(1+i)$ (b) $-(1+i)$
 (c) $\frac{-(1+i)}{\sqrt{2}}$ (d) $-\frac{1}{2}(1+i)$

Passage-VI

The equation $f(x) = a_0x^n + a_1x^{n-1} + a_2x^{n-2} + \ldots + a_{n-1}x + a_n = 0$ where $a_0, a_1, \ldots, a_{n-1}, a_n \in C$, the set of complex numbers, and $a_0 \neq 0$, is said to be an equation of degree n. An equation of degree n has exactly n roots. Let $\alpha_1, \alpha_2, \ldots, \alpha_n \in C$ be the n roots.

Then

$$f(x) = a_0(x-\alpha_1)(x-\alpha_2)\ldots(x-\alpha_n)$$

$$\alpha_1 + \alpha_2 + \ldots + \alpha_n = -\frac{a_1}{a_0}$$

$$\alpha_1\alpha_2\ldots\alpha_n = (-1)^n \frac{a_n}{a_0}$$

71. If $z_i, i = 1, 2, \ldots, 6$ are the vertices of the six sided regular polygon inscribed in a circle $|z| = 2$, then $\sum_{i=1}^{6} z_i^2$ is equal to
 (a) 0 (b) 1
 (c) −1 (d) None of these

72. If $\omega_k, k \in I; 0 \leq k \leq n-1$ are the nth roots of unity, then the maximum value of $(n!)^{1/n}\,\omega^{(n-1)/2}$ is
 (a) $\frac{n}{\omega - 1}$ (b) $\frac{1}{\omega - 1}$
 (c) $\frac{n-1}{\omega - 1}$ (d) None of these

Passage-VII

The equation of a straight line in the complex plane is given by $z\,\bar{a} + \bar{z}\,a + b = 0$

Where a is a constant complex number and 'b' is a constant real number

We can put z = x + iy. Let a = $\alpha + i\beta$, where, α, β are constant real numbers then the equation becomes $(x + iy)(\alpha - i\beta) + (x - iy)(\alpha + i\beta) + b = 0$

$$\Rightarrow \quad 2\alpha x + 2\beta y + b = 0 \text{ or } \alpha x + \beta y + \frac{b}{2} = 0$$

This equation is the Cartesian form of the line given by the equation the slope of the line given by equation (ii) is

$$-\frac{\alpha}{\beta} = -\frac{\text{Re}(a)}{\text{Im}(a)} = \frac{\text{Re}(a)}{\text{Im}(\bar{a})} = \frac{\frac{a+\bar{a}}{2}}{\frac{\bar{a}-a}{2i}} = i\frac{\bar{a}+a}{\bar{a}-a}$$

73. Two straight lines represented by complex equations $z\bar{a}_1 + \bar{z}a_1 + b_1 = 0 = 0$ and $z\bar{a}_2 + \bar{z}a_2 + b_2 = 0$ are parallel if
 (a) $\frac{a_1}{a_2}$ is purely real
 (b) $\frac{a_1}{a_2}$ is purely imaginary

(c) a_1a_2 is purely real

(d) None of these

74. Two straight lines are perpendicular if

(a) $a_1\bar{a}_2 - \bar{a}_1a_2 = 0$ (b) $a_1\bar{a}_1 - \bar{a}_2a_2 = 0$

(c) $a_1\bar{a}_2 + \bar{a}_1a_2 = 0$ (d) $a_1a_2 + \bar{a}_1\bar{a}_2 = 0$

Passage-VIII

Suppose z and w be two complex numbers such that $|z| \le 1$, $|w| \le 1$ and $|z - iw| = |z - i\bar{w}| = 2$. And $|z + w| \le |z| + |\bar{w}|$, answer the following questions

$a = \cos\alpha + i\sin\alpha, b = \cos\beta + i\sin\beta$

75. Which of the following is true for z and w

(a) $\text{Re}(z) = \text{Re}(w)$ (b) $I_m(z) = I_m(w)$

(c) $\text{Re}(z) = I_m(w)$ (d) None of these

76. Number of complex numbers z satisfying the above conditions is

(a) 1 (b) 2

(c) 4 (d) indeterminate

Passage-IX

ω ($\ne 1$) is an nth root of unity

77. Find the value of $1 + 2\omega + \ldots + n\,\omega^{n-1}$

(a) $\frac{n}{\omega - 1}$ (b) $\frac{n}{\omega + 1}$

(c) $\frac{n-1}{\omega - 1}$ (d) None of these

78. Find the value of $(2 - \omega)(2 - \omega^2) \ldots (2 - \omega^{n-1})$

(a) $\frac{n}{\omega - 1}$ (b) $\frac{n}{\omega + 1}$

(c) $\frac{n-1}{\omega - 1}$ (d) None of these

Passage-X

Let the complex numbers z_1, z_2, and z_3 represents the vertices A, B and C of a triangle ABC respectively, which is inscribed in the circle of radius unity and centre at origin. The internal bisector of the angle A meets the circumcircle again at the point P. Which is represented by the complex number z_4, and altitude from A to BC meets the circumcircle at Q, given by z_5.

79. Find the value of $\frac{z_2z_3}{z_4^2}$

(a) 1 (b) 1/2

(c) $\frac{1}{4}$ (d) None of these

80. Find the value of $\frac{z_2z_3}{z_1z_5}$

(a) 1 (b) 1/2

(c) $\frac{1}{4}$ (d) None of these

Passage-XI

Let α, β, γ be three real numbers such that $\alpha^2 + \beta^2 = \gamma - \gamma^2$ and $z = \frac{\alpha + i\beta}{1 - \gamma}$.

81. Find the value of $|z^2|$

(a) γ (b) $1 - \gamma$

(c) $\frac{\gamma}{1-\gamma}$ (d) $\frac{1-\gamma}{\gamma}$

82. Find $\alpha + \beta$

(a) $\frac{z + \bar{z}}{2(1 + |z|^2)}$ (b) $\frac{(\bar{z} - z)i}{2(1 + |z|^2)}$

(c) $\frac{(z - \bar{z}i)}{2(1 + |z|^2)}$ (d) none of these

IIT Corner

Passage-XII

Let A, B, C be there sets of complex numbers as defined below

$$A = \{z : Imz \ge 1\}$$

$$B = \{z : |z - 2 - i| = 3\}$$

$$C = \{z : \text{Re}(1 - i)z = \sqrt{2}\}$$

83. The number of elements in the set $A \cap B \cap C$ is **(2008)**

(a) 0 (b) 1

(c) 2 (d) ∞

84. Let z be any point in $A \cap B \cap C$. Then, $|z + 1 - i|^2 + |z - 5 - i|^2$ lies between **(2008)**

(a) 25 and 29 (b) 30 and 34

(c) 35 and 39 (d) 40 and 44

85. Let z be any point $A \cap B \cap C$ and let w be any point satisfying $|w-2-i| < 3$. Then, $|z|-|w|+3$ lies between **(2008)**

(a) – 6 and 3 (b) – 3 and 6

(c) – 6 and 6 (d) – 3 and 9

Passage-XIII

Let $S = S_1 \cap S_2 \cap S_3$, where $S_1 = \{z \in \mathbb{C} : |z| < 4\}$,

$$S_2 = \left\{z \in \mathbb{C} : Im\left[\frac{z-1+\sqrt{3}i}{1-\sqrt{3}i}\right] > 0\right\}$$

and $S_3 = \{z \in \mathbb{C} : Re\, z > 0\}$.

86. Area of S = **(JEE Adv. 2013)**

(a) $\frac{10\pi}{3}$ (b) $\frac{20\pi}{3}$

(c) $\frac{16\pi}{3}$ (d) $\frac{32\pi}{3}$

87. min $|1-3i-z| = \in S$ **(JEE Adv. 2013)**

(a) $\frac{2-\sqrt{3}}{2}$ (b) $\frac{2+\sqrt{3}}{2}$

(c) $\frac{3-\sqrt{3}}{2}$ (d) $\frac{3+\sqrt{3}}{2}$

(Integer Type)

88. If $a = \cos 2\alpha + i\sin 2\alpha, b = \cos 2\beta + i\sin 2\beta$, $c = \cos 2\gamma + i\sin 2\gamma$ and $d = \cos 2\delta + i\sin 2\delta$, $P = \sqrt{abcd} + \frac{1}{\sqrt{abcd}}$ then $\frac{\text{p}}{\cos(\alpha+\beta+\gamma+\delta)}$ is equal to

89. If $x = \omega - \omega^2 - 2$, then the value of $x^4 + 3x^3 + 2x^2 - 11x$ is

90. If $x + iy = \sum_{k=1}^{10}\left(\sin\frac{2\pi k}{11} - i\cos\frac{2\pi k}{11}\right)$ then the value of x+y is

91. If $\alpha = \cos\left(\frac{8\pi}{11}\right) + i\sin\left(\frac{8\pi}{11}\right)$, and $P = Re\left(\alpha + \alpha^2 + \alpha^3 + \alpha^4 + \alpha^5\right)$, then find the value of 6|P|

92. If $1, \omega, \omega^2, \ldots.\omega^8$ are the 9, 9th roots of unity and z_1 and z_2 are any two complex numbers, then $\frac{\left\{\sum_{k=0}^{8}|z_1 + \omega^k z_2|^2\right\}}{|z_1|^2 + |z_2|^2}$ is equal to

93. If 1, ω, ω^2 be the three cube roots of unity, then $(1+\omega)(1+\omega^2)(1+\omega^4)(1+\omega^8)\ldots$ to 2n factors is equal to

94. If $f(x) = x^4 - 8x^3 + 4x^2 + 4x + 39$ and f(3+ 2i) $= a + ib$, then $\frac{|a-b|}{24}$ is equal to

95. Let z_1 and z_2 be n^{th} roots of unity which subtend a right angle at the origin, then find the remainder when n divided by 4

96. Consider an ellipse having its foci at $A(z_1)$ and $B(z_2)$ in the Argand plane. If the eccentricity of the ellipse be 'e' and it is known that origin is an interior point of the ellipse, and $e \in \left(a, \frac{b|z_1 - z_2|}{c|z_1| + d|z_2|}\right)$ then the value of a + b + c + d is

97. Find the value of $e^{2mi\cot^{-1}p} \cdot \left(\frac{pi+1}{pi-1}\right)^m$

98. If $z = \frac{q+ir}{1+p}$, and $\frac{p+iq}{1+r} = \frac{1+iz}{1-iz}$, then $p^2 + q^2 + r^2$ equal to

99. Find the number of integral values of α for which the equation $z + \alpha|z-1| + 2i = 0$ has a solution

100. If points A, B, C represent the roots $\alpha, \beta+i, \beta-i$, such that $\alpha, \beta+i, \beta-i$ are three roots of $f(x) = 0$, and the roots of the equation $f'(x) = 0$ are either complex or equal, then find the square of maximum length of altitude through A

101. Find the sum of maximum and minimum distance of the curve $10(z\bar{z}) - 3i\{z^2 - (\bar{z})^2\} - 16 = 0$ from origin.

IIT Corner

102. If z is any complex number satisfying $|z-3-2i| \le 2$, then the minimum value of $|2z-6+5i|$ is **(2011)**

103. Let $\omega = e^{\frac{i\pi}{3}}$, and a, b, c, x, y, z be non-zero complex numbers such that **(2011)**

$a + b + c = x$

$a + b\omega + c\omega^2 = y$

$a + b\omega^2 + c\omega = z$

Then, the value of $\frac{|x|^2+|y|^2+|z|^2}{|a|^2+|b|^2+|c|^2}$ is

104. For any integer *k*, let $\alpha_k = \cos\left(\frac{k\pi}{7}\right) + i \sin\left(\frac{k\pi}{7}\right)$, where $i = \sqrt{-1}$. The value of the expression $\frac{\sum_{k=1}^{12}|\alpha_{k+1}-\alpha_k|}{\sum_{k=1}^{3}|\alpha_{4k-1}-\alpha_{4k-2}|}$ is

(JEE Adv. 2015)

(Assertion and Reason)

In the following questions, each question contains Statement I (Assertion) and Statement II (Reason). Each question has 4 choices (a), (b), (c) and (d) out of which only one is correct. The choices are

(a) Statement I is True, Statement II is True; Statement II is a correct explanation for Statement I.

(b) Statement I is True, Statement II is True; Statement II is not a correct explanation for Statement I.

(c) Statement I is True, Statement II is False.

(d) Statement I is False, Statement II is True.

105. **Statement I:** If $z + \frac{1}{z} = 1$ then

$$z^{4000} + \frac{1}{z^{4000}} = 1$$

Statement II: If $z + \frac{1}{z} = 1$ then $z^3 + \frac{1}{z^3} = -2$

106. **Statement I:** If z_1, z_2, z_3 are complex number representing the points A, B, C such that $\frac{2}{z_1} = \frac{1}{z_2} + \frac{1}{z_3}$. Then, circle through A, B, C such passes through origin.

Statement II: For a cyclic quadrilateral sum of opposite angles is π

107. **Statement I:** If $z + \frac{1}{z} = 1$ ($z \ne 0$ is a complex number), then the maximum value of $|z|$ is $\frac{\sqrt{5}+1}{2}$

Statement II: If $(|z|-a)(|z|-b) < 0$, then

$|z| \in (a,b)$

108. **Statement I:** If z_1 and z_2 are two complex numbers such that $|z_1| = |z_2| + |z_1 - z_2|$, then $\frac{z_1}{z_2}$ is purely real.

Statement II: $\arg(z) = 0 \Rightarrow z$ is purely real.

109. **Statement I:** If $|z| < \sqrt{2} - 1$, then $|z^{\dagger} + 2z\cos\alpha|$ is less than 8.

Statement II: $|z_1 + z_2| < |z_1| + |z_2|$

110. **Statement I:** If $x^2 + x + 1 = 0$, then the value of $(x + \frac{1}{x})^2 + (x^2 + \frac{1}{x^{\dagger}})^2 + \ldots + (x^{27} + \frac{1}{x^{27}})^2$ is 57.

Statement II: ω, ω^2 are the roots of equation $x^2 + x + 1 = 0$

111. **Statement I:** The equation $z\bar{z} + a\bar{z} + \bar{a}z + \lambda = 0$, where *a* is a complex number represents a circle in Argand plane, if λ is real and $|a|^2 > \lambda$

Statement II: The radius of the circle $z\bar{z} + \bar{a}z + a\bar{z} + = 0$ is $\sqrt{a\bar{a} - \lambda}$

112. **Statement I:** Let z_1 and z_2 are two distinct points in an Argand plane and a and b are real. If $a|z_1| = b|z_2|$, then $\frac{az_1}{bz_2} + \frac{bz_2}{az_1}$ is a point on the line segment [–2, 2] of the real axis.

Statement II: If arg $(z_1) = \theta$ and arg $(z_2) = \theta + \alpha$ and a and b are real then $\frac{az_1}{bz_2} + \frac{bz_2}{az_1} = 2\cos\alpha$

113. **Statement I:** The number of complex numbers z satisfying $|z|^2 + a|z| + b = 0 (a, b \in R)$ is at the most 2.

Statement II: A quadratic equation in which all the coefficients are non-zero can have at the most two roots.

114. **Statement I:** If a, b, c are non-zero real numbers and the equation $ax^2 + bx + c + i = 0$ has purely imaginary roots then $a = b^2c$.

Statement II: If the roots of the equation is complex conjugate of each other then sum of the roots may be zero.

(Matrix Match Type)

115.

A	If $\|z_1 - z_2\| = 5$ & $\|\|z - z_1\| - \|z - z_2\|\| = k$, where k is a real parameter less than 5 always represents	P	Portion of a straight Line
B	If $\|z - (3 + 2i)\| = \left\|z \cos\left(\frac{\pi}{4} - \arg(z)\right)\right\|$, then locus of z is	Q	Parabola
C	$\left\|z - 1 - \sin^{-1}\frac{1}{\sqrt{3}}\right\| + \left\|z - \cos^{-1}\frac{1}{\sqrt{3}} + \frac{\pi}{2}\right\| = 1$, then locus of z is	R	Hyperbola
D	Let 'z' be a complex number and 'a' be a real parameter such that $z^2 + az + a^2 = 0$, then locus of z is	S	Pair of straight lines

116.

A	$\|z - \cos^{-1}\cos 12\| - \|z - \sin^{-1}\sin 12\| = 8(\pi - 3)$	P	Straight Line
B	$z^2 + k_1 = i\|z_1\|^2 + k_2; k_2 \neq k_1 \in R - \{0\}$	Q	Rectangular Hyperbola
C	Locus of the point z satisfying the equation $Re(z^2) = Re(z + \bar{z})$	R	Intersection of hyperbola
D	Locus of the point z satisfying the equation $\left\|\frac{2z - i}{z + 1}\right\| = 2$, where $i = \sqrt{-1}$	S	Portion of a straight Line

117.

A	The complex numbers z is simultaneously satisfy the equations $\left\|\frac{z - 12}{z - 8i}\right\| = \frac{5}{3}, \left\|\frac{z - 4}{z - 8}\right\| = 1$, then the $Re(z)$ is	P	2
B	.If ω is a complex cube roots of unity, then $(1 - \omega + \omega^2)(1 - \omega^2 + \omega^4)(1 - \omega^4 + \omega^8)(1 - \omega^8 + \omega^{16})$ is equal to	Q	12
C	The value of $\left(\frac{1 + i}{\sqrt{2}}\right)^8 + \left(\frac{1 - i}{\sqrt{2}}\right)^8$ is equal to	R	16
D	If $\|z_1\| = 1, \|z_2\| = 2, \|z_3\| = 3$ and $\|9 z_1z_2 + 4z_1z_3 + z_2z_3\| = 12$ then $6\|z_1 + z_2 + z_3\|$ is equal to	S	6

118.

A	If ω_1 and ω_2 are the non-real complex cube roots of unity then $\omega_1^4 + \omega_2^4$ equals to	p	$2\omega_1\omega_2$
B	If ω_1 and ω_2 are the non-real complex cube roots of unity and $1 + z + z^3 + z^4 = 0$, then, $\|z\|$ equals to	Q	0
C	If ω and ω_2 be the non-real cube roots of unity and $\frac{1}{a+\omega} + \frac{1}{b+\omega} + \frac{1}{c+\omega} = \frac{2}{\omega_p^2}$ and $\frac{1}{a+\omega^2} + \frac{1}{b+\omega^2} + \frac{1}{c+\omega^2} = \frac{2}{\omega}$, where a, b, c are real then the value of $\frac{1}{a+1} + \frac{1}{b+1} + \frac{1}{c+1}$ is equal to:	R	$-\frac{1}{\omega_1\omega_2}$
D	If ω_1 and ω_2 are the non-real complex cube roots of unity then $\omega_1 - \omega_2 + \omega_1^2 - \omega_2^2 + \omega_1^3 - \omega_2^3$ equals to	S	$\frac{1}{\omega_1\omega_2}$

119.

A	If complex number z ($\neq 0$) satisfies the equation $\|iz - 1\| + \|z - i\| = 2$, then cos (arg z) equals to	P	–1/2
B	If z_1 and z_2 are two consecutive n^{th} roots of unity then $\cos\{\arg(\frac{z_1}{z_2})\}$ equals to	Q	0
C	If $z = \sqrt{-1-\sqrt{-1-\sqrt{\ldots\infty}}}$, then Re (z)	R	2
D	The locus of the centre of a circle which touches the circles $\|z - z_1\| = 2$ and $\|z - z_2\| = 4$, externally is a hyperbola if $\|z_1 - z_2\|$, is greater than	S	$\cos\left(\frac{2\pi}{n}\right)$

120.

A	If $z^2 + z + 1 = 0$, then the value of $(z + \frac{1}{z})† + (z^2 + \frac{1}{z†})^2 + (z^3 + \frac{1}{z‡})^2 + \ldots + (z^{24} + \frac{1}{z^{24}})$ is equal to	P	5
B	Let $\alpha \in R$ and z is a complex number and p and q be the real and imaginary parts of $(z - 1)(\cos\alpha - i\sin\alpha) + (z - 1)^{-1}(\cos\alpha + i\sin\alpha)$. If the locus of points on Argand diagram representing the complex number z such that q = 0 is a circle, and the radius of the circle is r then 12 r is	Q	12
C	An equilateral triangle has vertices at points $A\left(\frac{2}{\sqrt{3}}e^{\frac{i\pi}{2}}\right)$, $B\left(\frac{2}{\sqrt{3}}e^{\frac{-i\pi}{6}}\right)$ and $C\left(\frac{2}{\sqrt{3}}e^{\frac{-i5\pi}{6}}\right)$ If P (z) be any point on its incircle then $AP^2 + BP^2 + CP^2$ is equal to	R	2
D	If $\frac{a}{b-c} + \frac{b}{c-a} + \frac{c}{a-b} = 0$, where a, b, c are three distinct complex numbers, then the value of $\frac{a†}{(b-c)†} + \frac{b†}{(c-a)†} + \frac{c†}{(a-b)†}$ is equal to :	S	48

IIT CORNER

Each question contains statements given in two columns, which have to be matched. The statements in column-I are labeled A, B, C and D, while the statement in column – I are labeled p, q, r, s and t. Any given statement in column-I can have correct matching with ONE OR MORE statement(s) in column-II. The appropriate bubbles corresponding to the answers to these questions have to be darkened as illustrated in the following example:

If the correct matches are A-p, s and t; B-q and r; C-p and q; and D-s then the correct darkening of bubbles will look like the given.

121. $z \neq 0$ is a complex number **(1992 -2 Marks)**

Column I	**Column II**
(A) Re z = 0	(p) Re $z^2 = 0$
(B) Arg z = $\frac{\pi}{4}$	(q) Im $z^2 = 0$
	(r) Re z^2 = Im z^2

122. Match the statements in **Column I** with those in **Column II**. ***(2010)***

[**Note:** Here z takes value in the complex plane and Im z and Re z denote, respectively, the imaginary part and the real part of z.]

Column I	**Column II**
(A) The set of points z satisfying $\lvert z - i\lvert z\rvert\rvert = \lvert z + i\lvert z\rvert\rvert$ is contained in or equal to	(p) an ellipse with eccentricity $\frac{4}{5}$
(B) The set of points z satisfying $\lvert z+4\rvert + \lvert z-4\rvert = 10$ is contained in or equal to	(q) The set of points z satisfying Im z = 0
(C) If $\lvert w\rvert = 2$, then the set of points $z = w - \frac{1}{w}$, is contained in or equal to	(r) The set of points z satisfying $\lvert Imz\rvert \le 1$
(D) If $\lvert w\rvert = 1$, then the set of points $z = w + \frac{1}{w}$ is contained in or equal to	(s) The set of points z satisfying $\lvert Rez\rvert \le 2$
	(t) The set of points z satisfying $\lvert z\rvert \le 3$

123. Let $z_k = \cos\left(\frac{2k\pi}{10}\right) + i\sin\left(\frac{2k\pi}{10}\right)$; $k = 1, 2, ..., 9$.

(JEE Adv. 2014)

	List-I		**List-II**
P.	For each z_k there exists as z_j such that $z_k . z_j = 1$	**1.**	True
Q.	There exists a, k $\in$ {1, 2, ..., 9} such that $z_1.z = z_k$ has no solution z in the set of complex numbers	**2.**	False
R.	$z_k = \cos\left(\frac{2k\pi}{10}\right) + i\sin\left(\frac{2k\pi}{10}\right)$ equals	**3.**	1
S.	$1 - \sum_{k=1}^{9} \cos\left(\frac{2k\pi}{10}\right)$ equals equals	**4.**	2

	P	Q	R	S
(a)	1	2	4	3
(b)	2	1	3	4
(c)	1	2	3	4
(d)	2	1	4	3

(More Than One Correct MCQs)

124. The locus of the centre of a circle which touches the circle $|z - z_1| = a$ and $|z - z_2| = b$ externally (z, z_1 and z_2 are complexnumbers) may be

(a) an ellipse (b) a hyperbola
(c) a circle (d) a straight line

125. If $\dfrac{\tan\alpha - i(\sin\frac{\alpha}{2} + \cos\frac{\alpha}{2})}{1 + 2i\sin\frac{\alpha}{2}}$ is purely imaginary, then α is given by

(a) $\frac{\pi}{4}$ (b) $\frac{5\pi}{4}$
(c) $-\frac{\pi}{4}$ (d) $-\frac{5\pi}{4}$

126. If $z_1 = a + ib$ and $z_2 = c + id$ are complex numbersuch that $|z_1| = |z_2| = 1$ and $Re(z_1\overline{z_2}) = 0$, then the pair of complex number $a + ic = w_1$ and $b + id = w_2$ satisfies

(a) $|w_1| = |w_2| = 1$ (b) $|w_2| \neq |w_2|$
(c) $Re(w_1\bar{w}_2) = 0$ (d) $Re(w_1\bar{w}_2) = -1$

127. If $1, \alpha, \alpha^2, \ldots, \alpha^{n-1}$ are the n, nth roots of unity, and $S_n = \sum_{i=1}^{n-1} \frac{1}{2 - \alpha^i}$ then

(a) $S_{10} = \frac{2^{12} + 1}{2^9 - 1}$ (b) $S_{10} = \frac{2^{12} + 1}{2^{10} - 1}$
(c) $S_{12} = \frac{10(2^{11}) + 1}{2^{12} - 1}$ (d) $S_{12} = \frac{11(2^{11}) + 1}{2^{12} - 1}$

128. One vertex of the triangle of maximum area that can be inscribed in the curve $|z - 2i| = 2$, is $2 + 2i$, then

(a) Triangle is right angled
(b) Triangle is equilateral
(c) Other vertices are $-1 + i(2 \pm \sqrt{3})$
(d) Other vertices are $-1 \pm i(2 - \sqrt{3})$

129. If $z_1 = a_1 + ib_1$ and $z_2 = a_2 + ib_2$ are complex numbers such that $|z_1| = 1, |z_2| = 2$ and $Re(z_1 z_2) = 0$, then

(a) $\left|a_1 + \frac{ia_2}{2}\right| = 1$ (b) $|2b_1 + ib_2| = 1$
(c) $\left|a_1 - \frac{ia_2}{2}\right| = 1$ (d) $|2b_1 - ib_2| = 2$

130. If ω is a non-real cube root of unity, then which one of the following is correct about the value of $S_n = 1\cdot(2-\omega)(2-\omega^2) + 2\cdot(3-\omega)(3-\omega^2) + \ldots + (n-1)(n-\omega)(n-\omega^2)$

(a) $S_{20} = 36081$ (b) $S_{10} = 2016$
(c) $S_{31} = 216195$ (d) $S_{15} = 11011$

131. If z is a complex number satisfying $z + z^{-1} = 1$, then $z^n + z^{n-1}, n \in N$, has the value

(a) $2(-1)^n$ when n is a multiple of 3
(b) $(-1)^{n-1}$ when n is not a multiple of 3
(c) $(-1)^{n+1}$ when n is a multiple of 3
(d) 0, when n is not a multiple of 3

132. ABCD is a square, vertices being taken in the anti-clockwise sense. If vertex A is represented by the complex number z and the intersection of the diagonals is the origin, then which all statements is/are correct

(a) D represents the complex number $-iz$
(b) D represents the complex number $\overline{iz}$
(c) B represents the complex number $\overline{iz}$
(d) B represents the complex number iz

133. If α is a complex constant such that $\alpha z^2 + z + \bar{\alpha} = 0$ has a real root, then

(a) $|\alpha + \bar{\alpha}| = 1$
(b) $|\alpha + \bar{\alpha}| = 0$
(c) $|\alpha + \bar{\alpha}| = 2$
(d) the absolute value of the real root is 1

134. If z_0, z_1 represents points P, Q on the locus $|z-1|=1$ and the line segment PQ subtends an angle $\frac{\pi}{2}$ at the points $z=1$, then

(a) $z_1 = 1+i(z_0-1)$ (b) $-i = \frac{z_1-1}{z_0-1}$

(c) $z_1 = 1-i(z_0-1)$ (d) $z_1 = -i(z_0-1)$

135. If α is the fifth root of unity then

(a) $|1+\alpha+\alpha\dagger+\alpha\ddagger+\alpha^4| = 1$

(b) $|1+\alpha+\alpha\dagger+\alpha\ddagger| = 1$

(c) $|1+\alpha+\alpha\dagger| = 2\cos\frac{\pi}{5}$

(d) $|1+\alpha| > 2\cos\frac{\pi}{7}$

136. If $z = \sqrt{20i-21} + \sqrt{21+20i}$, then the principle value of arg z can be

(a) $\frac{\pi}{4}$ (b) $\frac{3\pi}{4}$

(c) $-\frac{\pi}{4}$ (d) $-\frac{3\pi}{4}$

137. P is the locus of points given by $|z-4-i|=2$. And Q is the locus of points given by $|z-3+i|$ $=3$, then

(a) $4+\frac{4}{\sqrt{5}}+i(1-\frac{2}{\sqrt{5}})$ is the complex number associated to both the sets P and Q

(b) There are two complex numbers satisfy both P and Q

(c) The maximum value of $|z_1-z_2|$ is $=5+\sqrt{5}$ here z_1 lies on P and z_2 on Q

(d) The minimum value of $|z_1-z_2|$ is $=5-\sqrt{5}$ here z_1 lies on P and z_2 on Q

138. One vertex of the triangle of maximum area that can be inscribed in the curve $|z-(1+i)|=\sqrt{10}$ is $-2+2i$, then the remaining vertices are

(a) $(\frac{5-\sqrt{3}}{2})+i(\frac{1+3\sqrt{3}}{2})$

(b) $(\frac{5-\sqrt{3}}{2})+i(\frac{1-3\sqrt{3}}{2})$

(c) $(\frac{5+\sqrt{3}}{2})+i(\frac{1-3\sqrt{3}}{2})$

(d) $(\frac{5+\sqrt{3}}{2})\,i(\frac{1+3\sqrt{3}}{2})$

139. A complex number z satisfies the equation $|z|^2 - 2iz + 2c(1+i) = 0$, where c is real. The values of c for which the above equation has at least one solution can be given by

(a) $c \in (-\infty, -1-\sqrt{2})$

(b) $c \in [-1-\sqrt{2}, -1+\sqrt{2}]$

(c) $c \in (-1+\sqrt{2}, \infty)$

(d) the number of integral values that c can assume is 3

140. If the lines $a\bar{z}+\bar{a}z+b=0$ and $c\bar{z}+\bar{c}z+d=0$ are mutually perpendicular, where a and c are non-zero complex numbers and b and d are real numbers, then

(a) $a\bar{a}+c\bar{c}=0$

(b) $a\bar{c}$ is purely imaginary

(c) $\arg(\frac{a}{c}) = \pm\frac{\pi}{2}$ (d) $\frac{a}{a} = -\frac{c}{c}$

141. If the complex number z satisfies $|iz+2|$ – Im (z), then

(a) minimum value of $|z|$ is 1 and locus of z is a parabola

(b) maximum value of $|z|$ is $\sqrt{3}$ and locus of z is a straight line

(c) locus of z is a parabola with focus (0, 2)

(d) maximum value of arg z is $\frac{3\pi}{4}$ and locus of z is a straight line

142. Let z_1, z_2 be two complex numbers represented by points on the circles $|z|=1$ and $|z|=2$ respectively then

(a) $|2z_1+z_2| \le 4$ (b) $|2z_1+z_2| \le 3$

(c) $\left|z_2+\frac{1}{z_1}\right| \le 3$ (d) $\left|z_2+\frac{1}{z_1}\right| \ge 3$

IIT CORNER

143. If $z_1 = a+ib$ and $z_2 = c+id$ are complex numbers such that $|z_1| = |z_2| = 1$ and Re($z_1\bar{z}_2$) = 0, then the pair of complex numbers $w_1 = a+ic$ and $w_2 = b+id$ satisfies **(1985)**

(a) $|w_1| = 1$ (b) $|w_2| = 1$

(c) Re($w_1\bar{w}_2$) = 0 (d) none of these

144. Let z_1 and z_2 be complex numbers such that $z_1 \neq z_2$ and $|z_1| = |z_2|$. If z_1 has positive real part and z_1 has negative imaginary part, then $\frac{z_1+z_2}{z_1-z_2}$ may be **(1986)**

(a) Zero (b) real and positive

(c) real and negative (d) purely imaginary

145. If let z_1 and z_2 two non-zero complex numbers such that $|z_1+z_2| = |z_1|+|z_2|$, then Arg z_1 – Arg z_2 is equal to **(1987)**

(a) $-\pi$ (b) $-\frac{\pi}{2}$

(c) 0 (d) $\frac{\pi}{2}$ (e) π

146. The value of $\sum_{k=1}^{6}\left(\sin\frac{2\pi k}{7} - i\cos\frac{2\pi k}{7}\right)$ is **(1988)**

(a) – 1 (b) 0

(c) – i (d) i (e) none

147. If ω is an imaginary cube root of unity, then $(1+\omega-\omega^2)^7$ equals **(1998)**

(a) 128ω (b) -128ω

(c) $128\omega^2$ (d) $-128\omega^2$

148. The value of the sum $\sum_{n=1}^{13}\left(i^n + i^{n+1}\right)$, where $i = \sqrt{-1}$, equals **(1998)**

(a) i (b) $i-1$

(c) $-i$ (d) 0

149. Let z_1 and z_2 be two distinct complex numbers and let z = $(1-i)z_1 + tz_2$ for some real t with 0 < t < 1. If Arg (w) denotes the principal argument of a non-zero complex number w, then **(2010)**

(a) $|z-z_1| + |z-z_2| = |z_1-z_2|$

(b) $\text{Arg}(z-z_1) = \text{Arg}(z-z_2)$

(c) $\begin{vmatrix} z-z_1 & \bar{z}-\bar{z}_1 \\ z_2-z_1 & \bar{z}_2-\bar{z}_1 \end{vmatrix} = 0$

(d) Arg $(z-z_1)$ = Arg (z_2-z_1)

150. Let w = $\frac{\sqrt{3}+i}{2}$ and P = $\{ w^n : n = 1,2,3,\ldots\}$, Further $H_1 = \left\{z \in C : Re z > \frac{1}{2}\right\}$

$H_2 = \left\{z \in C : Re z < -\frac{1}{2}\right\}$, where c is the set of all complex numbers.

If $z_1 \in P \cap H_1, z_2 \in P \cap H_2$ and O represents the origin, then $\angle z_1Oz_2$ = **(JEE Adv. 2013)**

(a) $\frac{\pi}{2}$ (b) $\frac{\pi}{6}$

(c) $\frac{2\pi}{3}$ (d) $\frac{5\pi}{6}$

151. Let a, $\notin$ b and $a^2 + b^2 \neq 0$.

Suppose $S = \left\{Z \in C : Z = \frac{1}{a+ibt}, + \in \mathbb{R}, t \neq 0\right\}$

, where $i = \sqrt{-1}$. If z = x+ iy and z $\notin$ S, then (x, y) lies on

(JEE Adv. 2016)

(a) the circle with radius $\frac{1}{2a}$ and centre $\left(\frac{1}{2a}, 0\right)$ for a > 0, b ≠ 0

(b) the circle with radius $-\frac{1}{2a}$ and centre $\left(-\frac{1}{2a}, 0\right)$ for a < 0, b ≠ 0

(c) the x-axis for a ≠ 0, b = 0

(d) the y-axis for a = 0, b ≠ 0

(Subjective Questions)

(IIT JEE)

152. Express $\frac{1}{1-\cos\theta+2i\sin\theta}$ in the form x + iy. **(1978)**

153. If x = $a+b$, $y = a\gamma + b\beta$ and $z = a\beta + b\gamma$ where γ and β are the complex cube roots of unity, show that xyz = $a^3 + b^3$. **(1978)**

154. If $x + iy = \sqrt{\frac{a+ib}{c+id}}$, prove that

$$\left(x^2 + y^2\right)^2 = \frac{a^2 + b^2}{c^2 + d^2}.$$ **(1979)**

155. It is given that n is an odd integer greater than 3, but n is not a multiple of 3. Prove that $x^3 + x^2 + x$ is a factor of $(x+1)^n - x^n - 1$. **(1980)**

156. Find the real values of x and y for which the following equation is satisfied

$$\frac{(1+i)x - 2i}{3+i} + \frac{(2-3i)y + i}{3-i} = i$$ **(1980)**

157. Let the complex number z_1, z_2 and z_3 be the vertices of an equailateral triangle. Let z_0 be the circumcentre of the triangle. Then prove that $z_1^2 + z_2^2 + z_3^2 = 3z_0^2$. **(1981)**

158. Prove that the complex numbers z_1, z_2 and the origin form an equilateral triangle only if $z_1^2 + z_2^2 - z_1 z_2 = 0$ **(1983)**

159. If $a_1, a_2 \ldots\ldots, a_{n-1}$ are the n roots of unity, then show that $(1 - a_1)(1 - a_2)(1 - a_3) \ldots \ldots (1 - a_{n-1}) = n$ **(1984)**

160. Show that the area of the triangle on the Argand diagram formed by the complex numbers z, iz and $z + iz$ is $\frac{1}{2}|z|^2.$ **(1986)**

161. Complex number z_1, z_2, z_3 are the vertices A, B, C respectively of an isosceles right angled triangle with right angle at C. Show that $(z_1 - z_2)^2 = 2(z_1 - z_3)(z_3 - z_2)$. **(1986)**

162. Let $Z_1 = 10 + 6i$ and $Z_2 = 4 + 6i.$ if Z is any complex number such that the argument of $\frac{Z - Z_1}{Z - Z_2}$ is $\frac{\pi}{4}$, then prove that $|Z - 7 - 9i| = 3\sqrt{2}.$ **(1990)**

163. If $iz^3 + z^2 - z + i = 0$, then show that $|z| = 1$. **(1995)**

164. If $|Z| \le 1, |W| \le 1,$ show that $|Z - W|^2 \le (|Z| - |W|)^2 + (Arg\, Z - Arg\, W)^2$ **(1995)**

165. Find all non-zero complex numbers Z satisfying $\bar{Z} = iZ^2$. **(1996)**

166. Let $\bar{b}z + b\bar{z} = c, b \neq 0$, be a line in the complex plane, when $\bar{b}$ is the complex conjugate of b. If a point z_1 is the reflection of a point z_2 through the line, then show that $c = \bar{z_1}b + z_2\bar{b}$. ***(1997)***

167. Let z_1 and z_2 be roots of the equation $z^2 + pz + q = 0$, where the coefficients p and q may be complex numbers. Let A and B represent z_1 and z_2 in the complex plane. If $\angle AOB = \alpha \neq 0$ and $OA = OB$, where O is the origin, prove that $p^2 = 4q\cos^2\left(\frac{\alpha}{2}\right)$. **(1997)**

168. For complex numbers z and w, proved that $|z|^2 w - |w|^2 z = z - w$ if and only if $z = w$ or $z\bar{w} = 1$. **(1999)**

169. Let a complex number $\alpha, \alpha \neq 1$, be a root of the equation $z^{p+q} - z^p - z^q + 1 = 0$, where p, q are distinct primes. Show that either $1 + \alpha + \alpha^2 + \ldots + \alpha^{p-1} = 0$ or $1 + \alpha + \alpha^2 + \ldots + \alpha^{q-1} = 0$ but not both together. **(2002)**

170. If z_1 and z_2 are two complex numbers such that $|z_1| < 1 < |z_2|$ then prove that $\left|\frac{1 - z_1\bar{z}_2}{z_1 - z_2}\right| < 1$. **(2003)**

171. Prove that there exist no complex number z such that $|z| < \frac{1}{3}$ and $\sum_{r=1}^{n} a_r z^r = 1$, where $|a_r| < 2$. **(2003)**

172. Find the centre and radius of circle given by $\left|\frac{z - \alpha}{z - \beta}\right| = k, k \neq 1$. Where, $z = x + iy, \alpha = \alpha_1 + i\alpha_2, \beta = \beta_1 + i\beta_2$ **(2004)**

173. If one of the vertices of the square circumscribing the circle $|z - 1| = \sqrt{2}$ is $2 + \sqrt{3}i$. Find the other vertices of the square. **(2005)**

Concept Eliminator (CE)

Olympiad Tonic

1. Let z_1, z_2, z_3 be distinct complex numbers such that $|z_1| = |z_2| = |z_3| = 5$ then find the minimum value of

$$\frac{1}{|z_1 - z_2||z_1 - z_3|} + \frac{1}{|z_2 - z_1||z_2 - z_3|} + \frac{1}{|z_3 - z_1||z_3 - z_2|}$$

2. Find all complex numbers z such that

$$|z - |z + 1|| = |z + |z - 1||$$

3. Let a, b, be non-zero complex numbers. Prove that the equation $az^3 + bz^2 + \bar{b}z + \bar{a} = 0$ has at least one root with absolute value equal to 1.

4. Let z be a complex number such that $z \in C / R$ and $Im\left(\frac{1 + z + z^2}{1 - z + z^2}\right) = 0$. Prove that $|z| = 1$.

5. Let z_1, z_2, z_3 be complex numbers that lie on $|z| = 1$ such that $z_1 + z_2 + z_3 \neq 0$ and $z_1^2 + z_2^2 + z_3^2 = 0$, then for all integers $n \geq 2$, find the all possible values of $|z_1^n + z_2^n + z_3^n|$

6. Let $n > 0$ be an integer and let z be a complex number such that $|z| = 1$, then prove that

$$n|1 + z| + |1 + z^2| + |1 + z^3| + \ldots + |1 + z^{2n}| + |1 + z^{2n+1}| \geq 2n$$

7. Let λ be a real number and let $n \geq 2$ be an integer then solve the equation

$$\lambda(\bar{z} + z^n) = i(\bar{z} - z^n)$$

8. Let z_1, z_2, z_3 be complex numbers such that

$$z_1 + z_2 + z_3 \neq 0 \text{ and } |z_1| = |z_2| = |z_3|$$

Prove that, $\text{Re}\left(\frac{1}{z_1} + \frac{1}{z_2} + \frac{1}{z_3}\right)$.

$$\text{Re}\left(\frac{1}{z_1 + z_2 + z_3}\right) \geq 0$$

9. For all complex numbers z with $|z| = 1$ find the minimum and maximum value of

$$|1 - z| + |1 + z^2|$$

10. Let a, b, c, z be complex numbers such that $|a| = |b| = |c|$ and $az^2 + bz + c = 0$. Then find the minimum and maximum value of $|z|$

$$\frac{\sqrt{5} - 1}{2} \leq |z| \leq \frac{\sqrt{5} + 1}{2}$$

11. Let $\omega = \cos\frac{2\pi}{2n+1} + i\sin\frac{2\pi}{2n+1} . n \geq 0$. and let $z = \frac{1}{2} + \omega + \omega^2 + \ldots + \omega^n$. Prove that: $\text{Im}(z^{2k}) = \text{Re}(z^{2k+1}) = 0$ for all $k \in \mathbb{N}$:

12. Let p, q be complex numbers such that $|p| + |q| < 1$. Prove that the moduli of the roots of the equation $z^2 + pz + q = 0$ are less than 1.

13. Let a, b, c be distinct complex numbers such that

$$(a - b)^7 + (b - c)^7 + (c - a)^7 = 0$$

Prove that a, b, c are the coordinates of the vertices of an equilateral triangle.

14. Three equal circles $C_1 (O_1 : r)$, $C_2 (O_2 : r)$ and $C_3 (O_3 : r)$ have a common point O. Circles C_1 and C_2, C_2 and C_3, C_3 and C_1 meet again at points A, B, C respectively. Prove that the circumradius of triangle A B C is equal to r.

15. Find the value of

$(1 + \omega)(1 + \omega^2)(1 + \omega^3)\ldots(1 + \omega^{1987})$ where ω is complex cube root of unity.

16. Let $\omega \neq 1$ be a cube root of unity then find the value of

$$(1 - \omega + \omega^2)(1 - \omega^2 + \omega^4)\ldots(1 - \omega^n + \omega^{2n})$$

17. Consider a complex number $z = \frac{2 + i}{2 - i}$. Prove that $|z| = 1$, but z is not an nth root of unity for any positive integer n.

18. Let U_n be the set of the nth roots of unity. Prove that

$$\prod_{\varepsilon \in U_n}\left(\varepsilon + \frac{1}{\varepsilon}\right) = \begin{cases} 0 \text{ if } n \equiv 0(mod\,4). \\ 2 \text{ if } n \equiv 1(mod\,4). \\ -4 \text{ if } n \equiv 2(mod\,4). \\ 2 \text{ if } n \equiv 3(mod\,4). \end{cases}$$

19. Let $z_1, z_2, \dots z_n$ be the coordinates of the vertices of a regular polygon then prove that $z_1^2 + z_2^2 + \dots + z_n^2 = z_1z_2 + z_2z_3 + \dots + z_nz_1$.
20. Let $n \geq 4$ and let $a_1, a_2, \dots, a_n$ be the coordinates of the vertices of a regular polygon. Prove that

$$a_1a_2 + a_2a_3 + \dots + a_na_1 = a_1a_3 + a_2a_4 + \dots + a_na_2$$

21. Let $z_1, z_2, \dots . z_n$ be complex numbers such that $|z_1| = \dots = |z_n| = 1$

$$z = \left(\sum_{k=1}^{n} z_k\right) \cdot \left(\sum_{k=1}^{n} \frac{1}{z_k}\right)$$

Prove that $\text{Im}(z) = 0$ and $0 \leq \text{Re}(z) \leq n^2$.

Direction for question number 22-23

Let $z_0, z_1, z_2, \dots, z_n$ be complex numbers such that $(k+1)z_{k+1} - i(n-k)z_k = 0$ for all $k \in \{0, 1, 2, \dots, n-1\}$

22. Find z_0 such that $z_0 + z_1 + \dots . + z_n = 2^n$.
23. For the value of z_0 determined above, prove that $|z_0|^2 + |z_1|^2 + \dots + |z_n|^2 < \frac{(3n+1)^n}{n!}$
24. Let $\omega = \cos\frac{2\pi}{2n+1} + i\sin\frac{2\pi}{2n+1} \cdot n \geq 0$ and let

$z = \frac{1}{2} + \omega + \omega^2 + \dots + \omega^n$.

Prove that: $(2z+1)^{2n+1} + (2z-1)^{2n+1} = 0$.

25. Let $z_1, z_2, \dots, z_n, \dots,$ be the coordinates of the vertices of a regular polygon with the circumcenter at the origin. Prove that there are $i, j, k \in \{1, 2, \dots, n\}$ such that $z_i + z_j = z_k$ if and only if 6 divides n.
26. Consider a polynomial with real coefficients, $f(x) = x^4 + ax^3 + bx^2 + cx + d$ and has and real roots. Prove that if $|f(i)| = 1$, then $a = b = c = d = 0$.
27. If $11z^{10} + 10iz^9 + 10iz - 11 = 0$, then prove that $|z| = 1$.[Putnam Mathematical Competition]
28. Prove that the average of the numbers k sin k° (k = 2, 4, 6, …, 180) is cot 1°.

[USA Mathematical Olympiad]

29. Let $n \geq 3$ be a positive integer and let $\varepsilon \neq 1$ be an nth root of unity then prove that $|1 - \varepsilon| > \frac{2}{n-1}$.
30. Let $z_1, z_2, \dots, z_n$ be distinct complex numbers such that $|z_1| = |z_2| = \dots |z_n|$ then prove that

$$\sum_{1 \leq i < j \leq n} \left|\frac{z_i + z_j}{z_i - z_j}\right|^2 \geq \frac{(n-1)(n-2)}{2}$$

Solutions

Concept Applicator (CA)–1

1. Since 315 = 4(78) + 3

$i^{315} = i^{4(78)+3} = i^{4(78)}\, i^3 = (i^4)^{78}\, i^3 = i^3 = -i$

$(\because i^4 = 1, i^3 = -i)$

2. $S = 1 + i + i^2 + i^3 + \ldots\ldots\ldots + i^{30}$

$= \frac{i^{31}-1}{i-1} = \frac{i^3-1}{i-1}$

Since $(i^{31} = (i^4)^7\, i^3 = i^3)$

$S = \frac{-i-1}{i-1} = \frac{(1+i)(1+i)}{(1-i)(1+i)}$

$= \frac{1+i^2+2i}{2} = \frac{2i}{2} = i \qquad (\because i^2 = -1)$

3. Let $S = i^n + i^{n+1} + i^{n+2} + i^{n+3}$

$= i^n(1 + i + i^2 + i^3)$

$= i^n(1 + i - 1 - i) \qquad (\because i^2 = -1 \text{ and } i^3 = -i)$

$= 0$

4. $z = \frac{1(-2-i)}{(-2+i)(-2-i)} = \frac{-2-i}{(-2)^2-i^2} = \frac{-2-i}{4+1} = \frac{-2-i}{5}$

$z = -\frac{2}{5} - i\frac{1}{5}$

5. Since the sum of any four consecutive power of i is zero.

$\therefore \sum_{n=1}^{13}(i^n + i^{n+1}) = \sum_{n=1}^{13} i^n + \sum_{n=1}^{13} i^{n+1}$

$= (i + i^2 + i^3 + \ldots + i^{13}) + (i^2 + i^3 + i^4 + \ldots + i^{14})$

$= (i + 0) + (i^2 + 0) = i - 1$

6. Here, $\frac{2^n}{(1+i)^{2n}} + \frac{(1+i)^{2n}}{2^n}$

$= \frac{2^n}{(1+i^2+2i)^n} + \frac{(1+i^2+2i)^n}{2^n}$

$= \frac{2^n}{(2i)^{2n}} + \frac{(2i)^n}{(2)^{2n}} = \frac{1}{i^n} + i^n = \frac{i^n}{i^{2n}} + i^n$

$= \frac{i^n}{(-1)^n} + i^n = i^n\left\{\frac{1}{(-1)^n} + 1\right\}$ or $i^n\{(-1)^n + 1\}$

7. As, $(1+i)^{6n} + (1-i)^{6n}$

$= (2i)^{3n} + (-2i)^{3n}$, as n is odd

$= (2i)^{3n} - (2i)^{3n} = 0$

8. $\frac{(1-i)^3}{1-i^3} = \frac{1-3i+3i^2-i^3}{1+i} = \frac{-2-2i}{1+i} = -2$

9. $\left(\frac{1+i}{1-i}\right)^n = \left(\frac{1+i}{1-i} \times \frac{1+i}{1+i}\right)^n = \left(\frac{1-1+2i}{1+1}\right)^n$

$= i^n = -1$, If $n = 2$

10. $\sqrt{-3-4i} = x + iy$

Square both sides

$-3 - 4i = x^2 - y^2 + i\,2xy$

$x^2 - y^2 = -3 \qquad \ldots$(i)

$2xy = -4 \qquad \ldots$(ii)

$(x^2 + y^2)^2 = (x^2 - y^2)^2 + (2xy)^2$

$= 3^2 + (-4)^2 = 5^2$

$x^2 + y^2 = 5 \qquad \ldots$(iii)

From eq. (i) and eq. (ii)

$2x^2 = 2$

So, $x = \pm 1$

$2y^2 = 8$

So, $y = \pm 2$

As from eq. (ii) $2xy = -4$

So, $xy = -2$

Thus, if $x > 0, y < 0$

and if $x < 0, y > 0$

$\sqrt{-3-4i} = 1 - 2i$ or $-1 + 2i$

Concept Applicator (CA)–2

1. Multiplicative inverse of z^2

$= \frac{1}{z^2} = \frac{1}{(1+i)^2} = \frac{1}{2i} = -\frac{i}{2}$

2. $\sqrt{2i} = \sqrt{(1+i)^2} = (1+i)$

$\Rightarrow \frac{1}{1+i} = \frac{1}{1+i} \times \frac{1-i}{1-i} = \frac{1}{2}(1-i)$

3. $\left|\frac{(3+2i)^2}{(4-3i)}\right| = \frac{|3+2i|^2}{|4-3i|} = \frac{\left(\sqrt{9+4}\right)^2}{\left(\sqrt{16+9}\right)} = \frac{13}{5}$

4. Let $z = \frac{3+4i}{4-5i} = \frac{3+4i}{4-5i} \times \frac{4+5i}{4+5i}$

$= \frac{12+15i+16i+20i^2}{16-25i^2} = \frac{12+31i-20}{16+25}$

$= -\frac{8}{41} + \frac{31}{41}i$

Then, $\bar{z} = -\frac{8}{41} - \frac{31}{41}i$ and

$|z| = \sqrt{\left(-\frac{8}{41}\right)^2 + \left(-\frac{31}{41}\right)^2} = \sqrt{\frac{64+961}{(41)^2}} = \frac{5}{\sqrt{41}}$

∴ Multiplicative inverse of

$z = \frac{\bar{z}}{|z|^2} = \frac{-\frac{8}{41} - \frac{31}{41}i}{\frac{25}{41}} = \frac{-8-31i}{25} = -\frac{8}{25} - \frac{31}{25}i$

5. $= \left|\frac{\beta - \alpha}{1 - \bar{\alpha}\beta}\right| = \left|\frac{\beta - \alpha}{\alpha\bar{\alpha} - \bar{\alpha}\beta}\right|$

$= \frac{|\beta - \alpha|}{|\alpha - \beta||\bar{\alpha}|} = \frac{|-1||\alpha - \beta|}{|\alpha - \beta||\alpha|} = 1$

6. $\frac{1+z}{1+\bar{z}} = \frac{z\bar{z} + z}{1+\bar{z}} \left(\because |z| = 1 \Rightarrow |z|^2 = 1 \Rightarrow z\bar{z} = 1\right)$

$= \frac{z(\bar{z}+1)}{1+\bar{z}} = z$

Option (a)

7. $w = \frac{1-iz}{z-i}$

As $|w| = 1 \Rightarrow |z-i| = |1-iz| = |z+i|$ (as $\frac{1}{i} = -i$)

$\Rightarrow$ z lies the right bisector of the segment connecting the point 1 and – 1. Thus 'z' lies on the real axis.

Option (b)

8. For every $a \in R, |a| = \sqrt{a^2}$

$\therefore |a|^2 = a^2$ Now, $(|x| - |y|)^2 \geq 0$

$\Rightarrow |x|^2 + |y|^2 - 2|x||y| \geq 0$

$\Rightarrow 2|x||y| \leq |x|^2 + |y|^2$

$\Rightarrow |x|^2 + |y|^2 + 2|x||y| \leq 2|x|^2 + 2|y|^2$

$\Rightarrow (|x| + |y|)^2 \leq 2(x^2 + y^2)$

$\Rightarrow (|x| + |y|)^2 \leq 2|z|^2$

$\therefore |x| + |y| \leq \sqrt{2}|z|$

So $k = \sqrt{2}$

Option (c)

9. $-2 + ixy = \overline{(x+y-i)}$

$-2 + ixy = x + y + i$

So, $xy = 1$

$x + y = -2$

$x + \frac{1}{x} = -2$

$x^2 + 2x + 1 = 0$

So, $(x+1)^2 = 0$, thus $x = -1$

$y = \frac{1}{x} = -1$

Hence, $x = -1, y = -1$

10. $\frac{1}{z} = \frac{1}{a-ib} - \frac{1}{a-ic} = \frac{i(b-c)}{(a-ib)(a-ic)}$

$\Rightarrow \frac{1}{|z|} = \frac{|b-c|}{\sqrt{a^2+b^2}\sqrt{a^2+c^2}}$

$\Rightarrow |z| = \frac{\sqrt{a^2+b^2}\sqrt{a^2+c^2}}{|b-c|}$

Concept Applicator (CA)–3

1. Given $z = 1 + \sqrt{3}\,i$

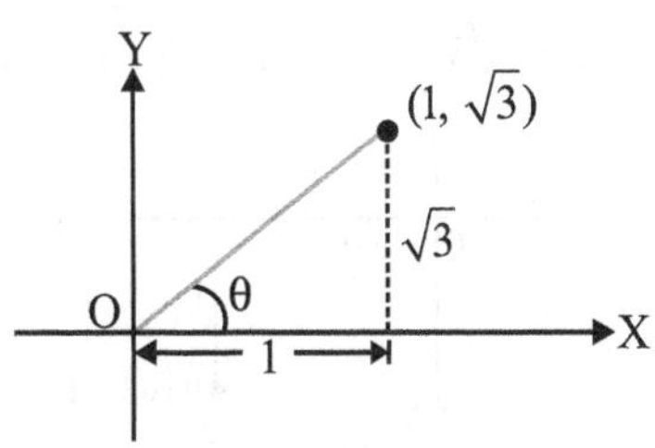

Modulus of z, $|z| = \sqrt{(1+3)} = 2$

$$\tan\theta = \frac{\sqrt{3}}{1} = \tan\frac{\pi}{3}$$

$$\theta = \frac{\pi}{3}$$

Argument of $z = \frac{\pi}{3}$

2. Let, $z = \sqrt{3} + i$, $|z| = \sqrt{(3+1)} = 2$

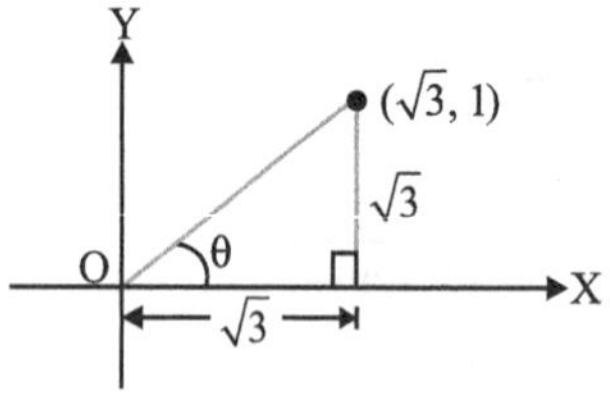

$$\tan\theta = \frac{1}{\sqrt{3}} = \tan\frac{\pi}{6}$$

So, $\theta = \frac{\pi}{6}$

Argument of $z = \frac{\pi}{6}$

3. Let, $z = -1 + \sqrt{3}\,i$, $|z| = \sqrt{(-1)^2 + (-\sqrt{3})^2} = 2$

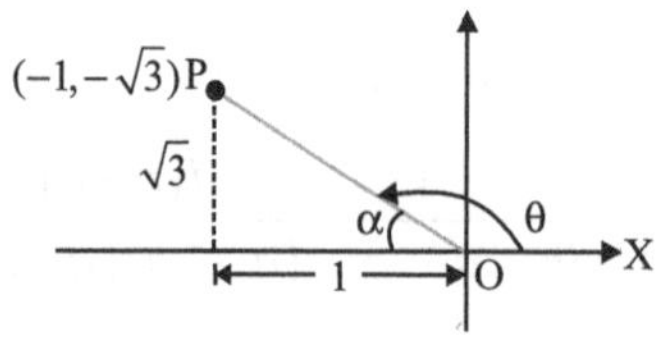

$$\tan\alpha = \frac{\sqrt{3}}{1} = \tan\frac{\pi}{3}$$

or $\alpha = \frac{\pi}{3}$

$$\theta = \pi - \alpha = \pi - \frac{\pi}{3} = \frac{2\pi}{3}$$

So, $|z| = 2$ and $\arg z = \frac{2\pi}{3}$

4. Given $z = \sqrt{3} - i$, so $|z| = \sqrt{(\sqrt{3})^2 + (1)^2} = 2$

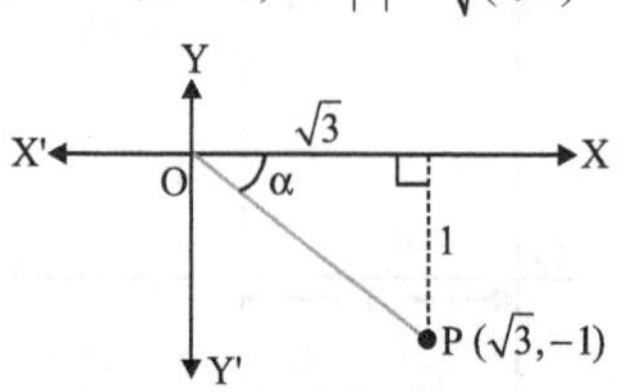

The complex number is in fourth quadrant.

$$\tan\alpha = \frac{1}{\sqrt{3}} = \tan\frac{\pi}{6}$$

So, $\alpha = \frac{\pi}{6}$,

Thus, principle argument, $\theta = -\frac{\pi}{6}$

Polar form of $z = 2\left(\cos\left(-\frac{\pi}{6}\right) + i\sin\left(-\frac{\pi}{6}\right)\right)$

$$= 2\left(\cos\frac{\pi}{6} - i\sin\frac{\pi}{6}\right)$$

5. Here, $z = -1 - i$ which lies in third quadrant.

$$|z| = \sqrt{(-1)^2 + (-1)^2} = \sqrt{2}$$

$$\tan\alpha = \frac{1}{1} = 1 \text{ so, } \alpha = \frac{\pi}{4}$$

$$\theta = -(\pi - \alpha)$$

$$= -\left(\pi - \frac{\pi}{4}\right)$$

$$= -\frac{3\pi}{4}$$

$$\arg z = -\frac{3\pi}{4}$$

Polar form of z, $\sqrt{2}\left(\cos\left(-\frac{3\pi}{4}\right) + i\sin\left(-\frac{3\pi}{4}\right)\right)$

6. $e^{iA}.e^{iB}.e^{iC}$

$$= (\cos A + i\sin A)(\cos B + i\sin B)(\cos C + i\sin C)$$

$$= \cos(A+B+C) + i\sin(A+B+C)$$

$$= \cos\pi + i\sin\pi = -1$$

7. $\text{amp}\,(z) = \text{amp}\left(\frac{\sqrt{3}+i}{\sqrt{3}-i}\right)$

$$= amp\left(\sqrt{3}+i\right) - amp\left(\sqrt{3}-i\right)$$

$$= tan^{-1}\frac{1}{\sqrt{3}} - tan^{-1}\left(\frac{-1}{\sqrt{3}}\right)$$

$$= \frac{\pi}{6} + \frac{\pi}{6} = \frac{\pi}{3}$$

8. $\text{amp}\,((x-1) + iy) = amp(x + i(y+3))$

$$\tan^{-1}\frac{y}{x-1} = \tan^{-1}\frac{y+3}{x}$$

$$\therefore \quad xy = (x-1)(y+3) \Rightarrow 3(x-1) = y$$

$\therefore \quad \dfrac{x-1}{y}=\dfrac{1}{3}$

9. $z=\dfrac{i(\sqrt{3}+i)^4}{4(1-\sqrt{3}i)^2}=\dfrac{i}{4}\cdot\dfrac{(2+2\sqrt{3}i)^2}{-2-2\sqrt{3}i}$

$=\dfrac{\sqrt{3}}{2}-\dfrac{i}{2}$

$\therefore \quad amp(z)=-\tan^{-1}\left(\dfrac{1}{\sqrt{3}}\right)=-\dfrac{\pi}{6}$

10. Let $z_1=r_1(\cos\theta_1+i\sin\theta_1)$, then $\left|\dfrac{z_1}{z_2}\right|=1$

$\Rightarrow \quad |z_1|=|z_2|$

$\Rightarrow |z_1|=|z_2|=r_1$

Now $\arg(z_1z_2)=0 \Rightarrow \arg(z_1)+\arg(z_2)=0$

$\Rightarrow arg(z_2)=-\theta_1$

There $z_2=r_1[\cos(-\theta_1)+i\sin(-\theta_1)]$

$=r_1(\cos\theta_1-i\sin\theta_1)=\overline{z}_1$

$\Rightarrow \overline{z}_2=(\overline{\overline{z}_1})=z_1$

$\Rightarrow \quad |z_2|^2=z_1z_2$

Concept Applicator (CA)–4

1. $|z_1+z_2+z_3|=|z_1-a+z_2-b+z_3-c+a+b+c|$

$\le |z_1-a|+|z_2-b|+|z_3-c|+|a+b+c|$

$\le a+b+c+a+b+c \ \le 2(a+b+c)$

Greatest value = 2 (a + b + c)

2. Let $|z_1|=|z_2|=r$

$\Rightarrow z_1=r(\cos\theta+i\sin\theta)$

and $z_2=r\left(\cos\left(\dfrac{\pi}{2}-\theta\right)+i\sin\left(\dfrac{\pi}{2}-\theta\right)\right)$

$\Rightarrow z_1z_2=r^2i$, which is purely imaginary.

3. Here $\dfrac{z_1}{z_2}=ik$ (purely imaginary) where $k\in R$.

$$\left|\dfrac{z_1-z_2}{z_1+z_2}\right|=\left|\dfrac{\dfrac{z_1}{z_2}-1}{\dfrac{z_1}{z_2}+1}\right|$$

$$=\dfrac{|ik-1|}{|ik+1|}=\dfrac{\sqrt{(k^2+1)}}{\sqrt{(k^2+1)}}=1$$

4. Given $x+iy=(a+ib)(b+ic)(c+id)(d+ie)$

Take modulus on both sides.

$|x+iy|=|(a+ib)(b+ic)(c+id)(d+ie)|$

$|x+iy|=|(a+ib)||(b+ic)||(c+id)||(d+ie)|$

$\sqrt{x^2+y^2}=$

$\sqrt{(a^2+b^2)}\sqrt{(b^2+c^2)}\sqrt{(c^2+d^2)}\sqrt{(d^2+e^2)}$

Squaring both sides.

$x^2+y^2=(a^2+b^2)(b^2+c^2)(c^2+d^2)(d^2+e^2)$.

5. $|\omega|=\left|\dfrac{1-iz}{z-i}\right|=\dfrac{|1-iz|}{|z-i|}=1$

$|1-iz|^2=|z-i|^2$

$(1-iz)\,\overline{(1-iz)}=(z-i)\,\overline{(z-i)}$

$(1-iz)(1+i\overline{z})=(z-i)(\overline{z}+i)$

$1-iz+i\overline{z}+z\overline{z}=z\overline{z}-i\overline{z}+iz+1$

$-iz+i\overline{z}=-i\overline{z}+iz$

$2iz=2i\overline{z}$

So, $z=\overline{z}$ thus z is purely real.

6. $z+\dfrac{1}{z}=2\cos\theta$

$z^2-2z\cos\theta+1=0$

So, $z^2-2\cos\theta\, z+\cos^2\theta=-1+\cos^2\theta$

$(z-\cos\theta)=i\sin\theta$ or $-i\sin\theta$

$z=\cos\theta+i\sin\theta$ or $z=\cos\theta-i\sin\theta$

$|z|=1$

Let $z=\cos\theta+i\sin\theta=e^{i\theta}$

$\left|\dfrac{z^{2n}-1}{z^{2n}+1}\right|=\left|\dfrac{e^{2in\theta}-1}{e^{2in\theta}+1}\right|=\left|\dfrac{e^{in\theta}-e^{-in\theta}}{e^{in\theta}+e^{-in\theta}}\right|$

$=\left|\dfrac{2i\sin(n\theta)}{2\cos(n\theta)}\right|=|i\tan(n\theta)|=|\tan(n\theta)|$

7. Let $z=\left[\dfrac{\cos\theta+i\sin\theta}{\sin\theta+i\cos\theta}\right]=\left[\dfrac{\cos\theta+i\sin\theta}{-i^2\sin\theta+i\cos\theta}\right]$

$=\dfrac{\cos\theta+i\sin\theta}{i(\cos\theta-i\sin\theta)}=\dfrac{e^{i\theta}}{ie^{-i\theta}}=\dfrac{1}{i}e^{2i\theta}$

$$z^{100} = \left(\frac{1}{i}e^{2i\theta}\right)^{100} = \frac{1}{i^{100}}e^{i200\theta}$$

$$= \frac{1}{\left(i^4\right)^{25}} (\cos 200\,\theta + i \sin 200\,\theta)$$

$$= \cos 200\,\theta + i \sin 200\,\theta$$

8. Let

$$z = \left[\frac{(\cos\theta + i\sin\theta)^{-1}(\cos 2\theta + i\sin 2\theta)^{-2}}{(\cos 3\theta + i\sin 3\theta)^{-3}(\cos 4\theta - i\sin 4\theta)^{-4}}\right]$$

$$= \frac{\left(e^{-i\theta}\right)^{-1}\left(e^{2i\theta}\right)^{-2}}{\left(e^{3i\theta}\right)^{-3}\left(e^{-4i\theta}\right)^{-4}}$$

$$= \frac{e^{i\theta}e^{-4i\theta}}{e^{-9i\theta}e^{-i16\theta}} = \frac{e^{-3i\theta}}{e^{-25i\theta}} = e^{i22\theta}$$

$$= \cos(22\theta) + i\sin(22\theta)$$

9. $\left(\sqrt{3}+i\right) = 2\left(\frac{\sqrt{3}}{2}+\frac{1}{2}i\right) = 2\left(\cos\frac{\pi}{6} + i\sin\frac{\pi}{6}\right)$

$$= 2\,e^{i\pi/6}$$

$$(1-i) = \sqrt{2}\left(\frac{1}{\sqrt{2}} - i\frac{1}{\sqrt{2}}\right) = \sqrt{2}e^{-i\pi/4}$$

$$z = \frac{\left(2e^{i\pi/6}\right)^{99}}{(\sqrt{2}e^{-i\pi/4})^{101}} = \frac{2^{99}}{\left(\sqrt{2}\right)^{101}}\frac{e^{\frac{i99\pi}{6}}}{e^{\frac{-i101\pi}{4}}}$$

$$= 2^{\left(99-\frac{101}{2}\right)}e^{\frac{i99\pi}{6}+\frac{i101\pi}{4}}$$

$$|z| = 2^{\frac{198-101}{2}} = (2)^{97/2}$$

Now calculate argument

$$\frac{99\pi}{6} + \frac{101\pi}{4} = \frac{33\pi}{2} + 25\pi + \frac{\pi}{4}$$

$$= 16\,\pi + \frac{\pi}{2} + 25\pi + \frac{\pi}{4}$$

$$= 16\,\pi + 24\pi + \pi + \frac{\pi}{2} + \frac{\pi}{4}$$

$$= 40\,\pi + +\pi + \frac{3\pi}{4} = 42\pi - \frac{\pi}{4}$$

$$\arg z = -\frac{\pi}{4}$$

and $|z| = (2)^{\frac{97}{2}}$

so $z = (2)^{\frac{97}{2}}\left(\cos\left(-\frac{\pi}{4}\right) + i\sin\left(-\frac{\pi}{4}\right)\right)$

10. Here $x = 1 + i$

$$x - 1 = i$$

$$(x-1)^2 = i^2$$

$$x^2 - 2x + 1 = -1$$

So, $x^2 - 2x + 2 = 0$

Multiplying by x

So, $x^3 - 2x^2 + 2x = 0$

$$x^3 + 8x^2 + 2x + 1 = (x^3 + 2x) + 8x^2 + 1$$

$$= 2x^2 + 8x^2 + 1 = 10x^2 + 1 \quad (\because x^2 = 2x - 2)$$

$$= 10(2x - 2) + 1$$

$$= 20x - 19$$

$$= 20(1+i) - 19 = 1 + 20i$$

Concept Applicator (CA)–5

1. As, $3^{49}(x+iy) = \left(\frac{3}{2}+\frac{\sqrt{3}}{2}i\right)^{100}$

$$= \left(i\sqrt{3}\left(\frac{1-i\sqrt{3}}{2}\right)\right)^{100}$$

$$\Rightarrow 3^{49}(x+iy) = i^{100}.3^{50}.(-\omega)^{100}$$

$$\Rightarrow 3^{49}(x+iy) = 3^{50}.\omega = 3^{50}.\left(-\frac{1}{2}+\frac{i\sqrt{3}}{2}\right)$$

$$\therefore x + iy = -\frac{3}{2} + \frac{3\sqrt{3}}{2}i$$

2. Here, $\frac{(1+i)^{2n} - (1-i)^{2n}}{\left(1+\omega^4-\omega^2\right)\left(1-\omega^4+\omega^2\right)}$

$$= \frac{\left\{(2i)^n - (-2i)^n\right\}}{\left\{\left(-2\omega^2\right)\left(-2\omega\right)\right\}}$$

$$= \frac{2^n i^n\left\{1-(-1)^n\right\}}{4} = \begin{cases} 0 & \text{, n is even} \\ 2^{n-2}i^n, & \text{n is odd} \end{cases}.$$

3. We have,

$$(16)^{1/4} = \left(2^4\right)^{1/4} = 2(1)^{1/4}$$

$$= 2(\cos 0 + i\sin 0)^{1/4}$$

$= 2\left\{\cos\frac{1}{4}(2k\pi + 0) + i\sin(2k\pi + 0)\right\}, k = 0, 1, 2, 3$

$= 2\times 1, 2\times i, 2\times(-1), 2\times(-i) = \pm 2, \pm 2i$

4. $x^2 - x + 1 = 0$

$\Rightarrow x = \frac{1\pm\sqrt{3}i}{2} = -\omega, -\omega^2$

$\therefore \sum_{n=1}^{5}(x^{2n} + \frac{1}{x^{2n}} + 2)$

$\Rightarrow \left(x^2 + \frac{1}{x^2} + 2\right) + \left(x^4 + \frac{1}{x^4} + 2\right)$

$+\left(x^6 + \frac{1}{x^6} + 2\right) + \left(x^8 + \frac{1}{x^8} + 2\right) + \left(x^{10} + \frac{1}{x^{10}} + 2\right)$

$\Rightarrow (\omega^2 + \omega^4 + \omega^6 + \omega^8 + \omega^{10})$

$+\left(\frac{1}{\omega^2} + \frac{1}{\omega^4} + \frac{1}{\omega^6} + \frac{1}{\omega^8} + \frac{1}{\omega^{10}}\right) + 10$

$\Rightarrow -1 - 1 + 10 = 8$

5. $z = -2 + 2\sqrt{3}i = 4\omega$

$z^{2n} + 2^{2n}z^n + 2^{4n} = 4^{2n}\omega^{2n} + 2^{2n}.4^n.\omega^n + 2^{4n}$

$= 4^{2n}\left(\omega^{2n} + \omega^n + 1\right)$

$= 0$, if n is not a multiple of 3

$= 3.4^{2n}$, if n is a multiple of 3

6. We know that

$\omega = \frac{(-1+\sqrt{-3})}{2}$ & $\omega^2 = \frac{(-1-\sqrt{-3})}{2}$

$\left(-1+\sqrt{-3}\right)^{62} + \left(-1-\sqrt{-3}\right)^{62}$

$= 2^{62}\omega^{62} + 2^{62}\left(\omega^2\right)^{62}$

$- 2^{62}\left[\left(\omega^3\right)^{20}\omega^2 + \left(\omega^3\right)^{41}\omega\right]$

$= 2^{62}\left[\omega^2 + \omega\right] = -2^{62}$

7. Given that

$z + \frac{1}{z} = 1$ or $\frac{z^2+1}{z} = 1$ or $z^2 - z + 1 = 0$

$\Rightarrow z = -\omega$ or $-\omega^2$

$\therefore z^{99} + \frac{1}{z^{99}} = (-1) + \frac{1}{(-1)} = -2$

8. Given expression

$\left(3 + 5\omega + 3\omega^2\right)^2 + \left(3 + 3\omega + 5\omega^2\right)^2$

$= \left\{3\left(1+\omega^2\right) + 5\omega\right\}^2 + \left\{3\left(1+\omega\right) + 5\omega^2\right\}^2$

$= \left(-3\omega + 5\omega\right)^2 + \left(-3\omega^2 + 5\omega^2\right)^2$

$= 4\omega^2 + 4\omega = -4$

9. $xyz = (a+b)\left(a\omega + b\omega^2\right)\left(a\omega^2 + b\omega\right)$

$= (a+b)\left(a^2 - ab + b^2\right) = a^3 + b^3$

10. $(1+\omega)^3 - \left(1+\omega^2\right)^3 = \left(-\omega^2\right)^3 - (-\omega)^3$

$= -\omega^6 + \omega^3 = -1 + 1 = 0$

Concept Applicator (CA)–6

1. $x^2 + y^2 + (4-3i)(x+iy) + (4+3i)(x-iy) + 5$

$= 0$

or $x^2 + y^2 + 8x + 6y + 5 = 0$

$\therefore$ Radius $= \sqrt{4^2 + 3^2 - 5} = \sqrt{20} = 2\sqrt{5}$

So it represents a circle with centre (–4, –3) and radius $2\sqrt{5}$

2. $|z-\omega|^2 + \left|z-\omega^2\right|^2 = \lambda$

$= \left|\omega - \omega^2\right|^2$

$= \left|\frac{-1+i\sqrt{3}}{2} - \frac{-1-i\sqrt{3}}{2}\right|^2 = \left|\frac{-1+i\sqrt{3}+1+i\sqrt{3}}{2}\right|^2$

$= \left|i\sqrt{3}\right|^2 = 3$

3. (b) $\therefore |z_1 - z_2| \ge \left||z_1| - |z_2|\right|$

$\therefore |z_2 - 3 - 4i| \ge \left||z_2| - |3+4i|\right| \Rightarrow 5 \ge \left||z_2| - |-5|\right|$

$\Rightarrow -5 \le |z_2| - 5 \le 5 \Rightarrow |z_2| \le 10 \Rightarrow -|z_2| \ge -10$

$\therefore |z_1| - |z_2| \ge 12 - 10 = 2$

But $|z_1 - z_2| \ge |z_1| - |z_2| \ge 2$

$\Rightarrow$ Minimum value of $|z_1 - z_2|$ is 2

ALTERNATIVE METHOD :

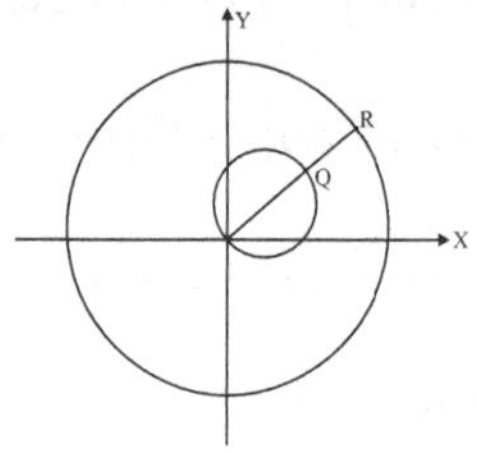

$|z_1| = 12 \Rightarrow z_1$ lies on circle of radius 12 and centre of origin $|z_2 - 3 - 4i| = 5 \Rightarrow z_2$ lies on circle of radius 5 and centre at (3, 4). Thus OR = 12, OQ = 10.

Now, minimum value of $|z_1 - z_2|$ = QR

= 12 – 10 = 2

4. $Im\left(\dfrac{2z+1}{iz+1}\right) = -2$

Let $z = x + iy$

$$\therefore \frac{2z+1}{iz+1} = \frac{2(x+iy)+1}{i(x+iy)+1} = \frac{(2x+1)+i(2y)}{(1-y)+i(x)}$$

$$\text{or } \frac{2z+1}{iz+1} = \frac{\{(2x+1)+i(2y)\}\{(1-y)-ix\}}{(1-y)^2+x^2}$$

$$= \frac{(2x+1)(1-y)+2xy+i\{2y(1-y)-x(2x+1)\}}{(1-y)^2+x^2}$$

Since, $Im\left(\dfrac{2z+1}{iz+1}\right) = -2$

we have, $\dfrac{-(2x^2+2y^2+x-2y)}{(1-2y+x^2+y^2)} = -2$

$\Rightarrow 2x^2 + 2y^2 + x - 2y = 2 - 4y + 2x^2 + 2y^2$

$\Rightarrow x + 2y = 2$, which is a straight line.

5. (c) We have, $z_1 - z_4 = z_2 - z_3$ or

$\dfrac{z_1+z_3}{2} = \dfrac{z_2+z_4}{2}$ i.e., the diagonals bisect each other.

$\therefore$ It is a parallelogram. Also, $amp\dfrac{z_4 - z_1}{z_2 - z_1} = \dfrac{\pi}{2}$

$\Rightarrow$ Angle at z_1 is a right angle.

$\therefore$ It is a rectangle and hence a cyclic quadrilateral.

6. (c) The vertices of the triangle are A (1, 0), B$\left(\dfrac{1}{\sqrt{2}}, \dfrac{1}{\sqrt{2}}\right)$ and vertices C (0, 1), now find the length of each side of triangle

$\therefore AB^2 = 2 - \sqrt{2}$

$BC^2 = 2 - \sqrt{2}$

$AC^2 = 1 + 1 = 2 \quad \therefore AB = BC$

7. (b) Vertices of triangle are (1,1), (–1, 1) (0, 2)

Required area

$$= \frac{1}{2}\begin{vmatrix} 1 & 1 & 1 \\ -1 & 1 & 1 \\ 0 & 2 & 1 \end{vmatrix} = \frac{1}{2}\{1-2-(-1-0)+(-2-0)\} = -1$$

$\therefore$ Area = 1

8. (a) Let $z = x + iy$ where $x, y \in R$

From the given information, $|iz - 1| + |z - i| = 2$

$\Rightarrow |ix - y - 1| + |x + iy - i| = 2$

$\Rightarrow \sqrt{(-y-1)^2 + x^2} + \sqrt{x^2 + (y-1)^2} = 2$

$\Rightarrow x^2 = 0$ or $x = 0$

$\therefore$ The locus of z is a straight line.

9. (a) We have, $z_1 = 1 + \sqrt{3}i$

$\dfrac{z_2}{z_1} = e^{i\frac{2\pi}{3}}, \dfrac{z_3}{z_1} = e^{-\frac{i2\pi}{3}}$

$\Rightarrow z_2 = -2$ and $z_3 = 1 - i\sqrt{3}$

10. (c) Let $z = x + iy$

Then, $\dfrac{z+2i}{z+4} = \dfrac{x+iy+2i}{x+iy+4} = \dfrac{x+i(y+2)}{(x+4)+iy}$

$$= \frac{[x+i(y+2)][(x+4)-iy]}{(x+4)^2+y^2}$$

$$= \frac{(x^2+4x+y^2+2y)+i(2x+4y+8)}{(x+4)^2+y^2}$$ Since,

$Re\left(\dfrac{z+2i}{z+4}\right) = 0$

$\Rightarrow x^2 + y^2 + 4x + 2y = 0$, which represents a circle with centre (–2, –1).

And radius is $\sqrt{2^2 + 1} = \sqrt{5}$

Exercise A

Concept Builder (CB)

1. (c) We have, $Im(z^2) = 4$

$\Rightarrow Im\left[(x^2 - y^2) + 2ixy\right] = 4$

[Putting $z = x + iy$]

$\Rightarrow \quad 2xy = 4$ or $xy = 2$

2. (b) $amp(z_1 z_2) = 0 \Rightarrow amp(z_1) + amp(z_2) = 0$

$\therefore amp(z_1) = -amp(z_2) = amp(\bar{z}_2)$

Since $|z_1| = |z_2|$, we get $|z_1| = |\bar{z}_2|$. So, $z_1 = \bar{z}_2$

Also, $z_1 z_2 = \bar{z}_2 z_2 = |z_2|^2 = 1$, because $|z_2| = 1$

3. (b) As, $(1+i)^3 + (1-i)^6$

$= (1+i)^3 + ((1-i)^2)^3$ since $(1-i)^2 = -2i$

$= (1+i)^3 + (-2i)^3$,

Now using $a^3 + b^3 = (a+b)(a^2 - ab + b^2)$

$= \{1 + i - 2i\}\{(1+i)^2 - (1+i)(-2i) + (-2i)^2\}$

$= (1-i)\{2i + 2i + 2i^2 + 4i^2\}$

$= (1-i)\{4i - 6\} = 4i - 6 - 4i^2 + 6i$

$= -2 + 10i = 2(-1 + 5i)$

4. (a) We have, $|z - 4i| = |z + 4i|$

$\Rightarrow |x + i(y-4)|^2 = |x + i(y+4)|^2$

$\Rightarrow x^2 + (y-4)^2 = x^2 + (y+4)^2$

$\Rightarrow 16y = 0; y = 0$, which is x–axis.

5. (d) We have, $Re(z^2) = 4$

$\Rightarrow \quad Re\left[(x^2 - y^2) + 2ixy\right] = 4$

[Putting $z = x + iy$]

$\Rightarrow x^2 - y^2 = 4$

which is a rectangular hyperbola.

6. (c) $(\sqrt{3} + i) = (a + ib)(c + id)$

Taking, argument on both sides

$\Rightarrow \tan^{-1}\frac{1}{\sqrt{3}} = \tan^{-1}\frac{b}{a} + \tan^{-1}\frac{d}{c}$

$\Rightarrow \tan^{-1}\frac{b}{a} + \tan^{-1}\frac{d}{c} = n\pi + \frac{\pi}{6}, n \in Z$

7. (a) Since $|z - 1| = 1$ so we can assume $z - 1 = e^{i\theta}$

$\Rightarrow z = 1 + \cos\theta + i\sin\theta$

$= 2\cos^2\frac{\theta}{2} + 2i\sin\frac{\theta}{2}\cos\frac{\theta}{2}$

$= 2\cos\frac{\theta}{2}\left(\cos\frac{\theta}{2} + i\sin\frac{\theta}{2}\right)$

$\therefore \arg(z) = \frac{\theta}{2} = \frac{1}{2}\arg(z-1)$

or $\arg(z-1) = 2\arg(z)$

8. (c) $|z - 1| = |z + i|$

$\Rightarrow z$ is perpendicular bisector of (1, 0) & (0, –1) hence a straight line which passes through origin.

9. (b) $i + \sqrt{3} = 2\left(\frac{\sqrt{3}}{2} + \frac{i}{2}\right) = 2\left(\cos\frac{\pi}{6} + i\sin\frac{\pi}{6}\right)$

10. (c) $|z_1 + z_2| = |z_1| + |z_2|$

$\Rightarrow \cos(\theta_1 - \theta_2) = 1$

$\Rightarrow \theta_1 - \theta_2 = 0$

$\Rightarrow \arg(z_1) = \arg(z_2)$

Smart Cracker

If $|z_1 + z_2| = |z_1| + |z_2|$, then Oz_1 and Oz_2 are co–linear. Here O is origin, so $\arg(z_1) = \arg(z_2)$

11. (d) $-1 + i\sqrt{3} = 2\left(-\frac{1}{2} + \frac{i\sqrt{3}}{2}\right)$

$= 2\left(\cos\frac{2\pi}{3} + i\sin\frac{2\pi}{3}\right) = 2e^{i2\pi/3}$

$\Rightarrow \theta = \frac{2\pi}{3}$ & $r = 2$

12. (a) $\left(\frac{1+i\sqrt{3}}{1-i\sqrt{3}}\right)^6+\left(\frac{1-i\sqrt{3}}{1+i\sqrt{3}}\right)^6$

$$=\left(\frac{\frac{1+i\sqrt{3}}{2}}{\frac{1-i\sqrt{3}}{2}}\right)^6+\left(\frac{\frac{1-i\sqrt{3}}{2}}{\frac{1+i\sqrt{3}}{2}}\right)^6$$

$$=\left(\frac{-\omega^2}{-\omega}\right)^6+\left(\frac{-\omega}{-\omega^2}\right)^6=2$$

13. (a) We know that $|z-z_1|=|z-z_2|$ represents a perpendicular bisector and given expression is in this format so it represents a straight line.

14. (a) Here,

$$\left(\frac{z_1}{z_2}\right)^2=\left(\frac{\sqrt{3}+i\sqrt{3}}{\sqrt{3}+i}\right)^2=\frac{3-3+6i}{3-1+2\sqrt{3}}=\frac{-3i\omega}{2}$$

$$\left(\frac{z_1}{z_2}\right)^{50}=\left(\left(\frac{z_1}{z_2}\right)^2\right)^{25}=\left(\frac{-3i\omega}{2}\right)^{25}$$

$$=-\left(\frac{3}{2}\right)^{25}i\omega=\left(\frac{3}{2}\right)^{25}\left(\frac{\sqrt{3}}{2}+\frac{i}{2}\right)$$

$\therefore \left(\frac{z_1}{z_2}\right)^{50}$ lies in I quadrant.

15. (b) We have, $(z-1)^n=i(z+1)^n$

$\therefore |z-1|^n=|i||z+1|^n$

$\Rightarrow |z-1|^n=|z+1|^n$

$\Rightarrow |z-1|=|z+1|$

$\Rightarrow |z-1|^2=|z+1|^2$

$\Rightarrow (x-1)^2+y^2=(x+1)^2+y^2$

$\Rightarrow 4x=0 \Rightarrow x=0$

$\therefore$ The roots lies on the y–axis

16. (a) $|z|+|z-1|\geq|z-z+1|\geq 1$

17. (a) $\frac{\left(\sin\frac{\pi}{8}+i\cos\frac{\pi}{8}\right)^8}{\left(\sin\frac{\pi}{8}-i\cos\frac{\pi}{8}\right)^8}$

$$=\frac{\sin\pi+i\cos\pi}{\sin\pi-i\cos\pi}=\frac{-i}{i}=-1$$

18. (b) $\frac{1+a}{2}=\frac{1}{2}\left(1+\cos\frac{4\pi}{3}+i\sin\frac{4\pi}{3}\right)$

$$=\frac{1}{2}\cdot 2\cos\frac{2\pi}{3}\left(\cos\frac{2\pi}{3}+i\sin\frac{2\pi}{3}\right)$$

$$=-\frac{1}{2}\left(\cos\frac{2\pi}{3}+i\sin\frac{2\pi}{3}\right)$$

$$\therefore \left(\frac{1+a}{2}\right)^{3n}=\left(-\frac{1}{2}\right)^{3n}\left(\cos\frac{2\pi}{3}+i\sin\frac{2\pi}{3}\right)^{3n}$$

$$=\frac{(-1)^n}{2^{3n}}$$

19. (d) $\frac{(\cos\theta+i\sin\theta)^4}{(i\cos\theta+\sin\theta)^5}$

$$=\frac{(\cos\theta+i\sin\theta)^4}{i^5}(\cos\theta+i\sin\theta)^5$$

$$=-i(\cos 9\theta+i\sin 9\theta)$$

$$=\sin 9\theta-i\cos 9\theta$$

20. (a) $z=\frac{1-i\sqrt{3}}{1+i\sqrt{3}}=\frac{(1-i\sqrt{3})^2}{1+3}=\frac{1-3-2\sqrt{3}i}{4}$

$$=-\frac{1}{2}-\frac{\sqrt{3}}{2}i=\cos 240^\circ+i\sin 240^\circ$$

$\therefore \arg(z)=240^\circ=\frac{4\pi}{3}$

Or principle argument is $\frac{-2\pi}{3}$

21. (c) Here, $a=\frac{1}{2}(\sqrt{3}+i)$, $r=\frac{1}{2}(\sqrt{3}+i)$

$\therefore\ T_n=ar^{n-1}=\left[\frac{1}{2}(\sqrt{3}+i)\right]^n$

Put $\frac{\sqrt{3}}{2}=r\cos\theta, \frac{1}{2}=r\sin\theta$ These give $r=1$ and $\theta=\frac{\pi}{6}$

$\therefore\ T_n=[r(\cos\theta+i\sin\theta)]^n$

$\Rightarrow |T_n|=1$

22. (a) $z_0^7=1$, so $z_0^6=\frac{1}{z_0}$

$$1+z_0+z_0^2+z_0^3+z_0^4+z_0^5+z_0^6=0$$

$$\log_2\left|1+z_0+\ldots+z_0^5-\frac{1}{z_0}\right|=\log_2\left|-z_0^6-z_0^6\right|$$

$= \log_2 |2z_0^6|$

$= \log_2 2 = 1$

23. (b) Let $z = x + iy$

Then, $\dfrac{z+2i}{z+2} = \dfrac{x+iy+2i}{x+iy+2} = \dfrac{x+(y+2)i}{(x+2)+iy}$

$= \dfrac{[x+(y+2)i][(x+2)-iy]}{(x+2)^2+y^2}$

$= \dfrac{(x^2+2x+y^2+2y)+i(2x+2y+4)}{(x+4)^2+y^2}$

Since, $Im\left(\dfrac{z+2i}{z+2}\right) = 0$

$\Rightarrow x + y + 2 = 0$

which represents a straight line.

24. (d) Given that $z_3 = \frac{1}{2}(z_1 + z_2)$ so z_3 is mid point of z_1 and z_2. Clearly, $|z_1| < |z_3| < |z_2|$

25. (b) Since, $|z_1| = |z_2| = |z_3|$, the origin is the circumcentre of the triangle.

∴ The triangle is equilateral so circum centre and centroid will coincide and so origin is the centroid.

$\therefore \dfrac{z_1 + z_2 + z_3}{3} = 0$

26. (c) $p^{1/3} = (-p)^{1/3}(-1)^{1/3}$

$= -(-p)^{1/3}, -(-p)^{1/3}\omega, -(-p)^{1/3}\omega^2$

$= -q, -q\omega, -q\omega^2$ where $q = (-p)^{1/3}$

Let $\alpha = -q, \beta = -q\omega, \gamma = -q\omega^2$

$\therefore \dfrac{x\alpha + y\beta + z\gamma}{x\beta + y\gamma + z\alpha} = \dfrac{-q(x + y\omega + z\omega^2)}{-q(x\omega + y\omega^2 + z)}$

$= \dfrac{x\omega^3 + y\omega^4 + z\omega^2}{x\omega + y\omega^2 + z} = \omega^2$

27. (c) We have $|z - 4i| + |z + 4i| = 10$

$\Rightarrow |z - (0 + 4i)| + |z - (0 - 4i)| = 10$

This represents an ellipse since, $|4i + 4i| < 10$

28. (d) Let $\overline{(\cos x - i \sin 2x)} = \sin x + i \cos 2x$

$\Rightarrow \cos x + i \sin 2x = \sin x + i \cos 2x$

$\therefore \cos x = \sin x$ ...(i)

and $\sin 2x = \cos 2x$ ...(ii)

$\Rightarrow \tan x = 1$ and $\tan 2x = 1$

which is impossible.

29. (b) Let $\arg(z) = \theta$ $\therefore \arg(w) = \pi - \theta$

$\therefore \overline{w} = -z$ i.e., $z = -\overline{w}$

30. (d) The vertices of triangle are i, ω, ω^2 and these points are

$A(0, 1), B\left(-\frac{1}{2}, \frac{\sqrt{3}}{2}\right), C\left(-\frac{1}{2}, -\frac{\sqrt{3}}{2}\right)$

∴ Triangle is as shown below,

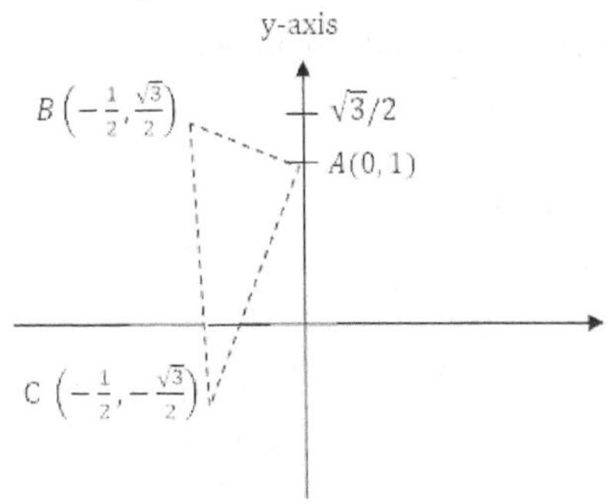

Area of triangle from figure is

$= \frac{1}{2} \cdot \frac{1}{2} \times |\sqrt{3}| = \frac{\sqrt{3}}{4}$ sq unit

31. (c) Given that $\dfrac{AB}{BC} = \sqrt{2}$ now consider the rotation about 'B', from rotation theorem we get,

$\dfrac{z_1 - z_2}{z_3 - z_2} = \dfrac{|z_1 - z_2|}{|z_3 - z_2|} e^{i\pi/4} = \dfrac{AB}{BC} e^{i\pi/4}$

$= \sqrt{2}\left(\dfrac{1}{\sqrt{2}} + \dfrac{i}{\sqrt{2}}\right) = 1 + i$

$\Rightarrow z_1 - z_2 = (1 + i)(z_3 - z_2)$

$\Rightarrow z_1 - (1 + i)z_3 = z_2(1 - 1 - i)$

$\Rightarrow iz_2 = -z_1 + (1 + i)z_3$

$\Rightarrow z_2 = iz_1 - i(1 + i)z_3 = z_3 + i(z_1 - z_3)$

32. (d) The given equation written as

$$\arg[z-(1+i)] = \begin{cases} \frac{3\pi}{4}, \text{ when } |z| \le |z-2| \\ -\frac{\pi}{4}, \text{ when } |z| > |z-2| \end{cases}$$

$\Rightarrow$ The locus is a set of two rays defined in two different domain

33. (b) Let

$f(z) = g(z)(z-i)(z+i) + az + b; a, b \in C$

$f(i) = i \Rightarrow ai + b = i$ ...(i)

$f(-i) = 1+i$

$\Rightarrow a(-i) + b = 1+i$ ...(ii)

Solving equations (i) and (ii), we get

$a = \frac{i}{2}, b = \frac{1}{2} + i$

Hence, required remainder

$= az + b = \frac{1}{2}iz + \frac{1}{2} + i$

34. (a) We have,

$$\frac{(1+i)x-2i}{3+i} + \frac{(2-3i)y+i}{3-i}$$

$$= \frac{\begin{Bmatrix} (3-i)(1+i)x - 2i(3-i) + \\ (3+i)(2-3i)y + i(3+i) \end{Bmatrix}}{3^2 - i^2} = i$$

or $(3-i)(1+i)x - 2i(3-i) + (3+i)(2-3i)y$
$+i(3+i) = i(9+1)$

Equating real and imaginary parts we get

$x = 3, y = -1$

So x + y = 2

35. (b) Let $z = x + iy$ where $x, y \in R$

$\therefore x^2 + y^2 \ne 0$ and $x = 0$

$\therefore z = 0 + iy = iy, y \ne 0 \Rightarrow z^2 = -y^2$

$\Rightarrow Im(z^2) = 0$

36. (b) We have, $(x+iy)^{1/3} = a + ib$

or $x + iy = (a+ib)^3$

$= a^3 + 3a^2(ib) + 3a(ib)^2 + (ib)^3$

$= a^3 + 3a^2bi + 3ab^2i^2 + i^3b^3$

$= a^3 + 3a^2bi - 3ab^2 - ib^3$

$= a(a^2 - 3b^2) + ib(3a^2 - b^2)$

Now equating real and imaginary part we get

$\therefore x = a(a^2 - 3b^2)$ and $y = b(3a^2 - b^2)$

$\Rightarrow \frac{x}{a} = a^2 - 3b^2$ and $\frac{y}{b} = 3a^2 - b^2$

$\therefore \frac{x}{a} + \frac{y}{b} = 4(a^2 - b^2)$

37. (a) $\frac{1-ix}{1+ix} = a - ib$

$\Rightarrow \frac{(1-ix)(1-ix)}{(1+ix)(1-ix)} = a - ib$

$\Rightarrow \frac{1-x^2-2ix}{1+x^2} = \frac{a-ib}{1}$

equating real and imaginary part we get

$\Rightarrow \frac{1-x^2}{1+x^2} = a$ and $\frac{2x}{1+x^2} = b$

Now, x can be written as

$$x = \frac{\frac{2x}{1+x^2}}{\frac{2}{1+x^2}} = \frac{\frac{2x}{1+x^2}}{\frac{1-x^2}{1+x^2}+1} = \frac{b}{1+a}$$

38. (d) Consider, $(|x| - |y|)^2 \ge 0$

$\Rightarrow |x|^2 + |y|^2 - 2|x||y| \ge 0$

$\Rightarrow 2|x||y| \le |x|^2 + |y|^2$

$\Rightarrow |x|^2 + |y|^2 + 2|x||y| \le 2|x|^2 + 2|y|^2$

$\Rightarrow (|x| + |y|)^2 \le 2(x^2 + y^2)$

$\Rightarrow (|x| + |y|)^2 \le 2|z|^2$

$\therefore |x| + |y| \le \sqrt{2}|z|$

$\lambda^2 = 2$

39. (b) $x^2 - 2x + 4 = 0$

$\Rightarrow x = 1 \pm \sqrt{3}i = 2\left(\frac{1}{2} \pm \frac{\sqrt{3}}{2}i\right) = -2\omega, -2\omega^2$

$\Rightarrow \alpha^6 + \beta^6 = (64\omega^6 + 64\omega^{12}) = 128$

40. (b) $|z+1|=|(z+4)-3|\le|z+4|+|-3|\le 3+3=6$

$\therefore$ Greatest value of $|z+1|=6$

Also, $|z+1|\ge|z+4|-|-3|\ge 0$

$\therefore 0\le|z+1|\le 6$

So $|z+1|$ can assume 7 numerical values.

41. (a) Given that, $x+iy=\dfrac{u+iv}{u-iv}$

Taking modulus both the sides

$$|z|=\frac{\sqrt{u^2+v^2}}{\sqrt{u^2+v^2}}=1$$

42. (c) Simplify, $\dfrac{1+i}{1-i}=i$ and $\dfrac{1-i}{1+i}=-i$

$$\therefore \left(\frac{1+i}{1-i}\right)^3-\left(\frac{1-i}{1+i}\right)^3=i^3-(-i)^3=x+iy$$

$\Rightarrow -i+i^3=x+iy$

$\Rightarrow -i-i=x+iy; x+iy=0-2i$

Hence, $(x,y)=(0,-2)$

43. (c) Taking modulus and squaring both sides, we get

$$|1+i|^2\cdot|1+2i|^2\cdot|1+3i|^2\ldots|1+ni|^2=|\alpha+i\beta|^2$$

$$\Rightarrow (1+1)\cdot(1+4)\cdot(1+9)\ldots(1+n^2)=\alpha^2+\beta^2$$

$$\Rightarrow 2\cdot 5\cdot 10\ldots(1+n^2)=\alpha^2+\beta^2$$

44. (a) We have, $P^2+Q^2=1$.

Let $P=\cos\theta, Q=\sin\theta$

$$\therefore \frac{1+P+iQ}{1+P-iQ}=\frac{1+\cos\theta+i\sin\theta}{1+\cos\theta-i\sin\theta}$$

$$=\frac{\cos\frac{\theta}{2}\left(\cos\frac{\theta}{2}+i\sin\frac{\theta}{2}\right)}{\cos\frac{\theta}{2}\left(\cos\frac{\theta}{2}-i\sin\frac{\theta}{2}\right)}=\frac{\left(\cos\frac{\theta}{2}+i\sin\frac{\theta}{2}\right)^2}{1}$$

$=\cos\theta+i\sin\theta=P+iQ$

45. (c) Let $z-1=r(\cos\theta+i\sin\theta)=re^{i\theta}$

$\therefore$ Given exp.

$$(z-1)(\cos\alpha-i\sin\alpha)+(z-1)^{-1}\times(\cos\alpha+i\sin\alpha)$$

$$=re^{i(\theta-\alpha)}+\frac{1}{r}e^{-i(\theta-\alpha)}$$

Since, imaginary part of given expression is zero, so $r\sin(\theta-\alpha)-\frac{1}{r}\sin(\theta-\alpha)=0$

$r^2-1=0\Rightarrow r^2=1$

$\Rightarrow r=1\Rightarrow |z-1|=1$ or $\sin(\theta-\alpha)=0$

$\Rightarrow \theta-\alpha=0\Rightarrow\theta=\alpha\Rightarrow \arg(z-1)=\alpha$

Concept Cracker (CC)

1. (a) From the given condition we can infer that $z, w, -iw$ and $i\bar{w}$ lie inside or on the boundary of the circle $|z|\le 1$. Since $|z+iw|=|z-i\bar{w}|=2$ we get $-iw, i\bar{w}$ are the other end points of the diameter having one end point as z. So,

$$-iw=i\bar{w}=-z.$$

Therefore, $|z+z|=2$ or $|z|=1$

but, $-z=-iw=i\bar{w}$

$\Rightarrow w=-iz$ and $\bar{w}=iz$

$\therefore \bar{w}=i\bar{z}$ and $\bar{w}=iz$

$\therefore i\bar{z}=iz\Rightarrow z=\bar{z}$

z is real.

and $|z|=1$, we get $z=1,-1$.

2. (a) Let $a+bi$ be a square root of $-2+2\sqrt{3}i$.

$\therefore (a+bi)^2=-2+2\sqrt{3}i$

$a^2+2abi+i^2b^2=-2+2\sqrt{3}i$

$\Rightarrow (a^2-b^2)+2abi=-2+2\sqrt{3}i$

Equating the real and imaginary part we get

$\therefore\ a^2-b^2=-2$ ...(i)

$2ab=2\sqrt{3}$ ...(ii)

Now, $(a^2+b^2)^2=(a^2-b^2)^2+4a^2b^2$

$=(-2)^2+(2\sqrt{3})^2=4+12=16$

$\therefore a^2+b^2=\sqrt{16}=4$ ...(iii)

From equations (i) and (iii), we get

$2a^2 = 2$ or $a^2 = 1$, $\therefore a = \pm 1$

$2b^2 = 6$ or $b^2 = 3$, $\therefore b = \pm\sqrt{3}$

either $a = 1, b = \sqrt{3}$ or $a = -1, b = -\sqrt{3}$

Hence, the two square root are $1+\sqrt{3}i$ and $-1-\sqrt{3}i$ i.e., $\pm(1+\sqrt{3}i)$

Smart Cracker

$$\sqrt{z} = \pm\left(\sqrt{\frac{|z|+a}{2}} + i\sqrt{\frac{|z|-a}{2}}\right)$$

$$= \pm\left(\sqrt{\frac{4-2}{2}} + i\sqrt{\frac{4+2}{2}}\right) = \pm\left(1+i\sqrt{3}\right)$$

3. (c) Since $1, \alpha_1, \alpha_2, \ldots, \alpha_{n-1}$ are the nth roots of unity so

$\therefore x^n - 1 = (x-1)(x-\alpha_1)(x-\alpha_2)\ldots(x-\alpha_{n-1})$

Putting $x = -1$, we get

$$(-1)^n - 1 = (-1-1)(-1-\alpha_1)(-1-\alpha_2)\ldots(-1-\alpha_{n-1})$$

$$\Rightarrow (-1)^n - 1 = (-1)^n(2)(1+\alpha_1)(1+\alpha_2)\ldots(1+\alpha_{n-1})$$

$$\Rightarrow (1+\alpha_1)(1+\alpha_2)\ldots(1+\alpha_{n-1}) = \frac{1}{2}\left\{1-(-1)^n\right\}$$

4. (b) Consider

$$|\alpha+\beta|^2 = (\alpha+\beta)(\overline{\alpha+\beta}) = (\alpha+\beta)(\bar{\alpha}+\bar{\beta})$$

$$= \alpha\bar{\alpha} + \beta\bar{\beta} + \bar{\alpha}\beta + \alpha\bar{\beta}$$

$$= |\alpha|^2 + |\beta|^2 + \bar{\alpha}\beta + \alpha\bar{\beta} \quad \ldots(i)$$

Now consider

$$|\alpha-\beta|^2 = (\alpha-\beta)(\bar{\alpha}-\bar{\beta}) = \alpha\bar{\alpha} + \beta\bar{\beta} - \bar{\alpha}\beta - \alpha\bar{\beta}$$

$$= |\alpha|^2 + |\beta|^2 - \bar{\alpha}\beta - \alpha\bar{\beta} \quad \ldots(ii)$$

On adding Equations (i) and (ii), we get

$$|\alpha|^2 + |\beta|^2 = \frac{1}{2}\left\{|\alpha+\beta|^2 + |\alpha-\beta|^2\right\}$$

5. (a) From the given information $\alpha^5 = 1$

$$\therefore \left|1+\alpha+\alpha^2+\alpha^{-2}-\alpha^{-1}\right|$$

$$= \left|1+\alpha+\alpha^2+\alpha^3-\alpha^4\right|$$

$$= \left|1+\alpha+\alpha^2+\alpha^3+\alpha^4-2\alpha^4\right|$$

$$= \left|\frac{1-\alpha^5}{1-\alpha} - 2\alpha^4\right| = \left|2\alpha^4\right| = 2|\alpha|^4 = 2\times 1 = 2$$

So $2^{\left|1+\alpha+\alpha^2+\alpha^{-2}-\alpha^{-1}\right|} = 4$

6. (a) $x + iy = 1 - t + i\sqrt{t^2+t+2}$,

$\Rightarrow x = 1-t, y = \sqrt{t^2+t+2}$,

In order to find the locus we have to eliminate 't' and find the relation between 'x' and 'y'

Since $x = 1 - t$ so $t = 1 - x$ and so $y^2 = t^2 + t + 2$

$$= (1-x)^2 + 1 - x + 2 = \left(x-\frac{3}{2}\right)^2 + \frac{7}{4}$$

or $y^2 - \left(x-\frac{3}{2}\right)^2 = \frac{7}{4}$, which is a hyperbola.

7. (b) Let $z = x + iy$

$$\therefore \sqrt{x^2+(y+1)^2} - \sqrt{x^2+(y-1)^2} = k \quad \ldots(i)$$

or $x^2 + (y+1)^2 - x^2 - (y-1)^2$

$$= k\cdot\left\{\sqrt{x^2+(y-1)^2} + \sqrt{x^2+(y-1)^2}\right\}$$

$$\therefore \sqrt{x^2+(y-1)^2} + \sqrt{x^2+(y+1)^2} = \frac{4y}{k} \ldots(ii)$$

From eqs (i) and (ii),

$$2\sqrt{x^2+(y+1)^2} = k + \frac{4y}{k}$$

$$4x^2 + 4y^2 + 8y + 4 = k^2 + \frac{16y^2}{k^2} + 8y$$

$$\Rightarrow 4x^2 + \left(4 - \frac{16}{k^2}\right)y^2 = k^2 - 4$$

For an hyperbola, $\frac{4k^2-16}{k^2} < 0 \Rightarrow k^2 - 4 < 0$

$\Rightarrow |k| < 2$

Smart Cracker

$|z-z_1| - |z-z_2| = k$ represents hyperbola if $|k| < |z_1 - z_2|$

Thus, $|z+i| - |z-i| = k$ represents hyperbola, if $|k| < |i+i|$ or $|k| < 2 \Rightarrow -2 < k < 2$

8. (c) Consider z = x + iy

$$\sqrt{x^2+(y-1)^2}+\sqrt{x^2+(y+1)^2}=k \quad \ldots(i)$$

or $x^2+(y-1)^2-x^2-(y+1)^2$

$$= k\left\{\sqrt{x^2+(y-1)^2}-\sqrt{x^2+(y+1)^2}\right\}$$

$$\therefore \sqrt{x^2+(y-1)^2}-\sqrt{x^2+(y+1)^2}=-\frac{4y}{k}\ldots(ii)$$

From eqs (i) and (ii), $2\sqrt{x^2+(y-1)^2}=-\frac{4y}{k}$

$$\Rightarrow 4x^2+\left(4-\frac{16}{k^2}\right)y^2=k^2-4$$

For an ellipse $= 4-\frac{16}{k^2}>0,\ k^2-4>0$

or $|k|>2$

Smart Cracker

$|z-z_1|+|z-z_2|=k$ represents ellipse if

$|k|>|z_1-z_2|$

Thus , $|z-i|+|z+i|=k$ represents ellipse if

$|k|>|i+i|$ or $|k|>|2i|$

$\therefore\ |k|>2$

9. (b) $|iz+z_0|=|i(z-i)-1+5+3i|$

$=|i(z-i)+4+3i|$

$\le |i||z-i|+|4+3i|$

$\le 1\cdot 2+5=7$

10. (a) Let $z=i^i$

$$log_e z = log_e i^i = i log_e i = i log_e e^{i\frac{\pi}{2}}$$

$$= i^2\frac{\pi}{2}log_e e = -\frac{\pi}{2}$$

$\therefore z=e^{-\pi/2}$

11. (b) If $|z|=|z-1|\Rightarrow |z|^2=|z-1|^2$

$\Rightarrow z\bar{z}=(z-1)(\bar{z}-1)$

$\Rightarrow z\bar{z}=z\bar{z}-\bar{z}-z+1$

$\therefore z+\bar{z}=1$

Again, if $|z|=|z+1|\Rightarrow |z|^2=|z+1|^2$

$\Rightarrow z\bar{z}=z\bar{z}+\bar{z}+z+1$

$\Rightarrow z+\bar{z}=-1$

$\Rightarrow |z+\bar{z}|=1$

12. (c) The triangle with vertices $A(z_1), B(z_2)$ and $C(z_3)$ is equilateral.

$\Rightarrow z_1^2+z_2^2+z_3^2=z_2z_3+z_3z_1+z_1z_2$

$\Rightarrow (z_1+z_2+z_3)^2=3(z_2z_3+z_3z_1+z_1z_2)\ \ldots$(i)

z_1, z_2, z_3 are roots of $x^3+3px^2+3qx+r=0$

$\therefore z_1+z_2+z_3=-3p$

and $z_2z_3+z_3z_1+z_1z_2=3q$

$\therefore$ Eq. (i) $\Rightarrow (-3p)^2=3(3q)\Rightarrow p^2=q$

13. (b) Given expression is

$$\sum_{k=2}^{n}(k-1)(k-\omega)(k-\omega^2)$$

$$=\sum_{k=2}^{n}(k-1)(k^2+k+1)=\sum_{k=2}^{n}(k^3-1)$$

$$=(2^3+3^3+\ldots+n^3)-(n-1)$$

$$=\left(\sum n^3-1^3\right)-n+1=\left(\frac{n(n+1)}{2}\right)^2-n$$

14. (a) If $\log_{\frac{1}{2}}\frac{|z|^2+2|z|+4}{2|z|^2+1}<0=\log_{\frac{1}{2}}1$

$$\Rightarrow \frac{|z|^2+2|z|+4}{2|z|^2+1}>\left(\frac{1}{2}\right)^0=1$$

or $|z|^2+2|z|+4>2|z|^2+1$

or $|z|^2-2|z|-3<0$

or $(|z|+1)(|z|-3)<0 \qquad \therefore\ |z|-3<0$

15. (c) $\frac{z-z_1}{z-z_2}=\frac{(x+iy)-(8+4i)}{(x+iy)-(6+4i)}$

$$=\frac{(x-8)+i(y-4)}{(x-6)+i(y-4)}$$

We have, $\arg\left(\frac{z-z_1}{z-z_2}\right)=\frac{\pi}{4}$

$$\Rightarrow \arg(z-z_1)-\arg(z-z_2)=\frac{\pi}{4}$$

$$\Rightarrow \tan^{-1}\frac{y-4}{x-8}-\tan^{-1}\frac{y-4}{x-6}=\frac{\pi}{4}$$

$$\Rightarrow \frac{2(y-4)}{x^2+y^2-14x-8y+64}=1$$

$$\Rightarrow x^2+y^2-14x-10y+72=0$$

$$\Rightarrow (x-7)^2+(y-5)^2=\left(\sqrt{2}\right)^2$$

$$\Rightarrow |z-(7+5i)|=\sqrt{2}$$

16. (a) Consider numerator, $|1-\bar{z}_1z_2|^2-|z_1-z_2|^2$

$$=(1-\bar{z}_1z_2)\overline{(1-\bar{z}_1z_2)}-(z_1-z_2)\overline{(z_1-z_2)}$$

$$(\because z\bar{z}=|z|^2)$$

$$=(1-\bar{z}_1z_2)\left(1-\overline{\bar{z}_1z_2}\right)-(z_1-z_2)(\bar{z}_1-\bar{z}_2)$$

$$\left(\because \overline{z_1-z_2}=\bar{z}_1-\bar{z}_2 \text{ and } \bar{1}=1\right)$$

$$=(1-\bar{z}_1z_2)(1-z_1\bar{z}_2)-(z_1-z_2)(\bar{z}_1-\bar{z}_2)$$

$$=1-z_1\bar{z}_2-\bar{z}_1z_2+z_1\bar{z}_1z_2\bar{z}_2-z_1\bar{z}_1+z_1\bar{z}_2$$

$$+\bar{z}_1z_2-z_2\bar{z}_2$$

$$=1+|z_1|^2|z_2|^2-|z_1|^2-|z_2|^2=\left(1-|z_1|^2\right)\left(1-|z_2|^2\right)$$

$$\frac{\left\{|1-\bar{z}_1z_2|^2-|z_1-z_2|^2\right\}}{\left\{\left(1-|z_1|^2\right)\left(1-|z_2|^2\right)\right\}}=1$$

17. (a) We know, $\cos z=\dfrac{e^{iz}+e^{-iz}}{2}$

$$=\frac{e^{i^2\log_e(2-\sqrt{3})}+e^{-i^2\log_e(2-\sqrt{3})}}{2}$$

$$=\frac{e^{\log_e(2-\sqrt{3})^{-1}}+e^{\log_e(2-\sqrt{3})}}{2}$$

$$=\frac{(2-\sqrt{3})^{-1}+(2-\sqrt{3})}{2}$$

$$=\frac{1}{2}\left[\frac{1}{2-\sqrt{3}}+\left(2-\sqrt{3}\right)\right]$$

$$=\frac{1}{2}\left[\frac{2+\sqrt{3}}{4-3}+\left(2-\sqrt{3}\right)\right]$$

$$=\frac{1}{2}\left[2+\sqrt{3}+2-\sqrt{3}\right]=2$$

18. (c) Given, $|z|=2$

Let $\alpha=5z-1 \Rightarrow |\alpha+1|=5|z|=5(2)=10$

$\Rightarrow |\alpha+1|=10$

$\therefore$ Locus of α, i.e., $5z-1$ is the circle having centre at -1 and radius 10. $\therefore a=10$.

Again, $z_1^2+z_2^2-2z_1z_2\cos\theta=0$

$$\Rightarrow \left(\frac{z_1}{z_2}\right)^2-2\frac{z_1}{z_2}\cos\theta+1=0$$

$$\Rightarrow \frac{z_1}{z_2}=\frac{2\cos\theta\pm\sqrt{4\cos^2\theta-4}}{2}=\cos\theta\pm i\sin\theta$$

$$\Rightarrow \left|\frac{z_1}{z_2}\right|=|\cos\theta\pm i\sin\theta|=1=\frac{a}{10}$$

$$\therefore |z_1|:|z_2|=a:10$$

19. (d) Slope of $\alpha\bar{z}+\bar{\alpha}z+1=0=0$ is

$$-\frac{\text{Coeff. of } \bar{z}}{\text{Coeff. of } z}=-\frac{\alpha}{\bar{\alpha}}$$

Slope of $\beta\bar{z}+\bar{\beta}z-1=0$ is $-\dfrac{\text{Coeff. of } \bar{z}}{\text{Coeff. of } z}=-\dfrac{\beta}{\bar{\beta}}$

The lines are perpendicular

$$\therefore \left(-\frac{\alpha}{\bar{\alpha}}\right)+\left(-\frac{\beta}{\bar{\beta}}\right)=0$$

$$\Rightarrow \alpha\bar{\beta}+\bar{\alpha}\beta=0$$

20. (d) $\tan\left[i\ln\left(\dfrac{a-ib}{a+ib}\right)\right]$

Let $a+ib=re^{i\theta}$, $r^2=a^2+b^2$

$$\Rightarrow a+ib=re^{i\theta},\ \tan\theta=\frac{b}{a}$$

$$\Rightarrow \frac{a-ib}{a+ib}=e^{-2i\theta}$$

$$iln\left(\frac{a-ib}{a+ib}\right)=iln\left(e^{-2i\theta}\right)=2\theta$$

$$\Rightarrow \tan[iln\left(\frac{a-ib}{a+ib}\right)]=\tan 2\theta$$

$$=\frac{2\tan 2\theta}{1-\tan^2\theta}=\frac{2b/a}{1-\dfrac{b^2}{a^2}}=\frac{2ab}{a^2-b^2}$$

21. (c) Let $z = x + iy$, then

$$z^2 + \bar{z}^2 = 2 \Rightarrow (x+iy)^2 + (x-iy)^2 = 2$$

$\Rightarrow x^2 - y^2 = 1$ which represent a hyperbola.

22. (d) We have, $|z-1| + |z+1| \le 4$

$$\Rightarrow \sqrt{(x-1)^2 + y^2} + \sqrt{(x+1)^2 + y^2} \le 4$$

where $z = x + iy$

$$\Rightarrow P^2 + Q^2 \le 4 \quad \ldots(i)$$

where, $P = \sqrt{(x-1)^2 + y^2}$

and $Q = \sqrt{(x+1)^2 + y^2}$

But $\left[(x-1)^2 + y^2\right] - \left[(x+1)^2 + y^2\right] = -4x$

i.e., $P^2 - Q^2 = -4x \quad \ldots(ii)$

Dividing eq. (ii) by (i), we get

$$\Rightarrow 2\sqrt{(x-1)^2 + y^2} \le 4 - x$$

$\Rightarrow 3x^2 + 4y^2 \le 12$ [squaring and simplifying]

or $\frac{x^2}{4} + \frac{y^2}{3} \le 1$

23. (c) Let A, B, C be the vertices of the equilateral triangle represented by the complex numbers z_1, z_2 and z_3 respectively.

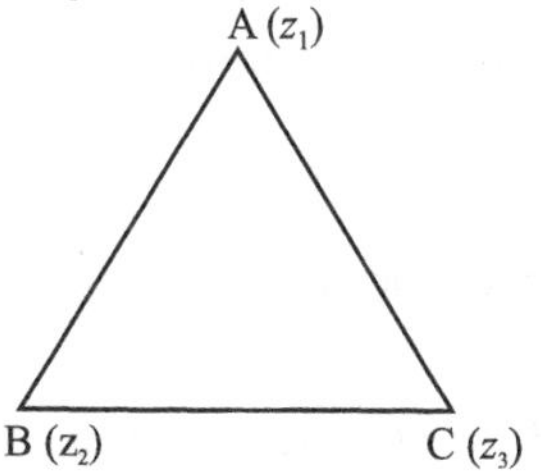

Since, ΔABC is equilateral.

$\therefore AB = BC = CA$

$\therefore |z_1 - z_2| = |z_2 - z_3| = |z_3 - z_1|$

$\therefore \left|\frac{z_1 - z_2}{z_3 - z_2}\right| = \left|\frac{z_2 - z_3}{z_1 - z_3}\right|$ [Each ratio = 1] ...(i)

Also, $\angle CBA = \angle ACB$

$\therefore \arg\left(\frac{z_1 - z_2}{z_3 - z_2}\right) = \arg\left(\frac{z_2 - z_3}{z_1 - z_3}\right) \quad \ldots(ii)$

From Eqs. (i) and (ii), we get $\frac{z_1 - z_2}{z_3 - z_2} = \frac{z_2 - z_3}{z_1 - z_3}$

or

$$z_1^2 - z_1z_2 - z_1z_3 + z_2z_3 = z_3z_2 - z_2^2 - z_3^2 + z_2z_3$$

or $z_1^2 + z_2^2 + z_3^2 - z_1z_2 - z_2z_3 - z_3z_1 = 0$

24. (d) $t_n = (n+1)\left(n + \frac{1}{\omega}\right)\left(n + \frac{1}{\omega^2}\right)$

$$= n^3 + n^2\left(\frac{1}{\omega^2} + \frac{1}{\omega} + 1\right) + n\left(1 + \frac{1}{\omega^2} + \frac{1}{\omega}\right) + 1$$

$$= n^3 + n^2\left(\omega + \omega^2 + 1\right) + n\left(\omega + \omega^2 + 1\right) + 1$$

$$= n^3 + 1$$

$$\therefore \quad s_n = \sum_{r=1}^{n} t_r = \sum_{r=1}^{n} (r^3 + 1) = \frac{n^2(n+1)^2}{4} + n$$

$S_{10} = 3025 + 10 = 3035$

$S_{20} = 44100 + 20 = 44120$

25. (d) $z^2 + az + a^2 = 0$

$\Rightarrow z = a\omega, a\omega^2$

(where 'ω' is non–real roots of unity)

$\Rightarrow$ locus of z is a pair of straight lines and $\arg(z) = \arg(a) + \arg(\omega)$ or $\arg(a) + \arg(\omega^2)$

$\Rightarrow \arg(z) = \pm\frac{2\pi}{3}$

Also, $|z| = |a||\omega|$ or $|a||\omega^2| \Rightarrow |z| = |a|$

All the three statements are correct.

26. (c) Given equations are $|z| - 4 = 0$

and $|z - i| - |z + 5i| = 0$

Let $z = x + iy$, these equations becomes

$|x + iy| = 4$ i.e., $x^2 + y^2 = 16 \quad \ldots(i)$

and $|x + iy - i| = |x + iy + 5i|$

or $x^2 + (y-1)^2 = x^2 + (y+5)^2$ i.e., $y = -2 \quad \ldots(ii)$

Putting $y = -2$ in Eq. (i), $x^2 + 4 = 16$

or $x = \pm 2\sqrt{3}$

hence, the complex numbers z satisfy the given equations are

$z_1 = (2\sqrt{3}, -2)$ and $z_2 = (-2\sqrt{3}, -2)$

that is, $z_1 = 2\sqrt{3} - 2i, z_2 = -2\sqrt{3} - 2i$

27. (b) $2\cos\theta = x + \frac{1}{x}, 2\cos\phi = y + \frac{1}{y}$

$\Rightarrow x^2 - 2x\cos\theta + 1 = 0$

$\Rightarrow x = \frac{2\cos\theta \pm \sqrt{4\cos^2\theta - 4}}{2}$

$\Rightarrow x = \cos\theta \pm i\sin\theta = e^{\pm i\theta}$

Similary $y = e^{\pm i\phi} \Rightarrow \frac{x}{y} + \frac{y}{x} = 2\cos(\theta + \phi)$

$x^m y^n + \frac{1}{x^m y^n} = 2\cos(m\theta + n\phi)$

$\frac{x^m}{y^n} + \frac{y^n}{x^m} = 2\cos(m\theta - n\phi)$

$xy + \frac{1}{xy} = 2\cos(\theta + \phi)$

28. (c) We have,

$z_k = 1 + a + a^2 + \ldots + a^k = \frac{1 - a^{k+1}}{1 - a}$

$\Rightarrow -z_k + \frac{1}{1-a} = \frac{a^{k+1}}{1-a}$

$\Rightarrow \left|z_k - \frac{1}{1-a}\right| = \frac{|a|^{k+1}}{|1-a|} < \frac{1}{|1-a|}$ (since $|a| < 1$)

$\therefore$ Vertices of the polygon $z_1, z_2, \ldots, z_n$ lie within the circle $\left|z - \frac{1}{1-a}\right| = \frac{1}{|1-a|}$

29. (a) Let $z = x + iy$

$\therefore |z| - z = 1 + 2i$

$\Rightarrow \sqrt{x^2 + y^2} - (x + iy) = 1 + 2i$

$\Rightarrow \left(\sqrt{x^2 + y^2} - x\right) + i(-y) = 1 + 2i$

$\Rightarrow \sqrt{x^2 + y^2} - x = 1$ and $y = -2$

If $y = -2, \sqrt{x^2 + 4} - x = 1$

$\Rightarrow x^2 + 4 = (1 + x)^2$

$\Rightarrow 2x = 3; x = \frac{3}{2} \therefore z = x + iy = \frac{3}{2} - 2i$

So $2|x - y| = 2|3/2 + 2| = 7$

30. (d) $|z_1| = |z_2|$

$Re(z_1) > 0, Im(z_2) < 0$

$Re\left(\frac{z_1 + z_2}{z_1 - z_2}\right) = \frac{1}{2}\left(\frac{z_1 + z_2}{z_1 - z_2} + \frac{\bar{z}_1 + \bar{z}_2}{\bar{z}_1 - \bar{z}_2}\right)$

$= \frac{1}{2}\left(\frac{(z_1 + z_2)(\bar{z}_1 - \bar{z}_2) + (\bar{z}_1 + \bar{z}_2)(z_1 - z_2)}{(z_1 - z_2)(\bar{z}_1 - \bar{z}_2)}\right)$

$$= \frac{1}{2}\left(\frac{z_1\bar{z}_1 - z_1\bar{z}_2 + z_2\bar{z}_1 - z_2\bar{z}_2 + z_1\bar{z}_1 + z_1\bar{z}_2 - z_2\bar{z}_1 - z_2\bar{z}_2}{|z_1 - z_2|^2}\right)$$

$= \frac{1}{2}\left(\frac{2|z_1|^2 - 2|z_2|^2}{|z_1 - z_2|^2}\right) = 0$

$\Rightarrow \left(\frac{z_1 + z_2}{z_1 - z_2}\right)$ is purely imaginary

31. (b) $\left|\frac{z_1 z - z_2}{z_1 z + z_2}\right| = k$

$\Rightarrow \left|\frac{z - \frac{z_2}{z_1}}{z + \frac{z_2}{z_1}}\right| = k$

Clearly, if $k \neq 0, 1$, then z would lie on a circle. If, $k = 1, z$ would lie on a perpendicular bisector of the line segment joining $\frac{z_2}{z_1}$ and $-\frac{z_2}{z_1}$. If $k = 0, z$ represents a point.

32. (a) Let $z = x + iy$ where $x, y \in R$

$\therefore \alpha + i\beta = \tan^{-1}(x + iy)$

$\Rightarrow \alpha - i\beta = \tan^{-1}(x - iy)$

$\Rightarrow 2\alpha = \tan^{-1}(x + iy) + \tan^{-1}(x - iy)$

$= \tan^{-1}\frac{2x}{1 - x^2 - y^2}$

$\therefore x^2 + y^2 + 2x\cot 2\alpha = 1$

33. (b) Let $a+ib=re^{i\theta}$ then $a-ib=re^{-i\theta}$

$$\therefore i\log\frac{a-ib}{a+ib}=i\log\left(e^{-2\theta i}\right)$$

$$=i(-2\theta i)\log e=2\theta$$

$$\therefore \cos\left(i\log\frac{a-ib}{a+ib}\right)=\cos 2\theta$$

$$=\frac{1-\tan^2\theta}{1+\tan^2\theta}=\frac{1-\frac{b^2}{a^2}}{1+\frac{b^2}{a^2}}=\frac{a^2-b^2}{a^2+b^2}$$

34. (a) Given that, $z_1\left(z_1^2-3z_2^2\right)=2$ ...(i)

$z_2\left(3z_1^2-z_2^2\right)=11$ ...(ii)

Now multiply equation (ii) by i and add it to equation (i), we get

$z_1^3-3z_2^2z_1+i\left(3z_1^2z_2-z_2^3\right)=2+11i$

$\Rightarrow (z_1-iz_2)^3=2+11i$ ...(iii)

Next, multiply equation (ii) by i and subtract it from equation (i), we get

$z_1^3-3z_2^2z_1-i\left(3z_1^2z_2-z_2^3\right)=2-11i$

$\Rightarrow \left(z_1-iz_2^2\right)^3=2-11i$ ...(iv)

Then multiply the results of equations (iii) and (iv), we get

$\left(z_1^2+z_2^2\right)^3=4+121$

$\Rightarrow z_1^2+z_2^2=5$

35. (c) Here, $\sqrt{1-c^2}=nc-1$

$\Rightarrow 1-c^2=n^2c^2-2nc+1$

$\therefore \frac{c}{2n}=\frac{1}{1+n^2}$ ...(i)

or $\frac{c}{2n}(1+nz)\left(1+\frac{n}{z}\right)=\frac{1}{1+n^2}\left\{1+n^2+n\left(z+\frac{1}{z}\right)\right\}$

$$=\frac{1}{1+n^2}\{1+n^2+n\cdot(2\cos\theta)\}$$

$$=\frac{\left(\left(1+n^2\right)+2n\cos\theta\right)}{\left(1+n^2\right)}$$

$$=1+\left(\frac{2n}{1+n^2}\right)\cos\theta \quad \text{[using Eq. (i)]}$$

$$=1+c\cos\theta$$

36. (c) We have,

$$|(x-2)+i(y-1)|=|z|\left|\frac{1}{\sqrt{2}}\cos\theta-\frac{1}{\sqrt{2}}\sin\theta\right|,$$

where

$$\theta=\arg(z),\sqrt{(x-2)^2+(y-1)^2}=\frac{1}{\sqrt{2}}|x-y|,$$

it represents distance from a fixed point (2, 1) is equal to distance from a fixed straight line $x-y=0$.

which is a parabola.

37. (c) We have,

$$\frac{1}{a+\omega}+\frac{1}{b+\omega}+\frac{1}{c+\omega}=2\omega^2=\frac{2}{\omega}$$

and $\frac{1}{a+\omega^2}+\frac{1}{b+\omega^2}+\frac{1}{c+\omega^2}=2\omega=\frac{2}{\omega^2}$

$\therefore$ ω and ω^2 are roots of the equation

$$\frac{1}{a+x}+\frac{1}{b+x}+\frac{1}{c+x}=\frac{2}{x}$$

Thus, when $x=1$, we get

$$\frac{1}{a+1}+\frac{1}{b+1}+\frac{1}{c+1}=\frac{2}{1}=2$$

38. (b) By definition,

$z_1=x_1+iy_1;\ z_2=x_2+iy_2;\ z_3=x_3+iy_3$

are collinear, if $\begin{vmatrix}x_1 & y_1 & 1\\ x_2 & y_2 & 1\\ x_3 & y_3 & 1\end{vmatrix}=0$

$$\Rightarrow \begin{vmatrix}a & b & 1\\ a' & b' & 1\\ a-a' & b-b' & 1\end{vmatrix}=0$$

$\Rightarrow ab'=a'b$

39. (a)Let $x^2+y^2=1$ where $z_1=i, z_2=1, z_3=-1$,

$\therefore z=-i$

$\Rightarrow zz_1+z_2z_3=1-1=0$

We will discuss more about this in "Concept Deviator".

40. (c) Given $\frac{z_2 - z_3}{z_1 - z_3} = \frac{1 - i\sqrt{3}}{2} = \frac{1}{2} - i\frac{\sqrt{3}}{2}$

$= \cos\left(-\frac{\pi}{3}\right) + i\sin\left(-\frac{\pi}{3}\right)$

$\Rightarrow \frac{z_2 - z_3}{z_1 - z_3} = \cos\frac{\pi}{3} + i\sin\frac{\pi}{3}$

$\Rightarrow \left|\frac{z_2 - z_3}{z_1 - z_3}\right| = 1$ and $\arg\left(\frac{z_2 - z_3}{z_1 - z_3}\right) = \frac{\pi}{3}$

If z_1, z_2, z_3 represent vertices A, B and C respectively of ΔABC,

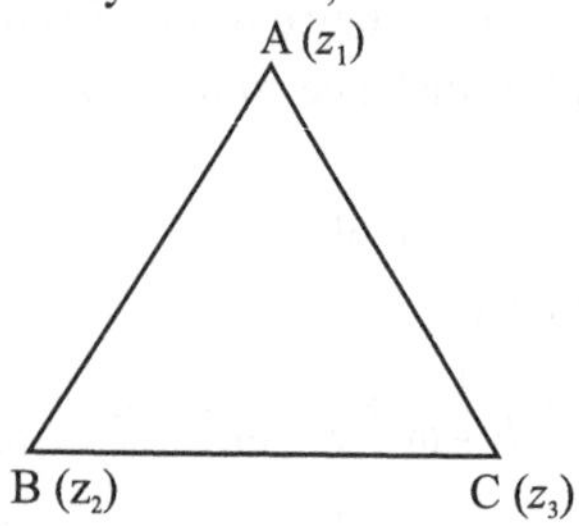

Then $\left|\frac{z_2 - z_3}{z_1 - z_3}\right| = 1$

$\Rightarrow |z_2 - z_3| = |z_1 - z_3|$

$\Rightarrow BC = AC$.

Also $\arg\left(\frac{z_2 - z_3}{z_1 - z_3}\right) = \frac{\pi}{3} \Rightarrow \angle BCA = \frac{\pi}{3}$

$\therefore$ ΔABC is equilateral.

41. (a) Given that $|\bar{z}\omega| = |\bar{z}||\omega| = |z||\omega| = |z\omega| = 1$

$\text{Arg}(\bar{z}\omega) = \arg(\bar{z}) + \arg(\omega)$

$= -\arg(z) + \arg\omega = -\frac{\pi}{2}$

$\therefore\ \bar{z}\omega = -i$

42. (d) Given equation is

$z^2 + az + b = 0; z_1 + z_2 = -a \ \&\ z_1 z_2 = b$

We know that for an equilateral triangle

$z_1^2 + z_2^2 + z_3^2 = z_1 z_2 + z_2 z_3 + z_3 z_1$

0, z_1, z_2 form an equilateral Δ

$\therefore\ 0^2 + z_1^2 + z_2^2 = 0.z_1 + z_1.z_2 + z_2.0$

$\Rightarrow z_1^2 + z_2^2 = z_1 z_2 \Rightarrow (z_1 + z_2)^2 = 3z_1 z_2$

$\therefore\ a^2 = 3b.$

43. (b) Given $\left(\frac{1+i}{1-i}\right)^x = 1$

$\Rightarrow \left[\frac{(1+i)^2}{(1+i)(1-i)}\right]^x = 1$

$\left(\frac{1 + i^2 + 2i}{1+1}\right)^2 = 1$

$\Rightarrow (i)^x = 1;$

$\therefore$ x = 4n; n $\in Z^+$

44. (c) arg zw = π

$\Rightarrow$ arg z + arg w = π ...(1)

$\bar{z} + i\bar{w} = 0$

$\Rightarrow \bar{z} = -i\bar{w}$

$\therefore\ z = iw$

$\Rightarrow$ arg z = $\frac{\pi}{2}$ + arg w

$\Rightarrow$ arg z = $\frac{\pi}{2} + \pi$ − arg z (from (1))

$\therefore$ arg z = $\frac{3\pi}{4}$

45. (a) $z^{\frac{1}{3}} = p + iq$

$\Rightarrow z = p^3 + (iq)^3 + 3p(iq)(p + iq)$

$\Rightarrow x - iy = p^3 - 3pq^2 + i(3p^2 q - q^3)$

$\therefore$ x = $p^3 - 3pq^2 \Rightarrow \frac{x}{p} = p^2 - 3q^2$

y = $q^3 - 3p^2 q \Rightarrow \frac{y}{q} = q^2 - 3p^2$

$\therefore\ \frac{x}{p} + \frac{y}{q} = -2p^2 - 2q^2$

$\therefore\ \frac{\left(\frac{x}{p} + \frac{y}{q}\right)}{(p^2 + q^2)} = -2$

46. (b) From the given equation $|z^2 - 1| = |z|^2 + 1$

$\Rightarrow |z^2 - 1|^2 = (z\bar{z} + 1)^2$

$\Rightarrow (z^2 - 1)(\bar{z}^2 - 1) = (z\bar{z} + 1)^2$

$\Rightarrow z^2\bar{z}^2 - z^2 - \bar{z}^2 + 1 = z^2\bar{z}^2 + 2z\bar{z} + 1$

$\Rightarrow z^2 + 2z\bar{z} + \bar{z}^2 = 0$

$\Rightarrow (z + \bar{z})^2 = 0$

$\Rightarrow z = -\bar{z}$

⇒ z is purely imaginary

47. (c) Given $|z_1 + z_2| = |z_1| + |z_2|$

⇒ z_1 and z_2 are collinear and are to the same side of origin; hence arg z_1 − arg z_2 = 0.

48. (c) As given $w = \dfrac{z}{z - \frac{1}{3}i}$

$\Rightarrow |w| = \dfrac{|z|}{\left|z - \frac{1}{3}i\right|} = 1$

$\Rightarrow |z| = \left|z - \frac{1}{3}i\right|$

⇒ z is equidistant from origin and from $\left(0, \frac{1}{3}\right)$

hence z lies on bisector of the line joining points, (0, 0) and (0, 1/3).

Hence z lies on a straight line.

49. (d) $\sum_{k=1}^{10}\left(\sin\frac{2k\pi}{11} + i\cos\frac{2k\pi}{11}\right)$

$= i\sum_{k=1}^{10}\left(\cos\frac{2k\pi}{11} - i\sin\frac{2k\pi}{11}\right)$

$= i\sum_{k=1}^{10} e^{-\frac{2k\pi}{11}} = i\left\{\sum_{k=0}^{10} e^{-\frac{2k\pi}{11}} - 1\right\}$

$= i\left[1 + e^{-\frac{2\pi}{11}i} + e^{-\frac{4\pi}{11}i} + \ldots 11 \text{ terms}\right] - i$

$= i\left[\dfrac{1 - \left(e^{-\frac{2\pi}{11}}\right)^{11}}{1 - e^{-\frac{2\pi}{11}i}}\right] - i = i\left[\dfrac{1 - e^{-2\pi i}}{1 - e^{-\frac{2\pi}{11}i}}\right] - i$

$= i \times 0 - i \qquad \left[\because e^{-2\pi i} = 1\right]$

$= -i$

50. (d) $z^2 + z + 1 = 0$

$\Rightarrow z = \omega$ or ω^2

So, $z + \dfrac{1}{z} = \omega + \omega^2 = -1$

$z^2 + \dfrac{1}{z^2} = \omega^2 + \omega = -1$,

$z^3 + \dfrac{1}{z^3} = \omega^3 + \omega^3 = 2$

$z^4 + \dfrac{1}{z^4} = -1,\ z^5 + \dfrac{1}{z^5} = -1,\ z^6 + \dfrac{1}{z^6} = 2$

∴ The given sum = 1 + 1 + 4 + 1+ 1+ 4 = 12

51. Since $|z + 4| \le 3$, so z lies on or inside the circle with centre (−4, 0) and radius 3 units.

Q. 51

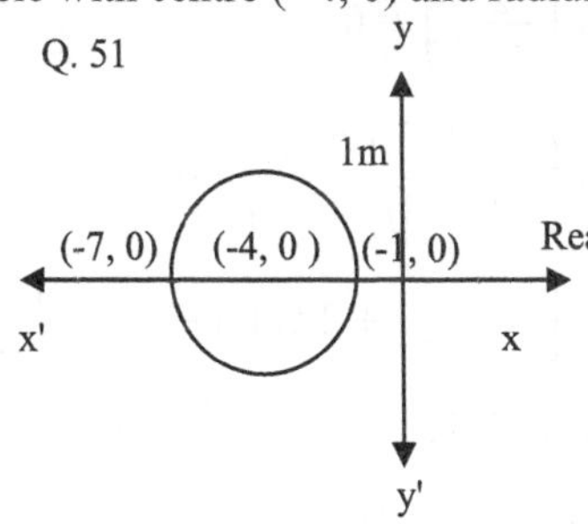

From the above diagram maximum value of $|z + 1|$ is 6

ALTERATION SOLUTION:

$|z + 1| = |z + 4 - 3| \le |z + 4| + |-3| \le |3| + |-3|$

$\Rightarrow |z + 1| \le 6$

$\Rightarrow |z + 1|_{max} = 6$

52. (c) $\left(\dfrac{1}{i - 1}\right) = \dfrac{1}{-1 - i} = \dfrac{-1}{i + 1}$

53. (a) Let z = $x + i$

$|z - 1| = |z + 1| (x - 1)^2 + y^2 = (x + 1)^2 + y^2$

⇒ Re z = 0

⇒ x = 0

$|z - 1| = |z - i| (x - 1)^2 + y^2 = x^2 + (y - 1)^2$

⇒ x = y

$|z + 1| = |z - i| (x + 1)^2 + y^2 = x^2 + (y - 1)^2$

Only (0, 0) will satisfy all conditions.

⇒ Number of complex number z = 1

54. (c) Given that real part of roots is 1

Let roots are 1 + $pi, 1 + q$

∴ Sum of roots.

= 1 + $pi + 1 + qi = -\alpha$ which is real

$\Rightarrow q = -p$ or root are

$= 1 + pi$ and $1 - pi$ product of roots

$= 1 + p^2 = \beta \in (1, \infty)$

$p \neq 0$ as roots are distinct.

55. (a) Let $z = x + iy$

$\therefore\ z^2 = x^2 - y^2 + 2ixy$

Now $\frac{z^2}{z-1}$ = is real

$\Rightarrow \text{Im}\left(\frac{z^2}{z-1}\right) = 0$

$\Rightarrow \text{Im}\left(\frac{x^2 - y^2 + 2ixy}{(x-1)+iy}\right) = 0$

$\Rightarrow \text{Im}\left[\left(x^2 - y^2 + 2ixy\right)(x-1) - iy\right] = 0$

$\Rightarrow 2xy(x-1) - y(x^2 - y^2) = 0$

$\Rightarrow y(x^2 - y^2 - 2x) = 0$

$\Rightarrow y = 0;\ x^2 + y^2 - 2x = 0$

$\therefore$ z lies either on real axis or on a circle through origin.

56. (c) Given $|z| = 1$, $\arg z = \theta$

As we know, $\bar{z} = \frac{1}{z}$

$\therefore \arg\left(\frac{1+z}{1+\bar{z}}\right) = \arg\left(\frac{1+z}{1+\frac{1}{z}}\right) = \arg(z) = \theta.$

Board Level Questions

Level 1

1. (b) The closest distance = length of the perpendicular from the origin on the line $a\bar{z} + \bar{a}z + a\bar{a} = 0$

$= \frac{a(0) + \bar{a}|0| + a\bar{a}}{2|a|} = \frac{|a|^2}{2|a|} = \frac{|a|}{2}$

2. (a) $|z - z_1| = |z - z_2| = |z - z_3|$

$\therefore$ z must be the mid–point of z_1 and z_2 and since : $|z - z_1| = |z - z_2|$ and z, z_1, z_2 are collinear.

$\Rightarrow$ z is circumcentre of Δ formed by z_1, z_2 and z_3

$\therefore \arg\left(\frac{z_3 - z_1}{z_3 - z_2}\right) = \pm\frac{\pi}{2}$ (neglecting –ve)

3. (c) We have, $3x^2 + px + 3 = 0, p > 0$

Let the roots be α, α^2

Sum of the roots $\alpha + \alpha^2 = -\frac{p}{3}$ and

Product of the roots $\alpha \cdot \alpha^2 = \frac{3}{3} = 1$

$\Rightarrow \alpha = 1, \omega, \omega^2$

$\therefore\ -\frac{p}{3} = 1 + 1$

$\Rightarrow p = -6$ (neglecting) or $p = -3\left(\omega + \omega^2\right)$

$\Rightarrow p = 3$

4. (b) Since z_1, z_2, z_3 are vertices of an equilateral triangle

$z_1^2 + z_2^2 + z_3^2 = z_2z_3 + z_3z_1 + z_1z_2$ ...(i)

Now, $(z_1 + z_2 + z_3)^2 = \sum z_1^2 + 2\sum z_2z_3$

$= \sum z_1^2 + 2\sum z_1^2 = 3\sum z_1^2$

Since, the triangle is equilateral, so the centroid and circumcentre coincide.

$\therefore z_0 = \frac{z_1 + z_2 + z_3}{3}$

$\therefore (z_1 + z_2 + z_3)^2 = 9z_0^2$

$\therefore 3\sum z_1^2 = 9z_0^2$ or $\sum z_1^2 = 3z_0^2$

5. (a) We have, $\frac{(\cos 75° + i\sin 75°)}{(\cos 30° + i\sin 30°)}$

$= \frac{\{\cos(75° - 30°) + i\sin(75° - 30°)\}}{\cos^2 30° + \sin^2 30°}$

$= (\cos 45° + i\sin 45°) = \frac{1}{\sqrt{2}}(1 + i)$

6. (d) Using de Moivre's theorem, the given expression

$= \frac{(\cos\theta + i\sin\theta)^{-14}(\cos\theta + i\sin\theta)^{-15}}{(\cos\theta + i\sin\theta)^{48}(\cos\theta + i\sin\theta)^{-30}}$

$= (\cos\theta + i\sin\theta)^{-47} = \cos 47\theta - i\sin 47\theta$

7. (a) We have, $|z-1|^2+|z+1|^2=4$

$\Rightarrow |(x-1)+iy|^2+|(x+1)+iy|^2=4$

$\Rightarrow (x-1)^2+y^2+(x+1)^2+y^2=4$

$\Rightarrow 2(x^2+1)+2y^2=4$

$\therefore$ The locus of P is $x^2+y^2=1$.

8. (d) We have, $\log_{1/2}\left(\frac{|z-1|+4}{3|z-1|-2}\right)>1$

$=\log_{1/2}\left(\frac{1}{2}\right)$

$\Rightarrow \frac{|z-1|+4}{3|z-1|-2}<\frac{1}{2}$ [since $\log_a x$ is a decreasing function if $a<1$]

$\Rightarrow 2|z-1|+8<3|z-1|-2$

$\Rightarrow |z-1|>10$

This is necessary but not sufficient condition.

9. (a) $z=(-i\omega)^5+(-i\omega^2)^5=i$

10. (d) We have, $\alpha=\omega$ and $\beta=\omega^2$

$(a+b+c)(a\omega+b\omega^2+c)(a\omega^2+b\omega+c)$

$=(a+b+c)(a^2+b^2+c^2-ab-bc-ca)$

$=a^3+b^3+c^3-3abc$

11. (b) Consider

$(a+b)+(a\omega+b\omega^2)+(a\omega^2+b\omega)$

$=a(1+\omega+\omega^2)+b(1+\omega+\omega^2)$

$=a\times 0+b\times 0=0$

Hence, from the relation $x^3+y^3+z^3-3xyz$

$=(x+y+z)(x^2+y^2+z^2-yz-zx-xy)=0$

So $x^3+y^3+z^3=3xyz$

Hence, $(a+b)^3+(a\omega+b\omega^2)^3+(a\omega^2+b\omega)^3$

$=3(a^3+b^3)$

12. (a) We have, $z=\frac{(1+\cos\theta+i\sin\theta)^5}{(\cos\theta+i\sin\theta)^3}$

$=\frac{\left\{2\cos^2\frac{\theta}{2}+2i\sin\frac{\theta}{2}\cos\frac{\theta}{2}\right\}^5}{\cos 3\theta+i\sin 3\theta}$

$=\frac{32\cos^5\frac{\theta}{2}\left\{\cos\frac{\theta}{2}+i\sin\frac{\theta}{2}\right\}^5}{\cos 3\theta+i\sin 3\theta}$

$=32\cos^5\left(\frac{\theta}{2}\right)\left\{\cos\frac{5\theta}{2}+i\sin\frac{5\theta}{2}\right\}\{\cos 3\theta-i\sin 3\theta\}$

$=32\cos^5\frac{\theta}{2}\left[\cos\left\{\frac{5\theta}{2}-3\theta\right\}+i\sin\left\{\frac{5\theta}{2}-3\theta\right\}\right]$

$=32\cos^5\frac{\theta}{2}\left[\cos\left(-\frac{\theta}{2}\right)+i\sin\left(-\frac{\theta}{2}\right)\right]$

$\text{Arg}(z)=-\frac{\theta}{2}, |z|=32\cos^5\frac{\theta}{2}$

13. (c) We have, $z^3+2z^2+2z+1=0$

$\Rightarrow (z+1)(z^2+z+1)=0$

Its roots are -1, ω and ω^2. Now substitute these values in the given equation

The root $z=-1$ does not satisfy the equation $z^{1985}+z^{100}+1=0$ but $z=\omega$ and $z=\omega^2$ satisfy it. Hence, ω and ω^2 are the common roots.

14. (a) We have,

$\frac{1}{z}=\frac{1}{\cos\theta+i\sin\theta}=\cos\theta-i\sin\theta$

$\therefore z^n=(\cos\theta+i\sin\theta)^n=\cos n\theta+i\sin n\theta$,

and $\frac{1}{z^n}=(\cos\theta-i\sin\theta)^n=\cos n\theta-i\sin n\theta$

Hence, $z^n+\frac{1}{z^n}=2\cos n\theta$ and

$z^n-\frac{1}{z^n}=2i\sin n\theta$

15. (b) We know that,

$(a-b)^2\ge 0, a^2+b^2\ge 2ab$ ...(i)

but $|z|=\sqrt{a^2+b^2}$; so from eq. (i), $|z|^2\ge 2ab$

$\therefore |z|^2+a^2+b^2\ge a^2+b^2+2ab$

or $|z|^2 + |z|^2 \geq (a+b)^2$; $\quad \therefore \ 2|z|^2 \geq (a+b)^2$

$\therefore \ \sqrt{2}|z| \geq a+b$ as $|z|$ is positive

$\therefore \ |z| \geq \frac{1}{\sqrt{2}}(a+b)$

16. (b) We have,

$$\frac{z^{2n}-1}{z^{2n}+1} = \frac{(\cos\theta + i\sin\theta)^{2n}-1}{(\cos\theta + i\sin\theta)^{2n}+1}$$

$$= \frac{\cos 2n\theta + i\sin 2n\theta - 1}{\cos 2n\theta + i\sin 2n\theta + 1}$$

(Using de Moivre's Theorem)

$$= \frac{(1-2\sin^2 n\theta) + 2i\sin n\theta\cos n\theta - 1}{(2\cos^2 n\theta - 1) + 2i\sin n\theta \cos n\theta + 1}$$

$$= \frac{i\sin n\theta\cos n\theta + i^2\sin^2 n\theta}{\cos^2 n\theta + i\sin n\theta\cos n\theta}\left(\text{since } i^2 = -1\right)$$

$$= \frac{i\sin n\theta(\cos n\theta + i\sin n\theta)}{\cos n\theta(\cos n\theta + i\sin n\theta)} = i\tan n\theta$$

So put n= 10 the result is $i \tan 10\theta$

17. (b) Given,

$$z_1 z_2 z_3 \ldots \infty = \left(\cos\frac{\pi}{2} + i\sin\frac{\pi}{2}\right)\left(\cos\frac{\pi}{2^2} + i\sin\frac{\pi}{2^2}\right)\left(\cos\frac{\pi}{2^3} + i\sin\frac{\pi}{2^3}\right)\ldots$$

$$= \cos\left(\frac{\pi}{2} + \frac{\pi}{2^2} + \frac{\pi}{2^3} + \ldots\right) + i\sin\left(\frac{\pi}{2} + \frac{\pi}{2^2} + \frac{\pi}{2^3} + \ldots\right) \ldots \text{(i)}$$

Since $\frac{\pi}{2} + \frac{\pi}{2^2} + \frac{\pi}{2^3} + \ldots$ is a GP, whose common ratio is $r = \frac{1}{2}$ and the first term is $a = \frac{\pi}{2}$.

Thus, $\frac{\pi}{2} + \frac{\pi}{2^2} + \frac{\pi}{2^3} + \ldots = \frac{a}{1-r} = \frac{\frac{\pi}{2}}{1-\frac{1}{2}} = \pi$

$\therefore$ From eq. (i),

$$z_1 z_2 z_3 \ldots \infty = \cos\pi + i\sin\pi = -1$$

18. (c) Consider, $\omega^{3p} + \omega^{3p+1} + \omega^{3r+2}$

$$= \omega^{3p} + \omega^{3q}\cdot\omega + \omega^{3r}.\omega^2$$

$$= (\omega^3)^p + (\omega^3)^q\cdot\omega + (\omega^3)^r\cdot\omega^2$$

$$= 1 + \omega + \omega^2 = 0 \qquad \left[\text{since } \omega^3 = 1\right]$$

19. (b) We have,

$$|z| = \left|z + \frac{2}{z} - \frac{2}{z}\right| \leq \left|z + \frac{2}{z}\right| + \frac{2}{|z|}$$

$$\Rightarrow |z| \leq 2 + \frac{2}{|z|}$$

$$\Rightarrow |z|^2 \leq 2|z| + 2$$

$$\Rightarrow |z|^2 - 2|z| + 1 \leq 1 + 2$$

$$\Rightarrow (|z|-1)^2 \leq 3$$

$$\Rightarrow -\sqrt{3} \leq |z| - 1 \leq \sqrt{3}$$

$$\Rightarrow 1-\sqrt{3} \leq |z| \leq 1+\sqrt{3}$$

So maximum value of $|z|$ is $1+\sqrt{3}$

20. (d) Let $z_1 = 1-3i$, $z_2 = 4+3i$ and $z_3 = 3+i$. Then, z_3 divides the line segment joining z_1 and z_2 in the ratio 2 : 1 internally. So, the points z_1, z_2, z_3 are collinear.

Level 2

21. We have, $|z_1 + z_2| = \left|\frac{1}{z_1} + \frac{1}{z_2}\right| = \left|\frac{z_1+z_2}{z_1 z_2}\right|$

$$\Rightarrow |z_1 + z_2|\left(1 - \frac{1}{|z_1 z_2|}\right) = 0 \qquad \Rightarrow |z_1 z_2| = 1$$

22. Given Expression

$$= (1+i)^{n_1} + (1-i)^{n_1} + (1+i)^{n_2} + (1-i)^{n_2}$$

$$= 2^{\frac{n_1}{2}}\left(\frac{1}{\sqrt{2}} + i\frac{1}{\sqrt{2}}\right)^{n_1} + 2^{n_1/2}\left(\frac{1}{\sqrt{2}} - i\frac{1}{\sqrt{2}}\right)^{n_1}$$

$$+2^{\frac{n_2}{2}}\left(-\frac{1}{\sqrt{2}} + i\frac{1}{\sqrt{2}}\right)^{n_2} + 2^{n_2/2}\left(\frac{1}{\sqrt{2}} - i\frac{1}{\sqrt{2}}\right)^{n_2}$$

$$= 2^{\frac{n_1}{2}}\left\{\left(\cos\frac{\pi}{4} + i\sin\frac{\pi}{4}\right)^{n_1} + \left(\cos\frac{\pi}{4} - i\sin\frac{\pi}{4}\right)^{n_1}\right\}$$

$$+2^{\frac{n_2}{2}}\left\{\left(\cos\frac{\pi}{4} + i\sin\frac{\pi}{4}\right)^{n_2} + \left(\cos\frac{\pi}{4} - i\sin\frac{\pi}{4}\right)^{n_2}\right\}$$

$$= 2^{\frac{n_1}{2}}\left\{\cos\frac{n_1\pi}{4} + i\sin\frac{n_1\pi}{4} + \cos\frac{n_1\pi}{4} - i\sin\frac{n_1\pi}{4}\right\}$$

$$+2^{\frac{n_2}{2}}\left\{\cos\frac{n_2\pi}{4} + i\sin\frac{n_2\pi}{4} + \cos\frac{n_2\pi}{4} - i\sin\frac{n_2\pi}{4}\right\}$$

$$= 2^{\frac{n_1}{2}}\cdot 2\cos\frac{n_1\pi}{4} + 2^{\frac{n_2}{2}}\cdot 2\cos\frac{n_2\pi}{4}$$

real i.e., n_1 n_2 are any two positive integers

23. (b) If z is a root of $(z-1)^n = 2\omega(z+1)^n$, then

$$\left(\frac{z-1}{z+1}\right)^n = 2\omega \Rightarrow \left|\frac{z-1}{z+1}\right|^n = |2\omega| = 2$$

$$\Rightarrow \left|\frac{z-1}{z+1}\right| = 2^{1/n}$$

Since $2^{1/n} > 1$, we get z lies on a circle.

24. Given that $z = \dfrac{3}{2+\cos\theta + i\sin\theta}$

$$\frac{9}{|z|^{\dagger}} = (2+\cos\theta)^{\dagger} + \sin^{\dagger}\theta = 5 + 4\cos\theta$$

And $\dfrac{3}{z} + \dfrac{3}{\bar{z}} = 4 + 2\cos\theta$

Eliminating θ we get $\dfrac{9}{|z|^{\dagger}} - 6\left(\dfrac{1}{z} + \dfrac{1}{\bar{z}}\right) = -3$

$$|z|^2 - 2(\bar{z} + z) = -3$$

25. Given that $\sqrt{x+iy} = \dfrac{a+ib}{c+id}$

$$\Rightarrow x + iy = \left(\frac{a+ib}{c+id}\right)^2$$

$$\Rightarrow |x+iy| = \left|\frac{(a+ib)^2}{(c+id)^2}\right| = \left|\frac{a+ib}{c+id}\right|^2$$

$$\Rightarrow \sqrt{x^2+y^2} = \frac{a^2+b^2}{c^2+d^2} = 2$$

$$\Rightarrow x^2 + y^2 = 4$$

26. We have $x^4 - 1 = (x-1)(x-x_1)(x-x_2)(x-x_3)$

or $(x-x_1)(x-x_2)(x-x_3) = \dfrac{x^4-1}{x-1}$

Put $x = \omega$ & ω^2, we get

$$(\omega - x_1)(\omega - x_2)(\omega - x_3) = \frac{\omega^4 - 1}{\omega - 1}$$

$$= \frac{\omega - 1}{\omega - 1} = 1$$

and $(\omega^2 - x_1)(\omega^2 - x_2)(\omega^2 - x_3) = \dfrac{\omega^8 - 1}{\omega^2 - 1} = 1$

So, $\dfrac{(\omega^2 - x_1)(\omega^2 - x_2)(\omega^2 - x_3)}{(\omega - x_1)(\omega - x_2)(\omega - x_3)} = 1$

27. Consider

$$|a + bw + cw^2|^2$$

$$= (a + bw + cw^2)(a + c\bar{w} + c\bar{w}^2)$$

$$= (a + bw + cw^2)(a + bw^2 + cw)$$

$$= a^2 + b^2 + c^2 - bc - ca - ab$$

$$= \frac{1}{2}[(b-c)^2 + (c-a)^2 + (a-b)^2] \geq 3$$

As a, b, c are distinct integers the least value of

$$(b-c)^2 + (c-a)^2 + (a-b)^2 \geq 6$$

28. Given that $|z_1 + z_2|^2 = |z_1|^2 + |z_2|^2$

$$\Rightarrow (z_1 + z_2)(\bar{z}_1 + \bar{z}_2) = |z_1|^2 + |z_2|^2$$

$$\Rightarrow |z_1|^2 + z_2\bar{z}_1 + z_1\bar{z}_2 + |z_2|^2 = |z_1|^2 + |z_2|^2$$

$$\Rightarrow z_2\bar{z}_1 + z_1\bar{z}_2 = 0 \Rightarrow = \bar{z}_1 z_2 = -z_1\bar{z}_2$$

So $-z_1\bar{z}_2$ is purely imaginary and statement (i) is correct

$z_1/\bar{z}_2$ is purely imaginary and statement (ii) is incorrect

$z_1\bar{z}_2 + \bar{z}_1 z_2 = 0$ and statement (iii) is correct

$\Rightarrow \dfrac{z_1}{z_2}\,|z_2|^2$ is purely imaginary.

Since $|z_2|^2$ is real, we get z_1/z_2 is purely imaginary.

As z_1/z_2 is purely imaginary, $\arg(z_1/z_2) = \pm\dfrac{\pi}{2}$

Thus, 0, z_1, z_2 forms a right triangle.

So statement (iv) is incorrect

29. $\arg\left(\dfrac{z - z_1}{z_2 - z}\right) = \dfrac{\pi}{2} \Rightarrow \text{Re}\left(\dfrac{z - z_1}{z_2 - z}\right) = 0$

$$\Rightarrow \frac{z - z_1}{z_2 - z} + \frac{\bar{z} - \bar{z}_1}{\bar{z}_2 - \bar{z}} = 0$$

$$\Rightarrow (z - z_1)(\bar{z}_2 - \bar{z}) + (z_2 - z)(\bar{z} - \bar{z}_1) = 0$$

$$\Rightarrow z(\bar{z}_1 + \bar{z}_2) + \bar{z}(z_1 + z_2) - 2z\bar{z} - (z_1\bar{z}_2 + z_2\bar{z}_1)$$

$$= 0$$

$$\Rightarrow z\ \bar{z} - (5+i)z - (5-i)\bar{z} + 21 = 0$$

$$\Rightarrow |z - (5-i)| = \sqrt{5}$$

30. $|z_1 + z_2 + z_3| = |(z_1 - 1) + (z_2 - 2) + (z_3 - 3) + 6|$

$$< |z_1 - 1| + |z_2 - 2| + |z_3 - 3| + 6$$

$$< 1 + 2 + 3 + 6 = 12$$

31. For $x \neq 1$, we have $1 + x + x^2 + \ldots + x^n$

$$= \frac{x^{n+1} - 1}{x - 1}$$

Differentiating w. r. t. x, we get

$0 + 1 + 2x + \ldots + nx^{n-1}$

$$= \frac{(x-1)(n+1)x - (x^{+1} - 1).1}{(x-1)\dagger}$$

$\Rightarrow 1 + 2x + \ldots + nx^{n-1}$

$$= \frac{(n+1)x}{x-1} - \frac{x^1 - 1}{(x-1)^2}$$

Multiplying both sides by x, we get

$\Rightarrow x + 2x^2 + \ldots + nx^n$

$$= \frac{(n+1)x^{n+1}}{x-1} - \frac{x^{n+2} - x}{(x-1)^2}$$

Differentiating w. r. t. x, we get

$1 + 4x + \ldots + n^2 x^{n-1}$

$$= \left(\frac{(n+1)^2 x^n}{x-1} - \frac{(n+1)x^{n-1}}{(x-1)^2}\right)$$

$$- \left(\frac{(n+2)x^{n+1} - 1}{(x-1)^2} - \frac{2(x^{n+1} - x)}{(x-1)^3}\right)$$

Putting $x = \omega$, we get

$1 + 4\omega + \ldots\ldots\ldots + n^2\omega^{n-1} =$

$$\frac{(n+1)^2}{\omega - 1} - \frac{(2n+3)\omega - 1}{(\omega-1)^2} + \frac{2(\omega^2 - \omega)}{(\omega-1)^3}$$

$$(\because \omega^n = 1)$$

$$= \frac{(n+1)^2(\omega - 1) - (2n+3)\omega + 1 + 2\omega}{(\omega-1)^2}$$

$$= \frac{(n^2 + 2n + 1 - 2n - 3 + 2)\omega - n^2 - 2n - 1 + 1}{(\omega-1)^2}$$

$$= \frac{n^2\omega - n^2 - 2n}{(\omega-1)^2}$$

ALTERNATIVE METHOD:

$S_n = 1 + 4\omega + 9\omega^2 + \ldots. + n^2\omega^{n-1}$

$$\omega S = \omega + 4\omega^2 + \ldots + (n-1)^2\omega^{n-1} + n\dagger\omega^n$$

$$(1-\omega)S = 1 + 3\omega + 5\omega^2 + \ldots + (2n-1)\omega^{n-1} - n\dagger\omega^n \quad \ldots(1)$$

Let $S'_n = 1 + 3\omega + 5\omega^2 + \ldots + (2n-1)\omega^{n-1}$

$$\omega S_n = \omega + 3\omega\dagger + \ldots + (2n-3)\omega^{n-1} + (2n-1)\omega^n$$

$$(1-\omega)S'_n = 1 + 2\omega + 2\omega\dagger + \ldots + 2\omega^{n-1} - (2n-1)\omega$$

$$= 1 - (2n-1)\omega^n + \frac{2\omega(1 - \omega^{n-1})}{1 - \omega}$$

$\therefore$ From (1)

$$(1-\omega)S_n = \frac{1 - (2n-1)\omega}{1 - \omega} + \frac{2\omega(1 - \omega^{n-1})}{(1-\omega)^2 - n\dagger}$$

$$\Rightarrow S_n = \frac{2(-n+1)}{(1-\omega)\dagger} + \frac{2\omega - 2}{(1-\omega)\ddagger} - \frac{n\dagger}{1-\omega}$$

$$(\because \omega^n = 1)$$

$$S_n = \frac{n\dagger\omega - n\dagger - 2n}{(\omega - 1)\dagger}$$

32. Let $z = x + iy$, where $x, y \in R$

$\Rightarrow x^2 + y^2 - 2i(x + iy) + 2c(a + i) = 0$

$\Rightarrow x^2 + y^2 + 2y + 2ac = 0$ and $-2x + 2c = 0$

$\therefore y^2 + 2y + c^2 + 2ac = 0$ and $x = c$

Since $y \in R$, we have $\Delta \geq 0$

$4 - 4(c^2 + 2ac) \geq 0$

$\Rightarrow c^2 + 2ac - 1 \leq 0$

The above inequality must be true $\forall c \in R$ (given)

$\therefore \Delta \leq 0 \Rightarrow 4a^2 + 4 \leq 0$ which is impossible for any $a \in I$.

Hence, number of intetgral values of a for which it is true is zero.

33. We know that

$$z_1^2 + z_2^2 + z_3^2 - z_1z_2 - z_2z_3 - z_3z_1 = 0,$$

or

$$a^2 - 1 + 2ai + 1 - b^2 + 2bi + 0 - a + b - i - abi = 0$$

equating real & imaginary part, we get

$\therefore$ $a^2 - b^2 - a + b = 0$ and $2a + 2b - ab - 1 = 0$

$\Rightarrow$ $a = b$ and $2a + 2b - ab - 1 = 0$

($\because$ $a + b = 1$ does not give real solution)

$\therefore$ $a = b$ and $a^2 - 4a + 1 = 0$

$a = b = 2 - \sqrt{3}$ ($\because$ $a < 1, b < 1$)

34. $\dfrac{2}{z_1} = \dfrac{1}{z_2} + \dfrac{1}{z_3}$

$$\Rightarrow \frac{1}{z_1} - \frac{1}{z_2} = \frac{1}{z_3} - \frac{1}{z_1}$$

$$\Rightarrow \frac{z_2 - z_1}{z_1z_2} = \frac{z_1 - z_3}{z_1z_3}$$

$$\Rightarrow \frac{z_2 - z_1}{z_3 - z_1} = -\frac{z_2}{z_3}$$

$$\Rightarrow \arg\left(\frac{z_2 - z_1}{z_3 - z_1}\right) = \arg\left(-\frac{z_2}{z_3}\right) = \arg\left(\frac{z_2}{z_3}\right) \pm \pi$$

$$\Rightarrow arg\left(\frac{z_2 - z_1}{z_3 - z_1}\right) = arg\left(\frac{z_2 - 0}{z_3 - 0}\right) \pm \pi$$

$\Rightarrow \angle CAB = \angle COB \pm \pi$

$\Rightarrow \angle CAB - \angle COB = \pm \pi$

$\Rightarrow$ Points *O, A, B, C* are concyclic.

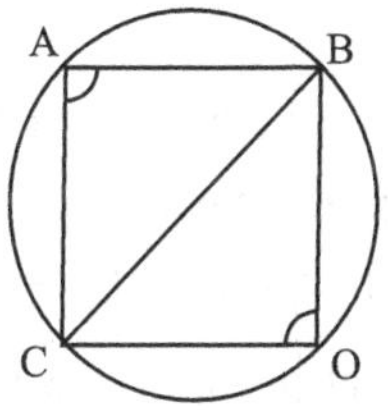

35. We have

$$z = \frac{1}{2}(\sqrt{3} - i) = -\frac{1}{2}i(1 + i\sqrt{3}) = i\omega^2$$

where $\omega \neq 1$ is a cube of unity.

$\therefore$ $z^{89} = (i\,\omega^2)^{89} = i^{89}\,\omega^{178} = i\omega$

Also, $i^{97} = i^{96}\,i = i$

Thus, $(z^{89} + i^{97})^{94} = (i\omega + i^{94} = [i(-\omega^2)]^{94}$

$= -\omega^2$

Also, $z^n = i^n \omega^{2n}$

$\therefore$ The given equation becomes

$\omega^2 = i^n\,\omega^{2n}$

$\Rightarrow i^n\,\omega^{2n-2} = -1$

This is possible if n is in the form of 4k + 2 and 2n – 2 is in the form of 3p.

The smallest value of n is 10.

36. $|z + \bar{z}| + |z - \bar{z}| = 2$

$\Rightarrow |2\,\text{Re}(z)| + |2Im(z)i| = 2$

$\Rightarrow |x| + |y| = 1$

Which represents a square.

Also $|z + i| + |z - i| = 2$ represents a line segment

Two complex numbers i and –i will satisfy these condition

37. We have $z = \dfrac{b + ic}{1 + a}$

$$\Rightarrow \bar{z} = \frac{b - ic}{1 + a}$$

$$\Rightarrow z + \bar{z} = \frac{2b}{1 + a},\ z - \bar{z} = \frac{2ic}{1 + a},$$

And $z\bar{z} = \dfrac{b^2 + c^2}{(1 + a)^2} = \dfrac{1 - a^2}{(1 + a)^2} = \dfrac{1 - a}{1 + a}$

We have

$$\frac{1 + iz}{1 - iz} = \frac{1 + iz}{1 - iz} \times \frac{1 + i\bar{z}}{1 + i\bar{z}} = \frac{1 + i(z + \bar{z}) - z\bar{z}}{1 - i(z - \bar{z}) + z\bar{z}}$$

$$= \frac{1 + \dfrac{2bi}{1 + a} - \dfrac{1 - a}{1 + a}}{1 + \dfrac{2c}{1 + a} + \dfrac{1 - a}{1 + a}} = \frac{1 + a + 2bi - 1 + a}{1 + a + 2c + 1 - a}$$

$$= \frac{2(a + ib)}{2(1 + c)} = \frac{a + ib}{1 + c}$$

38. If α satisfies given equation, then $\alpha \neq 0$ for otherwise, $\sin\theta_4 = 3$.

$$3 = |3| = \left|\sin\theta_1 z^3 + \sin\theta_2 z^2 + \sin\theta_3 z + \sin\theta_4\right|$$

$$\leq |\sin\theta_1||z|^3 + |\sin\theta_2||z|^2 + |\sin\theta_3||z| + |\sin\theta_4|$$

$$\leq |z|^3 + |z|^2 + |z| + 1$$

$$\leq 1 + |z| + |z|^2 + |z|^3 + |z|^4 + \ldots \leq \frac{1}{1 - |z|}$$

$$\Rightarrow 1 - |z| < 1/3 \Rightarrow |z| > 2/3$$

So $|\alpha| > 2/3$

39. We have $\log_{(0.8)}|z+1| > \log_{(0.8)}|z-1|$

$\Rightarrow |z+1| < |z-1|$ $\quad [0 < 0.8 < 1]$

$\Rightarrow (x+1)^2 + y^2 < (x-1)^2 + y^2$

[where z = x + iy]

$\Rightarrow 4x < 0$ or $x < 0$ or $R(z) < 0$

Exercise B
Concept Deviator (CD)
(Single Correct MCQs)

1. (d) The roots of $x^2 + x + 1 = 0$ are ω and ω^2,

So $h(\omega) = 0$ and $h(\omega^2) = 0$ also $w^3 = 1$

$\Rightarrow w\,f(1) + w^2 g(1) = 0$

and $w^2 f(1) + wg(1) = 0$

$\Rightarrow f(1) = g(1) = 0$

$\therefore\ h(1) = f(1) + g(1) = 0$

Hence all the three statements are correct.

2. (a) Let S = $|z_2 - z_3|^2 + |z_3 - z_1|^2 + |z_1 - z_2|^2$

Let α = minimum value of S

And β = maximum value of S

Here S ≥ 0 and S = 0 for $z_1 = z_2 = z_3$.

So minimum value of $|z_2 - z_3|^2 + |z_3 - z_1|^2 + |z_1 - z_2|^2$ is 0

We know that $|a-b|^2 = |a|^2 + |b|^2 - \bar{a}b - a\bar{b}$,

so $|z_2 - z_3|^2 + |z_3 - z_1|^2 + |z_1 - z_2|^2$

$= 6 - (z_2\bar{z}_3 + \bar{z}_2 z_3 + z_3\bar{z}_1 + \bar{z}_3 z_1 + z_1\bar{z}_2 + \bar{z}_1 z_2)$

$= 6 - \bar{z}_1(z_2 + z_3) - \bar{z}_2(z_3 + z_1) - \bar{z}_3(z_1 + z_2)$ ...(i)

$|z_1 + z_2 + z_3|^2 = |z_1|^2 + |z_2|^2 + |z_3|^2 + \bar{z}_1(z_2 + z_3) + \bar{z}_2(z_3 + z_1) + \bar{z}_3(z_1 + z_2)$

$= \bar{z}_1(z_2 + z_3) + \bar{z}_2(z_3 + z_1) + \bar{z}_3(z_1 + z_2) \geq -3$ ...(ii)

From equation (i) and (ii), we get

$|z_2 - z_3|^2 + |z_3 - z_1|^2 + |z_1 - z_2|^2 \leq 9$

Maximum value of $|z_2 - z_3|^2 + |z_3 - z_1|^2 + |z_1 - z_2|^2$ is 9 when

$z_1 = 1 \quad z_2 = \omega \quad z_3 = \omega^2$

where $\omega \neq 1$ is a cube root of unity.

Required sum is 0+9 =9

3. (c) Given that $z_1 + z_2 + z_3 + z_4 = 0$

$\Rightarrow \dfrac{z_1 + z_2}{2} = -\dfrac{z_3 + z_4}{2}$

$\dfrac{z_1 + z_2}{2}$ represents mid point E of AB

And $\dfrac{z_3 + z_4}{2}$ represents mid point F of DC

ABCD is a parallelogram inscribed in a circle and hence a rectangle

Then ABCD is necessarily rectangle, so statement (i) and (iii) are correct

4. (a) Let α (purely imaginary) be a root of the given equation then

$\alpha = -\bar{\alpha}$. Also $a\alpha^2 + b\alpha + c = 0$ ...(i)

or, $\overline{a\alpha^2 + b\alpha + c} = \bar{0}$

$\Rightarrow \bar{a}\bar{\alpha}^2 + \bar{b}\bar{\alpha} + \bar{c} = 0$

$\Rightarrow \bar{a}\alpha^2 - \bar{b}\alpha + \bar{c} = 0$ ...(ii)

$[\because z = -\bar{z}]$

Solving (i) and (ii) simultaneously

$\dfrac{\alpha^2}{b\bar{c} + c\bar{b}} = \dfrac{\alpha}{c\bar{a} - a\bar{c}} = \dfrac{1}{-a\bar{b} - \bar{a}b}$

Eliminating α, we get

$(b\bar{c} + c\bar{b})(a\bar{b} + \bar{a}b) + (c\bar{a} - a\bar{c})^2 = 0$

5. (a) Given that $|z-1| + |z+3| \leq 8$

$\therefore$ z lies inside or on the ellipse whose foci an (1, 0) and (–3, 0) and vertices are (–5, 0) and (3, 0). Clearly the minimum and maximum values of $|z-4|$ are 1 and 9 respectively. Thus,

$1 \leq |z-4| \leq 9$.

6. (b) The equation is $(3 + 4i)^x = 5^x$

$\Rightarrow \left(\frac{3}{5} + \frac{4}{5}i\right)^x = 1$

$\Rightarrow (\cos\alpha + i\sin\alpha)^x = 1$,

where $\cos\alpha = \frac{3}{5}$ and $\sin\alpha = \frac{4}{5}$

If x is integer, then the equation is
$\cos\alpha x + i\sin\alpha x = 1$

$\Rightarrow \cos\alpha x = 1$

and $\sin\alpha x = 0 \Rightarrow \alpha x = 2n\pi, n \in I$

$\Rightarrow x = \frac{2n\pi}{\alpha}$

Clearly, only integral value of x can be x = 0.

7. (c) $(i - z)(1 + 2i) + (1 - iz)(3 - 4i) = 1 + 7i$
$\Rightarrow i - z - 2 - 2iz + 3 - 3iz - 4i - 4z = 1 + 7i$
$\Rightarrow z + iz + 2i = 0$
$\Rightarrow (x - y) + i(x + y + 2) = 0$ [put z = x + iy]
$\Rightarrow x - y = 0$ and $x + y + 2 = 0$
$\Rightarrow x = y = -1$
$\therefore z = -1 - i$

8. (c) Since α is a complex nth root of unity
So $1 + \alpha + \alpha^2 + \ldots + \alpha^{n-1} = 0$

Now, $\sum_{r=0}^{n-1} |z_1 + \alpha^r z_2|^2$

$= \sum_{r=0}^{n-1} (z_1 + \alpha^r z_2)(\bar{z}_1 + \bar{\alpha}^r \bar{z}_2)$

$= \sum_{r=0}^{n-1} |z_1|^2 + z_1\bar{z}_2 \sum_{r=0}^{n-1} \bar{\alpha}^r + z_2\bar{z}_1 \sum_{r=0}^{n-1} \alpha^r$

$+ \sum_{r=0}^{n-1} |z_2|^2 |\alpha|^{2r}$

$= n(|z_1|^2 + |z_2|^2)$

9. (c) Given that

$$\frac{1}{a+\omega} + \frac{1}{b+\omega} + \frac{1}{c+\omega} + \frac{1}{d+\omega} = \frac{1}{\omega} \quad \ldots(i)$$

Taking conjugate we get,

$$\frac{1}{a+\omega\dagger} + \frac{1}{b+\omega\dagger} + \frac{1}{c+\omega\dagger} + \frac{1}{d+\omega\dagger} = \frac{1}{\omega\dagger} \quad \ldots(ii)$$

Subtract equation (ii) from equation (i) we get

$$\Sigma\left(\frac{1}{a+\omega} - \frac{1}{a+\omega\dagger}\right) = \frac{1}{\omega} - \frac{1}{\omega\dagger}$$

$$\Rightarrow (\omega^2 - \omega)\,\Sigma\frac{1}{(a+\omega)(a+\omega^2)} = \omega\dagger - \omega$$

$$\Rightarrow \Sigma\frac{1}{a^2 - a + 1} = 1$$

$$\text{So } \Sigma\frac{3}{a^2 - a + 1} = 3$$

10. (b) The angles of ΔDEF are $\pi - 2A, \pi - 2B$ and $\pi - 2C$, respectively $\frac{z_3 - z_1}{z_2 - z_1}$ is purely real

$\Rightarrow \arg\left(\frac{z_3 - z_1}{z_2 - z_1}\right) = 0$ or π

$\angle EDF = 0$ or π

$\therefore \pi - 2A = 0$ or $\pi \Rightarrow A = \frac{\pi}{2}$

11. (a) We have $\left|z - \frac{4}{z}\right| \ge \left||z| - \left|\frac{4}{z}\right|\right|$

$\Rightarrow \left||z| - \left|\frac{4}{z}\right|\right| \le 2$

$\Rightarrow -2 \le |z| - \frac{4}{|z|} \le 2$

$\Rightarrow |z|^2 + 2|z| - 4 \ge 0$ and $|z|^2 - 2|z| - 4 \le 0$

$\Rightarrow |z| \le -\sqrt{5} - 1$ or $|z| \ge \sqrt{5} - 1$

and $1 - \sqrt{5} \le |z| \le 1 + \sqrt{5}$

$\Rightarrow \sqrt{5} - 1 \le |z| \le \sqrt{5} + 1$

(Common value of $|z|$ from above solutions)

$\therefore$ Least value of $|z| = \sqrt{5} - 1$ and greatest value of $|z| = \sqrt{5} + 1$.

So integral values that $|z|$ can assume is 2 and 3

12. (a) Let, z = x + iy, then

arg (z – 3i) = arg (x + iy 3i) – $\frac{3\pi}{4}$

$\Rightarrow x < 0, y - 3 > 0$ $\left(\because \frac{3\pi}{4} \text{ is in II quadrant}\right)$

and $\frac{y-3}{x} = \tan\frac{3\pi}{4} = -1$

$\Rightarrow y = -x + 3 \;\forall\; x < 0$ and $y > 3$ …(i)

Now consider arg (2z + 1 – 2i)

$= \arg[(2x + 1) + i(2y - 2)] = \frac{\pi}{4}$

$\Rightarrow 2x + 1 > 0, 2y - 2 > 0$ ($\because \frac{\pi}{4}$ is in I quadrant) and $\frac{2y-2}{2x+1} = \tan\frac{\pi}{4} = 1 \Rightarrow 2y-2 = 2x+1$

$\Rightarrow y = x + \frac{3}{2} \ \forall x > -\frac{1}{2}, y > 1$ ...(ii)

Straight line (i) and (ii) do not intersect in the given domain so there exist no point of intersection.

13. (d) $\cos\frac{2\pi k}{11} + i\sin\frac{2\pi k}{11}$ represents 11th roots of unity for k = 0, 1, 2, …., 10

∴ Their sum = 0

$$\Rightarrow \sum_{k=1}^{10}(\cos\frac{2\pi k}{11} + i\sin\frac{2\pi k}{11}) = 0$$

$$\Rightarrow 1 + \sum_{k=1}^{10}(\cos\frac{2\pi k}{11} + i\sin\frac{2\pi k}{11}) = 0$$

$$\Rightarrow -i\sum_{k=1}^{10}(\cos\frac{2\pi k}{11} + i\sin\frac{2\pi k}{11}) = i$$

$$\Rightarrow \sum_{k=1}^{10}(\sin\frac{2\pi k}{11} - i\cos\frac{2\pi k}{11}) = i$$

14. (a) From the rotation theorem,

$$\frac{z_1 - 0}{\bar{z}_1 - 0} = \frac{|z_1 - 0|}{|\bar{z}_1 - 0|} e^{2i\pi/n} = e^{2\pi i/n}$$

$$\Rightarrow \quad \frac{z_1}{\bar{z}_1} = e^{2\pi i/n} \quad ...(i)$$

Let, $z_1 = re^{i\theta}$, then

$$\therefore \quad \frac{z_1}{\bar{z}_1} = \frac{re^{i\theta}}{re^{-i\theta}} = e^{2\pi i/n} \quad \therefore \theta = \frac{\pi}{n}$$

Now, $\frac{\text{Im}(z_1)}{\text{Re}(z_1)} = \frac{\sqrt{3}-1}{\sqrt{3}+1}$

$$\Rightarrow \frac{r\sin(\pi/n)}{r\cos(\pi/n)} = \frac{\sqrt{3}-1}{\sqrt{3}+1}$$

$$\Rightarrow \tan\frac{\pi}{n} = \tan\frac{\pi}{12}$$

$$\therefore \quad n = 12$$

15. (b) Let, ω and ω^2 represent the points A and B and P is the variable point represented by z in the Argand diagram

Then AP = $|z-\omega|$; BP = $|z-\omega^2|$

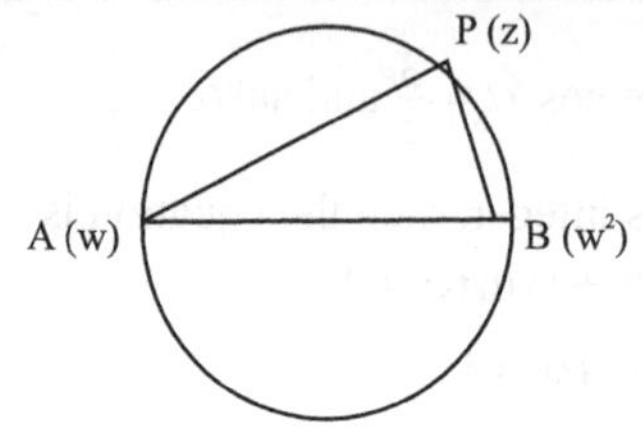

$\because \angle APB = \frac{\pi}{2}; \therefore AP^2 + BP^2 = AB^2$

$$\Rightarrow |z-\omega|^2 + |z-\omega^2|^2 = |\omega-\omega^2|^2 = 3$$

16. (a) Let

$$\frac{1}{|z_2 - z_3|} = \frac{2}{|z_3 - z_1|} = \frac{3}{|z_1 - z_2|} = \lambda(say)$$

$$\Rightarrow 1 = \lambda|z_2 - z_3|, 2 = \lambda|z_3 - z_1|, 3 = \lambda|z_1 - z_2|$$

$$\Rightarrow 1 = \lambda^2|z_2 - z_3|^2 = \lambda^2(z_2 - z_3)(\bar{z}_2 - \bar{z}_3)$$

$$\therefore \frac{1}{(z_2 - z_3)} = \lambda^2(\bar{z}_2 - \bar{z}_3)$$

Similarly, $\frac{4}{(z_3 - z_1)} = \lambda^2(\bar{z}_3 - \bar{z}_1)$ and

$$\frac{9}{|z_1 - z_2|} = \lambda^2(\bar{z}_1 - \bar{z}_2)$$

$$\frac{1}{(z_2 - z_3)} + \frac{4}{(z_3 - z_1)} + \frac{9}{(z_1 - z_2)} = 0$$

17. (b) Let the centre of the circle be z_0 and radius r. Then its equation is

$|z - z_0| = r$ .... (i)

The circle (i) touches the circle $|z - z_1| = a$ externally

∴ Distance between centres = sum of radii

$\Rightarrow |z_0 - z_1| = a + r$ ...(ii)

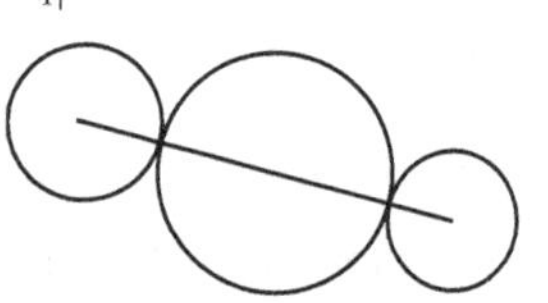

Similarly, $|z_0 - z_2| = b + r$ ...(iii)

From equations (ii), and (iii)

$|z_0 - z_1| - |z_0 - z_2| = a - b$

∴ z_0 lies on the curve $|z - z_1| - |z - z_2| = a - b$, which represents a hyperbola.

18. (b) Let z = x + i y, then $|z|^2 + |z-3|^2 + |z-6i|^2$

$= 3x† + 3y† + 45 - 6x - 12y$

$= 3\left[(x-1)^2 + (y-2)^2 + 10\right]$

$\therefore$ Minimum value $= 3 \times 10 = 30$, which occurs when x = 1, y = 2.

Smart Cracker
$\lvert z - z_1\rvert^2 + \lvert z - z_2{}^2\rvert^2 + \lvert z - z_3\rvert^2$ attains the least value if z coincides with centroid of triangle formed by z_1, z_2, z_3.

19. (a) $\because$ ω is an imaginary *n*th root of unity

$\therefore \omega^n = 1$ and $1 + \omega + \omega^2 + \ldots + \omega^{n-1} = 0$

Let $S = \sum_{r=1}^{n} (ar + b)\, \omega^{r-1}$

$= a\sum_{r=1}^{n} r\omega^{r-1} + b\sum_{r=1}^{n} \omega^{r-1}$

$= a\{1 + 2\omega + 3\omega^2 + 4\omega^3 + \ldots + n\omega^{n-1}\}$
$+ b\{1 + \omega + \omega^2 + \omega^3 + \ldots + \omega^{n-1}\}$ ....(i)

As $1 + 2\omega + 3\omega^2 + \ldots + n\omega^{n-1}$ is an A.G.P.,

So $1 + 2\omega + 3\omega^2 + \ldots. + n\omega^{n-1}$

$= \dfrac{1 + \omega + \omega† + \ldots\ldots\ldots + \omega^{n-1}}{1-\omega} - \dfrac{n\omega}{1-\omega}$

Hence,

$S = a\left\{\dfrac{1 + \omega + \omega† + \ldots\ldots\ldots + \omega^{n-1}}{1-\omega} - \dfrac{n\omega^n}{1-\omega}\right\}$

$+ b\{1 + \omega + \omega† + \ldots + \omega^{n-1}\}$

$= a\left\{\dfrac{0}{1-\omega} - \dfrac{n}{1-\omega}\right\} + 0 = \dfrac{na}{\omega - 1}$

Now put n = 10, so required value is

$\dfrac{na}{\omega-1} = \dfrac{10a}{\omega-1}$

20. (c) Since z_1, z_2, z_3 lie on a circle with centre origin and radius 1 unit so $|z_1| = |z_2| = |z_3| = 1$

Also given that $\dfrac{z_1^2}{z_2 z_3} + \dfrac{z_2^2}{z_3 z_1} + \dfrac{z_3^2}{z_1 z_2} = -1$

Multiply by $z_1 z_2 z_3$ we get

$z_1^3 + z_2^3 + z_3^3 = -z_1 z_2 z_3$

or $-4z_1 z_2 z_3 = z_1^3 + z_2^3 + z_3^3 - 3z_1 z_2 z_3$

$= (z_1 + z_2 + z_3)\left(z_1^2 + z_2^2 + z_3^2 - z_2 z_3 - z_3 z_1 - z_1 z_2\right)$

$= (z_1 + z_2 + z_3)\left[(z_1 + z_2 + z_3)^2 - 3(z_2 z_3 + z_3 z_1 + z_1 z_2)\right]$

Now let $z = z_1 + z_2 + z_3$ then

$z^3 - 3z(z_2 z_3 + z_3 z_1 + z_1 z_2) + 4z_1 z_2 z_3 = 0$

$z^3 = z_1 z_2 z_3\left[3z\left(\dfrac{1}{z_1} + \dfrac{1}{z_2} + \dfrac{1}{z_3}\right) - 4\right]$

$= z_1 z_2 z_3\left[3z(\bar{z}_1 + \bar{z}_2 + \bar{z}_3) - 4\right]$

$= z_1 z_2 z_3\left[3z\bar{z} - 4\right]$

$\Rightarrow |z|^3 = |z_1||z_2||z_3|\left|(3|z|^2 - 4)\right|$

$\Rightarrow \rightarrow |z|^3 - \left|(3|z|^2 - 4)\right| = 0$

Now, we have two cases

Case (i): If $|z| \geq 2/\sqrt{3}$

$|z|^3 - 3|z|^2 + 4 = 0$

$\Rightarrow (|z| - 2)\left(|z|^2 - |z| - 2\right) = 0$

$\Rightarrow (|z| - 2)^2(|z| + 1) = 0$

$\Rightarrow |z| - 2 = 0$ or $|z| = 2$

Case (ii): If $|z| < 2/\sqrt{3}$, then

$|z|^3 + 3|z|^2 - 4 = 0$

$\Rightarrow (|z| - 1)\left(|z|^2 + 4|z| + 4\right) = 0$

$\Rightarrow |z| - 1 = 0 \Rightarrow |z| = 1.$

Required sum is 1 + 2 = 3

21. (a) We have $\left(\dfrac{1+ia}{1-ia}\right)^4 = z$ and $|z| = 1$ so assume

$z = \cos\theta + i\sin\theta$

Let $\left(\dfrac{1+ia}{1-ia}\right)^4 = \cos\theta + i\sin\theta$

$\Rightarrow \dfrac{1+ia}{1-ia} = (\cos\theta + i\sin\theta)^{1/4}$

$= (\cos(2k\pi + \theta) + i\sin(2k\pi + \theta))^{1/4}$, $k \in Z$

$= \cos\left(\dfrac{2k\pi + \theta}{4}\right) + i\sin\left(\dfrac{2k\pi + \theta}{4}\right)$, $k = 0, 1, 2, 3$

$\therefore \dfrac{1+ia}{1-ia} = \cos\alpha + i\sin\alpha$, where $\alpha = \dfrac{2k\pi+\theta}{4}$

$\Rightarrow \dfrac{2}{2ia} = \dfrac{\cos\alpha + i\sin\alpha + 1}{\cos\alpha + i\sin\alpha - 1}$

$= \dfrac{2\cos^2\frac{\alpha}{2} + 2i\sin\frac{\alpha}{2}\cos\frac{\alpha}{2}}{-2\sin^2\frac{\alpha}{2} + 2i\sin\frac{\alpha}{2}\cos\frac{\alpha}{2}}$

$= \dfrac{2\cos\frac{\alpha}{2}(\cos\frac{\alpha}{2} + i\sin\frac{\alpha}{2})}{2i\sin\frac{\alpha}{2}(\cos\frac{\alpha}{2} + i\sin\frac{\alpha}{2})} = \dfrac{1}{i}\cot\dfrac{\alpha}{2}$

$\therefore ia = i\tan\dfrac{\alpha}{2} \Rightarrow a = \tan\dfrac{\alpha}{2}$

$\therefore a = \tan\dfrac{2k\pi+\theta}{8}$, k = 0, 1, 2, 3 = $\tan\dfrac{\theta}{8}$,

$\tan\left(\dfrac{\pi}{4}+\dfrac{\theta}{8}\right), \tan\left(\dfrac{\pi}{2}+\dfrac{\theta}{8}\right), \tan\left(\dfrac{3\pi}{4}+\dfrac{\theta}{8}\right)$

$\therefore$ All roots are real and distinct.

22. (b) $|z_1+z_2|^2 + |z_2+z_3|^2 + |z_3+z_1|^2$

$= 2(|z_1|^2 + |z_2|^2 + |z_3|^2) + (z_1\bar{z}_2 + \bar{z}_1 z_2 + z_2\bar{z}_3 + \bar{z}_2$
$+ z_2\bar{z}_3 + z_3\bar{z}_1 + \bar{z}_3 z_1)$

$= 24 + (z_1\bar{z}_2 + \bar{z}_1 z_2 + z_2\bar{z}_3 + \bar{z}_2 z_3 + z_3\bar{z}_1 + \bar{z}_3 z_1)$ ...(i)

Also, $|z_1+z_2+z_3|^2 \geq 0$

$\Rightarrow z_1\bar{z}_2 + \bar{z}_1 z_2 + z_2\bar{z}_3 + \bar{z}_2 z_3$
$+ z_3\bar{z}_1 + \bar{z}_3 z_1 \geq -12$

$\therefore |z_1+z_2|^2 + |z^2+z^2|^2 + |z_3+z_2|^2 \geq 12$

so required minimum value is 12

23. (a) Given that

$2|z-z_1| = |z-z_2| + |z-z_3|$ ...(i)

$2|z-z_2| = |z-z_1| + |z-z_3|$ ...(ii)

$2|z-z_3| = |z-z_1| + |z-z_2|$ ...(iii)

Now, subtracting (i) from (ii),

$2(|z-z_2| - |z-z_1|) = |z-z_1| - |z-z_2|$

$\Rightarrow |z-z_1| = |z-z_2|$

Similarly solving other pairs of equations and combining the results,

$|z-z_1| = |z-z_2| = |z-z_3|$

$\Rightarrow$ z is the circumcentre of Δ ABC.

24. (a) Given complex number is

$\dfrac{(4+3i)(1-2i)}{(1+2i)(1-2i)} = \dfrac{4+3i-8i+6}{1+4} = 2-i$

And given straight line is $iz = \bar{z}$

$i(x+iy) = x - iy$

$ix - y = x - iy$

$x + y - i(x+y) = 0$

$x + y = 0.$

Reflection is $(1-2i)$.

25. (d) Given equation is $x^4 - 2x^2 + 4 = 0$

$\Rightarrow (x^2-1)^2 = -3$

$\Rightarrow x^2 = 1 \pm i\sqrt{3}$

$= 2\left[\cos\dfrac{\pi}{3} \pm i\sin\dfrac{\pi}{3}\right]$

Roots are $\sqrt{2}e^{\frac{i\pi}{6}}, -\sqrt{2}e^{\frac{i\pi}{6}}, \sqrt{2}e^{\frac{-i\pi}{6}} - \sqrt{2}e^{\frac{-i\pi}{6}}$

They lie on a circle radius $\sqrt{2}$, and the diagonals intersect at an angle $\dfrac{\pi}{3}$, so they represent the vertices of an inscribed rectangle.

26. (b) Let z = x + iy

$\therefore$ the given equation of the curve can be written as

$$2|2x| = 32 + \left(\frac{2y}{i}\right)^2$$

$\Rightarrow \quad 4|x| = 32 + (-4y^2)$

$\Rightarrow \quad |x| = 8 - y^2$

$\Rightarrow \quad y^2 = -|x| + 8 = \begin{cases} x+8, x<0 \\ -x+8, x\geq 0 \end{cases}$

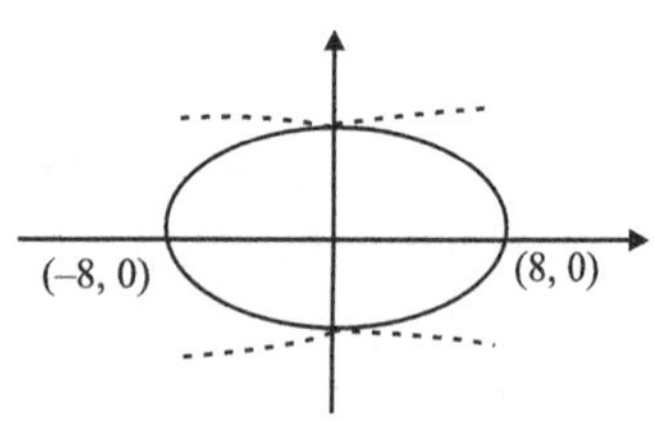

From the above graph, the maximum distance between the two can be 16 units.

27. (d) The given equation is $z^2 = \bar{z}^3$

$\Rightarrow |z|^2 = |\bar{z}|^3$

$\Rightarrow |z|^2 = |z|^3$

$\Rightarrow |z| = 0$ or $|z|^2 = 1$

$\Rightarrow |z| = 0$ or $|z| = 1$

Case (i) If $|z| = 0$, then $z = 0$

Case (ii) If $|z| = 1$, then $\bar{z} = \frac{1}{z}$ and the equation becomes

$z^2 = \frac{1}{z^3}$

$\Rightarrow z^5 = 1$ This equation has 5 roots.

So, there are 6 complex numbers satisfying the given equation .

28. (d) Given that: $\left|\frac{z-25}{z-1}\right| = 5$

$\Rightarrow |z-25|^2 = 25|z-1|^2$

$\Rightarrow |z|^2 - 25z - 25\bar{z} + 625$

$= 25\{|z|^2 - z - \bar{z} + 1\}$

$\Rightarrow 24|z|^2 = 600$

$\Rightarrow |z|^2 = 25$

$\Rightarrow |z| = 5$

29. (d) $|z-1| \le 1$ represent the interior and boundary of the circle with centre at $1 + 0i$ and radius 1 and $|z-2| = 1$ represent circle with centre $2 + 0i$ and radius 1.

Clearly the points z satisfying $|z-1| \le 1$ and $|z-2| = 1$ lie on a portion of circle

$\therefore OA \le |z| \le OC\ (= OD)$

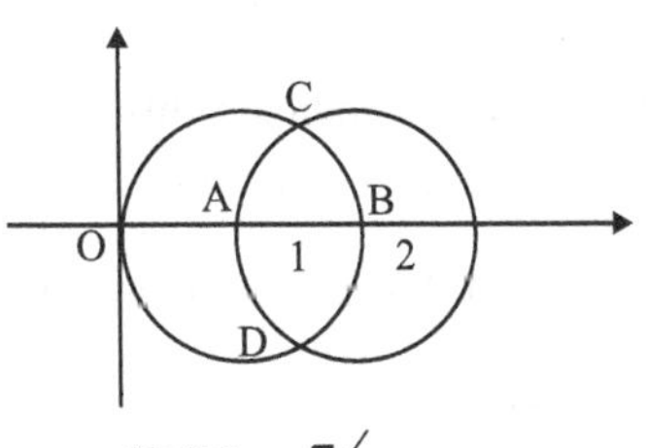

As $\angle OCB = \pi/2$,

$OC^2 = OB^2 - BC^2 = 4 - 1 = 3$

$\Rightarrow OC = \sqrt{3}$

Thus, $|z|^2 \le 3$

$1 \le |z| \le \sqrt{3}$

30. (b) $(x-1)^3 + 8 = 0$

$\Rightarrow (x-1)^3 = -8 = (-2)^3$

$\Rightarrow x - 1 = -2$ or -2ω or $-2\omega^2$

$\Rightarrow x = -1, 1 - 2\omega, 1 - 2\omega^2$

31. (d) Consider $\frac{1+i}{1-i}$, multiplying numerator and denominator by $(1+i)$ we get

$$\frac{1+i}{1-i} = \frac{(1+i)^2}{(1-i)(1+i)} = \frac{1-1+2i}{2} = i$$

Now $i^n = 1$

$\Rightarrow$ the smallest positive integral value of n should be 4.

32. (a) Given that $|z - 5i| = |z + 5i|$

$\Rightarrow$ z remains equidistant from $z_1 = 5i$ and $z_2 = -5i$.

$\therefore$ z lies on perpendicular bisector of line segment joining z_1 and z_2. which is clearly the axis of x i.e. $y = 0$

33. (b) Since $\left(\frac{\sqrt{3}}{2} + \frac{i}{2}\right) = -i\left(\frac{-1}{2} + \frac{i\sqrt{3}}{2}\right) = i\omega$

$$\frac{\sqrt{3}}{2} - \frac{i}{2} = i\left(\frac{-1}{2} - \frac{i\sqrt{3}}{2}\right) = i\omega^2$$

$$\therefore\ z = (-i\omega)^5 + (i\omega^2)^5 = -i\omega^2 + i\omega$$

$$= i(\omega - \omega^2) = i(i\sqrt{3}) = -\sqrt{3}$$

$\Rightarrow Re(z) < 0$ and $Im(z) = 0$

34. (d) $|z-4| < |z-2|$

$\Rightarrow |(x-4) + iy| < |(x-2) + iy|$

$\Rightarrow (x-4)^2 + y^2 < (x-2)^2 + y^2$

$\Rightarrow -8x + 16 < -4x + 4 \quad \Rightarrow 4x - 12 > 0$

$\Rightarrow x > 3 \Rightarrow \text{Re}(z) > 3$

Alternative Solution:

Since $|z-4|$ represent distance of z from point (4,0) and $|z-4|^2$ represent distance of z from point (2, 0)

Now $|z-4| < |z-2|$ which means point is closer to 4 than 2 i.e $x > 3$ or $\text{Re}(z) > 3$

35. (b) $|\omega| = 1 \Rightarrow \left|\frac{1-zi}{z-i}\right| = 1$

$\Rightarrow |1-iz| = |z-i|$

$\Rightarrow |-i||z+i| = |z-i|$

$\Rightarrow |z-(-i)| = |z-i|$

$\Rightarrow$ z is equidistant from (0, – 1) and (0, 1)

$\Rightarrow$ z lies on perpendicular bisector of (0, –1) and (0, 1), i.e., x–axis, i.e., y=0

$\therefore$ *z* lies on real axis.

36. (b) If vertices of a parallelogram are z_1, z_2, z_3, z_4 then as diagonals bisect each other.

$\therefore \frac{z_1+z_3}{2} = \frac{z_2+z_4}{2}$

$\Rightarrow z_1 + z_3 = z_2 + z_4$

37. (b) Let *ABC* be the Δ with vertices *abc* and *PQR* be the Δ with vertices u, v, w.
Then c = (1 – r) *a* + *rb*

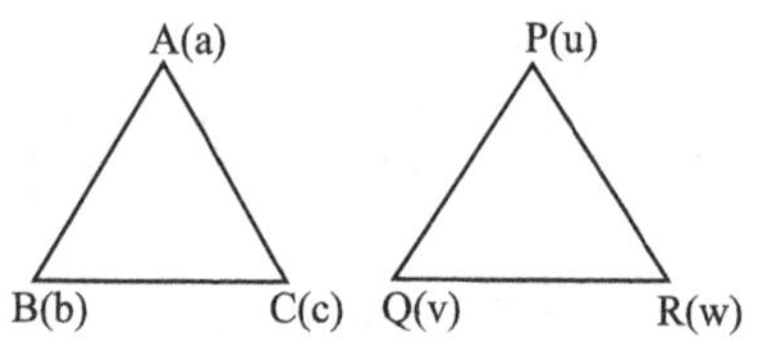

$\Rightarrow c - a = r(b-a)$

$\Rightarrow \frac{c-a}{b-a} = r$ ...(i)

$\Rightarrow w = (1-r)u + rv$

$\Rightarrow \frac{w-u}{v-u} = r$ ...(ii)

From (i) and (ii) $\left|\frac{c-a}{b-a}\right| = \left|\frac{w-u}{v-u}\right|$ and arg $\left(\frac{c-a}{b-a}\right) = \arg\left(\frac{w-u}{v-u}\right)$

$\Rightarrow \frac{AC}{AB} = \frac{PR}{PQ}$ and $\angle CBA = \angle RPQ$

$\Rightarrow \Delta ABC \sim \Delta PQR$

ALTERNATIVE SOLUTION:

Two triangles wih vertices a, b, c and u, v, w are simialy if

$$\begin{vmatrix} a & u & 1 \\ b & v & 1 \\ c & w & 1 \end{vmatrix} = 0$$

Here we have

$$\begin{vmatrix} a & u & 1 \\ b & v & 1 \\ c & w & 1 \end{vmatrix}$$

$$= \begin{vmatrix} a & u & 1 \\ b & v & 1 \\ c-(1-r)a-rb & w-(1-r)u-rw & 1-(1-r)-r \end{vmatrix}$$

$\left[R_3 \to R_3 - (1-r)R_1 - rR_2\right]$

But given that

$c = (1-r)a + rb$ and $w = (1-r)u + rw$

$\therefore$ Determinant = 0

$\Rightarrow$ The two Δ's are similar.

38. (b) $(1+\omega)^7 = A + B\omega$

$\Rightarrow \left(-\omega^2\right)^7 = A + B\omega \left(\because 1+\omega+\omega^2 = 0\right)$

$\Rightarrow -\omega^{14} = A + B\omega$

$\Rightarrow -\omega^2 = A + B\omega$ $\left(\because \omega^3 = 1\right)$

$\Rightarrow 1 + \omega = A + B\omega$

$\Rightarrow A = 1, B = 1$

39. (d) $\because |z| = |\omega|$ and arg $z = \pi - \arg\omega$

Let $\omega = re^{i\theta}$ *then* $z = re^{i(\pi-\theta)}$

$\Rightarrow z = re^{i\pi}.e^{-i\theta}$

$= \left(re^{-i\theta}\right)(cos\pi + i\,sin\pi) = \bar{\omega}(-1) = -\bar{\omega}$

40. (c) Given that $|z+i\omega| = |z-i\bar{\omega}|$

$\Rightarrow |z-(-i\omega)| = |z-(-\overline{i\omega})|$

$\Rightarrow$ *z* lies on perpendicular bisector of the line segment joining $(-i\omega)$ and $(-\overline{i\omega})$, which is real axis, $(-i\omega)$ and $(-\overline{i\omega})$ being mirror images of each other.

$\therefore$ Im(z) = 0.

If z = x then $|z| \le 1$

$\Rightarrow x^2 \le 1 \Rightarrow 0 \le x \le 1$

41. (d) $(1+i)^{n_1} + \left(1+i^3\right)^{n_1} + \left(1+i^5\right)^{n_2} + \left(1+i^7\right)^{n_2}$

$= (1+i)^{n_1} + (1-i)^{n_1} + (1+i)^{n_2} + (1-i)^{n_2}$

Using $1+i=\sqrt{2}(\cos\pi/4+i\sin\pi/4)$

and $1-i=\sqrt{2}(\cos\pi/4-i\sin\pi/4)$

Then

$$=(\sqrt{2})^{n_1}\left[\cos\frac{n_1\pi}{4}+i\sin\frac{n_1\pi}{4}\right]$$

$$+(\sqrt{2})^{n_1}\left[\cos\frac{n_1\pi}{4}-i\sin\frac{n_1\pi}{4}\right]+$$

$$(\sqrt{2})^{n_2}\left[\cos\frac{n_2\pi}{4}+i\sin\frac{n_2\pi}{4}\right]+$$

$$(\sqrt{2})^{n_2}\left[\cos\frac{n_2\pi}{4}-i\sin\frac{n_2\pi}{4}\right]$$

$$=(\sqrt{2})^{n_1}\left[2\cos\frac{n_1\pi}{4}\right]+(\sqrt{2})^{n_2}\left[2\cos\frac{n_2\pi}{4}\right]$$

= real number irrespective of the values of $n_1>0$ and $n_2>0$

42. (c) We know that $\omega=-\frac{1}{2}+\frac{i\sqrt{3}}{2}$

$$4+5(\omega)^{334}+3(\omega)^{365}=4+5\omega+3\omega^2$$

$$=1+2\omega+3(1+\omega+\omega^2)=1+(-1+i\sqrt{3})=i\sqrt{3}$$

43. (a) $\arg(-z)-\arg(z)=\arg\left(\frac{-z}{z}\right)=\arg(-1)=\pi$

44. (a) $|z_1|=|z_2|=|z_3|=1$ (given)

Now, $|z_1|=1\Rightarrow|z_1|^2=1\Rightarrow z_1\bar{z}_1=1$

Similarly $z_2\bar{z}_2=1$, $z_3\bar{z}_3=1$

Now, $\left|\frac{1}{z_1}+\frac{1}{z_2}+\frac{1}{z_3}\right|=|\bar{z}_1+\bar{z}_2+\bar{z}_3|=1$

$\Rightarrow|\overline{z_1+z_2+z_3}|=1$

$\Rightarrow|z_1+z_2+z_3|=1$

45. (d) Let $z=(1)^{1/n}=(\cos 2k\pi+i\sin 2k\pi)^{1/n}$

$$z=\cos\frac{2k\pi}{n}+i\sin\frac{2k\pi}{n},k=0,1,2,\ldots,n-1.$$

Let $z_1=\cos\left(\frac{2k_1\pi}{n}\right)+i\sin\left(\frac{2k_1\pi}{n}\right)$

and $z_2=\cos\left(\frac{2k_2\pi}{n}\right)+i\sin\left(\frac{2k_2\pi}{n}\right)$

be the two values of z. s.t. they subtend $\angle$ of 90° at origin.

$$\therefore\ \frac{2k_1\pi}{n}-\frac{2k_2\pi}{n}=\pm\frac{\pi}{2}$$

$\Rightarrow 4(k_1-k_2)=\pm n$

As k_1 and k_2 are integers and $k_1\neq k_2$.

$\therefore$ n = 4k, k $\in$ 1

46. (c) $\frac{z_1-z_3}{z_2-z_3}=\frac{1-i\sqrt{3}}{2}$

$$\arg\left(\frac{z_1-z_3}{z_2-z_3}\right)=\arg\left(\frac{1-i\sqrt{3}}{2}\right)$$

$\Rightarrow\arg(\cos(-\pi/3)+i\sin(-\pi/3))$

$\Rightarrow$ angle between z_1-z_3 and z_2-z_3 is 60°.

and $\left|\frac{z_1-z_3}{z_2-z_3}\right|=\left|\frac{1-i\sqrt{3}}{2}\right|$

$\Rightarrow\left|\frac{z_1-z_3}{z_2-z_3}\right|=1$

$\Rightarrow|z_1-z_3|=|z_2-z_3|$

So in triangle two sides are equal and one angle is 60°. so it is an equilateral triangle.

47. (b) We know that,

$$|z_1-z_2|\geq||z_1|-|z_2|| \quad \ldots(1)$$

Here $|z_1|=12$ and $|z_2-3-4i|=5$

but $|z_2-(3+4i)|\geq||z_2|-|3+4i||$

$\Rightarrow 5\geq|z_2|-5$

$\Rightarrow|z_2|\leq 10$

Also from (i) $|z_1-z_2|$ will have least value when $|z_2|$ has greatest value i.e.10

$\therefore|z_1-z_2|\geq 12-10=2$

Thus min. value of $|z_1-z_2| is 2$.

ALTERNATIVE SOLUTION:

$|z_1|=12$

$\Rightarrow z_1$ lies on a circle with centre (0, 0) and radius 12 units, and $|z_2-3-4i|=5$

$\Rightarrow z_2$ lies on a circle with centre (3, 4) and radius 5 units.

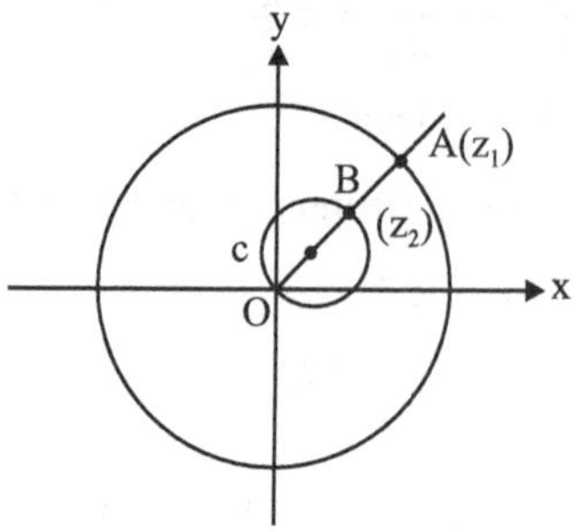

Distance between the two centers is 5 units so required minimum distance is 12–(5+5) = 2.

48. (a) $\omega = \dfrac{z-1}{z+1}$

From componendo and dividendo

$\Rightarrow z = \dfrac{1+\omega}{1-\omega}; |z| = |1$

$\Rightarrow \left|\dfrac{1+\omega}{1-\omega}\right| = 1$

$\Rightarrow |\omega+1| = |\omega-1|$

$\Rightarrow \omega$ is equidistant from (1, 0) and (–1, 0) and hence must lie on perpendicular bisector of line segment joining (1, 0) and (–1, 0) i.e. imaginary axis $\Rightarrow \omega$ is purely imaginary $\Rightarrow \text{Re}(\omega) = 0$.

49. (b) $\left(1+\omega^2\right)^n = \left(1+\omega^4\right)^n$

$\Rightarrow (-\omega)^n = (1+\omega)^n = (-\omega^2)^n \Rightarrow \omega^n = 1 \Rightarrow n = 3$

50. (a) Here.

$AB = AC = AD = 2$

$\therefore$ BCD is an arc of a circle with centre at A and redius 2. Shaded region is outer part of this sector ABCDA.

$\therefore$ For any point z on arc BCD

$|z-(-1)| = 2$

Then for shaded region, $|z+1| > 2$ ...(i)

For shaded region we also have

$-\pi/4 < arg(z+1) < \pi/4$

or $|arg(z+1)| < \pi/4$ ...(ii)

From relation (i) and (ii) we have z :

$|z+1| > 2$ and $|\arg(z+1)| < \pi/4$

51. (b) Given that a, b, c are integers not all equal, and ω is cube root of unity $\neq 1$, so we can take

$\omega = \dfrac{-1+i\sqrt{3}}{2}, \&\omega^2 = \dfrac{-1-i\sqrt{3}}{2}$

Consider $\left|a+b\omega+c\omega^2\right|$

$= \left|a+b\left(\dfrac{-1+i\sqrt{3}}{2}\right)+\left(\dfrac{-1-i\sqrt{3}}{2}\right)\right|$

$= \left|\left(\dfrac{2a-b-c}{2}\right)+i\left(\dfrac{b\sqrt{3}-c\sqrt{3}}{2}\right)\right|$

$= \dfrac{1}{2}\sqrt{(2a-b-c)^2+3(b-c)^2}$

$= \dfrac{1}{2}\sqrt{4a^2+b^2+c^2-4ab+2bc-4ac +3b^2+3c^2-6bc}$

$= \sqrt{a^2+b^2+c^2-ab-bc-ca}$

$= \sqrt{\dfrac{1}{2}\left[(a-b)^2+(b-c)^2+(c-a)^2\right]}$

R.H.S. will be min. when a = b = c, but as per given condition we cannot take a = b = c

$\therefore$ The min value is obtained when any two are zero and third is a minimum magnitude integer i.e. 1.

Thus b = c = 0, a = 1 gives us the minimum value 1.

52. (b) Operating $R_1 + R_2 + R_3$, we get

$$\begin{vmatrix} 3 & 0 & 0 \\ 1 & -1-\omega^2 & \omega^2 \\ 1 & \omega^2 & \omega^4 \end{vmatrix} = 3[-\omega-1-\omega] = 3\left(\omega^2-\omega\right)$$

53. (d) $\because \dfrac{w-wz}{1-z}$ is purely real, we know that if z is purely real then $z = \bar{z}$

$\therefore \left(\overline{\dfrac{w-\bar{w}z}{1-z}}\right) = \left(\dfrac{w-\bar{w}z}{1-z}\right)$

$\Rightarrow \dfrac{\bar{w}-w\bar{z}}{1-\bar{z}} = \dfrac{w-\bar{w}z}{1-z}$

$\Rightarrow \bar{w} - \bar{w}z - w\bar{z} + wz\bar{z} = w - w\bar{z} - \bar{w}z + \bar{w}z\bar{z}$

$\Rightarrow w - \bar{w} = (w - \bar{w})|z|^2$

$\Rightarrow \; |z|^2 = 1$ ($\because$ w = $\alpha + i\beta$ and $\beta \neq 0$)

$\Rightarrow |z| = 1$ also given $z \neq 1$

$\therefore$ The required set is

$\{z : |z| = 1, z \neq 1\} = 3\omega(\omega - 1)$

54. (d) $\overline{OP} = \overline{OA} + \overline{AP}$

$\Rightarrow \quad \overline{OP} = \overline{OA} + \overline{OB}$

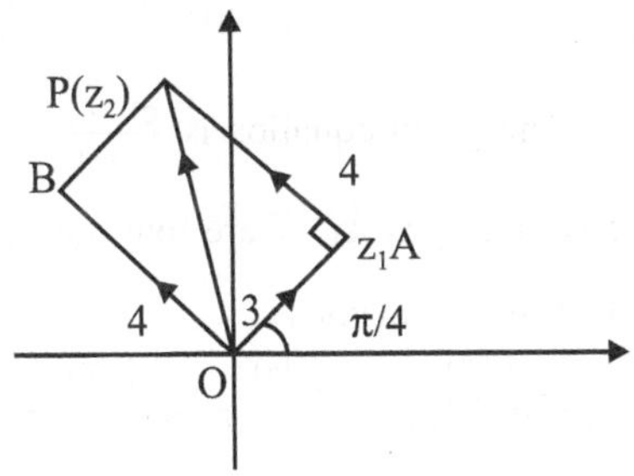

$\Rightarrow \overline{OP} = 3e^{i\pi/4} + 4e^{i(\pi/2 + \pi/4)}$

$= 3e^{i\pi/4} + 4e^{i\pi/2}.e^{i\pi/4}$

$= 3e^{i\pi/4} + 4ie^{i\pi/4} = e^{i\pi/4}(3 + 4i)$.

55. (d) Given $|z| = 1$ and $z \neq \pm 1$

Consider $\dfrac{1}{1 - z^2} = \dfrac{z}{z\bar{z} - z^2}$

$[\because \; |z| = 1 \Rightarrow |z|^2 = z\bar{z} = 1]$

$= \dfrac{1}{\bar{z} - z}$ = purely imaginary number

$\therefore \; \dfrac{z}{1 - z^2}$ must lie on y–axis.

56. (d) The initial position of point is $Z_0 = 1 + 2i$

$\therefore \; Z_1 - (1 + 5) + (2 + 3)i = 6 + 5i$

Now Z_1 is moved through a distance of $\sqrt{2}$ units in the direction $i + j$.(i.e. by $1 + i$).

$\therefore$ It becomes $Z_1' = Z_1 + (1 + i) = 7 + 6i$

Now OZ_1' is rotated through an angle $\dfrac{\pi}{2}$ in anticlockwise direction therefore

$$Z_2 = iZ_1' = -6 + 7i$$

57. (d) $z = \cos\theta + i\sin\theta$

$\Rightarrow z^{2m-1} = (\cos\theta + i\sin\theta)^{2m-1}$

$= \cos(2m-1)\theta + i\sin(2m-1)\theta$

$\therefore$ Im (z^{2m-1}) = $\sin(2m-1)\theta$

$\therefore \; \sum_{m=1}^{15} \text{Im}(z^{2m-1}) = \sum_{m=1}^{15} \sin(2m-1)\theta$

$= \sin\theta + \sin 3\theta + \sin 5\theta + \ldots +$ upto 15 terms

$= \dfrac{\sin\left[15\left(\frac{20}{2}\right).\sin[\theta + 14 \times \theta]\right]}{\sin\theta}$

$\left[u \sin g \sin\alpha + \sin(\alpha + \beta) + i\sin(\alpha + 2\beta) + \ldots n \text{ terms} = \dfrac{\sin\left(n\beta/2.\sin\left[\alpha + (n-1)\beta/2\right]\right)}{\sin(\beta/2)}\right]$

$= \dfrac{\sin 15\theta.\sin 15\theta}{\sin\theta} = \dfrac{\sin 30^\circ.\sin 30^\circ}{\sin 2^\circ}$

$= \dfrac{1}{4\sin 2^\circ}$

58. (a) given z = $x + iy$ where z and y are integer

Also $z\bar{z}^3 + \bar{z}z^3 = 350$

$\Rightarrow |z|^2(\bar{z}^2 + z^2) = 350$

$\Rightarrow (x^2 + y^2)(x^2 - y^2) = 175$

$\Rightarrow (x^2 + y^2)(x^2 - y^2) = 25 \times 7 \qquad \ldots(i)$

or $(x^2 + y^2)(x^2 - y^2) = 35 \times 5 \qquad \ldots..(ii)$

$\because$ x and y are integers,

$\therefore \; x^2 + y^2 = 25$ and $x^2 - y^2 = 7$

$\Rightarrow x^2 = 16$ and $y^2 = 9$

$\Rightarrow x \pm 4$ and $y = \pm 3$

$\therefore$ Vertices of rectangle are (4, 3), (4, –3), (–4, –3), (–4, 3).

So, area of rectangle = 8 × 6 = 48 sq. units Now from equation (ii)

or $x^2 + y^2 = 35$ and $x^2 - y^2 = 5$

$\Rightarrow x^2 = 20$, which is not possible for any integral value of x.

59. (d) $\because$ $Im(z) \neq 0 \Rightarrow z$ is non real

and equation $z^2 + z + (1-a) = 0$

will have non real roots, if $D < 0$

$\Rightarrow 1 - 4(1-a) < 0 \Rightarrow 4a < 3 \Rightarrow a < \frac{3}{4}$

$\therefore$ a can not take the value $\frac{3}{4}$

60. (c) As α lies on the circle

$(x - x_0)^2 + (y - y_0)^2 = r^2$

$\therefore \ |\alpha - z_0|^2 = r^2$

$\Rightarrow (\alpha - z_0)(\bar{\alpha} - \bar{z}_0) = r^2$

$\Rightarrow \alpha\bar{\alpha} - \alpha\bar{z}_0 - \bar{\alpha}z_0 + z_0\bar{z}_0 = r^2$

$\Rightarrow |\alpha|^2 + |z_0|^2 - \alpha\bar{z}_0 - \bar{\alpha}z_0 = r^2 \quad ...(i)$

Also $\frac{1}{\bar{\alpha}}$ lies on the circle $(x - x_0)^2 + (y - y_0)^2$

$= 4r^2$

$\therefore \ \left|\frac{1}{\bar{\alpha}} - z_0\right|^2 = 4r^2$

$\Rightarrow \left(\frac{1}{\bar{\alpha}} - z_0\right)\left(\frac{1}{\alpha} - \bar{z}_0\right) = 4r^2$

$\Rightarrow \frac{1}{\alpha\bar{\alpha}} - \frac{z_0}{\alpha} - \frac{\bar{z}_0}{\bar{\alpha}} + z_0\bar{z}_0 = 4r^2$

$\Rightarrow \frac{1}{|\alpha|^2} - \frac{z_0\bar{\alpha}}{|\alpha|^2} - \frac{\bar{z}_0\alpha}{|\alpha|^2} + |z_0|^2 = 4r^2$

$\Rightarrow 1 + |\alpha|^2|z_0|^2 - z_0\bar{\alpha} - \bar{z}_0\alpha = 4r^2|\alpha|^2 \quad$ (ii)

Subtracting eq^n (i) from (ii) we get

$1 - |\alpha|^2 + |z_0|^2(|\alpha|^2 - 1) = r^2(4|\alpha|^2 - 1)$

or $(|\alpha|^2 - 1)(|z_0|^2 - 1) = r^2(4|\alpha|^2 - 1)$

Using $|z_0|^2 = \frac{r^2 + 2}{2}$ we get

$$(|\alpha|^2 - 1)\frac{r^2}{2} = r^2(4|\alpha|^2 - 1)$$

$\Rightarrow |\alpha|^2 - 1 = 8|\alpha|^2 - 2 \Rightarrow |\alpha| = \frac{1}{\sqrt{7}}$

(Passage Type)

For (61–62)

Given that $\frac{a}{b} + \frac{b}{c} + \frac{c}{b} - 1$

$\Rightarrow \frac{cis\alpha}{cis\beta} + \frac{cis\beta}{cis\gamma} + \frac{cis\gamma}{cis\alpha} = 1$

Where cis represents $\cos\theta + i\sin\theta$

$\Rightarrow cis(\alpha - \beta) + cis(\beta - \gamma) + cis(\gamma - \alpha) = 1$

Equating the real part of both sides

$\Rightarrow \cos(\alpha - \beta) + \cos(\beta - \gamma) + \cos(\gamma - \alpha) = 1$

Similarly equating imaginary parts we get

$4(\sin(\alpha - \beta) + \sin(\beta - \gamma) + \sin(\gamma - \alpha)) = 0$

61. (a)

62. (d)

63. (a, b) The given equation is $\frac{z^5 - 1}{z - 1} = 0$ which means that z_1, z_2, z_3, z_4 are four out of five 5th roots of unity except 1.

So $z_1 = e^{\left(\frac{i2\pi}{5}\right)}, z_2 = e^{\left(\frac{i4\pi}{5}\right)}, z_3 = e^{\left(\frac{i6\pi}{5}\right)}, z_4 = e^{\left(\frac{i8\pi}{5}\right)}$

Then, $z_1^4 = \left(e^{\left(\frac{i2\pi}{5}\right)}\right)^4 = e^{\left(\frac{i8\pi}{5}\right)} = z_4$

$z_2^4 = \left(e^{\left(\frac{i4\pi}{5}\right)}\right)^4 = e^{\left(\frac{i16\pi}{5}\right)} = e^{\left(\frac{i6\pi}{5}\right)} = z_3$

$z_3^4 = \left(e^{\left(\frac{i6\pi}{5}\right)}\right)^4 = e^{\left(\frac{i24\pi}{5}\right)} = e^{\left(\frac{i4\pi}{5}\right)} = z_2$

$z_4^4 = \left(e^{\left(\frac{i8\pi}{5}\right)}\right)^4 = e^{\left(\frac{i32\pi}{5}\right)} = e^{\left(\frac{i2\pi}{5}\right)} = z_1$

So $z_1^4 + z_2^4 + z_3^4 + z_4^4 = z_4 + z_3 + z_2 + z_1$

$P = z_1^4 + z_2^4 + z_3^4 + z_4^4 + 1^4 = 0$

$\Rightarrow \left|\sum_{i=1}^{4} z_i^4\right| = 1$

$Q = z_1^5 + z_2^5 + z_3^5 + z_4^5 = 4$

$\Rightarrow \left|\sum_{i=1}^{4} z_i^5\right| = 4$

64. (b, c) Since $z^4 + z^3 + z^2 + z + 1$

$= (z - z_1)(z - z_2)(z - z_3)(z - z_4)$

Putting $z = -2$ both the sides and we get

$$\sum_{i=1}^{4}(z_i+2)=11$$

Now consider $|z_1+z_2|=\sqrt{2+2\cos 144^\circ}$ for minimum $=2\cos 72^\circ=\frac{\sqrt{5}-1}{2}$ whose greatest integer is 0.

65. (b, c) Since origin is the centre of the polygon and $z_0, z_1, \ldots, z_{n-1}$ represent the affixes of $A_1, \ldots, A_n$, such that $z_0 = 1, z_1=\alpha,\quad z_2=\alpha^2,\ldots,$ $z_{n-1}=\alpha^{n-1}$, where $\alpha = e^{\frac{i2\pi}{n}}$

Now, $|A_1A_r|^2 = |\alpha^r - 1|^2 = |1-\alpha^r|^2$

$$= \left|1-\cos\frac{2r\pi}{n} - i\sin\frac{2r\pi}{n}\right|^2$$

$$= \left(1-\cos\frac{2r\pi}{n}\right)^2 + \left(\sin\frac{2r\pi}{n}\right)^2$$

$$= 2-2\cos\frac{2r\pi}{n}$$

$$\therefore \sum_{r=2}^{n}|A_1A_r|^2 = \sum_{r=2}^{n}\left(2-2\cos\frac{r\pi}{n}\right)$$

$$=2(n-1)-2\left\{\cos\frac{2\pi}{n}+\cos\frac{4\pi}{n}+\ldots+\cos\frac{2(n-1)\pi}{n}\right\}$$

$$=2(n-1)-2(-1)=2n$$

$$\therefore S_n = |A_1A_2|^2+|A_2A_3|^2+\ldots+|A_1A_n|^2 = 2n$$

66. (a, b, c, d) Now consider

$$P_n = |A_1A_2||A_1A_3|\ldots|A_1A_n|$$

$$=|1-\alpha||1-\alpha^2||1-\alpha^3|\ldots|1-\alpha^{n-1}|$$

$=\left|(1-\alpha)(1-\alpha^2)(1-\alpha^3)\ldots(1-\alpha^{n-1})\right|$. Since, $1,\alpha,\alpha^2,\ldots,\alpha^{n-1}$ are the roots of $z^n-1=0$

$$\Rightarrow (z-1)(z-\alpha)(z-\alpha^2)\ldots(z-\alpha^{n-1}) = z^n-1$$

$$\Rightarrow (z-\alpha)(z-\alpha^2)\ldots(z-\alpha^{n-1})$$

$$=\frac{z^n-1}{z-1}=1+z+z^2+\ldots+z^{n-1}$$

Substituting $z=1$, we have

$$(1-\alpha)(1-\alpha^2)\ldots(1-\alpha^{n-1})=n$$

$$P_n = |1-\alpha||1-\alpha^2|\ldots|1-\alpha^{n-1}|=n$$

For (67 – 68)

Consider equation $x^6-6x+20=0$ it has no real roots, because $f'(x)=0$ at $x=1$, and $f(1)=15>0$

There are no real roots of the equation

$$z^6-6z+20=0$$

If $x+iy$ is a root, then $x-iy$ is also a root.

Let the roots be $x_1\pm iy_1, x_2\pm iy_2, x_3\pm iy_3$

Sum of the roots $=2(x_1+x_2+x_3)=0$

$\Rightarrow x_1+x_2+x_3=0$

$\Rightarrow$ One of x_1, x_2, x_3 is negative and other two are positive or vice–versa

$\Rightarrow$ The number of roots in each of the quadrant is either 1 or 2.

67. (c)
68. (b)
69. (b) Given that points are $A\left(2e^{\frac{i\pi}{4}}\right), B\left(2e^{11i\pi/12}\right)$ and $C(2e^{-5i\pi/12})$ so ABC is an equilateral triangle.

$$AP^2+BP^2+CP^2$$

$$=|z-z_1|^2+|z-z_2|^2+|z-z_3|^2$$

$$=3|z|^2+|z_1|^2+|z_2|^2+|z_3|^2-z(\bar{z}_1+\bar{z}_2+\bar{z}_3)$$
$$-\bar{z}(z_1+z_2+z_3)$$

since ΔABC is equilateral.

Since $|z_1|=|z_2|=|z_3|=2$

And $\frac{z_1+z_2+z_3}{3}=\frac{\bar{z}_1+\bar{z}_2+\bar{z}_3}{3}=0$

Also, $|z|=1$

$\therefore\ AP^2+BP^2+CP^2=3\times1+12=15$

70. (a) Consider the diagram as per the given condition

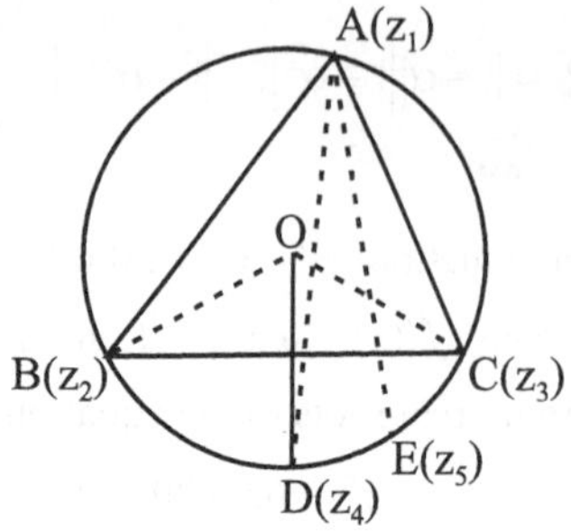

$\angle BOD = \angle COD = A$

$\angle COE = 2\angle CAE = 2\left(\frac{\pi}{2} - C\right)$

$\therefore \frac{z_3}{z_5} = e^{i(\pi - 2C)}$

Also, $\angle AOB = 2C$

$\Rightarrow \frac{z_2}{z_1} = e^{i2C}$

$\Rightarrow \frac{z_3 z_2}{z_1 z_5} = e^{i\pi} = -1$

$z_5 = \frac{-z_2 z_3}{z_1} = -\sqrt{2}(1+i)$

71. (a) Here, $z_1 + z_2 + z_3 + \ldots + z_6 = 0$ .... (i)

$z_2 = z_1 e^{(i2\pi)/n}$ if $e^{(i2\pi)/n} = \alpha$

$\Rightarrow z_2 = z_1 \alpha$

Similarly,

$z_3 = z_1\alpha^2, z_4 = z_2\alpha^3, z_5 = z_2\alpha^4, z_6 = z_1\alpha^5$

Squaring and adding and then using Eq. (i), we get $z_1^2 + z_2^2 + z_3^2 + z_4^2 + z_5^2 + z_6^2 = 0$

72. (b) Since $AM \geq GM$, so

$$\frac{1 + 2\omega + 3\omega^2 + 4\omega^3 + \ldots + n\omega^{n-1}}{n} \geq \left\{ n!\omega^{\frac{n(n-1)}{2}} \right\}^{\frac{1}{n}}$$

Now, $S = 1 + 2\omega + 3\omega^2 + 4\omega^3 + \ldots + n\omega^{n-1}$ ...(i)

$\omega S = \omega + 2\omega^2 + 3\omega^3 + \ldots + (n-1)\omega^{n-1} + n\omega^n$

$(1-\omega)S = 1 + \omega + \omega^2 + \ldots + \omega^{n-1} - n\omega^n$ ...(ii)

$\Rightarrow S = \frac{0 - n(1)}{1 - \omega}$

$\Rightarrow S = \frac{n}{\omega - 1}$,

Required maximum value $= \frac{1}{\omega - 1}$

73. (a) The lines are parallel if their slopes are equal, so

$i\left(\frac{\bar{a}_1 + a_1}{\bar{a}_1 - a_1}\right) = i\left(\frac{\bar{a}_2 + a_2}{\bar{a}_2 - a_2}\right)$

$\Rightarrow \bar{a}_1\bar{a}_2 + a_1\bar{a}_2 - \bar{a}_1 a_2 - a_1 a_2$

$= \bar{a}_1\bar{a}_2 + \bar{a}_1 \; a_2 - a_1\bar{a}_2 - a_1 a_2$

$\Rightarrow a_1\bar{a}_2 = \bar{a}_1 \; a_2 \qquad \Rightarrow \frac{a}{a_2} = \left(\frac{\bar{a}_1}{\bar{a}_2}\right)$

$\Rightarrow \frac{a_1}{a_2}$ is purely real

74. (c) The straight line are perpendicular if

$i\left(\frac{\bar{a}_1 + a_1}{\bar{a}_1 - a_1}\right) . i\left(\frac{\bar{a}_2 + a_2}{\bar{a}_2 - a_2}\right) = -1$

$\Rightarrow \bar{a}_1\bar{a}_2 + \bar{a}_1 \; a_2 + a_1\bar{a}_2 + a_1 a_2$

$= \bar{a}_1\bar{a}_2 - \bar{a}_1 a_2 - a_1\bar{a}_2 + a_1 a_2$

$\Rightarrow \bar{a}_1 \; a_2 + a_1\bar{a}_2 = 0$

75. (d) Let $z = x + iy$ and $\omega = \alpha + i\beta$

Now $|z + i\omega| = 2 \Rightarrow (z + i\bar{\omega})(\bar{z} - i\bar{\omega}) = 4$

$|z|^2 + |\omega|^2 + i\omega\bar{z} - i\bar{\omega}z = 4$

$\Rightarrow i\omega\bar{z} - i\bar{\omega}z = 2$ ...(i)

and $|z - i\bar{\omega}| = 2$

$\Rightarrow (z - i\bar{\omega})(\bar{z} + i\omega) = 4$

$\Rightarrow |z|^2 + |\omega|^2 + i\omega z - i\bar{\omega}\bar{z} = 4$

$\Rightarrow i\omega z - i\bar{\omega}\bar{z} = 2$ ...(ii)

Add (i) and (ii),

$\Rightarrow i(\omega - \bar{\omega})(z + \bar{z}) = 4$

$\Rightarrow i(2i\beta)(2x) = 4$

$\Rightarrow \beta x = -1$ ...(iii)

From equations (i) and (ii)

$\Rightarrow i\ (\omega + \bar{\omega})(z - \bar{z}) = 0 \Rightarrow \alpha y = 0$ ...(iv)

So either $\alpha = 0$ or $y = 0$.

If $y = 0$, then $x^2 + y^2 = 1 \Rightarrow x = \pm 1 \Rightarrow z = 1$ or -1

If $\alpha = 0$, then $\alpha^2 + \beta^2 = 1 \Rightarrow \beta = \pm 1 \Rightarrow \omega = \pm i$

So, $I_m(z) = \text{Re}(\omega) = 0$

76. (b) As $z = \pm 1$, so two values of z can be obtained.

77. (a) 1, $\omega, \omega^2 \ldots\ldots\omega^{n-1}$ are the nth roots of unity.

$\Rightarrow 1 + \omega + \omega^2 + \ldots + \omega^{n-1} = 0$

Now let $s = 1 + 2\omega + 3\omega^2 + \ldots\ldots + n\omega^{n-1}$

$\omega s = \omega + 2\omega^2 + \ldots\ldots + (n-1)\omega^{n-1} + n\omega^n$

$(1 - \omega)s = 1 + [\omega + \omega^2 + \ldots +$

$\omega^{n-1}] - n\omega^n = -n \Rightarrow s = \dfrac{n}{\omega - 1}$

78. (d) Again $x^n - 1 = (x - 1)(x - \omega)(x - \omega^2) \ldots (x - \omega^{n-1})$

$\Rightarrow (x - \omega)(x - \omega^2) \ldots (x - \omega^{n-1}) = \dfrac{x^n - 1}{x - 1}$

Put $x = 2$

$\Rightarrow (2 - \omega)(2 - \omega^2) \ldots (2 - \omega^{n-1}) = 2^n - 1$

79. Consider the diagram as per the given condition

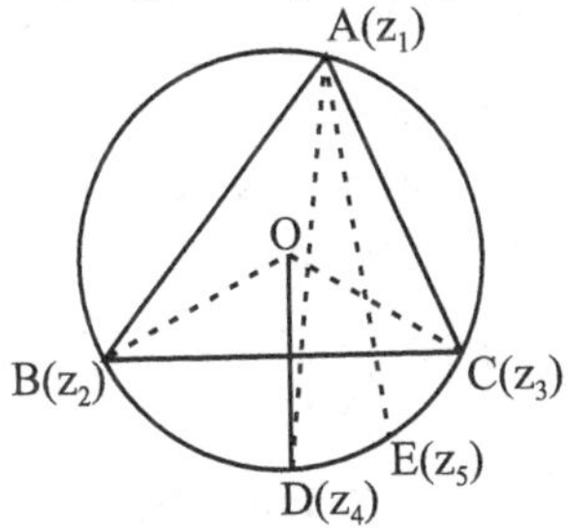

$\angle BOD = \angle COD = A$

$\therefore \dfrac{z_4 - 0}{z_2 - 0} = e^{iA}$

and $\dfrac{z_3 - 0}{z_4 - 0} = e^{iA}$

$\therefore z_4^2 = z_2 z_3 \Rightarrow \dfrac{z_2 z_3}{z_4^2} = 1$

80. (d) $\angle COE = 2\angle CAE$

$= 2\left(\dfrac{\pi}{2} - C\right) \quad \therefore \dfrac{z_3}{z_5} = e^{i(\pi - 2C)}$

Also, $\angle AOB = 2C$

$\Rightarrow \dfrac{z_2}{z_1} = e^{i2C}$

$\Rightarrow \dfrac{z_3 z_2}{z_1 z_5} = e^{i\pi} = -1$

81. (c) Given that $z = \dfrac{\alpha + i\beta}{1 - \gamma}$.

so $|z|^2 = \dfrac{\alpha^2 + \beta^2}{(1-\gamma)^2} = \dfrac{\gamma - \gamma^2}{(1-\gamma)^2} = \dfrac{\gamma}{1-\gamma}$

82. (d) We know that

$|z|^2 = \dfrac{\gamma}{1-\gamma} = -1 + \dfrac{1}{1-\gamma}$

$\Rightarrow 1 + |z|^2 = \dfrac{1}{1-\gamma}$

$\Rightarrow \gamma = \dfrac{|z|^2}{1 + |z|^2}$

Now, $\dfrac{2\alpha}{1-\gamma} = z + \bar{z}$

$\Rightarrow 2\alpha = (z + \bar{z})(1 - \gamma) = \dfrac{z + \bar{z}}{1 + |z|^2}$

$\Rightarrow \alpha = \dfrac{z + \bar{z}}{2(1 + |z|^2)}$

Now consider $\dfrac{2i\beta}{1-\gamma} = z - \bar{z}$

$\Rightarrow 2i\beta = (z - \bar{z})(1 - \gamma) = \dfrac{z - \bar{z}}{1 + |z|^2}$

$\Rightarrow \beta = \dfrac{(\bar{z} - z)i}{2(1 + |z|^2)}$

So $\alpha + \beta =$

$$\frac{z + \bar{z}}{2(1+|z|^2)} + \frac{(\bar{z} - z)i}{2(1+|z|^2)} = \frac{\{z(1-i) + \bar{z}(1+i)\}}{\{2(1+|z|^2)\}}$$

For (83–85)

Given that $A = \{z : Imz \geq 1\} = \{(x, y) : y \geq 1\}$

So A is the set of all points lying on or above the line $y = 1$.

$$B = \{z : |z - 2 - i| = 3\} = \{(x, y) : (x-2)^2 + (y-1)^2 = 9\}$$

$\Rightarrow B$ is the set of all points lying on the boundary of the circle with centre (2, 1) and radius 3.

$C = \{z : \text{Re}[(1-i)z] = \sqrt{2}\} = \{(x, y) : x + y = \sqrt{2}\}$

$\Rightarrow C$ is the set of all points lying on the straight line $x + y = \sqrt{2}$.

Graphically, the three sets are represented as shown below:

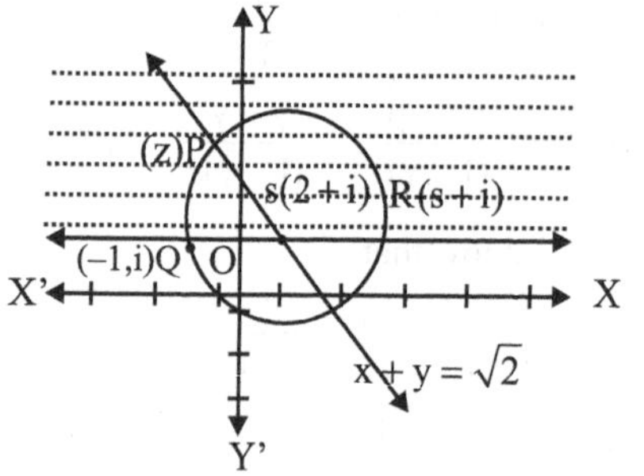

83. (b) From graph $A \cap B \cap C$ consist of only one point P [the common point of the region $y \geq 1, (x-2)^2 + (y-1)^2 = 9$ and $x + y = \sqrt{2}$]

$\therefore n(A \cap B \cap C) = 1$

84. (c) Since z is a point of $A \cap B \cap C$

$\Rightarrow$ z represents the point of P

$\therefore |z + 1 - i|^2 + |z - 5 - i|^2$

$\Rightarrow |z - (-1 + i)|^2 + |z - (5 + i)|^2$

$\Rightarrow PQ^2 + PR^2 = QR^2 = 6^2 = 36$

which lies between 35 and 39

85. (d) Given that $|w - 2 - i| < 3$

$\Rightarrow$ distance between w and $2 + i$ i.e. is smaller than 3.

$\Rightarrow$ w is a point lying inside the circle with centre $(2+i)$ and radius 3.

$\Rightarrow$ distance between z (i.e. the point P) and w should be smaller than 6 (the diameter of the circle)i.e. $|z - w| < 6$

But we know that $||z| - |w|| < |z - w|$

$\Rightarrow ||z| - |w|| < 6$

$\Rightarrow -6 < |z| - |w| < 6, or -3 < |z| - |w| + 3 < 9$

86. (b) Area of shaded region

$= \frac{\pi}{4} \times 4^2 + \frac{\pi \times 4^2 \times 60^\circ}{360^\circ}$

$= 4\pi + \frac{8\pi}{3} = \frac{20\pi}{3}$

87. (c) $S_1 : x^2 + y^2 < 16$

$S_2 : Im\left[\frac{(x-1) + i(y + \sqrt{3})}{1 - i\sqrt{3}}\right] > 0$

$\Rightarrow \sqrt{3}(x-1) + (y + \sqrt{3}) > 0 \Rightarrow y + \sqrt{3}x > 0$

$S_3 : x > 0$

Then S : $S_1 \cap S_2 \cap S_3$ is as shown in the figure given below,

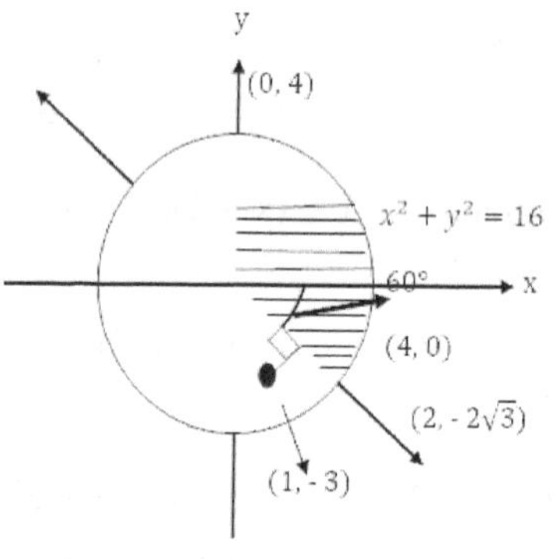

$\therefore$ min $|1 - 3i - z|$ = min distance between z and $(1, -3)$ $z \in s$

clearly (from figure) minimum distance between $z \in s$ and $(1, -3)$ from line $y + x\sqrt{3} = 0$ i.e. $\left|\frac{\sqrt{3} - 3}{\sqrt{3} + 1}\right| = \frac{3 - \sqrt{3}}{2}$.

(Integer type)

88. **(2)** We have, $abcd = \cos(2\alpha + 2\beta + 2\gamma + 2\delta) + i\sin(2\alpha + 2\beta + 2\gamma + 2\delta)$

$\therefore \sqrt{abcd} = [\cos(2\alpha + 2\beta + 2\gamma + 2\delta) + i\sin(2\alpha + 2\beta + 2\gamma + 2\delta)]^{1/2}$

or $\sqrt{abcd}$

$= \cos(\alpha + \beta + \gamma + \delta) + i\sin(\alpha + \beta + \gamma + \delta)$ ...(i)

$\therefore \frac{1}{\sqrt{abcd}}$

$= \cos(\alpha+\beta+\gamma+\delta) - i\sin(\alpha+\beta+\gamma+\delta)$ ...(ii)

On adding Eqs. (i) and (ii), we obtain

$$\sqrt{abcd} + \frac{1}{\sqrt{abcd}} = 2\cos(\alpha+\beta+\gamma+\delta)$$

$$\frac{p}{\cos(\alpha+\beta+\gamma+\delta)} = 2$$

89. **(7)** We have, $x = \omega - \omega^2 - 2$ or $x + 2 = \omega - \omega^2$

Squaring, $x^2 + 4x + 4 = \omega^2 + \omega^4 - 2\omega^3$

$= \omega^2 + \omega^3 \cdot \omega - 2\omega^3 = \omega^2 + \omega - 2$

$= -1 - 2 = -3$

$\Rightarrow x^2 + 4x + 7 = 0$

Dividing $x^4 + 3x^3 + 2x^2 - 11x$ by $x^2 + 4x + 7$,

we get, $x^4 + 3x^3 + 2x^2 - 11x - 6$

$$= (x^2 + 4x + 7)(x^2 - x - 1) + 7$$

$\Rightarrow (0)(x^2 - x - 1) + 1 = 0 + 1 = 7$

90. **(1)** We have,

$$\sum_{k=1}^{10}\left(\sin\frac{2\pi k}{11} - i\cos\frac{2\pi k}{11}\right)$$

$$= \sum_{k=1}^{10}\left(-i^2\sin\frac{2\pi k}{11} - i\cos\frac{2\pi k}{11}\right)$$

$$= -i\sum_{k=1}^{10}\left(\cos\frac{2\pi k}{11} + i\sin\frac{2\pi k}{11}\right)$$

$$= -i\sum_{k=1}^{10} e^{i\frac{2\pi k}{11}}$$

$= -i$ (sum of 11th roots of unity – 1)

$= -i(0-1) = i$

So x+ y = 0+1 =1

91. **(3)** We have, $\alpha = \cos\left(\frac{8\pi}{11}\right) + i\sin\left(\frac{8\pi}{11}\right) = e^{\frac{8\pi}{11}i}$

$\text{Re}(\alpha + \alpha^2 + \alpha^3 + \alpha^4 + \alpha^5)$

$$= \frac{\alpha+\alpha^2+\alpha^3+\alpha^4+\alpha^5+\bar\alpha+\bar\alpha^2+\bar\alpha^3+\bar\alpha^4+\bar\alpha^5}{2}$$

$$= \frac{-1+\left(\begin{matrix}1+\alpha+\alpha^2+\alpha^3+\alpha^4+\alpha^5 \\ +\bar\alpha+\bar\alpha^2+\bar\alpha^3+\bar\alpha^4+\bar\alpha^5\end{matrix}\right)}{2}$$

$= \frac{-1+0}{2}$ (sum of 11, 11th roots ....)

$= -\frac{1}{2}$ so 6|P| = 3

92. **(9)** Lets generalize it for n roots of unity

Since, $1, \omega, \omega^2, \ldots \omega^{n-1}$ are the n, nth roots of unity

$\therefore \sum_{k=0}^{n-1}\omega^k = 0$ and $\sum_{k=0}^{n-1}\bar\omega^k = 0$

Now,

$$\sum_{k=0}^{n-1}|z_1 + \omega^k z_2|^2$$

$$= \sum_{k=0}^{n-1}(z_1 + \omega^k z_2)\{\bar z_1 + (\bar\omega)^k \bar z_2\}$$

$$= \sum_{k=0}^{n-1}\left[z_1\bar z_1 + z_1\bar z_2(\bar\omega)^k + \bar z_1 z_2\omega^k + z_2\bar z_2(\omega)^k(\bar\omega)^k\right]$$

$$= \sum_{k=0}^{n-1}|z_1|^2 + \sum_{k=0}^{n-1} z_1\bar z_2(\bar\omega)^k + \sum_{k=0}^{n-1}\bar z_1 z_2\omega^k + \sum_{k=0}^{n-1}|z_2|^2$$

$$= n|z_1|^2 + 0 + 0 + n|z_2|^2 = n(|z_1|^2 + |z_2|^2)$$

$$\frac{\left\{\sum_{k=0}^{8}|z_1 + w^k z_2|^2\right\}}{|z_1|^2 + |z_2|^2} = n$$

Here n =9

93. **(1)** We have, $(1+\omega)(1+\omega^2)(1+\omega^4)(1+\omega^8)\ldots$ to $2n$ factors

$= (1+\omega)(1+\omega^2)(1+\omega^3\cdot\omega)(1+\omega^6\cdot\omega^2)\ldots$ to $2n$ factors

$= (1+\omega)(1+\omega^2)(1+\omega)(1+\omega^2)\ldots$ to $2n$ factors [since $\omega^3 = \omega^6 = \ldots =$]

$= [(1+\omega)(1+\omega)\ldots$ to n factors$]$

$[(1+\omega^2)(1+\omega^2)$ to n factors$]$

$$= (1+\omega)^n(1+\omega^2)^n = [(1+\omega)(1+\omega^2)]^n$$

$$= (1+\omega+\omega^2+\omega^3)^n = (0+1)^n = 1$$

$[\text{since } 1+\omega+\omega^2 = 0, \omega^3 = 1]$

Smart Cracker
Since total number of factors is 2n so let us assume that n=1 then just solve $(1+\omega)(1+\omega^2)=1$ so 1 is the answer

94. **(9)** Let $x = 3 + 2i$

$\Rightarrow (x-3)^2 = 4i^2 = -4$

$\Rightarrow x^2 - 6x + 13 = 0$

Thus, $x^4 - 8x^3 + 4x^2 + 4x + 39$

$= (x^2 - 6x + 13)(x^2 - 2x - 21) - 96x + 312$

$\therefore f(3+2i) = 0(x^2 - 2x - 21) - 96(3+2i) + 312$

$= 24 - 192i = a + ib$

$\therefore a = 24$ and $b = -192$ then $\frac{|a-b|}{24} = \frac{216}{24} = 9$

95. **(0)** The *n, nth* root of unity are given by $cis\left(\frac{2k\pi}{n}\right), k = 0,1,2,\ldots,n-1$

Let $z_1 = cis\left(\frac{2k_1\pi}{n}\right), and\ z_2 = cis\left(\frac{2k_2\pi}{n}\right)$

where k_1 *and* k_1 are distinct integers from the set $\{0,1,2,\ldots,n-1\}$

As z_1 *and* z_2 subtend a right angle at the origin, therefore,

$$\left|\frac{2k_1\pi}{n} - \frac{2k_2\pi}{n}\right| = \frac{\pi}{2}, k_1 \neq k_2$$

$\Rightarrow \frac{2\pi}{n}|k_1 - k_2| = \frac{\pi}{2} \Rightarrow 4|k_1 - k_2| = n$

So, $n = 4$ (a positive integer)

So n is a multiple of 4 and required remainder is 0

96. **(3)** If $P(z)$ be any point on the ellipse. Then equation of the ellipse is,

$$|z - z_1| + |z - z_2| = \frac{|z_1 - z_2|}{e} \quad \ldots(i)$$

$\therefore$ For $P(z)$ to lie inside ellipse, we have

$$|z - z_1| + |z - z_2| < \frac{|z_1 - z_2|}{e}$$

It is given that origin is an interior point of the ellipse.

$\Rightarrow |0 - z_1| + |0 - z_2| < \frac{|z_1 - z_2|}{e}$

$\Rightarrow e \in \left(0, \frac{|z_1 - z_2|}{|z_1| + |z_2|}\right)$

So a + b + c + d = 0 + 1 + 1 + 1 = 3

97. **(1)** Let $\cot^{-1} p = \theta$, then $p = \cot\theta$

$\therefore\ e^{2mi\cot^{-1}p} \cdot \left(\frac{pi+1}{pi-1}\right)^m = e^{2mi\theta} \cdot \left(\frac{i\cot\theta + 1}{i\cot\theta - 1}\right)^m$

$= e^{2mi\theta} \cdot \left(\frac{i(\cot\theta - i)}{i(\cot\theta + 1)}\right)^m = e^{2mi\theta} \cdot \left(\frac{\cot\theta - i}{\cot\theta + i}\right)^m$

$= e^{2mi\theta} \cdot \left(\frac{\cos\theta - i\sin\theta}{\cos\theta + i\sin\theta}\right)^m = e^{2mi\theta} \cdot \left(\frac{e^{-i\theta}}{e^{i\theta}}\right)^m$

$= e^{2mi\theta}\left(e^{-2i\theta}\right)^m = e^0 = 1$

98. **(1)** We have, $z = \frac{q+ir}{1+p}$; $\therefore iz = \frac{-r+iq}{1+p}$

$\therefore$ By componendo and dividendo,

$\frac{1+iz}{1-iz} = \frac{1+p-r+iq}{1+p+r-iq}$

$\therefore \frac{p+iq}{1+r} = \frac{1+iz}{1-iz}$ if $\frac{p+iq}{1+r} = \frac{1+p-r+iq}{1+p+r-iq}$

or $p(1+p+r) + q^2 + i\{q(1+p+r) - pq\}$

$= (1+r)(1+p-r) + iq(1+r)$

$\Rightarrow p(1+p+r) + q^2 = (1+r)(1+p-r)$ and $(1+p+r) - pq = q(1+r)$, (this is obviously true)

$\therefore$ The condition is

$p(1+p+r) + q^2 = (1+r)(1+p-r)$

or $p + p^2 + pr + q^2 = 1 + p - r + r + pr - r^2$

or $p^2 + q^2 + r^2 = 1$

99. **(3)** Let $z = x + iy$

We have, $z + \alpha|z - 1| + 2i = 0$

$\Rightarrow x + i(y+2) + \alpha\sqrt{(x-1)^2 + y^2} = 0$

Equating real and imaginary parts

$y + 2 = 0 \therefore y = -2$

and $x+\alpha\sqrt{(x-1)^2+4}=0$

$\therefore x^2=\alpha^2\left(x^2-2x+5\right)$

or $\left(1-\alpha^2\right)x^2+2\alpha^2x-5\alpha^2=0$

Since x is real.

$\therefore D=B^2-4AC\geq 0$

$\Rightarrow 4\alpha^4+20\alpha^2\left(1-\alpha^2\right)\geq 0$

$\Rightarrow -4\alpha^4+5\alpha^2\geq 0$

$\Rightarrow 4\alpha^2\left(\alpha^2-\frac{5}{4}\right)\leq 0$

$\Rightarrow -\frac{\sqrt{5}}{2}\leq\alpha\leq\frac{\sqrt{5}}{2}$

Integral values of a are –1, 0 and 1

100. **(3)** Maximum length of altitude from A is AD $=|\alpha-\beta|$

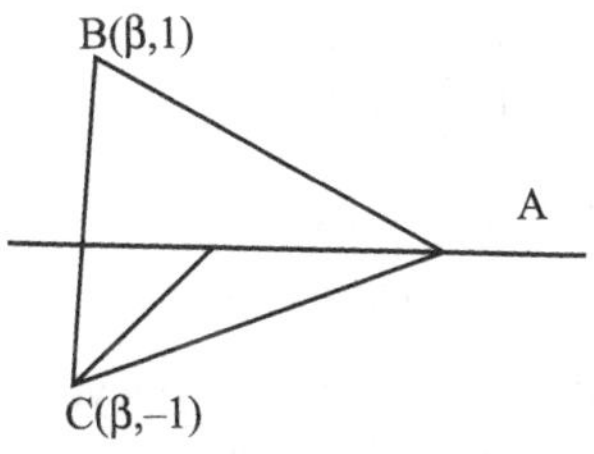

Since $\alpha, \beta+i, \beta-i$ are three roots of f(x) =0

so, let f(x) = $k(x-\alpha)(x-\beta-i)(x-\beta+i)$

$=k(x-\alpha)\left((x-\beta)^2+1\right)$

f'(x)= $k\left((x-\beta)^2+1\right)+k(x-\alpha)(2(x-\beta))$

$\Rightarrow$ f'(x) = $k\left(3x^2-2(\alpha+2\beta)x+\left(\beta^2+2\alpha\beta+1\right)\right)$

$\Delta=4\ (\alpha+2\beta)^2-4.3\left(\beta^2+1+2\alpha\beta\right)$

$\Delta=4\left\{\alpha^2+\beta^2-2\alpha\beta-3\right\}$

$\Delta=4\left\{(\alpha-\beta)^2-3\right\}$

For equal or complex roots, we have

$(\alpha-\beta)^2-3\leq 0$

$\Rightarrow |\alpha-\beta|\leq\sqrt{3}$, where $|\alpha-\beta|=AD$.

Hence maximum length of altitude AD is $\sqrt{3}$.

And square of that length is 3

101. **(3)** The given equation is

$$10(z\bar{z})-3i\left\{z^2-(\bar{z})^2\right\}-16=0$$

or $10(x^2+y^2)-3i(2x)(2iy)-16=0$

$\Rightarrow 5(x^2+y^2)+6xy-8=0$ …….. (1)

Let $(r\cos\theta, r\sin\theta)$ be a point on (1), then

$5r^2+6r^2\sin\theta\cos\theta-8=0\Rightarrow r^2=\frac{8}{5+3\sin 2\theta}$

Clearly, $1\leq r^2\leq 4 \quad\Rightarrow 1\leq|r|\leq 2$

So minimum distance is 1 and maximum distance is 2 and its sum is 2 + 1 = 3

102. **(5)** Consider $|z-3-2i|\leq 2$

Which represents interior points of a circle with centre (3, 2) and radius 2.

Now $|2z-6+5i|=2\left|z-\left(3-\frac{5}{2}i\right)\right|$

= 2 × distance of z from p (where Z lies in or on the circle)

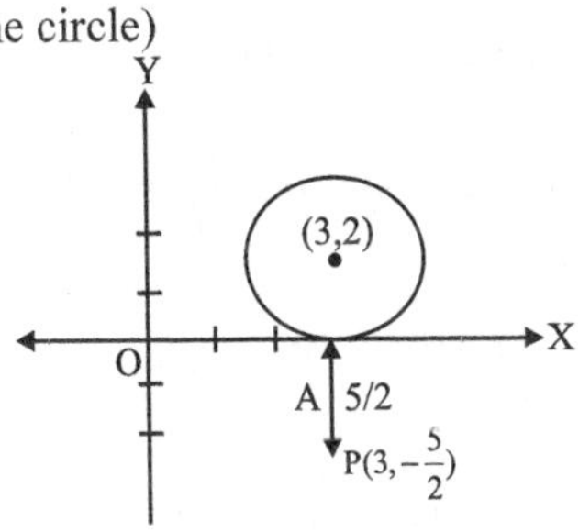

Also min distance of z from $P=\frac{5}{2}$

$\therefore$ Minimum value of $|2z-6+5i|=5$

ALTERNATIVE:

$\left|z-3+\frac{5}{2}i\right|=\left|z-3-2i+2i+\frac{5}{2}i\right|$

$=\left|(z-3-2i)+\frac{9}{2}i\right|$

103. **(3)** The expression may not attain integral value for all a, b, c. If we consider a = b = c then x = 3a, y = $a\left(1+\omega+\omega^2\right)=a\left(1+i\sqrt{3}\right)$

Z = $a\left(1+\omega+\omega^2\right)=a\left(1+i\sqrt{3}\right)$

$\Rightarrow \quad |x|^2+|y|^2+|z|^2=9|a|^2+4|a|^2+4|a|^2$

$=17|a|^2$

$$\Rightarrow \frac{|x|^2+|y|^2+|z|^2}{|a|^2+|b|^2+|c|^2}=\frac{17}{3}$$

(which is not an integer)

Note: However if $\omega = e^{i\left(2\frac{\pi}{3}\right)}$, then the value of expression can be evaluated as follows

$$\frac{|x|^2+|y|^2+|z|^2}{|a|^2+|b|^2+|c|^2}=\frac{x\bar{x}+y\bar{y}+z\bar{z}}{|a|^2+|b|^2+|c|^2}$$

$$=\frac{(a+b+c)(\bar{a}+\bar{b}+\bar{c})+(a+b\omega+c\omega^2)(\bar{a}+\bar{b}\omega^2+\bar{c}\omega)+(a+b\omega^2+c\omega)(\bar{a}+\bar{b}\omega+\bar{c}\omega^2)}{|a|^2+|b|^2+|c|^2}$$

$$=\frac{3|a|^2+3|b|^2+3|c|^2+(a\bar{b}+\bar{a}b+b\bar{c}+\bar{b}c+a\bar{c}+\bar{a}c)(1+\omega+\omega^2)}{|a|^2+|b|^2+|c|^2}$$

$= 3\ (\because 1+\omega+\omega^2=0)$

104. **(4)** $\alpha_k=\cos\frac{k\pi}{7}+i\sin\frac{k\pi}{7}=e^{\frac{i\pi k}{7}}$

$\alpha_{k+1}-\alpha_k$

$=e^{\frac{i\pi(k+1)}{7}}-e^{\frac{i\pi k}{7}}=e^{\frac{i\pi k}{7}}(e^{i\pi/7}-1)$

$|\alpha_{k+1}-\alpha_k|=|e^{i\pi/7}-1|$

$\Rightarrow \sum_{k=1}^{12}|\alpha_{k+1}-\alpha_k|=12|e^{i\pi/7}-1|$

Similarly $\sum_{k=1}^{3}|\alpha_{4k-1}-\alpha_{4k-2}|=3|e^{i\pi/7}-1|$

$$\therefore \frac{\sum_{k=1}^{12}|\alpha_{k+1}-\alpha_k|}{\sum_{k=1}^{3}|\alpha_{4k-1}-\alpha_{4k-2}|}=4$$

(Assertion and Reason)

105. (d) Solution: $z+\frac{1}{z}=1$

$\Rightarrow z^2-z+1=0 \qquad \therefore z=-\omega,-\omega^2$

Now, for

$z=-\omega, p=(-\omega)^{4000}+\frac{1}{(-\omega)^{4000}}=\omega+\frac{1}{\omega}=-1$

so statement I is false.

Similarly $z^3+\frac{1}{z^3}=(-\omega)^3+\frac{1}{(-\omega)^3}=-2$

so statement II is true.

106. (b) Statement II is a fact and hence correct.

Now consider statement I–

$\frac{2}{z_1}=\frac{1}{z_2}+\frac{1}{z_3}\Rightarrow \frac{1}{z_1}-\frac{1}{z_2}=\frac{1}{z_3}-\frac{1}{z_1}$

$\Rightarrow \frac{z_2-z_1}{z_2z_1}=\frac{z_1-z_3}{z_1z_3} \qquad \Rightarrow \frac{z_2-z_1}{z_3-z_1}=-\frac{z_2}{z_3}$

$\Rightarrow \arg\left(\frac{z_2-z_1}{z_3-z_1}\right)=\arg\left(-\frac{z_2}{z_3}\right)$

$\Rightarrow \arg\left(\frac{z_2-z_1}{z_3-z_1}\right)=\pi-\arg\left(\frac{z_2}{z_3}\right)$

$=\pi-\arg\left(\frac{z_2-0}{z_3-0}\right)$

$\Rightarrow \angle CAB=\pi-\angle COB$

$\Rightarrow \angle CAB+\angle COB=\pi$

$\Rightarrow$ Points O, A, B, C are concyclic.

107. (a) Statement II is a fact and correct

Now consider statement I

$|z|=\left|z+\frac{1}{z}-\frac{1}{z}\right|\le\left|z+\frac{1}{z}\right|+\frac{1}{|z|}\le 1+\frac{1}{|z|}$

$\Rightarrow |z|^2-|z|-1\le 0$

$\Rightarrow |z|\in\left[\frac{1-\sqrt{5}}{2},\frac{1+\sqrt{5}}{2}\right]$

But as $|z|>0$, we have, $|z|\in\left(0,\frac{1+\sqrt{5}}{2}\right]y$

$\Rightarrow$ Maximum value of $|z|$ is $\frac{1+\sqrt{5}}{2}$

108. (a) Statement II is true.

Now consider statement I,

$|z_1|=|z_2|+|z_1-z_2|$

$\therefore (|z_1-z_2|)^2=(|z_1|-|z_2|)^2$

$\Rightarrow |z_1|^2+|z_2|^2-2|z_1||z_2|\cos(\theta_1-\theta_2)$

$= |z_1|^2 + |z_2|^2 - 2|z_1||z_2|$

$\Rightarrow \cos(\theta_1 - \theta_2) = 1$

$\Rightarrow \theta_1 - \theta_2 = 0$

$\Rightarrow \arg(z_1) - \arg(z_2) = 0$

$\Rightarrow \arg\left(\frac{z_1}{z_2}\right) = 0 \Rightarrow \frac{z_1}{z_2}$ is purely real.

$\Rightarrow \text{Im}\left(\frac{z_1}{z_2}\right) = 0$

Statement I and Statement II are true and Statement II is correct explanation of Statement I.

109. (a) Statement II is a fact and hence correct

Consider statement II:–

$\left|z^2 + 2z\cos\alpha\right| < \left|z^2\right| + \left|2z\cos\alpha\right| < \left|z^2\right| + 2|z| < 8$

110. (d) Statement II is correct

Now consider Statement I

$x + \frac{1}{x} = -1,\ x^2 + \frac{1}{x^2} = -1,$

$x^3 + \frac{1}{x^3} = 2,\ x^4 + \frac{1}{x^4}$

$= x + \frac{1}{x},\ x^5 + \frac{1}{x^5} = -1,\ x^6 + \frac{1}{x^6} = 2$ etc.

$$\Rightarrow \left(x + \frac{1}{x}\right)^2 + \left(x^2 + \frac{1}{x^2}\right)^2 + \left(x^3 + \frac{1}{x^3}\right)^2$$

$$+ \left(x^4 + \frac{1}{x^4}\right)^2 + \left(x^5 + \frac{1}{x^5}\right)^2 + \left(x^6 + \frac{1}{x^6}\right)^2$$

$$+ \left(x^7 + \frac{1}{x^7}\right)^2 + \ldots + \left(\left(x^3\right)^9 + \frac{1}{\left(x^3\right)^9}\right)^2$$

$= (1 + 1 + 4) + (1 + 1 + 4) + (1 + 1 + 4) + \ldots 9$ times $= 6 \times 9 = 54$.

111. (a) Statement II is correct.

Statement I–

The equation can be rewritten as

$z\bar{z} + \bar{a}z + a\bar{z} + a\bar{a} = a\bar{a} - \lambda$

$\Rightarrow (z + a)\left(\overline{a + z}\right) = a\bar{a} - \lambda$

$\Rightarrow |z + a| = \sqrt{a\bar{a} - \lambda}$

Since, $a\bar{a}$ is real, λ should be real.

$\sqrt{a\bar{a} - \lambda}$ represents radius of the circle.

Hence, Statement I and Statement II both are True.

112. (a) Consider Statement I:– $|z_1| = r_1$ and $|z_2| = r_2$

Then $z_1 = r_1 e^{i\theta}$ & $z_2 = r_2 e^{i(\theta+\alpha)}$

$$\frac{az_1}{bz_2} + \frac{bz_2}{az_1} = \frac{ar_1e^{i\theta}}{br_2e^{i(\theta+\alpha)}} + \frac{br_2e^{i(\theta+\alpha)}}{ar_1e^{i\theta}}$$

$$= e^{-i\alpha} + e^{i\alpha} = 2\cos\alpha \in [-2, 2]$$

113. (c) Consider Statement I $|z| = \frac{-a \pm \sqrt{a^2 - 4b}}{2}$,

if $a^2 > 4b$ and $|z|$ is a positive number c, then

$|z| = c$

$\Rightarrow z = c(\cos\theta + i\sin\theta)$

$\Rightarrow$ Infinite complex numbers satisfy for $b \le 0$ given equation. So statement I is false.

Now consider statement II :– A quadratic can have more than two roots if all coefficients are zero.

114.(c) Statement II is incorrect as sum of the roots must be zero

Now consider statement I

$ax^2 + bx + c + i = 0$

$$x = \frac{-b \pm \sqrt{b^2 - 4a(c + i)}}{2a} = \frac{-b \pm (p + iq)}{2a} (say)$$

Now, $\frac{-b \pm p}{2a} = 0 \quad \Rightarrow b = \pm p$

So, $\sqrt{b^2 - 4ac - 4ai} = p + iq$

$\Rightarrow b^2 - 4ac = p^2 - q^2$ and $2pq = -4a$

Eliminating p and q we get $\rightarrow a = b^2c$ since all coefficient are not real so the statement II is not true

(Matrix Match Type)

115. $A \rightarrow R,\ B \rightarrow Q,\ C \rightarrow P,\ D \rightarrow S$

A. We know that $\left||z - z_1| - |z - z_2|\right| = k$ represents a hyperbola, if $|z_1 - z_2| > k$.

Here $|z_1 - z_2| = 5$ so given expression represents a hyperbola

$A \to R$

B. Let $\arg(z) = \theta$ then given equation is

$$|(x-3)+i(y-2)| = |z|\left|\frac{1}{\sqrt{2}}\cos\theta + \frac{1}{\sqrt{2}}\sin\theta\right|$$

$$\Rightarrow \sqrt{(x-3)^2+(y-2)^2} = \frac{1}{\sqrt{2}}|x+y|.$$

This is equivalent to locus of points whose distance from point (3, 2) and from straight line x+y =0 is equal, which represent a parabola with focus (3, 2) and directrix x+y=0

B→Q

C. Given,

$$\left|z-1-\sin^{-1}\frac{1}{\sqrt{3}}\right| + \left|z-\cos^{-1}\frac{1}{\sqrt{3}}+\frac{\pi}{2}\right| = 1$$

Since, $1+\sin^{-1}\frac{1}{\sqrt{3}}+\cos^{-1}\frac{1}{\sqrt{3}}-\frac{\pi}{2} = 1$

$\therefore |z-z_1|+|z-z_2| = |z_1-z_2|$

$\Rightarrow$ Locus of z is line segment joining z_1 and z_2

$C \to P$

D. Since $z^2+az+a^2=0$

$\Rightarrow z = a\omega, a\omega^2$ (where 'ω' is non–real cube root of unity)

$\Rightarrow$ Locus of z is a pair of straight lines

$D \to S$

116. $A \to S$, $B \to R$, $C \to Q$, $D \to P$

A. Given,

$$\left||z-\cos^{-1}\cos 12| - |z-\sin^{-1}\sin 12|\right| = 8(\pi-3)$$

Since, $|\cos^{-1}\cos 12 - \sin^{-1}\sin 12| = 8(\pi-3)$

$\therefore$ Locus of z is portion of a line joining z_1 and z_2 except the segment between z_1 and z_2.

$A \to S$

B. $z^2 - i|z_1|^2 = k_2 - k_1$

$\therefore x^2-y^2+2ixy-i_1\lambda_1^2 = \lambda_2$ let $|z_1| = \lambda_1$

$\Rightarrow x^2-y^2=\lambda_2$ and $xy = \frac{\lambda_1^2}{2}$

$\therefore$ Locus of z is point of intersection of hyperbola.

$B \to R$

C. Put $= x+iy$.

$\therefore \text{Re}(x+iy)^2 = \text{Re}(x+iy+x-iy)$

$$x^2-y^2=2x$$

or $x^2-y^2-2x=0$

Rectangular hyperbola, eccentricity $=\sqrt{2}$

C→Q

D. Since $\left|\frac{2z-i}{z+i}\right| = 2$

$$\Rightarrow \left|\frac{z-\frac{i}{2}}{z+1}\right| = \frac{2}{2} = 1 \qquad \Rightarrow \left|\frac{z-\frac{i}{2}}{z+1}\right| = 1$$

$\Rightarrow \left|z-\frac{i}{2}\right| = |z+1|$ *ie,* straight line.

D → P

117. $A \to S$, $B \to R$, $C \to P$, $D \to Q$

A. Here, $\left|\frac{z-4}{z-8}\right| = 1$

$\Rightarrow \left|\frac{x-4+iy}{x-8+iy}\right| = 1$

$\Rightarrow (x-4)^2+y^2 = (x-8)^2+y^2$

$\Rightarrow x = 6$

With $x=6$, $\left|\frac{z-12}{z-8i}\right| = \frac{5}{3}$

$\Rightarrow y^2-25y+136=0$

$\Rightarrow 9(36+y^2) = 25\left[36+(y-8)^2\right]$

$\Rightarrow y = 17, 8$

Thus, the required numbers are

$$z = 6+17i, 6+8i$$

Hence, the value of Re(z) is 6.

$A \to S$

B. We have,

$$(1-\omega+\omega^2)(1-\omega^2+\omega^4)(1-\omega^4+\omega^8)(1-\omega^8+\omega^{16})$$

$$=(1+\omega^2-\omega)(1-\omega^2+\omega)(1-\omega+\omega^2)(1-\omega^2+\omega)$$

[Using $\omega^4 = \omega^3 \cdot \omega = \omega; \omega^8 = \left(\omega^3\right)^2 \cdot \omega^2; \omega^{16}$

$= \left(\omega^3\right)^5 \cdot \omega = \omega$ and $\omega^3 = 1$]

$= \left(-\omega - \omega\right)\left(-\omega^2 - \omega^2\right)\left(-\omega - \omega\right)\left(-\omega^2 - \omega^2\right)$

$= \left(-2\omega\right)\left(-2\omega^2\right)\left(-2\omega\right)\left(-2\omega^2\right) = 16 \cdot \omega^6$

$= 16\left(\omega^3\right)^2 = 16\left(1\right)^2 = 16$

B→R

C. Solution: We have, $\left(\frac{1+i}{\sqrt{2}}\right)^8 + \left(\frac{1-i}{\sqrt{2}}\right)^8$

$= \left[\cos\frac{\pi}{4} + i\sin\frac{\pi}{4}\right]^8 + \left[\cos\frac{\pi}{4} - i\sin\frac{\pi}{4}\right]^8$

$= \cos 2\pi + i\sin 2\pi + \cos 2\pi - i\sin 2\pi$

$= 2\cos 2\pi = 2\left(1\right) = 2$

[By de Moivre's theorem]

C→P

D. $z_1\overline{z_1} = 1, z_2\overline{z_2} = 4, z_3\overline{z_3} = 9$

$\therefore \left|9z_1z_2 + 4z_1z_3 + z_2z_3\right| = 12$

$= \left|z_1z_2z_3\overline{z_3} + z_1z_3z_2\overline{z_2} + z_2z_3z_1\overline{z_1}\right| = 12$

$\Rightarrow \left|z_1z_2z_3\right|\left|\overline{z_1} + \overline{z_2} + \overline{z_3}\right| = 12$

$\Rightarrow 6\left|z_1 + z_2 + z_3\right| = 12$

D → Q

118. *A→R, B→S, C→P, D → Q*

A. Let $\omega_1 = \omega$ and $\omega_2 = \omega^2$ then

$\omega_1^4 + \omega_2^4 = \omega^4 + \omega^8 = \omega + \omega^2 = -1 = -\frac{1}{\omega_1\omega_2}$

A→R,

B. $1 + z + z^3 + z^4 = 0$

$\Rightarrow (z + 1)^2.(1 - z + z^2) = 0$

$\Rightarrow z = -1, -\omega, -\omega^2$

$\Rightarrow |z| = 1 = \frac{1}{\omega_1\omega_2}$

B→S

C. The given relation can be written as

$\frac{1}{a+\omega} + \frac{1}{b+\omega} + \frac{1}{c+\omega} = \frac{2}{\omega}$

and $\frac{1}{a+\omega^2} + \frac{1}{b+\omega^2} + \frac{1}{c+\omega^2} = \frac{2}{\omega^2}$

$\Rightarrow \omega$ and ω^2 are roots of

$\frac{1}{a+x} + \frac{1}{b+x} + \frac{1}{c+x} = \frac{2}{x}$

$\Rightarrow \frac{3x^2 + 2\left(a+b+c\right)x + bc + ca + ab}{\left(a+x\right)\left(b+x\right)\left(c+x\right)} = \frac{2}{x}$

$\Rightarrow x^3 - \left(bc + ca + ab\right)x - 2abc = 0$ …… (i)

Two roots of the equation (i) are ω and ω^2. Let the third root be α, then

$\alpha + \omega + \omega^2 = 0 \Rightarrow \alpha = -\omega - \omega^2 = 1$.

$\therefore \alpha = 1$ will satisfy equation

$\Rightarrow \frac{1}{a+1} + \frac{1}{b+1} + \frac{1}{c+1} = 2 = 2\omega_1\omega_2$

C→P,

D. Given $\omega_1 - \omega_2 + {\omega_1}^2 - {\omega_2}^2 + {\omega_1}^3 - {\omega_2}^3 = (\omega_1 + {\omega_1}^2 + {\omega_1}^3) - (\omega_2 + {\omega_2}^2 + {\omega_2}^3) = 1 - 1 = 0$

D → Q

119. *A →Q, B→S, C→P, D → R*

A. $\left|iz - 1\right| + \left|z - i\right| = 2$

$\Rightarrow \left|z + i\right| + \left|z - i\right| = 2$ and $\left|i - \left(-i\right)\right| = 2$

So, z lies on the line segment joining the points i and – i and hence arg(z) = $\frac{\pi}{2}$ or $-\frac{\pi}{2}$.

So cos(arg(z)) = 0

A→Q,

B. (C) $z_1 = e^{\frac{2qr}{n}i}$ and $z_2 = e^{\frac{2pr}{n}i}$ where q, p ∈ Z and 0 ≤ q, p ≤ n

$\therefore \arg\left(\frac{z_1}{z_2}\right) = 2\left(q - p\right)\frac{\pi}{n}$, which is a multiple of $\frac{2\pi}{n}$

For consecutive vertices q–p =1 so required value is $\cos\left(\frac{2\pi}{n}\right)$

B→S

C. $z = \sqrt{-1-\sqrt{-1-\sqrt{\ldots\infty}}} = \sqrt{-1-z}$

$\Rightarrow z^2 + z + 1 = 0 \qquad \Rightarrow \frac{-1 \pm i\sqrt{3}}{2}$

Re(z) = –1/2

C→P

D. The required condition of being hyperbola is $|z_1 - z_2| > |2-4| = 2$

D → R

120. *A→S, B→Q, C→P, D→R*

A. Solving $z^2 + z + 1 = 0$

$\Rightarrow z = \frac{-1 \pm i\sqrt{3}}{2}$ or $z = \omega \; or \; \omega^2$

$(z + \frac{1}{z})^2 + (z^2 + \frac{1}{z^2})^2 + (z^3 + \frac{1}{z^3})^2 + \ldots + (z^{24} + \frac{1}{z^{24}}) = (\omega + \frac{1}{\omega})^2 + (\omega^2 + \frac{1}{\omega^2})^2 + (\omega^3 + \frac{1}{\omega^3})^2 + \ldots + (\omega^{24} + \frac{1}{\omega^{24}}) = 48$

A→S,

B. Let z = x + iy

$\Rightarrow p + qi = (x-1+iy)(\cos\alpha - i\sin\alpha) + ((x-1)+iy)^{-1}(\cos\alpha + i\sin\alpha)$

$= (x-1)\cos\alpha + y\sin\alpha + i\{y\cos\alpha - (x-1)\sin\alpha\} + \frac{(x-1)-iy}{(x-1)^2+y^2}(\cos\alpha + i\sin\alpha)$

$= (x-1)\cos\alpha + y\sin\alpha + i\{y\cos\alpha - (x-1)\sin\alpha\} + \frac{1}{(x-1)^2+y^2}\{(x-1)\cos\alpha + y\sin\alpha + i(x-1)\sin\alpha - y\cos\alpha\}$

$\therefore q = y\cos\alpha - (x-1)\sin\alpha + \frac{(x-1)\sin\alpha - y\cos\alpha}{(x-1)^2+y^2}$

$= \frac{y\cos\alpha - (x-1)\sin\alpha}{(x-1)^2+y^2}\{(x-1)^2 + y^2 - 1\}$

Given that, q = 0

Either $y\cos\alpha - (x-1)\sin\alpha = 0$

or $(x-1)^2 + y^2 = 1 \Rightarrow (x-1)^2 + y^2 = 1$

It represent a circle with radius r =1 so 12r=12

B→Q,

C. Given that $A\left(\frac{2}{\sqrt{3}}e^{\frac{i\pi}{2}}\right)$, $B\left(\frac{2}{\sqrt{3}}e^{\frac{-\pi}{6}}\right)$ and $C\left(\frac{2}{\sqrt{3}}e^{\frac{-5\pi}{6}}\right)$

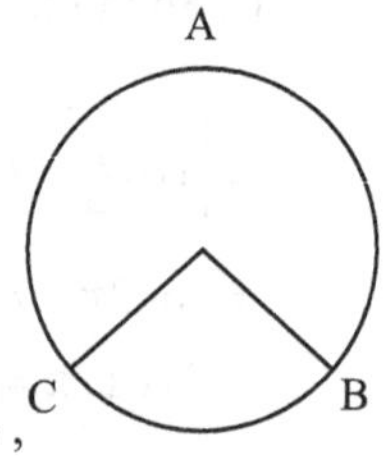

So A is $\frac{2}{\sqrt{3}}i$,

B is $\frac{2}{\sqrt{3}}\left(\frac{\sqrt{3}}{2} - \frac{i}{2}\right) = 1 - \frac{i}{\sqrt{3}}$,

C is $-1 - \frac{i}{\sqrt{3}}$.

Radius of incircle is $\frac{1}{\sqrt{3}}$, so any point P on incircle is

$z = \frac{1}{\sqrt{3}}(\cos\alpha + i\sin\alpha)$

Now, $AP^2 + BP^2 + CP^2$

$= \left(\frac{1}{\sqrt{3}}\cos\alpha\right)^2 + \left(\frac{1}{\sqrt{3}}\cos\alpha - 1\right)^2 + \left(\frac{1}{\sqrt{3}}\cos\alpha + 1\right)^2 + \left(\frac{1}{\sqrt{3}}\sin\alpha - \frac{2}{\sqrt{3}}\right)^2 + \left(\frac{1}{\sqrt{3}}\sin\alpha + \frac{1}{\sqrt{3}}\right)^2 + \left(\frac{1}{\sqrt{3}}\sin\alpha + \frac{1}{\sqrt{3}}\right)^2 = 5$

C→ R

D. Given that

$\frac{a}{b-c} + \frac{b}{c-a} + \frac{c}{a-b} = 0$

$\Rightarrow \left(\frac{a}{b-c} + \frac{b}{c-a} + \frac{c}{a-b}\right)^2 = 0$

$$\Rightarrow \frac{a^2}{(b-c)^2}+\frac{b^2}{(c-a)^2}+\frac{c^2}{(a-b)^2}$$

$$+2\left[\frac{ab}{(b-c)(c-a)}+\frac{bc}{(c-a)(a-b)}+\frac{ca}{(a-b)(b-c)}\right]=0$$

$$\Rightarrow \frac{a^2}{(b-c)^2}+\frac{b^2}{(c-a)^2}+\frac{c^2}{(a-b)^2}$$

$$+2\left[\frac{ab(a-b)+bc(b-c)+ca(c-a)}{(b-c)(c-a)(a-b)}\right]=0 \ldots(1)$$

But ab(a – b) + bc(b – c) + ca(c – a) = – (b – c)(c – a)(a – b)

$$\Rightarrow \frac{a^2}{(b-c)^2}+\frac{b^2}{(c-a)^2}+\frac{c^2}{(a-b)^2}+2(-1)=0$$

$$\Rightarrow \frac{a^2}{(b-c)^2}+\frac{b^2}{(c-a)^2}+\frac{c^2}{(a-b)^2}=2$$

D→ R

121. (A) → (q)

$z \neq 0$ Let z = $a+ib$

Re(z) = 0 ⇒ z = $ib \Rightarrow z^2=-b^2$

$\therefore\ Im(z)^2 = 0$

∴ (A) → (q)

Arg $z=\frac{\pi}{4} \Rightarrow a=b \Rightarrow$ z = $a+ia$

$z^2 = a^2 - a^2 + 2ia^2$

$z^2 = 2ia^2 \Rightarrow \text{Re}\ (z)^2 = 0$

∴ (B) → (p)

(A) → (q, r), (B) → (p) ; (C) → (p, r, t) ; D → (q, r, s, t)

122. Consider (A):– $|z-i|z||=|z+i|z||$

⇒ z is equidistant from two points (0, $|z|$) ≤ 1

(A) → (q, r)

Consider (B):– Sum of distant of z from two fined points (–4, 0) and (4, 0) is 10 which is greater than 8.

∴ z traces an elipes with $2a=10$ and $2ae=8$

$\Rightarrow e = 4/5$

(B) → p

Consider (C)

Let $\omega = 2(\cos\theta + i\sin\theta)$

then z =

$$\omega-\frac{1}{\omega}=2(\cos\theta+i\sin\theta)-\frac{1}{2}(\cos\theta-i\sin\theta)$$

$$\Rightarrow x+iy=\frac{3}{2}\cos\theta+i\frac{5}{2}\sin\theta$$

Here $|z|=\sqrt{\frac{9+25}{4}}=\sqrt{\frac{34}{4}}\leq 3$ and $|R_e(z)|\leq 2$

also $x=\frac{3}{2}\cos\theta, y=\frac{5}{2}\sin\theta \Rightarrow \frac{4x^2}{9}+\frac{4y^2}{25}=1$

Which is an ellipse with $e=\sqrt{1-\frac{9}{25}}=\frac{4}{5}$

(C) → (p, s, t)

Consider (D)

Let $\omega = \cos\theta + i\sin\theta$ then

$$z = 2\cos\theta \Rightarrow Imz = 0$$

Also $|z|\leq 3$ and $|Im(z)|\leq 1, \|R_e(z)|\leq 2|$

(D) → (q, r, s, t)

123. (c) (P) → (1) : $z_k=\cos\frac{2k\pi}{10}+i\sin\frac{2k\pi}{10}$, $k = 1$ to 9

$\therefore\ z_k = e^{i\frac{2k\pi}{10}}$

Now $z_k.z_j = 1 \Rightarrow z_j = \frac{1}{z_k} = e^{-i\frac{2k\pi}{10}} = \overline{z_k}$

We know if z_k is 10th root of unity so will be $\overline{z}_k$.

∴ For every z_k, there exist $z_i = \overline{z}_k$

Such that $z_k.z_j = z_k.\overline{z}_k = 1$

Hence the statement is true.

(Q) → (2) $z_1 = z_k \Rightarrow z = \frac{z_k}{z_1}$ for $z_1 \neq 0$

∴ We can always find a solution to $z_1.z = z_k$

Hence the statement is false.

(R) → (3) : We know

$$z^{10}-1=(z-1)(z-z_1)....(z-z_9)$$

$$\Rightarrow (z-z_1)(z-z_2).....(z-z_9)=\frac{z^{10}-1}{z-1}$$

$= 1 + z + z^2 + ... z^9$

For z = 1 we get

$$(1-z_1)(1-z_2).....(1-z_9)=10$$

$$\therefore \frac{|1-z_1||1-z_2|.....|1-z_9|}{10}=1$$

(S) → (4) : 1, Z_1, Z_2, ... Z_9 are 10th roots of unity.

$\therefore Z^{10} - 1 = 0$

From equation $1 + Z_1 + Z_2 + ... + Z_9 = 0$

Re (1) + Re (Z_1) + Re (Z_2) + ... + Re(Z_9) = 0

⇒ Re (Z_1) + Re (Z_2) + ... Re(Z_9) = – 1

$$\Rightarrow \sum_{K=1}^{9}\cos\frac{2k\pi}{10}=-1 \Rightarrow 1-\sum_{K=1}^{9}\cos\frac{2k\pi}{10}=2$$

Hence (c) is the correct option.

(More Than One Correct MCQs)

124. (b, d) Let $A(z_1), B(z_2)$ be the centre of given circles and P be the centre of the variable circle which touches given circles externally, then $|AP| = a + r$ and $|BP| = b + r$, where r is the radius of the variable circle. On subtraction, we get $|AP| - |BP| = |a - b|$, a constant. Hence, locus of P is

(i) right bisector $[AB]$ if $a = b$

(ii) a hyperbola if $|a - b| < |AB| = |z_2 - z_1|$

(iii) an empty set if $|a - b| > |AB| = |z_2 - z_1|$

(iv) Set of all points on line AB except those which lie between A and B if $|a - b| = |AB| = 0$.

125. (a. b) A complex number is purely imaginary if its real part is zero

$$Re\left(\frac{\tan\alpha - i\left(\sin\frac{\alpha}{2} + \cos\frac{\alpha}{2}\right)}{1 + 2i\sin\frac{\alpha}{2}}\right) = 0$$

$$Re\left(\frac{\{\tan\alpha - i\left(\sin\frac{\alpha}{2} + \cos\frac{\alpha}{2}\right)\}\left(1 - 2i\sin\frac{\alpha}{2}\right)}{1 + 4sin^2\frac{\alpha}{2}}\right) = 0$$

$$Re\left\{\frac{\tan\alpha - 2\sin\frac{\alpha}{2}\left(cos\frac{\alpha}{2} + sin\frac{\alpha}{2}\right) - i\left(\tan\alpha . 2\sin\frac{\alpha}{2} + \sin\frac{\alpha}{2} + \cos\frac{\alpha}{2}\right)}{1 + 4sin^2\frac{\alpha}{2}}\right\} = 0$$

$$\Rightarrow \tan\alpha = 2\sin^2\frac{\alpha}{2} + \sin\alpha$$

$$\Rightarrow \frac{\sin\alpha}{\cos\alpha} = \sin\alpha + 1 - \cos\alpha$$

$$\Rightarrow \sin\alpha = \sin\alpha . \cos\alpha + \cos\alpha - \cos^2\alpha$$

$$\Rightarrow \sin\alpha(1 - \cos\alpha) = \cos\alpha(1 - \cos\alpha)$$

$$\Rightarrow \sin\alpha = \cos\alpha, \cos\alpha = 1$$

$$\Rightarrow \alpha = n\pi + \frac{n}{4}, \alpha = 2n\pi (n \in I)$$

126. (a, c) Given $z_1 = a + ib, z_2 = c + id$

$$|z_1| = |z_2| = 1 \Rightarrow a^2 + b^2 = c^2 + d^2 = 1$$

$$w_1 = a + ic, w_2 = b + id$$

$$z_1\overline{z_2} = (a + ib)(c - id)$$

$$= (ac + bd) + i(bc - ad)$$

$$As, Re(z_1\overline{z_2}) = 0 \Rightarrow ac + bd = 0 \Rightarrow ac = -bd$$

$$w_1\overline{w_2} = (a + ic)(b - id)$$

$$= (ab + cd) + i(bc - ad)$$

We have $a^2 + b^2 = c^2 + d^2$

$$\Rightarrow a^2 - c^2 = d^2 - b^2$$

$$\Rightarrow a^2 - c^2 + 2i\,ac = d^2 - b^2 - 2i\,bd\,(as\,ac = -bd)$$

$$\Rightarrow (a + ic)^2 = (d - ib)^2$$

$$\Rightarrow a + ic = (d - ib) \text{ or } -d + ib$$

$$\Rightarrow a = d \text{ and } c = -b \text{ or } a = -d, b = c$$

$$\Rightarrow c^2 + d^2 = b^2 + d^2,\ a^2 + c^2 = a^2 + b^2$$

$$\Rightarrow a^2 + c^2 = 1, b^2 + d^2 = 1$$

$$\Rightarrow |w_1| = |w_2| = 1 \text{ also } ab + cd = -cd + cd = 0$$

$$Re(w_1\overline{w_2}) = 0$$

127. (b, c) Let, $1, \alpha, \alpha^2, \ldots, \alpha^{n-1}$ are the n, nth roots of unity,

$$\therefore x^n - 1 = (x - 1)(x - \alpha)(x - \alpha^2)\ldots(x - \alpha^{n-1})$$

$$\Rightarrow \log(x^n - 1) = \log(x - 1) + \log(x - \alpha) + \log(x - \alpha^2) + \ldots + \log(x - \alpha^{n-1})$$

On differentiating both side w.r.t. 'x', we get

$$\frac{nx^{n-1}}{x^n - 1} = \frac{1}{x - 1} + \frac{1}{x - \alpha} + \frac{1}{x - \alpha^2} + \ldots + \frac{1}{x - \alpha^{n-1}}$$

Put $x = 2$, we get

$$\frac{n2^{n-1}}{2^n - 1} = \frac{1}{1} + \frac{1}{2 - \alpha} + \frac{1}{2 - \alpha^2} + \ldots + \frac{1}{2 - \alpha^{n-1}}$$

$$\therefore \frac{n \cdot 2^{n-1}}{2^n - 1} - 1 = \sum_{i=1}^{n-1}\frac{1}{2 - \alpha^i}$$

Hence, $S_n = \sum_{i=1}^{n-1} \frac{1}{2-\alpha^i} = \frac{n \cdot 2^{n-1} - 2^n + 1}{2^n - 1}$

$= \frac{(n-2)2^{n-1}+1}{2^n - 1}$

Now put n= 10 we get

$S_{10} = \sum_{i=1}^{9} \frac{1}{2-\alpha^i} = \frac{8(2^9)+1}{2^{10}-1} = \frac{2^{12}+1}{2^{10}-1}$

Now put n= 12 we get

$S_{12} = \sum_{i=1}^{11} \frac{1}{2-\alpha^i} = \frac{10(2^{11})+1}{2^{12}-1}$

128. (b, c) Since area is maximum so, the inscribed triangle is equilateral

Now from rotation theorem

$\Rightarrow \frac{z_2 - z_0}{z_1 - z_0} = e^{i\frac{2\pi}{3}}, \frac{z_3 - z_0}{z_1 - z_0} = e^{-i\frac{2\pi}{3}}$

$\Rightarrow z_2 = -1 + i(2+\sqrt{3})$

and $z_3 = -1 + i(2-\sqrt{3})$

129. (a, c, d) From the given information

$a_1^2 + b_1^2 = 1, a_2^2 + b_2^2 = 4 \text{ and } a_1 a_2 = b_1 b_2$

$a_2^2 + b_2^2 = 4a_1^2 + 4b_1^2$

$(a_2 + 2ia_1)^2 = (2b_1 + ib_2)^2$

$a_2 = \pm 2b_1$

$\Rightarrow \left|a_1 + \frac{ia_2}{2}\right|^2 = \left|a_1 - \frac{ia_2}{2}\right|^2$

$= a_1^2 + b_1^2 = 1$

and $|2b_1 + ib_2|^2 = |2b_1 - ib_2|^2 = 4b_1^2 + b_2^2 = 4$

130. (a, b, c, d) $(n-1)(n-\omega)(n-\omega^2)$

$= (n-1)\{n^2 - (\omega+\omega^2)n + \omega^3\}$

$= (n-1)(n^2+n+1) = n^3 - 1$, etc

$\therefore$ Series $= \sum_{n=1}^{n-1}(n^3 - 1)$

$= \{1^3 + 2^3 + 3^3 + \ldots + (n-1)^3\} - (n-1)$

$= \frac{n^2(n-1)^2}{4} - n + 1$

$S_{20} = \frac{(20^2)(19^2)}{4} - 20 + 1 = 36081$

$S_{10} = \frac{(10^2)(9^2)}{4} - 10 + 1 = 2016$

$S_{31} = \frac{(31^2)(30^2)}{4} - 31 + 1 = 216195$

$S_{15} = \frac{(15^2)(14^2)}{4} - 15 + 1 = 11011$

131. (a, b) $z + z^{-1} = 1 \Rightarrow z^2 - z + 1 = 0$

$\therefore z = \frac{1 \pm \sqrt{3}i}{2} = -\omega, -\omega^2$

$\therefore z^n + z^{-n} = (-\omega)^n + (-\omega)^{-n}$

or $(-\omega^2)^n + (-\omega^2)^{-n}$

$= (-1)^n \omega^n + (-1)^{-n} \frac{1}{\omega^n}$

or $(-1)^n \omega^{2n} + (-1)^{-n} \cdot \frac{1}{\omega^{2n}}$

$= (-1)^n \left(\omega^n + \frac{1}{\omega^n}\right)$

or $(-1)^n \left(\omega^{2n} + \frac{1}{\omega^{2n}}\right)$

$= (-1)^n \left(\omega^n + \frac{1}{\omega^n}\right)$,

beacuse $\omega^{3n} = 1$

$= (-1)^n \cdot (\omega^n + \omega^{2n})$

$= (-1)^n \cdot (1+1)$ if n is a multiple of 3

or $(-1)^n \cdot (-1)$ if n is not a multiple of 3

132. (a, d) $OB = OD = OA = |z|$

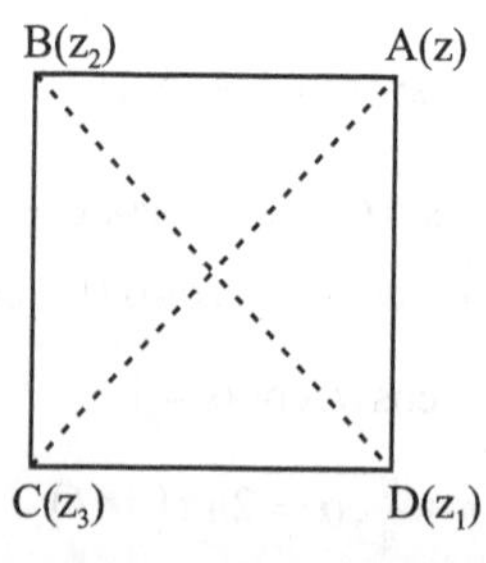

If amp $z=\theta$, then amp $z_1=\theta-\frac{\pi}{2}$,

amp $z_2=\theta+\frac{\pi}{2}$ and $z=|z|\{\cos\theta+i\sin\theta\}$

$$\therefore z_1=|z|\left\{\cos\left(\theta-\frac{\pi}{2}\right)+i\sin\left(\theta-\frac{\pi}{2}\right)\right\}$$

$$=|z|\{\sin\theta-i\cos\theta\}$$

$$=|z|(-i)(\cos\theta+i\sin\theta)=-iz$$

$$z_2=|z|\left\{\cos\left(\theta+\frac{\pi}{2}\right)+i\sin\left(\theta+\frac{\pi}{2}\right)\right\}=iz$$

133.(a, d) Let $z=k$, be a real root.

Then, $\alpha k^2+k+\bar{\alpha}=0$ .... (i)

let $\alpha=p+iq$

Then, $(p+iq)k^2+k+p-iq=0$

$\Rightarrow pk^2+k+p=0$ and $qk^2-q=0$

$\Rightarrow k=\pm1(\text{since } q\neq 0)$

$\therefore$ Eq. (i) $\Rightarrow \alpha\pm1+\bar{\alpha}=0$

$|\alpha+\bar{\alpha}|=1$

Also, $|k|=1$

134. (a, b, c) $|z_1-1|=1, |z_0-1|=1 \therefore \left|\frac{z_1-1}{z_0-1}\right|=1$

$|z-1|=1$

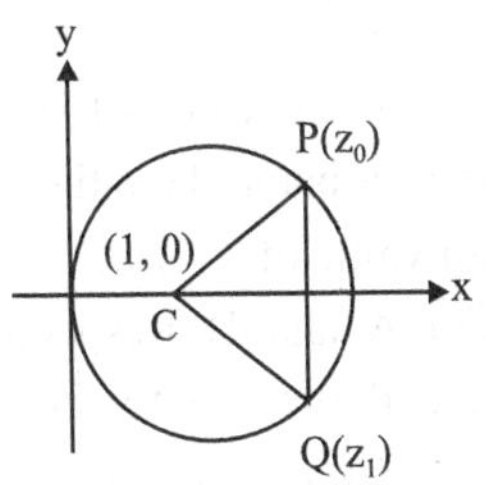

$$\angle QCP=\frac{\pi}{2}\Rightarrow \text{amp}\left(\frac{z_1-1}{z_0-1}\right)=\frac{\pi}{2},-\frac{\pi}{2}$$

$$\therefore \frac{z_1-1}{z_0-1}=1\cdot\left(\cos\frac{\pm\pi}{2}+i\sin\frac{\pm\pi}{2}\right)=i,-i$$

$$\therefore z_1-1=(z_0-1)i, z_1-1=-(z_0-1)i$$

135. (b, c, d) Since α^5 is the fifth root of unity so $\alpha^5=1$, and

$$a=\cos\frac{2\pi}{5}+i\sin\frac{2\pi}{5}$$

and $1+a+a^2+a^3+a^4=0$

$$\therefore |1+a+a^2+a^3|=|-a^4|=|a|^4=1$$

Also, $|1+a+a^2|=|-a^3(1+a)|$

$$=|-a^3||1+a|=|1+a|$$

$$=\left|1+\cos\frac{2\pi}{5}+isin\frac{2\pi}{5}\right|$$

$$=\left|2\cos\frac{\pi}{5}\left(\cos\frac{\pi}{5}+isin\frac{\pi}{5}\right)\right|=2\cos\frac{\pi}{5}$$

So $|1+a|=|1+a+a^2|=2\cos\frac{\pi}{5}>2\cos\frac{\pi}{7}$

136. (a, b, c, d) Consider $\sqrt{20i-21}=\pm(2+5i)$

And $\sqrt{21+20i}=\pm(5+2i)$

Now we have following cases-

Case (i) $z=7+7i\Rightarrow \arg z=\frac{\pi}{4}$

Case (ii) $z=-3+3i\Rightarrow \arg(z)=\frac{3\pi}{4}$

Case (iii) $z=-7-7i\Rightarrow \arg z=-\frac{3\pi}{4}$

Case (iv) $z=3-3i\Rightarrow \arg z=-\frac{\pi}{4}$

137. (a, b, c) Equation of Q

$(x-3)^2+(y+1)^2=9$

$x^2+y^2-6x+2y+1=0$ ...(i)

Center (3, -1) and radius = 3

Equation of curve P

$(x-4)^2+(y-1)^2=4$

$\Rightarrow x^2+y^2-8x-2y+13=0$ ...(ii)

Center (4, 1) and radius = 2

Distance between the centers is

$$(4-3)^2+(1+1)^2=\sqrt{5}$$

So these two circles intersect at two points

Solving P and Q we get equation of common chord.

$2x+4y-121=0$

$x = 6 - 2y$ ...(iii)

Solving equation (i) and (iii) we get

$y = 1 - \frac{2}{\sqrt{5}}, x = 4 + \frac{4}{\sqrt{5}}$

138. (b, d) The given curve is a circle with center z_0 (1,1) and radius $2\sqrt{2}$ and for maximum area the triangle is equilateral

$$\arg\left(\frac{z_2 - z_0}{z_1 - z_0}\right) = \frac{2\pi}{3} \Rightarrow \arg\left(\frac{z_2 - (1+i)}{-3+i}\right) = \frac{2\pi}{3}$$

$$\Rightarrow z_2 - (1+i) = e^{\frac{i2\pi}{3}}(-3+i)$$

$$\Rightarrow z_2 = \left(\frac{-1+i\sqrt{3}}{2}\right)(-3+i) + 1 + i$$

$$= \frac{(-1+i\sqrt{3})(-3+i) + 2 + 2i}{2}$$

$$\Rightarrow z_2 = \frac{\left(5-\sqrt{3}\right) + i(1-3\sqrt{3})}{2}$$

Similarly $\arg\left(\frac{z_3 - z_0}{z_1 - z_0}\right) = -\frac{2\pi}{3}$

$$z_3 = e^{\frac{i2\pi}{3}}(-3+i) + (1+i)$$

$$\Rightarrow z_3 = -\left(\frac{1+i\sqrt{3}}{2}\right)(-3+i) + (1+i)$$

$$= \frac{-\left(\left(-3-\sqrt{3}\right) + i\left(1-3\sqrt{3}\right)\right) + 2(1+0)}{2}$$

$$\Rightarrow z_2 = \frac{\left(5+\sqrt{3}\right) + i(1+3\sqrt{3})}{2}$$

139. (b, d) Let z = $x + iy$ then the equation is

$$x^2 + y^2 - 2i(x+iy) + 2c(1+i) = 0$$

$$\Rightarrow (x^2 + y^2 + 2y + 2c) + i(2c - 2x) = 0$$

$$\Rightarrow x^2 + y^2 + 2y + 2c = 0 \text{ and } x = c$$

$$\Rightarrow c^2 + y^2 + 2y + 2c = 0$$

$$\Rightarrow y = -1 \pm\sqrt{1-2c-c^2}$$

$$\because y \in R \;\Rightarrow\; 1 - 2c - c^2 \geq 0$$

$$\Rightarrow c^2 + 2c - 1 \leq 0 \Rightarrow -1-\sqrt{2} \leq c \leq -1+\sqrt{2}$$

140. (b, c, d) Let $a = a_1 + ia_2$ and $c = c_1 + ic_2$ then

Slope of the line

$a\bar{z} + \bar{a}z + b = 0$ is $-\frac{a_1}{a_2}$ and slope of the line

$c\bar{z} + \bar{c}z + d = 0$ is $-\frac{c_1}{c_2}$

Since the two lines are perpendicular to each other

So, $\left(-\frac{a_1}{a_1}\right) \times \left(-\frac{c_1}{c_2}\right) = -1 \Rightarrow a_1c_1 + a_2c_2 = 0$

$$\Rightarrow \left(\frac{a+\bar{a}}{2}\right)\left(\frac{c+\bar{c}}{2}\right) + \left(\frac{a-\bar{a}}{2i}\right)\left(\frac{c-\bar{c}}{2i}\right) = 0$$

$$\Rightarrow a\bar{c} + \bar{a}c = 0$$

$\therefore$ $a\bar{c}$ is purely imaginary. Also

$\frac{a}{c} = -\left(\frac{\bar{a}}{\bar{c}}\right) \Rightarrow \frac{a}{c}$ is also purely imaginary

$$\Rightarrow \arg\left(\frac{a}{c}\right) = \pm\frac{\pi}{2}$$

141. (a, c) $|iz + 2| = Im(z) \Rightarrow |z - 2i| = Im(z)$

$\Rightarrow$ Distance of z from 2i = distance of z from real axis

$\Rightarrow$ Locus of z is parabola having focus at (0, 2) and

Directrix at x – axis so its cartestan equation is $x^2 = 4(y-1)$

Clearly minimum value of |z | = 1 (Distance from vertex of parabola to the origin)

Maximum value of |z| is ∞.

Also point A and B Respectively represent minimum and maximum values of argument [*OA* and *OB* represent tangents from origin whose equations are $y = \pm x$]

142. (a, c) $|2z_1 + z_2| \leq |2z_1| + |z_2| = 2|z_1| + |z_2| = 2 \times 1 + 2 = 4$

Again $\left|z_2 + \frac{1}{z_1}\right| \leq |z_2| + \left|\frac{1}{z_1}\right| = 2 + \left|\frac{1}{z_1}\right| = 2 + \frac{1}{1}$

$= 3$

143. (a, b, c) Given that If $z_1 = a + ib$ and $z_2 = c + id$ and $|z_1| = |z_2| = 1$

$\therefore$ Let $z_1 = \cos\theta + i\sin\theta = a + ib$ and

$z_2 = \cos\phi + i\sin\phi = c + id$

$\Rightarrow a = \cos\theta; b = \sin\theta;$

$c = \cos\phi; d = \sin\phi;$

Also $w_1 = a + ic = \cos\theta + i\cos\phi$

$w_2 = b + id = \sin\theta + i\sin\phi$

Now consider Re $(z_1\bar{z}_2) = 0$

$\Rightarrow \text{Re}\left[(\cos\theta + i\sin\theta)(\overline{\cos\phi + i\sin\phi})\right] = 0$

$\Rightarrow \text{Re}\left[(\cos\theta + i\sin\theta)(\cos\phi - i\sin\phi)\right] = 0$

$\Rightarrow \text{Re}\left[\cos(\theta - \phi) + i\sin(\theta - \phi)\right] = 0$

$\Rightarrow \cos(\theta - \phi) = 0 \Rightarrow \theta - \phi = \pi/2$

$\Rightarrow \theta = \pi/2 + \phi$

$\therefore w_1 = \cos(\pi/2 + \phi) + i\cos\phi$

$= -\sin\phi + i\cos\phi;$

$w_2 = \sin(\pi/2 + \phi) + i\sin\phi = \cos\phi + i\sin\phi$

$\therefore |w_1| = 1; |w_2| = 1$

Also

Re $(w_1\bar{w}_2)$

$= \text{Re}\left[(-\sin\phi + i\cos\phi)(\cos\phi - i\sin\phi)\right]$

$\Rightarrow \text{Re}[(-\sin\phi\cos\phi + \sin\phi\cos\phi) + i(\cos^2\phi + \sin^2\phi)]$

$\Rightarrow \text{Re}(w_1\bar{w}_2) = 0$

144. (a, d) Let $z_1 = a + ib,\ a > 0$ and $b \in R; z_2 = c + id, d < 0, c \in R$

then $|z_1| = |z_2| \Rightarrow a^2 + b^2 = c^2 + d^2$

$= a^2 - c^2 = d^2 - b^2$ ...(i)

Now, $\dfrac{z_1 + z_2}{z_1 - z_2} = \dfrac{(a+c) + i(b+d)}{(a-c) + i(b-d)}$

$$= \frac{\left[(a^2 - c^2) + (b^2 - d^2)\right] + i\left[(a-c)(b+d) - (a+c)(b-d)\right]}{(a-c)^2 + (b-d)^2}$$

$$= \frac{i\left[(a-c)(b+d) - (a+c)(b-d)\right]}{(a-c)^2 + (b-d)^2}$$

[Using (1)]

= Purely imaginary number or zero in case $a + c = b + d = 0.$

145. (c) $|z_1 + z_2| = |z_1| + |z_2|$

$\Rightarrow \cos(\theta_1 - \theta_2) = 1$

$\Rightarrow \theta_1 - \theta_2 = 0$

$\Rightarrow \arg(z_1) = \arg(z_2)$

Smart Cracker
If $\lvert z_1 + z_2\rvert = \lvert z_1\rvert + \lvert z_2\rvert$ then oz_1 and oz_2 are co-linear here O is origin, so $\arg(z_1) = \arg(z_2)$

146. (d) Let z = $\cos\dfrac{2\pi}{7} + i\sin\dfrac{2\pi}{7}$

Using DeMoiver's theorem, we get

$z^k = \cos\dfrac{2\pi k}{7} + i\sin\dfrac{2\pi k}{7}$

Now, $\sum_{k=1}^{6}\left(\sin\dfrac{2\pi k}{7} - i\cos\dfrac{2\pi k}{7}\right)$

$= \sum_{k=1}^{6}(-i)\left(\cos\dfrac{2\pi k}{7} + i\sin\dfrac{2\pi k}{7}\right)$

$= (-i)\sum_{k=1}^{6} z^k = -iz\dfrac{(1 - z^6)}{1 - z} = -i\left(\dfrac{z - z^7}{1 - z}\right)$

$= (-i)\left(\dfrac{z - 1}{1 - z}\right) =$

[Using $z^7 = \cos 2\pi + i\sin 2\pi = 1$]

$= i\left(\dfrac{1 - z}{1 - z}\right) = i$

147. (d) Given $(1 + \omega + \omega^2)^7 = (-\omega^2 - \omega^2)^7$

$(-2)^7(\omega^2)^7 = -128\omega^{14} = -128\omega^2$

148.(b) $\sum_{i=1}^{13}\left(i^n + i^{n+1}\right) = \sum_{i=1}^{13} i^n (1+i)$

$= (1+i)\sum_{i=1}^{13} i^n$

This form a G.P.

Sum of G.P. $= i(1+i)\dfrac{\left(1-i^{13}\right)}{1-i} = i-1$

149.(a, c, d) Given that $z = (1-t)z_1 + t z_2$ where $0 < t < 1$

$\Rightarrow z = \dfrac{(1-t)z_1 + t z_2}{(1-t)+t}$

$\Rightarrow$ z divides the line joining of z_1 and z_2 internally in the ratio t: (1-t).

$\therefore$ z_1, z and z_2 are collinear

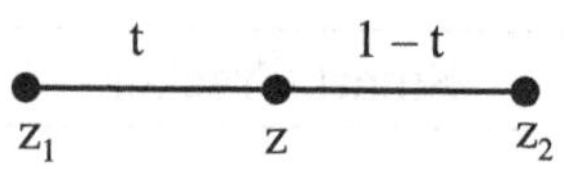

$\Rightarrow |z - z_1| + |z - z_2| = |z_1 - z_2|$

Also $z = (1-t)z_1 + t z_2$

$\Rightarrow \dfrac{z - z_1}{z_2 - z_1} = t =$ purely real number

$\therefore \arg\left(\dfrac{z - z_1}{z_2 - z_1}\right) = 0$

$\Rightarrow \arg(z - z_1) = \arg(z_2 - z_1)$

Also $\dfrac{z - z_1}{z_2 - z_1} = t$

$\Rightarrow \dfrac{\bar{z} - \bar{z}_1}{\bar{z}_2 - \bar{z}_1} = t \Rightarrow \dfrac{z - z_1}{z_2 - z_1} = \dfrac{\bar{z} - \bar{z}_1}{\bar{z}_2 - \bar{z}_1}$

$\Rightarrow (z - z_1)(\bar{z}_2 - \bar{z}_1) = (\bar{z} - \bar{z}_1)(z_2 - z_1)$

$\Rightarrow \begin{vmatrix} z - z_1 & \bar{z} - \bar{z}_1 \\ z_2 - z_1 & \bar{z}_2 - \bar{z}_1 \end{vmatrix} = 0$

150. (c, d) $w = \dfrac{\sqrt{3}+i}{2} = \cos\dfrac{\pi}{6} + i\sin\dfrac{\pi}{6}$

and $w^n = \cos\dfrac{n\pi}{6} + i\sin\dfrac{n\pi}{6}$

$\therefore$ P contains all those points which lies on unit circle and have argument $\dfrac{\pi}{6}, \dfrac{2\pi}{6}, \dfrac{3\pi}{6}$ and so on. As $z_1 \in P \cap H_1$ and $z_2 \in P \cap H_2$, therefore z_1 and z_2 can have possible positions as shown in the figure.

Re z $< -\dfrac{1}{2}$ Re z $> -\dfrac{1}{2}$

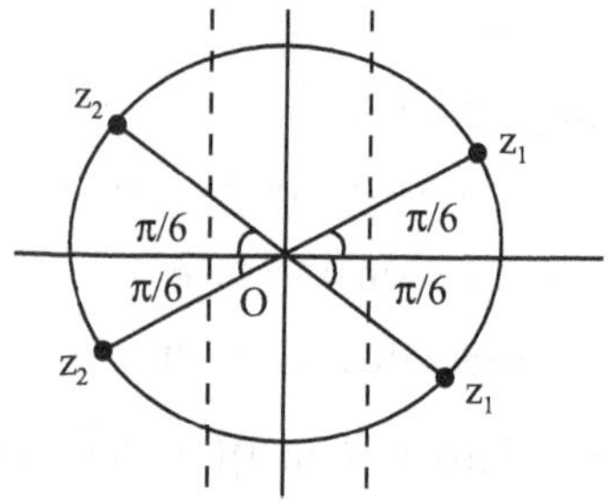

$\therefore \angle z_1 O z_2$ can be $\dfrac{2\pi}{3}$ or $\dfrac{5\pi}{6}$.

151. (a, c, d) $z = \dfrac{1}{a + ibt} = x + iy$

$\Rightarrow \quad x + iy = \dfrac{a - ibt}{a^2 + b^2t^2}$

$\Rightarrow \quad x = \dfrac{a}{a^2 + b^2t^2},\ y = \dfrac{-bt}{a^2 + b^2t^2}$

$\Rightarrow \quad x^2 + y^2 = \dfrac{1}{a^2 + b^2t^2} = \dfrac{x}{a}$

$\Rightarrow \quad x^2 + y^2 - \dfrac{x}{a} = 0$

$\therefore$ Locus of z is a circle with centre $\left(\dfrac{1}{2a}, 0\right)$ and radius $= \dfrac{1}{2|a|}$ irrespective of 'a' +ve or –ve

Also for b = 0, a ≠ 0, we get, y = 0

$\therefore$ locus is x-axis

and for a = 0, b ≠ 0 we get x = 0

$\Rightarrow$ locus is y-axis.

$\therefore$ a, c, d are the correct options.

(Subjective Questions)

IIT JEE

152. $\dfrac{1}{1 - \cos\theta + 2i\sin\theta}$

$= \dfrac{1}{2\sin^2\theta/2 + 4i\sin\theta/2\cos\theta/2}$

$$= \frac{1}{2\sin\theta/2}\left[\frac{\sin\theta/2 - 2i\cos\theta/2}{(\sin\theta/2 + 2i\cos\theta/2)(\sin\theta/2 - 2i\cos\theta/2)}\right]$$

$$= \frac{1}{2\sin\theta/2}\left[\frac{\sin\theta/2 - 2i\cos\theta/2}{\left(\sin^2\theta/2 + 4\cos^2\theta/2\right)}\right]$$

$$= \frac{2}{2\sin\theta/2}\left[\frac{\sin\theta/2 - 2i\cos\theta/2}{5 + 3\cos\theta}\right]$$

$$= \left(\frac{1}{5 + 3\cos\theta}\right) + \frac{1}{\sin\theta/2}\left(\frac{-2\cos\theta/2}{5 + 3\cos\theta}\right)i$$

153. Given that β and γ are the complex cube roots of unity so, let $\beta = \omega$ and $\gamma = \omega^2$ then $\omega + \omega^2 + 1 = 0$ and $\omega^3 = 1$.

Now consider

$xyz = (a+b)(a\omega^2 + b\omega)(a\omega + b\omega^2)$

$= (a+b)(a^2\omega^3 + ab\omega^4 + ab\omega^2 + b^2\omega^3)$

$= (a+b)(a^2 + ab\omega + ab\omega^2 + b^2)$ (Since $\omega^3 = 1$)

$= (a+b)(a^2 + ab(\omega + \omega^2) + b^2)$

$= (a+b)(a^2 - ab + b^2)$ (Since $\omega + \omega^2 = -1$)

$= a^3 + b^3$ Proved.

154. Given $x + iy = \sqrt{\dfrac{a+ib}{c+id}}$ Squaring both the sides we get

$\Rightarrow (x+iy)^2 = \dfrac{a+ib}{c+id}$ ...(i)

Now taking conjugate on both sides, we get

$(x-iy)^2 = \dfrac{a-ib}{c-id}$ ...(ii)

Multipluing (i) and (ii)

$\left(x^2+y^2\right)^2 = \dfrac{a^2+b^2}{c^2+d^2}$

155. Given that n is an odd integer > 3 and n is not a multiple of 3.

Let $p(x) = (x+1)^n - x^n - 1$

and $q(x) = x^3 + x^2 + x$

$q(x) = x(x^2 + x + 1) = x(x-\omega)(x-\omega^2)$

where ω and ω^2 are cube roots of unity.

Clearly $0, \omega, \omega^2$ are zeros of the polynominal $q(x)$.

Now, $p(0) = 1^n - 0 - 1 = 0$

$\Rightarrow$ 0 is a zero of (x).

$p(\omega) = ((\omega+1)^n - \omega^n - 1)$

$\Rightarrow \left(-\omega^2\right)^n - \omega^n - 1$

$= -\omega^{2n} - \omega^n - 1$ [$\because n$ is odd]

$= -\left(\omega^n + \omega^{2n}\right) - 1$

$= -(-1) - 1$

$\left[\because \omega^n + \omega^{2n} + 1 = 0 \text{ if } n \neq 3m\right] = 0$

$\Rightarrow \omega$ is a zero of $p(x)$.

Now consider $p(\omega^2) = (\omega^2+1)^n - (\omega^2)^n - 1$

$\Rightarrow (-\omega)^n - 2\omega^n - 1$

$= -\omega^n - \omega^{2n} - 1 = -\left(1 + \omega^n + \omega^{2n}\right) = 0$

[for $n \neq 3m, m \in N$]

$\Rightarrow \omega^2$ is a zero of $p(x)$.

$\because 0, \omega, \omega^2$ are zeroes of $p(x)$.

$\therefore x, x-\omega, x-\omega^2$ are factors of $p(x)$.

$\Rightarrow x(x-\omega)(x-\omega^2)$ is a factor of $p(x)$.

$\Rightarrow x^3 + x^2 + x$ is a factor of $p(x)$.

156. $\dfrac{(1+i)x - 2i}{3+i} + \dfrac{(2-3i)y + i}{3-i} = i$

$\Rightarrow (4+2i)x - 6i - 2 + (9-7i)y + 3i - 1 = 10i$

$\Rightarrow (4x + 9y - 3) + (2x - 7y - 3)i = 10i$

$\Rightarrow 4x + 9y - 3 = 0$ and $2x - 7y - 3 = 10$

on solving above two equations, we get $x = 3, y = -1$

157. Let the vertices of ΔABC be represented by z_1, z_2 and z_3, and its circumcentre by z_0.

Applying Rotation theorem in ΔABC, about A and B we get

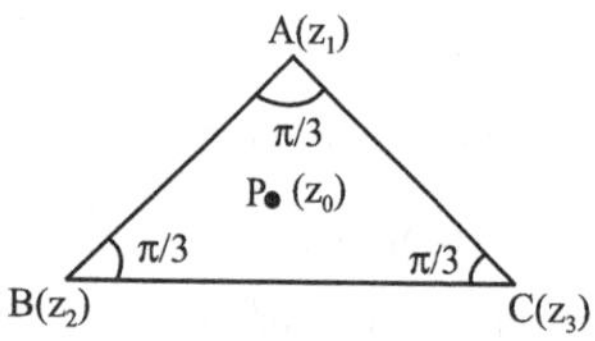

$z_3 - z_1 = (z_2 - z_1)e^{i\pi/3}$

and $z_1 - z_2 = (z_3 - z_2)\, e^{i\pi/3}$

Dividing we get $\frac{z_3 - z_1}{z_1 - z_2} = \frac{z_2 - z_1}{z_3 - z_2}$

$\Rightarrow (z_3 - z_1)(z_3 - z_2) = -(z_2 - z_1)^2$

$\Rightarrow z_3^2 - z_2z_3 - z_1z_3 + z_1z_2 + z_2^2 + z_1^2 - 2z_1z_2 = 0$

$\Rightarrow z_1^2 + z_2^2 + z_3^2 = z_1z_2 + z_2z_3 + z_3z_1 \quad \ldots(i)$

Since in an equilateral triangle circumcentre coincides with centroid. Therefore

$z_0 = \frac{z_1 + z_2 + z_3}{3}$

Or $9\, z_0^2 = z_1^2 + z_2^2 + z_3^2 + 2z_1z_2 + 2z_2z_3 + 2z_3z_1$

$\Rightarrow \frac{9z_0^2 - (z_1^2 + z_2^2 + z_3^2)}{2} = z_1z_2 + z_2z_3 + z_3z_1 \quad \ldots(ii)$

From equation (i) and (ii) we get,

$z_1^2 + z_2^2 + z_3^2 = \frac{9\, z_0^2 - (z_1^2 + z_2^2 + z_3^2)}{2}$

$\Rightarrow 2(z_1^2 + z_2^2 + z_3^2) + (z_1^2 + z_2^2 + z_3^2) = 9\, z_0^2$

$\Rightarrow 3(z_1^2 + z_2^2 + z_3^2) = 9\, z_0^2$

$\Rightarrow z_1^2 + z_2^2 + z_3^2 = 3\, z_0^2$ Proved

158. We know that if z_1, z_2, z_3 form an equilateral triangle then

$z_1^2 + z_2^2 + z_3^2 = z_1z_2 + z_2z_3 + z_3z_1$

putting $z_3 = 0$

we get $z_1^2 + z_2^2 = z_1z_2$ or $z_1^2 + z_2^2 - z_1z_2 = 0$

Proved

159. 1, $a_1, a_2, \ldots a_{n-1}$ are the n roots of unity. Or roots of equation $x^n - 1 = 0$

So

$x^n - 1 = (x-1)(x-a_1)(x-a_2)\ldots.(x-a_{n-1}) \quad \ldots(i)$

$\Rightarrow \frac{x^n - 1}{x-1} = (x-a_1)(x-a_2)\ldots.(x-a_{n-1}) \quad \ldots(ii)$

LHS of above equation is the sum

$1 + x + x^2 + \ldots\ldots. + x^{n-1}$

$\therefore$ We get

$1 + x + x^2 + \ldots\ldots. + x^{n-1}$

$= (x-a_1)(x-a_2)\ldots.(x-a_{n-1})$

Putting x = 1,

1+1+1+n times $= (1-a_1)(1-a_2)\ldots(1-a_{n-1})$

$\Rightarrow (1-a_1)(1-a_2)\ldots(1-a_{n-1}) = n$

160. Let $A = z = x + iy,\ \ B = iz = -y + ix,$

$C = z + iz = (x - y + i(x + y)$

Now, area of

$$\Delta ABC = \frac{1}{2}\begin{vmatrix} x & y & 1 \\ -y & x & 1 \\ x-y & x+y & 1 \end{vmatrix}$$

Operating $R_2 - R_1, R_3 - R_1$, we get

$$\Delta ABC = \frac{1}{2}\begin{vmatrix} x & y & 1 \\ -y-x & x-y & 0 \\ -y & x & 0 \end{vmatrix}$$

$= \frac{1}{2}|x(-y-x) + y(x-y)|$

$= \frac{1}{2}|-xy - x^2 + xy - y^2| = \frac{1}{2}|-x^2 - y^2|$

$= \frac{1}{2}|x^2 + y^2| = \frac{1}{2}|z|^2$ Hence proved.

161. $\because BC = AC$ and $\angle C = \frac{\pi}{2}$, we have by rotation

about C in anti clockwise sense

$CB = CAe^{i\pi/2}$

$\Rightarrow z_2 - z_3 = (z_1 - z_3)e^{i\pi/2}$

$\Rightarrow z_2 - z_3 = i(z_1 - z_3)$

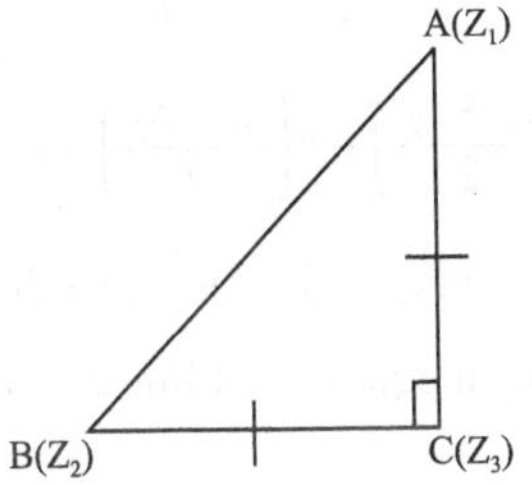

Squaring both sides, we get

$(z_2 - z_3)^2 = -(z_1 - z_3)^2$

$\Rightarrow z_2^2 + z_3^2 - 2z_2z_3 = -z_1^2 - z_3^2 + 2z_1z_3$

$\Rightarrow z_1^2 + z_2^2 - 2z_1z_2$

$= 2z_1z_3 + 2z_2z_3 - 2z_3^2 - 2z_1z_2$

$\Rightarrow (z_1 - z_2)^2 = 2\left[(z_1z_3 - z_3^2) - (z_1z_2 - z_2z_3)\right]$

$\Rightarrow (z_1 - z_2)^2 = 2(z_1 - z_3)(z_3 - z_2)$

162. We are given that $z_1 = 10 + 6i$ and $z_2 = 4 + 6i$

Also $\arg\left(\frac{z - z_1}{z - z_2}\right) = \frac{\pi}{4}$

$\Rightarrow \arg(z - z_1) - \arg(z - z_2) = \frac{\pi}{4}$

$\Rightarrow \arg((x + iy) - (10 + 6i) - \arg((x + iy) - (4 + 6i)) = \frac{\pi}{4}$

$\Rightarrow \arg[(x - 10) + i(y - 6)] - arg[(x - 4) + i(y - 6)] = \frac{\pi}{4}$

$\Rightarrow \tan^{-1}\left(\frac{y - 6}{x - 10}\right) - \tan^{-1}\left(\frac{y - 6}{x - 4}\right) = \frac{\pi}{4}$

$\Rightarrow \tan^{-1}\left(\frac{\frac{y-6}{x-10} - \frac{y-6}{x-4}}{1 + \frac{(y-6)^2}{(x-4)(x-10)}}\right) = \frac{\pi}{4}$

$\Rightarrow \frac{(x-4)(y-6) - (x-10)(y-6)}{(x-4)(x-10) + (x-6)^2} = tan\frac{\pi}{4} = 1$

$\Rightarrow (x - 4 - x + 10)(y - 6)$

$(x-4)(x-10) + (x-6)^2$

$\Rightarrow 6y - 36 = x^2 + y^2 - 14x - 12y + 40 + 36$

$\Rightarrow x^2 + y^2 - 14x - 18y + 112 = 0$

$\Rightarrow (x^2 + 14x + 49) + (y^2 - 18y + 81) = 18$

$\Rightarrow (x - 7)^2 + (y - 9)^2 = (3\sqrt{2})^2$

$\Rightarrow |(x + iy) - (7 - 9i)| = 3\sqrt{2}$

$\Rightarrow |z - (7 + 9i)| = 3\sqrt{2}$

163. Dividing both the side of the equation by i we get

$z^3 - iz^2 + iz + 1 = 0$

or $z^2(z - i) + i(z - i) = 0$ as $1 = -i^2$

or $((z - i)(z^2 + i) = 0 \quad \therefore z = i \text{ or } z^2 = -i$

$\therefore |z| = |i| = 1 \text{ or } |z^2| = |z|^2 = |-i| = 1$

$\Rightarrow |z| = 1$

Hence in either case $|z| = 1$

164. Let $Z = r_1(\cos\theta_1 + i\sin\theta_1)$

and $W = r_2(\cos\theta_2 + i\sin\theta_2)$

We have $|Z| = r_1, |W| = r_2$, Arg $Z = \theta_1$

and Arg $W = \theta_2$

Since $|Z| \le 1, |W| \le 1$, so $r_1 \le$ and $r_2 \le 1$

Consider $Z - W = (r_1\cos\theta_1 - r_2\cos\theta_2)$

$+ i(r_1\sin\theta_1 - r_2\sin\theta_2)$

$|Z - W|^2$

$= (r_1cos\theta_1 - r_2\cos\theta_2)^2 + (r_1sin\theta_1 - r_2\,sin\theta_2)^2$

$= r^2_1cos^2\theta_1 + r^2_2\cos^2\theta_2 - 2r_1r_2cos\theta_1cos\theta_2$

$+ r^2_1sin^2\theta_1 + r^2_2sin^2\theta_2 - 2r_1r_2sin\theta_1sin\theta_2$

$= r^2_1(cos^2\theta_1 + sin^2\theta_1) +$

$r^2_2(\cos^2\theta_2 + sin^2\theta_2) - 2r_1r_2$

$(\cos\theta_1\cos\theta_2 + \sin\theta_1\sin\theta_2)$

$= r_1^2 + r_2^2 - 2r_1r_2\cos(\theta_1 - \theta_2)$

$= (r_1 - r_2)^2 + 2r_1r_2[1 - \cos(\theta_1 - \theta_2)]$

$= (r_1 - r_2)^2 + 4r_1r_2\sin^2\left(\frac{\theta_1 - \theta_2}{2}\right)$

$= |r_1 - r_2|^2 + 4r_1r_2\left|\sin\left(\frac{\theta_1 - \theta_2}{2}\right)\right|^2$

$\le |r_1 - r_2|^2 + 4\left|\sin\left(\frac{\theta_1 - \theta_2}{2}\right)\right| \qquad [\because r_1, r_2 \le 1]$

But $|\sin\theta| \le \forall|\theta|\ for\ \theta \in R$

Therefore, $|Z - W|^2 \le |r_1 - r_2|^2 + 4\left|\frac{\theta_1 - \theta_2}{2}\right|^2$

$\le |r_1 - r_2|^2 + |\theta_1 - \theta_2|^2$

Thus $|Z - W|^2 \le (|Z| - |W|)^2 + (\text{Arg}Z - \text{Arg}W)^2$

165. Let $z = x + iy$ then $\bar{z} = iz^2$

$\Rightarrow x - iy = i(x^2 - y^2 + 2ixy)$

$\Rightarrow x - iy = i(x^2 - y^2) - 2xy$

$\Rightarrow x(1 + 2y) = 0;\ x^2 - y^2 + y = 0$

$\Rightarrow x = 0$ or $y = -\frac{1}{2}$ $\Rightarrow x = 0$ or $y = 0, 1$

or $y = -\frac{1}{2}, x = \pm\frac{\sqrt{3}}{2}$

For non zero complex number z

$x = 0, y = 1;$

$x = \frac{\sqrt{3}}{2},\ y = -\frac{1}{2}; x = \frac{\sqrt{3}}{2}, y = -\frac{1}{2}$

$\therefore z = i, \frac{\sqrt{3}}{2} - \frac{i}{2}, -\frac{\sqrt{3}}{2} - \frac{i}{2}$

166.

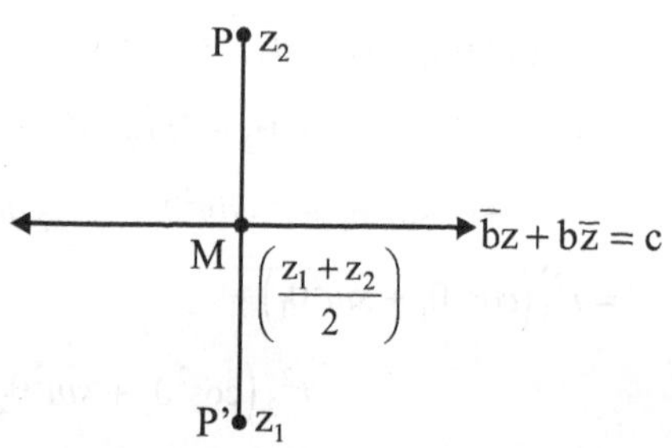

Since z_1 is a reflection of point z_2, through the line $\bar{b}z + b\bar{z} = c$ ...(i)

Therefore the mid point of z_1 and z_2 i.e. $\frac{z_1 + z_2}{2}$ must lie on straight line (i), so

$\bar{b}\left(\frac{z_1 + z_2}{2}\right) + b\left(\overline{\frac{z_1 + z_2}{2}}\right) = c$

$\Rightarrow \bar{b}\left(\frac{z_1 + z_2}{2}\right) + b\left(\frac{\bar{z}_1 + \bar{z}_2}{2}\right) = c$

$\Rightarrow (\bar{b}z_1 + \bar{b}z_2) + (b\bar{z}_1 + b\bar{z}_2) = 2c$ ...(ii)

Now given equation of line can also be written as

$b(x - iy) + \bar{b}(x + iy) = c$

or $(b + \bar{b})x - i(b - \bar{b})y = c$

$\Rightarrow$ slope of line $= \frac{b + \bar{b}}{i(b - \bar{b})} = m_1$ (say)

Also slope of line joining $z_1(x_1 + iy_1)$ and $z_2(x_2 + iy_2)$ is

$m_2 = \frac{y_2 - y_1}{x_2 - x_1} = \frac{\frac{z_2 - \bar{z}_2}{2i} - \frac{z_1 - \bar{z}_1}{2i}}{\frac{z_2 + \bar{z}_2}{2} - \frac{z_1 + \bar{z}_1}{2}}$

$= \frac{z_2 - \bar{z}_2 - z_1 + \bar{z}_1}{i(z_2 + \bar{z}_2 - z_1 - \bar{z}_1)}$

As the above two lines are perpendicular to each other

$\therefore m_1 \times m_2 = -1$

$\Rightarrow \frac{b + \bar{b}}{i(b - \bar{b})} \times \frac{z_2 - \bar{z}_2 - z_1 + \bar{z}_1}{i(z_2 + \bar{z}_2 - z_1 - \bar{z}_1)} = -1$

$\Rightarrow \frac{b + \bar{b}}{(b - \bar{b})} \times \frac{z_2 - \bar{z}_2 - z_1 + \bar{z}_1}{z_2 + \bar{z}_2 - z_1 - \bar{z}_1} = 1$

$\Rightarrow bz_2 - b\bar{z}_2 - bz_1 + b\bar{z}_1 + \bar{b}z_2 - \bar{b}\bar{z}_2 - \bar{b}z_1 + \bar{b}\bar{z}_1$

$= bz_2 + b\bar{z}_2 - bz_1 - b\bar{z}_1 - \bar{b}z_2 - \bar{b}\bar{z}_2 + \bar{b}z_1 + \bar{b}\bar{z}_1$

$\Rightarrow \bar{b}z_1 + b\bar{z}_2 = \bar{b}z_2 + b\bar{z}_1$

Substituting this value in equation (ii) we get

$b\bar{z}_2 + b\bar{z}_1 + \bar{b}z_2 + b\bar{z}_1 = 2c$

$\Rightarrow \bar{b}z_2 + b\bar{z}_1 = c$

$\Rightarrow \bar{z}_1 b + z_2\bar{b} = c$ Hence proved.

ALTERNATIVE SOLUTION:

Given that z_1 is the reflection of z_2 through the line $\bar{b}z + b\bar{z} = c$ ...(ii)

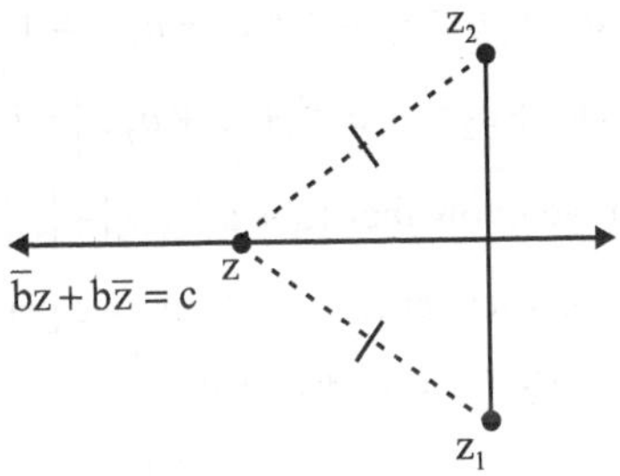

so for any arbitrary point z on the line, we must have

$|z - z_1| = |z - z_2|$

$\Rightarrow |z - z_1|^2 = |z - z_2|^2$

$\Rightarrow (z - z_1)(\overline{z - z_1}) = (z - z_2)(\overline{z - z_2})$

$\Rightarrow (z - z_1)(\bar{z} - \bar{z}_1) = (z - z_2)(\bar{z} - \bar{z}_2)$

$\Rightarrow (\bar{z}_2 - \bar{z}_1)z + (z_2 - z_1)\bar{z} + z_1\bar{z}_1 - z_2\bar{z}_2 = 0$ ...(ii)

The above equation represents the given line. Thus equation (i) and (ii) must be identical.

$\therefore$ From equations (i) and (ii) we get

$$\frac{\bar{b}}{\bar{z}_2 - \bar{z}_1} = \frac{b}{z_2 - z_1} = \frac{c}{z_2\bar{z}_2 - z_1\bar{z}_1} = k(say)$$

$\Rightarrow \bar{b} = k(\bar{z}_2 - \bar{z}_1),\ b = k(z_2 - z_1)$

$c = k(z_2\bar{z}_2 - z_1\bar{z}_1)$

Now,

$\bar{z}_1 b + z_2\bar{b} = \bar{z}_1(k(z_2 - z_1)) + z_2 k(\bar{z}_2 - \bar{z}_1)$

$= k(\bar{z}_1 z_2 - z_1\bar{z}_1 + z_2\bar{z}_2 - z_2\bar{z}_1)$

$= k(z_2\bar{z}_2 - z_1\bar{z}_1) = c$ Hence proved.

167. Given equation is $z^2 + pz + q = 0$

Using sum and product of roots

$$z_1 + z_2 = -p, z_1 z_2 = q$$

Now by rotation through α in anticlockwise direction $z_2 = z_1 e^{i\alpha}$ ...(i)

$$\frac{z_2}{z_1} = \frac{e^{i\alpha}}{1} = \frac{\cos\alpha + i\sin\alpha}{1}$$

Add 1 in both sides to get $z_1 + z_2 = -p$

$$\therefore \frac{z_1 + z_2}{z_1} = \frac{1 + \cos\alpha + i\sin\alpha}{1}$$

$$= 2\cos\frac{\alpha}{2}\left[\cos\frac{\alpha}{2} + i\sin\frac{\alpha}{2}\right]$$

or $\dfrac{(z_1 + z_2)}{z_1} = 2\cos\dfrac{\alpha}{2} e^{i\alpha/2}$

On squaring $(z_1 + z_2)^2 = 4\cos^2(\alpha/2) z_1^2 .\ e^{i\alpha}$

$= 4\cos^2\dfrac{\alpha}{2} z_1^2 \cdot \dfrac{z_2}{z_1} = 4\cos^2\dfrac{\alpha}{2} z_1 z_2$

or $p^2 = 4q\cos^2\dfrac{\alpha}{2}$

168. Given that z and w are two complex numbers. We need to prove $|z|^2 w - |w|^2 z = z - w \Leftrightarrow$ $z = w$ or $z\bar{w} = 1$

Let us consider

$|z|^2 w - |w|^2 z = z - w$ ...(i)

$\Rightarrow z(1 + |w|^2) = w(1 + |z|^2)$

$\Rightarrow \dfrac{z}{w} = \dfrac{1 + |z|^2}{1 + |w|^2}$ = a real number

$\Rightarrow \overline{\left(\dfrac{z}{w}\right)} = \dfrac{z}{w}$

$\Rightarrow \dfrac{\bar{z}}{\bar{w}} = \dfrac{z}{w}$

$\Rightarrow \bar{z}w = z\bar{w}$ ...(ii)

Now from equation (i).

$z\bar{z}w - w\bar{w}z = z - w$

$z(\bar{z}w - 1) - w(w\bar{z} - 1) = 0$

$z(z\bar{w} - 1) - w(z\bar{w} - 1) = 0$

(Using equation(ii))

$\Rightarrow (z\bar{w} - 1)(z - w) = 0 \Rightarrow z\bar{w} = 1$ or $z = w$

Conversely if $z = w$ then

L.H.S. of

(i) $= |w|^2 w - |w|^2 w = 0$ or $= w - w = 0$

Also if $z\bar{w} = 1$ then

L.H.S. of (i) $= z\bar{z}w - w\bar{w}z$

$= z\bar{z}w - w\bar{w}z = z - w =$ R.H.S. Hence proved.

169. The given equation can be written as

$$\left(z^p - 1\right)\left(z^q - 1\right) = 0$$

$\therefore z = (1)^{1/p}$ or $(1)^{1/p}$ ...(i)

Where p and q are distinct prime numbers.

So both the equations will have distinct roots and as $z \neq 1$, both will not be simultaneously zero for any value of z given by equations in (1)

Also $1 + \alpha + \alpha^2 + \ldots + \alpha^{p-1} = \dfrac{1-\alpha^p}{1-\alpha} = 0 (\alpha \neq 1)$

or $1 + \alpha + \alpha^2 + \ldots + \alpha^p = \dfrac{1-\alpha^q}{1-\alpha} = 0 (\alpha \neq 1)$

Because of (i) either $\alpha^p = 1$ and $\alpha^q = 1$ but not both simultaneously as p and q distinct primes.

170. Given that $|z_1| < 1 < |z_2|$

Then $\left|\dfrac{1 - z_1\bar{z}_2}{z_1 - z_2}\right| < 1$ is true

if $|1 - z_1\bar{z}_2| < |z_1 - z_2|$ is true

if $|1 - z_1\bar{z}_2|^2 < |z_1 - z_2|^2$ is true

if $(1 - z_1\bar{z}_2)\overline{(1 - z_1\bar{z}_2)} < (z_1 - z_2)\overline{(z_1 - z_2)}$ is true

if $(1 - z_1\bar{z}_2)(1 - \bar{z}_1 z_2) < (z_1 - z_2)(\bar{z}_1 - \bar{z}_2)$

if $1 - z_1\bar{z}_2 - \bar{z}_1 z_2 + z_1\bar{z}_1 \; z_2\bar{z}_2 <$

$z_1\bar{z}_1 - z_1\bar{z}_2 - \bar{z}_1 z_2 + z_2\bar{z}_2$ is true

if $1 + |z_1|^2 |z_2|^2 < |z_1|^2 + |z_2|^2$ is true

if $(1 - |z_1|^2)(1 - |z_2|^2) < 0$ is true.

which is obviously true as $|z_1| < 1 < |z_2|$

$\Rightarrow |z_1|^2 < 1 < |z_2|^2$

$\Rightarrow \left|1 - |z_1|^2\right| > 0$ and $\left(1 - |z_2|^2\right) < 0$

Hence proved.

171. Let us consider, $\sum_{r=1}^{n} a_r z^r = 1$ where $|a_r| < 2$

$\Rightarrow a_1 z + a_2 z^2 + a_3 z^3 + \ldots + a_n z^n = 1$

$\Rightarrow |a_1 z + a_2 z^2 + a_3 z^3 + \ldots + a_n z^n| = 1$ ...(i)

But we know that $|z_1 + z_2| \leq |z_1| + |z_2|$

$\therefore$ Using its generalized from, we get

$|a_1 z + a_2 z^2 + a_3 z^3 + \ldots + a_n z^n|$

$$\leq |a_1 z| + |a_2 z^2| + \ldots + |a_n z^n|$$

$\Rightarrow 1 \leq |a_1||z| + |a_2||z^2| + |a_3||z^3| + \ldots + |a_n||z^n|$

But given that $|a_r| < 2 \forall r = 1(1)^n$

$\therefore 1 < 2\left[|z| + |z|^2 + |z|^3 + \ldots + |z|^n\right]$

(Since $|z^n| = |z|^n$)

$\Rightarrow 1 < 2\left[\dfrac{|z|\left(1 - |z|^n\right)}{1 - |z|}\right]$

$\Rightarrow 2\left[\dfrac{|z| - |z|^{n+1}}{1 - |z|}\right] > 1$

$\Rightarrow 2\left[|z| - |z|^{n+1}\right] > 1 - |z|$

$(\because 1 - |z| > 0 \; as \; |z| < 1/3)$

$\Rightarrow \left[|z| - |z|^{n+1}\right] > \dfrac{1}{2} - \dfrac{1}{2}|z|$

$\Rightarrow \dfrac{3}{2}|z| > \dfrac{1}{2} + |z|^{n+1}$

$\Rightarrow |z| > \dfrac{1}{3} + \dfrac{2}{3} \; |z|^{n+1}$

$\Rightarrow |z| > \dfrac{1}{3}$

which is a contradiction as given that $|z| < \dfrac{1}{3}$

$\therefore$ There exist no such complex number.

172. The given circle is $\left|\frac{z-\alpha}{z-\beta}\right| = k, k \neq 1$

where $z = x+iy, \alpha = \alpha_1 + i\alpha_2, \beta = \beta_1 + i\beta_2$

Equation of circle in standard form

$z\bar{z} - \bar{a}z + a\bar{a} - r^2 = 0 \qquad \ldots(i)$

where a is the centre and r is the radius.

Now write given equation of circle into its standard form

$$\left|\frac{z-\alpha}{z-\beta}\right| = k$$

$$\Rightarrow |z-\alpha| = k|z-\beta|$$

Squaring on both sides, we get

$$|z-\alpha|^2 = k^2|z-\beta|^2$$

$$\Rightarrow (z-\alpha)\left(\overline{z-\alpha}\right) = k^2(z-\beta)\left(\overline{z-\beta}\right)$$

$$\Rightarrow (z-\alpha)(\bar{z}-\bar{\alpha}) = k^2(z-\beta)(\bar{z}-\bar{\beta})$$

$$\Rightarrow (1-k^2)z\bar{z} - (\alpha - k^2\beta)\bar{z} - (\bar{\alpha} - k^2\bar{\beta})z + \alpha\bar{\alpha} - k^2\beta\bar{\beta} = 0$$

$$\Rightarrow z\bar{z} - \left(\frac{\alpha-k^2\beta}{1-k^2}\right)\bar{z} - \left(\frac{\bar{\alpha}-k^2\bar{\beta}}{1-k^2}\right)z + \left(\frac{\alpha\bar{\alpha}-k^2\beta\bar{\beta}}{1-k^2}\right) = 0$$

$$\Rightarrow z\bar{z} - \left(\frac{\alpha-k^2\beta}{1-k^2}\right)\bar{z} - \left(\overline{\frac{\alpha-k^2\beta}{1-k^2}}\right)z + \left(\frac{\alpha\bar{\alpha}-k^2\beta\bar{\beta}}{1-k^2}\right) = 0 \quad \ldots(ii)$$

On comparing (i) and (ii), we get

$$a = \frac{u - k'\beta}{1-k^2} \text{ and } a\bar{a} - r^2 = \frac{\alpha\alpha - k^2\beta\bar{\beta}}{1-k^2}$$

$$\therefore Centre\,is\,a = \frac{\alpha-k^2\beta}{1-k^2} \text{ and r is given by}$$

$$\left(\frac{\alpha-k^2\beta}{1-k^2}\right)\left(\overline{\frac{\alpha-k^2\beta}{1-k^2}}\right) - r^2 = \frac{\alpha\bar{\alpha}-k^2\beta\bar{\beta}}{1-k^2}$$

$$\Rightarrow r^2 = \left(\frac{\alpha-k^2\beta}{1-k^2}\right)\left(\frac{\bar{\alpha}-k^2\bar{\beta}}{1-k^2}\right) - \frac{\alpha\bar{\alpha}-k^2\beta\bar{\beta}}{1-k^2}$$

$$= \frac{\alpha\bar{\alpha} - k^2\alpha\bar{\beta} - k^2\bar{\alpha}\beta + k^4\beta\bar{\beta} - \alpha\bar{\alpha} + k^2\beta\bar{\beta} + k^2\alpha\bar{\alpha} - k^4\beta\bar{\beta}}{(1-k^2)^2}$$

$$= \frac{k^2\alpha\bar{\alpha} + k^2\beta\bar{\beta} - k^2\alpha\bar{\beta} + k^2\bar{\alpha}\beta}{(1-k^2)^2}$$

$$= \frac{k^2\left[\alpha\bar{\alpha} - \beta\bar{\beta} - \alpha\bar{\beta} - \bar{\alpha}\beta\right]}{(1-k^2)^2} = \frac{k^2(\alpha-\beta)(\bar{\alpha}-\bar{\beta})}{(1-k^2)^2}$$

$$= \frac{k^2(\alpha-\beta)\left(\overline{\alpha-\beta}\right)}{(1-k^2)^2} = \frac{k^2|\alpha-\beta|^2}{(1-k^2)^2}$$

$$\therefore Radius\,r = \frac{k}{|1-k^2|}\left\|\alpha-\beta\right\|$$

173 . The given circle is $|z-1| = \sqrt{2}$ where $z_0 = 1$ the centre of the circle is (1,0) and radius $\sqrt{2}$.

z_1 is one of the vertex of square inscribed in the given circle.

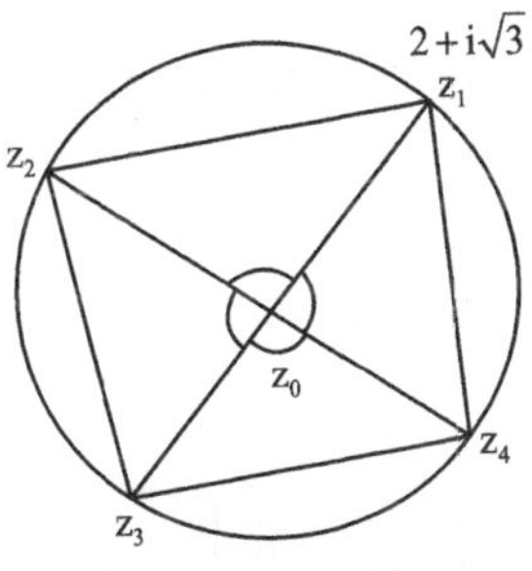

Now z_2 can be obtained by rotating z_1 by an $\angle 90°$ in anticlockwise sense, about centre z_0.

Thus, $z_2 - z_0 = (z_1 - z_0)e^{i\pi/2}$

or $z_2 - 1 = (2 + i\sqrt{3} - 1)i$

$\Rightarrow z_2 = i - \sqrt{3} + 1$

$z_2 = (1-\sqrt{3}) + i$

Again rotating $z_2\,by$ $90°$ about z_0 we get

$z_3 - z_0 = (z_2 - z_0)i$

$\Rightarrow z_3 - 1 = \left[(1-\sqrt{3}) + i - 1\right]i = -\sqrt{3}$ i-1

$\Rightarrow z_3 = -i\sqrt{3}$

and similarly $1 = \left(-i\sqrt{3} - 1\right)i = \sqrt{3} - i$

$\Rightarrow z_4 = \left(\sqrt{3} + 1\right) - i$

So the remaining vertices are

$\left(1 - \sqrt{3}\right) + i, -i\sqrt{3}, \left(\sqrt{3} + 1\right) - i$

Concept Eliminator (CE)

Olympiad Tonic

1. Let z_1, z_2, z_3 are represented by points A, B and C on argand diagram.

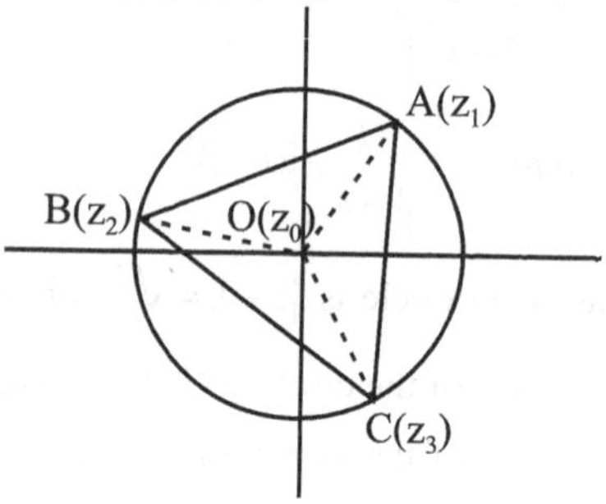

From rotation theorem we get

$$\frac{z_0 - z_1}{z_2 - z_1} = \left|\frac{z_0 - z_1}{z_2 - z_1}\right| e^{i\left(\frac{\pi}{2} - c\right)} = \frac{5}{c} e^{i\left(\frac{\pi}{2} - c\right)}$$

And $\frac{z_3 - z_1}{z_0 - z_1} = \left|\frac{z_3 - z_1}{z_0 - z_1}\right| e^{i\left(\frac{\pi}{2} - B\right)} = \frac{b}{5} e^{i\left(\frac{\pi}{2} - B\right)}$

From above two results we get

$$\frac{z_1^2}{(z_1 - z_2)(z_1 - z_3)} = \frac{25}{bc}\left(e^{i(B-C)}\right)$$

The following identity is easy to verify

$$\frac{z_1^2}{(z_1 - z_2)(z_1 - z_3)} + \frac{z_2^2}{(z_2 - z_1)(z_2 - z_3)} + \frac{z_3^2}{(z_3 - z_1)(z_3 - z_2)}$$

$$= \frac{25}{abc}\left(ae^{i(B-C)} + be^{i(C-A)} + ce^{i(A-B)}\right) = 1.$$

Because, in a triangle

$a\sin(B - C) + b\sin(C - A) + c\sin(A - B) = 0$

and

$a\cos(B - C) + b\cos(C - A) + c\cos(A - B) = \frac{abc}{R^2}$

Putting the modulus we get

$$1 = \left|\sum \frac{z_1^2}{(z_1 - z_2)(z_1 - z_3)}\right| \le \sum \frac{|z_1|^2}{|z_1 - z_2||z_1 - z_3|}$$

$$= R^2 \sum \frac{1}{|z_1 - z_2||z_1 - z_3|}$$

$$\frac{1}{|z_1 - z_2||z_1 - z_3|} + \frac{1}{|z_2 - z_1||z_2 - z_3|} + \frac{1}{|z_3 - z_1||z_3 - z_2|} \ge \frac{1}{R^2} = \frac{1}{25}$$

So required minimum value is $\frac{1}{25}$

2. Given equation is $| z - | z + 1 || = | z + | z - 1 ||$
It is true if and only if

$| z - | z + 1 ||^2 = | z + | z - 1 ||^2$.

i. e., $(z - | z + 1 |) . (\bar{z} - | z + 1 |)$

$= (z + | z - 1 |) . (\bar{z} + | z - 1 |)$.

or

$z.\bar{z} - (z + \bar{z}) | z + 1 | + | z + 1 |^2$

$= z . \bar{z} + (z + \bar{z}) . | z - 1 | + | z - 1 |^2$.

or $| z + 1 |^2 - | z - 1 |^2$

$= (z + \bar{z}) . (| z + 1 | + | z - 1 |)$.

i. e., $(z + 1)(\bar{z} + 1) - (z - 1)(\bar{z} - 1)$

$= (z + \bar{z}) . (| z + 1 | + | z - 1 |)$.

or $2(z + \bar{z}) = (z + \bar{z}) . (| z + 1 | + | z - 1 |)$

i. e ., $z + \bar{z} = 0$

Or $| z + 1 | + | z - 1 | = 2$.

z is any real number with $-1 \le z \le 1$.

The equation $z + \bar{z} = 0$ has the solutions $z = b$ i, $b \in R$. Hence, the solutions to be equation are

$\{b i : b \in R\} \cup \{a \in R : a \in [-1, 1]\}$.

3. Given equation is $az^3 + bz^2 + \bar{b}z + \bar{a} = 0$
We can observe that if z is a root of the equation, then $\frac{1}{\bar{z}}$ is also a root of the equation.

So, if z_1, z_2, z_3 are the roots of the equation, then $\frac{1}{\bar{z}_1}, \frac{1}{\bar{z}_2}, \frac{1}{\bar{z}_3}$ are the same roots, not necessarily in the same order. Now we have two cases

Case (i) If $z_k = \frac{1}{\bar{z}_k}$ for some k = 1, 2, 3:

then $|z_k|^2 = z_k\bar{z}_k = 1$ and hence proved .

Case (ii) if $z_k \neq \frac{1}{\bar{z}_k}$ for all k = 1, 2, 3,

Then we may consider without loss of generality that

$z_1 = \frac{1}{\bar{z}_2}, z_2 = \frac{1}{\bar{z}_3}, z_3 = \frac{1}{\bar{z}_1}$.

The first two equalities yield $z_1 . \bar{z}_2 . z_2 . \bar{z}_3 = 1$. Hence $|z_1| . |z_2|^2 . |z_3| = 1$. On the other hand, $z_1 z_2 z_3 = -\frac{\bar{a}}{a}$, so $|z_1||z_2||z_3| = 1$. It follows that $|z_2| = 1$, proved.

4. Given expression is $\frac{1+z+z^2}{1-z+z^2} = 1 + \frac{2z}{1-z+z^2}$

$\in$ R if and only if $\frac{z}{1-z+z^2} \in$ R.

That is, $\frac{1-z+z^2}{z} = \frac{1}{z} - 1 + z \in$ R.

Or $z + \frac{1}{z} \in$ R.

Or $z + \frac{1}{z} = \bar{z} + \frac{1}{\bar{z}}$,

i. e., $(z - \bar{z})(1 - |z|^{2}) = 0$.

So $z = \bar{z}$ or $|z| = 1$.

But z is not a real number, So $|z| = 1$.

5. Let $s_1 = z_1 + z_2 + z_3$,

$s_2 = z_1z_2 + z_2z_3 + z_3z_1$,

And $s_3 = z_1z_2z_3$

Now consider the cubic equation

$z^3 - s_1z^2 + s_2z - s_3 = 0$

It has three roots z_1, z_2, z_3 .

Given that $z^2_1 + z_2^2 + z^2_3 = 0$, we have

$s^2_1 = 2 s_2$ ...(i)

And also,

$s_2 = s_3\left(\frac{1}{z_1} + \frac{1}{z_2} + \frac{1}{z_3}\right) = s_3\left(\bar{z}_1 + \bar{z}_2 + \bar{z}_3\right)$

$= s_3 . \bar{s}_1$ ...(ii)

From equations (i) and (ii) $s_1^2 = 2s_3 . \bar{s}_1$

and consequently,

$|s_1|^2 = 2|s_3| . |\bar{s}_1| = 2|s_1|$.

Because $s_1 \neq 0$, we have $|s_1| = 2$, so $s_1 = 2\lambda$ with $|\lambda| = 1$.

From equations (i) and (ii) it also follows that

$s_2 = \frac{1}{2}s_1^2 = 2\lambda^2$ and $s_3 = \frac{s_2}{\bar{s}_1} = \frac{2\lambda^2}{2\bar{\lambda}} = \lambda^3$

The equation with roots z_1, z_2, z_3 becomes

$z^3 - 2\lambda z^2 + 2\lambda^2 z - \lambda^3 = 0$.

This is equivalent to $(z - \lambda)(z^2 - \lambda z + \lambda^2) = 0$.

The roots are $\lambda, \lambda\omega, -\lambda\omega^2$,

where $\varepsilon = \frac{1}{2} + i\frac{\sqrt{3}}{2}$.

Then

Without loss of generality we may assume that $z_1 = \lambda, z_2 = \lambda\omega, z_3 = -\lambda\omega^2$. Using the relations $\omega^2 - \omega + 1 = 0$ and $\omega^3 = -1$,

So, $E_n = |z_1^n + z_2^n + z_3^n| = |\lambda^n + \lambda^n\omega^n + (-1)^n\lambda^n\omega^{2n}|$

$= |1 + \omega^n + (-1)^n\omega^{2n}|$.

Since $E_{k+6} = E_k$ for all integers k and that equalities

$E_0 = 3, \quad E_1 = 2, E_2 = 0, E_3 = 1, E_4 = 0, E_5 = 2$.

$E_n = |z_1^n + z_2^n + z_3^n| \in \{0, 1, 2, 3\}$

ALTERNATE SOLUTION

It is clear that z_1^2, z_2^2, z_3^2 are distinct. Otherwise, if, for example, $z_1^2 = z_2^2$, then $1 = |z_3^2| = |-(z_1^2 + z_2^2)| = 2|z^2_1| = 2$, a contradiction .

From $z_1^2 + z_2^2 + z_3^2 = 0$ it follows that z_1^2, z_2^2, z_3^2 are the coordinates of the vertices of an equilateral triangle . Hence we may assume that $z_2^2 = \omega z_1^2$ and $z_3^2 = \omega^2 z_1^2$, where $\omega^2 + \omega + 1 = 0$. Because $z_2^2 = \omega^4 z_1^2$ and $z_3^2 = \omega^2 z_1^2$ it follows that $z_2 = \pm\omega^2 z_1$ and $z_3 = \pm\omega z_1$. Then $|z_1^n + z_2^n + z_3^n| = |(1 + (\pm\omega)^n + (\pm\omega^2)^n) z_1^n|$ $= |1 + (\pm\omega^2)^n| \in \{0, 1, 2, 3\}$

6. Consider LHS

$n|1 + z| + |1 + z^2| + |1 + z^3| + \ldots + |1 + z^{2n}| + |1 + z^{2n+1}|$

$= \sum_{k=1}^{n} (|1+z|+|1+z^{2k+1}|) + \sum_{k=1}^{n} |1+z^{2k}|$

$\geq \sum_{k=1}^{n} |z - z^{2k+1}| + \sum_{k=1}^{n} |1+z^{2k}|$

$= \sum_{k=1}^{n} (|z||1-z^{2k}| + |1+z^{2k}|)$

$= \sum_{k=1}^{n} (|1-z^{2k}| + |1+z^{2k}|)$

$\geq \sum_{k=1}^{n} |1-z^{2k}+1+z^{2k}| = 2n.$

Hence Proved.

7. The equation can be re-written as

$$z^n(\lambda+i) = \bar{z}(-\lambda+i).$$

Taking the absolute values of both sides of the equation,

$|z|^n = |\bar{z}| = |z|$, hence $|z| = 0$ or $|z| = 1$.

If $|z| = 0$, then $z = 0$ which satisfies the equation.

If $|z| = 1$ then $\bar{z} = \frac{1}{z}$ and the equation may be rewritten as $z^{n+1} = \frac{-\lambda+i}{\lambda+i}$.

Because $\left|\frac{\lambda+i}{\lambda+i}\right| = 1$, there exists $t \in [0, 2\pi)$ such that $\frac{\lambda+i}{\lambda+i} = \cos t + i\sin t$.

Then $z_k = \cos\frac{t+2k\pi}{n+1} + i\sin\frac{t+2k\pi}{n+1}$ for $k = 0, 1, \ldots, n$ are the other solutions to the equation (besides $z = 0$).

8. Let $r = |z_1| = |z_2| = |z_3| > 0$. Then $z_1\bar{z}_1 = z_2\bar{z}_2 = z_3\bar{z}_3 = r^2$ and $\frac{1}{z_1}+\frac{1}{z_2}+\frac{1}{z_3}$

$= \frac{\bar{z}_1+\bar{z}_2+\bar{z}_3}{r^2} = \frac{\overline{z_1+z_3+z_3}}{r^2}.$

now consider $\frac{1}{z_1+z_2+z_3} = \frac{\overline{z_1+z_3+z_3}}{|z_1+z_2+z_3|^{\dagger}}$

and, $\text{Re}\left(\frac{1}{z_1}+\frac{1}{z_2}+\frac{1}{z_3}\right) . \text{Re}\left(\frac{1}{z_1+z_2+z_3}\right)$

$= \text{Re}\left(\frac{\overline{z_1+z_3+z_3}}{r^2}\right) . \text{Re}\left(\frac{\overline{z_1+z_3+z_3}}{|z_1+z_2+z_3|^{\dagger}}\right)$

$= \frac{\left(Re\left(\overline{z_1+z_3+z_3}\right)\right)^{\dagger}}{r^{\dagger}|z_1+z_2+z_3|^{\dagger}} \geq 0.$ Hence proved

9. Let $z = \cos t + i\sin t$ Then

$|1-z|$

$= \sqrt{(1-\cos t)^2 + \sin^2 t} = \sqrt{2-2\cos t} = 2\left|\sin\frac{t}{2}\right|$

and $|1+z^2|$

$= \sqrt{(1+\cos 2t)^2 + \sin^2 2t} = \sqrt{2+2\cos 2t}$

$= 2|\cos t| = 2\left|1-2\sin^2\frac{t}{2}\right|.$

So $|1-z|+|1+z^2| = 2\left|\sin\frac{t}{2}\right| + 2\left|1-2\sin^2\frac{t}{2}\right|$

$= |a| + |1-2a^2|$, for $a = \sin\frac{t}{2} \in [-1, 1]$

In the given range [-1, 1] the required minimum value is $\frac{1}{\sqrt{2}}$ and maximum value is 2

10. Let $r = |a| = |b| = |c| > 0$.

Then $|az^2| = |-bz-c| \leq |b||z| + |c|$.

Hence $r|z^2| \leq r|z| + r$. it follows that $|z^2| - |z| - 1 \leq 0$. so $|z| \leq \frac{1+\sqrt{5}}{2}$.

Now consider

$|c| = |-az^2 - bz| \leq |a||z|^2 + |b||z|$

$|z|^2 + |z| - 1 \geq 0.$

Thus $|z| \geq \frac{\sqrt{5}-1}{2}$

So $\frac{\sqrt{5}-1}{2} \leq |z| \leq \frac{\sqrt{5}+1}{2}$

11. We have

$\omega^{2n+1} = 1$ and $1 + \omega + \omega^2 + \ldots + \omega^{2n} = 0.$

Then $\frac{1}{2} + \omega + \omega^2 + \ldots + \omega^n + \omega^n(\omega + \omega^2 + \ldots + \omega^n) + \frac{1}{2} = 0.$

Or $z + \omega^n(z - \frac{1}{2}) + \frac{1}{2} = 0.$

Hence $z = \frac{1}{2} \cdot \frac{\omega^n - 1}{\omega^n + 1}.$

we have $\bar{z} = \frac{1}{2}\frac{\frac{1}{\omega^n}-1}{\frac{1}{\omega^n}+1} = -z.$

Since, $\bar{z} = -z$, so, $\text{Re}(z) = 0$.

Hence I m $(z^{2k}) = \text{Re}(z^{2k+1}) = 0$ for all $k \in N$
Hence Proved

12. Because $z_1 + z_2 = -p$ and $z_1 z_2 = q$, and given inequality $|p| + |q| < 1$
so $|z_1 + z_2| + |z_1 z_2| < 1$.
But $||z_1| - |z_2|| \le |z_1 + z_2|$,
hence $|z_1| - |z_2| + |z_1||z_2| - 1 < 0$
if and only if $(1 + |z_2|)(|z_1| - 1) < 0$
and $|z_2| - |z_1| + |z_2||z_1| - 1 < 0$
if and only if $(1 + |z_1|)(|z_2| - 1) < 0$.
Or, $|z_1| < 1$ and $|z_2| < 1$, hence proved.

13. Let $x = a - b$, $y = b - c$, $z = c - a$. Because $x + y + z = 0$ and $x^7 + y^7 + z^7 = 0$,
So $(x + y)^7 - x^7 - y^7 = 0$.
This is equivalent to $7xy(x + y)(x^2 + xy + y^2)^2 = 0$.
But $xyz \ne 0$, so $x^2 + xy + y^2 = 0$, i.e., $x^3 = y^3$,
From symmetry, $x^3 = y^3 = z^3$,
hence $|x| = |y| = |z| =$ or $|a - b| = |b - c| = |c - a|$ so triangle is equilateral

14. Consider the complex plane with origin at point O and let z_1, z_2, z_3 be the coordinates of the centres O_1, O_2, O_3, respectively.
Then the points A, B, C have the coordinates
$z_1 + z_2, z_2 + z_3, z_3 + z_1$,
hence $AB = |(z_1 + z_2) - (z_2 + z_3)|$
$= |z_1 - z_3| = O_1 O_3$.
Similarly, $BC = O_1O_2$ and $AC = O_2O_3$, hence triangles ABC and $O_1O_2O_3$ are congruent. And so, their circumradii are equal.
But $OO_1 = OO_2 = OO_3 = r$, the circumradius of triangle $O_1O_2O_3$ and ABC is equal to r,

15. Since $\omega^3 = 1$ and $\omega^2 + \omega + 1 = 0$ and
$1987 = 662 \times 3 + 1$
Hence, $(1+\omega)(1+\omega^2)(1+\omega^3)\ldots(1+\omega^{1987})$

$$= \sum_{k=0}^{661}\left[(1+\omega^{3k+1})(1+\omega^{3k+2})(1+\omega^{3k+3})\right](1+\omega^{1987})$$

$$= \sum_{k=0}^{661}\left[(1+\omega)(1+\omega^2)(1+1)\right](1+\omega)$$

$$= (1+\omega)\left[2(1+\omega+\omega^2+\omega^3)\right]^{662}$$

$$= (1+\omega)\left[2^{662}\right] = 2^{662}(-\omega^2) = 2^{661}(1+i\sqrt{3})$$

16. Since Let $\omega \ne 1$ be a cube root of unity then $1 + \omega + \omega^2 = o$
So $1 - \omega + \omega^2 = -2\omega$ & $1 + \omega - \omega^2 = -2\omega^2$

$$\text{Then, } 1 - \omega^n + \omega^{2n} = \begin{cases} 1 & \text{if } n \equiv 0 \pmod 3 \\ -2 & \text{if } n \equiv 1 \pmod 3 \\ -2 & \text{if } n \equiv 2 \pmod 3 \end{cases}$$

So the product of any three consecutive factors of the given product equals
$1 \cdot (-2\omega) \cdot (-2\omega^2) = 2^2$.
Therefore

$$(1-\omega+\omega^2)(1-\omega^2+\omega^4)\ldots(1-\omega^n+\omega^{2n})$$

$$= \begin{cases} 2^{\frac{2n}{3}} & \text{if } n \equiv 0 \pmod 3 \\ -2^{2\left(\frac{n}{3}\right)+1}\omega & \text{if } n \equiv 1 \pmod 3 \\ 2^{2\left(\frac{n}{3}\right)+2} & \text{if } n \equiv 2 \pmod 3 \end{cases}$$

17. Obviously $|z| = 1$. Assume by contradiction that there is an integer $n \ge 1$ such that $z^n = 1$
Then $(2 + i)^n = (2 - i)^n$,
So $(2 - i)^n = (2 + i)^n = ((2 - i) + 2i)^n$

$$= (2-i)^n + \binom{n}{1}(2-i)^{n-1}2i + \ldots + \binom{n}{n-1}(2-i)$$

$(2i)^{n-1} + (2i)^n$ Or $(2i)^n = (-2 + i)$

$$\left[\binom{n}{1}(2i-1)^{n-2}2i + \ldots + \binom{n}{n-1}(2i)^{n-1}\right]$$

$= (-2 + i)(a + bi)$ With $a, b \in \mathbb{Z}$.
Now by taking modulus we get
$2^n = 5(a^2 + b^2)$, a contradiction and hence proved.

18. Consider the polynomial
$f(x) = X^n - 1 = \prod_{\varepsilon \in U_n}(X - \varepsilon)$.

$$\prod_{\varepsilon \in U_n}\left(\varepsilon + \frac{1}{\varepsilon}\right) = \prod_{\varepsilon \in U_n}\frac{\varepsilon^2+1}{\varepsilon}$$

$$= \frac{\prod_{\varepsilon \in U_n}(\varepsilon+i)(\varepsilon-i)}{\prod_{\varepsilon \in U_n}\varepsilon}$$

$$= \frac{\prod_{\varepsilon \in U_n}(i+\varepsilon)\prod_{\varepsilon \in U_n}(-i+\varepsilon)}{(-1)^n f(0)}$$

$$= \frac{f(-1).f(i)}{(-1)^{n-1}} = \frac{\left[(-i)^n - 1\right]\left(i^n - 1\right)}{(-1)^{n-1}}$$

If $n \equiv 0 \pmod 4$, then $i^n = 1$

and $\prod_{\varepsilon \in U_n}\left(\varepsilon + \frac{1}{\varepsilon}\right) = 0,$

If $n \equiv 1 \pmod 2$, then $(-1)^{n-1} = 1$ and

$\prod_{\varepsilon \in U_n}\left(\varepsilon + \frac{1}{\varepsilon}\right) = (-i^n - 1)(i^n - 1)$

$= -(i^{2n} - 1) = -((-1)^n - 1) = -(-1 - 1) = 2.$

If $n \equiv 2 \pmod 4$, then $(-1)^{n-1} = -(i^{2n} - 1) = -1,$

$(-i)^n = i^n = i^2 = -1,\ i^n = -1,$

hence $\prod_{\varepsilon \in U_n}\left(\varepsilon + \frac{1}{\varepsilon}\right) = \frac{(-1-1)(-1-1)}{-1} = -4.$

If $n \equiv 3 \pmod 4$, then $(-10^{n-1} = 1$ and

$\prod_{\varepsilon \in U_n}\left(\varepsilon + \frac{1}{\varepsilon}\right) = (-i^n - 1)(i^n - 1)$

$= (i^3 - 1)(-i^3 - 1) = -(i^6 - 1)$

$= -((-1)^3 - 1) = 2.$

To sum up the above result we get

$$\prod_{\varepsilon \in U_n}\left(\varepsilon + \frac{1}{\varepsilon}\right) = \begin{cases} 0 & \text{if } n \equiv 0 \pmod 4. \\ 2 & \text{if } n \equiv 1 \pmod 4. \\ -4 & \text{if } n \equiv 2 \pmod 4. \\ 2 & \text{if } n \equiv 3 \pmod 4. \end{cases}$$

Hence proved.

19. Without loss of generality we may assume that the centre of the polygon is the origin of the complex plane.

Let $z_k = z_1\varepsilon^{k-1}$, where

$\varepsilon = \cos\frac{2\pi}{n} + i\sin\frac{2\pi}{n}$, k = 1, …, n.

The right hand side is equal to $z_1z_2 + z_2z_3 + \ldots + z_nz_1 = \sum_{k=1}^{n} z_k z_{k+1}$

$$= \sum_{k=1}^{n} z_1^2\varepsilon^{2k-1} = z_1^2\varepsilon.\ \frac{1-\varepsilon^{2n}}{1-\varepsilon^2} = 0.$$

Now consider $z_1^2 + z_2^2 + \ldots + z_n^2 = \sum_{k=1}^{n} z_1^2$

$$= \sum_{k=1}^{n} z_1^2\ \varepsilon^{2k-2} = z_1^2\frac{1-\varepsilon^{2n}}{1-\varepsilon^2} = 0$$

So LHS = RHS Proved

20. Without loss of generality we can assume that the centre of the polygon is the origin in complex plane and $a_k = a_1\varepsilon^{k-1}$, k = 1 … n, where $\varepsilon = \cos\frac{2\pi}{n} + i\sin\frac{2\pi}{n}$.

Consider LHS $a_1a_2 + a_2a_3 + \ldots + a_na_1$

$$= a_1^2\sum_{k=1}^{n}\varepsilon^{2k-1} = a_1^2\varepsilon\frac{1-\varepsilon^{2n}}{1-\varepsilon^2} = 0$$

Now consider RHS

$$a_1^2\sum_{k=1}^{n}\varepsilon^{2k} = a_1^2\varepsilon^2\frac{1-\varepsilon^{2n}}{1-\varepsilon^2} = 0.$$

So LHS = RHS Proved

21. Note that $\overline{z_k} = \frac{1}{z_k}$ for all k = 1, …, n.

$$\overline{z} = \left(\sum_{k=1}^{n}\overline{z_k}\right)\left(\sum_{k=1}^{n}\frac{1}{\overline{z_k}}\right)$$

$$= \left(\sum_{k=1}^{n}\frac{1}{z_k}\right)\left(\sum_{k=1}^{n}z_k\right) = z.$$

So z is a real number and Im(z) =0

Let $z_k = \cos a_k + i\sin a_k$, where a_k are real numbers. For $k = \overline{1,n}$. Then

$$z = \left(\sum_{k=1}^{n}\cos a_k + i\sum_{k=1}^{n}\sin a_k\right)\left(\sum_{k=1}^{n}\cos a_k - i\sum_{k=1}^{n}\sin a_k\right)$$

$$= \left(\sum_{k=1}^{n}\cos a_k\right)^2 + \left(\sum_{k=1}^{n}\sin a_k\right)^2 \geq 0.$$

Now consider $z = \left(\sum_{k=1}^{n}\cos^2 a_k + i\sum_{k=1}^{n}\sin^2 a_k\right)$

$$+ 2\sum_{1\le i<j\le n}(\cos\alpha_i\cos\alpha_j + \sin\alpha_i\sin\alpha_j)$$

$$\le n + 2\sum_{1\le i<j\le n}\cos(\alpha_i - \alpha_j) \le n + 2\binom{n}{2}$$

$$= n + 2\frac{n(n-1)}{2} = n^2. \text{ Hence proved.}$$

22. From induction we can say that $z_k = i^k\binom{n}{k}z_0$, for all $k \in \{0, 1, \ldots, n\}$.

Then $z_0 + z_1 + \ldots + z_n = 2^n$ if and only if $z_0(1 + i)^n = 2^n$. i. e., $z_0 = (1 - i)^n$

23. From the above result we have

$|z_0|^2 + |z_1|^2 + \ldots + |z_n|^2$

$= |z_0|^2 \left(\binom{n}{0}^2 + \binom{n}{1}^2 + \ldots + \binom{n}{n}^2\right)$

$= |z_0|^2 \cdot \binom{2n}{n}$

$= 2^n \cdot \binom{2n}{n} = \dfrac{2^n}{n!} 2n(2n-1) \ldots (n+1)$

$< \dfrac{2^n}{n!}\left(\dfrac{2n(2n-1)\ldots(n+1)}{n}\right)^n$

$= \dfrac{(3n+1)^n}{n!}$ [From A M – G M inequality]

24. We have $\omega^{2n+1} = 1$

and $1 + \omega + \omega^2 + \ldots + \omega^{2n} = 0$.

Then $\frac{1}{2} + \omega + \omega^2 + \ldots + \omega^n + \omega^n(\omega + \omega^2 + \ldots + \omega^n) + \frac{1}{2} = 0$. Or $z + \omega^n(z - \frac{1}{2}) + \frac{1}{2} = 0$.

Or $2z + 1 = -\omega^n(2z - 1)$.

Since $\omega^{2n+1} = 1$. we obtain $(2z+1)^{2n+1}$.

And hence $(2z+1)^{2n+1} + (2z-1)^{2n+1} = 0$

25. Let $\varepsilon = \cos\dfrac{2\pi}{n} + i\sin\dfrac{2\pi}{n}$.

Then $z_P = z_1 . \varepsilon^{P-1}$, for all $p = \overline{1, n}$.

Given condition $z_i + z_j = z_k$ if and only if $1 + \varepsilon^{j-i} = \varepsilon^{k-i}$, i.e.,

$2\cos\dfrac{(j-i)\pi}{n}\left[\cos\dfrac{(j-i)\pi}{n} + i\sin\dfrac{(j-i)\pi}{n}\right]$

$= \cos\dfrac{2(k-i)\pi}{n} + i\sin\dfrac{2(k-i)\pi}{n}$,

From above result we can deduce $\dfrac{(j-i)\pi}{n} = \dfrac{\pi}{3} = \dfrac{2(k-i)\pi}{n}$, i.e, $n = 6(k-i) = 3(j-i)$,

hence 6 divides n.

Now, if 6 divides n, let

$i = 1$, $j = \frac{n}{3} + 1$, $k = \frac{n}{6} + 1$ and we have

$z_i + z_l = z_k$ Proved.

26. Let x_1, x_2, x_3, x_4 be the real roots of the given polynomial f(x). Then

$f(x) = (x - x_1)(x - x_2)(x - x_3)(x - x_4)$

and $|f(i)| = \sqrt{1 + x_1^2} . \sqrt{1 + x_2^2} . \sqrt{1 + x_3^2} . \sqrt{1 + x_4^2}$.

Given that $|f(i)| = 1$, which is possible only when $x_1 = x_2 = x_3 = x_4 = 0$ and so $a = b = c = d = 0$, Proved.

27. The equation can be rewritten as

$z^9 = \dfrac{11 - 10iz}{11z + 10iz}$.

Now assume that $z = a + b\,i$, and taking modulus both the sides we get

$|z^9| = \dfrac{|11 - 10iz|}{|11z + 10i|}$

$= \dfrac{\sqrt{11^2 + 220b + 10^2(a^2 + b^2)}}{\sqrt{11^2(a^2 + b^2) + 220b + 10^2}}$.

Let f (a, b) and g (a, b) denote the numerator and denominator of the right hand side. Now take two cases

Case (i) If $|z| > 1$, then $a^2 + b^2 > 1$, so g (a, b) > f (a, b), leading to $|z^9| < 1$, a contradiction hence this case is ruled out.

Case (ii) If $|z| < 1$, then $a^2 + b^2 < 1$, so g (a, b) < f (a, b), Or $|z^9| > 1$, again a contradiction so this case is also not possible.

Hence $|z| = 1$.

28. Let $z = \cos t + i\sin t$. Then

$z + 2z^2 + \ldots + nz^n$

$= (z + \ldots + z^n) + (z^2 + \ldots + z^n) + \ldots z^n$

$= \dfrac{1}{z-1}[(z^{n+1} - z) + (z^{n+1} - z^2) + \ldots + (z^{n+1} - z^n)]$

$= \dfrac{nz^{n+1}}{z-1} - \dfrac{nz^{n+1} - z}{(z-1)^2}$

From this we can deduce

$\sum_{k=1}^{n} k\cos kt$

$$= \frac{(n+1)\sin\frac{(2n+1)t}{2}}{2\sin\frac{t}{2}} - \frac{1-\cos(n+1)t}{4\sin^2\frac{t}{2}} \quad \ldots(i)$$

$$\sum_{k=1}^{n} k\sin kt = \frac{\sin(n+1)t}{4\sin^2\frac{t}{2}} - \frac{n\cos\frac{(2n+1)t}{2}}{2\sin\frac{t}{2}} \quad \ldots(ii)$$

From this

2 sin 2° + 4 sin 4° + … +178 sin 178° = 2(sin 2° + 2 sin 2 · 2° +…+89 sin 89 · 2°

$$= 2\left(\frac{\sin 90.2°}{4sin^2 1°} - \frac{90\cos 179°}{2\sin 1°}\right)$$

$$= -\frac{90\cos 179°}{\sin 1°} = 90\cot 1°.$$

Required average is

$$\frac{1}{90}(2\sin 2° + 4\sin 4° + \ldots + 178\sin$$
$$+178° + 180\sin 180°) = \cot 1°$$

Proved.

29. Since $\varepsilon \neq 1$ be an nth root of unity so

$$\varepsilon^n - 1 = (\varepsilon - 1)(\varepsilon^{n-1} + \ldots + \varepsilon + 1)$$

But $\varepsilon \neq 1$ be an nth root of unity, so $\varepsilon^{n-1} + \ldots + \varepsilon + 1 = 0$.

Or $(\varepsilon^{n-1} - 1) + \ldots + (\varepsilon - 1) = -n$,

i.e., $(\varepsilon - 1)$

$$\left[\varepsilon^{n-2} + 2\varepsilon^{n-3} + .. + (n-2)\varepsilon + (n-1)\right] = -n.$$

Taking modulus both the sides we get

$$n = |\varepsilon - 1|\left|\varepsilon^{n-2} + 2\varepsilon^{n-3} + .. + (n-1)\right|$$

$$\leq |\varepsilon - 1|\left(\left|\varepsilon^{n-2}\right| + 2|\varepsilon|^{n-3} + .. + (n-1)\right)$$

Therefore

$$n \leq |1-\varepsilon|(1 + 2 + .. + (n-1)) = |1-\varepsilon|\frac{n(n-1)}{2},$$

i.e., we find the inequality $|1-\varepsilon| \geq \frac{2}{n-1}$.

Please Note:- Equality is not possible since the geometric images of 1, ε, …, ε^{n-1} are not collinear.

30. Let us assume that the points $A_1, A_2, \ldots, A_n$ with coordinates $z_1, z_2, \ldots, z_n$. The polygon $A_1, A_2, \ldots, A_n$ is inscribed in the circle with center at origin and radius $R = |z_1|$. The coordinate of the midpoint A_{ij} of the segment $[A_i A_j]$ is equal to $\frac{z_i + z_j}{2}$, for $1 \leq i < j \leq n$.

Hence, $|z_i + z_j|^2 = 4OA_{ij}^2$ and $|z_i + z_j|^2 = A_i A_j^2$.

But, $4OA_{ij}^2 = 4R^2 - A_i A_j^2$.

$$\text{The sum} \sum_{1\leq i<j\leq n}\left|\frac{z_i + z_j}{z_i - z_j}\right|^2 = \sum_{1\leq i<j\leq n}\frac{4OA_{ij}^2}{A_iA_j^2}$$

$$= \sum_{1\leq i<j\leq n}\frac{4R^2 - A_iA_j^2}{A_iA_j^2}$$

$$= 4R^2\sum_{1\leq i<j\leq n}\frac{1}{A_iA_j^2} - \binom{n}{2}.$$

From AM – HM inequality

$$\sum_{1\leq i<j\leq n}\frac{1}{A_iA_j^2} \geq \frac{\left(\binom{n}{2}\right)^2}{\sum_{1\leq i<j\leq n}A_iA_j^2}.$$

Since $\sum_{1\leq i<j\leq n} A_iA_j^2 \leq n^2.R^2$,

$$\text{Or} \sum_{1\leq i<j\leq n}\left|\frac{z_i + z_j}{z_i - z_j}\right|^2 \geq$$

$$4R^2\frac{\left(\binom{n}{2}\right)^2}{\sum_{1\leq i<j\leq n}A_iA_j^2} - \binom{n}{2} \geq \frac{4\left(\binom{n}{2}\right)^2}{n^2} - \binom{n}{2}$$

$$= \frac{\left(4\binom{n}{2} - n^2\right)\cdot\binom{n}{2}}{n^2}$$

$$= \frac{(n-1)(n-2)}{2},$$

Hence Proved